Creep in Structures

International Union of Theoretical and Applied Mechanics

M. Życzkowski (Ed.)

Creep in Structures

4th IUTAM Symposium, Cracow, Poland
September 10-14, 1990

Springer-Verlag
Berlin Heidelberg New York
London Paris Tokyo
Hong Kong Barcelona Budapest

Professor Dr. hab. Michał Życzkowski

Head, Institute of Mechanics and Machine Design
Cracow University of Technology
31-155 Cracow, ul. Warszawska 24
Poland

ISBN-13: 978-3-642-84457-7 e-ISBN-13: 978-3-642-84455-3
DOI: 10.1007/978-3-642-84455-3

Library of Congress Cataloging-in-Publication Data
Creep in structures : 4th symposium, Cracow, Poland, September 10-14, 1990
M. Życzkowski (ed.). (IUTAM-Symposia)
"International Union of Theoretical and Applied Mechanics."
Series statement from data sheet.
ISBN-13: 978-3-642-84457-7
1. Materials--Creep--Congresses. 2. Structural analysis (Engineering)--Congresses.
3. Viscoplasticity--Congresses.
I. Życzkowski, Michał. II. International Union of Theoretical and Applied Mechanics.
III. Series: IUTAM-Symposien.
TA418.22.C716 1991

Softcover reprint of the hardcover 1st edition 1991

61/3020 543210 – Printed on acid-free paper

Preface

There is a tradition to organize IUTAM Symposia "Creep in Structures" every ten years: the first Symposium was organized by N.J. Hoff in Stanford (1960), the second one by J. Hult in Göteborg (1970), and the third one by A.R.S. Ponter in Leicester (1980).

The fourth Symposium in Cracow, September 1990, gathered 123 participants from 21 countries and reflected rapid development of the theory, experimental research and structural applications of creep and viscoplasticity, including damage and rupture. Indeed, the scope of the Symposium was broad, maybe even too broad, but it was kept according to the tradition. Probably the chairman of "Creep in Structures V" in the year 2000 (if organized at all) will be forced to confine the scope substantially.

Participation in the Symposium was reserved for invited participants, suggested by members of the Scientific Committee. Total number of suggestions was very large and the response - unexpectedly high. Apart from several papers rejected, as being out of scope, over 100 papers were accepted for presentation. A somewhat unconventional way of presentation was introduced to provide ample time for fruitful and well prepared discussions: besides general lectures (30 minutes each), all the remaining papers were presented as short introductory lectures (10 minutes) followed by a 1-hour poster discussion with the authors and then by a general discussion. Such an approach made it possible to present general ideas orally, and then to discuss all the papers through and through.

Opening of the Symposium on Monday, September 10, 1990, in Assembly Hall of the oldest university in Poland (Uniwersytet Jagielloński), was devoted to professor Giuseppe Bernasconi (Politecnico di Milano), active member of the Scientific Committee, who passed away on October 18, 1989. His photo was demonstrated and his short scientific career was presented. Then, professor Władysław Muszyński, the Rector of the Cracow University of Technology, presented the host institution, professor Maria Szmyd-Dormus played Szymanowski and Chopin, and four general lectures followed. All the remaining lectures took place in Hotel pod Kopcem, beautifully located on top of a hill. On Tuesday all the participants were invited to the Town Hall by the Lord Mayor of Cracow, professor Jacek Woźniakowski, and on Thursday the conference dinner was organized in Ball Room of the Wieliczka salt mine.

This volume does not contain all the papers presented. Much to regret of the Scientific Committee some manuscripts were not submitted, and some were not included because of lack of space; however, they are shown in the Programme enclosed. Finally, the volume contains 81 papers divided into three almost equal parts: constitutive equations, combined loadings; damage, creep crack growth, creep rupture; structures, analytical and numerical methods, optimal design. General lectures are followed by the remaining papers in the alphabetical order. Full addresses of the participants are enclosed as well.

Finally, it is my great pleasure to express my cordial thanks to all members of the Scientific Committee, to all members of the Local Organizing Committee, to all supporting institutions (all names enclosed separately), to my wife Teresa Życzkowska, the organizer of the Ladies' Program, to my secretaries Jadwiga Deręgowska and Alicja Cioch, and to over 20 collaborators from the Institute of Mechanics and Machine Design, Cracow University of Technology, for their help which cannot be overestimated. Thanks are also due to the Springer-Verlag for publishing this volume which will certainly be useful both for researchers and designers.

Michał Życzkowski
December 1990

Scientific Committee

Prof. M. Życzkowski - Chairman

Prof. G. Bernasconi (Italy)

Prof. H. Besseling (The Netherlands)

Prof. J. Betten (Germany)

Prof. J. Hult (Sweden)

Prof. F.A. Leckie (USA)

Prof. J. Lemaître (France)

Prof. S. Murakami (Japan)

Prof. Yu.V. Nemirovsky (USSR)

Prof. A.R.S. Ponter (United Kingdom)

Prof. F.G. Rammerstorfer (Austria)

Prof. S.A. Shesterikov (USSR)

Prof. Z. Sobotka (Czechoslovakia)

Local Organizing Committee

Prof. M. Życzkowski - Chairman

Dr. E. Cegielski - Secretary

Dr. J. Bielski - Secretary

Prof. M. Chrzanowski

Prof. S. Piechnik

Prof. J. Skrzypek

Prof. Z. Waszczyszyn

Dr. A. Wróblewski - Technical Editor

Mr. A. Ganczarski

Sponsoring Institutions

International Union of Theoretical and Applied Mechanics

Politechnika Krakowska (The Cracow University of Technology)

Uniwersytet Jagielloński (The Jagiellonian University)

Komitet Mechaniki Polskiej Akademii Nauk (Committee of Mechanics, Polish Academy of Sciences)

List of Participants

AINSWORTH, R.A.	Central Electricity Generating Board, Gloucestershire, BERKELEY GL13 9PB, **UNITED KINGDOM**
ALI-ZADEH, A.N.	5 Chkalov str. apar.70, BAKU, **AZERBAIJAN SSR**
ALTENBACH, H.	WB Werkstofftechnik, Technische Universität, Postfach 124, 3010 MAGDEBURG, **GERMANY**
ANDERSON, R.G.	19 The Leylands, West Beach, LYTHAM Lancashire, FY8 5QS, **UNITED KINGDOM**
BALKE, H.	Zentralinstitut für Festkörperphysik und Werkstofforschung, Akademie der Wissenschaften der DDR, Helmholtzstr. 20, 8027 DRESDEN, **GERMANY**
BATHE, K.J.	Deptartment of Mechanical Engineering, Massachusetts Institute of Technology, CAMBRIDGE, MA 02139, **USA**
BENALLAL, A.	Laboratoire de Mécanique et Technologie, E.N.S. de Cachan, 61, Avenue du Président Wilson, 94230 CACHAN, **FRANCE**
BERTRAM, A.	Bundesanstalt für Materialforschung und -prüfung, Unter den Eichen 87, D-1000 BERLIN 45, **GERMANY**
BESSELING, J.F.	Faculty of Mechanical Engng., TU Delft, P.O.Box 5033, 2600 GA DELFT, **THE NETHERLANDS**
BETTEN, J.	Rhein.-Westf. Techn. Hochschule Aachen, Templergraben 55, D-5100 AACHEN, **GERMANY**
BIELSKI, J.	Instytut Mechaniki i PKM, Politechnika Krakowska, ul. Warszawska 24, 31-155 KRAKÓW, **POLAND**
BOCHENEK, B.	Instytut Mechaniki i PKM, Politechnika Krakowska, ul. Warszawska 24, 31-155 KRAKÓW, **POLAND**
BODNAR, A.	Instytut Mechaniki Budowli, Politechnika Krakowska, ul. Warszawska 24, 31-155 KRAKÓW, **POLAND**
BOYLE, J.T.	Dept. of Engineering, University of Strathclyde, 75 Montrose Street, GLASGOW G1 1XJ, **UNITED KINGDOM**
BRÜLLER, O.	Institut für Mechanik A, Technische Universität, 8000 MÜNCHEN 2, **GERMANY**
BYKOVTSEV, G.I.	Institute for Automation, 5 Radio str., 690032 VLADIVOSTOK, **USSR**
CAILLETAUD, G.	École Nationale Supérieure des Mines de Paris, B.P.87, 91003 EVRY CEDEX, **FRANCE**
CESCOTTO, S.	Institut du Génie Civil, Université de Liège, Quai Banning 6, B-4000 LIEGE, **BELGIUM**
CEGIELSKI, E.	Instytut Mechaniki i PKM, Politechnika Krakowska, ul. Warszawska 24, 31-155 KRAKÓW, **POLAND**
CHRZANOWSKI, M.	Instytut Mechaniki Budowli, Politechnika Krakowska, ul. Warszawska 24, 31-155 KRAKÓW, **POLAND**

COGNARD, J.-Y.	Laboratoire de Mécanique et Technologie, E.N.S. de Cachan, 61, Avenue du Président Wilson, 94230 CACHAN, **FRANCE**
DING, J.L.	Dept. of Mechanical Engng., Washington State Univ., PULLMAN, WA 99164, **USA**
DOROSZ, S.	Zakład Mechaniki Ośrodków Ciągłych, Instytut Podstawowych Problemów Techniki PAN, ul. Świętokrzyska 21, 00-049 WARSZAWA, **POLAND**
DULMAN, V.M.	Vilnius Civil Engineering Institute, 232055 VILNIUS, **Lithuanian SSR**
EDELSTEIN, W.S.	Dept. of Mathematics, Illinois Inst. of Technology, CHICAGO, Ill.60616, **USA**
FINNIE, I.	Dept. of Engineering, University of California, BERKELEY, Cal. 94720, **USA**
FISCHER, F.D.	Inst. für Mechanik, Montanuniversität Leoben, Franz-Josef-Str. 18, A-8700 LEOBEN, **AUSTRIA**
GANCZARSKI, A.	Instytut Mechaniki i PKM, Politechnika Krakowska, ul. Warszawska 24, 31-155 KRAKÓW, **POLAND**
GIESSEN, van der, E.	Dept. of Mechanical Engng., University of Technology, 2638 CD DELFT, **THE NETHERLANDS**
GOKHFELD, D.A.	Chelabinsky Politekhn. Inst., Prospekt Lenina 76, 454 044 CHELABINSK, **USSR**
GOLUB, V.P.	Institute of Mechanics, Academy of Sciences Ukr. SSR, Nesterov str.3, 252 057 KIEV-57, **USSR**
GUMMERT, P.	1. Institut für Mechanik, Technische Universität Berlin, Straße des 17 Juni 135, D-1000 BERLIN 12, **GERMANY**
HADDAD, Y.M.	Dept. of Mechanical Engng., University of Ottawa, 770 King Edward, OTTAWA, Ontario K1N 6N5, **CANADA**
HALL, F.R.	The University of Birmingham, Edgbaston, BIRMINGHAM B15 2TT, **UNITED KINGDOM**
HAYHURST, D.R.	Dept. of Mechanical Engng., University of Sheffield, Mappin Street, SHEFFIELD S1 3JD, **UNITED KINGDOM**
HEEMANN, U.	Institut für Tieflagerung, Gesellschaft für Strahlen- und Umweltforschung, Theodor-Heuss-Str.4, D-3300 BRAUNSCHWEIG, **GERMANY**
HUANG, N.C.	Dept. of Aerospace and Mech., Univ. of Notre Dame, NOTRE DAME, IN 46556, **USA**
HYDE, T.H.	Dept. of Engineering, The University, NOTTINGHAM NG7 2RD, **UNITED KINGDOM**
IGARI, T.	Nagasaki Research and Development Center, Mitsubishi Heavy Industries, Akunoura-machi 1-1, NAGASAKI 850-91, **JAPAN**
INOUE, T.	Dept. of Mechanical Engng., Kyoto University, Yoshida-Honmachi, Sakyo-ku, KYOTO, 606 **JAPAN**

ISHIKAWA, H. — Dept. of Mechanical Engng., Hokkaido University, Kita-Juhsanjo, Kita-ku, SAPPORO, 060 **JAPAN**

JAKOWLUK, A. — Katedra Mechaniki Stosowanej, Politechnika Białostocka, ul. Czysta 11, 15-463 BIAŁYSTOK, **POLAND**

KAWAI, M. — Inst. of Engng. Mechanics, University of Tsukuba, TSUKUBA, 305 **JAPAN**

KIENZLER, R. — Fraunhofer-Institut für Werkstoffmechanik, Wöhlerstrasse 11, D-7800 FREIBURG, **GERMANY**

KITAMURA, T. — Dept. of Mechanical Engng., Kyoto University, Yoshida-Honmachi, Sakyo-ku, KYOTO, 606 **JAPAN**

KNAUSS, W.G. — Graduate Aeronautical Lab., California Institute of Technology, PASADENA, Calif. 91009, **USA**

KOWALCZYK, K. — Instytut Mechaniki i PKM, Politechnika Krakowska, ul. Warszawska 24, 31-155 KRAKÓW, **POLAND**

KOWALEWSKI, Z. — Zakład Mechaniki Ośrodków Ciągłych, Instytut Podstawowych Problemów Techniki PAN, ul. Świętokrzyska 21, 00-049 WARSZAWA, **POLAND**

KOZŁOWSKI, R. — Instytut Materiałoznawstwa, Politechnika Krakowska, ul. Warszawska 24, 31-155 KRAKÓW, **POLAND**

KRASIŃSKI, M. — Instytut Mechaniki i PKM, Politechnika Krakowska, ul. Warszawska 24, 31-155 KRAKÓW, **POLAND**

KREMPL, E. — Dept. of Mechanical Engng., Rensselaer Polytechnic Institute, TROY, N.Y.12181, **USA**

KRUCH, S. — Office National d'Etudes et de Recherches Aérospatiales, F-92320 CHATILLON, **FRANCE**

KRUŻELECKI, J. — Instytut Mechaniki i PKM, Politechnika Krakowska, ul. Warszawska 24, 31-155 KRAKÓW, **POLAND**

KUCIEL, S. — Instytut Mechaniki i PKM, Politechnika Krakowska, ul. Warszawska 24, 31-155 KRAKÓW, **POLAND**

KURTYKA, T. — Instytut Mechaniki i PKM, Politechnika Krakowska, ul. Warszawska 24, 31-155 KRAKÓW, **POLAND**

LECKIE, F.A. — Dept. of Mechanical Engng., Univ. of California, SANTA BARBARA, Calif. 93106, **USA**

LEMAITRE, J. — Laboratoire de Mécanique et Technologie, E.N.S. de Cachan, 61, Avenue du Président Wilson, 94230 CACHAN, **FRANCE**

LIS, Z. — Instytut Technologii i Konstr. Bud., Politechnika Poznańska, ul. Piotrowo 5, 60-965 POZNAŃ, **POLAND**

LITEWKA, A. — Instytut Technologii i Konstr. Bud., Politechnika Poznańska, ul. Piotrowo 5, 60-965 POZNAŃ, **POLAND**

LOKOSHTCHENKO, A.M. — Institute of Mechanics, Moscow University, 119 899 MOSKVA, **USSR**

LUIJERINK, J. — Dept. of Civil Engng., University of Technology, Stevinweg 1, 2628 CD DELFT, **THE NETHERLANDS**

ŁACZEK, S. — Instytut Mechaniki i PKM, Politechnika Krakowska, ul. Warszawska 24, 31-155 KRAKÓW, **POLAND**

MANG, H.	Institut für Festigkeitslehre, TU Wien, Karlsplatz 13/202, A-1040 WIEN, **AUSTRIA**
MATOUSEK, J.	Jahodova 2888/15, 106 00 PRAHA 10, **CZECHOSLOVAKIA**
MIĆUNOVIĆ, M.V.	Mašinski Fakultet, Univerzitet u Kragujevcu, Sestre Janjić 6, 34000 KRAGUJEVAC, **YUGOSLAVIA**
MRÓZ, Z.	Zakład Mechaniki Ośrodków Ciągłych, Instytut Podstawowych Problemów Techniki PAN, ul. Świętokrzyska 21, 00-049 WARSZAWA, **POLAND**
MUC, A.	Instytut Mechaniki i PKM, Politechnika Krakowska, ul. Warszawska 24, 31-155 KRAKÓW, **POLAND**
MURAKAMI, S.	Dept. of Mechanical Engng., Nagoya University, Furo-cho, Chikusa-ku, NAGOYA, 464 **JAPAN**
MURZEWSKI, J.	Instytut Mat. i Konstr. Bud., Politechnika Krakowska, ul. Warszawska 24, 31-155 KRAKÓW, **POLAND**
NEMIROVSKY, Yu.V.	Institut Mekhaniki, Sibirskoe Otdelenie A.N. SSSR, 630090 NOVOSIBIRSK, **USSR**
NEUMEISTER, J.	Division of Solid Mechanics, Chalmers University of Technology, S-41296 GÖTEBORG, **SWEDEN**
OBERAIGNER, E.R.	Inst. für Mechanik, Montanuniversität Leoben, Franz-Josef-Str. 18, A-8700 LEOBEN, **AUSTRIA**
OHJI, K.	Dept. of Mechanical Engng., Osaka University, Yamadaoka 2-1, SUITA, OSAKA, 565 **JAPAN**
OHNAMI, M.	Dept. of Mechanical Engng., Ritsumeikan University, Tojiin Kita-machi 28-1, Kita-ku, KYOTO, 603 **JAPAN**
OHNO, N.	Dept. of Mechanical Engng., Nagoya University, Furo-cho, Chikusa-ku, NAGOYA, 464-01 **JAPAN**
ONARAN, K.	Dept. of Civil Engineering, Istanbul Technical University, ISTANBUL, **TURKEY**
OSTOJA-STARZEWSKI, M.	Michigan State University, EAST LANSING, **USA**
OSTROWSKI, L.	Inst. für Mechanik, Technische Universität Hannover, Appelstrasse 11, D-3000 HANNOVER 1, **GERMANY**
PALEY, M.	Dept. of Solid Mechanics, Tel-Aviv University, RAMAT - AVIV 69978, **ISRAEL**
PENNY, R.K.	Dept of Mechanical Engng., University of Cape Town, RONDENBOSCH 7700, **SOUTH AFRICA**
PIECHNIK, S.	Instytut Mechaniki Budowli, Politechnika Krakowska, ul. Warszawska 24, 31-155 KRAKÓW, **POLAND**
PINEAU, A.	École Nationale Supérieure des Mines de Paris, B.P.87, 91003 EVRY CEDEX, **FRANCE**
PIQUES, R.	École Nationale Supérieure des Mines de Paris, B.P.87, 91003 EVRY CEDEX, **FRANCE**
PONTER, A.R.S.	University of Leicester, University Road, LEICESTER LE1 7RH, **UNITED KINGDOM**

POTAPOV, V.D.	Moskovsky Inst. Inzhenerov Zheleznodor. Transporta, MOSKVA, **USSR**
PRIEST, R.H.	Nuclear Electric PLC, Bridgewater Road, BRISTOL BS13 8AN, **UNITED KINGDOM**
PYRZ, R.	Dept. of Mechanical Engng., University of Aalborg, Pontoppidanstraede 101, DK-9220 AALBORG, **DENMARK**
RAMMERSTORFER, F.G.	Inst. für Leichtbau und Flugzeugbau, TU Wien, Gusshausstrasse 27-29/317, A-1040 WIEN, **AUSTRIA**
RIVLIN, R.S.	Center for the Appl. of Mathematics, Lehigh Univ., BETHLEHEM, PA 18015, **USA**
RYSZ, M.	General Engng. Dept., University of Puerto Rico, MAYAGUEZ, Puerto Rico 00708, **USA**
SADAKOV, O.S.	Chelabinsky Politekhn. Inst., Prospekt Lenina 76, 454 044 CHELABINSK, **USSR**
SANECKI, H.	Instytut Mechaniki i PKM, Politechnika Krakowska, ul. Warszawska 24, 31-155 KRAKÓW, **POLAND**
SANZHAROVSKY, R.S.	Leningradsky Inzhenerno-Stroitelny Institut, LENINGRAD, **USSR**
SHESTERIKOV, S.A.	Institute of Mechanics, Moscow University, 119 899 MOSKVA, **USSR**
SHEVTCHENKO, Y.N.	Institute of Mechanics, Academy of Sciences Ukr. SSR, Nesterov str.3, 252 057 KIEV-57, **USSR**
SKRZYPEK, J.	Instytut Mechaniki i PKM, Politechnika Krakowska, ul. Warszawska 24, 31-155 KRAKÓW, **POLAND**
SOBOTKA, Z.	CSAV, Ustav Matematiky, Švédskà 10, 15000 PRAHA, **CZECHOSLOVAKIA**
STAMM, H.	Joint Research Centre, Inst. for Advanced Materials, 21020 ISPRA (VARESE), **ITALY**
SZUWALSKI, K.	Instytut Mechaniki i PKM, Politechnika Krakowska, ul. Warszawska 24, 31-155 KRAKÓW, **POLAND**
SZYSZKOWSKI, W.	Dept. of Mechanical Engng., Univ. of Saskatchewan, SASKATOON SK S7N OWO, **CANADA**
TAKEZONO, S.	Dept. of Energy Engng., Toyohashi University, Tempaku-cho, TOYOHASHI, 440 **JAPAN**
TOWNLEY, C.H.A.	National Power Nuclear, Berkeley Nuclear Laboratories, Gloucestershire, BERKELEY, GL13 9PB, **UNITED KINGDOM**
TRĄMPCZYŃSKI, W.	Zakład Mechaniki Ośrodków Ciągłych, Instytut Podstawowych Problemów Techniki PAN, ul. Świętokrzyska 21, 00-049 WARSZAWA, **POLAND**
TROJNACKI, A.	Instytut Mechaniki i PKM, Politechnika Krakowska, ul. Warszawska 24, 31-155 KRAKÓW, **POLAND**
TYLIKOWSKI, A.	Instytut Budowy Maszyn, Politechnika Warszawska, ul. Narbutta 84, 02-524 WARSZAWA, **POLAND**

VINOGRADOV, A.M.	Dept. of Mechanical Engng., University of Calgary, 2500 University Drive N.W., CALGARY, ALBERTA T2N 1N4, **CANADA**
WALCZAK, J.	ADINA R + D. INC, 71 Elton Avn., WATERTOWN, MA 02172, **USA**
WANG, REN	Dept. of Mechanics, Peking University, BEIJING 100 871, **CHINA**
WANIEWSKI, M.	Zakład Mechaniki Ośrodków Ciągłych, Instytut Podstawowych Problemów Techniki PAN, ul. Świętokrzyska 21, 00-049 WARSZAWA, **POLAND**
WASZCZYSZYN, Z.	Instytut Mechaniki Budowli, Politechnika Krakowska, ul. Warszawska 24, 31-155 KRAKÓW, **POLAND**
WATANABE, O.	Inst. of Engng. Mechanics, University of Tsukuba, Ibaraki 305, TSUKUBA, **JAPAN**
WEBER, H.	Institut für Verfahrenstechnik und Mechanik, Universität Karlsruhe, Postfach 6380, D-7500 KARLSRUHE 1, **GERMANY**
WRÓBLEWSKI, A.	Instytut Mechaniki i PKM, Politechnika Krakowska, ul. Warszawska 24, 31-155 KRAKÓW, **POLAND**
YANG, T.Q.	Dept. of Mechanics, Huazhong University of Science and Technology, WUHAN 430074, **CHINA**
YANG, YEONG-BIN	Dept. of Civil Engng., Nat. Taiwan University, TAIPEI, **CHINA (TAIWAN) 10764**
YUMASHEVA, M.A.	Institute of Mechanics, Moscow University, 119 899 MOSKVA, **USSR**
ZABORSKI, A.	Instytut Mechaniki Budowli, Politechnika Krakowska, ul. Warszawska 24, 31-155 KRAKÓW, **POLAND**
ZADOYAN, M.A.	Institut Mekhaniki, AN Arm.SSR, Prospekt Marshala Bagramiana 24b, 375019 EREVAN, **Arm. SSR**
ZHANG, WEN	Dept. of Math. and Mechanics, Fudan University, SHANGHAI, 201903 **CHINA**, temp. address: room 5-021, Massachusetts Institute of Technology, CAMBRIDGE, MA 02139, **USA**
ZIELIŃSKI, A.	Instytut Mechaniki i PKM, Politechnika Krakowska, ul. Warszawska 24, 31-155 KRAKÓW, **POLAND**
ZOLOCHEVSKY, A.	str. R.Eidemana 15/104, 310118 KHARKOV,118, **USSR**
ŻUCHOWSKI, R.	Instytut Materiałoznawstwa i Mech. Techn., Politechnika Wrocławska, ul. Smoluchowskiego 26, 50-370 WROCŁAW, **POLAND**
ŻYCZKOWSKI, M.	Instytut Mechaniki i PKM, Politechnika Krakowska, ul. Warszawska 24, 31-155 KRAKÓW, **POLAND**

Scientific Programme

Monday, September 10, morning

OPENING OF THE SYMPOSIUM

M. ŻYCZKOWSKI, Chairman

W. MUSZYŃSKI, Rector of the Cracow University of Technology

SESSION 1: **General Lectures**
Chairman: R.S. Rivlin   Assistant chairman: K. Szuwalski

A.R.S. PONTER: *Creep and rupture of structures subjected to variable loading*

D.A. GOKHFELD, O.S. SADAKOV: *An unified mathematical model for plasticity and creep under variable-repeated loading*

SESSION 2: **General Lectures**
Chairman: R.K. Penny   Assistant chairman: K. Szuwalski

K.J. BATHE, M. KOJIĆ, J. WALCZAK: *Advances in finite element methods for elasto-plastic and creep analysis*

M. ŻYCZKOWSKI: *Optimal structural design under creep conditions*

Monday, September 10, afternoon

SESSION 3: **Viscoplasticity**
Chairman: M. Ohnami   Assistant chairman: J. Krużelecki

A. BENALLAL, J. LEMAITRE: *Localization phenomena in thermo-elasto-plasticity and thermo-elastoviscoplasticity* (general lecture)

E. KREMPL: *Modeling of short- and long-term creep using the viscoplasticity theory based on overstress*

O. WATANABE, S.N. ATLURI: *Some unification of creep theories based on the internal time concept*

P. GUMMERT: *A stress-strain relationship (constitutive equation) for non-linear and compressible viscoelastic materials*

G. CAILLETAUD, P. PILVIN: *Intergranular and transgranular hardening in viscoplasticity*

M.V. MIĆUNOVIĆ: *A viscoplasticity theory of irradiated materials*

T. KURTYKA, M. ŻYCZKOWSKI: *A theory of viscoplasticity allowing for distortional hardening*

Session 4: **Plates, Wedges**
Chairman: S. Cescotto   Assistant chairman: A. Zieliński

Z. SOBOTKA: *Bending of the orthotropic viscoelastic plates with thermal effects*

H. ALTENBACH: *Modelling of viscoelastic behaviour of plates*

A. BODNAR, M. CHRZANOWSKI: *A non-unilateral damage in creeping plates*

T. IGARI, K. SETOGUCHI, S. NOMURA: *Prediction of macroscopic and local creep behaviour of perforated plates*

M.A. ZADOYAN: *The stress state of the composite wedge in stationary creep*

Tuesday, September 11, morning

Session 5: **Analytical and Numerical Methods**
Chairman: F.G. Rammerstorfer   Assistant chairman: S. Łaczek

J.N.C. GUERREIRO, A.F.D. LOULA, J.T. BOYLE: *Finite element methods in stress analysis for creep* (general lecture)

R.K. PENNY: *Contributions to robust methods of creep analysis*

S. CESCOTTO: *Creep analysis in a large strain context by unified constitutive equations*

W. SZYSZKOWSKI, P.G. GLOCKNER: *A numerical method of nonlinear viscoelastic structural analysis*

J.-Y. COGNARD, P. LADEVEZE: *The large time increment method applied to cycling loadings*

W.S. EDELSTEIN: *Some analytic results pertaining to stress redistribution in a class of structures undergoing primary or secondary creep*

Session 6: **Constitutive Equations, Transformations**
Chairman: P. Gummert   Assistant chairman: T. Kurtyka

T. INOUE: *Thermo-mechanical simulation of some engineering processes incorpolating solidification*

F.D. FISCHER, E.R. OBERAIGNER: *Diffusional transformation and creep in steels*

K. ONARAN: *Nonlinear constitutive equations for creep of structural materials*

G.I. BYKOVTSEV: *Constitutive equations of transient metal creep*

YEONG-BIN YANG, L.J. LEU: *Constitutive law and force recovery for large strain problems*

V.P. GOLUB: *Theory of creep and long-term strength of isotropic strengthening media*

O. BRÜLLER: *Some new aspects of the nonlinear characterization of creep and stress relaxation of polymers*

Tuesday, September 11, afternoon

SESSION 7: **Creep-Fatigue Interaction**
Chairman: E. Krempl Assistant chairman: B. Bochenek

K. OHJI, S. KUBO: *Fatigue crack propagation behavior under creep conditions* (general lecture)

R. OHTANI, T. KITAMURA, N. TADA: *Characterization of creep-fatigue failure process based on stochastic behaviour of small crack initiation and growth*

D.A. MILLER, R.H. PRIEST: *The assessment of creep-fatigue initiation of crack growth*

D. BESDO, L. OSTROWSKI: *On the creep-ratchetting of AlMgSi 0.5 at elevated temperature; experimental investigations*

T. INOUE, S. IMATANI, M. TANAKA: *A viscoplastic constitutive relationship and the FEM implementation of stress/strain concentration in notched component*

SESSION 8: **Cavities, Voids, Microcracks**
Chairman: D.R. Hayhurst Assistant chairman: A. Ganczarski

A. M. LOKOSHTCHENKO: *Investigation of damage of metals by a method of resistivity measurements*

H.A. BAHR, H. BALKE, W. POMPE, S. WERDIN: *Nonstationary nucleation of cavities at grain boundaries*

E. van der GIESSEN, V. TVERGAARD: *On microcracking due to cavitation and grain boundary sliding in creeping polycrystals*

E. STEIN, U. HEEMANN: *Transient creep and creep fracture in rock salt structure*

J. BESSON, M. ABOUAF, F. MAZEROLLE, P. SUQUET: *Compressive creep tests on porous ceramic notched specimens*

Wednesday, September 12, morning

SESSION 9: **Creep Buckling**
Chairman: N.C. Huang Assistant chairman: A. Trojnacki

W.G. KNAUSS, T.M. MINAHEN: *Creep buckling in viscoelastic materials* (general lecture)

S.A. SHESTERIKOV, M.A. YUMASHEVA: *On a flow potential for non-linearly viscous medium, taking into account instantaneous plasticity*

S.A. SHESTERIKOV: *Formulation of a creep equation and stability problems*

A.M. VINOGRADOV: *Nonlinear instability induced by creep*

M. PALEY, J. ABOUDI: *Creep buckling of inelastic plates*

J. MURZEWSKI: *Creep buckling of steel columns in fire temperatures*

SESSION 10: **Creep Damage and Rupture in Materials**
Chairman: H. Ishikawa Assistant chairman: K. Szuwalski

H.J. BÖHM, F.G. RAMMERSTORFER: *The development of residual microstresses during processing and loading of metal matrix composites*

A. LITEWKA, Z. LIS: *Creep rupture of metals under tri-axial state of stress*

M. OSTOJA-STARZEWSKI: *Percolation models as a basis of material failure*

H. STAMM, U. von ESTORFF: *Determination of creep damage in an austenitic steel*

Y.M. HADDAD: *A microstructural approach to the creep response of fibrous systems*

C. ALBERTINI, M. MONTAGNANI, M.V. MIĆUNOVIĆ: *Uniaxial viscoplastic experiments of predamaged AISI 316H*

J. LUIJERINK: *Creep of concrete during load-controlled experiments*

Wednesday, September 12, afternoon

SESSION 11: **Constitutive Equations, Combined Loadings**
Chairman: F.D. Fischer Assistant chairman: A. Wróblewski

J. BETTEN: *Applications of tensor functions in creep mechanics* (general lecture)

R.S. RIVLIN: *Is objectivity an axiom ?*

N. OHNO, T. MIZUNO, H. KAWAJI, H. KAWAI: *Effects of slip and lattice rotation on combined tension-torsion creep of a directionally solidified alloy*

J.L. DING, W.N. FINDLEY: *Creep deformation under variable complex loadings*

W. TRĄMPCZYŃSKI, Z. MRÓZ: *Anisotropic creep hardening rule for metals and its application to non-proportional loading*

H. WEBER: *Characterization of nonlinear viscoelastic materials for a special class of loading histories*

M. WANIEWSKI: *Effects of prior plastic strain path shapes on creep behaviour of incompressible materials*

Z. KOWALEWSKI: *The influence of deformation history on creep of pure copper*

SESSION 12: **Lifetime of Structures**
Chairman: I. Finnie Assistant chairman: K. Kowalczyk

R.A. AINSWORTH: *Creep life estimates for defective structures*

I.W. GOODALL, R.A. AINSWORTH: *An assessment procedure for the high temperature response of structures*

C.H.A. TOWNLEY: *Time dependent structural mechanics - an engineer's view of research requirements*

R.G. ANDERSON: *The verification of the solutions of large problems of discontinuous creeping structures*

R.S. SANZHAROVSKY: *Creep of statically indeterminate reinforced concrete structures*

Thursday, September 13, morning

SESSION 13: **Creep Crack Growth**
Chairman: Ren Wang   Assistant chairman: S. Kuciel

S. MURAKAMI, M. MIZUNO: *Mechanical modeling of irradiation creep and its application to the analysis of creep crack growth* (general lecture)

F.R. HALL, D.R. HAYHURST: *Creep continuum damage mechanics studies of size effects, creep crack growth, and weldments*

N.C. HUANG: *Crack growth in linearly viscoelastic media*

T.H. HYDE, J.J. WEBSTER, A.C. CHAMBERS: *Mixed-mode creep crack growth*

R. PIQUES, A. PINEAU: *Global and local approaches to creep crack initiation and creep crack growth*

R. KIENZLER, T. HOLLSTEIN: *Theoretical, experimental and numerical investigations of creep crack growth*

SESSION 14: **Optimization under Creep Conditions**
Chairman: S.A. Shesterikov   Assistant chairman: B. Bochenek

F.A. LECKIE: *Composite design against creep* (general lecture)

Yu. V. NEMIROVSKY: *Problems of creep rupture and optimization of structures under creep conditions*

K. SZUWALSKI: *Structures optimal with respect to ductile rupture time vs. structures of uniform strength*

A. GANCZARSKI, J. SKRZYPEK: *Optimal shape of prestressed disks against brittle rupture under unsteady creep conditions*

M. RYSZ: *Optimal shaping of a cross section of cylindrical shell under combined loadings with respect to creep stability*

A. WRÓBLEWSKI: *Optimal design of circular and annular plates with respect to creep buckling*

Thursday, September 13, afternoon

SESSION 15: **Shells, Tubes**
Chairman: W.S. Edelstein   Assistant chairman: J. Krużelecki

J.F. BESSELING: *Finite axisymmetric creep deformation of shells of revolution* (general lecture)

I. FINNIE: *Stress concentrations and failure due to dissimilar creep properties in weld metal and base metal*

H. WALTER, H. MANG: *Nonlinear finite element analysis of prestressed concrete shells considering creep, shrinkage and aging of concrete*

Yu.N. SHEVTCHENKO: *Elastic-plastic deformation of structure elements taking into account the damage of material at creep*

S. TAKEZONO, K. MIGITA, K. TANI: *The creep of anisotropic multi-layered moderately thick shells of revolution*

A.A. ČYRAS, V.M. DULMAN: *Mathematical models of the stress and strain analysis problem under creep conditiions*

SESSION 16: **Cyclic Loadings**
Chairman: K. Onaran   Assistant chairman: M. Krasiński

M. OHNAMI, M. SAKANE, T. HASEBE: *Creep-fatigue life and cyclic constitutive relation of MAR-M247 ds superalloy at high temperature*

H. ISHIKAWA: *A hybrid constitutive model for cyclic, and creep deformation*

V.N. MADUDIN, O.S. SADAKOV: *Generalized models of structures for cyclic plastic and creep analysis*

R. ŻUCHOWSKI: *Analysis of deformation process in thin-walled tubes under thermal cyclic conditions*

Friday, September 14, morning

SESSION 17: **Composites**
Chairman: Yu. V. Nemirovsky   Assistant chairman: A. Muc

J.M. NEUMEISTER: *Creep rupture in fibre bundles with hardening-softening fibre material*

R. PYRZ: *A hybrid model for creep and damage in woven composite materials*

SESSION 18: **Dynamics**
Chairman: Yu. V. Nemirovsky   Assistant chairman: H. Sanecki

T.Q. YANG, Z.W. YANG, REN WANG: *Dynamic response of a viscoelastic circular plate on a viscoelastic half space foundation*

V.D. POTAPOV: *On the stability of viscoelastic structures under a stochastic excitation*

A. TYLIKOWSKI: *Stability and bounds on motion of viscoelastic columns with imperfections and time-dependent forces*

WEN ZHANG: *Dynamic behaviour and creep instability of high spinning visco-elastic deformable body*

Friday, September 14, afternoon

SESSION 19: **Creep Damage and Rupture of Structures**
Chairman: H. Mang Assistant chairman: A. Ganczarski

J.L. CHABOCHE, S. KRUCH, P.M. LESNE: *A new damage approach for creep crack growth prediction*

J. MATOUŠEK: *Evaluation of creep damage in some major components of steam turbine*

S. KUBO, K. OHJI: *Small-scale creep prediction of notch stress and strain under creep conditions*

S. PIECHNIK, A. ZABORSKI: *Stress-assisted corrosion in visco-elastic material*

SESSION 20: **Constitutive Equations, Anisotropy**
Chairman: Z. Sobotka Assistant chairman: T. Kurtyka

M. KAWAI: *Anisotropic creep behaviour induced by plastic deformation*

A. BERTRAM, J. OLSCHEWSKI, M. ZELEWSKI: *Anisotropic creep modelling for f.c.c. single crystals*

O.S. SADAKOV, V.N. MADUDIN, M.V. APAICHEV: *On the state equations for anisotropic elasto-visco-plastic bodies*

A. JAKOWLUK, E. MIELESZKO, M. PLEWA: *Description of creep anisotropy and reduced stress*

A. ZOLOCHEVSKY: *Creep of isotropic and anisotropic materials with different behaviour in tension and compression*

CLOSURE OF THE SYMPOSIUM:

M. ŻYCZKOWSKI, Chairman

Contents

Part I:
Constitutive Equations, Combined Loadings

Applications of Tensor Functions in Creep Mechanics

J. Betten

Department of Mathematical Models in Materials Science
Technical University Aachen
D-5100 Aachen, Templergraben 55, Germany

Summary
This paper will provide a short survey of some recent advances in the mathematical modelling of materials behaviour under creep conditions. The mechanical behaviour of anisotropic solids requires a suitable mathematical modelling. The properties of tensor functions with several argumenttensors constitute a rational basis for a consistent mathematical modelling of complex material behaviour.

This paper presents certain principles, methods, and recent succesful applications of tensor functions in solid mechanics. The rules for specifying irreducible sets of tensor invariants and tensor generators for material tensors of rank two and four are also discussed.

Furthermore, it is very important that the scalar coefficients in constitutive and evolutional equations are determined as functions of the integrity basis and experimental data. It is explained in detail that these coefficients can be determined by using tensorial interpolation methods. Some examples for practical use are discussed.

Finally, we have carried out our own experiments in order to examine the validity of the mathematical modelling.

1. Introduction

Over the last decade much effort has been devoted to the elaboration of phenomenological theories describing the relation between force and deformation in bodies of materials which obey neither the linear laws of the classical theories of elasticity nor the hydrodynamics of viscous fluids.

Material laws and constitutive theories are the fundamental bases for describing the mechanical behaviour of materials under multi-axial states of stress involving creep and creep

M. Zyczkowski (Ed.)
Creep in Structures
IUTAM Symposium Cracow/Poland 1990

rupture. The tensor function theory has become a powerful tool [1,2] for solving such complex problems.

The mechanical behaviour of anisotropic solids (which are materials with oriented internal structures produced by forming processes and manufacturing procedures or induced by permanent deformation) requires a suitable mathematical modelling. The properties of tensor functions with several argumenttensors constitute a rational basis for a consistent mathematical modelling of complex material behaviour [3,4].

In creep mechanics one can differentiate between three stages: the *primary*, *secondary*, and *tertiary* creep stage. These terms correspond to a decreasing, constant, and increasing creep strain-rate respectively, and were introduced by *Andrade* in 1910.

In order to describe the creep behaviour of metals in the primary stage *tensorial non-linear constitutive equations* involving the *strain-hardening hypothesis* are proposed in [5]. Based upon these general relations, the primary creep behaviour of a thin-walled circular cylindrical shell subjected to internal pressure are also analysed in [5]. The creep buckling of cylindrical shells subjected to internal pressure and axial compression was investigated in [6] by considering *tensorial non-linearities* and *anisotropic primary creep*.

In the following, the secondary creep stage of isotropic and anisotropic solids in a state of multi-axial stress will first be discussed. Creep deformations of metals usually remain unaffected if hydrostatic pressure is superimposed. In order to describe the secondary creep behaviour of isotropic materials some authors use a *creep potential* [7,8,9], which is a scalar-valued function of *Cauchy's stress tensor*. One can show that the *creep potential theory* is compatible with the *tensor function theory* provided the material is *isotropic*, and additional conditions are fulfilled [10]. However, the creep potential hypothesis only furnishes restricted forms of constitutive equations and, therefore, has only limited justification if the material is *anisotropic*.

The tertiary creep phase is considered in the other part of this paper. In this phase, the creep process is accompanied by the formation of microscopic cracks on the grain boundaries, so that damage-accumulation occurs. In some cases voids are caused by a given stress history and, therefore, they are distributed anisotropically amongst the grain boundaries. Thus, the mechanical behaviour will be *anisotropic* and it is therefore necessary to investigate this kind of anisotropy by introducing appropiately defined *anisotropic damage tensors* into constitutive equations [11,12].

Problems of *creep damage* have been investigated by many authors. Very extensive surveys into recent advances in *damage mechanics* are given in [13,14], for instance. Further contributions to the theory of *Continuum Damage Mechanics* should be mentioned, e.g. [15-27].

2. Secondary Creep Behaviour

When investigating the secondary creep behaviour one can presume the existence of a creep potential. The *creep potential hypothesis* is based upon the *principle of maximum dissipation rate*, from which one obtains the *flow rule*

$$d_{ij} = \dot{\lambda}\partial F/\partial\sigma_{ij} \tag{2.1}$$

by applying *Lagrange's method* in connection with a *creep condition* F = const. as a subsidiary condition, where d_{ij} are the cartesian components of the so-called *rate-of-deformation tensor* $\underset{\sim}{d}$ and σ_{ij} are the cartesian components of *Cauchy's stress tensor*. The tensor $\underset{\sim}{d}$ is not to be confused with the *material time derivative* $\dot{\underset{\sim}{\varepsilon}}$ of the *infinitesimal strain tensor* $\underset{\sim}{\varepsilon}$ as described in detail in [24]. Only in cases of small displacement gradients and small velocity gradient tensors do we find $d_{ij} \approx \varepsilon_{ij}$. Based upon the *hyptheses of the equivalent dissipation rate*, the *Lagrange multiplier* $\dot{\lambda}$ has been determined for isotropic materials in [8] and for anisotropic materials in [28] by taking the *Norton-Bailey creep law* into account. For isotropic materials the creep potential $F = F(\sigma_{ij})$ is a scalar-valued tensor function only of the *Cauchy stress tensor*. It is evident from the theory of isotropic tensor functions that the creep potential can be expressed as a single-valued function of the *integrity basis*, the elements of which are the *irreducible basic invariants* $S_\upsilon = \mathrm{tr}\, \underset{\sim}{\sigma}^\upsilon$, $\upsilon = 1,2,3$, of the stress tensor. Thus, by considering the form $F = F(S_1, S_2, S_3)$ and using the classical flow rule (2.1), one immediately obtains the constitutive equation

$$d_{ij} = \dot{\Lambda}[(\partial F/\partial S_1)\delta_{ij} + 2(\partial F/\partial S_2)\sigma_{ij} + 3(\partial F/\partial S_3)\sigma_{ij}^{(2)}] \ . \qquad (2.2)$$

Instead of the *creep potential hypothesis*, one can represent the rate-of-deformation tensor as a symmetric *tensor-valued function* of one second-order argument tensor:

$$d_{ij} = f_{ij}(\underset{\sim}{\sigma}) = \varphi_0\delta_{ij} + \varphi_1\sigma_{ij} + \varphi_2\sigma_{ij}^{(2)} \ , \qquad (2.3)$$

where φ_0, φ_1, φ_2 are scalar-valued functions of the integrity basis S_1, S_2, S_3.

If one compares the *minimum polynomial representation* (2.3) with the result (2.2) based upon the *creep potential hypothesis*, one arrives at the following identities:

$$\varphi_0 = \dot{\Lambda}\partial F/\partial S_1 \ , \qquad \varphi_1 = 2\dot{\Lambda}\partial F/\partial S_2 \ , \qquad \varphi_2 = 3\dot{\Lambda}\partial F/\partial S_3 \ . \qquad (2.4)$$

By eliminating the creep potential F in (2.4) one achieves the following sufficient and necessary condition

$$(3\frac{\partial\varphi_1}{\partial S_3} - 2\frac{\partial\varphi_2}{\partial S_2})\varphi_0 + (\frac{\partial\varphi_2}{\partial S_1} - 3\frac{\partial\varphi_0}{\partial S_3})\varphi_1 + (2\frac{\partial\varphi_0}{\partial S_2} - \frac{\partial\varphi_1}{\partial S_1})\varphi_2 = 0 \ , \qquad (2.5)$$

as has already been pointed out in more detail in [10].

As one can see from the results given above, the theory of the creep potential is *compatible* with the tensor function theory provided the condition (2.5) has been fulfilled. That is why this condition has been called the *condition of compatibility* in [10].

The creep potential hypothesis is *not* compatible with the tensor function theory for more complicated examples than those mentioned above. It furnishes only restricted forms of constitutive equations for *anisotropic materials* in particular, even if a general creep potential has been assumed. Consequently, the classical flow rule (2.1) must be modified, and for each such modification *conditions of compatibility* must be found. Some examples are discussed in [10], for instance a case in which oriented solids can be described by a *material tensor of rank two*. In such cases the creep potential is a scalar-valued function of two symmetric second-order tensors, $F = F(\sigma_{ij}, A_{ij})$, the integrity basis of which consists of ten irreducible elements. A more general case is also described in [10] by using a *material tensor of rank four*.

It must be noted that, from the physical point of view [29], the assumption of a creep potential has only limited justification, especially in the anisotropic case and in the tertiary creep stage. Fortunately, constitutive equations can be represented in full as tensor-valued functions.

For example, the tensor function theory has been applied in [30-33] in order to study the effect of *plastic prestrain* on the secondary creep behaviour. Plastic prestrain can produce *anisotropy* in materials which are initially isotropic. This anisotropy greatly influences the creep behaviour of such materials. *Tensorial non-linear constitutive equations* for this secondary creep behaviour are proposed in [33], based upon the representation theory of tensor functions. The induced anisotropy is characterized by introducing a second rank *prestrain tensor* as an additional argument tensor of the constitutive equations. Numerical results are compared with our own creep tests on plastic prestrained specimens of INCONEL 617 at 950°C.

A *simplified theory* has been developed in [9,28] in order to study anisotropic effects. This theory is based upon the isotropic concept (2.2) and (2.3), where the *Cauchy stress tensor* σ_{ij} is replaced by the *mapped stress tensor*

$$\tau_{ij} = \beta_{ijkl}\sigma_{kl} \quad . \tag{2.6}$$

According to this assumption the anisotropy of the material is entirely involved in the fourth-rank tensor β_{ijkl}. For example, in [9] the *orthotropic* case is considered, and the

Poynting-effect is taken into account. Furthermore, the influence of the *second-order effect* and of the *anisotropy* on the creep behaviour of a thin-walled tube subjected to internal pressure has been investigated in [9,34].

3. Tensorial Generalization of Norton's Creep Law

The well-known *Norton creep law* is assumed to be valid for the secondary creep stage and is characterized by the uni-axial relation

$$d/d_0 = (\sigma/\sigma_0)^n \quad \text{or} \quad d = K\sigma^n , \qquad (3.1a,b)$$

where d is the strain rate, σ the uni-axial true stress, and d_0, σ_0, n, K are constants. Some authors [35] are losing faith in *Norton's law* since they feel that their new *θ Projection Concept* provides a far more comprehensive description of creep behaviour for design. In this new approach, normal creep curves are envisaged as the sum of a decaying primary and an accelerating tertiary stage, i.e., the *secondary stage* is merely the period of ostensibly constant rate observed when the decay in the creep rate during the primary stage is offset by the gradual acceleration caused by tertiary processes.

This concept neglects the secondary component and may be valid for some special materials, e.g. $\frac{1}{2}$ Cr $\frac{1}{2}$ Mo $\frac{1}{4}$ V, as has been discussed in detail in [35]. However, an extended secondary creep stage can be observed for many materials. Thus, in spite of the discussion in [35], it is very important that *Norton's* law be generalized to multi-axial states of stress. This can be achieved by using an *isotropic tensor function* (2.3). The main problem then is to determine the scalar coefficients φ_0, φ_1, φ_2 in (2.3) as functions of the experimental data (K, n) in (3.1b) and of the integrity basis, the elements of which are the irreducible invariants of the *Cauchy stress tensor* $\underset{\sim}{\sigma}$.

Alternatively, we can represent the constitutive equation in the form

$$d_{ij} = f_{ij}(\underset{\sim}{\sigma}') = \varphi_0\delta_{ij} + \varphi_1\sigma'_{ij} + \varphi_2\sigma'^{(2)}_{ij} \,, \tag{3.2}$$

where $\sigma'_{ij} := \sigma_{ij} - \sigma_{kk}\delta_{ij}/3$ are the cartesian components of the stress deviator $\underset{\sim}{\sigma}'$. Note that the scalar coefficients φ_0, φ_1, φ_2 in (3.2) differ from the corresponding scalars in (2.3). In the special case of *incompressible* behaviour ($d_{kk} = 0$) we derive from (3.2):

$$3\varphi_0 + \varphi_2\sigma'^{(2)}_{kk} = 0 \Rightarrow \varphi_0 = - 2\varphi_2 J'_2/3 \tag{3.3}$$

with the quadratic invariant $J'_2 = \sigma'_{ik}\sigma'_{ki}/2$ of the stress deviator, so that the constitutive equation (3.2) is reduced to the form

$$d_{ij} = \varphi_1\sigma'_{ij} + \varphi_2\sigma''_{ij} \tag{3.4}$$

containing the traceless tensors

$$\sigma'_{ij} = \partial J'_2/\partial\sigma_{ij} \quad \text{and} \quad \sigma''_{ij} = \partial J'_3/\partial\sigma_{ij} \tag{3.5a,b}$$

with the cubic invariant $J'_3 = \sigma'_{ij}\sigma'_{jk}\sigma'_{ki}/3$ of the stress deviator.

Tensorial interpolation methods have been developed by the author in an earlier paper [36] in order to determine the scalar coefficients φ_0, φ_1, φ_2 in tensor functions such as (3.2). These methods are very useful for engineering applications as has been illustrated in detail in [2,24,37], which discuss some examples. One of these examples is the tensorial generalization of *Norton's power law* (3.1a,b) by way of the tensor function (3.2) in which the scalar coefficients have been determined by using a tensorial interpolation method:

$$\varphi_0 = \frac{1}{9}(1 - 8\upsilon + 6\upsilon n)K\sigma^n \,, \tag{3.6a}$$

$$\varphi_1 = \frac{2}{3}(1 + \upsilon + \frac{3}{2}\upsilon n)K\sigma^{n-1} \,, \tag{3.6b}$$

$$\varphi_2 = (1 + \upsilon - 3\upsilon n)K\sigma^{n-2} \,, \tag{3.6c}$$

where K,n are the experimental data taken from (3.1b), while υ is the transverse contraction ratio.

Assuming incompressibility (3.4) and neglecting tensorial nonlinearity $\{\varphi_2 = 0 \Rightarrow \varphi_0 = 0$ and $\varphi_1 = (1+\upsilon)K\sigma^{n-1}\}$, then starting from (3.2) we arrive at the simplified constistutive equation

$$d_{ij} = 3K\sigma^{n-1}\,\sigma'_{ij}/2\ , \tag{3.7a}$$

if we assume that $\upsilon = 1/2$ for isochoric behaviour. The result (3.7a) is identical to the constitutive equation proposed by *Leckie* and *Hayhurst* [38]. If we insert the *v. Mises equivalent stress* $\sigma = \sqrt{3J_2'}$ into (3.7a), we arrive at the constitutive equation

$$d_{ij} = 3K(3J_2')^{(n-1)/2}\,\sigma'_{ij}/2 \tag{3.7b}$$

used by *Odquist* and *Hult* [39].

The equivalent stress σ in (3.6a,b,c) can be determined as a function of the stress invariants by applying the *hypothesis of the equivalent dissipation rate* $\dot{D} := \sigma_{ij}d_{ji} = \sigma d$, where $\dot{D}$ is defined as the *rate of dissipation of creep energy*. This problem has been expounded in more detail in [40]. The result is:

$$\sigma^3 + A\sigma^2 + B\sigma + C = 0\ , \tag{3.8}$$

where the abbreviations

$$A = -(1 - 8\upsilon + 6\upsilon n)J_1/9\ , \tag{3.9a}$$

$$B = -4(1 + \upsilon + 3\upsilon n/2)J_2'/3\ , \tag{3.9b}$$

$$C = -(1 + \upsilon - 3\upsilon n)(3J_3' + 2J_1J_2'/3) \tag{3.9c}$$

have been used. Thus, the scalar coefficients (3.6a,b,c) are functions of the irreducible invariants

$$J_1 = \sigma_{kk} , \quad J_2' = \sigma_{ik}'\sigma_{ki}'/2 , \quad J_3' = \sigma_{ij}'\sigma_{jk}'\sigma_{ki}'/3 \tag{3.10a,b,c}$$

and of experimental data (K,n,υ):

$$\varphi_\alpha = \varphi_\alpha(J_1, J_2', J_3' ; K,n,\upsilon) , \qquad \alpha = 0,1,2 . \tag{3.11}$$

This statement is compatible with the representation theory of tensor-valued functions (3.2) in which the coefficients φ_α, $\alpha = 0,1,2$, are scalar-valued functions of the integrity basis (3.10).

In the case of incompressible behaviour ($\upsilon = 1/2$), the first invariant J_1 has no invluence. The cubic equation (3.8) then takes the reduced form

$$\sigma^3 + B^*\sigma + C^* = 0 \tag{3.12}$$

with the abbreviations $B^* = -(2 + n)J_2'$ and $C^* = 9(n - 1)J_3'/2$ depending on the irreducible invariants (3.10b,c) of the stress deviator.

Further examples are discussed in [40] and in [24,37] the damage state in the tertiary creep stage has also been included.

4. Tertiary Creep Behaviour

The tertiary creep phase is accompanied by the formation of microscopic cracks on the grain boundaries, so that damage-accumulation occurs.

In a uni-axial tension specimen, material deterioration can be described by introducing an additional variable ω or, alternatively, $\psi = 1 - \omega$ into constitutive equations, i.e. the strain rate d can be expressed as $d = f(\sigma,\omega)$, or $d = f(\sigma,\psi)$, where σ is the uni-axial stress [7,41,42]. The material parameters ω and ψ describe the current *damage* state and the *continuity* of the material respectively. The parameter of *continuity*, ψ, represents that fraction of the cross-sectional area which is not occupied by either voids or internal fissures. The *net-stress* acting over the cross-section of the uni-axial specimen is then $\hat{\sigma} = \sigma/\psi$. When $\psi = 1$ the material is in its virgin undamaged state, and when $\psi = 0$, the material can no longer sustain any load. In the latter case the constitutive equation would be required in order to approach an infinite strain rate. Furthermore, it is assumed that the damage rate $\dot{\omega}$, or alternatively the rate of continuity change $\dot{\psi}$, is also governed by the uni-axial stress and by the current state of continuity, i.e. $\dot{\omega} = g(\sigma,\omega)$, or $\dot{\psi} = - g(\sigma,\omega)$.

The forms of the functions f and g have been discussed in detail by many scientists, for instance by *Rabotnov* [7], *Chrzanowski* [43], *Leckie* and *Ponter* [44], *Leckie* and *Hayhurst* [38], *Goel* [45], *Hayhurst et al.* [46]. Often the forms

$$d/d_0 = (\sigma/\sigma_0)^n/(1-\omega)^m \ , \qquad \dot{\omega}/\dot{\omega}_0 = (\sigma/\sigma_0)^\nu/(1-\omega)^\mu \qquad (4.1a,b)$$

are used, where $n \geq \nu$, m, μ, d_0, $\dot{\omega}_0$, and σ_0 are constants. The undamaged case ($\omega = 0$) hereby leads to *Norton's power law* (3.1), which is assumed to be valid for the secondary creep stage, while the creep rate d approaches infinity as ω approaches 1.

Integrating the *kinetic equation* (4.1b) under the initial condition $\omega(t = 0) = 0$ and inserting the result into (4.1a), one arrives at the relation [47]

$$\frac{d}{d_0} = \left(\frac{\sigma}{\sigma_0}\right)^n \left[1 - k \left(\frac{\sigma}{\sigma_0}\right)^\nu \dot{\omega}_0 t\right]^{-m/k} \quad \text{with } k = 1 + \mu \ . \qquad (4.2)$$

A further integration leads to the tertiary creep strain

$$\varepsilon_t = a[1 - (1 - bt)^{1-c}]/b(1 - c) \ , \qquad (4.3)$$

if we take the initial condition $\varepsilon_t(0) = 0$. The abbreviations

$$a = d_0(\sigma/\sigma_0)^n \ , \qquad b = k(\sigma/\sigma_0)^\nu \dot{\omega}_0 \ , \qquad c = m/k \qquad (4.4a,b,c)$$

have been introduced in (4.3).

Since creep rupture is characterised by $\omega = 1$ or $d \to \infty$, we can immediately find the time to rupture from (4.2):

$$t_r = 1/k(\sigma/\sigma_0)^{\upsilon}\,\dot{\omega}_0 \ . \tag{4.5}$$

If, for convenience sake, the constants m and μ are taken to be equal to the parameters n and υ respectively, the relations (4.1a,b) can then be simplified to

$$d = K\hat{\sigma}^{n} \ , \qquad \dot{\omega} = L\hat{\sigma}^{\upsilon} \ , \tag{4.6a,b}$$

where $\hat{\sigma} = \sigma/(1 - \omega)$ is interpreted as the *net-stress* acting over the current cross-sectional area of a uni-axial specimen. Thus, the simplification (4.6a,b) can be called the *net-stress concept*. From (4.1a,b) and (4.6a,b) we read: $K = d_0/\sigma_0^n$ and $L = \dot{\omega}_0/\sigma_0^{\nu}$. One immediately arrives at the first relation (4.6a) from *Norton's law* (3.1b) if we replace the nominal stress σ by the net-stress $\hat{\sigma}$. Furthermore, a *tensorial generalization* of (4.6a,b) can be achieved in a very similar manner to that described in chapter 3, where the *Norton creep law* has been generalized.

Because of the simplifications $m = n$ and $\mu = \upsilon$, which lead to (4.6a,b), the creep rupture time (4.5) takes the form

$$t_r = 1/(1 + \upsilon)L\sigma^{\upsilon} \ , \tag{4.7}$$

where the nominal stress σ can be interpreted as the actual stress at the beginning of the tertiary creep stage ($\omega = 0$), i.e., considering *Norton's law* (3.1b) and starting from (4.7) we arrive at the formula

$$d_{min}^{\upsilon/n}\, t_r = K^{\upsilon/n}/L(1 + \upsilon) \ . \tag{4.8a}$$

The quantity d_{min} in (4.8a) is the steady-state or minimum creep-rate. Assuming $\upsilon = n$, one arrives at the *Monkman-Grant* [48] relationship

$$d_{min}\, t_r = K/L(1 + n) = \text{const.} \tag{4.8b}$$

Thus, the *net-stress concept* (4.6a,b) with identical exponents, $\upsilon = n$, is compatible with the model of *Monkman-Grant*. The justification of this model has, for example, been analysed in [49,50,51,52].

Under certain conditions, the grain boundaries in polycrystals slide during creep deformation. *Edward* and *Ashby* illustrate in [50] that this sliding can be accommodated in various ways: elastically, by diffusion, or by non-uniform creep or plastic flow of the grains themselves. In other cases, holes or cracks appear at the grain boundaries and grow until they link, leading to an *intergranular creep fracture*. When fracture is of this sort, the *Monkman-Grant rule* can be approximatively confirmed. Often, however, the *Monkman-Grant product*, $d_{min}t_r$, is proportional to the *strain-to-rupture*, ε_r, as has been observed in [49,52] or in our own experiments.

Because of its microscopic nature, *damage* generally has an *anisotropic* character even if the material was originally isotropic. The fissure orientation and length cause anisotropic macroscopic behaviour. Therefore, damage in an isotropic or anisotropic material which is in a state of multi-axial stress can only be described in a tensorial form.

When generalizing the uni-axial concept (4.1a,b) or (4.6a,b), constitutive equations and anisotropic growth equations are expressed as the tensor-valued functions

$$d_{ij} = f_{ij}(\underset{\sim}{\sigma},\underset{\sim}{\omega}) \ , \qquad \overset{\circ}{\omega}_{ij} = g_{ij}(\underset{\sim}{\sigma},\underset{\sim}{\omega}) \ , \tag{4.9a,b}$$

respectively, where $(\overset{\circ}{\ })$ denotes the *Jaumann derivative*, $\underset{\sim}{\sigma}$ is the *Cauchy stress tensor*, and $\underset{\sim}{\omega}$ represents an appropriately defined *damage tensor*.

Damage tensors are constructed, for instance, in [11,23]. Furthermore, we also refer to the work of *Murakami* and *Ohno* [15]. They assumed that damage accumulating in the process of creep can be expressed through a symmetric tensor of rank two.

Rabotnov [53] has also introduced a symmetric second-order tensor of damage and defined a symmetric *net-stress tensor* $\hat{\underset{\sim}{\sigma}}$ by way of a linear transformation

$$\sigma_{ij} = \Omega_{ijkl}\hat{\sigma}_{kl} \ , \tag{4.10}$$

where the fourth-order tensor $\underset{\sim}{\Omega}$ is assumed to be symmetric.

However, in [12] it has been pointed out in more detail that the fourth-order tensor in (4.10) is only symmetric with reference to the first index pair ij, but not to the second, kl. Thus, the net-stress tensor is *not* symmetric in a case of anisotropic damage. The net-stress tensor can be decomposed into a symmetric part and into an antisymmetric one, where only the symmetric part is equal to the net-stress tensor indroduced by *Robotnov* [53], as shown in [12].

Starting from a third-order skew-symmetric tensor of continuity to represent area vectors (*bivectors*) of *Cauchy's tetrahedron* in a damaged state, one finally arrives at a second-order damage tensor which has the *diagonal form* with respect to the rectangular *Cartesian* coordinate system under consideration, as has been pointed out in detail in [11].

The symmetric tensor-valued functions (4.9a,b) are valid for an *isotropic material* in an *anisotropic damage state*. Furthermore, one must differentiate between *anisotropic damage growth* and the *initial anisotropy* resulting from a forming process, for instance, rolling. Constitutive equations and anisotropic damage growth equations are then represented by expressions such as

$$d_{ij} = f_{ij}(\underset{\sim}{\sigma},\underset{\sim}{\omega};\underset{\sim}{A}) \quad \text{and} \quad \mathring{\omega}_{ij} = g_{ij}(\underset{\sim}{\sigma},\underset{\sim}{\omega};\underset{\sim}{A}) \tag{4.11a,b}$$

respectively, where $\underset{\sim}{A}$ is a fourth-order constitutive tensor with components A_{pqrs} characterizing the anisotropy from, for example rolling, i.e. the anisotropy of the material in its undamaged state.

A general representation of (4.11a), and similarly (4.11b) is given through a linear combination

$$\underset{\sim}{d} = \sum_{\alpha} \varphi_{\alpha}\, {}^{\alpha}\underset{\sim}{G} \quad \text{or} \quad d_{ij} = \sum_{\alpha} \varphi_{\alpha}\, {}^{\alpha}G_{ij} \,, \tag{4.12}$$

where the $\underset{\sim}{G}$'s are symmetric *tensor generators* of rank two involving the argument tensors $\underset{\sim}{\sigma}$, $\underset{\sim}{\omega}$, $\underset{\sim}{A}$. Some possible methods in arriving at such tensor generators have been discussed in [1,11,12], for instance. The coefficients φ_{α} in (4.12) are scalar-valued functions of the integrity basis associated with the representation (4.12). They must also contain experimental data measured in uni-axial creep tests. The main problems are: to construct an irreducible set of tensor generators and to determine the scalar coefficients involving experimental data.

A further aim is to represent the constitutive equation (4.12) in *canonical form*

$$d_{ij} = {}^0H_{ijkl}\delta_{kl} + {}^1H_{ijkl}\sigma_{kl} + {}^2H_{ijkl}\sigma_{kl}^{(2)}, \qquad (4.13)$$

where ${}^0H_{ijkl},\ldots,{}^2H_{ijkl}$ are the cartesian components of fourth-order tensor-valued functions ${}^0\underset{\sim}{H},\ldots,{}^2\underset{\sim}{H}$ depending on the *damage tensor* $\underset{\sim}{\omega}$ and the *anisotropy tensor* $\underset{\sim}{A}$. The canonical form (4.13) is a representation in three terms, which are the contributions of zero, first, and second orders in the stress tensor $\underset{\sim}{\sigma}$ influenced by the functions ${}^0\underset{\sim}{H}$, ${}^1\underset{\sim}{H}$, and ${}^2\underset{\sim}{H}$, respectively.

It may be impossible to find a canonical form (4.13) for all types of anisotropy. However, for the most important kinds of anisotropy, namely *transversely isotropic* and *orthotropic behaviour*, the constitutive equation (4.12) can be expressed in the *canonical form* (4.13) as has been illustrated in [1,10,12], for instance.

When formulating constitutive equations such as (4.11a) one has to take the following into account: the undamaged case ($\underset{\sim}{\omega} \to \underset{\sim}{0}$) immediately leads to the secondary creep stage, while the *rate-of-deformation tensor* $\underset{\sim}{d}$ approaches infinity as $\underset{\sim}{\omega}$ approaches $\underset{\sim}{\delta}$. In view of polynomial representations of constitutive equations it is convenient to use the tensor

$$D_{ij} := (\delta_{ij} - \omega_{ij})^{(-1)} = \psi_{ij}^{(-1)} \qquad (4.14)$$

as an argument tensor instead of the tensorial damage variable $\underset{\sim}{\omega}$. Thus, expressions such as $d_{ij} = f_{ij}(\underset{\sim}{\sigma},\underset{\sim}{D},\underset{\sim}{A})$ must be taken into consideration. Some possible representations of such functions have been discussed in detail in [1,11,12]. These representations may be too complicated for practical use. Therefore, we have to look for simplified representations. As an example, we suggest the following *simplified* constitutive equation

$$d_{ij} = f_{ij}(\underset{\sim}{t},\underset{\sim}{\tau}) = \frac{1}{2} \sum_{\upsilon,\mu=0}^{2} \Phi_{(\upsilon,\mu)} [t_{ik}^{(\upsilon)} \tau_{kj}^{(\mu)} + \tau_{ik}^{(\mu)} t_{kj}^{(\upsilon)}] , \qquad (4.15)$$

where the linear transformations

$$t_{ij} = D_{ijpq}\sigma_{pq} = t_{ji} \quad \text{with} \quad D_{ijpq} := \frac{1}{2} (D_{ip}D_{jq} + D_{iq}D_{jp}) \qquad (4.16)$$

and

$$\tau_{ij} = A_{ijpq}\sigma_{pq} = \tau_{ji} \qquad (4.17)$$

have been introduced [11]. We can see that the first linear transformation (4.16) considers the *anisotropic damage state,* while the second one expresses the *initial anisotropy* of the material. The main problem now is to determine the scalar coefficients $\Phi_{(\upsilon,\mu)}$ in (4.15) as functions of the integrity basis and experimental data. This can be achieved by using the *interpolation method* developed in [1,36] and applied in [1,24,37].

The tensor $\underset{\sim}{t}$ defined in (4.16) is called the *pseudo-net-stress* tensor. This tensor is symmetric even in the anisotropic damage case. The non-symmetric property of the *actual net-stress tensor* $\hat{\underset{\sim}{\sigma}}$ in (4.10) is a disadvantage, and this tensor is awkward to use in constitutive equations.

In recent years much effort has been devoted to the elaboration of *evolutional equations* such as (4.11b). The results are discussed in [54]. For the sake of simplicity, we can once again use the linear transformations (4.16) and (4.17), i.e., as in (4.15), we have to represent the tensor-valued function $\mathring{D}_{ij} = g_{ij}(\underset{\sim}{t},\underset{\sim}{\tau})$, where $\mathring{D}_{ij}$ is the *Jaumann derivative* of the tensor (4.14).

5. Comparisons with our own Experiments

As has been pointed out in more detail by *Avula* [55], "the validity of a model should not be judged by mathematical rationality alone; nor should it be judged purely by empirical validation at the cost of mathematical and scientific principles. A combination of rationality and empiricism (logic and pragmatism) should be used in the validation".

"Experimental observations and measurements are generally accepted to constitute the backbone of physical sciences and engineering because of the physical insight they offer to the scientist for formulating the theory. The concepts that are developed from observations are used as guides for the design of new eyperiments, which in turn are used for validation of the theory. Thus, experiments and theory have a hand-in-hand relationship" [55].

However, it must be noted that experimental results can differ greatly from the reality just like a bad mathematical model [56].

Apart from experimental data taken from literature, the author and his coworkers have also taken results from their own experimental measurements in order to examine the validity of a mathematical model, for instance, experiments by *Waniewski* [32,57], *Betten* and *Waniewski* [33,34,58,59], *Betten, Breitbach* and *Waniewski* [60], to name but a few. Further experiments are in preparation [47].

References

1. Betten, J.: Tensorrechnung für Ingenieure. Stuttgart: B.G. Teubner-Verlag 1987.

2. Betten, J.: Recent Advances in Mathematical Modelling of Materials Behaviour. Presented at the Seventh Intern. Conf. on Mathematical and Computer Modelling, Chicago 1989, published in the proceedings.

3. Boehler, J.P. (ed.): Applications of Tensor Functions in Solid Mechanics. Wien, New York: Springer-Verlag 1987.

4. Betten, J.: Elastizitäts- und Plastizitätslehre. Braunschweig, Wiesbaden: Vieweg-Verlag 1985, 2nd edition 1986.

5. Betten, J.; Borrmann, M.; Butter, T.: Materialgleichungen zur Beschreibung des primären Kriechverhaltens innendruckbeanspruchter Zylinderschalen aus isotropem Werkstoff. Ingenieur-Archiv 60 (1989) 99-109.

6. Betten, J.; Butters, T.: Rotationssymmetrisches Kriechbeulen dünnwandiger Kreiszylinderschalen im primären Kriechbereich. Forschung im Ingenieurwesen 56 (1990) 84-89.

7. Rabotnov, Yu.N.: Creep Problems in Structural Members. Amsterdam, London: North-Holland 1969.

8. Betten, J.: Zur Verallgemeinerung der Invariantentheorie in der Kriechmechanik. Rheol. Acta 14 (1975) 715-720.

9. Betten, J.: Creep Theory of Anisotropic Solids. Journal of Rheology 25 (1981) 565-581, presented at the Golden Jubilee Meeting of The Society of Rheology, Boston, MA, 1979.

10. Betten, J.: The Classical Plastic Potential Theory in Comparison with the Tensor Function Theory. Engineering Fracture Mechanics 21 (1985) 641-652, presented at the Int. Symposium "Plasticity Today", Udine 1983.

11. Betten, J.: Damage Tensors in Continuum Mechanics. Journal de Mécanique théorique et appliquée 2 (1983) 13-32, presented at Euromech Colloquium 147 on Damage Mechanics, Paris-VI, Cachan, 1981.

12. Betten, J.: Net-stress Analysis in Creep Mechanics. Ingenieur Archiv 52 (1982) 405-419, presented at the "Second German-Polish Symposium on Inelastic Solids and Structures", Bad Honnef 1981.

13. Bodner, S.R.; Hashin, Z. (eds.): Mechanics of Damage and Fatigue. New York ... Toronto: Pergamon Press 1986.

14. Krajcinovic, D.; Lemaitre, J. (eds.): Continuum Damage Mechanics. Wien, New York: Springer-Verlag 1987.

15. Murakami, S.; Ohno, N.: A Continuum Theory of Creep and Creep Damage, in: Ponter, A.R.S.; Hayhurst, D.R. (eds.) "Creep in Structures 1980". Berlin, Heidelberg, New York: Springer-Verlag 1981, 422-444.

16. Murakami, S.: Notion of Continuum Damage Mechanics and its Application to Anisotropic Creep Damage Theory. J. of Engin. Materials and Technology 105 (1983) 99-105.

17. Murakami, S.; Sawczuk, A.: A unified Approach to Constitutive Equations of Inelasticity based on Tensor Function Representations. Nucl. Engin. and Design. 65 (1981) 33-47.

18. Murakami, S.; Sanomura, Y.; Saitoh, K.: Formulation of Cross-Hardening in Creep and its Effect on the Creep Damage Process of Copper. J. of Engin. Materials and Technology 108 (1986) 167-173.

19. Murakami, S.: Progress of Continuum Damage Mechanics. JSME Intern. Journal 30 (1987) 701-710.

20. Chaboche, J.L.: Anisotropic Creep Damage in the Framework of Continuum Damage Mechanics. Nucl. Engng. Des. 79 (1984) 304-319.

21. Chrzanowski, M.: Use of the Damage Concept in Describing Creep-Fatigue Interaction under Prescribed Stress. Int. J. mech. Sci. 18 (1976) 69-73.

22. Krajcinovic, D.: Constitutive Equations for Damaging Materials. J. Appl. Mech. 50 (1983) 355-360.

23. Betten, J.: Damage Tensors and Tensors of Continuity in Stress Analysis. Z. Naturforschung 38a (1983) 1383-1390.

24. Betten, J.: Applications of Tensor Functions to the Formulation of Constitutive Equations Involving Damage and Initial Anisotropy. Engng. Frac. Mech. 25 (1986) 573-584, presented at the IUTAM Symposium on Mechanics of Damage and Fatigue, Haifa and Tel Aviv, Israel, 1985.

25. Onat, E.T.: Representation of Mechanical Behavior in the Presence of Internal Damage. Engng. Frc. Mech. 25 (1986) 605-614, presented at the IUTAM Symposium on Mechanics of Damage and Fatigue, Haifa and Tel Aviv, Israel, 1985.

26. Litewka, A.; Morzyńska, J.: Theoretical and Experimental Study of Fracture for a Damaged Solid. Res Mechanica 27 (1989) 259-272.

27. Litewka, A.; Hult, J.: One Parameter CDM Model for Creep Rupture Prediction. Eur. J. Mech., A / Solids, 8 (1989) 185-200.

28. Betten, J.: Representation of Constitutive Equations in Creep Mechanics of Isotropic and Anisotropic Materials, in: Ponter, A.R.S.; Hayhurst, D.R. (eds.) "Creep in Structures 1980". Berlin, Heidelberg, New York: Springer-Verlag 1981, 179-201.

29. Rice, J.R.: On the Structure of Stress-Strain Relations for Time-Dependent Plastic Deformations in Metals. Trans. ASME, J. Appl. Mech. 37 (1970) 728-737.

30. Sawczuk, A.; Anisimowicz, M.: Tensor Functions Approach to Creep Laws after Prestrain, in: Ponter, A.R.S.; Hayhurst, D.R. (eds.) "Creep in Structures 1980". Berlin, Heidelberg, New York: Springer-Verlag 1981, 220-232.

31. Sawczuk, A.; Trampczyński, W.A.: A Theory of Anisotropic Creep after Plastic Pre-Straining. Int. J. Mech. Sci. 24 (1982) 647-653.

32. Waniewski, M.: A simple law of steady-state creep for material with anisotropy induced by plastic prestraining. Ing.-Archiv 55 (1985) 368-375.

33. Betten, J.; Waniewski, M.: Einfluß der plastischen Anisotropie auf das sekundäre Kriechverhalten inkompressibler Werkstoffe. Rheol. Acta 25 (1986) 166-174.

34. Betten, J.; Waniewski, M.: Multiaxial creep behaviour of anisotropic materials. Arch. Mech. 41 (1989) 679-695, presented at the Fourth Polish-German Symposium on Mechanics of Inelastic Solids and Structures, Mogilany, Poland, 1987.

35. Brown, S.G.R.; Evans, R.W.; Wilshire, B.: Exponential Descriptions of normal Creep Curves. Scripta Metallurgica 20 (1986) 855-860.

36. Betten, J.: Interpolation Methods for Tensor Functions, in: Avula, J.R. et al. (eds.) Mathematical Modelling in Science and Technology. New York: Pergamon Press 1984, 52-57.

37. Betten, J.: Generalization of nonlinear material laws found in experiments to multi-axial states of stress. Eur. J. Mech., A / Solids, 8 (1989) 325-339, presented at Euromech Colloquium 244 on Experimental Analysis of Nonlinear Problems in Solid Mechanics, Poznan, Poland, 1988.

38. Leckie, F.A.; Hayhurst, D.R.: Constitutive Equations for Creep Rupture. Acta Metallurg. 25 (1977) 1059-1070.

39. Odquist, F.K.G.; Hult, J.: Kriechfestigkeit metallischer Werkstoffe. Berlin, Göttingen, Heidelberg: Springer-Verlag 1962.

40. Betten, J.: Beitrag zur tensoriellen Verallgemeinerung einachsiger Stoffgesetze. Z. angew. Math. Mech. (ZAMM) 66 (1986) 577-581.

41. Kachanov, L.M.: On the time to failure under creep conditions (in Russian). Izv. Akad. Nauk USSR Otd. Tekh. Nauk 8 (1958) 26-31.

42. Kachanov, L.M.: Introduction to Continuum Damage Mechanics. Dordrecht, Boston, Lancaster: Martinus Nijhoff Publishers 1986.

43. Chrzanowski, M.: The Description of Metallic Creep in the Light of Damage Hypothesis and Strain Hardening. Krakow: Diss. hab., Politechnika Krakowska 1973.

44. Leckie, F.A.; Ponter, A.R.S.: On the State Variable Description of Creeping Materials. Ing. Archiv 43 (1974) 158-167.

45. Goel, R.P.: On the Creep Rupture of a Tube and a Sphere. J. Appl. Mech., Trans. ASME 43 (1975) 625-629.

46. Hayhurst, D.R.; Trampczynski, W.A.; Leckie, F.A.: Creep Rupture and Non-Proportinal Loading. Acta Metall. 28 (1980) 1171-1183.

47. Betten, J.; El-Magd, E.; Meydanli, S.C.: Anisotropic damage groth under multi-axial stress (theory and experiments), in preparation.

48. Monkman, F.C.; Grant, N.J.: An Empirical Relationship between Rupture Life and Minimum Creep Rate in Creep-Rupture Tests. Proc. ASTM 56 (1956) 593-620.

49. Ilschner, B.: Hochtemperatur-Plastizität. Berlin, Heidelberg, New York: Springer-Verlag 1973.

50. Edward, G.H.; Ashby, M.F.: Intergranular Fracture during Power-Law Creep. Acta Metall. 27 (1979) 1505-1518.

51. Evans, H.E.: Mechanisms of Creep Rupture. London / New York: Elsevier Appl. Sci. Publishers 1984.

52. Riedel, H.: Fracture at High Temperatures. Berlin /.../ Tokyo: Springer-Verlag 1987.

53. Rabotnov, Y.N.: Creep Rupture, in: Hetenyi, M.; Vincenti, H. (eds.) Appl. Mech. Conf., Stanford University 1968, 342-349.

54. Betten, J.; Meydanli, S.C.: Materialgleichungen, DFG-Report 1990, not yet published.

55. Avula, X.J.R.: Mathematical Modelling. Encyclopedia of Physical Science and Technology 7 (1987) 719-728.

56. Betten, J.: Die Traglasttheorie der Statik als mathematisches Modell. Schweizerische Bauzeitung 91 (1973) 6-9.

57. Waniewski, M.: The influence of direction and value of plastic prestrain on steady-state creep rate using the combined isotropic-kinematic hardening rule. Rozpr. Inz. 23 (1984) 4.

58. Betten, J.; Waniewski, M.: Biaxial Tension Creep Test of Rolled Sheet-Metals. Arch. Mech. 43 (1991), in press, presented at III Sympozjum nt. Zagadnień Pelzania Materialow, Bialystok 1989.

59. Betten, J.; Waniewski, M.: Stress-Path Influence on Multiaxial Creep Behaviour due to Multiple Load Changes. Presented at the 4th Intern. Conf. on Creep and Fracture, Swansea 1990, published in the proceedings.

60. Betten, J.; Breitbach, G.; Waniewski, M.: Multiaxial Anisotropic Creep Behaviour of Rolled Sheet-Metals. Z. angew. Math. Mech. (ZAMM) 70 (1990) 371-379.

A Unified Mathematical Model for Plasticity and Creep Under Variable-Repeated Loading

D.A. GOKHFELD and O.S. SADAKOV

Cheliabinsk Politechnical Institute, USSR

Summary

In this paper emphasis is made on the concept of rheonomic nature of inelastic deformation of any type. It is shown that the both processes (creep and plasticity) can get a common descriptions without any inconsistency by the structural model employing (Masing, Besseling and others). Some experimental data concerning the problem are discussed.

Introduction

The inelastic strain is usually divided into time-independent (scleronomic) and time-dependent (rheonomic) parts. This components are determined independently using corresponding mathematical models of specific structures. Naturally, by this approach the interaction between the two processes is disregarded. Meanwhile, the experiments show that this interaction is often rather essential, in particular in cyclic loading conditions.

The necessity of separate plastic and creep analysis usually is grounded due to the existence of definite differences between the corresponding slip mechanisms. But any analogies in regularities of the both processes set us thinking about their unity. Besides, it is rather complicated to separate plastic and viscous deformations in experiments, especially at non-isothermal loading conditions.

The main problem consists in determination if a state equation could be suggested to reflect actual creep curves as well as stress-strain diagrams. Some attempts to solve the problem were not successful until now.

The structural model of media; general statements

In accordance with the model any volume element is supposed to be a set of subelements (SEs) having equal values of strain, temperature, elastic modulus and thermal expansion coefficient. The SEs are perfectly viscous, i.e. their inelastic strain rate depends upon individual stress and temperature only. As SEs have different rheological characteristics the inelastic strain and stress distributions within a body element are nonuniform and vary along with the deformation process.

The considered version of the structural model [1-3] concerns materials which have cyclically stable deformation behaviour initially or after a few preliminary loading cycles.

M. Zyczkowski (Ed.)
Creep in Structures
IUTAM Symposium Cracow/Poland 1990

It is supposed that the SE rheological functions (RFs) are mutually similar:

$$\dot{p}^{(k)} = \Phi_k(r^{(k)}, T) = \Phi(r^{(k)}/z_k, T) \ . \tag{1}$$

Here $\dot{p}^{(k)}$, $r^{(k)}$ are the SE inelastic strain rate and elastic strain; $k = 1, 2, \ldots, N$ - subelement number; z_k - identifying multipliers which have ensemble average $< z_k >= 1$.

$$\dot{p}_s = \Phi(r, T) \ , \tag{2}$$

where $\dot{p}_s$ is secondary creep rate.

The assumption (1) plays a governing role as it reduces the number of RFs. Its legitimacy for various alloys was confirmed by experimental verification.

Elastic strain r is used in relations (1), (2) instead of corresponding stress σ because by temporary elimination of the Young's modulus temperature dependence some important regularities can be revealed.

The identification of the model consists of two basic functions determination: the RF and the inhomogeneity function (IF) are to be find using the necessary test data. The latter function is connected with the shape of strain-stress diagram.

The both types of inelastic deformation can be reflected by the model depending on adopted shape of the RF. Its four types are shown in Fig. 1: case 3 which is typical for a

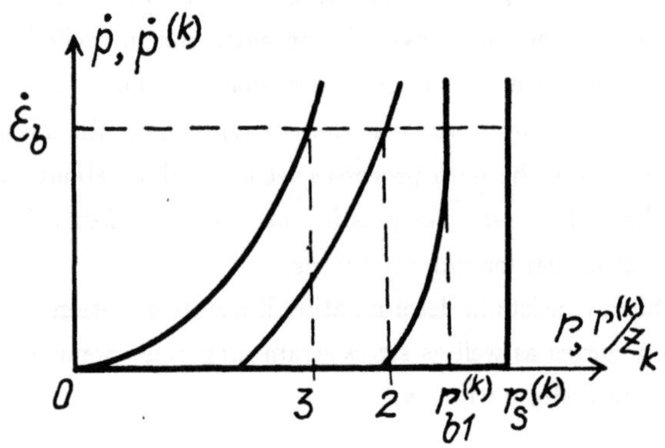

Figure 1: Types of rheological function

perfectly rheonomic material; analogous shape 2 reflecting the zone below the creep limit. The case 0 illustrates the degenerated RF:

$$\begin{aligned} \dot{p}_s &= \ 0 \quad \text{at } r < r_b \ , \\ \dot{p}_s &= \ \tfrac{0}{0} \quad \text{at } r = r_b \ . \end{aligned} \tag{3}$$

It reflects deformation behaviour of perfectly plastic medium. Shape 1 includes both types of inelastic regimes.

It must be stressed that at operating temperature levels the RFs of actual structural alloys have a specific form. They approach the polygonal line OAB as an asymptote.

Similarity principle (SP)

Under proportional loading conditions taking into account the mentioned character of the RFs the structural model relations can be reduced to a state equation formulated in macro-terms directly [1,3]. A new system of macroscopic parameters connected with current values r, ε and their magnitudes r_i, ε_i in the turning points of loading trajectory are introduced. At any time the last turning point remaining in the material "memory" is characterized by values r_ν, ε_ν. Then the new coordinates are

$$r_* = r - r_\nu \ , \quad \varepsilon_* = \varepsilon - \varepsilon_\nu \ ; \tag{4}$$

the following state parameters have to be introduced:

$$C = r_*/\varepsilon_* \tag{5}$$

and θ which is defined by the parametric equation

$$r_* = \theta f(\varepsilon_*/\theta) \ . \tag{6}$$

The latter one can be both positive and negative. θ can be easily determined by use of a proper geometrical construction. The function f is the basic stress-strain diagram ($r = f(\varepsilon)$). Finally, let us introduce another parameter

$$\kappa = \theta + \Sigma \ , \tag{7}$$

where $\Sigma = \sum_{i=1}^{\nu} \theta_i$ is constant at any loading stage, i.e. between the subsequent reversals. It is adopted that the values of r_ν, ε_ν, Σ at initial loading stage become zero.

The state equation is:

$$\dot{p} = \Phi(\kappa r_b, T)[1 - K(C)] \ , \tag{8}$$

where r_b is the maximum value of $f(\varepsilon)$, $K = df/d\varepsilon$ - tangent modulus for a point of the $f(\varepsilon)$ determined by a secant modulus C. Integration of the eq. (8) for constant values of $\dot{\varepsilon}$, T leads to

$$\kappa = \Phi^0(\dot{\varepsilon}, T)/r_b \ . \tag{9}$$

Φ^0 - function inverse to Φ with respect to the first argument. As it follows from the relations (6), (7), all the stress-strain diagrams for various constant $\dot{\varepsilon}$ and T are relatively similar to coordinate origin if the turning points are superposed. It can be shown that $\max r_*$ (6) tends to the value at which secondary creep rate is equal to the prescribed strain rate ($\dot{p}_s = \dot{\varepsilon}$).

Turning points emerge when reversals (sign changes) of either strain rate $\dot{\varepsilon}$ or secant modulus rate $\dot{C} = dC/dt$ occur. The latter case is connected with a rapid temperature fall (or strain rate increase).

Rheological functions of structural alloys

Three different methods of RF determination can be used [1-3]. The first one follows from its definition and needs a set of secondary creep curves. The second one is based on relation $\dot{p}_s = \dot{\varepsilon}$ at $r \to \kappa r_b$. This time a set of stress-strain diagrams at various strain rates is the necessary experimental information (at T=const). Eq. 5 opens the most effective way to solve the problem. It can be done by common utilization of stress-strain diagram and primary creep (or relaxation) curve.

As it was found, the structural alloys RF usually can be approximated by a polygonal line with a sufficient accuracy, if logarithmic (in some cases semilogarithmic) coordinates are used. Section I of the broken line (Fig. 2) which is characterized by maximum

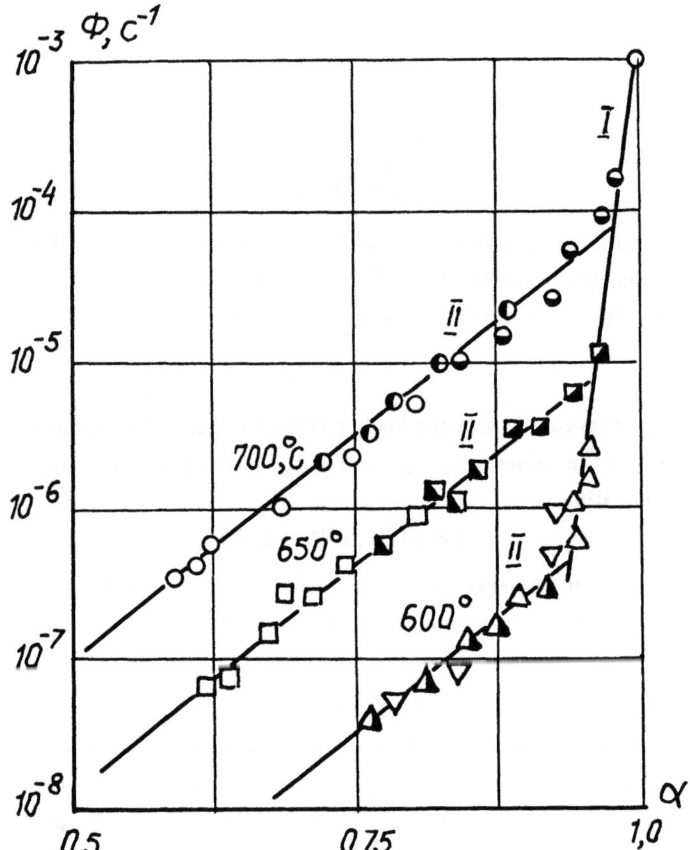

Figure 2: Rheological function of an austenic steel

steepness corresponds to a deformation mechanism which is close to the plastic one. Slope of the second section is essentially less, it corresponds to typical viscous deformation processes.

As it is seen, the slope of the broken line sections remain practically constant for various constant values of temperature if the relative x - coordinate is used.

On the structural model adequacy

Rather good correlation of the data obtained when principally different methods of RF determination are used (Fig. 2) can be considered as convincing experimental verification of the model. For the latter purpose various special tests were realized as well. The experiment illustrated by Fig. 3a, consists of two analogous tests. They include loading

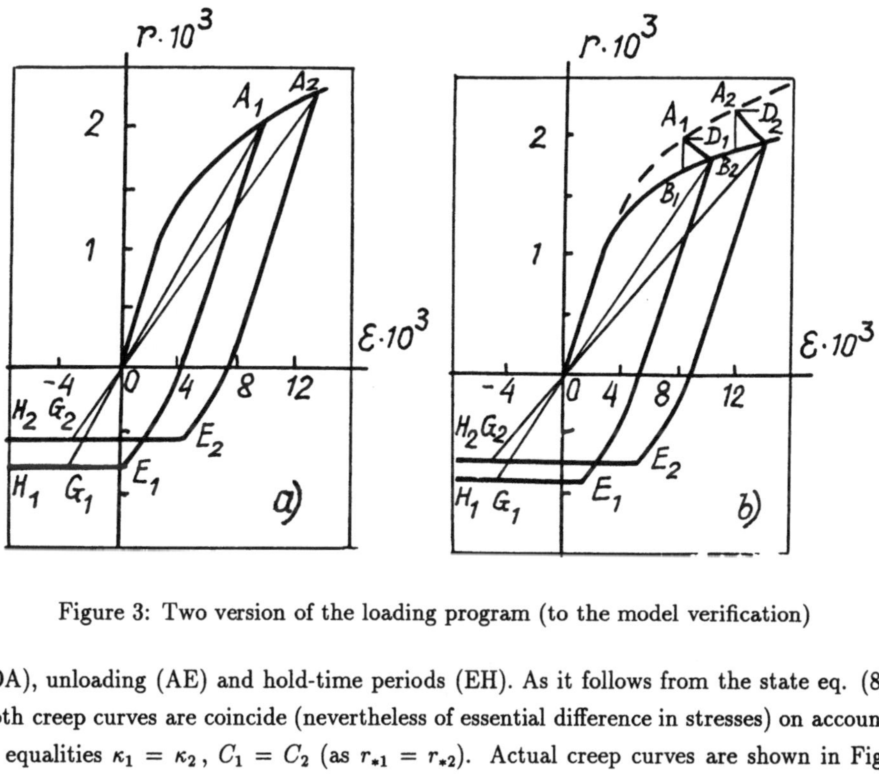

Figure 3: Two version of the loading program (to the model verification)

(OA), unloading (AE) and hold-time periods (EH). As it follows from the state eq. (8) both creep curves are coincide (nevertheless of essential difference in stresses) on account of equalities $\kappa_1 = \kappa_2$, $C_1 = C_2$ (as $r_{*1} = r_{*2}$). Actual creep curves are shown in Fig. 4. Coincidence is violated when any state point passes across the position G: at this moment the corresponding turning point A is excluded from the material memory (the general "memory rules" are not discussed here, see [1-3]).

An analogous effect is expected if the both turning points A are displaced to positions on another but common stress-strain curve (Fig. 3b). The hold-time periods AD in our test were continued until creep deformation practically stopped. The experiments (nickel alloy, T=700 C) confirm the prognosis related to the second loading programm as well.

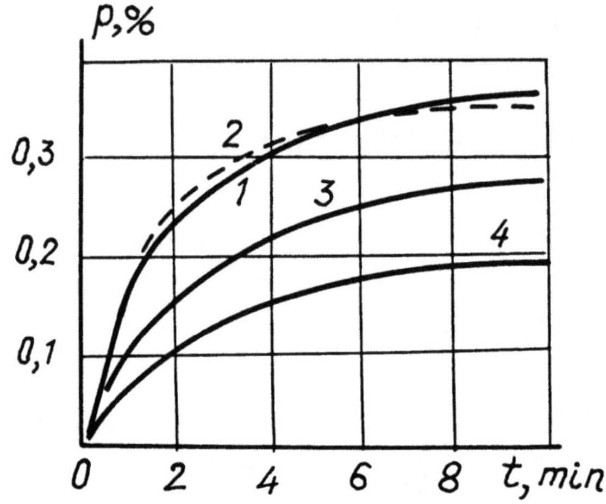

Figure 4: Creep curves: 1,2 - after the turning points; 3,4 - initial (without deformation reversals)

Conclusion

The structural model presents a common description of the rheonomic and scleronomic processes without any contradiction. The both processes can be considered as time-dependent ones. The existence of different deformation regimes can be taken into account, if necessary. In particular, such necessity appears at life-time determination due to low-cycle fatigue damage accumulation.

Thus the structural model is sufficiently "flexible" to meet various demands. In any case interaction between two inelastic regimes is taken into account automatically.

References

[1] Gokhfeld, D.A., Sadakov, O.S.: *Plasticity and creep in structural elements under repeated loading*, Mashinostrojenije, Moscow 1984 (in Russian)

[2] Gokhfeld, D.A., Sadakov, O.S., Martynenko, M.E.: *On history dependence of stress-strain diagrams and creep curves under variable repeated loading*, In: Trans. 5th Int. Conf. on SMiRT, W. Berlin, 1979, vol. L, L1/7, 1-8

[3] Gokhfeld, D.A., Kononov, K.M., Poroshin, V.B., Sadakov, O.S.: *Couple mathematical models for cyclic inelastic deformation and damage accumulation processes*, In: Trans. 10th Int. Conf. on SMiRT, Anaheim (USA), 1989, vol. L, 19-24

Anisotropic Creep Modeling for f.c.c. Single Crystals

A. BERTRAM, J. OLSCHEWSKI, M. ZELEWSKI, R. SIEVERT
Federal Institute for Materials Research and Testing (BAM), Berlin, FRG

Summary

The one-dimensional behavior of single crystal superalloys at high temperatures under constant and cyclic creep conditions is described by means of a 4-parameter rheological model based on linear viscoelasticity. Tertiary creep is taken into account by reducing the effective cross section by means of an additional damage parameter. Tensile creep tests have been used for the identification of the material constants by a non-linear optimization procedure. For the generalization to three dimensions, a complete tensor-representation of cubic material symmetry is given. It contains twelve (temperature dependent) material parameters. Some results by finite element analysis will be presented.

Introduction

Within the last decade, single crystal superalloys have become more and more important for blading within the hot section of gas turbines. Because of their f.c.c. crystal structure the material behavior is highly orientation dependent and, hence, must be treated as (cubic) anisotropic. At high temperatures the inelastic properties under constant and cyclic loads are of paramount interest for the design of blades. Creep experiments show us that, after a short phase of primary creep, the secondary or steady creep phase almost immediately turns over into the overlinear phase of tertiary creep. Because of the absence of grain boundaries in single crystals, the creep deformation can reach about 20% so that nonlinear geometrical effects are to be considered at that stage, although the physical nonlinearities due to creep damage are predominant.

One-dimensional theory

As starting point for our analysis we choose a four-parameter model from linear viscoelasticity which consists of two elasticities and two viscosities (Fig.1). The governing differential equation in the stress σ and the tension ε which completely describes the behavior of the model is

$$\ddot{\sigma} + a_1 \dot{\sigma} + a_2 \sigma = a_3 \ddot{\varepsilon} + a_4 \dot{\varepsilon} , \qquad (1)$$

where the four (positive) material constants are given by

$$a_1 = C / D + C / L + K / L , \quad a_2 = (C K) / (L D) , \quad a_3 = C + K , \quad a_4 = C K / D \qquad (2)$$

M. Zyczkowski (Ed.)
Creep in Structures
IUTAM Symposium Cracow/Poland 1990

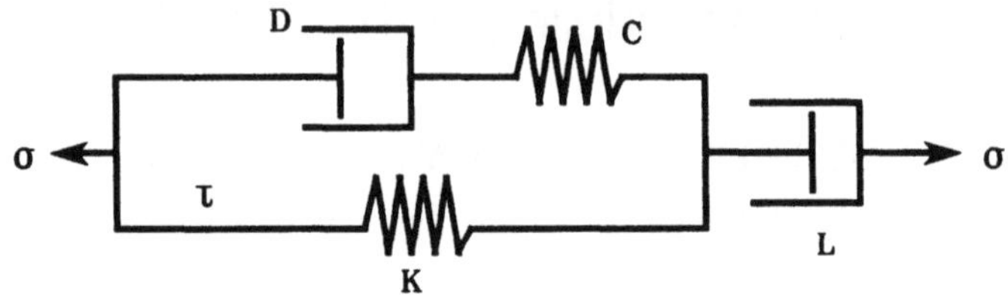

Fig. 1. Rheological model.

By introducing the stress in the lower branch of the model as an internal variable

$$\tau(t) := K \int_0^t \{ \varepsilon(s)^{\cdot} - \sigma(s) \, / \, L \} \, ds \tag{3}$$

we obtain the evolution equations for the model

$$\tau^{\cdot} = c_1 \sigma^{\cdot} + c_2 (\sigma - \tau)$$

$$\varepsilon^{\cdot} = c_3 \sigma^{\cdot} + c_4 \sigma + c_5 \tau \tag{4}$$

such that the constants c_i are determined by (5)

$$c_1 = K \, / \, (C + K) \, , \quad c_2 = c_1 C \, / \, D \, , \quad c_3 = 1 \, / \, (C + K) \, , \quad c_4 = c_3 a_1 \, , \quad c_5 = - c_3 C \, / \, D.$$

The two evolution equations (4) are appropiate for numerical integration and may serve for describing the primary and secondary creep behavior in constant and cyclic creep tests as shown in Figs. 2, 3. The asymptotic behavior as $t \to \infty$ is linear, being governed by the viscosity L. Hence the range of this simple model is limited to short creep intervals.

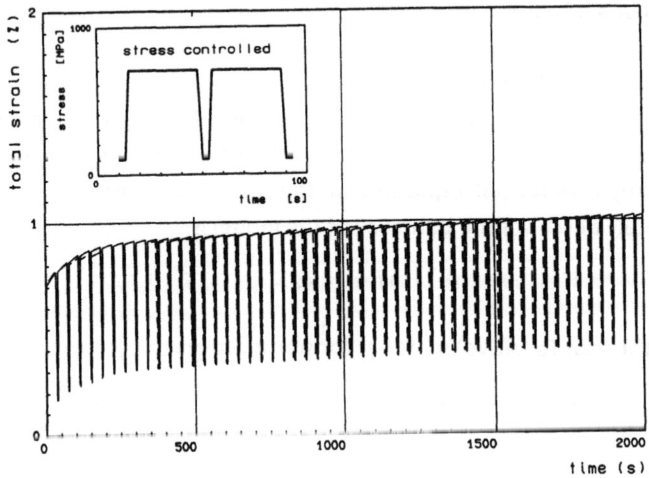

Fig. 2. Cyclic creep with hold times. (dashed line: experiment)

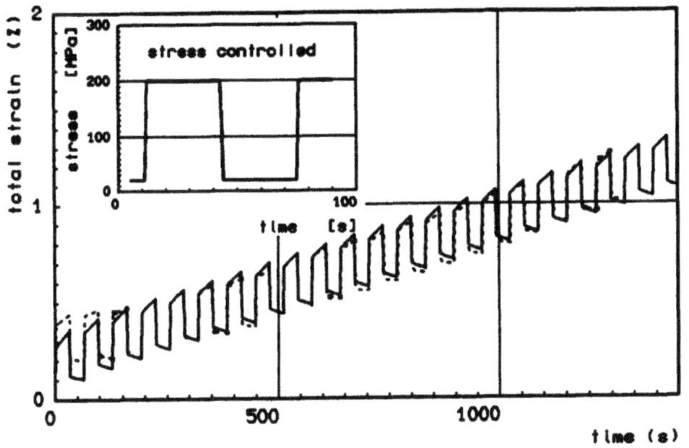

Fig. 3. Cyclic creep with hold times at 980 C.

Damage modeling

If we are interested in the long range creep behavior till final failure, we introduce a damage variable δ, which may be interpreted as the loss of effective cross section as suggested by KACHANOV and RABOTNOV (see KRAJCINOVIC / LEMAITRE [1])

$$A_e = A_o \, (1 - \delta), \tag{6}$$

where A_o is the apparent cross section area, A_e is the effective (or fictitious) area which initially equals A_o .However, this is by no means the only interpretation of δ . In analogy to (4) we assume the evolution function of the damage parameter

$$\dot{\delta} = c_6 \, \sigma + c_7 \, \dot{\sigma} + c_8 \, \tau + c_9 \, \delta \tag{7}$$

In Fig. 4 a set of experimental results of constant creep tests is plotted. The points of measurements are marked by symbols. All of these tests were performed at a temperature of 760 °C by an INSTRON tensile testing machine on cylindrical samples of the nickel base superalloy SRR99 with orientations near to [1 0 0] . Note the large scatter of experimental results at the same stress level, which is typical for creep tests.

As a first step, these data were used to determine the constants in (4) and (7) by a nonlinear optimization technique which minimizes the difference between measured and calculated deformation in each of these points for each test seperately. It turned out that the dominant term in (7) is the first one, so that a reduced form

$$\dot{\delta} = c_6 \, \sigma \tag{8}$$

gives similar results. The curves in Fig. 4 show these calculated creep curves which prove the capability of the model to describe creep behavior in the full range of these tests.

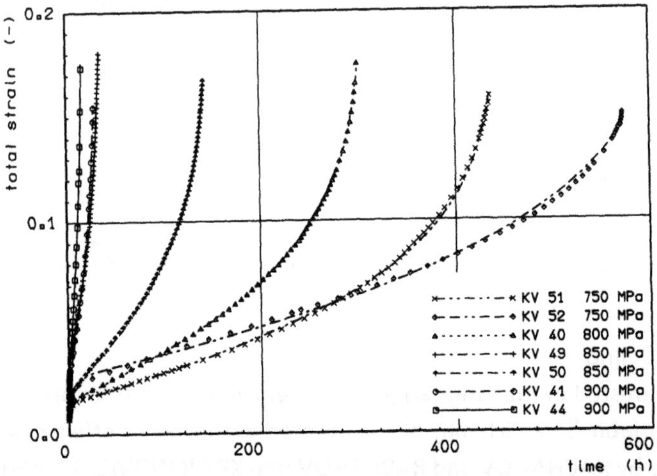

Fig. 4. Creep tests at 760 C. Experiment and calculation.

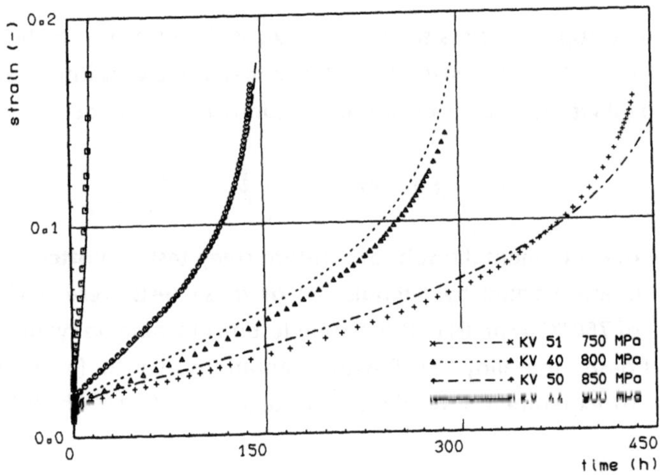

Fig. 5. Global adjustment of creep tests KV51, 40, 44, 50

In order to do a global identification, i. e. to give a single set of material constants that describe all of the creep tests in the given range, we suggest the following procedure. First, select one test as a master curve for the others (in our case we chose KV51) and identify C, K, D, L, and c_6 by fitting this single test. Second, introduce a linear time transformation by

$$t^* = t / \alpha \tag{9}$$

where t is the real time, t^* is an artificial or internal time, and α is a time scaling factor, that depends on the initial stress σ_0. In our case the linear form

$$\alpha = (k_1 \sigma_0 + k_2) \tag{10}$$

gives satisfying results. We now solve the evolution equations (4) with respect to the internal time t^* and afterwards transform the solution to the real time by means of (9) and (10). The results are given in Fig.5 together with a selected set of creep tests. The parameter list is given in Table 1.

C	K	D	L
GPa	GPa	GPa h	GPa h
50.52	49.64	57.69	6775.

c_6	k_1	k_2
$(\text{GPa h})^{-1}$	$(\text{GPa})^{-1}$	-
1.448	- 6.366	5.775

Table 1. Material constants for SRR99 at 760 C

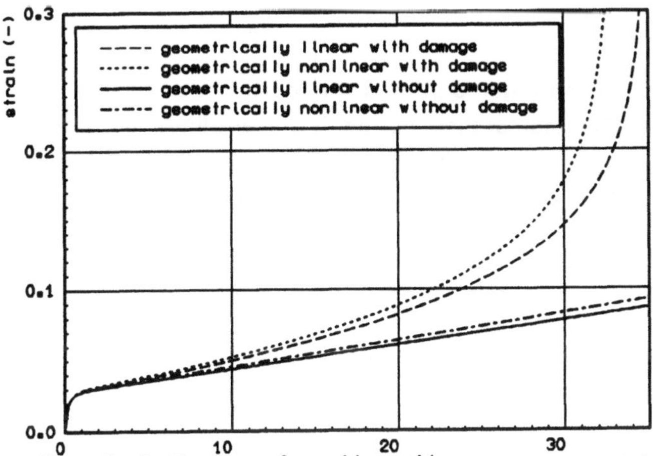

Fig. 6. Influence of nonlinearity.

In order to investigate the influence of geometrical nonlinear effects, a comparison between geometrical linear and nonlinear calculations (assuming isochoric creep deformations) is given in Fig. 6 with and without damage for a single set of material parameters. The difference between the two approaches is remarkable for larger deformations but small if compared with the effect of damage. Thus, we can conclude that the effect of geometrical linearization can be equally included by an appropriate choice of damage constants (as being done in Figs. 4 and 5).

Three-dimensional theory

Three dimensional creep laws are needed not only to predict the creep behavior in arbitrary orientations, but, moreover, for inhomogeneous tests such as on notched bars. In literature only few suggestions have been made to include anisotropy in three-dimensional creep laws (see, e. g., MURAKAMI / SAWCZUK [2] , BETTEN [3] and [4]).

We obtain the three-dimensional constitutive model by generalizing the one-dimensional one in analogy to the three-dimensional linear elastic law which is well known since about one century for this crystal class (see VOIGT [5]). In doing so, let the vectors $\mathbf{e}_i$ describe the three orthogonal crystal axes. We introduce three projections (fourth order tensors) by

$$\begin{aligned} \mathbf{P}_1 &= \tfrac{1}{3}\, \mathbf{I} \otimes \mathbf{I} \\ \mathbf{P}_2 &= \mathbb{I} - \textstyle\sum_i (\mathbf{e}_i \otimes \mathbf{e}_i \otimes \mathbf{e}_i \otimes \mathbf{e}_i) \\ \mathbf{P}_3 &= \mathbb{I} - \mathbf{P}_1 - \mathbf{P}_2 \end{aligned} \tag{11}$$

Here, $\mathbf{I}$ is the second order identity, $\mathbb{I}$ the fourth order identity, and $\otimes$ the tensor product. The advantage of these notions is that any linear transformation between symmetric tensors that fullfills cubic anisotropy is a linear combination of these three projections. For example, the cubic form of HOOKE´s law is

$$\mathbf{S} = \mathbf{C}[\mathbf{E}] \qquad \text{with} \qquad \mathbf{C} = \textstyle\sum_i c_i\, \mathbf{P}_i\,, \tag{12}$$

where $\mathbf{S}$ is the stress tensor and $\mathbf{E}$ the linear strain tensor. If we use linear laws like (12) for each element in our rheological model, we obtain the three dimensional cubic generalization of (1)

$$\mathbf{S}^{\cdot\cdot} + \mathbf{A}_1[\mathbf{S}^{\cdot}] + \mathbf{A}_2[\mathbf{S}] = \mathbf{A}_3[\mathbf{E}^{\cdot\cdot}] + \mathbf{A}_4[\mathbf{E}^{\cdot}] \tag{13}$$

where $\mathbf{A}_j$, $j = 1,\ldots, 4$, are fourth order material tensors and, thus, linear combinations like

$$\mathbf{A}_j := d_{1j}\,\mathbf{P}_1 + d_{2j}\,\mathbf{P}_2 + d_{3j}\,\mathbf{P}_3 = \sum_i d_{ij}\,\mathbf{P}_i \tag{14}$$

with

$$d_{i1} = C_i / D_i + C_i / L_i + K_i / L_i \,,\; d_{i2} = (C_i K_i) / (L_i D_i)\,,$$
$$d_{i3} = C_i + K_i \,,\; d_{i4} = C_i K_i / D_i \;,\; i = 1,2,3 \tag{15}$$

with $3 \times 4 = 12$ real constants d_{ij}. For example, the cubic form of the evolution equation is

$$\mathbf{T}^{\cdot} = \sum_i \mathbf{P}_i \left[f_i \mathbf{S}^{\cdot} + g_i (\mathbf{S} - \mathbf{T})\right]$$
$$\mathbf{E}^{\cdot} = \sum_i \mathbf{P}_i \left[h_i \mathbf{S}^{\cdot} + p_i \mathbf{S} + q_i \mathbf{T}\right]. \tag{16}$$

Here $\mathbf{T}$ is the tensor of the internal variables defined by

$$\mathbf{T}(t) := \sum_i K_i\,\mathbf{P}_i \left[\int_0^t \{\mathbf{E}(s)^{\cdot} - \mathbf{S}(s) / L_i\}\, ds\right] \tag{17}$$

and the real constants are determined by (18)

$$f_i = K_i / (C_i + K_i)\,,\; g_i = f_i C_i / D_i \,,\; h_i = 1 / (C_i + K_i)\,,\; p_i = h_i d_{i1} \,,\; q_i = -h_i C_i / D_i$$

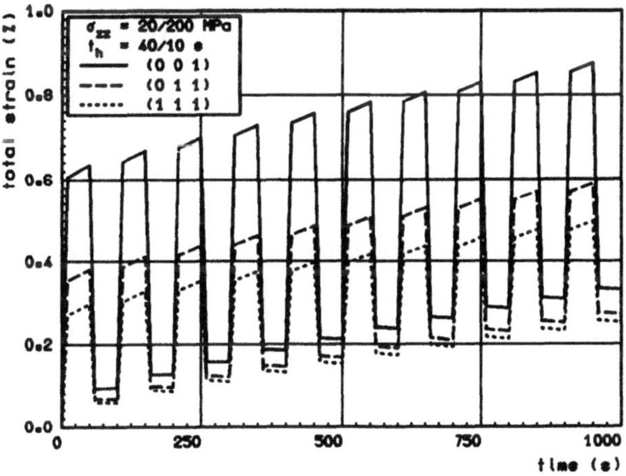

Fig. 7. Cyclic creep with different orientations

It is easy to make (16) explicit in terms of stress increments, if one prefers deformations as independent variables. These versions are appropriate for implementation into finite-element-codes. The results shown in Fig. 7 have been obtained by using ADINA [6] for tensile cyclic creep test of different orientations. The twelve material constants were only roughly fitted to experimental results. In BERTRAM et al. [7] applications to notched specimens are shown. A three-dimensional version of the damage equation will not be given here.

References

1. Krajcinovic, D.; Lemaitre, J.: Continuum damage mechanics. Wien, New York: Springer-Verlag (1987).

2. Murakami, S.; Sawczuk, A.: A unified approach to constitutive equations of inelasticity based on tensor function representations. Nucl. Engn. Des. 65 (1981) 33-47.

3. Betten, J.: Materialgleichungen zur Beschreibung des sekundären und tertiären Kriechverhaltens anisotroper Stoffe. Z. ang. Math. Mech. 64, 6 (1984) 211-220.

4. Betten, J.: Representation of initial and deformation induced anisotropy in creep mechanics. In: Advances in plasticity 1989. Khan, A. S.; Tokuda, M. (eds.) Oxford: Pergamon Press (1989) 677-681.

5. Voigt, W.: Lehrbuch der Kristallphysik. Leipzig, Berlin: Teubner (1910).

6. Bathe, K.-J.: ADINA, a finite element program for automatic dynamic incremental nonlinear analysis. Report AE 84-1. ADINA Eng. Inc., Watertown MA, USA (1984).

7. Bertram, A.; Olschewski, J.; Sievert, R.; Zelewski, M.: Constitutive modeling of the creep behaviour of single crystals with applications to notched specimens. Proc. 3. Int. Conf. Const. Laws for Eng. Mat., Tucson (1991) (to appear).

Acknowledgement

This report was supported by the *Bundesminister für Forschung und Technologie* under grant O3 M 3005 C4.

On the Creep-Ratchetting of AlMgSi 0.5 at Elevated Temperature; Experimental Investigations

DIETER BESDO, LESZEK OSTROWSKI
Institut für Mechanik, University of Hannover, Appelstr.11, 3 Hannover 1, RFG.

Introduction

Combined creep-ratchetting experiments according to Krempl and Ruggles were carried out with thin-walled AlMgSi 0,5 tubes at two temperatures 100°C and 150°C. Characteristic values are: maximum circumferential tension σ_R in zero-to-tension experiments and cyclic time $2*T_R$.
For history I ratchetting stress σ_{RAT} is determined by the strain controlled tensile test up to $\Delta\varepsilon$ = 0,7 %. For history II σ_{RAT} is determined by the relaxation that follows the tensile test. In history III the tensile test is limited by σ_{RAT} , that corresponds to the value of the relaxation experiments. In a certain range σ_{RAT} and T_R have a strong influence on the accumulated ratchetting strain ε_{RAT} .The viscose part of stress in this case forms the urging power for ratchetting deformations as it does with stainless-steel -type 304. Owing to a stepwise decrease of σ_{RAT} as well as T_R the viscose character of ratchetting rapidly declines until negative values can be reached.
The chosen strain-rate $\dot{\varepsilon} = 5 * 10^{-6}$ sec^{-1} leads to an instable material behaviour. The tension-to-fracture curve shows a negative stress sensitivity.

Material, test specimens.

The investigated specimen (Geometry see Fig.1) has following chemical composition: Si-0.5 , Fe-0.18 , Cu-0.003 , Ti-0.004 , Mg-0.5 , Al-98.81 % . The thin-walled tubes are warm aged at 150°C for 20 hours before the actual experiments.
The servo-hydraulic experimental arrangement, which was constructed by the second author, has two separated hydraulic circles (inner pressure, axial loading). Warming-up is done by a coupled resistance-

M. Zyczkowski (Ed.)
Creep in Structures
IUTAM Symposium Cracow/Poland 1990

induction-heating. At a nominal temperature of 100°C there was a tolerance of ± 1 °C in the zone of measurement, at 150°C $\pm$ 2°C were measured.
All strains and stresses of the experiments are technical circumferential-tensions and -strains computed by standard methods.
Radial deformations are measured by three inductive displacement-pickups, that were distributed every 120° on the tube circumference.
Another inductive displacement-pickup determines axial deformations, which serve for an estimation of disturbing bending moments.
All experiments are controlled by a c't 68000 processor of a GWK-real -time computer. The arising data were digitalised for further interpretation and simultaneously plotted.

Testing Program

The idea of the test method was taken from Ruggles and Krempl. The set of experiments is schematically shown in Fig.2a. Three different test histories were performed.

OA_0 -strain controlled loading of a virigin specimen for history I and II to a strain of 0,7 % with a strain rate $\dot{\varepsilon} = 5*10^{-6}sec^{-1}$. That determines the upper bound of the circumferential tensions σ_{RAT} for ratchetting tests. For history II σ_{RAT} is decreased by the following relaxation, Fig.2b. The same σ_{RAT} are taken for history III, that runs without pre-deformation up to $\Delta\varepsilon$ = 0,7 %. Essential features of the tension-to-fracture curve are shown in Fig.3. A negative tension sensitivity appears for both temperatures. Although inner pressure was controlled in steps of 0.003 MPa, the P-L zone occurs from certain deformations up. This leads to simultaneous elastic unloading.

AAi - optional, only at history II. Relaxation time T_{REL} is 2.5 , 5.0, 10 and 60 min. Relaxation time leads to a variation of the maximum stresses σ_{Ai} for the following ratchetting σ_{RAT}, which characterizes the distance to a quasi-static equilibrium for the viscose material.

AiB - for all 3 histories strain-controlled unloading with $\dot{\varepsilon}_{AiB} = -\dot{\varepsilon}_{OA}$.

BC - ratchetting; determined by the number of cycles (500 for all histories), cycle time $2*T_R$ and maximum stress σ_{RAT}. Two values for rise-time (T_R) were assumed (T_R = 25 sec or 250 sec). σ_{RAT} is: $\sigma_{RAT} = \sigma_A$ for history I, $\sigma_{RAT} = \sigma_{Ai}$ for histories II and III.

For history III the part OA is limited by σ_{Ai} (determined from the relaxation tests). The different σ_{Ai} lead to a variation of $\dot{\sigma}_{RAT}$. The total time for BC is with T_R= 25 sec ⇒ 500*25*2 sec and with T_R = 250 sec ⇒ 500*250*2 sec.

CDEF- creep test, CD-stress-controlled loading with $\dot{\sigma}_{CD} = \dot{\sigma}_{RAT}$ and $\sigma_D = \sigma_{RAT} = \sigma_{CR}$), creep for 1 hour (CF) and stress controlled unloading with $\sigma_{EF} = -\sigma_{RAT}$ up to zero (F).

FGH - strain controlled zero-to-tension loading FG with $\dot{\varepsilon}_{FG} = \dot{\varepsilon}_{OA}$ up to $\varepsilon_G = \varepsilon_F + \Delta\varepsilon_{OA}$,then immediate unloading with $\dot{\varepsilon}_{GH} = -\dot{\varepsilon}_{OA}$ up to zero. Stress in G is on the one hand compared with σ_{Ao}, on the other hand determines the maximum tension of the following step.

HI - slow, quasi-static stress-controlled loading up to $\sigma_{MAX} = \sigma_G$. Time for HI is assumed to be 2*30 min. Here the further possibility of the material for viscose behaviour after preceding deformations is examined (deformation history).

Test results, conclusion

The performed experiments with 60 specimen revealed a versatile character of material-behaviour.

* ratchetting

Characteristic results (of some tubes) are summarized in table I. It should be taken into account, that for history III also the variation of ε_{FG} was introduced. For constitutive modelling three kinds of behaviour can be distinguished :

- viscose behaviour:

  For large σ_{RAT} (more than $0.85*\sigma_{Ao}$ at 150°C and more than $0.97*\sigma_{Ao}$ at 100°C) the material behaves significantly time-dependent.

  In the special case of $\sigma_{RAT} = \sigma_{Ao}$ for T_R= 250 sec the tubes burst after a certain number of cycles ($\varepsilon_{BURST} \cong 8\%$), but for T_R= 25 sec ε_{RAT} takes its maximum value (both history I).

- plastic domain:

  By a stepwise decrease of σ_{RAT} the strain-rate sensitivity rapidly declines. A wide range of tension with time-independent increase of irreversible deformations follows the relatively small zone of viscose tensions.

  Amounts of $\Delta\varepsilon_{RAT}$ compared to those of the viscose zone are extremely small.

- Further decrease of σ_{RAT} (after 1h relaxation) leads to the new phenomenon of negative ratchetting. Experiments with ε_{OA}= 0.7% (compared to history III considered as pre-deformation) and suitable combinations of σ_{RAT} and $\dot{\sigma}_{RAT}$ directly after unloading (B) show a sucsessive reduction of total ε until a state of saturation is reached. For history III dynamic recovery is far smaller. A transient transition from zero to negative strains and then backwards to positive (though very small) strains could also be observed.

* creep:

All three zones of creep were observed. Fig. 7 shows the relation between ε_{RAT} and ε_{CR}. In the range of viscose ratchetting there is a quasi-linear connection between them. Understandably ε_{CR} increases at large $\dot{\sigma}_{RAT}$. For history III ε_{CR} is smaller than the corresponding ε_{CR} at same σ_{RAT} of history II.In general: creep and ratchetting are qualitatively alike (large σ_{RAT}) and the material shows strain- and time-softening.

*FGH:

This step should clarify, if the material was hardened by preceding deformation O-F. Surprisingly σ_G is determined by the tension-to-fracture curve with negative stress-sensitivity discribed above. It seems, that the tension -to-fracture curve is an objective characteristics of the material.

*HI: That means, that the slow step HI to proove the possibility (of the material) of further viscose behaviour is unnecessary. Results here show time- and strain-independent $\Delta\varepsilon_{HI}$.

The experiments performed shall serve as a base for identification of parameters for the material law proposed by first author .

References :

Ruggles, M.B ; Krempl, E. : The Influence of Test Temperature on the Ratchetting Behavior of Type 304 Stainless Steel. J. of Eng. Tech. ASME., 111 (1989), P. 378-383.
O'Donell, R.G. ; Parker, B.A. : Constitutive Relations Describing Flow Behaviour in Aluminium Alloys. Res Mechanica 9 (1986), P. 17-33.
Nouailhas, D. : Prediction of Ratchetting and Creep Effects at Room Temperature on 304 Stainless Steel . Res Mechanica 30 (1990), P. 223-242.
Besdo, D. : Endochronic Constitutive Law for Finite Deformations Presentend in the Space of Strains. Proc. Int. Conf. Const. Laws for Eng. Mat., Ed. by Fan Jinghong Sumio Murakami, Aug. 11-13,(1989), P. 699–702.
Besdo, D.; Ostrowski, L. : Cyclic Plasticity with Creep Interaction. In 28th Polish Solid Mechanics Conference. (Kozubnik 1990), S. 18-23.

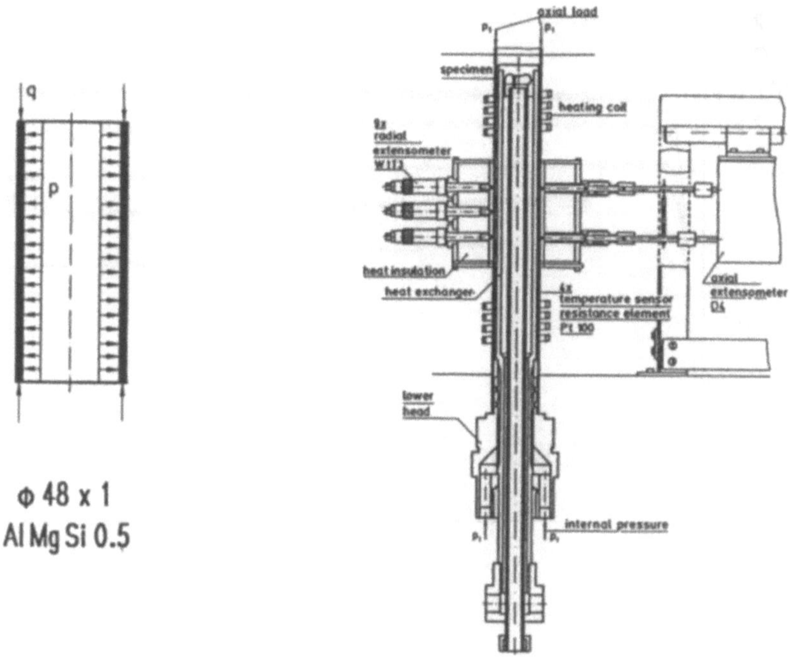

Fig. 1 Geometry of specimen . Servo — hydraulic test equipment.

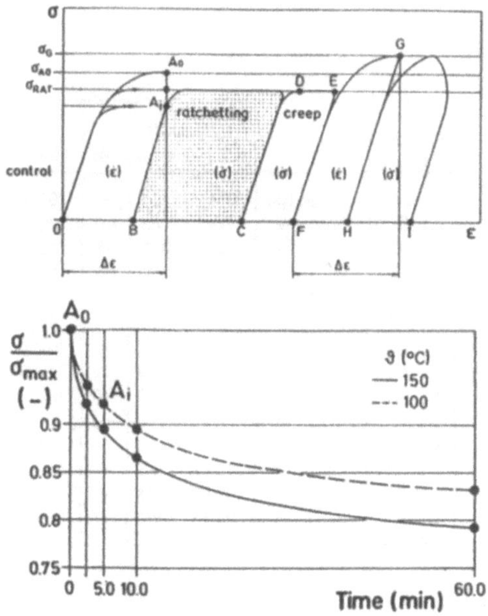

c)

t_{REL} (MIN)	σ_{RAT}	
	θ = 150 °C	100 °C
—	162.79	194.55
2.5	149.71	182.86
5.0	146.43	179.99
10.0	140.21	173.40
60.0	129.67	162.04

Fig. 2 a) loading pattern , b) relaxation stresses for histories II, III , c)stress values for relaxation time t_{REL}= 2.5 , 5.0 , 10.0 and 60.0 min.

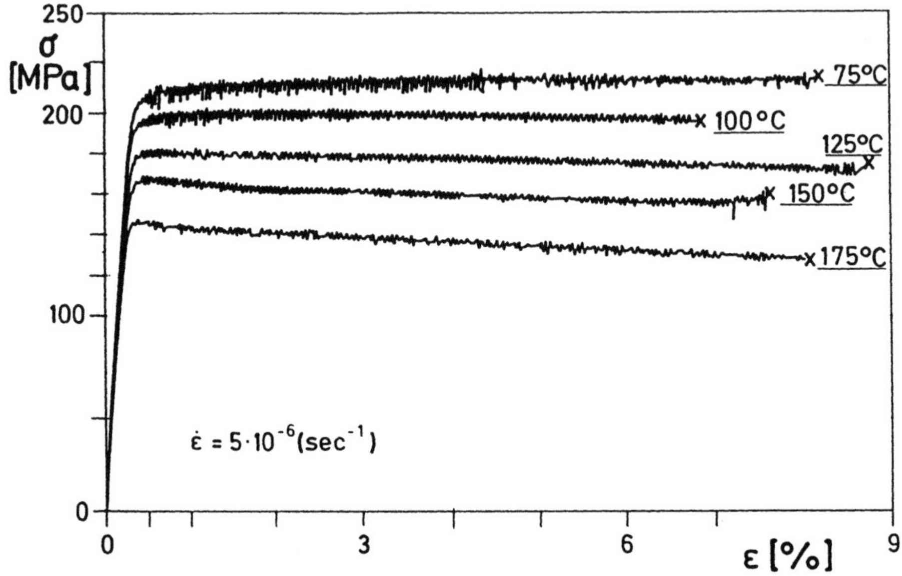

Fig. 3 Tension - to - fracture curves, temperature influence. The negative stress sensitivity for $\dot{\varepsilon} = 5*10^{-6}$ sec^{-1} for AlMgSi 0,5.

Fig	Tube	Hist.	θ [°C]	ε_{OA} [%]	T_R [sec]	t_{REL} [min]	σ_{RAT} [MPa]	$\Delta\varepsilon_{FG}$ [%]	$\Delta\varepsilon_{RAT}$ [%]	$\Delta\varepsilon_{CR}$ [%]	σ_G [MPa]	ε_{HI} [%]
4a	50	I	150	0.7	25	—	162.8	.70	0.3658	4.5647	156.92	0.163
b	07	II	-"-	-"-	25	60	129.7	-"-	-0.0260	0.0404	158.31	0.043
c	37	II	-"-	-"-	250	5	146.4	-"-	1.5000	1.7757	148.05	0.041
d	13	II	-"-	-"-	250	60	129.7	-"-	0.0656	0.0653	152.79	0.054
5a	61	I	100	-"-	25	—	194.5	-"-	0.1282	3.6740	198.99	0.060
b	06	II	-"-	-"-	25	60	162.0	-"-	-0.0026	0.0136	198.98	0.033
c	64	II	-"-	-"-	250	5	179.9	-"-	0.0475	0.1155	196.38	0.041
d	12	II	-"-	-"-	250	60	162.0	-"-	-0.0120	0.0114	193.07	0.035
6a	60	III	150	.2592	250	—	146.4	.70	0.3553	0.2456	157.63	0.085
b	52	III	-"-	.2511	25	—	140.2	.25	0.0444	0.1458	155.87	0.039
c	57	III	-"-	.3042	25	—	153.0	.70	0.0493	0.5099	160.95	0.100
d	54	III	-"-	.2496	250	—	140.2	.25	0.2130	0.1409	154.00	0.034

Tab. I Experimental results for all three histories.

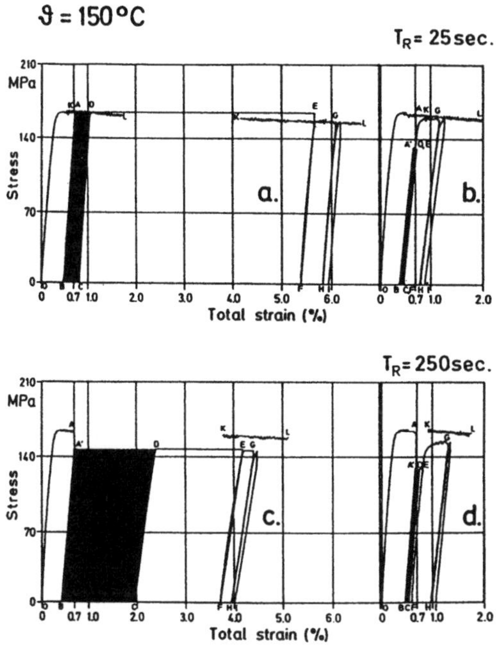

Fig. 4 (Tab. I) Experimental results for $\Theta = 150^0$ C, history I (a) and history II (b–d).
Specimens :
a) 50, b) 07, c) 37, d) 13.

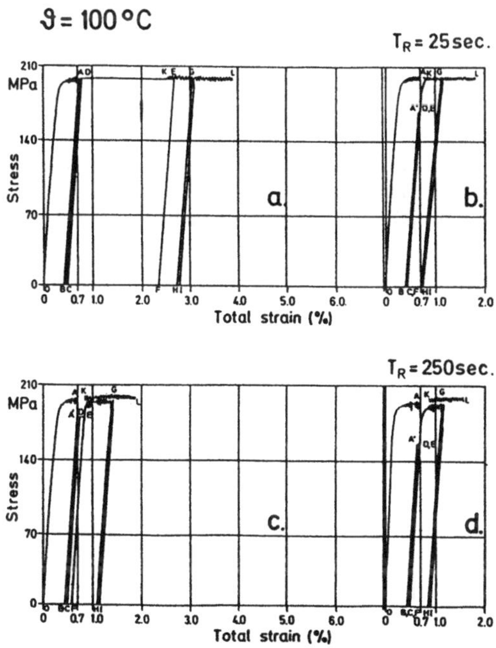

Fig. 5 (Tab. I) Experimental results for $\Theta = 100^0$ C, history I (a) and history II (b–d).
Specimens :
a) 61, b) 06, c) 64, d) 12.

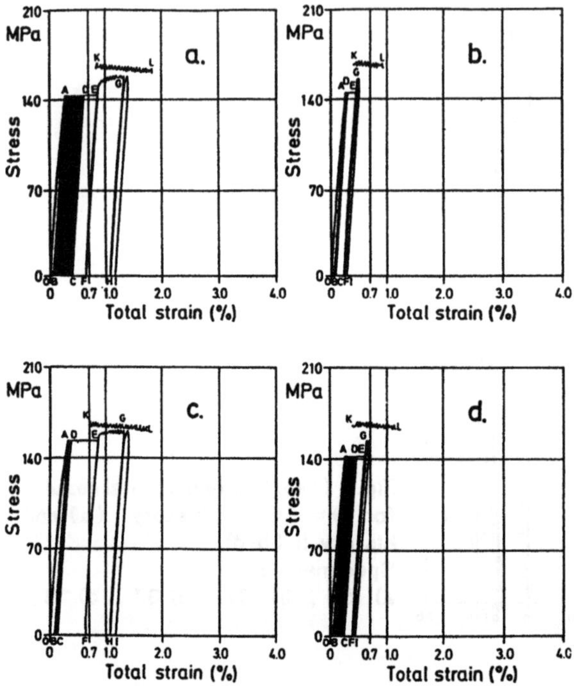

Fig. 6 (Tab. I) Experimental results for $\Theta = 150°C$, history III (a - d). Specimens : a) 60, b) 52, c) 57, d) 54 .

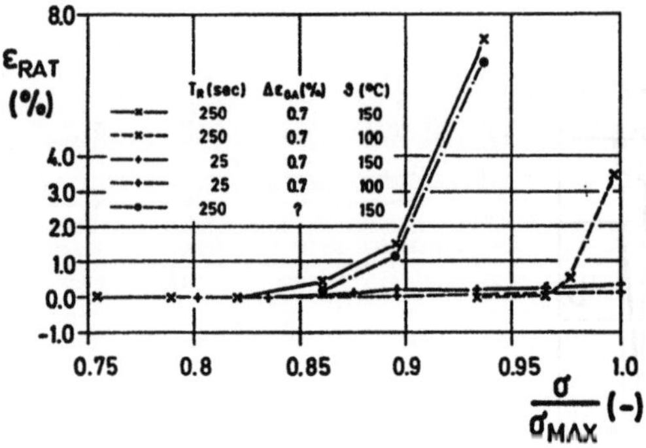

Fig. 7a Correlation between ratchetting strain ε_{RAT} and σ_{RAT} after relaxation .

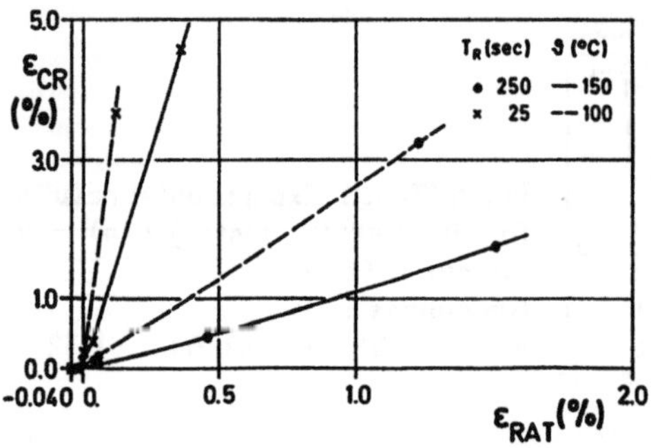

Fig. 7b Correlation between ratchetting and creep for $\theta = 100°C$ and $150°C$, history I - II for different σ_{RAT}.

Compressive Creep Tests on Porous Ceramic Notched Specimens

J. Besson [1] and M. Abouaf [2]
Ecole Nationale des Mines de Paris, Centre P.M. Fourt
BP87 91003 Evry Cedex France

F. Mazerolle and P. Suquet
CNRS/LMA, 31 Chemin J. Aiguier
13402 Cedex 9 Marseille France

Abstract

Porous cylindrical notched specimens are tested in compression in order to generate densification gradients which are measured using an X–ray scanner. These measurements are compared with good agreement with numerical results obtained by Finite Element Simulation. From a set the calculations made using different rheological behaviours, it is believed that accurate *in–situ* measurement of densities by X–ray tomography can be used to analyse the rheology of porous materials.

Introduction

Powder processing has been used for over ten years in industrial applications, essentially to produce high quality metallic or ceramic parts mainly by using Hot Isostatic Pressing (HIP) which is now a widely used process. However it is well known that the final products differ in scale but also in shape ; this can be caused by stiff encapsulation cans [1, 2] or by densification gradients generated by low heat transfer [3]. It could therefore be very useful to be able to forecast the behaviour of the can and the powder during densification. This can be done using mechanical constitutive equations that allow for the simulation of the HIP process [1, 2, 4]. It remains however very difficult to validate these constitutive laws using HIP tests (canning problems, monitoring of displacement...). The purpose of this work is to provide an experimental procedure to generate large density gradients inside cylindrical notched samples and to measure these gradients using an X–ray scanner. It is then possible to compare experimental results with numerical simulation and to check the validity of the constitutive laws used for the calculations.

---

[1] Present address : University of California Santa Barbara, Materials Department, Santa Barbara 93106 CA, USA
[2] Present address : CICE SA, 63 rue Beaumarchais, 93104 Montreuil, France

M. Zyczkowski (Ed.)
Creep in Structures
IUTAM Symposium Cracow/Poland 1990

1. Experimental procedure

Tests specimens were produced using a fine grained alumina powder * of high purity (Al_2O_3 content> 99.98%). Initial grain size is about 0.25 μm. Cylinders (diameter 30 mm) of various relative densities were hot–pressed in a vacuum furnace under a load of 20 MPa and a temperature variyng betwen 1250 and 1300°C using graphite dies and pistons. These cylinders are then machined to produce a notch with a radius of 6.66 mm. The smaller diameter of the sample is equal to 16.67 mm (*cf.* fig. 1).

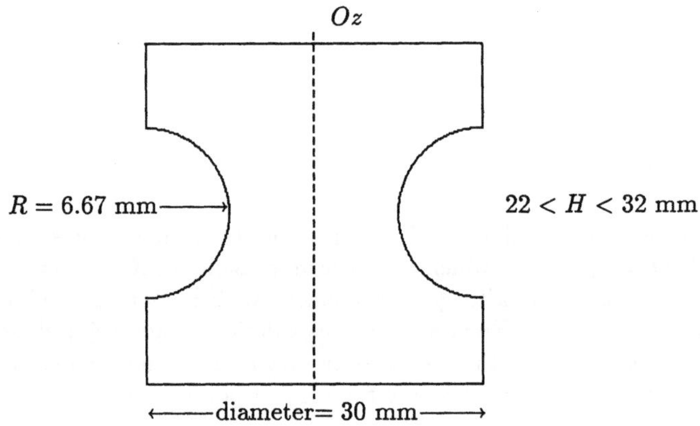

Fig 1 : Geometry of notched samples

The notched samples are then pressed again using a temperature between 1270 and 1310°C and an axial load varying between 7000 and 10000 N ; such values correspond to a stress of 10 to 15 MPa on an homogeneous cylinder having a diameter of 30 mm. Four samples were produced and pressed (*cf.* table 1 for characteristics).

sample	initial height (mm)	initial relative density	final height (mm)	Δh (mm)
1	28.24	0.823	26.48	1.76
2	25.46	0.897	23.68	1.78
3	31.10	0.744	28.16	2.94
4	24.18	0.923	22.58	1.60

Table 1 : Tested notched samples

* Criceram HR8

Tested notched specimens were then examinated using an X–ray scanner, which allows for the measurement of densities inside the ceramic part. The following precedure was used : the X–ray device measures a parameter called Hounsfield density (HD with $-1000 < \text{HD} < +1000$) which is representative of X–ray absorption by the matter ; this parameter can be related to the relative density of the sample using specimens having a known uniform density. The relation between the relative density ρ and the Hounsfield density is linear (the experimental data and the fitted curve are plotted on fig. 2) :

$$\rho = 0.5707 + 5.64\,10^{-4}\,\text{HD} \qquad \text{for} \quad 0.7 \leq \rho \leq 1.0 \tag{1}$$

Density maps of the sample in a vertical plane (parallel to Oz) and in an horizontal plane (perpendicular to Oz) can then be obtained. Tests show that measurements are more accurate when using a perpendicular section ; this is probably due to the axisymmetric distribution of density about the Oz axis. The image has the thickness of 3 mm and is defined by square pixels having a length of 0.7 mm.

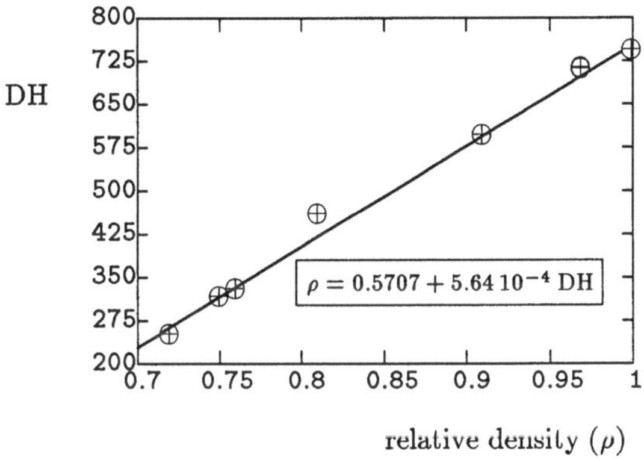

Fig 2 : Hounsfield density vs. relative density : (⊕) experimental data, (—) fitted curve

The same procedure was used by relating the Vickers micro-hardness (under a load of 2 kg) to the relative density. The minimum size of the Vickers indentations is equal to 45 μm and is therefore much bigger than the pores remaining in the material, so that the measurement is representative of an average value of density. Relative density and micro–hardness ($\text{H}_\text{V}^{2\text{kg}}$) are related by the following formula (*cf.* fig 3)

$$\rho = 0.629 + 1.769\,10^{-2}\,\text{H}_\text{V}^{2\text{kg}} \qquad \text{for} \quad 0.7 \leq \rho \leq 1.0 \tag{2}$$

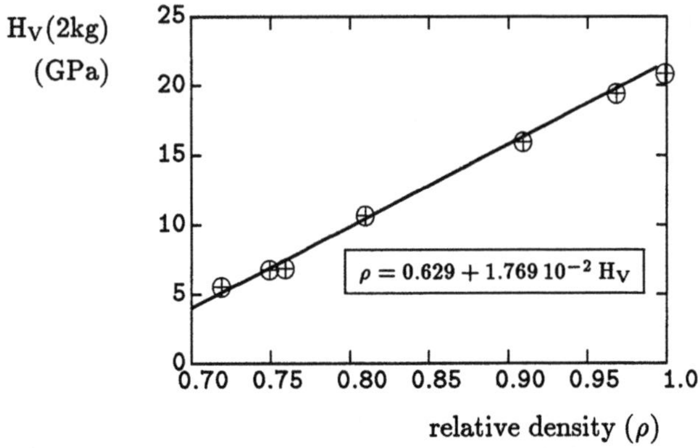

Fig 3 : Vickers micro–hardness (2 kg) vs. relative density : (⊕) experimental data, (—) fitted curve

2. Constitutive equations for porous materials

In this study, the mechanical behaviour of the porous material will be described using constitutive equations allowing for volume changes during visco–plastic flow which are an extension of the J_2 Mises theory to porous materials [1, 4, 5, 6] The equivalent stress σ_{eq}, which includes the effect of pressure, is defined as

$$\sigma_{eq}^2 = 3cJ_2 + fI_1^2$$
$$\text{with} \quad \begin{cases} I_1 = \mathrm{Tr}(\tilde{\sigma}) = \sum_i \sigma_{ii} & \text{first invariant} \\ J_2 = \frac{1}{2}\tilde{s} : \tilde{s} & \text{second invariant} \\ \tilde{s} = \tilde{\sigma} - \frac{1}{3}I_1\tilde{1} & \text{deviator of the stress tensor} \end{cases} \tag{3}$$

where $\tilde{\sigma}$ is the Cauchy stress tensor and $\tilde{1}$ the unitary second order tensor. c and f are two functions representing the stress localisation induced by the porosity, which *a priori* depend only on the relative density. For full density $c = 1$ and $f = 0$ so that the previous equations are equivalent to the definition of the Mises equivalent stress. Normality rule is further assumed ; using the following definition of the equivalent strain rate ($\dot{\epsilon}_{eq}$) : $\rho\dot{\epsilon}_{eq}\sigma_{eq} = \tilde{\sigma} : \dot{\tilde{\epsilon}}$, the visco–plastic strain rate tensor ($\dot{\tilde{\epsilon}}$) is given by

$$\dot{\tilde{\epsilon}} = \rho\frac{\dot{\epsilon}_{eq}}{\sigma_{eq}}\left(\frac{3}{2}c\tilde{s} + fI_1\tilde{1}\right) \tag{4}$$

$\dot{\epsilon}_{eq}$ and σ_{eq} are related using the creep law of the material. Hot–pressing and sinter–forging tests have shown that the creep law can be expressed as follows [2]

$$\frac{1}{\dot{\epsilon}_{eq}} = \frac{1}{\dot{\epsilon}_{\mathrm{diff}}} + \frac{1}{\dot{\epsilon}_{ir}}$$
$$\text{with} \quad \begin{cases} \dot{\epsilon}_{\mathrm{diff}} = \dfrac{A_{\mathrm{diff}}\exp(-Q_{\mathrm{diff}}/RT)}{TG^3}\sigma_{eq} & \text{diffusion} \\ \dot{\epsilon}_{ir} = \dfrac{A_{\mathrm{ir}}\exp(-Q_{\mathrm{ir}}/RT)}{G}\sigma_{eq}^2 & \text{interface reaction control} \end{cases} \tag{5}$$

where G is the grain size, T the absolute temperature and R the gas constant. Creep constants in the previous equations are the following ones

A_{diff}	$= 6.15\,10^{-9}$	$s^{-1}.MPa^{-1}.m^3.K$
A_{ir}	$= 121$	$s^{-1}.MPa^{-2}m$
Q_{diff}	$= 419$	$kJ.mol^{-1}$
Q_{ir}	$= 495$	$kJ.mol^{-1}$
$0.35\ \mu m < G < 0.50\ \mu m$		

The deformation mechanism is diffusion at grain boundaries (Coble creep [7]) or diffusion accomodated superplasticity [8] which can be controlled by interface reactions (*i.e.* production and annihilation kinetics of vacancies at grain boundaries) [9]. The slower mechanism controls the deformation rate.

The two functions c and f are plotted on fig. 4 for densities between 0.75 and 1.0. An interesting feature is that the ratio f/c remains constant and close to 1/2 for densities up to 0.9. This means for example, that during forging of a porous specimen the radius will not change ; this observation is in good agreement with preliminary results reported by Casagrande *et al.*[10].

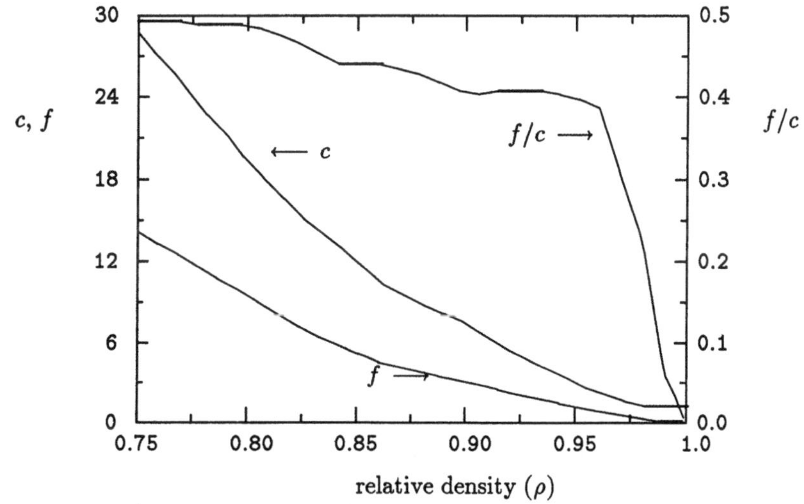

Fig 4 : functions c, f and f/c vs. relative density

3. Comparison between finite element computation and experimental data

Finite element computation of the notched specimens under load were carried out using the program Zébulon developed by the Centre des Matériaux de l'École des Mines de Paris. Algorithms used for the computation were presented elsewhere [1, 2]. The computation is carried out using the axisymmetry of the problem, with a constant load on the top of the specimen and constant and uniform temperature, all other surfaces are free. The computation is stopped when the calculated height is equal to the observed one (see table 1). Calculated times necessary to obtain the given displacement differ from the observed ones by a factor between 1/2 and 2. Comparison between measurement (X–ray scanner and micro–hardness) and computation are presented for sample 3 on fig. 5,6 and 7. The same kind of agreement between calculation and experiment is obtained with sample 1. In samples 2 and 4 the density gradients are too small to be accurately measured. In addition the X–ray examination of sample 4 before machining shows a density gradient between the top and the bottom of the sample which is probably due to temperature gradient in the furnace or wall friction in the hot–pressing die. The maximum density difference between computation and measurement is 0.03. The scanner measurements are inaccurate near the free surfaces due to the thickness of the X–ray beam which is both in the sample and in the air.

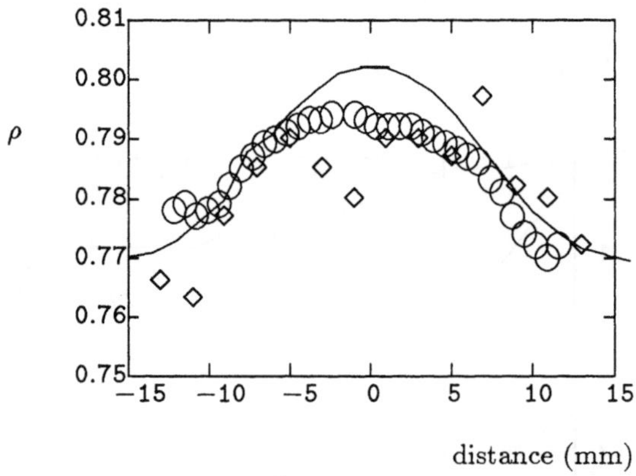

Fig 5 : sample 3 : radial density profile in the upper part of the notched specimen (—) FEM calculation, (◇) Vickers measurement (○) X–ray scanner measurement

4. Effect of rheological parameters on density gradients

Simulation were carried out in order to check the effect of the rheology of the porous material on the density gradients in the powder. Two parameters have been studied : the stress exponent n with $\dot{\epsilon}_{eq} = A\sigma_{eq}^n$ and the ratio f/c which roughly characterises the densification/shear behaviour. The following hypothesis have been used

$$\begin{aligned} c &= 1 + 160(1-\rho) \\ f &= (c-1)/\kappa \end{aligned} \tag{6}$$

The calculated structures have the same geometry than the one used for the experiments ; the height is equal to 30 mm and the height variation is 3 mm. The loading

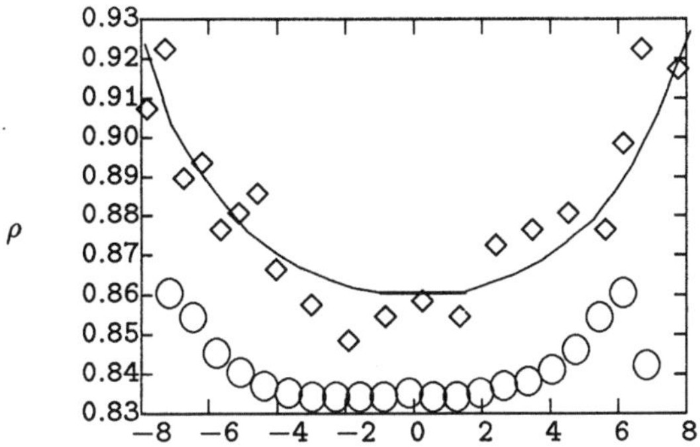

Fig 6 : sample 3 : radial density profile in the notched region of the specimen (—) FEM calculation, (◇) Vickers measurement (○) X–ray scanner measurement

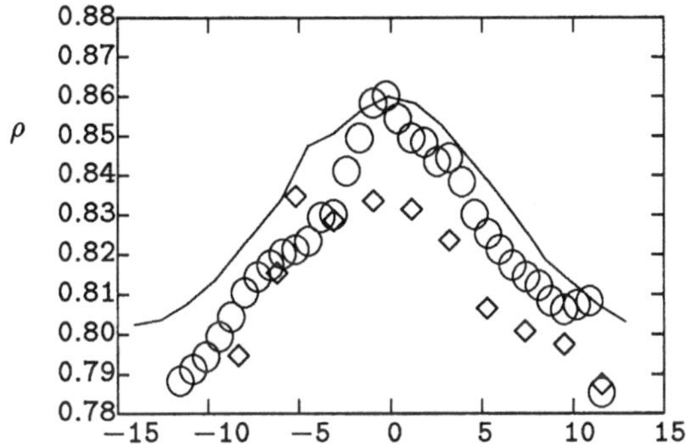

Fig 7 : sample 3 : axial density profile along the Oz axis of the notched specimen (—) FEM calculation, (◇) Vickers measurement (○) X–ray scanner measurement

conditions are also the same. Various calculations have been carried out with $\kappa = 2, 8$ and $n = 1, 3, 9$. Various density maps are shown on fig. 7. For a fixed κ, a high n produces higher density gradients ; and for fixed n a low κ produces the same effect. It is also noted that the maximum relative density in the sample is always observed at the notch tip. Calculations show that stresses in the notched section slightly depends on κ or n after a sufficiently high deformation. Both shear ($\sqrt{3J_2}$) and pressure ($I_1/3$) are maximum at the notch tip. Comparison between theses maps and accurate measurement using X–ray tomography could provide a rapid estimation of the rheological behaviour of the porous body.

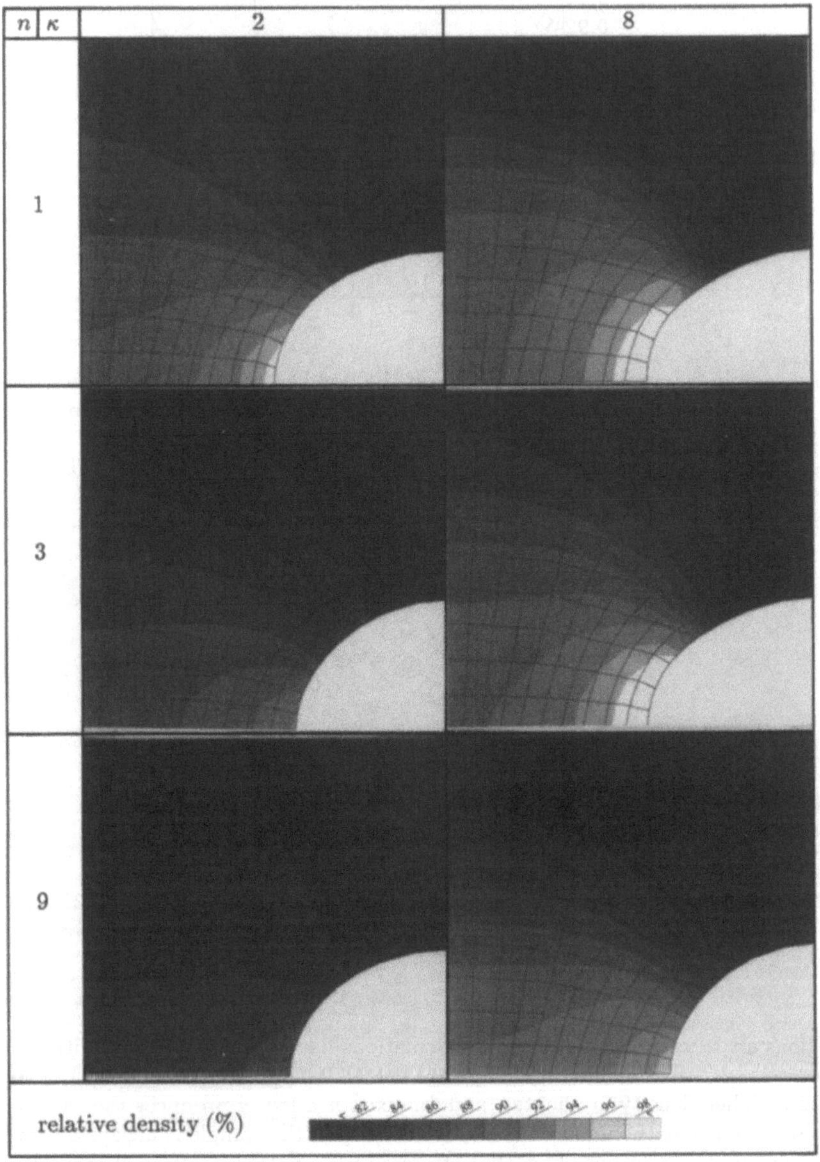

Fig 8 : Comparison of density maps for various n and κ for a given height variation

Conclusion

Notched alumina samples have been tested under compression in order to generate density gradients in the structure. Measurements of these gradients using an X–ray scanner and Vickers indentations have been compared to numerical results obtained using finite element computation. A continuum description of ceramic powders has been used to represent the behaviour of alumina. Such an approach can be useful to check the constitutive equations of the porous medium since the parameters of these equations are obtained using tests that generate homogeneous deformations (*i.e.* HIP, hot–pressing or sinter–forging), or to quickly determine these parameters by comparing observed density maps with calculated ones.

References

[1] Abouaf M., Chenot J.L., Raisson G. and Bauduin P., Finite Element Simulation of Hot–Isostatic Pressing of Metal Powders *Int. J. for Numerical Methods in Engineering, Vol. 25, No. 1, pp191–212* (1988)

[2] Besson J. and Abouaf M., Finite Element Simulation of Hot–Isostatic Pressing of Ceramic Powders *paper presented at the* 2nd *Int. Conf. on HIP : Theory and Application, Gaithersburg June 1989*(1989)

[3] Li W., Ashby M.F. and Easterling K.E., On Densification and Shape Change during Hot Isostatic Pressing, *Acta Met., Vol. 35, No. 12, pp2831–2842* (1987)

[4] Nohara A., Soh T. and Nakagawa T., Numerical Simulation of the Hot–Isostatic Pressing Process *Int. J. for Numerical Methods in Engineering, Vol. 25, No. 1, pp213–225* (1988)

[5] Green R.J., A Plasticity Theory for Porous Solids *Int. J. mech. Sci., Vol. 14, pp215–224,* (1972)

[6] Shima S. and Oyane M., Plasticity Theory for Porous Metals *Int. J. Mech. Sci., Vol. 18, pp285–291* (1976)

[7] Coble R.L., Diffusion Sintering in the Solid State, *Kinetics of High–Temperature Processes, ed. Kingery W.D., pp147–162* (1959)

[8] Ashby M.F. and Verrall R.A., Diffusion Accomodated Flow and Superplasticity *Acta Met., Vol. 21, No. 2, pp149–163* (1973)

[9] Arzt E., Ashby M.F. and Verrall R.A., Interface Controlled Diffusional Creep *Acta Met., Vol. 31, No. 12, pp1977–1989* (1983)

[10] Casagrande A., Zok F. and Evans A.G., private communication (1990)

Some New Aspects of the Nonlinear Characterization of Creep and Stress Relaxation of Polymers

O.S. BRÜLLER

Department A of Mechanics
Technical University of Munich

Summary

Based on the well-known mechanical models made of springs and dashpots, the linear viscoelastic behavior of materials can be described by exponential series with a limited number of terms. Since the models for creep (increasing function) and stress relaxation (decreasing function) are different, the series are also of different type. The approach can be extended to the nonlinear viscoelastic range by introducing some load dependent nonlinearity functions. In the present paper it is shown that both, creep and stress relaxation can be represented by the same type of exponential series. Further, based on experimental data, it is shown that the instantaneous response of the nonlinear polymeric material to the sudden loading step is always linear elastic. Finally, a way to reduce considerably the experimental efforts needed for the material characterization is shown.

Linear and Nonlinear Viscoelasticity

Simple stress relaxation of polymeric materials in the linear viscoelastic range can be described easily by using the generalized Maxwell model, shown in Fig. 1 [1,2].

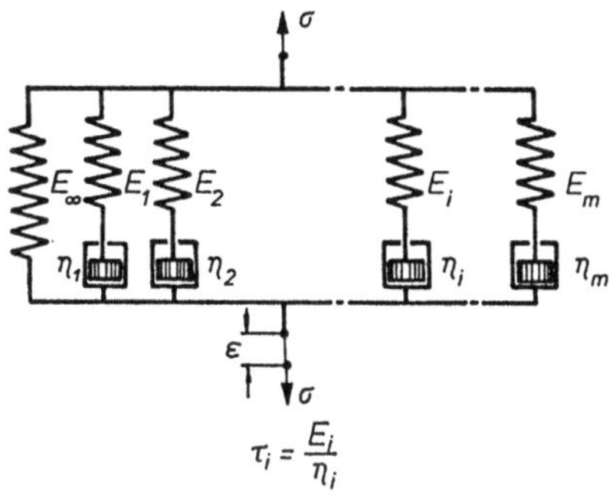

Fig. 1: The generalized Maxwell model

M. Zyczkowski (Ed.)
Creep in Structures
IUTAM Symposium Cracow/Poland 1990

On the other hand, the creep behavior of a polymer in the linear range can be described by the Kelvin model (Fig. 2).

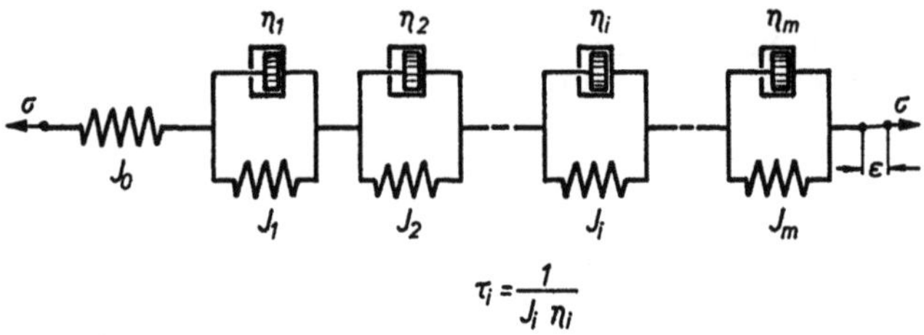

Fig. 2: The generalized Kelvin model

From those models on can derive easily the relations for:

a) The relaxation modulus (from the Maxwell model)

$$E(t) = E_\infty + \Delta E(t) \tag{1}$$

with:

$$\Delta E(t) = \sum_{i=1}^{m} E_i \exp(\frac{-t}{\tau_i}) \tag{2}$$

b) The creep compliance (from the Kelvin model)

$$J(t) = J_0 + \Delta J(t) \tag{3}$$

with:

$$\Delta J(t) = \sum_{i=1}^{m} J_i[1 - \exp(\frac{-t}{\tau_i})] \tag{4}$$

To describe the behavior in the nonlinear viscoelastic range the above given relation may be modified by introducing some load dependent nonlinearity functions. One obtains:

$$E_n(t, \epsilon_0) = h_\infty(\epsilon_0)\, E_\infty + h_t(\epsilon_0)\, \Delta E(t) \tag{5}$$

and:

$$J_n(t, \sigma_0) = g_0(\sigma_0)\, J_0 + g_t(\sigma_0)\, \Delta J(t) \tag{6}$$

In all these relations the symbols represent:

ϵ_0 = constant strain
σ_0 = constant stress
t = time
h_∞, h_t = strain-dependent nonlinearity factors
g_0, g_t = stress-dependent nonlinearity factors
E_∞ = equilibrium value of the linear stress-relaxation modulus
$\Delta E(t)$ = transient term of the linear stress-relaxation modulus
J_0 = instantaneous value of the linear creep compliance
$\Delta J(t)$ = transient term of the linear creep compliance
E_i, J_i = linear material parameters
τ_i = relaxation or retardation times

In order to describe the creep or stress-relaxation of a nonlinear viscoelastic material it is necessary to know the numerical values of all the parameters of Eqs. (1) - (6). A procedure for the determination of this data on the basis of a series of long-term experiments is presented in [3].

It should be mentioned that the structure of the mathematical relations for the relaxation modulus and the creep compliance are different: while in Eq. (1) the constant term represents the equilibrium (infinite) value of the relaxation modulus, in Eq. (3) it reflects the instantaneous compliance (i.e. the response of the material to the sudden loading step). It is clear that only the instantaneous compliance (or its inverse, the initial modulus) can be determined experimentally.

It should also be pointed out here that in the linear viscoelastic range the following identities hold:

$$h_\infty = h_t = 1 \qquad g_0 = g_t = 1 \tag{7}$$

Experimental Results

Results obtained from long-term experiments under isothermal loading conditions with several thermoplastic polymers, such as PMMA, PVC, PS, ABS, POM and ABS under uniaxial loading and pure shear (torsion of thin walled cylindrical specimens) have shown that the above mentioned characterization leads to an accurate description of the nonlinear viscoelastic behavior of those materials. Moreover, more complicated loading histories can be predicted on this basis, as for example sudden periodic loading and unloading [4].

As an example the creep behavior of PVC in pure shear is depicted in Fig. 1. A very good agreement between experiments and computed data can be seen. Fig. 2 shows the course of the nonlinearity functions vs. applied stress.

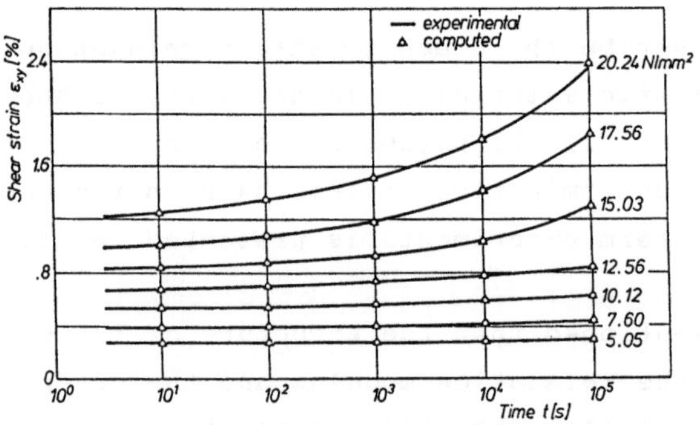

Fig. 1. Creep curves of PVC in shear

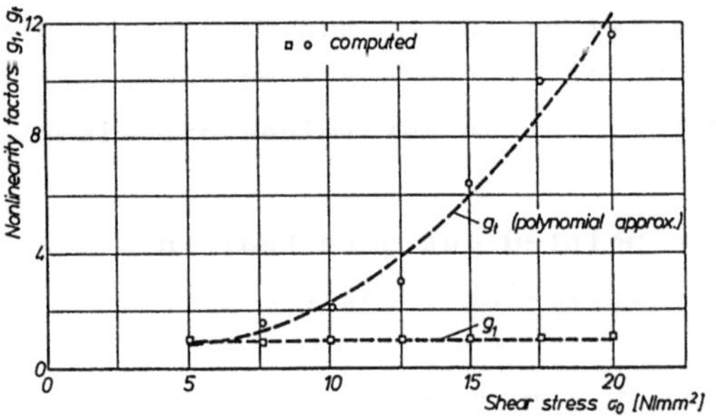

Fig. 2: Nonlinearity factors of PVC in shear creep

In this graph one can observe that the nonlinearity function of the instantaneous response of the material is practically equal to one in the entire range of applied stresses, even in the nonlinear range. This has been confirmed with all the tested materials in shear as well as in uniaxial tension.

Consequences for the Material Characterization

The fact that the instantaneous material response to a loading step is linear elastic has major consequences for the material characterization, namely only one nonlinearity function is needed to describe the material behavior. Accordingly, the linear elastic loading step is described by the parameter wich can now be recognized as beeing the instantaneous elastic compliance, i.e.:

$$J_0 \equiv J_{el} \tag{8}$$

Consequently, in the case of stress-relaxation the instantaneous response of the material must also be linear elastic. From Eqs. (1) and (2) one obtains:

$$h_\infty E_\infty + h_t \Delta E(0) = h_\infty E_\infty + h_t \sum_{i=1}^{m} E_i = E_{el} \tag{9}$$

with:

$$E_{el} = \frac{1}{J_{el}} \tag{10}$$

Eq. (9) means that between the two nonlinearity functions of the stress-relaxation representation a linear relationship must exist. This was confirmed with all the tested materials as shown for example in [3].

With this important result some additional relations can be obtained. If the instantaneous part of the relaxation modulus, given by Eq. (9):

$$h_\infty E_\infty + h_t \sum_{i=1}^{m} E_i \tag{11}$$

is added to and subtracted from the modulus itself:

$$E_n(t,\epsilon_0) = h_\infty E_\infty + h_t \sum_{i=1}^{m} E_i \exp(\frac{-t}{\tau_i}) \tag{12}$$

one obtains:

$$\begin{aligned} E_n(t,\epsilon_0) &= h_\infty E_\infty + h_t \sum_{i=1}^{m} E_i - h_\infty E_\infty - h_t \sum_{i=1}^{m} E_i \\ &+ h_\infty E_\infty + h_t \sum_{i=1}^{m} E_i \exp(\frac{-t}{\tau_i}) \end{aligned} \tag{13}$$

which can be rewritten as:

$$E_n(t,\epsilon_0) = E_{el} - h_t \sum_{i=1}^{m} E_i \,[1 - \exp(\frac{-t}{\tau_i})] \tag{14}$$

This new form of the stress-relaxation modulus has exactly the same structure as the expression for the creep compliance, with exception of the minus sign in front of the time-dependent term. It is based, like the relation for the creep compliance, on the initial modulus (which is generally known) instade of the equilibrium modulus (which can not be determined). Consequently, the nonlinear behavior under stress-relaxation loading can be described (like the creep compliance) by only one nonlinearity function, namely the nonlinearity function which defines the transient, time-dependent reliance of the relaxation modulus upon the applied strain.

On the other hand one should note that by using a similar procedure, the nonlinear creep compliance could be expressed on the basis of the eqilibrium (infinite) compliance, in the form (which has no practical advantages):

$$J_n(t,\sigma_0) = g_\infty J_\infty - g_t \sum_{i=1}^{m} J_i \exp(\frac{-t}{\tau_i}) \tag{15}$$

An Example

In Fig. 3 experimental stress-relaxation curves of POM in simple tension are shown. In the same figure theoretical values obtained on the basis of Eq. (14) are also represented (for details on the computational procedure see [3]). A good agreement between experiment and theory can be observed.

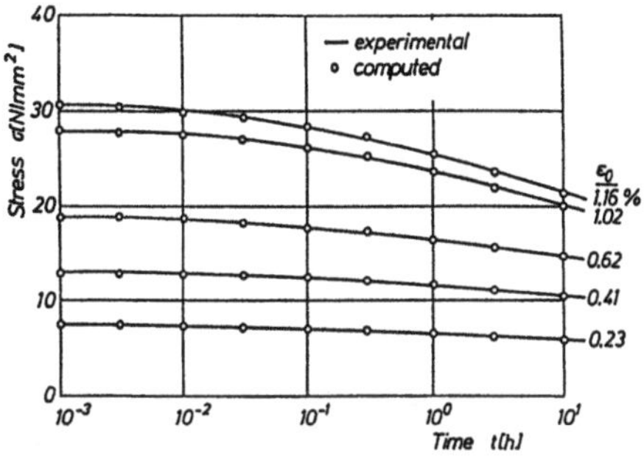

Fig. 3: Stress relaxation curves of POM in tension

The only one existent nonlinearity function (the time dependent one) used in this computation is depicted in Fig. 4. It should be noted, that here, as well as in all other cases of stress-relaxation investigated by the author, the deviation from linear viscoelastic behavior is practically linear.

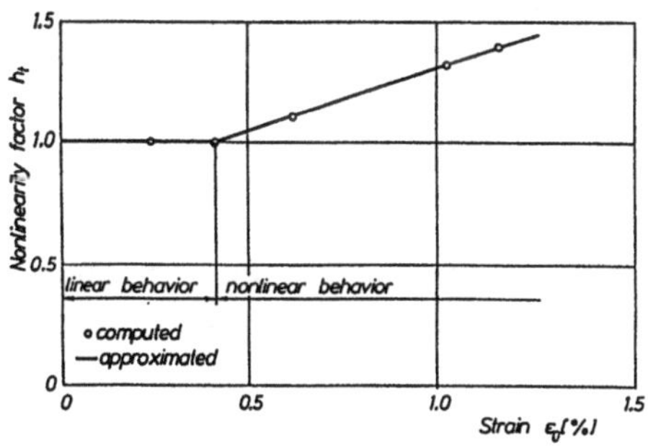

Fig. 4: The nonlinearity function of POM in tensile relaxation

On the basis of the simplified relations presented here, it is sufficient to conduct only a reduced number of experiments over the whole time-range for which the characterization is needed. From these tests, the linear parameters could be computed, if the nonlinearity function of the time dependent term would be known. Since this nonlinearity function is valid for the entire considered time-range of the experiments, it is sufficient to perform a series of tests at different loading levels in a very limited time-interval. They allow the numerical determination of the needed nonlinearity function, and subsequently, that of the linear parameters.

As an example, the stress-relaxation of ABS is considered. The experimental data is illustrated by five relaxation curves at strain values between 0.3% and 1.2% up to 300 hours (Fig. 5).

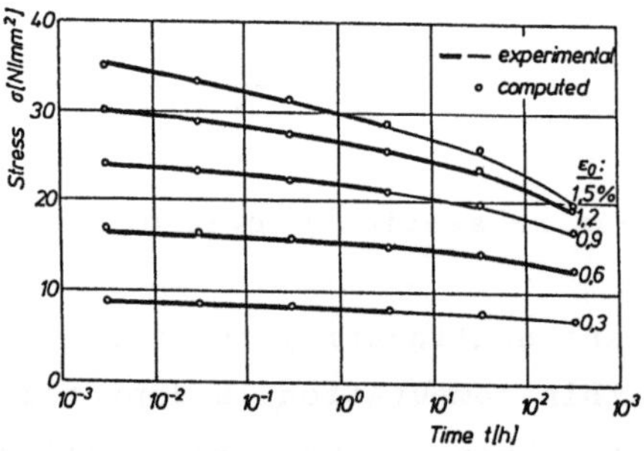

Fig. 5: Stress-relaxation curves of ABS

The experimental data used for the material characterization were taken only from the bulk lines represented in Fig. 5. For the determination of the nonlinearity factor, all the curves in a limited range up to 3 hours were used; for the computation of the linear parameters, the curves at 0.6% and 1.2% strain up to 300 hours were considered. In this way about 60% of the experimental time have been saved.

References

1. Williams, M.L.: Structural Analysis of Viscoelastic Materials. AAIA J. 2 (1964) 785-808.

2. Schapery, R.A.: On the Characterization of Nonlinear Viscoelastic Materials. Polym. Eng. Sci. 9 (1969) 295-310.

3. Brüller, O.S.: On the Nonlinear Characterization of the Long-Term Behavior of Polymeric Materials. Polym. Eng. Sci. 27 (1987) 144-148.

4. Brüller, O.S.: On the Nonlinear Response of Polymers to Periodical Sudden Loading and Unloading. Polym. Eng. Sci. 25 (1985) 604-607.

Constitutive Equations of Nonstationary Metal Creep

G.I. BYKOVTSEV

Institute for Automation and Control Processes,
Vladivostok, USSR

Summary

Kinematic equations of creep theory have been proposed by Yu.N. Rabotnov in [1]. Attempts of many researches to state constitutive equations for structural parameters caused discrepancy between theoretical and experimental results in the case of variable step load. We propose to use experimental reload data to define the structural parameters. In building this theory we use and extend ideas in [2,3].

Assumptions on creep phenomena in a specimen under torsion and off-loading

The theory of nonstationary metal creep is based on the assumption that creep curves under constant stress can be approximated by the expression

$$p(t) = V(\sigma)t + a(\sigma)(1 - e^{-mt}) , \tag{1}$$

where p is the creep strain, σ is the stress, $V(\sigma)$ and $a(\sigma)$ are functions of σ defined by experiment, and m is a constant that is independent of stress. In [3,4] it is shown that the proposed relationship and experimental data are in good coincidence for metal.

If on the stationary curve segment, stress is suddenly relieved, then reset strain is observed, which terminates in a time interval equal to load time. The maximum magnitude of the reset strain

$$p(t_1) - p(\infty) = C(\sigma) , \tag{2}$$

is independent of the stress-relief time t_1 on the stationary curve segment and is a function of initial stress σ. The relationship was verified in V.I. Gorelov's experiments for many metals.

Constitutive equations for simple tension

We will construct the simplest version of the kinetic creep theory to describe exactly the experimental relationship (1) and (2).

M. Zyczkowski (Ed.)
Creep in Structures
IUTAM Symposium Cracow/Poland 1990

Kinetic theory equations with one structural parameter will take the form

$$\begin{aligned}\dot{p} &= a_1(\sigma)p + b_1(\sigma)q + C_1(\sigma) \ , \\ \dot{q} &= a_2(\sigma)p + b_2(\sigma)q + C_2(\sigma) \ . \end{aligned} \tag{3}$$

Eqs. (3) for fixed stress are linear. The basic nonlinear effects are described by using $a_i(\sigma)$ and $b_i(\sigma)$. To describe relationship (1) and (2), equations (3) become

$$\begin{aligned}\dot{p} &= q(\sigma)C^{-1}(\sigma)q + V(\sigma) + ma(\sigma) \ , \\ \dot{q} &= -mg - m^2C(\sigma) \ . \end{aligned} \tag{4}$$

The initial state $p(0) = q(0) = 0$. For $\sigma = 0$, the equality $a(0)C^{-1}(0) = 1$ is hold.

As the parameter q is introduced in a rather formal way, it can be replaced by any parameter r which is a linear function of p and q:

$$r = d_1 p + d_2 q \ , \quad d_i = const \ . \tag{5}$$

It is natural that the transformation (5) does not change the meaning of physical phenomena under description. We will use the relationship (4) in our research. These relationship include three experimental functions $a(\sigma)$, $b(\sigma)$, $C(\sigma)$ and a constant m. To construct these functions we have to make several experiments for measuring creep curves for different values of σ and estimating the magnitude of reset deformation on the stationary curve segment. Figure 1 illustrates the geometrical meaning of these quantities. It should

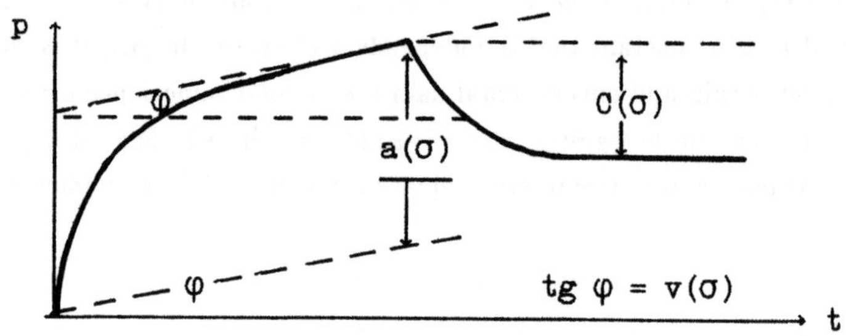

Figure 1:

be noted that $a(\sigma)$, $V(\sigma)$, $C(\sigma)$ are estimated from most reliable measurements made on the stationary curve segment.

The comparison of the results obtained by integrating (4) for various step changes of σ with experimental measurements on ZI437B, 40HSN2MWF and 5HNIM steels is made in [3,4,6]. At the beginning creep curves for five fixed stress values have been built. Using the curves, we have determined $a(\sigma)$, $V(\sigma)$ and m as well as $C(\sigma)$ associated with the stationary curve segment obtained after relieving the load.

The steel specimens have further been strained up to creep strain p_0 associated with the nonstationary curve segment. The load has been then relieved and the reset strain has been estimated. We have determined the experimental relationship $p_0 - p(\infty) = f(p_0)$ in this way.

Two specimens of ZI437B steel have been strained under fixed stress values σ_1 and σ_2 up to creep strain p_1. Then we have reloaded then on to $\sigma = 0.5(\sigma_1 + \sigma_2)$ and measured creep strain at various instants of time.

Finally, we have investigated the specimens under various step load with four load switches.

In all cases we have observed a good coincidence between experimental data and the results of integration of (4), except for the cases where multiple zero-reloads take place. The experimental creep curves obtained after two zero-reloaded switches lie higher than theoretical ones.

Constitutive equations at three-dimensional stressed conditions

For complex stressed conditions, generalized relationship are proposed in [3] in the form:

$$\begin{aligned} \dot{p}_{ij} &= a(J)C^{-1}(J)q_{ij} + \frac{3}{2}[V(J) + ma(J)]J^{-1}S_{ij} \, , \\ \dot{q}_{ij} &= -mq_{ij} - \frac{3}{2}m^2C(J)J_{-1}S_{ij} \, , \end{aligned} \tag{6}$$

where p_{ij} is the creep strain tensor, $S_{ij} = \sigma_{ij} - \frac{1}{3}\sigma_{kk}\delta_{ij}$ is the deviator stress tensor, q_{ij} is the kinetic parameter tensor, and J is the stress measure which will further take from $J = (\frac{3}{2}S_{ij}S_{ij})^{1/2}$.

The selection of the constitutive equations in the form (6) is stipulated by the following considerations:

1. For simple tension, relationship (6) become (4).

2. According to (6), initial torsion has no effect on following tension. This fact has been obtained experimentally by V.S. Namestnikov [7].

3. If a specimen creeps for some time at $\sigma_{12} = \tau$ and is reloaded onto $\sigma_{12} = -\tau$, then, according (6), the creep velocity at $\sigma_{12} = -\tau$ is greater than one for $\sigma_{12} = \tau$ at the instant of load switch. This fact has been observed by V.S. Namestnikov.

4. For simple load when stress σ_{ij} varies as some parameter, the creep velocity $\dot{p}_{ij}$ is proportional to deviator stress tensor S_{ij}. This relationships are observed in many experiments and stated in [4,5].

5. For reloads at complex stressed conditions, the creep velocity vector, according to (6), departs from tensor S_{ij} in the direction of increased load and then approaches

the deviator stress tensor S_{ij} asymptotically. This fact has been observed by experiment, described in [7,10] and, finally, stated by O.V. Sosnin [11].

The quantitative comparison of results, followed from (6), with experimental data obtained for specimens of 40HSN2MWF and 5HNIM steels is made in [3].

We will consider possible ways to simplify relationships (6). Using the results obtained from creep and reset curves for 40HN2MWF, 3HW40F and 5HNIM steels, we have draw a conclusion that ratios $a(\sigma)/V(\sigma)$ and $C(\sigma)/V(\sigma)$ are independent of stress values [5], i.e. $a(\sigma) = \alpha V(\sigma)$, $C(\sigma) = \beta V(\sigma)$.

It should be noted that relationships (2) follow from (4) only if $a(\sigma)/C(\sigma) = 1$. And if we assume that the experimental relationships are hold at $\sigma = 0$, then $\alpha = \beta$.

The equations (6) then take the form

$$\begin{aligned} \dot{p}_{ij} &= q_{ij} + \frac{3}{2}(1 + m\alpha)V(J)J^{-1}S_{ij} \\ \dot{q}_{ij} &= -mq_{ij} - \frac{3}{2}m^2\alpha V(J)J^{-1}S_{ij} \end{aligned} \tag{7}$$

Given a load trajectory S_{ij}, the integration of (9) results in the integral relationships:

$$p_{ij} = \frac{3}{2}\int_0^t \left[1 + m\alpha e^{m(\tau - t)}\right] V[J(\tau)]J^{-1}(\tau)S_{ij}(\tau)d\tau \tag{8}$$

The physical meaning of the structural parameters

We will consider the physical meaning of parameters q_{ij}. Assume that, on straining along some load trajectory, their values p_{ij}^0 and q_{ij}^0 are reached at time t_1. If $t = t_1$, zero-reload occurs. If $t > t_1$, then $S_{ij} = 0$ and relationships (6) takes the form:

$$\begin{aligned} \dot{p}_{ij} &= q_{ij} \, , \\ \dot{q}_{ij} &= -mq_{ij} \, . \end{aligned} \tag{9}$$

The solution of (11) is

$$\begin{aligned} p_{ij} &= q_{ij}^0 m^{-1}\left[1 - e^{-m(t-t_1)}\right] + p_{ij}^0 \, , \\ q_{ij} &= q_{ij}^0 e^{-m(t-t_1)} \, . \end{aligned} \tag{10}$$

For reversible creep strain, when values of t are sufficiently large, we get

$$p_{ij}^0 - p_{ij}(\infty) = -q_{ij}^0 m^{-1} \, .$$

Thus, the parameter q_{ij} introduced has a clear physical meaning. Tensor q_{ij} is proportional to reversible creep strain at any time.

Conclusion

We have proposed relationships (1) and (2) for creep and reset strains, respectively. On assuming the constitutive equations (3) for the kinetic creep theory, we have obtained, from (1) and (2), the constitutive equations (4) which include three experimental functions $a(\sigma)$, $V(\sigma)$, $C(\sigma)$ and a constant m, independent of stress. The geometric meaning of the introduced functions is illustrated in Fig. 1. The extension of (4) onto the three-dimensional stressed state in the form (6) has allowed us to describe the fundamental relationships observed during the experimental study of various metals. The structural parameters represent reversible creep strains.

References

[1] Rabotnov, Yu.N.: *On the equations of state for creep*, Progress in Appl. Mech., Macmillian, New York 1963, 307-315

[2] Astaphev, W.I.: *Description of creep-hardening with the help of a tensor internal variable*, Izv. AN SSSR, MTT **2**, (1987), 132-140 (in Russian)

[3] Bereznaia, O.I., Bykovtsev, G.I., Gorelov, W.I.: *Derivation of kinetic equations of the creep theory*, Izv. AN SSSR, MTT **1**, (1988), 147-157 (in Russian)

[4] Bykovtsev, G.I., Gorelov, W.I.: *On a certain constitutive law in metal creep*, Dokl. AN SSSR (DAN SSSR), **5**, (1983), 273, 1080-1082 (in Russian)

[5] Gorelov, W.I.: *Investigation of creep and recovery properties in metals*, Izv. AN SSSR, MTT **6**, (1987), 137-142 (in Russian)

[6] Bykovtsev G.I., Gorelov, W.I.: *Phenomenologic derivation of kinetic equations of creep theory*, Dokl. AN SSSR (DAN SSSR), **1**, (1985), 283, 83-84 (in Russian)

[7] Namestnikov, W.S.: *On creep caused by a changeable load in conditions of complex stress state*, Izv. AN SSSR, OTN **10**, (1957), 83-85 (in Russian)

[8] Namestnikov, W.S.: *Direct and indirect torsion in creep conditions*, PMTF, **1**, (1960), 121-122 (in Russian)

[9] Namestnikov, W.S.: *Technological creep of pure copper in conditions of complex stress state*, Izv. AN SSSR, OTN **4**, (1957), 127-130 (in Russian)

[10] Johnson, A.E., Henderson, J., Mathur, V.P.: *Creep under changing complex creep system*, Engineer. **5350**, (1985), 206, 209-216

[11] Sosnin O.W.: *On orientation of strain hardening in creep*, Izv. AN SSSR, MTT **3**, (1970), 120-124 (in Russian)

Creep of Metals Under Variable Complex Loadings

J.L. DING[1] and W.N. FINDLEY[2]

1. Department of Mechanical and Materials Engineering, Washington State University, Pullman, WA 99164-2920, USA
2. Division of Engineering, Brown University, Providence, RI 02912, USA

Summary
In recent years, numerous material models and experimental observations on the inelastic behavior of metals at elevated temperatures have been proposed and reported. Despite their various forms, the models can in general fit into some common structures. Identification of such structures is essential in the development of constitutive equations. It is also far more important than finding a specific relation for a specific material. The same is also true in the experimental work. In order to systematically characterize the material behavior or validate the models, a structured approach is necessary. In the current paper, instead of reviewing the details of the available theoretical and experimental work, it is attempted to summarize the current progress in terms of some basic theoretical structures and experimental methodology. Examples derived from these structures and their capability for predicting the creep deformation of metals under variable complex stresses are discussed. Finally emphasis of future research on creep under nonproportional loadings is suggested.

Formulation of the Constitutive Equations

The initial condition for most of the viscoplastic constitutive equations proposed so far is an isotropic material. In the early models, the formulation was usually based on the strain hardening approach in which the current inelastic strain rate is a function of the applied stress and the accumulated inelastic strain. Modifications have to be made for this approach in order to describe the possible softening, damage and the deformation induced anisotropy. Furthermore, using accumulated strain to characterize the inelastic state of a material has also been a debatable issue.

To improve the above deficiencies, in most of the advanced constitutive equations, a set of internal state variables, instead of the accumulated strain, has been adopted to characterize the inelastic state, and the deformation history is described by the evolution of the selected state variables. As far as the modeling of the inelastic deformation at elevated temperatures is concerned, usually the hardening or softening which is isotropic in nature is modeled through a scalar state variable R, sometimes called drag stress; and the deformation induced anisotropy is through a tensor variable $\underset{\sim}{\alpha}$, sometimes called back stress. A damage state variable could also be included to characterize the possible damaged state and the coupling effect with the inelastic deformation. The nature of the

M. Zyczkowski (Ed.)
Creep in Structures
IUTAM Symposium Cracow/Poland 1990

damage variable has been discussed by Cocks and Leckie [1], and Murakami [2] etc. Since damage is a variable which can essentially be characterized with the same approaches as for others, it will be excluded in most of the following discussions.

The general structure for modeling the inelastic material behavior by internal variables was studied by Rice [3], and Onat and Leckie [4] among others. The formulation starts with a potential function, usually free energy ψ which is a function of elastic strain ϵ^e, temperature T, and internal variables ξ. Partial differentiation with respect to strain and internal variables yields respectively the stress and the thermodynamic forces conjugate to the internal variables. Consideration of the first and second thermodynamic laws in addition to the basic kinematic such as the material objectivity then leads to certain constraints on the possible structures of the constitutive equations. Furthermore, assuming $\dot{\xi}$ is only a function of the thermodynamic force associated with ξ and the present state of the material, Rice [3] showed that it was possible to find a dissipative potential function (Ω) from which the inelastic strain rate $(\dot{\epsilon}^{VP})$, could be derived, i.e.,

$$\dot{\underset{\sim}{\epsilon}}^{VP} = \partial\Omega/\partial\underset{\sim}{\sigma} \quad \text{at fixed inelastic state} \tag{1}$$

where $\underset{\sim}{\sigma}$ is the stress tensor. Based on similar arguments, it can be shown that the evolution rule of the internal variables may be derived from the same potential. For example, see the work by Ponter and Leckie [5]. For an initially isotropic material, the flow potential is usually assumed to be a function of the second invariant of $(\underset{\sim}{\sigma}' - \underset{\sim}{\alpha})/R$ with the initial value of $\underset{\sim}{\alpha}$ equal to zero, where $\underset{\sim}{\sigma}'$ is stress deviator. Taking partial differentiation of Ω with respect to stress then yields a strain rate in the following form:

$$\dot{\underset{\sim}{\epsilon}}^{VP} = \lambda(\underset{\sim}{\sigma}' - \underset{\sim}{\alpha}) \tag{2}$$

where λ is again a scalar function of the second invariant of $(\underset{\sim}{\sigma}' - \underset{\sim}{\alpha})/R$. If $\underset{\sim}{\alpha}$ can be excluded, eqn. (2) is then reduced to the classical J_2 flow theory for isotropic materials.

The evolution rules for the state variables are usually assumed in the following forms:

$$\dot{\underset{\sim}{\alpha}} = h_\alpha \dot{\underset{\sim}{\epsilon}}^{VP} - r_\alpha \underset{\sim}{\alpha}, \tag{3}$$

$$\dot{R} = h_R \mid\mid \dot{\underset{\sim}{\epsilon}}^{VP} \mid\mid - r_R, \tag{4}$$

where h_α and h_R are the hardening functions, r_α and r_R are the softening or recovery functions, and $\mid\mid \dot{\underset{\sim}{\epsilon}}^{VP} \mid\mid$ is the effective magnitude of the strain rate tensor. Examples of the material models which fit into the above structure can be found in the work by Bodner and Partom [6], Miller [7] etc. Eqns. (2)-(4) are apparently in a differential form. Development of the advanced constitutive equations can also follow an integral approach where the deformation history is represented as a hereditary integral. Both approaches can theoretically be interrelated to each other. The integral can be defined with respect to either an extrinsic time, e.g., the multiple integral representation proposed by Rivlin and others which was originally applied to describe the nonlinear viscoelastic behavior of plastics and polymers, see the book by Findley et al., [8]; or an intrinsic time, e.g., the endochronic viscoplastic theory proposed by Valanis [9]. A more detailed discussion on the multiple integral representation will be given in a later section.

It should also be mentioned that nearly all the viscoplastic models treat the time-dependent inelastic strain as nonrecoverable. However, it has been observed that strain recovery does occur when the stress is released. In the vicous-viscoelastic model proposed by Findley and Lai [10], the inelastic strain was therefore further decomposed into a recoverable, i.e., viscoelastic and a nonrecoverable, i.e., viscoplastic component.

Experimental Characterization of the Constitutive Equations

Most of the experimental work in the area of constitutive modeling was performed under uniaxial stress state. Accordingly determination of material constants and test of models were limited to uniaxial stress condition too. From the physical point of view, although uniaxial tests can generate some data on the evolution of the material properties, they provide little information on how the hardening, softening, and damage progress in general, e.g., whether it is in an isotropic or anisotropic manner. From the mathematical point of view, when applied to uniaxial stress condition, the tensorial constitutive equations are actually reduced to scalar equations. In this case, since the tensor variable, such as $\underset{\sim}{\alpha}$ has been reduced to a scalar, it's really just another material parameter similar to R. Adding more material parameters in a constitutive equation can of course better describe the material behavior. However, the significance of one material parameter $\underset{\sim}{\alpha}$, which is a second rank tensor, as compared to another, R which is a zero rank tensor, can really not be distinguished. One of the fundamental differences between the tensor and the scalar variables is that $\underset{\sim}{\alpha}$, like $\underset{\sim}{\dot{\epsilon}}^{VP}$, is a directional quantity. To characterize the directional quantity, information on both magnitude and direction is needed. It thus appears that in order to systematically characterize the inelastic material behavior, the dimension of the material test has to be extended to at least two, e.g., combined tension and torsion test, so that at least two tensor components can be independently measured. Characterization of two components can lead to the construction of a vector in a two-dimensional space which can provide the information not only on the magnitude but also the direction of a tensor at least in a two-dimensional space.

In a recent work by Ding and Lee [11], creep tests under nonproportional loadings were suggested as a tool to systematically characterize the inelastic material behavior at elevated temperatures. They were especially useful for studying the directional characteristics of the hardening, softening or damage, or equivalently the significance of the tensor variables in a constitutive equation. The experimental program starts with a creep test under constant biaxial stress state. After the first loading step, the loading direction is abruptly changed to another direction. Both the direction and the magnitude of the inelastic strain rate are monitored during the test. The material response in the first loading step is used to characterize the initial inelastic state and those in the subsequent loading steps along different loading directions are used to study the possible deformation induced anisotropy. The underlying idea for using creep test to identify the tensor variable $\underset{\sim}{\alpha}$ is that since the stress is constant during a creep test, the direction of the creep rate direction should be governed by the current value of $\underset{\sim}{\alpha}$. In other words, study of the creep direction in a creep test should provide important information on identifying $\underset{\sim}{\alpha}$. This approach can be easily appreciated with eqn. (2) which clearly shows the dependence of creep direction on $\underset{\sim}{\alpha}$ when the stress is constant.

In addition to the characterization of the degree of anisotropy, multiaxial creep tests may also be used to study the directional characteristics of the evolution of the tensor variable $\underset{\sim}{\alpha}$. This can be illustrated by using eqns. (2) to (4) as an example. As discussed by Ding and Lee [11], eqn. (2) implies a self-consistent relation for $\dot{\underset{\sim}{\alpha}}$, which can be obtained by differentiating eqn. (2) with respect to time, i.e.,

$$\ddot{\underset{\sim}{\epsilon}}^{VP} = \lambda\dot{\underset{\sim}{\Sigma}} + \dot{\lambda}\,\underset{\sim}{\Sigma} \tag{5}$$

where $\underset{\sim}{\Sigma} = \underset{\sim}{\sigma}' - \underset{\sim}{\alpha}$. If the applied stress is constant as in a creep test, a simple manipulation will lead to eqn. (6)

$$\dot{\underset{\sim}{\alpha}} = \frac{\dot{\lambda}}{\lambda^2}\dot{\underset{\sim}{\epsilon}}^{VP} - \frac{1}{\lambda}\ddot{\underset{\sim}{\epsilon}}^{VP} \tag{6}$$

which indicates that the direction of $\dot{\underset{\sim}{\alpha}}$ can be related to the directions of $\dot{\underset{\sim}{\epsilon}}^{VP}$ and $\ddot{\underset{\sim}{\epsilon}}^{VP}$. Of course, eqn. (6) should theoretically be consistent with the evolution rule for $\underset{\sim}{\alpha}$, i.e., eqn. (3), under the constant stress loading condition.

For 2618-T61 aluminum alloy tested at 200 C, it was found by Ding and Lee [11] that for a virgin specimen the creep direction was always consistent with the applied stress direction. When the loading direction changed, the creep direction also rapidly approached the applied stress direction after a short transition period. Furthermore, it was found that the directions of $\dot{\underset{\sim}{\epsilon}}^{VP}$ are $\ddot{\underset{\sim}{\epsilon}}^{VP}$ are essentially parallel to each other which implied that $\dot{\underset{\sim}{\alpha}}$ should also be in the same direction as $\dot{\underset{\sim}{\epsilon}}^{VP}$ according to eqn. (6), and the directional softening term in eqn. (3) which depended on the current value of $\underset{\sim}{\alpha}$ may be negligible. The consistent coaxiality between $\underset{\sim}{\sigma}$ and $\dot{\underset{\sim}{\epsilon}}^{VP}$ and between $\dot{\underset{\sim}{\alpha}}$ and $\dot{\underset{\sim}{\epsilon}}^{VP}$ seems to suggest that the hardening or softening may progress in an isotropic manner, or equivalently the role of the tensor variable in the constitutive equation for the 2618 aluminum may not be significant. This is certainly an important piece of information. However, it can hardly be obtained without a systematic experimental investigation. Once the individual roles of the state variables can be determined, the rest of the work related to the characterization of a constitutive equation becomes quite straightforward.

Although the above discussion was mainly based on eqns. (2) to (4), it is believed that similar approach can be applied to more complicated equations.

Experimental Simulation of Creep of Metals Under Variable Complex Loading

The ultimate goal of a constitutive equation is to accurately predict the behavior of a structure under the possible service conditions. Creep experiments under variable complex loadings will not only improve the understanding of the material behavior under these conditions but also provide a valuable data base for evaluating the theories. The relevant work before 1980 was reviewed by Findley at al., [12]. More recently an extensive experimental study on creep of 2618 aluminum alloy was performed by Ding and Findley [13-15]. The stress histories examined can be grouped into: stepwise varied multiaxial proportional loadings in which the magnitude of stresses varied in steps but the loading directions were kept constant; stepwise varied nonproportional loadings including cyclic loadings; and axial stress relaxation combined with torsional creep and controlled strain rate tests both in tension and in torsion [16]. Similar work was also performed on 304 stainless steel by Cho and Findley [17-19].

Theoretical Simulation of Creep of Metals Under Variable Complex Loadings

Three tentative viscoplastic models formulated with different approaches were developed to simulate the inelastic behavior observed for 2618 aluminum. One is based on the isotropic strain hardening approach which is of the following form:

$$\dot{\epsilon}_{ij}^{VP} = f(\tau_{max})m\left[\frac{\frac{1}{2}\gamma_{max}}{\tau_{max}f(\tau_{max})}\right]^{\frac{m-1}{m}}\sigma'_{ij} \tag{7}$$

The function f and the constant m were determined from the creep test data under constant stress [20]. Use of maximum shear stress (τ_{max}) and maximum shear strain (γ_{max}) instead of the conventional effective stress and strain was based on the observation that the Tresca relation fitted the experimental data better than the Von Mises relation.

The second is formulated with one tensor state variable, namely,

$$\dot{\epsilon}_{ij}^{VP} = C\sinh(\| \Sigma_{ij} \|)\Sigma_{ij}/ \| \Sigma_{ij} \|, \tag{8}$$

$$\dot{\alpha}_{ij} = h(\| \alpha_{ij} \|, \| \dot{\epsilon}_{ij}^{VP} \|)\dot{\epsilon}_{ij}^{VP}, \tag{9}$$

where $\| \Sigma_{ij} \|, \| \alpha_{ij} \|$, and $\| \dot{\epsilon}_{ij}^{VP} \|$ are the effective magnitudes of the corresponding tensors which are again defined in terms of the Tresca relation. Constant C and function h were also determined from the creep test data [20].

The third one is actually the Bodner-Partom's model formulated with one scalar state variable. It has the following form:

$$\dot{\epsilon}_{ij}^{VP} = D_o\exp[-\frac{1}{2}\left(\frac{R^2}{3J_2}\right)^n]J_2^{-\frac{1}{2}}\sigma'_{ij}, \tag{10}$$

$$\dot{R} = p(R_1 - R)\dot{W}_p - q(R - R_o)^d, \tag{11}$$

where $\dot{W}_p$ is the plastic work rate, J_2 stands for the second invariant of the sress deviator, and D_o, n, R_o, R_1, p, q and d are the material constants determined from the constant strain rate tests [16].

As mentioned earlier, the viscoplastic models usually treat the inelastic strain as nonrecoverable. In order to take the strain recovery into consideration, the viscous-viscoelastic model mentioned earlier was also formulated. In the work by Ding and Findley [13-15], the viscoplastic component was modeled with eqns. (7), and (8) and (9); and the viscoelastic one by a third order multiple integral formulation which is of the following form for a uniaxial stress condition:

$$\begin{aligned}
\epsilon^{VE}(t) = &\int_0^t \phi_1(t-\tau_1)\dot{\sigma}(\tau_1)d\tau_1 \\
&+ \int_0^t \int_0^t \phi_2(t-\tau_1, t-\tau_2)\dot{\sigma}(\tau_1)\dot{\sigma}(\tau_2)d\tau_1 d\tau_2 \\
&+ \int_0^t \int_0^t \int_0^t \phi_3(t-\tau_1, t-\tau_2, t-\tau_3)\dot{\sigma}(\tau_1)\dot{\sigma}(\tau_2)\dot{\sigma}(\tau_3)d\tau_1 d\tau_2 d\tau_3 \qquad (12)
\end{aligned}$$

When applied to variable stress states, a modified superposition principle is usually used to simplify the integral. Other details can be found in the book by Findley et al. [8]. It should also be mentioned that in the work by Cho and Findley [17-19] and Findley and Lai [10], both inelastic strain components were modeled with the multiple integral formulation combined with a creep limit, a stress below which creep was negligible.

The details of the simulation can be found in the aforementioned references. Generally speaking, from the comparison between the viscoplastic and the viscous-viscoelastic models, it was found that the decomposition of the inelastic strain into viscoelastic and viscoplastic components was useful in accounting for not only the strain recovery but also the transient creep behavior upon reloadings and set-up loadings. As far as the three viscoplastic models are concerned, it was found that for proportional loadings, either uniaxial or multiaxial, the differences among the predictions from all three models were insignificant. However, for nonproportional loadings, they all had limitations on describing some of the features observed for the current aluminum alloy with the one formulated with one tensor state variable, i.e., eqns. (8) and (9), the worst. Examples of deficiencies were found in the effects of precreep in tension on the subsequent creep in torsion or vice versa, synergistic effects due to the addition of stress components, cyclic softening effects and the strongly nonlinear material behavior at high stress levels. The fact that eqns. (8) and (9) gave the poorest prediction may not be surprising based on the earlier observation that the evolution of the hardening or softening of the current aluminum may progress in an isotropic manner. Part of the deficiencies of eqn. (7) is apparently its negligence of the softening effects. For Bodner-Partom's model, it was found that with some modification of the softening function, a single scalar state variable seemed to be able to reasonably describe the observed material behavior [16].

Discussion

The above comparison again demonstrates one of the key points illustrated in the current presentation, namely, without a systematic experimental methodology, blindly curve fitting a proposed equation to some arbitrary experimental data could lead to serious errors in structural analysis. For example, even though the material constants in a constitutive equation based on a tensor state variable can be determined by curve fitting to the creep test data, the essence of the equation itself could be unjustifiable.

Another important observation from the above simulation study is that the inelastic behavior under proportional loadings, whether it is multiaxial or not, may not be discriminative enough to distinguish the characteristics of different types of constitutive equations. The emphasis of future research should probably be put on creep under nonproportional loadings which will provide valuable information for not only characterizing but also evaluating a constitutive equation. Modelling of the macroscopic behavior under these conditions, such as some of the features mentioned above, may have to rely on better understanding of the inherent deformation mechanisms. Perhaps some damage parameter may need to be included in the model even though distinguishing the damage from other softening mechanisms, e.g., aging, and modeling are still open questions.

Conclusion

Overall, it seems that significant progress has been made in the past decade on both experimental characterization and theoretical modelling of the inelastic behavior of metals at elevated temperatures. However, understanding and modelling the creep under nonproportional loadings remain to be the research challenges for the next decade.

References

1. Cocks, A.C.F.; Leckie, F.A.: Creep constitutive equations for damaging materials. Advances in Applied Mechanics, 25 (1987), 239-294.
2. Murakami, S.: Mechanical modeling of material damage. ASME J. Appl. Mech., 55 (1988), 280-286.
3. Rice, J.R.: On the structure of stress-strain relations for time-dependent deformation in metals. ASME J. Appl. Mech., 37 (1970), 728-737.
4. Onat, E.T.: Leckie, F.A.: Representation of mechanical behavior in the presence of changing internal structure. ASME J. Appl. Mech., 55 (1988), 1-10.
5. Ponter, A.R.S.; Leckie, F.A.: Constitutive relationships for time dependent deformation of metals. ASME J. Eng. Mat. & Tech., 98 (1976), 47-51.
6. Bodner, S.R.: Partom, Y.: Constitutive equations for elastic viscoplastic strain ahrdening materials. ASME J. Appl. Mech., 42 (1975) 385-389.
7. Miller, A.K.: An inelastic constitutive model for monotonic, cyclic and creep deformation. ASME J. Eng. Mat. & Tech., 98 (1976), 97-113.

8. Findley, W.N.; Lai, J.S.; Onaran, K.: Creep and relaxation of nonlinear viscoelastic materials. New York: Dover Publications, Inc. 1989.

9. Valanis, K.C.: A theory of viscoplasticity without a yield surface. Arch. Mech., 23(4) (1971), 517-551.

10. Findley, W.N.; Lai, J.S.: Creep and recovery of 2618 aluminum alloy under combined stress with a representation by a viscous-viscoelastic model. ASME J. Appl. Mech., 45 (1978), 507-514.

11. Ding, J.L.; Lee, S.R.: Development of viscoplastic constitutive equation through biaxial material testing. Exp. Mech., 25, (1987), 239-294.

12. Findley, W.N.;Cho, U.W.; Ding, J.L: Creep of metals and plastics under combined stresses, a review. ASME, J. Engr. Mat. & Tech., 101 (1979), 365-368.

13. Ding, J.L.; Findley, W.N.: Multiaxial creep of 2618 aluminum under proportional loading steps. ASME J. Appl. Mech., 51, (1984), 133-140.

14. Ding, J.L.; Findley, W.N.: Nonporoportional loading steps in multiaxial creep of 2618 aluminum. ASME J. Appl. Mech., 52, (1985), 621-628.

15. Ding, J .L.; Findley, W.N.: Simultaneous and mixed stress relaxation in tension and creep in torsion of 2618 aluminum. ASME J. Appl. Mech., 53, (1986), 529-535.

16. Lee, S.R.; Ding, J .L.: Viscoplastic constitutive modeling with one scalar state variable. Int. J. of Plasticity, 5 (1989), 617-637.

17. Cho, U.W.; Findley, W.N.: Creep and plastic strains of 304 stainless steel at 593 C under step stress changes, considering aging. ASME J. Appl. Mech., 49, (1982), 297-304.

18. Cho, U.W.; Findley, W.N.: Creep and plastic strains under side steps of tension and torsion for 304 stainless steel at 593 C. ASME J. Appl. Mech., 50 (1983), 580-586.

19. Cho, U.W.; Findley, W.N.: Creep and plastic strains under stress reversal in torsion with and without simultaneous tension for 304 stainless steel at 593 C. ASME J. Appl. Mech., 50, 1983, 587-592.

20. Ding, J.L.; Findley, W.N.: 48 hour multiaxial creep and creep recovery of 2618 aluminum alloy at 200 C. ASME J. Appl. Mech. 51, (1984), 125-132.

Theory of Creep and Long-Term Strength of Insotropic Hardening Media

V.P. GOLUB
Institute of Mechanics,
Academy of Sciences of Ukrainian SSR, Kiev, USSR

Summary

The theory of creep and long-term strength is founded on the principle of similarity of isochrones. The family of isochrones includes also the instantaneous stress-strain diagram. Special features of creep are determined by the character and intensity of strain-hardening of the medium. In particular, it is shown that the creep of linearly hardened materials is non-stationary, while in materials with power hardening all stages occur. A generalization of the theory on non-stationary loading and on complex stressed state is presented. A comparison with experiment is given.

Introduction

The use of modern methods of construction of constitutive relations in the theory of creep has allowed to obtain a broad range of relations - from approximating functions to non-linear integral equations. However the relations are derived from only a part of needed physical laws. The diversity of phenomenologic observations on various phenomena and micromechanism does not allow to generalize them on all materials and all loadings conditions.

The classical approach is based on the use of the hypothesis of "Mechanical equation of state"

$$F(\sigma, \varepsilon^C, \dot{\varepsilon}^C, t, \theta) = 0 , \tag{1}$$

which represents the relation between stress σ, strain ε^C, creep strain rate $\dot{\varepsilon}^C$, time t and temperature θ. The specific application of the function F is obtained on the basis of ageing and hardening concepts [1] and of the damage concept [2]. The three-dimensional creep law is formulated similarly as the Prandtl-Reuss equation

$$d\varepsilon_{ij}^C = \frac{3}{2}\frac{d\varepsilon^C}{\sqrt{3J_2}} s_{ij} , \quad (i,j = 1,2,3) , \tag{2}$$

where s_{ij} - the stress deviator, J_2 - second invariant of the stress deviator.

In the viscoelasticity the integral relations of Boltzmann-Volterra 2,3 are used

$$\varepsilon^C = \int_0^t \frac{\partial f(\sigma(\tau), t-\tau)}{\partial t} d\tau , \tag{3}$$

which allow to compute the creep deformation at every arbitrary loading history.

M. Zyczkowski (Ed.)
Creep in Structures
IUTAM Symposium Cracow/Poland 1990

In all existing approaches all unknown functions of creep are determined according to experimental graphs of creep in coordinates $(\varepsilon^C - t)$. These graphs are considered as fundamental properties of material, allowing to make generalizations. Actually, however, the generalizations are valid only for the initial stage of creep, where similarity conditions are satisfied. Numerous attempts to complicate equations (1) and (3) with the purpose of adequate description of the creep process as a whole prevent the practical applications of such theories.

Another approach to construction of the creep theory may be formulated as follows:

1. Fundamental properties of an isotropic creep medium are prescribed by isochronous diagrams in the plane $(\varphi(\varepsilon^r) - \varepsilon^r)$, which are similar at all stages. Coefficient of similarity has the structure of time operator.

2. The family of similar isochrones includes the instantaneous stress-strain diagram which is isochronic at $t = 0$. This allows to identify the propagation of creep process on the basis of the instantaneous stress-strain diagram. The initial state equation has the form

$$\Phi\left(\sigma, \varphi_0(\varepsilon^s, \dot{\sigma}), K(t), \theta\right) = 0 , \tag{4}$$

where φ_0 - function of instantaneous loading which depends on the stress rate $\dot{\sigma}$; $K(t)$ - coefficient of similarity.

Basic constitutive equations

The following relation represents the similarity of isochronic creep diagrams $\varphi_t(\varepsilon^r)$, which include also the diagram of instantaneous strain $\varphi_0(\varepsilon^s)$

$$\varphi_0(\varepsilon^s) = \varphi_t(\varepsilon^r)[1 + G(t)] . \tag{5}$$

The operator $G(t)$ will be chosen in such a manner that the condition

$$\varphi_t \big|_{t=0} = \varphi_0(\varepsilon^s) = \sigma \tag{6}$$

is satisfied. Here and above $\varepsilon^r = \varepsilon^C + \varepsilon_0$ - rheonomic strain ($\varepsilon_0 = \sigma/E$); ε^s - scleronomic strain, prescribed by diagram of instantaneous strain; σ - applied stress. In (5) it is also assumed $K(t) = 1 + G(t)$. The graphic interpretation of assumed notation is shown in Fig. 1a

It follows from the condition of similarity (5) that each creep isochrone may be reduced on the coordinate φ to the diagram of instantaneous deformation, for example, by

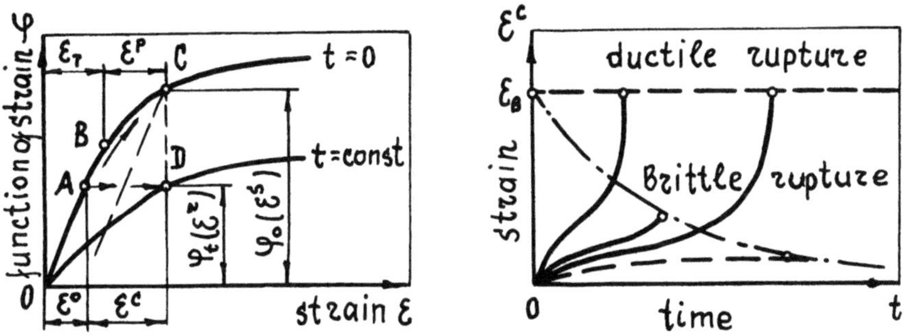

Figure 1: Isochronical creep diagrams (a) and creep fracture scheme (b)

adjustment of points D and C at fixed value of ε. In this case from (5) taking into account (6) we obtain for active processes

$$\varphi_0(\varepsilon^r) = [1 + G(t)]\sigma \ , \tag{7}$$

where it is assumed $\varepsilon^r = \varepsilon^C$. The criterion of activity in (7) is the increase of ε^r. The creep process is considered in plane $(\varphi - \varepsilon)$ as the alternation of active $(d\sigma > C\,; dt = 0)$ and of the passive $(d\sigma = 0\,; dt > 0)$ regimes of change the function φ_0.

We differentiate then both parts of the equation (7) with respect to t, its left hand side we multiply and divide by $d\varepsilon^r$. As a result the new relation is obtained for the coordinate system $(\varepsilon^r - t)$

$$\frac{d\varphi_0(\varepsilon^r)}{d\varepsilon^r}\frac{d\varepsilon^r}{dt} = \frac{d}{dt}\{\sigma\,[1 + G(t)]\} \ . \tag{8}$$

The solution of this equation with respect to the rheonomic strain rate $\dot{\varepsilon}^r$ allows to represent the constitutive equation of creep in the form

$$\frac{d\varepsilon^r}{dt} = \frac{d}{dt}\{\sigma\,[1 + G(t)]\}\left[\frac{d\varphi_0(\varepsilon^s)}{d\varepsilon^s}\right]^{-1} . \tag{9}$$

Here the value $d\varphi_0/d\varepsilon^C$ corresponds to tangent modulus $g_0(\varepsilon^s)$ and characterizes the intensity of the initial strain-hardening of the medium. As a result the creep strain rate is prescribed by equation (9) as the function of tangent modulus. As it is known the value $g_0(\varepsilon^s)$ is changing within the limits from E to 0, and for $\dot{\varepsilon}^C$ following estimations hold

$$\left.\frac{d\varepsilon^C}{dt}\right|_{g_0(\varepsilon^s)\to 0} \to 0 \ , \qquad \text{and} \qquad \left.\frac{d\varepsilon^C}{dt}\right|_{g_0(\varepsilon^s)\to 0} \to \infty \ . \tag{10}$$

These equations show that the equation (9) describes all characteristic stages of creep, from non-stationary to accelerated.

The second relation of the condition (10) actually determines the occurrence of mechanical instability in the creep process and may be applied as fracture criterion. From the diagram of instantaneous deformation it follows

$$g_0(\varepsilon^s) = \left.\frac{d\varphi_0(\varepsilon^s)}{d\varepsilon^s}\right|_{\varepsilon^s \to \varepsilon_B} \to 0 \,, \tag{11}$$

where ε_B - strain corresponding to the lose of stability of the instantaneous deformation. We integrate (9), taking into account (11) and, for the creep fracture time t_R following solution is obtained

$$\int_0^{t_R} \frac{\partial[1+G(t)]}{\partial t} \left[\frac{d\varphi_0(\varepsilon^s)}{d\varepsilon^s}\right] dt = \frac{\varepsilon_B}{\sigma} \,. \tag{12}$$

The graphic interpretation of this solution is shown in Fig. 1b. For deformations of fracture decreasing in time the solution is constructed with the use of the non-linear approximating function.

Creep and fracture of hardening materials

We consider the application of the constitutive equations obtained to the solution of problems of creep prediction and of long-term strength at stationary and stepwise loading. As the subject of investigation we choose materials with linear and power hardening. Methods of specification of unknown functions and of determination of coefficients are presented in [4].

Material with linear hardening will be described by instantaneous stress-strain relation in the form

$$\varphi_0(\varepsilon^s) = \begin{cases} E\varepsilon^e & (0 \le \varepsilon^e \le \varepsilon_T); \\ E\varepsilon_T + E^*\varepsilon^p & (0 \le \varepsilon^p \le \varepsilon_B - \varepsilon_T) \,, \end{cases} \tag{13}$$

where $\varepsilon^e, \varepsilon^p$ - elastic and plastic components of strain; E, E^* - modulus of elasticity and modulus of linear hardening; ε_T - strain corresponding to yield point σ_T. Upon substitution of (13) into (9) we obtain the expression for the creep strain rate $\dot{\varepsilon}^C$

$$\frac{d\varepsilon^C}{dt} = \begin{cases} aB\frac{\sigma}{E}\frac{1}{t^{1-B}} \,, \\ aB\frac{\sigma}{E^*}\frac{1}{t^{1-B}} \,. \end{cases} \tag{14}$$

Integrating (14), taking into account the initial conditions $\varepsilon^r = \varepsilon^0$ at $t = 0$, for creep strain ε^C we obtain

$$\varepsilon^C = \begin{cases} \frac{a}{E}\sigma t^B \,, \\ \frac{\sigma}{E}(1+at^B) + \frac{\sigma_T}{E}\left(1 - \frac{E}{E^*}\right) - \frac{\sigma}{E} \,. \end{cases} \tag{15}$$

Here it is assumed $G(t) = at^B$ with $a > 0$, $B > 1$. It is clear that the creep of linearly hardening materials can be non-stationary only.

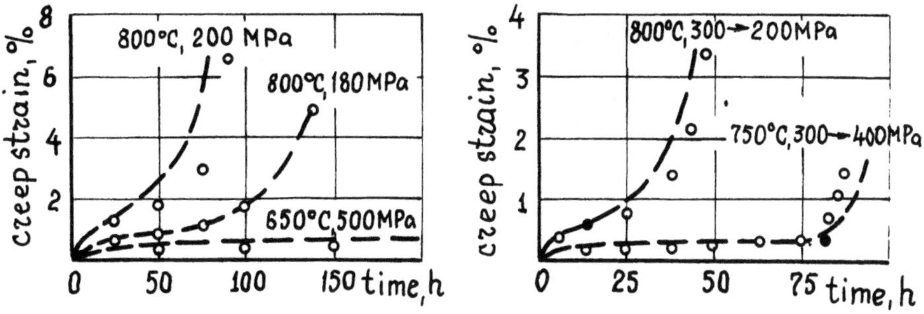

Figure 2: Theoretical and experimental creep curves of heatresistant alloy at constant (a) and onestage (b) loading

Materials with power hardening will be described by the instantaneous stress-strain relation in the form

$$\varphi_0(\varepsilon^s) = \begin{cases} E\varepsilon^e & (0 \le \varepsilon^e \le \varepsilon_T); \\ E\varepsilon_T + C(\varepsilon^p)^m & (0 \le \varepsilon^p \le \varepsilon_B - \varepsilon_T) , \end{cases} \tag{16}$$

where C , m - indexes of power hardening. In this case, by analogy with (14) and (15) for the creep strain rate $\dot{\varepsilon}^C$ we obtain

$$\frac{d\varepsilon^C}{dt} = \begin{cases} aB\frac{\sigma}{E}\frac{1}{t^{1-B}} \\ aB\left(\frac{1}{mC}\right)^{\frac{1}{m}}\frac{\sigma}{m}\left[\sigma(1+at^B) - E\varepsilon_r\right]^{\frac{1-m}{m}}\frac{1}{t^{1-B}} . \end{cases} \tag{17}$$

and for creep strain ε^C

$$\varepsilon^C = \begin{cases} \frac{\sigma}{E} \\ \left(\frac{1}{mC}\right)^{\frac{1}{m}}\left[\sigma(1+at^B) - \sigma_r\right]^{\frac{1}{m}} + \frac{\sigma_r - \sigma}{E} . \end{cases} \tag{18}$$

It is easy to prove [4] that in materials with power hardening both the stationary and the accelerated creep is occur. In Fig. 2a the results of computations for stationary loading, carried out with the use of Eqs. (15) and (18) are compared with experimental data. Solid line denotes here the portion of creep graph, used for determination of coefficients a and B.

In accordance with equation (9) the creep at stationary loadings is identified with passive (by increment of time) change of function φ_0. The change of function φ_0 at stepwise loadings is realized by increment of stress for fixed moment of time. As a result, the nonstationary creep is interpreted as the process of alternating the active and passive change of function φ_0 - owing to the stress increment and the time increment.

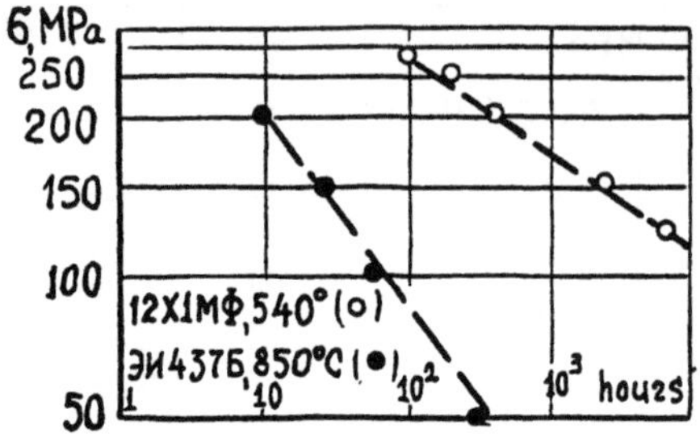

Figure 3: Lifetime of heat-resistant alloy for ductile fracture

At stepwise increase of the load, by integrating (9) for the region of non-stationary creep the relation holds

$$\varepsilon^C = \varepsilon_1^s + \frac{2}{E}\int_{\tau_1}^{t-\tau_1} \frac{\partial}{\partial t}\left[\sigma_2(1+at^B)\right] dt \,. \tag{19}$$

Here τ_1 - time of beginning change; σ_2 - stress at the second stage of loading. In Fig. 2b some results of computation of non-stationary creep are compared with experiment.

For evaluation of the long-term strength we consider now the conditions when accelerated creep develops. Taking into account the second relation (18) following relation is obtained for the time to fracture t_R from (12)

$$t_R = \left(\frac{1}{a}\right)^{\frac{1}{B}} \left[\frac{(\varepsilon_B - \varepsilon_T)^m - \frac{\sigma_T}{C}}{\frac{1}{C}} - 1\right]^{\frac{1}{B}} \,. \tag{20}$$

In Fig. 3 the results of computations carried out according to equation (2) are compared with experimental data. For determination of coefficients results of fracture tests were not used. The model (20) is adequate to the model of viscous failure.

The three-dimensional theory is constructed on the basis of uniform deformation graph.

References

[1] Davenport, C.C: *Correlation of creep an relaxation properties of copper*, Tran ASME Dec. (1936) 733-743.

[2] Rabotnov, Yu.N.: *Creep of structural elements*, Moscow: Nauka 1966 (in Russian).

[3] Christensen, R.: *Introduction to the theory of viscoelasticity*, Moscow: Mir 1974 (in Russian).

[4] Golub, V.P.: *On the theory of creep isotropic initially hardening media*, Prikl. Mekh. **25** 2 (1989) 90-100 (in Russian)

A Constitutive Equation for Non-Linear, Compressible Viscoelastic Materials with Fading Memory

PETER R.W. GUMMERT
TECHNICAL UNIVERSITY OF BERLIN
INSTITUTE FOR MECHANICS

SUMMARY

Numerous proposals have been made for time-dependent stress-strain relationships of viscoelastic materials in order to describe creep and relaxation and their application to structures. These differ or they are even contradictory in their assumptions, results, validity and compactness.
Basing on the principles of Rational Mechanics for simple, isotropic solids, using representation theorems for the functionals and expansions about theorems for the functionals and expansions about the "rest-history" and formulating a constraint condition for the compressibility a generalized constitutive equation is derived for geomtrical and physical nonlinear behaviour of materials, in which other proposed relations are included as special cases.

CONTENTS

1. FUNDAMENTALS

Starting from principles of Rational Mechanics using the

1. Principle of determinism
2. Principle of neighbourhood resp. local action
3. Principle of frame – indifference resp. of material objectivity
4. Principle of material symmetry (isotropic solid)

The following stress–strain relation is obtained first:

$$\mathbf{S}(x,t) = \mathop{\mathfrak{F}}_{0}^{\infty}[\mathbf{C}_t(s) \cdot \mathbf{B}(t)] \qquad \text{or} \qquad (*)$$

or

$$\mathbf{S}(x,t) = \mathbf{g}(\mathbf{B}(t)] + \mathop{\mathfrak{G}}_{0}^{\infty}[\mathbf{G}(s) \cdot \mathbf{B}(t)] \qquad (**)$$

M. Zyczkowski (Ed.)
Creep in Structures
IUTAM Symposium Cracow/Poland 1990

whereas $\mathbf{C}_t(s)$ ist the relative left and $\mathbf{B}(t)$ the instantaneous right CAUCHY-Tensor with t as presence time and $\tau = t-s$ as past time.
Approximation for the expansion about the "rest-history" ($0 \lesssim \epsilon \lesssim 1$):

$$\mathbf{F}(t) = \mathbf{I} + \epsilon\, \mathbf{H}(t) \qquad (\text{for } \epsilon = 1 \;:\; \mathbf{F} = \mathbf{I} + \mathbf{H})$$

$$\begin{aligned}
\mathbf{C}_t(s) &= \mathbf{F}_t^T(s) \cdot \mathbf{F}_t(s) \text{ with } \mathbf{F}_t(s) = \mathbf{F}(\tau) \cdot \mathbf{F}^{-1}(t)\\
&= [\,\mathbf{F}(\tau) \cdot \mathbf{F}^{-1}(t)]^T \cdot [\,\mathbf{F}(\tau) \cdot \mathbf{F}^{-1}(t)]\\
&= \mathbf{F}^{-T}(t) \cdot \mathbf{F}^T(\tau) \cdot \mathbf{F}(\tau) \cdot \mathbf{F}^{-1}(t)\\
&= [\,\mathbf{I} + \epsilon\, \mathbf{H}(t)]^{-T} \cdot [\,\mathbf{I} + \epsilon\, \mathbf{H}(\tau)]^T \cdot [\,\mathbf{I} + \epsilon\, \mathbf{H}(\tau)] \cdot [\,\mathbf{I} + \epsilon\, \mathbf{H}(t)]^{-1}\\
&= [\,\mathbf{I} - \epsilon\, \mathbf{H}^T(t) + \dots] \cdot [\,\mathbf{I} + \epsilon\, \mathbf{H}^T(\tau)] \cdot [\,\mathbf{I} + \epsilon\, \mathbf{H}(\tau)] \cdot [\,\mathbf{I} - \epsilon\, \mathbf{H}(t) + \dots]\\
&= \mathbf{I} + \epsilon\, [\,\mathbf{H}(\tau) + \mathbf{H}^T(\tau) - \mathbf{H}(t) - \mathbf{H}^T(t)]\\
&\quad + \epsilon^2\, [\,\mathbf{H}(\tau) + \mathbf{H}^T(\tau) - \mathbf{H}(t) - \mathbf{H}^T(t)]\\
&\quad + \epsilon^3\, [\,\mathbf{H}(\tau) + \mathbf{H}^T(\tau) - \mathbf{H}(t) - \mathbf{H}^T(t)] + \dots
\end{aligned}$$

Linearization in ϵ and introduction of the difference displacement gradient history and $\tau = t - s$ leads to

$$\mathbf{C}_t(s) = \mathbf{I} + \epsilon\, [\,\underbrace{\mathbf{H}(t-s) - \mathbf{H}(t)}_{\mathbf{H}_d(s)} + \underbrace{\mathbf{H}^T(t-s) - \mathbf{H}^T(t)}_{\mathbf{H}_d^T(s)}]$$

$$\boxed{\mathbf{C}_t(s) = \mathbf{I} + \epsilon\, (\,\mathbf{H}_d(s) + \mathbf{H}_d^T(s))} \qquad (1)$$

$$\begin{aligned}
\mathbf{B}(t) = \mathbf{F}(t) \cdot \mathbf{F}^T(t) &= [\,\mathbf{I} + \epsilon\, \mathbf{H}(t)] \cdot [\,\mathbf{I} + \epsilon\, \mathbf{H}(t)]^T\\
&= \mathbf{I} + \epsilon\, (\,\mathbf{H}(t) + \mathbf{H}^T(t)) + \epsilon^2\, \mathbf{H} \cdot \mathbf{H}^T
\end{aligned}$$

$$\boxed{\mathbf{B}(t) = \mathbf{I} + \epsilon\, (\mathbf{H} + \mathbf{H}^T)} \qquad (2)$$

(1) and (2) in (*) have the result:

$$\boxed{\mathbf{S}(x,t) = [(\,\mathbf{I} + \epsilon\, (\mathbf{H}_d(s) + \mathbf{H}_d^T(s)) \cdot (\,\mathbf{I} + \epsilon\, (\,\mathbf{H}(t) + \mathbf{H}^T(t))]} \qquad (3)$$

Using right GREEN Tensors $G^r = G = \frac{1}{2}[C - I]$ it is obtained:

$$G = \frac{1}{2}[F^T \cdot F - I] = \frac{1}{2}[(I + \epsilon H)^T \cdot (I + \epsilon H) - I]$$

$$= \frac{1}{2}[I + \epsilon(H + H^T) + \epsilon^2 H^T \cdot H - I]$$

$$\left\| \begin{array}{ll} G & = \frac{\epsilon}{2}[H + H^T] \\ G_d(s) & := \frac{\epsilon}{2}[H_d(s) + H_d(s)^T] \end{array} \right\|$$

$$\begin{aligned} S(x,t) &= \mathfrak{F}[(I + 2G_d(s)) \cdot (I + 2G(t))] \\ &= \mathfrak{F}[I] + \frac{\partial}{\partial\epsilon}\mathfrak{F}[(I + 2\epsilon G_d(s)) \cdot (I + 2G(t))] \Big|_{\epsilon=0} \end{aligned} \tag{4}$$

with the definition of the "weak" derivative of a functional

$$\delta\mathfrak{F}[u; v] := \lim_{\epsilon \to 0} \frac{\mathfrak{F}[u + \epsilon v] - \mathfrak{F}[u]}{\epsilon} = \frac{\partial}{\partial\epsilon}\left[\mathfrak{F}[u + \epsilon v]\right]_{\epsilon=0}$$

$$S(x,t) = \mathfrak{F}[I] + \delta\mathfrak{F}[G(t); G_d(s)] + 0(\epsilon^2)$$

If $\mathfrak{F}[I] = S^0 = 0$, that means that the reference configuration is defined as stress-less, we get

$$S(x,t) = \delta\mathfrak{F}[G(t); G_d(s)] = \mathfrak{G}[G(t); G_d(s)]$$

The variation $\delta\mathfrak{F}$ of a functional $\mathfrak{F}$ is always linear in the increment. That means that the expansion about the"rest-history" leads to a linear increment $G_d(s)$ and therefore the functional is linear in $G_d(s)$:

$$S = \overset{\infty}{\underset{0}{\mathfrak{G}}}[G(t);\ G_d(s)] \tag{5}$$

Aequivalent expression using equation (**) and expansion about the "rest-history" leads to

$$S = \mathfrak{G}[G(t); G_d^t(s)] = \mathfrak{G}[G(t), 0] + \delta\,\mathfrak{G}[G(t); G_d(s)]_0$$

since $G_d(s) = G(t-s) - G(t)\Big|_{s=0} = G(t) - G(t) = 0$

$$\mathbf{S}(x,t) = \mathbf{h}(\mathbf{G}(t)) + \mathfrak{h}[\mathbf{G}(t);\ \mathbf{G}_d(s)] \quad \text{(linear in } \mathbf{G}_d(s)\text{)} \qquad (6)$$

For $\mathbf{h}(\mathbf{G}(t))$ a tensor–polynom is taken up to order two (CAYLEY–HAMILTON)

$$\mathbf{h}(\mathbf{G}(t)) = \psi_0 \mathbf{I} + \psi_1 \mathbf{G}(t) + \psi_2 \mathbf{G}^2 \quad \text{where} \quad \psi_i = \psi_i(G_I, G_{II}, G_{III})$$

For $\mathfrak{h}(\mathbf{G}(t);\ \mathbf{G}_d(s))$ the representation is taken by theorem of RIESS

$$\delta\,\mathfrak{G}[\mathbf{G}(t);\ \mathbf{G}_d(s)] = \mathfrak{h}[\mathbf{G}(t);\ \mathbf{G}_d(s)] = \int_{s=0}^{\infty} [\ \overset{(4)}{\mathbf{H}}(s, \mathbf{G}(t)) \cdot\cdot\ \mathbf{G}_d(s)]\ ds$$

whereas $\overset{(4)}{\mathbf{H}}$ is a tensor of rank four depending on past–time s and $\mathbf{G}(t)$. So we get from (6)

$$\mathbf{S}(x,t) = \sum_{i=0}^{2} \psi_i\, \mathbf{G}^i + \int_{s=0}^{\infty} \underbrace{(\mathbf{H}(s, \mathbf{G}) \cdot\cdot\ \mathbf{G}_d)}_{\mathbf{J}}\ ds \qquad (7)$$

2. THEORY OF N–TH ORDER

Assuming the following representation for $\mathbf{J}$ under consideration of the linearity in $\mathbf{G}_d(s)$

$$\mathbf{J}(s, \mathbf{G}, \mathbf{G}_d) = [f_0(\ )\mathbf{I} + f_1(\ )\mathbf{G} + f_2(\ )\mathbf{G}^2] \cdot [\ g_0(\ )\mathbf{I} + g_1(\ )\mathbf{G}_d(s) + \mathbf{0}]$$

we obtain

$$\begin{aligned} \mathbf{J}(s, \mathbf{G}, \mathbf{G}_d) &= F_0 \mathbf{I} + F_1 \mathbf{G} + F_2 \mathbf{G}^2 \\ &+ F_{31} \mathbf{G}_d(s) + F_{32} \left[\mathbf{G}_d(s) \cdot \mathbf{G}(t) + \mathbf{G}(t) \cdot \mathbf{G}_d(s)\right] \\ &+ F_{33} \left[\mathbf{G}_d(s) \cdot \mathbf{G}^2(t) + \mathbf{G}^2(t) \cdot \mathbf{G}_d(s)\right] \end{aligned}$$

whereas

$$\begin{aligned} F_0 &= \Phi_{01}\mathrm{tr}[\mathbf{G}_d(s)] + \Phi_{02}\mathrm{tr}[\mathbf{G}_d(s) \cdot \mathbf{G}(t)] + \Phi_{03}\mathrm{tr}[\mathbf{G}_d(s) \cdot \mathbf{G}^2(t)] \\ F_1 &= \Phi_{11}\mathrm{tr}[\mathbf{G}_d(s)] + \Phi_{12}\mathrm{tr}[\mathbf{G}_d \cdot \mathbf{G}] + \Phi_{13}\mathrm{tr}[\mathbf{G}_d \cdot \mathbf{G}^2] \\ F_2 &= \Phi_{21}\mathrm{tr}[\mathbf{G}_d(s)] + \Phi_{22}\mathrm{tr}[\mathbf{G}_d \cdot \mathbf{G}] + \Phi_{23}\mathrm{tr}[\mathbf{G}_d \cdot \mathbf{G}^2] \end{aligned}$$

and $$F_{3i} = \Phi_{3i} \qquad \text{(since linear in } G_d(s)) , \quad \Phi_i = \Phi_i(G_I, G_{II}, G_{III}, s)$$

It follows

$$S = \psi_0 I + \psi_1 G + \psi_2 G^2 + \int \Big\{ [F_0 I + F_1 G + F_2 G^2 + F_{31} G_d(s) + F_{32} [G_d(s) \cdot G + G \cdot G_d(s)] + F_{33} [G_d(s) \cdot G^2 + G^2 \cdot G_d(s)] \Big\} ds$$

$$\boxed{\begin{aligned} S = & \ \psi_0 I + \psi_1 G + \psi_2 G^2 \\ & + \int_{s=0}^{\infty} \Big\{ [\Phi_{01} \mathrm{tr}(G_d(s)) + \Phi_{02} \mathrm{tr}(G_d(s) \cdot G) + \Phi_{03} \mathrm{tr}(G_d(s) \cdot G^2)] I \\ & + [\Phi_{11} \mathrm{tr}(G_d(s)) + \Phi_{12} \mathrm{tr}(G_d(s) \cdot G) + \Phi_{13} \mathrm{tr}(G_d(s) \cdot G^2)] \ G \\ & + [\Phi_{21} \mathrm{tr}(G_d(s)) + \Phi_{22} \mathrm{tr}(G_d(s) \cdot G) + \Phi_{23} \mathrm{tr}(G_d(s) \cdot G)] \ G^2 \\ & + [\Phi_{31} G_d(s) + \Phi_{32}(G_d(s) \cdot G(t) + G(t) \cdot G_d(s)) + \\ & + \Phi_{33}(G_d(s) \cdot G^2 + G^2 \cdot G_d(s))] \Big\} ds \qquad (8) \end{aligned}}$$

whereas $\Phi_{01} \dots \Phi_{33}$: 12 material functions for the memory part with $\Phi_{ij} = \Phi_{ij}(G_I, G_{II}, G_{III}, s)$

and ψ_1, ψ_2, ψ_3 : 3 material-functions for the spontaneous part with $\psi_i = \psi_i(G_I, G_{II}, G_{III})$

3. CONSISTENT THEORY OF 2ND-ORDER

By introducing a (spectral-)norm for the deformation tensors and defining an order of magnitude:

$$\boxed{\epsilon = \sup \ \| H \| \ ; \quad \| H \| = \mathrm{tr} \sqrt{[H \cdot H^T]} \qquad (9)}$$

a classification of all terms in (8) becomes possible:

$$G = H + H^T + H^T \cdot H = 0(\epsilon) = H + H^T + 0(\epsilon^2)$$

$$G_d(s) = 0(\epsilon) \qquad G^2 \cdot G_d(s) = 0(\epsilon^3)$$

$$G_d(s) \cdot G = 0(\epsilon^2) \qquad \mathrm{tr}\, G,\ \mathrm{tr}\, G_d = 0(\epsilon)$$

$$G^2 = 0(\epsilon^2) \qquad \mathrm{tr}\,(G \cdot G_d) = \mathrm{tr}\,(G_d \cdot G) = 0(\epsilon^2)$$

$$\mathrm{tr}\,(G_d \cdot G^2) = 0(\epsilon^3)$$

This way a consistent theory of order two, neglecting all terms of $0(\epsilon^3)$, can be generated

$$\begin{aligned} S(x,t) = \; & \psi_0 I + \psi_1 G(t) + \psi_2 G^2(t) \\ & + \int_0^\infty \Big\{ [\Phi_{01} \mathrm{tr}\,(G_d) + \Phi_{02} \mathrm{tr}\,(G_d \cdot G)]\, I \\ & + [\Phi_{11} \mathrm{tr}\,(G_d)]\, G(t) \\ & + [\Phi_{31} G_d(s) + \Phi_{32}(G_d(s) \cdot G(t) + G(t) \cdot G_d(s))] \Big\} ds \\ & + 0(\epsilon^3) \end{aligned} \qquad (10)$$

containing 8 material functions

$$\psi_i = \psi_i(G_I, G_{II}, G_{III}) \qquad i = 0,1,2$$

$$\Phi_{ij} = \Phi_{ij}(G_I, G_{II}, G_{III}, s) \qquad ij = 01,02,11,31,32$$

Note:

if the terms of order ϵ would be taken only, a consistent theory of order one with geometrical nonlinear behaviour would be the result (MAXWELL, KELVIN, BURGER and higher viscoelastic solids are included)

$$S = \psi_0(G_I) I + \psi_{10} G + \int_0^\infty [\Phi_{10} \mathrm{tr} G_d(s) I + \Phi_{31} G_d(s)]\, ds + 0(\epsilon^2)$$

$\psi_{10}, \Phi_{01}, \Phi_{31}$ constants, $\psi_0 = \psi_\infty \mathrm{tr}(G)$

4. CONSTRAINT CONDITION

In order to reduce the number of material functions and/but to avoid the assumption of incompressible materials (det $\mathbf{F} = 1$) a constraint condition for "weak" compressible matters is considered:

$$\det \mathbf{F} = III_F = 1 + \Delta \qquad (11)$$

Since the relation between the 3rd invariant of **F** and the invariants of **H** is:

$$\det \mathbf{F} = III_F = 1 + I_H + \underline{II_H + III_H} \quad , \text{ it follows}$$

$$I_H = \tfrac{1}{2}[\, I_H^2 - H_I^2 \,] = III_H = O(\epsilon^3), \qquad \operatorname{tr} \mathbf{H} = O(\epsilon^{3/2})$$

$$\operatorname{tr} \mathbf{G} = \operatorname{tr}[\underbrace{\mathbf{H}} + \underbrace{\mathbf{H}^T} + \underbrace{\mathbf{H}^T \cdot \mathbf{H}}]$$

$$= \quad O(\epsilon^{3/2}) \qquad O(\epsilon^3)$$

$$\operatorname{tr} \mathbf{G},\ \operatorname{tr} \mathbf{G}_d(s) = O(\epsilon^{3/2}) \qquad \mathbf{G}^2,\ \mathbf{G}_d \cdot \mathbf{G},\ \mathbf{G} \cdot \mathbf{G}_d = O(\epsilon^2)$$

$$\operatorname{tr} \mathbf{G}_d \cdot \mathbf{G},\ \operatorname{tr} \mathbf{G} \cdot \mathbf{G}_d = O(\epsilon^3) \qquad [\, \operatorname{tr} \mathbf{G}_d\,] \cdot \mathbf{G} = O(\epsilon^{5/2})$$

under the constraint condition $\det \mathbf{F} = 1 + I_H$ for weak compressible material the constitutive equation becomes:

$$\begin{aligned} \mathbf{S}(x,t) &= \psi_0 \mathbf{I} + \psi_1 \mathbf{G}(t) + \psi_2 \mathbf{G}^2(t) \\ &+ \int_{s=0}^{\infty} \Big\{ \Phi_{01} \operatorname{tr}(\mathbf{G}_d(s))\, \mathbf{I} + \Phi_{31} \mathbf{G}_d(s) \\ &\quad + \Phi_{32}[\mathbf{G}_d(s) \cdot \mathbf{G}(t) + \mathbf{G}(t) \cdot \mathbf{G}_d(s)] \Big\} ds \end{aligned} \qquad (12)$$

with only 6 material functions

$$\Phi_{01} = \Phi_0 ; \quad \Phi_{31} = \Phi_1 ; \quad \Phi_{32} = \Phi_2$$

$$\psi_i = \psi_i(I_G, II_G, III_G) ; \qquad \Phi_i = \Phi_i(I_G, II_G, III_G, s)$$

5. CONSTITUTIVE EQUATION

As the functional is expanded about the rest-history the material-functions can be expanded the same way:

$$\Phi_{ij}(G_I, G_{II}, G_{III}, s) = \Phi_{ij}(0,0,0,s) + \frac{\partial \Phi_{ij}}{\partial G_I}\bigg|_{0,s} \cdot (G_I - 0) + \frac{\partial \Phi_{ij}}{\partial G_{II}}\bigg|_{0,s} (G_{II} - 0) +$$

$$\left.\frac{\partial \Phi_{ij}}{\partial G_{III}}\right|_{0,s} \cdot (G_{III} - 0) + \frac{1}{2}\left\{ \left.\frac{\partial \Phi_{ij}}{\partial G_I}\right|_{0,s} G_I^2 + \right.$$

$$\left. \frac{\partial \Phi_{ij}}{\partial G_{II}^2} G_{II}^2 + \frac{\partial^2 \Phi_{ij}}{\partial G_I \partial G_{II}} + \cdots + \frac{\partial^2 \Phi}{\partial G_{II} \partial G_{III}} + \cdots\cdots \right\}$$

As at least all Φ are linked with $\operatorname{tr} \mathbf{G}_d(s)$ in sense of a theory of order two the expansion is considered to the first derivative only

$$\Phi_{ij}(G_I, G_{II}, G_{III}, s) = \Phi_{i0}(0,0,0,s) + \Phi_{i1} \cdot G_I + 0\,(\epsilon^2)$$

$$= \Phi_{i0}(s) + \Phi_{i1}(s)\,(\operatorname{tr} \mathbf{G})$$

Consequently the material functions become functions of s only. The number of them can be reduced to the number three, since the spontaneous part is always the initial value of the entire historical part. These three functions are to determine from experiments on multiaxial testing machine with tension and/or torsion and internal pressure of tubes. So finally we obtain

$$\mathbf{S}(x,t) = \Phi_0(G_I, G_{II})\,\mathbf{I} + \Phi_1(G_I)\,\mathbf{G} + \Phi_{20}\mathbf{G}^2$$

$$+ \int_{s=0}^{\infty} \Big[\Phi_0(s) \operatorname{tr} \mathbf{G}_d(s)\,\mathbf{I} + \Phi_1(s)\,\mathbf{G}_d(s) + \Phi_2(s)[\mathbf{G}_d(s) \cdot \mathbf{G} + \mathbf{G} \cdot \mathbf{G}_d(s)] \Big]\, ds \tag{13}$$

This work bases on publications of

TRUESDELL (1954)
NOLL ('57, '64, '72, '75)
COLEMAN ('64, '67)
GURTIN, OWEN, VALANIS, CARLSON
FREUDENTHAL, PRAGER, HILL, GEIRINGER
RIVLIN, ERINGEN
PARKUS, FLÜGGE, BECKER
TROSTEL, HAUPT, I. MÜLLER
GUMMERT ('71, '76, '80, '83, '86) and others

Transient Creep in Rock Salt Structure

U. Heemann[†1], E. Stein[†2]

Introduction and Abstract:

It is projected by the German governement to deposit radioactive waste in a great salt dome in northern Germany (Lower saxonia). Because of the necessity of guaranteeing the safety of the repository over long lasting periods of time, a very good knowledge of the behaviour of the used materials, especially the host rock salt and the probable backfilling, crushed salt, is indespensible. This is seen to be a basis of technical security calculations in underground dam constructions, backfilling and the behaviour of the host rock. An extended paper of our former research on this field has been presented in [SWM87]. The presented theory is founded on basic physical mechanisms of dislocations and developed to equations describing dislocation multiplication and annihilation. From these a complete theory of transient creep in rock salt can be deduced which is compared with experimental results and shows very good agreement.

Mean Free Path Length of Dislocations

An important basis of the here presented theory is the calculation of the mean free path length of dislocations gliding under the action of stress, but interacting with other dislocations. The probability of mutual capture of a pair of dislocations of opposite burgers vectors can be calculated by the ratio of capture distance u_c over mean dislocation distance d which can be described by the acting stress τ_{eff} and the square root of the dislocation density ρ (see fig. 1).

$$P(capture) = P_C = 2\frac{u_c}{d} = 2\frac{D\sqrt{\rho/2}}{4\tau_{eff}}, \tag{1}$$

$$u_c = \frac{D}{4\tau_{eff}}, \qquad d = \frac{1}{\sqrt{\rho/2}}.$$

By a simple statistical aproach [HEE89] the mean free path length of screw dislocations results as

$$L_T = \sum_{n=0}^{\infty} P_C(1-P_C)^n(n+1)d = \frac{1}{\beta u_c \rho} = \frac{1}{\beta}\frac{4\tau_{eff}}{D\rho}, \qquad \beta = 2 \tag{2}$$

In case of screw dislocations the argumentation is equivalent except for a different factor $\beta \approx 4$ because of their higher mobility. For the interaction of a mobile edge dislocation

[1] Gesellschaft für Strahlen- und Umweltforschung, Institut für Tieflagerung

[2] Institut für Baumechanik und Numerische Mechanik, Universität Hannover

M. Zyczkowski (Ed.)
Creep in Structures
IUTAM Symposium Cracow/Poland 1990

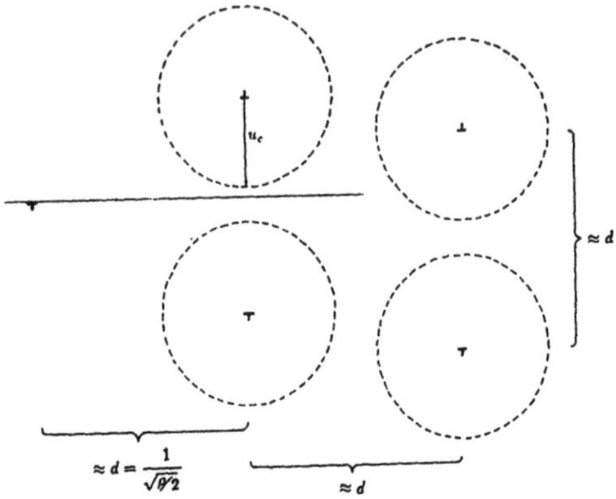

fig. 1: Passage of a mobile dislocation through an ensemble of mobile dislocations (total density ρ, mobile themselves or sessile) with capture radius u_c and mean distance d ($d \approx 1/\sqrt{\rho/2}$).

with a dislocation dipole it can be seen from fig. 2 that the gliding dislocation has either the possibility to pass the dipole, build a tripole or being exchanged by release of another dislocation. In a numerical analysis the capture probability for tripole building was found to be of definitely the same form as in eq. (1) thus leading to an equivalent result for the path length with a smaller factor $\beta \approx 1/3$.

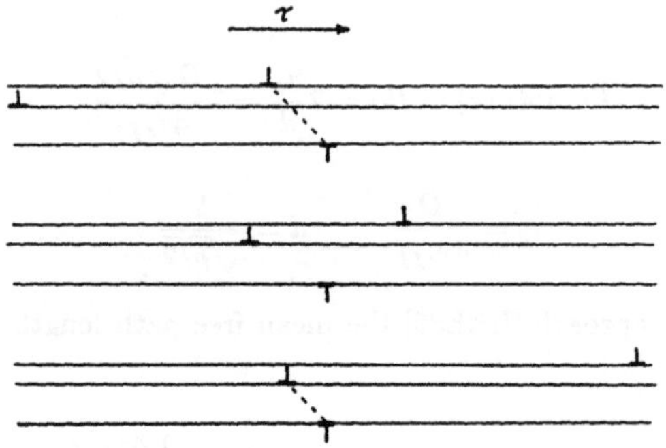

fig. 2: Interaction of a free dislocation with a dipole, here resulting in an exchange of the mobile and a sessile dislocation.

So the description is equivalent for screw, edge and dipole dislocations except the different factor β and it is supposed that this is right for all other reactions of mobile dislocations with higher multipoles.

Dislocation Multiplication

Screw dislocations have the ability to change their plane of glide (cross slip), thus giving them a higher mobility and furthermore, by switching back into their old glide plane (see fig. 3), pulling out new loops of dislocations so increasing the length and number of dislocations. But the cross slip distance has to be greater than the capture distance to enable them to overcome their mutual interaction. The probability for a screw dislocation to make a double cross slip step of more than a distance h has been described by Wiedersich [WIE62] by the probability q for the first cross slip and the probability p for switching back.

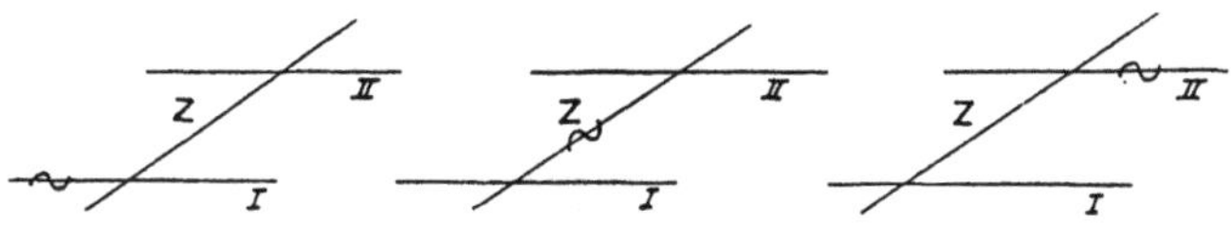

fig. 3: Cross slip of a screw dislocation from plane I to plane II via an intermediate plane Z.

$$Q = q \exp -(\frac{ph}{b}) \tag{3}$$

Orlov et al [OPS75] calculated them to be determined by the density of the forest dislocations and the effective stress

$$q = \chi\sqrt{c}\,\rho_f\,, \qquad p \approx k\tau_{eff}\,. \tag{4}$$

c is the density of point defects, κ and k are constants, and ρ_f is the density of forest dislocations. Inserting (4) and the capture distance for minimum slip length h, the probability Q results

$$Q(h \geq u_c) = q \exp -(\frac{pu_c}{b}) = \kappa\sqrt{c} \exp -(\frac{kD}{4b})\,\rho_f\,. \tag{5}$$

The rate of generation of new loops now can easily be found to be

$$\dot{n} = Q\mu v = Q_o \rho_f\, \mu v. \tag{6}$$

The screw dislocation density is simply connected to the number of loops and the glide distance of the edge dislocations which definitely determines the length of the screws

$$\mu = 4nL_T\,. \tag{7}$$

Differentiating n and inserting in eq. 6 gives the rate of dislocation density increase

$$\dot{\mu} = 4L_T\dot{n} = 4\frac{4\tau_{eff}}{\beta D\rho}Q_o\rho_f\mu v = \kappa\mu v\tau_{eff}\left(\frac{\rho_f}{\rho}\right). \tag{8}$$

The forest dislocations and edge dislocations can be assumed to be proportional to one another. So this equation yields to be identical to the empirical equation of Peisker, Haasen and Alexander [PHA62] and others which is found to be right in several different materials such as metalls, ceramics and ionic crystals except that they used total dislocation density.

Evolution Equations for Dislocation Densities and Creep

The annihilation of the screws is simply related to their mean free path length

$$\dot{\mu}^- = -\mu\frac{v}{L_S} = -\beta_1\frac{\mu^2 vD}{4\tau_{eff}} = -\gamma_1\frac{\mu^2 v}{\tau_{eff}}. \tag{9}$$

γ_1 is a constant. So the whole evolution equation is the sum of the last two ones.

$$\dot{\mu} = \dot{\mu}^+ + \dot{\mu}^- = \kappa_1\mu v\tau_{eff} - \gamma_1\frac{\mu^2 v}{\tau_{eff}}. \tag{10}$$

With every new loop also edge dislocations are generated and so their generation rate $\dot{\rho}$ is directly proportional to that of the screws.

Two pure glide mechanisms are supposed to be responsible for the annihilation of the edge dislocations, the first one stemming from the reaction of an orthogonal dislocation with the dipole, leaving only the catalyst dislocation, or the reaction of the dipole with an oblique dislocation which has been suggested by Strunk and Strecker [STS84]. Here only the reaction with the orthogonal dislocation has been taken into account, leading to an annihilation probability proportional to both dislocation densities, η denoting the orthogonal dislocations, and again inverse proportional to the stress.

$$\dot{\rho} = \kappa_2\mu v\tau_{eff} - \gamma_2\frac{\rho\eta u}{\tau_{eff}} \tag{11}$$

The second mechanism [STS84] is a little bit different, but should be described in a quite similar way.

Because first trials with a singleparametric equation for the whole dislocation density already gave quite good results, the evolution of the orthogonal or secondary dislocations has been assumed to be well described by just the same law.

$$\dot{\eta} = \kappa_3\eta u\tau_{eff} - \gamma_3\frac{\eta^2 u}{\tau_{eff}} \tag{12}$$

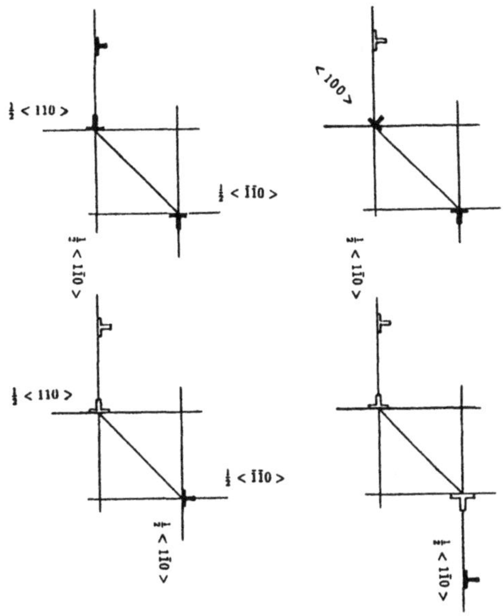

fig. 3: A dislocation dipole is dissolved by a "catalyst" - dislocation of an orthogonal glide system by combination and pure glide of the associated dislocations (without climb). In b) the burgers vectors add to to a joint one $\frac{1}{2} < 110 > +\frac{1}{2} < 1\bar{1}0 >=< 100 >$, in c) by the reaction $< 100 > +\frac{1}{2} < \bar{1}\bar{1}0 >= \frac{1}{2} < 1\bar{1}0 >$ the previous "catalyst" - dislocation is released again.

This has to be combined with an equation similar to that of Orowan, combining creep rate, dislocation densities and their velocities respectively

$$\dot{\varepsilon} = \phi b(\mu v + \eta u). \tag{13}$$

The velocities have been deduced by simple comparison with experimental creep rates [BLV88]

$$v = v_o \exp\left(\frac{-Q}{kT}\right)\tau_{eff} \sinh\left\langle\frac{\gamma\tau_{eff}}{(1-z)mkT}\right\rangle, \tag{14a}$$

$$u = u_o \exp\left(\frac{-Q}{kT}\right)\tau_{eff} \sinh\left\langle\frac{\gamma\tau_{eff}}{(1-z)mkT}\right\rangle. \tag{14b}$$

τ_{eff} is given by

$$\tau_{eff} = \tau - \alpha bG\sqrt{\rho_t}, \quad \rho_t = \mu + \rho + \eta. \tag{15}$$

But there is still one effect left to be incorporated into the model. When gliding dislocations are stopped at orthogonal dislocations, their retardation time till cutting determines their mean velocity.

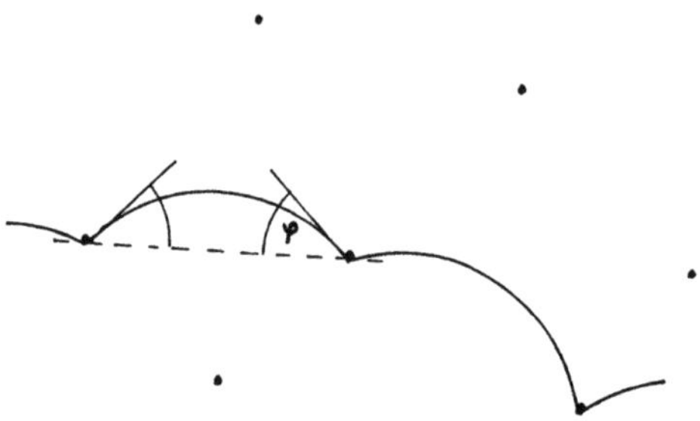

fig. 5: Bowing out of dislocations under stress at pointlike barriers (e.g. orthogonal dislocations).

$$\sin\varphi = \frac{\tau_{eff}}{bG\sqrt{\rho_f}}\,. \tag{16}$$

But when the sin of the bowing angle φ is of the order of 1 then the dislocations can bypass the hindrances and recombine behind, and so the velocity increases drastically which can be described – purely pragmatically – by a power of the right term of eq (16).

$$v = v_o'\left(\frac{\tau_{eff}}{bG\sqrt{\rho_t}}\right)^p \exp\left(\frac{-Q}{kT}\right)\tau_{eff}\sinh\langle\frac{\gamma\tau_{eff}}{(1-z)mkT}\rangle \tag{17a}$$

$$u = u_o'\left(\frac{\tau_{eff}}{bG\sqrt{\rho_t}}\right)^p \exp\left(\frac{-Q}{kT}\right)\tau_{eff}\sinh\langle\frac{\gamma\tau_{eff}}{(1-z)mkT}\rangle \tag{17b}$$

The above evolution equations have been compared with the empirical results of Blum [BLV88] on creep rate by solving the differential equations for a set of parameters and changing only three of them by a computer such to optimize the fit of theoretical equations and experimental results. A plot of the gained creep curve ($\log\dot{\varepsilon}$ over ε, fig. 6) for primary transient creep, both theoretical and empirical, nearly shows no difference.

The resulting parameters are in very good aggreement with experimental results of several different groups [PHA02, HAA64, CMN71, SMS76]. The evolution of the dislocation densities can only qualitatively be judged to be confident because of lack of data.

In order to check these equations also at strain hardening for constant strain rate it has been necessary to raise the value of the exponent p (see fig. 7). The failure of the specimen at the right part of the curve of course can not be reproduced from the above equations.

The necessity for the use of a new set of parameters is attributed to the purely pragmatical choose of the power term for the effect of bypassing of orthogonal dislocations and

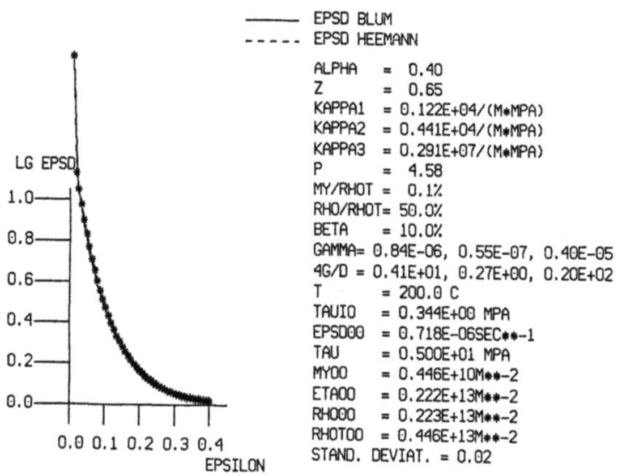

fig. 6: Comparison of the solution of equations (15) to (19) and (30) with the curve of Blum [BLV88](the stars mark the points of comparison) in uniaxial calculation.

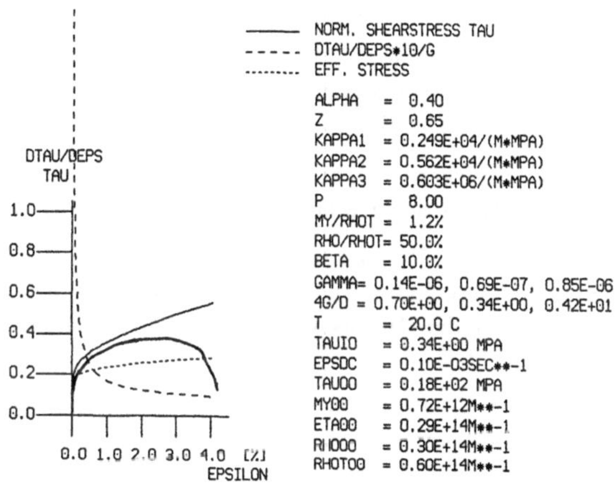

fig. 7: Hardening curve, calculated at room temperature and $\dot{\varepsilon} = 10^{-4}\text{sec}^{-1}$. Also enclosed into the plot is the experimental curve of an own test on creep fracture, for uniaxial loading.

recombining behind. A better theoretical description – especially regarding the effect of dislocation cutting on the velocity – is supposed to give much better results with only one set of parameters.

Aknowledgement:
This work has been sponsored by the Ministry of Research and Technology of the Federal Republic of Germany, conducted in the order of the Federal Institute of Geosciences and Natural Resources (BGR).

Citations

BLV88 W. Blum, S. Vogler: final report on the research - project KWA 5502/8, "Geotechnisches Verhalten verschiedener Salzgesteine" - Teilprojekt "Kriechen von natürlichem Steinsalz", University of Erlangen, 1988

CMN71 W.A. Coghlan, R.A. Menezes, W.D. Nix: A Study of High Temperature Creep in LiF Single Crystals. Philosophical Magazine **23**, 1515-1530, 1971

HAA64 P. Haasen: Dislocation Mobility and Generation. Discussions of the Faraday Society **38**, 191-200, 1964

HEE89 U. Heemann: Transientes Kriechen und Kriechbruch im Steinsalz. Doctoral thesis, Hannover, 1989

OPS75 A.N. Orlov, V.N. Perevesentsev, B.I. Smirnov Dependence of the Cross-Slip Parameters on Temperature, Stress, and Defect Structure of Crystals. physica status solidi (a) **32**, 35-42, 1975

PHA62 E. Peissker, P. Haasen, H. Alexander: Anisotropic Plastic Deformation of Indium Antimonide. Philosophical Magazine, Ser. 8, **7**, 1279-1303, 1962

SMS76 B.I. Smirnov, T.V. Samoĭlova: Influence of various Factors on Multiplikation of Screw Dislocations during Deformation of Crystals. Soviet Physics of Solid State **18** (6), 105-116, 1976

STS84 H.P. Strunk, H. Strecker: Annihilation of Edge Dislocation Dipoles by a Glide Mechanism. Materials Science and Engineering, **62**, 265-269, 1984

SWM87 E. Stein, D.W. wetjen, R. Mahnken, U. Heemann: Theoretische und numerische Aspekte der Berechnung thermo-mechanischer Kriechvorgänge im Steinsalz. Abhandlungen der Braunschweigischen Wissenschaftlichen Gesellschaft, **XXXIX**, 61-121, 1987

WIE62 H. Wiedersich: A Quantitative Theory for the Dislocation Multiplication during the Early Stages of Formation of Glide Bands. Journal of Applied Physics **33** (3), 854-858, 1962

A Hybrid Constitutive Model for Cyclic, and Creep Deformation

H. Ishikawa Professor
Department of Mechanical Engineering II
Hokkaido University
Sapporo 060 JAPAN

Summary

This paper shows the fact that back stress in cyclic plasticity plays an important role to describe the creep and relaxation behavior of SUS304 type stainless steel at room temperature following cyclic preloading if the evolution of kinematic back stress is simulated from the hybrid constitutive equations. Consequently, the experimental results of creep during cyclic loading could be described by the so-called Bailey-Norton law, if the effective stress is defined as stress measured from the current center of yield surface, i.e., kinematic back stress.

Testing Equipments and Procedure

The specimens used in this work have the gauge length of 50mm, and the inner and outer radii of the tube of 20mm and 23mm, respectively. The specimens were made of type 304 stainless steel subjected to heat treatment at 1070 ℃, its chemical compositions were 0.06C, 0.51Si, 1.06Mn, 0.028P, 0.012S, 8.81Ni, 18.38Cr, and Fe balance in weight percent. Young's modulus at room temperature was equal to E=198GPa, while sheering modulus G=73.5GPa.

A servocontrolled axial-torsion testing machine (Shimadzu EHF-EB10) together the Shimadzu 4825 controller and a computer (NEC PC-9801VM) was used for computerized testing and data acquisition accomplished by a program written in Machine Word. Strains measured using two strain gauges applied on diametrically opposite sides of the specimens. The axial force were measured using a load cell incorporated in the machine. The strain controlled mode was used for cyclic plasticity and relaxation tests, while the stress controlled mode for cyclic plasticity and creep test. With the aid of the computer system, the constant strain rate $\dot{\varepsilon} = 5\times10^{-4}$ (1/sec) or the constant stress rate $\dot{\sigma} = 10$ (MPa/sec) was maintained during plastic loading, and the resulting stress-strain responses were recorded on a

M. Zyczkowski (Ed.)
Creep in Structures
IUTAM Symposium Cracow/Poland 1990

floppy-disk by the computer. A data point was recorded whenever the strain changed by 0.0025%, or whenever the stress changed by 5 MPa, respectively.

Creep behavior at the same stress level during uniaxial loading

A stress-strain diagram of a specimen which was subjected to creep tests during uniaxial loading with the strain rate $\dot{\varepsilon}=5\times10^{-4}$ (1/sec) is given in Fig.1. Creep strains AB(=0.29%) and EF(=0.01%) are obtained after same time intervals of 300 sec under the same stress level of 250 MPa. The difference of creep strains between AB and EF may be caused by a structural change in the internal variable during unloading DE. What kind of internal variable governs this phenomenon? It should be worth seeking for a constitutive modeling of cyclic viscoplasticity. The similar problem concerned will be discussed in the following section. In addition, it is important to note that upon a further tensile under strain controlled loading with the same strain rate $\dot{\varepsilon}=5\times10^{-4}$ (1/sec), BCD, or FGH, the material continues to harden as in a regular tensile test, and the dashed extension emanating from A or D in Fig.1 is the curve for an uninterrupted tensile test. This phenomenon seems to give the important suggestion on an interaction between plastic strain and creep strain.

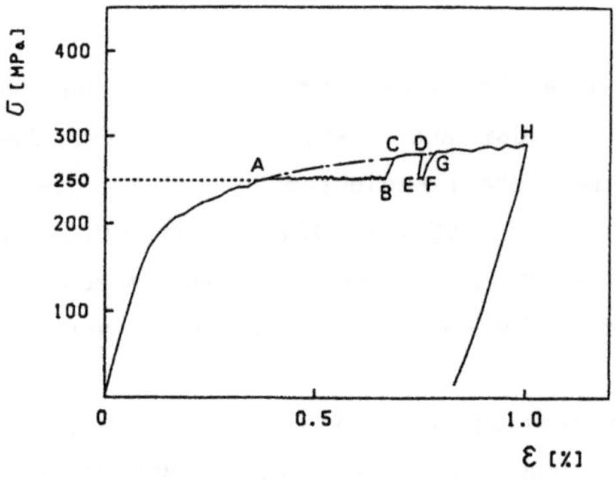

Fig.1 Creep behavior at the same stress level during uniaxial loading

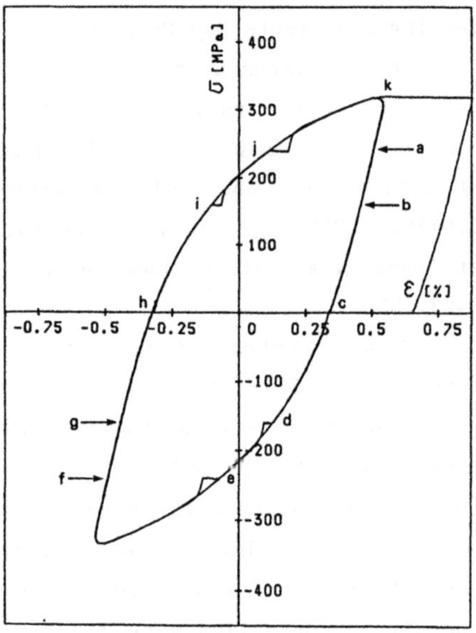

Fig.2 Intermittent creep periods of 300sec during cycling between ±0.55%

Creep and Relaxation during Cyclic loading

The experimental results were obtained in a quasi-cyclic-steady state. Namely, specimens were subjected to stress-controlled cycling with the stress rate $\dot{\sigma}$=10 (MPa/sec) at the strain amplitude of $\Delta\bar{\varepsilon}$ =1.1%, until the hysteresis loops were stable. Creep periods were then introduced around the loops and the creep behavior was obtained.

Figure 2 shows the intermittent creep periods of 300sec duration during strain cycling between ±0.55%. After stabilization of cyclic straining with the 5th cycle, the creep periods start at the point a. All consecutive creep-hold periods are designated by alphabet letters ending at k. In spite of the same stress levels at a and j, b and i, d and g, and at e and f, creep in the nearly straight sections of the hysteresis loop is very small or nonexistent. At c and h, which are points of zero stress, the magnitude of strain decreases by a small but noticeable amount. Due to the conventional creep theory, the creep rate depends on stress and time in the time-hardening theory or on stress and creep strain in the strain-hardening one. The results shown in Fig.2, however, are at variance with these theories. The additional observation of the results in Fig.2 shows that at the maximum stress level of the cycle the greatest amount of creep is found which subsequently subsides and then increases in the opposite direction as the curvature of the stress-strain curve increases. Therefore, a point of zero creep must exist by continuity arguments. However, this point of zero creep can not very well defined only from Fig.2, as pointed out by Kujawski et al.[1].

It is very reasonable that the origin of the stress space should move due to the plastic deformation as presented in many constitutive models for cyclic plasticity proposed presently; namely, the current center of yield surface, i.e., kinematic back stress evolved during cyclic preloading. The physical instinct shows that the point of zero creep should corresponds to this current center of yield surface. Nevertheless, the point of zero creep can not defined yet from an experiment because back stress is an internal variable. The author has succeed in defining the evolution of back stress from the computer simulation based on his constitutive modeling of cyclic plasticity [2,3]. Especially it was a very interesting fact that the shape of the equi-plastic-strain surfaces, among which the yield surface was confined to the surface determined from proof strain of 50μm/m, could be described by a regular ellipse if it were probed at simulated back stress, while the equi-plastic-strain surface has a deformed shape with a rounded nose and a flatten back at the zero stress state unloaded completely [2].

A brief summary of the mathematical structure of the hybrid constitutive model is reproduced from [2]. Combining the Mises type stress-hardening with the kinematic hardening, the following yield function (1) is chosen for cyclically stable materials at temperature T. The associated flow low is derived from the normality of the plastic strain increments to the yield surface and is given by equation (2). Using Ziegler type of assumption as the evolution equation of the center of the loading surface, the motion of the center is given by equation (3).

$$f = \frac{1}{2} C_{ijkl} (\sigma_{ij} - \sigma^c_{ij})(\sigma_{kl} - \sigma^c_{kl}) - \frac{1}{3} R^2(\kappa ,T) = 0 \tag{1}$$

$$d\varepsilon^p_{ij} = \frac{3d\bar{\varepsilon}^p}{2\,R} C_{ijkl} (\sigma_{kl} - \sigma^c_{kl}) \tag{2}$$

$$d\sigma^c_{ij} = \frac{3}{2\,R^2} [C_{klmn} (\sigma_{mn} - \sigma^c_{mn}) d\sigma_{kl}] (\sigma_{ij} - \sigma^c_{ij}) - (\frac{dR}{dW^p})(\sigma_{ij} - \sigma^c_{ij}) d\bar{\varepsilon}^p \tag{3}$$

In equations (1)-(3), C_{ijkl} is the plastic-deformation-induced anisotropy coefficient tensor of 4th rank, and σ_{ij} is stress. σ^c_{ij} is kinematic back stress, giving the translation of the yield surface. R is isotropic flow stress, describing the increase or decrease of the size of the yield surface, and κ is the hardening or softening parameter. Moreover, $d\varepsilon^p_{ij}$ and $d\bar{\varepsilon}^p$ are the plastic strain and the equivalent plastic strain increments.

All the stress-strain curves are assumed to be represented by the modified Ramberg-Osgood law (4).

$$\varepsilon_{ij} - \varepsilon^c_{ij} = \frac{\sigma_{ij} - \sigma^c_{ij}}{E} [1 + K \{ \frac{\bar{\sigma}}{\sigma_o} \}^m] \tag{4}$$

where E is Young's modulus, K is material constant, and $\bar{\sigma}$ is the equivalent stress. The quantities $\sigma_{o(n)}$ and $m_{(n)}$ in equation (4) are the reference stress at the proof strain ε_o =500 μm/m, and the exponent of hardening or softening after the (n-1)th inversion of loading, and they depend on the modified plastic work according to equation (5)-(6). R can be provided by equation (7) following equation (5).

$$\sigma_o = \sigma_{o(\infty)} [1 \mp \alpha \exp \{ - \frac{W^p_{(n-1)}}{W_o} \}] \tag{5}$$

$$m = m_{(\infty)} [1 \mp \beta \exp \{ - \frac{W^p_{(n-1)}}{W_1} \}] \tag{6}$$

$$R = R_{(\infty)} [1 \mp \lambda \exp \{ - \frac{W^p}{W_2} \}] \tag{7}$$

$\sigma_{o(\infty)}$, α, W_o, $m_{(\infty)}$, β, W_1, $R_{(\infty)}$, λ, and W_2 in equations (5) ~ (7) are constants prescribed by the history of loading or effected by a strain path memory. W^p in equation (7) is the accumulated plastic work which is equal to $W^p=W^p_{(n-1)}+\Delta W^p$ with the modified plastic work increment ΔW^p during loading in the current stage of deformation.

In the previous work [3], the yield function f in equation (1) was confirmed to be an adequate form for cyclic plasticity and the associated flow law given by equation (2) was also appreciated to be acceptable from the normality of the plastic strain increment vector to the subsequent yield surface. In addition, applicability of the Ramberg-Osgood law to the subsequent loading after cyclic loading is confirmed in [2]. $\sigma_{o(1)}$=223MPa, $\sigma_{o(2)}$=250MPa, $m_{(1)}$=6.7, and $m_{(2)}$=3.7 can be evaluated directly from the experimental results, among which the stabilized stress-strain loop is shown in Fig.2. Figure 3 shows the initial yield surface defined by a von Mises effective strain of 50 μm/m and determined by following 12 radial stress paths away from the origin of stress-space, alway returning to the origin after yielding was observed. The coordinates of stress-space were selected so that the curve described by the von Mises yield formulation is

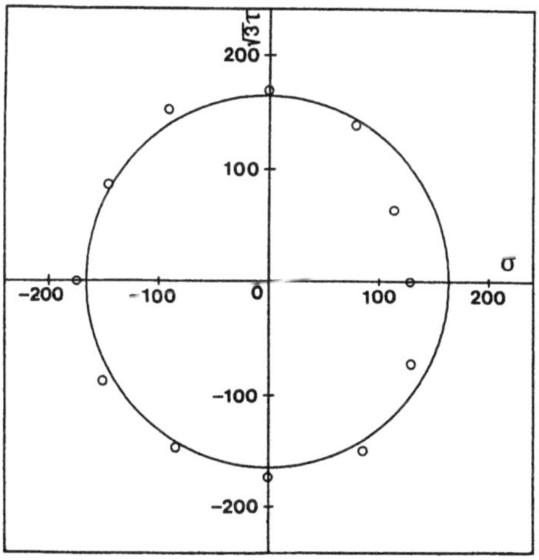

Fig.3 Initial yield surface

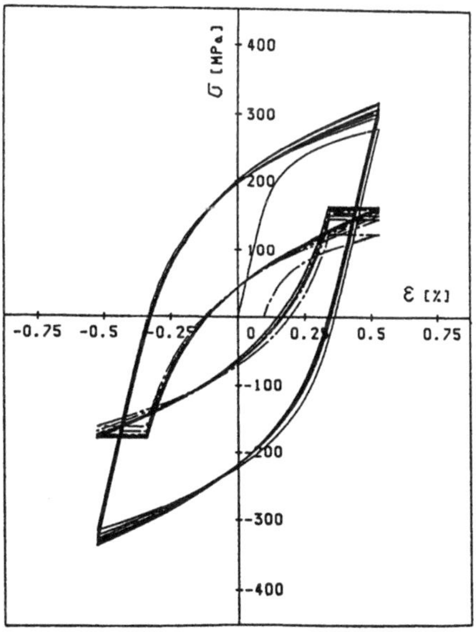

Fig.4 Cyclic tension-compression loading with strain amplitude of 1.1%

a circle. As a result, the von Mises formulation gives a good description of this initial yielding, and the anisotropy of the specimen cannot be appreciated initially. Then the nonzero coefficient tensors in equation (1) take $C_{1111}=2/3$ and $C_{1212}=1/2$. After several trials with the computer(NEC PC-9801VM), the simulated stress-strain curves shown by the solid lines in Fig.4 are depicted using $\sigma_{o(\infty)}=R_{(\infty)}=275$MPa , $\alpha=\lambda=0.07$, $W_o/\sigma_{o(\infty)}=W_2/\sigma_{o(\infty)}=0.10$, $m_{(\infty)}=3.3$, $\beta=0.13$ and $W_1/\sigma_{o(\infty)}=0.028$. The movement of the center of the yield surfaces during cyclic loading is traced by the dash-dot lines in Fig.4. The value of kinematic back stress during compression in the stabilized stress-strain loop is 160MPa. The stress at the point b in Fig.2 corresponds to this back stress, and therefore, the effective stress is defined in this work as stress measured from the points b or g, corresponding compression side of the stress-strain loop or tension one, respectively.

The creep curves corresponded to Fig.2 are given in Fig.5, where the absolute values of total strain measured from the point b or g is plotted during 300sec. The numerals in Fig.5 represent the effective stress. The solid lines show the creep along the tension side of the stress-strain loop, while the broken ones along the compression side. Two creep curves under the same stress level have a good agreement. This observation shows

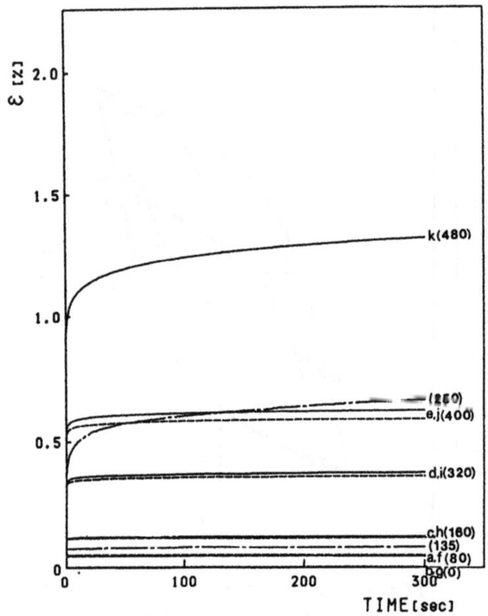

Fig.5 Total strain versus time for creep tests depicted in Figs.1 and 2

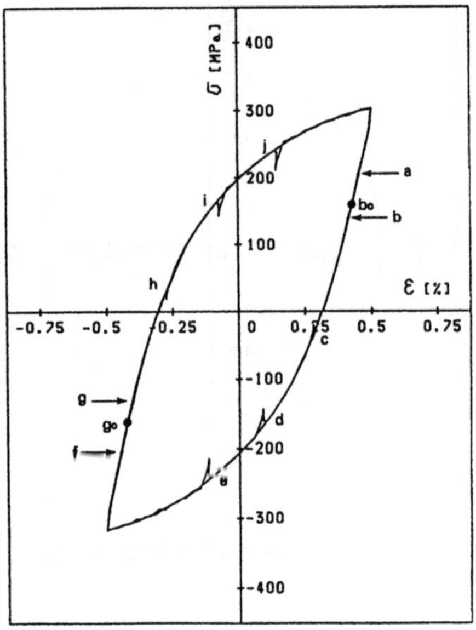

Fig.6 Intermittent relaxation periods of 300 sec during cycling between ± 0.5%

that the creep-rate depends on stress and time if stress is measured from the current center of loading surface as the conventional creep theory. Therefore, the evolution of back stress during cyclic preloading plays the important role to prescribe the following creep behavior. The creep curves corresponded to Fig.1 are also shown in Fig.5 by dash-dot lines. the numerals on the dash-dot lines represent also the effective stresses measured from the current centers of the stress space. The clear difference between creep during monotonic loading shown by dash-dot line and that during cyclic loading shown by solid or broken line represents the effect of cyclic plasticity, or fatigue on following creep behavior.

From the experimental results of creep during cyclic loading, the following conventional creep law has been obtained,

$$\varepsilon^c = 1.90\times10^{-5}\sigma^{1.72}t^{0.0181} \tag{8}$$

where σ is the effective stress measured from the current center of yield surface, and t is time in second. This equation (8) is the so-called Bailey-Norton law.

The results for intermittent relaxation tests shown in Fig.6 confirm the results obtained in creep. At the completion of the 20th cycle of

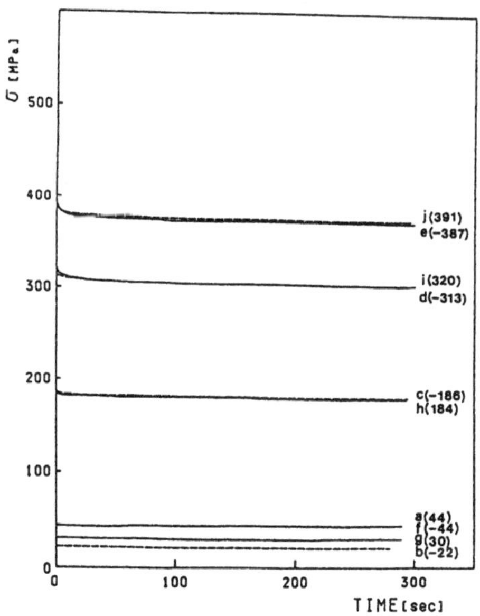

Fig.7 Stress versus time for relaxation tests depicted in Fig.6

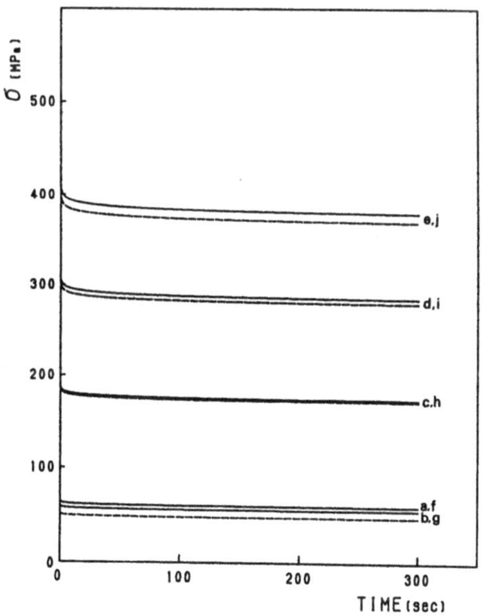

Fig.8 Prediction of relaxation of stress

strain controlled cycling between ±0.5% with the strain rate $\dot{\varepsilon}=5\times10^{-4}$ (1/sec), relaxation periods for 300 sec were introduced starting at a through j. The relaxation curves corresponded to Fig.6 are given in Fig.7. In Fig.7 the absolute values of stress measured from b_o or g_o is plotted during 300 sec, where b_o and g_o correspond to the current centers of yield surfaces such as the points b and g in Fig.2. The numerals in Fig.7 represent the effective stress at the commencement of stress relaxation. The solid lines show the relaxation along the tensile side of the stress-strain loop, while the broken ones along the compression side. The prediction of the relaxation of stress with use of Eq.(8) is shown in Fig.8. Comparison between the experimental results in Fig.7 and the numerical prediction in Fig.8 shows a good agreement. Therefore, the results for intermittent relaxation tests reconfirm the results obtained in creep.

Conclusion

To aim at a formulation of the unified constitutive model of cyclic plasticity and creep, the concept of effective stress which is defined in this paper as stress measured from the current center of yield surface, is employed to explain the intermittent creep or relaxation period after cyclic prestraining. Then the creep or relaxation behavior around the hysteresis loop of cyclic plasticity can be described by a conventional theory of creep. However, the cyclic plasticity in out-of-phase loading as mechanical ratchetting should be carefully examined to construct an unified theory.

References

1 Kujawski, D., Kallianpur, V., and Krempl , E., "An Experimental Study of Uniaxial Creep, Cyclic Creep and Relaxation of AISI Type 304 Stainless Steel at Room Temperature," J. Mech. Phys. Solids, Vol.28, 1980, pp.129-140.

2 Ishikawa, H., and Sasaki, K., "Stress-Strain Relations of SUS304 Stainless Steel After Cyclic Preloading," ASME Journal of Engineering Materials and Technology, Vol.111, 1989, pp.417-423.

3 Ishikawa, H., and Sasaki, K., "Yield Surfaces of SUS304 under Cyclic Loading," ASME Journal of Engineering Materials and Technology, Vol.110, 1988, pp.364-371.

Anisotropic Creep Behaviour Induced by Plastic Deformation

M. KAWAI

Institute of Engineering Mechanics
University of Tsukuba, Tsukuba 305, JAPAN.

ABSTRACT
A rate-dependent constitutive model which describes deformation induced anisotropy is proposed. The model falls in the category of the unified models. In the first place, we describe an inelastic strain-rate in terms of an isotropic tensor function of the deviatoric stress tensor and a second rank internal variable tensor. Based on its polynomial representation, an effective stress characterized by the internal variable is defined. Then, a simple expression of the internal state is postulated which accounts for a combined state of isotropic and anisotropic hardenings. The evolutional rate of the internal state is prescribed in terms of the effective inelastic strain rate which is defined analogous to the effective stress. Finally, the model behavior is demonstrated in the case of combined creep and plasticity.

INTRODUCTION

Anisotropy in strain-hardening of metals is intrinsic to inelastic deformation. The anisotropy developed, in turn, has a significant influence on subsequent inelastic deformation behaviors [1-5]. In constitutive modellings, therefore, descriptions of the *state* and the *evolution* of material anisotropy is characteristically important.

In most phenomenological rate-dependent formulations of inelasticity, the *effective stress concept* is adopted: a physical driving force of inelastic deformation is not the externally applied (deviatoric) stress $\mathbf{s}$ itself but a modified (effective) stress $\mathbf{s}^*$ which macroscopically reflects the current internal state (microstructure, damage, etc.). Many models proposed so far have a similar skeletal structure [6]. Inelastic strain rate $\dot{\mathbf{e}}$ is commonly expressed as

$$\dot{\mathbf{e}} = F\mathbf{s}^* \tag{1}$$

where F is a strain-rate scalar function of the effective stress and other variables. The form of the rate function F is influenced by the physical phenomena to be described and by the ease of material identification. The inelastic flow direction is parallel to the effective stress. In order to describe an anisotropic (kinematic) hardening, the effective stress $\mathbf{s}^*$ is usually defined by the difference between the applied stress $\mathbf{s}$ and an internal stress (back stress) $\boldsymbol{\xi}$: $\mathbf{s}^* = \mathbf{s} - \boldsymbol{\xi}$. When only the kinematic hardening variable is used, however, it is found that the creep strain under

M. Zyczkowski (Ed.)
Creep in Structures
IUTAM Symposium Cracow/Poland 1990

stress reversals is overestimated [1]. Therefore, the incorporation of some constraint on the kinematic hardening variable or a replacement with another effective stress including both aspects of anisotropic and isotropic hardenings are required [7, 8]. Such effective stresses accounting for the isotropic and anisotropic hardenings can be defined, for example, in terms of $\mathbf{s}^{**} = \mathbf{s}^* /D$ [9, 10], or $\|\hat{\mathbf{s}}^{**}\| = \langle \|\mathbf{s}^*\| - D \rangle$ [11], where D represents a drag stress D and $\|.\|$ denotes the magnitude of a quantity. In more general context, we may consider the effective stress defined in terms of a linear transformation of the applied stress: $\tilde{\mathbf{s}}^{**} = \Omega \mathbf{s}$ [4, 12].

The effective stress, in turn, characterizes the internal hardening variables. The effective stress $\mathbf{s}^*$ uniquely corresponds to a single internal variable $\xi = \mathbf{s} - \mathbf{s}^*$. In the case of $\mathbf{s}^{**} = (\mathbf{s} - \xi)/D$, however, the inversion is not unique, since two independent internal variables are integrated into one effective stress. The multiplied form makes it still more difficult to split into the expected isotropic and anisotropic hardenings. Even if we are concerned with only a simple monotonic proportional deformation, the use of latter effective stress requires one to identify the evolution of isotropic and anisotropic hardenings; in order to identify the anisotropy, we need at least reversed loading tests, and more precisely multiaxial non-proportional tests. Thus, taking ease of material identification into consideration, a simple expression of the effective stress will be useful which has a unique relation to the associated internal variables representing isotropic and anisotropic hardenings.

In the present study, we discuss a simple representation of the effective stress which incorporates a combined effect of isotropic and anisotropic hardenings. The emphasis is placed on a description of the anisotropic internal-state associated with the effective stress. An evolutional equation for the internal state is also formulated. The internal state and its evolution are described in a simple connection with the inelastic strain and inelastic strain rate. The integrated model falls in the category of unified constitutive models. The model behavior is demonstrated in a simple case of combined creep and plasticity, together with a comparison with the actual material behaviors [13].

ANISOTROPIC CONSTITUTIVE MODEL

Effective Stress

Consider a polynomial representation of inelastic strain rate [2, 3]: $\dot{\mathbf{e}} = \mathbf{f}(\mathbf{s}, \mathbf{p}) = a_i \mathbf{G}^i$, where $\mathbf{s}$ and $\mathbf{p}$ are the deviatoric Cauchy stress tensor and a second rank symmetric deviatoric tensor, respectively. The coefficients a_i are arbitrary scalar functions of the joint invariants of $\mathbf{s}$ and $\mathbf{p}$, and $\mathbf{G}^i$ irreducible tensor generators. Selecting only linear terms of the arguments, we have a reduced expression

$$\dot{\mathbf{e}} = a_1 \{\mathbf{s} - k_1 \mathbf{p} - k_2 \Sigma(\mathbf{p}) \mathbf{s}\} \tag{2}$$

where $a_3/a_1 = - k_1$, $a_5/a_1 = - k_2$, $\Sigma(\mathbf{p})\mathbf{s} = \mathbf{J}(\mathbf{sp} + \mathbf{ps})$, and $\mathbf{J}$ denotes a deviator generator tensor of fourth rank. The above equation proposes that $a_1 = F$, and an expression of the effective stress

$$\mathbf{s}^* = \{\mathbf{J} - k_2\Sigma(\mathbf{p})\}\mathbf{s} - k_1\mathbf{p} \tag{3}$$

As regards the effective stress, an anisotropic effect of the Bauschinger type is expressed only via kinematic hardening variable $\mathbf{p}$. The term $k_2\Sigma(\mathbf{p})\mathbf{s}$ may be used to account for a deviation from the von Mises flow rule. Also, it defines a magnification tensor of fourth rank $\mathbf{M} = \mathbf{J} - k_2\Sigma(\mathbf{p})$. Assuming that $k_2 = 0$, eq. (3) coincides with the expression used by Sawczuk and Trampczynski [3]. In the following discussion, we put aside the term $k_2\Sigma(\mathbf{p})\mathbf{s}$ for future application.

Anisotropic Internal State

In the first place, we consider a monotonic proportional loading for initially isotropic materials and postulate that a current state of the induced anisotropy, which is invisible, is expressed in terms of a *state equation* : $\mathbf{p} = \widehat{\mathbf{p}}(\mathbf{s}, \mathbf{e})$, where $\widehat{\mathbf{p}}$ denotes an isotropic tensor function and $\mathbf{e}$ the inelastic strain tensor. Based on the polynomial representation, an expression similar to eq. (2) is obtained: $\mathbf{p} = b_1\mathbf{s} + b_3\mathbf{e} + b_5\Sigma(\mathbf{e})\mathbf{s}$, where the coefficients b_i are scalar-valued functions of the joint invariants of $\mathbf{s}$ and $\mathbf{e}$.

Let us consider a simple form;

$$\mathbf{p} = B\mathbf{m} + C\mathbf{n} \tag{4}$$

where $\mathbf{m} = \mathbf{e}/\|\mathbf{e}\|$ and $\mathbf{n} = \mathbf{s}/\|\mathbf{s}\|$. The coefficients B and C, in general, are scalar functions of joint invariants of $\mathbf{s}$ and $\mathbf{e}$. We consider an initially isotropic material subjected to monotonic proportional deformation. In this case, $\mathbf{p} = (B + C)\mathbf{m}$, since $\mathbf{m}$ is coaxial with $\mathbf{n}$. If we denote the magnitude of $\mathbf{p}$ (i.e., a total strain hardening) in terms of H, then the coefficients B and C are expressible as

$$B = \mu H, \qquad C = (1 - \mu)H \tag{5}$$

where μ is a scalar variable.

Introducing two kinds of new variables:

$$\alpha = B\mathbf{m}, \qquad \rho = C \tag{6}$$

then eq. (4) is rewritten as

$$\mathbf{p} = \alpha + \rho\mathbf{n} \tag{7}$$

If ρ is non-negative (i.e., $\mu \leq 1$), the internal stress state $\mathbf{p}$ is identified in terms of a point on the surface

$$g(\mathbf{p};\ \alpha, \rho) = (3/2)^{1/2} \|\mathbf{p} - \alpha\| - \rho = 0 \tag{8}$$

The above description of anisotropic state is characteristic of the present model. The partition of the internal state is a hypothesis. Substitution of eq.(7) into eq. (3) results in a particular form of the effective stress

$$\mathbf{s}^* = (\|\mathbf{s}\| - k_1\rho)\mathbf{n} - k_1\alpha - k_2\{\Sigma(\alpha) + \rho\Sigma(\mathbf{n})\}\mathbf{s} \tag{9}$$

This effective stress $\mathbf{s}^*$ reflects both aspects of kinematic and isotropic hardenings, which depends on the parameter μ. Let us assume $k_1 = 1$ and $k_2 = 0$. When $\mu = 1$, $\mathbf{s}^* = \mathbf{s} - \alpha$; the model is in agreement with the pure kinematic hardening model. When $\mu = 0$, on the other hand, $\mathbf{s}^* = (\|\mathbf{s}\| - \rho)\mathbf{n}$, which corresponds to the pure isotropic model.

Assuming $k_2 = 0$, the magnitude of the effective stress $\|\mathbf{s}^*\|$ is in agreement with the overstress expression $\|\hat{\mathbf{s}}^{**}\| = \langle\|\mathbf{s}^*\| - D\rangle$ of the viscoplastic formulation [11] under monotonic proportional loading conditions. However, it is not so in general cases. Note that in the present framework we do not postulate the equi-dissipative potential surface within which the inelastic strain rate disappears. Furthermore, by the use of the present effective stress in eq. (1), the direction of the inelastic flow is influenced by the isotropic hardening variable ρ as well as the kinematic hardening variable α. In these respects as well as the state definition mentioned above, the present modelling is essentially different from others proposed so far.

Evolution of Internal State

We have defined an anisotropic internal state generated by a prior inelastic straining. In order to complete the present modelling, we have to prescribe the evolution of the current internal state. Considering again a monotonic proportional deformation, we can assume from eqs. (6) and (7) the following rate expressions:

$$\dot{\mathbf{p}} = \dot{\alpha} + \dot{\rho}\mathbf{n} \tag{10}$$

$$\dot{\alpha} = \mu H_1 \dot{\mathbf{e}}, \qquad \dot{\rho} = (1 - \mu)H_1\|\dot{\mathbf{e}}\| \tag{11}$$

Taking into account $\dot{\mathbf{p}} = H_1\dot{\mathbf{e}}$, and $\dot{\mathbf{p}} \bullet \mathbf{n} = \|\dot{\mathbf{p}}\| = H_1\dot{\mathbf{e}} \bullet \mathbf{n} = H_1\|\dot{\mathbf{e}}\|$, it is postulated that

$$\|\dot{\mathbf{p}}\| = H_1[\mu\|\dot{\mathbf{e}}\| + (1 - \mu)\|\dot{\mathbf{e}}\|] = H_1[\dot{\varepsilon}_{k(h)} + \dot{\varepsilon}_{i(h)}] \tag{12}$$

In eq. (12), it turns out that we assume a fictitious decomposition of the inelastic strain rate into two parts: a kinematic-hardening strain rate $\dot{\mathbf{e}}_{k(h)} = \mu\dot{\mathbf{e}}$ and an isotropic-hardening strain rate $\dot{\varepsilon}_{i(r)} = (1 - \mu)\|\dot{\mathbf{e}}\|$. The parameter μ describes a ratio of the inelastic strain contribution to the two types of hardenings. For incorporation of steady-state behaviors, furthermore, the evolutional equations for $\mathbf{p}$, α, and ρ are modified by including recovery terms. Assuming the strain hardening-recovery format of the Bailey-Orowan type [6, 9-11], we replace the time rate of $\mathbf{p}$ by

$$\dot{\mathbf{p}} = H_1\dot{\mathbf{e}} - R\mathbf{p} \tag{13}$$

The expression may be deduced from the polynomial representation of $\dot{\mathbf{p}} = \mathbf{h}[\dot{\mathbf{e}}, \mathbf{s}, \mathbf{p}(\mathbf{s}, \mathbf{e})]$. Here, substituting eq. (7) into eq. (13) and using eq. (12), we obtain

$$\dot{\alpha} = H_1\dot{\mathbf{e}}_{k(h)} - R\alpha, \qquad \dot{\rho} = H_1\dot{\varepsilon}_{i(h)} - R\rho \tag{14}$$

Analogous to eq. (12), we define here the recovery strain rates: a kinematic-recovery strain rate $\dot{\mathbf{e}}_{k(r)} = (R/H_1)\alpha$ and an isotropic-recovery strain rate $\dot{\varepsilon}_{i(r)} = (R/H_1)\rho$. Then, we consider the effective inelastic strain rates of kinematic and isotropic types:

$$\dot{\mathbf{e}}_k{}^* = \dot{\mathbf{e}}_{k(h)} - \dot{\mathbf{e}}_{k(r)}, \qquad \dot{\varepsilon}_i{}^* = \dot{\varepsilon}_{i(h)} - \dot{\varepsilon}_{i(r)} \tag{15}$$

In view of eq. (15), we have an alternative expression of eq. (13)

$$\dot{\mathbf{p}} = H_1\dot{\mathbf{e}}^* \tag{16}$$

where $\dot{\mathbf{e}}^* = \dot{\mathbf{e}} - \dot{\mathbf{e}}_{(r)} = \dot{\mathbf{e}}_k{}^* + \dot{\varepsilon}_i{}^*\mathbf{n}$, and $\dot{\mathbf{e}}_{(r)} = (R/H_1)\mathbf{p}$. The strain rate $\dot{\mathbf{e}}^*$ will be termed the (total) *effective inelastic strain rate*. We may consider slightly different evolutions which include different recovery strain rates: $\dot{\mathbf{e}}'_{k(r)} = \nu\dot{\mathbf{e}}_{(r)}$; $\dot{\varepsilon}'_{i(r)} = (1 - \nu)\|\dot{\mathbf{e}}_{(r)}\|$, where ν is an additional parameter. It is noted that a partition of the inelastic strain (rate) is also adopted in the Hart model [14]. However, the rheological structure of the present model is different [14].

APPLICATION TO COMBINED DEFOMATION OF CREEP AND PLASTICITY

In order to demonstrate the model behavior, a simple numerical calculation is made for the creep behavior after plastic deformation and compared with the experimental results [13]. Our

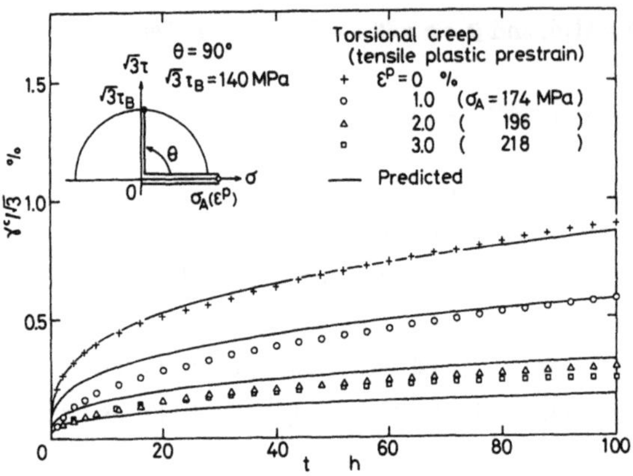

Fig.1. Subsequent torsional creep after tensile plastic prestraining ($\theta = 90°$)

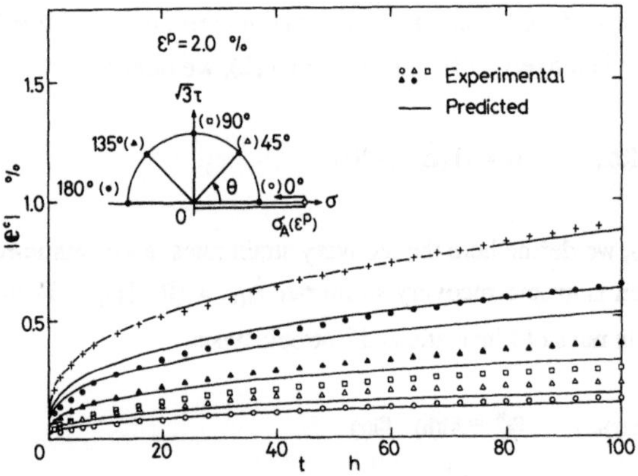

Fig.2. Subsequent creep curves in different directions after tensile plastic prestraining

discussion here is limited to almost instantaneous responses immediately after the onset of the subsequent creep deformation. Thus, we treat the internal variables as constant. For the sake of simplicity, we consider the rate function F

$$F = (3/2)mA^{1/m}\chi^{(m-1)/m}[\sigma_e^*]^{n/m-1}, \qquad \dot{\chi} = d_e \tag{17}$$

where $\sigma_e^* = (3/2)^{1/2}\|s^*\|$, and $d_e = (2/3)^{1/2}\|\dot{e}\|$. This corresponds to a modified version of the rate equation derived from the Bailey-Norton creep law: $\varepsilon = At^m\sigma^n$. The material constants involved in the Bailey-Norton equation were first determined based on the uniaxial creep curves

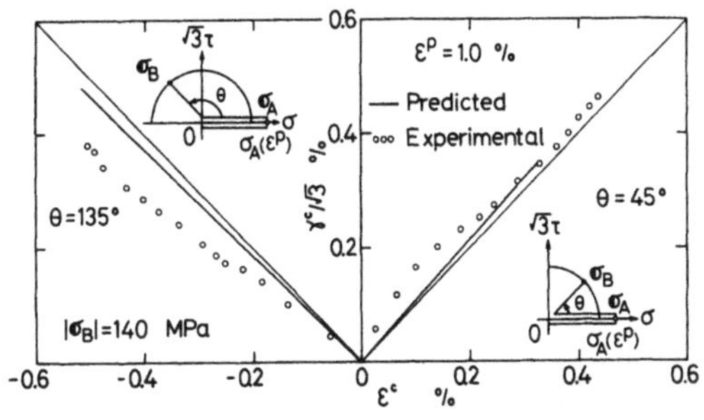

Fig. 3. Creep strain trajectories for $\theta = 45°$ and $135°$

of virgin materials: $A = 7.42 \times 10^{-16}$ MPa$^{-5.8}$ h$^{-0.31}$, $m = 0.31$, $n = 5.8$. Then, the change in axial internal stress p was identified using the subsequent creep curves in the same and opposite directions of the prior plastic straining, where we employed the uniaxial relation $\varepsilon = At^m(\sigma - p)^n$ at $t = 100$ h. The total hardening H was approximated by a linear relation $H = H_1(\|e\| - H_2)$, where $H_1 = 2188.11$ MPa, $H_2 = 0.28 \times 10^{-2}$, $c = 0.4$.

Figure 1 shows the predictions of subsequent torsional creep behaviors (θ =90°) in solid lines, together with the experimental results in symbols. Figure 2 shows the comparison of the subsequent creep curves for five different angles θ after plastic straining of $\varepsilon^p = 2.0$ %, where the crucial symbol is the torsional creep of virgin material. In Fig. 3, the creep strain paths are shown for $\theta = 45°$ and $135°$ after tensile plastic straining of $\varepsilon^p = 1.0$ %. As a whole, the model reproduced the trend of plastic strain induced creep anisotropy observed in the experiments. The non-collinearity between the applied stress and the creep strain rate vectors was underestimated.

CONCLUDING REMARKS

We discussed a form of effective stress and proposed a description of the associated internal state and its evolution involving both aspects of isotropic and anisotropic hardenings. The evolution was directly connected with inelastic straining. The present model falls in the category of the rate-dependent unified constitutive models. The model feature is a fictitious decomposition of inelastic strain (rate) into kinematic and isotropic parts. Further study is required especially on a physical characterization of the parameters μ and ν. A good selection of the functions F and R is important for ease of material identification. Power functions are manageable: $F = K\|s^*\|^{N-1}$, $R = L\|p\|^{M-1}$. Auxiliary rules are additionally required to apply the present model to more complex loading conditions as an unified model.

ACKNOWLEDGEMENT

The author is grateful to Professor S. Murakami of Nagoya University for his useful discussions. The financial support provided in part by the Ministry of Education and the University of Tsukuba is gratefully acknowledged.

REFERENCES

1. Ohashi. Y.; Ohno, N.; Kawai, M.: Evaluation of creep constitutive equations for type 304 stainless steel under repeated multiaxial loading. ASME J. Eng. Mat. Tech. 104 (1982) 159-164.

2. Sawczuk, A.; Anisimowicz, M.: Tensor functions approach to creep laws after prestrain. In: Ponter, A. R.S.; Hayhurst, D. R. (eds.) Creep in structures. Berlin, Heidelberg, New York: Springer-Verlag (1981) 220-232.

3. Sawczuk, A.; Trampczynski, W. A.: A theory of anisotropic creep after plastic prestraining. Int. J. Mech. Sci. 24 (1982) 647-653.

4. Waniewski, M.: A simple law of steady-state creep for material with anisotropy introduced by plastic prestraining. Ing. Arch. 55 (1985) 368-375.

5. Trampczynski, W. A.: The influence of cold work on the creep of copper under biaxial states of stress. Act. metall. 30 (1982) 1035-1041.

6. Krieg, R. D.: Numerical integration of some new unified plasticity-creep formulations. Proc. Int. Conf. SMiRT-4 (1977) M6/4*.

7. Murakami, S.; Ohno, N.: A constitutive equation of creep based on the concept of a creep-hardening surface. Int. J. Solids Struc. 18 (1982) 597-609 .

8. Mroz, Z.; Trampczynski, W. A.: On the creep-hardening rule for metals with a memory of maximal prestress. Int. J. Solids Struc. 20 (1984) 467-486.

9. Miller, A.: An inelastic constitutive model for monotonic, cyclic, and creep deformation. ASME J. Eng. Mat. Tech. 98 (1976) 97-113.

10. Krieg, R. D.; Swearengen, J. C.; Rohde, R. W.: A physically-based internal variable model for rate-dependent plasticity. In: Chang, T. Y.; Krempl, E. (eds.) Inelastic behavior of pressure vessel and piping components. PVP-PB-028 ASME Press (1978) 15-28.

11. Chaboche, J. L.: Viscoplastic constitutive equations for the description of cyclic and anisotropic behaviour of metals. Bull. Acad. Pol. Sci. Ser. Sci. Tech. 25 (1977) 33-42.

12. Betten, J.: Representation of constitutive equations in creep mechanics of isotropic and anisotropic materials. In: Ponter, A. R. S.; Hayhurst, D. R. (eds.) Creep in structures. Berlin, Heidelberg, New York: Springer-Verlag (1981) 179-201.

13. Ohashi, Y.; Kawai, M.; Momose, T.: Effects of prior plasticity on subsequent creep of type 316 stainless steel at elevated temperature. ASME J. Eng. Mat. Tech. 108 (1986) 68-74.

14. Hart, E. W.: Constitutive relations for the nonelastic deformation of metals. ASME J. Eng. Mat. Tech. 98 (1976) 193-202.

The Influence of Deformation History on Creep of Pure Copper

Z. KOWALEWSKI

Institute of Fundamental Technological Research
Polish Academy of Sciences
ul. Świętokrzyska 21, 00-049 Warsaw, Poland

Summary

A series of combined stress tests in tension-torsion space was performed on thin-walled tubes of M1E commercially pure copper. Interaction between plastic and creep deformations was studied experimentally with respect to creep behaviour subsequent to plastic prestraining. In combined plasticity-creep tests, specimens were first prestrained plastically at the room temperature in axial tensile or pure torsional directions, and were subsequently subjected to constant stress creep under various biaxial states with the two identical effective stress σ_i=31 and 41 [MPa] at 573K. Results for virgin material were utilized to verify of the basic assumptions of the quasi-linear creep theory. The stationary creep rates either for virgin copper or the same material plastically prestrained have been described using a modified Norton's law.

1. Introduction

Most engineering structures operate under variable and non-proportional loading conditions and some parts are subjected to prestrain during the life-time. Also manufacturing processes introduce certain predeformation in the final products and in consequence may effect on their mechanical properties. It is important from practical point of view to know the influence of such prestrain on different high temperature properties such as strain rate, life-time, ductility and fracture.

Deformation at room temperature prior to creep testing has been found from previous investigations to have either a beneficial effect on the creep of some materials, e.g. Inconel MA754 [1], copper [2,3], or a detrimental effect for other materials, e.g. Nimonic 80A [4,5], Al-2.5%Cu-1.2%Mg [6]. This was manifested by the decrease or increase of the secondary creep rate, respectively. In some materials the prior cold work have no significant influence, e.g. 1/2Cr-1/2Mo-1/4V [7].

Prior room temperature deformation can change the creep ductility and time-to-fracture [1,7,8]. In most cases prestraining was found to cause marked reductions in ductility and time-to-fracture, e.g. 1/2Cr-1/2Mo-1/4V [7], Inconel X-750 [8], Nimonic 80A [4,5,9]. Since the plastic prestrain has a various effect on creep behaviour depending on the material, in order to obtain a better understanding of this problem further systematic investigations are needed.

M. Zyczkowski (Ed.)
Creep in Structures
IUTAM Symposium Cracow/Poland 1990

2. Experimental procedure

The material used was a commercially pure copper, which was received as drawn rods of 45 [mm] diameter. The rods have been annealed before machining at 673K for two hours with subsequent furnace cooling. To check the initial isotropy of the material, preliminary simple tension tests were carried out on small cylindrical specimens cut out from a drawn rod in different directions. From these tests the copper was found to be initially isotropic. Data for the whole programme of the initial tests have been given ealier [10].

Creep specimens in the form of thin-walled tubes with the 25 [mm] outer diameter, 1.5 [mm] wall thickness, 40 [mm] gauge length and 140 [mm] total length were used.

The creep experiments were performed with the use of the high-temperature, combined loading creep testing machine [10,11]. The measuring technique corresponded to those presented in [12].

Experimental programme consisted of creep tests under biaxial stress state with various combinations of tension and torsion (σ_{12}/σ_{11}=0; 0.33; 0.5; ∞, where σ_{11} - tension stress, σ_{12} - shear stress) at 573K for pure copper in the virgin state and for the same material plastically prestrained at room temperature. Creep tests for virgin copper were realized for three levels of effective stress of Huber-Mises type: $\sigma_i = (\sigma_{11}^2 + 3\sigma_{12}^2)^{1/2}$=31; 41 and 45 [MPa], whereas prestrained material were examined for two values of σ_i equal to 31 and 41 [MPa]. The effect of cold work on creep behaviour was investigated according to the following procedure. Each tubular specimen was proportionally deformed up to 5% of plastic prestrain by uniaxial tension or pure torsion at room temperature and then unloaded. Subsequently, each specimen was heated uniformly at the test temperature 573K for 24 hours prior to testing and then was subjected to a selected biaxial loading at the constant stress level.

3. Results and discussion

3.1 Results for virgin copper

Experimental creep data for virgin material proved anisotropy. Steady creep rates under the same stress level for uniaxial tension creep tests were significantly higher than obtained for pure torsion creep tests. For example, if σ_i=41 [MPa] then for tension creep test $D_i=22.1\times10^{-5}$ [1/h] whereas for torsion creep test $D_i=9\times10^{-5}$ [1/h] (D_i - effective steady creep rate of Huber-Mises type). Experimentally obtained secondary creep rates are presented in Fig.1.

Time needed to reach the minimum creep rate depends on the direction of loading stress vector $\theta_\sigma = arctan\sqrt{3}\sigma_{12}/\sigma_{11}$. It is seen in Fig.2, that if the θ_σ angle increases

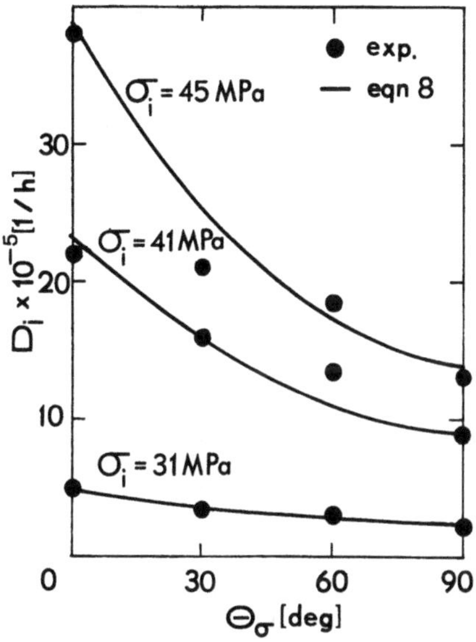

Fig.1. Secondary creep rates for virgin copper obtained for different stress states

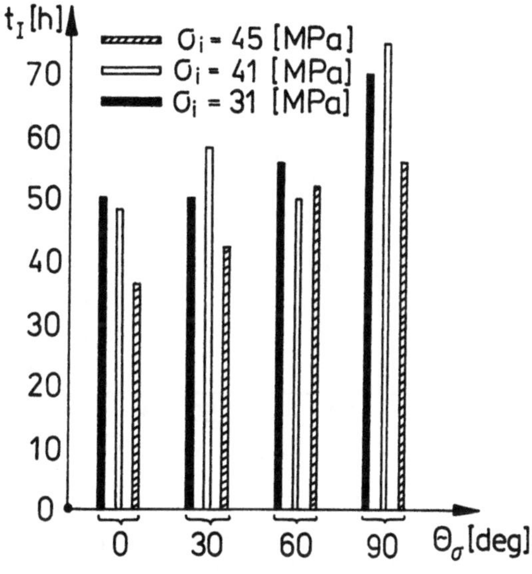

Fig.2. Duration of primary creep versus θ_σ

than the duration of primary creep is longer.

3.1.1 Analysis of the basic assumptions of the quasi-linear creep theory

Quasi-linear creep theory expressed by the equation in the form

$$D_{ij} = 3/2F(\sigma_i)G(t)S_{ij}, \tag{1}$$

where D_{ij}, S_{ij} denote creep strain rate and stress deviators, respectively, and $F(\sigma_i)G(t) = D_i(t)/\sigma_i$, is supported on the two following general assumptions:

1) - similarity of the stress and creep strain rate deviators,

2) - creep rate is a function of the second invariant of stress deviator.

In the equation (1) a dependence between the creep rate and stress tensors is linear, however, nonlinearity is included in the expression $F(\sigma_i)G(t)$.

In the case of stationary creep and under the first assumption the general creep law (1) takes the form

$$D_{ij} = CS_{ij}. \tag{2}$$

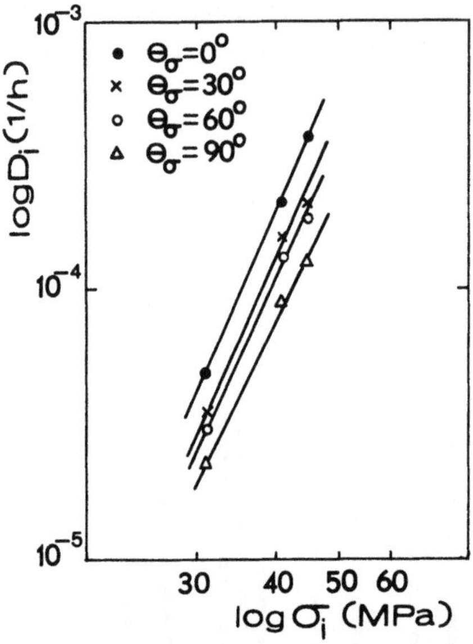

Fig.3. Plots of effective secondary creep rates for virgin copper versus effective stresses

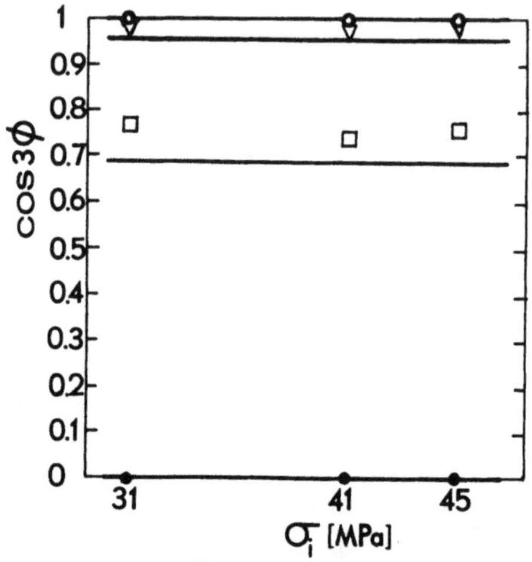

Fig.4. Checking of proportionality of the strain rate and stress deviators (∘ - $\theta_\sigma = 0^0$, ▽ - $\theta_\sigma = 30^0$, □ - $\theta_\sigma = 60^0$, • - $\theta_\sigma = 90^0$)

Moreover, it is further assumed that the scalar multiplier C depends solely on the second invariant of the stress deviator $C = C(J_2)$ (it corresponds to the second assumption). The creep law (2) can thus be written in terms of the effective stress and strain rate

$$D_i = f(J_2)\sigma_i, \tag{3}$$

where

$$D_i = \left[\tfrac{2}{3} D_{ij} D_{ij}\right]^{1/2}, \qquad \sigma_i = \left[\tfrac{3}{2} S_{ij} S_{ij}\right]^{1/2},$$

or in the more useful dimensionless form

$$\frac{D_i}{D_o} = \left[\frac{\sigma_i}{\sigma_o}\right]^n, \tag{4}$$

where n, D_o, σ_o denote the material constants.

An analysis of experimental data regarding creep in plane stress state for some materials indicates that the creep law (2) is too restrictive and should not be directly extended to the multiaxial stress states [13,14,15,16]. To check this point of view let us first consider, on the basis of the own programme of tests, the experimental evidence regarding the relation (3). In Fig.3 the experimentally obtained effective strain rates are plotted against the effective stresses for different stress states, and thus at different values of the remaining stress invariants. If relations (3) or (4) correct, the lines concerning various stress states should be coincided.

To discuss the applicability of the relation (2) and to answer the question whether the deviators of strain rate and stress are proportional, it is sufficient to check the corresponding Lode parameters expressed in the following form

$$\Phi_S = \frac{1}{3} arccos \left[\frac{3\sqrt{3} J_3}{2(J_2)^{3/2}} \right], \tag{5}$$

$$\Phi_D = \frac{1}{3} arccos \left[\frac{3\sqrt{3} I_3}{2(I_2)^{3/2}} \right]. \tag{6}$$

In the case of coaxiality of the respective deviators

$$\Phi_S = \Phi_D. \tag{7}$$

In Fig.4 the results are given concerning the tests to be performed. The components of the strain rate and stress deviators are not proportional, except the case of uniaxial tension $cos 3\Phi_D = 1$ and that of pure torsion $cos 3\Phi_D = 0$. The experimental points for biaxial tests are located decidedly above the line corresponding to $cos\Phi_D$ calculated for the applied state of stress employing the tensorially linear creep law (2). Thus, it may be concluded that the assumptions of quasi-linear creep theory are not satisfied under multiaxial stress states. Similar conclusion can be reached when considering results published by other investigators [13,14,15,16].

3.2 Results for prestrained copper

In Fig.5 the comparison of creep curves for virgin and prestrained copper is presented. It shows that the creep process under constant stress is generally affected by prior plastic strain at the room temperature. Cold work preceding the creep at 573K causes hardening effect expressed by significant decrease of the minimum creep rate (compare the results plotted in Fig.1 with those in Figs.6.7). The increase of the creep resistance is observed in the creep experiments of copper prestrained either by uniaxial tension or by pure torsion. The greatest hardening effect during creep was observed in the case of coaxiality of plastic prestraining and creep directions. The hardening effect is observed also in the cases when prestraining direction is different than that of the loading in the subsequent creep test.

Analysis of the creep curves presented in Fig.5 shows that plastic prestrain drastically reduces the duration of primary creep stage for both considered directions of plastic prestrain. In order to obtain better presentation of this effect comparison of the results for virgin and prestrained material is showed in Fig.8.

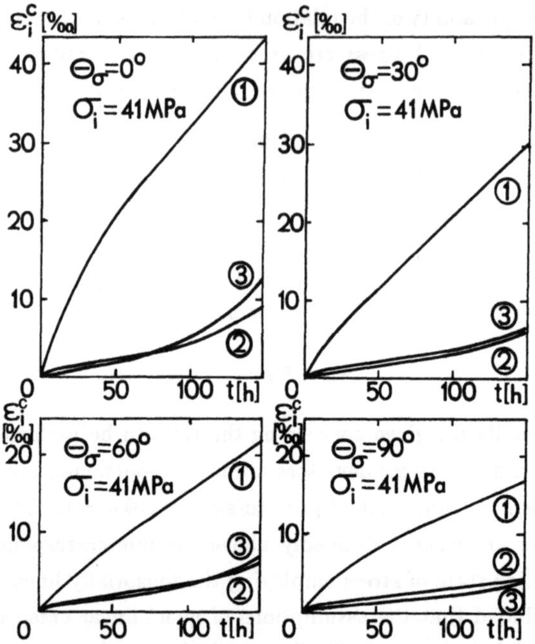

Fig.5. Comparison of creep curves for virgin copper (1) and the same material plastically prestrained by tension (2), and by torsion (3)

As shown in Fig.5, also duration of the secondary creep decreases significantly.

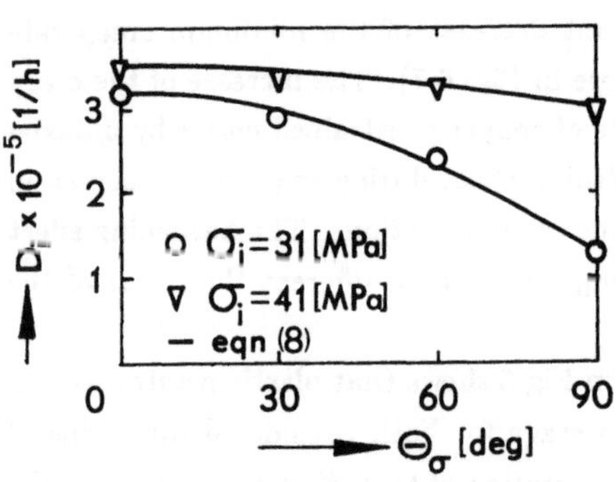

Fig.6. Secondary creep rates for copper prestrained by tension versus θ_σ

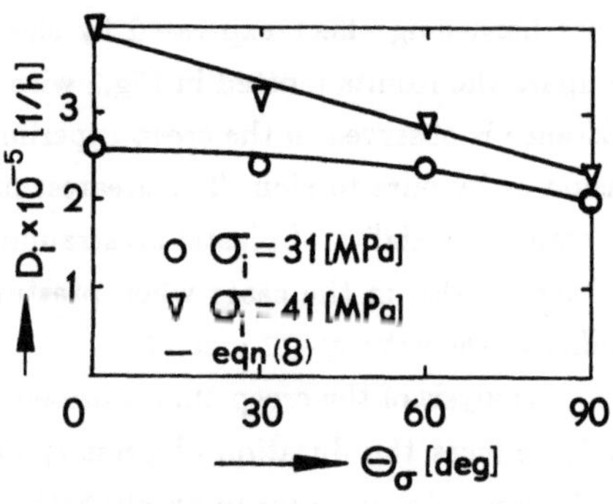

Fig.7. Secondary creep rates for copper prestrained by torsion versus θ_σ

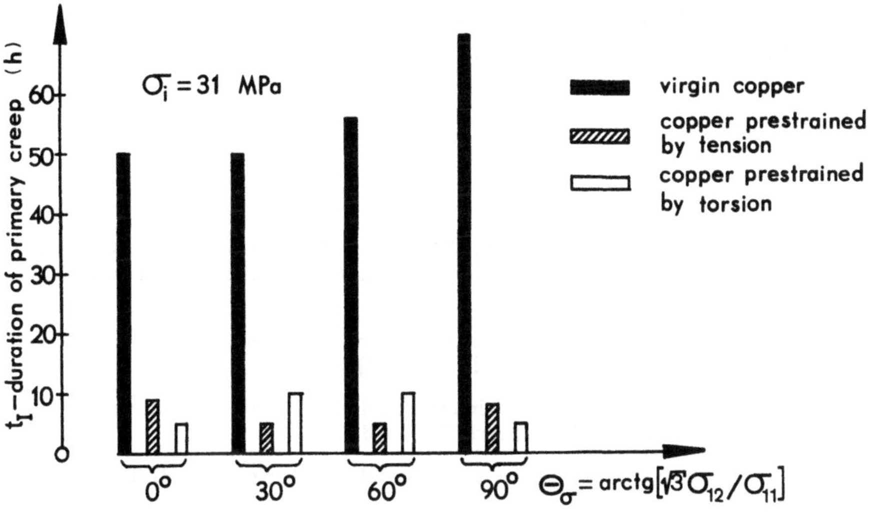

Fig.8. Duration of primary creep for virgin and prestrained copper

4. Description of secondary creep by modified Norton's law

The Norton's law in the form (4) describes with sufficient accuracy experimental data in the case of uniaxial creep tests whereas under combined stress states it gives large discrepancies. In previous papers [10, 11] the modification of this law was proposed and discussed in detail. The equation (4) of the steady stage creep in the modified form is

$$D_i = D_o(\theta_\sigma) \left[\frac{\sigma_i}{\sigma_o}\right]^{n(\theta_\sigma)} \tag{8}$$

where $D_o(\theta_\sigma)$ and $n(\theta_\sigma)$ are material parameters. Modification of the equation (4) is based on the assumption that material parameters may be expressed as certain functions depending on the direction (θ_σ) of the loading stress vector. In Figs.1,6,7 the experimental secondary creep rates are compared with those computed according to the equation (8).

5. Conclusions

The effect of predeformation on the creep behaviour was investigated. It was found that room temperature plastic prestraining caused creep strengthening of copper at 573K. This was connected with considerable decrease of the secondary creep rate. Cold work reduced significantly duration of the first and second creep stages.

The analysis of experimental data regarding creep in plane stress state proved that the quasi-linear creep law is too restrictive.

The steady creep rates have been described with good accuracy using the modified Norton's law.

References

1. Marlin, R.T.; Cosandey, F.; Tien, J.K.: The effect of predeformation on the creep and stress rupture of an oxide dispersion strengthened mechanical alloy. Met.Trans.A 11A (1980) 1771.
2. Trąmpczyński, W.A.: The influence of cold work on the creep of copper under biaxial states of stress. Acta Metall. 30 (1982) 1035.
3. Waniewski, M.: The influence of direction and value of plastic prestrain on creep in metals. (in Polish), IFTR Reports 34 (1983).
4. Dyson, B.F.; Rodgers, M.J.: Prestrain, cavitation and creep ductility. Metal Sci. 8 (1974) 261.
5. Dyson, B.F.; Loveday, M.S; Rodgers, M.J.: Grain boundary cavitation under various states of applied stress. Proc.R.Soc.Lond.A 349 (1976) 245.
6. Wilson, R.N.: The influence of 3% prestrain on the creep strength of Al-2.5%Cu-1.2%Mg alloys at 150°C. J.Inst.Met. 101 (1973) 188.
7. Tipler, H.J.; Varma, R.K.: The effect of prior room temperature deformation on creep, rupture and cavitation of 1/2Cr-1/2Mo-1/4V steels of commercial and high purity. 3-rd Int.Conf.Mech.Beh.Mat. (1979) 321
8. Pandey, M.C.; Mukherjee, A.K; Taplin, D.M.R.: Prior deformation effects on creep and fracture in inconel alloy X-750. Met.Trans.A 15A (1984) 1437.
9. Hayhurst, D.R.; Trąmpczyński, W.A.; Leckie, F.A.: Creep rupture under nonproportional loading. Acta Metall. 28 (1980) 1171.
10. Kowalewski, Z.: The influence of plastic anisotropy on the creep of metals under complex stress states. (in Polish), IFTR Reports 36 (1987).
11. Kowalewski, Z.: The surface of constant rate of energy dissipation under creep and its experimental determination. Arch.Mech. 39 (1987) 445.
12. Trąmpczyński, W.; Kowalewski, Z.: A tension-torsion testing technique. Proc. Symp.: Techniques for multiaxial creep testing. Elsevier Applied Science. London and New York (1986) 79.
13. Namestnikov, V.S.: On creep under constant combined loading. (in Russian), Izv. AN SSSR, OTN No 4 (1957) 141.
14. Oding, I.A.; Tulakow, G.A.: Creep of an austenitic steel under combined stress. (in Russian), Izv. AN SSSR, OTN No 1 (1958) 3.
15. Anisimowicz, M.: Studies on dynamic creep of alloys in plane stress state. (in Polish), IFTR Reports 38 (1978).
16. Sawczuk, A.; Anisimowicz, M.: Tensor functions approach to creep laws after prestrain. Proc. of 3-rd IUTAM Symp.: Creep in Structures. Springer-Verlag (1981), 220.

Modeling of Recovery of State Using the Viscoplasticity Theory Based on Overstress

E. KREMPL and P. S. MAJORS

Mechanics of Materials Laboratory
Rensselaer Polytechnic Institute
Troy, N.Y. 12180–3590, USA

Summary

Unified constitutive models, which do not postulate separate constitutive equations for creep and plasticity but consider all inelastic deformation rate dependent, have been applied to the modeling of transient and cyclic short–term deformation behavior. Their application to modeling long–term creep has been demonstrated in principle and hinges on the introduction of recovery terms in the growth laws for the state variables. After postulating a set of phenomenological indicators of recovery the viscoplasticity theory based on overstress (VBO) for small strain is introduced and the growth law for each of two state variables is augmented by a recovery term. Hypothetical, but realistic material properties are postulated and numerical experiments are performed to simulate the tests which were previously designated as indicators of significant recovery. The results show that each state variable has specific effects and that recovery terms in both state variables are needed to reproduce the significant phenomena.

Introduction

"Unified" constitutive equations which do not separately postulate constitutive equations for creep and plasticity but consider all inelastic deformation as rate dependent have been proposed during the last twenty years or so. They have been applied to short–term transient loadings in uniaxial and biaxial settings with considerable success. These theories typically contain one or more state variables which are endowed with a growth law, typically a first order nonlinear differential equation which is homogeneous of degree one in the rates. They contain "hardening" and "recall" terms to model the short–term inelastic deformation under uniaxial and multiaxial conditions.

At high homologous temperatures diffusion processes can be active which tend to counteract the effects of inelastic deformation. These processes are termed recovery of state or simply recovery. Their importance increases with increasing temperature; they take time to develop and even proceed when inelastic deformation has ceased. To model recovery of state so called recovery terms are introduced into the growth laws for the state variables following the Bailey–Orowan format. They are independent of rate so that their effects become important at long times.

To the knowledge of the authors no systematic study has been performed indicating how the presence of recovery might affect the real,

M. Zyczkowski (Ed.)
Creep in Structures
IUTAM Symposium Cracow/Poland 1990

macroscopic inelastic deformation behavior and how these so observed phenomena can be reproduced in the unified models. The purpose of this short paper is to propose a number of phenomenological indicators of the presence of significant recovery of state and to show how they can be reproduced within the context of the small strain viscoplasticity theory based on overstress (VBO). Space does not permit an extensive review and comparison with other theories.

Phenomenological Indicators of Recovery

As a guide for continuum modeling, indicators of the presence of significant recovery effects are necessary. The authors are not aware that such indicators have been explicitly presented. Some of the evidences to be presented below have been implicitly used to ascertain the presence of recovery. Others have their origin in the predictions of VBO without a recovery of state term and their comparisons with experimental results obtained at temperatures where recovery of state is anticipated to occur. All evidence is based on the uniaxial state of stress and the examples cited below are not considered to be exhaustive. Other indicators could be found but the ones cited have a discriminating quality which helps in the formulation of the model.

Creep in the Quasi Linear Elastic Region

Experimental evidence for ferritic steels and other alloys indicates that secondary and tertiary creep can be observed at stress levels which are in the quasi elastic region of the stress–strain diagram. Also secondary creep is explained in the materials science literature as the balance between hardening due to inelastic deformation and recovery of state.

Influence of Rest Periods at Zero Stress after Inelastic Deformation

This is the most frequently cited example of the influence of recovery of state. After loading and unloading, a rest period is introduced at point A in Fig. 1. If the subsequent stress–strain diagram is observed to depend on the time at rest at point A, recovery of state is said to have taken place. The details of the evidence depend on the control conditions (stress or strain control). As an example some negative strain can be observed while the specimen is at zero load in load control; a small increase in stress may be found if the specimen is in strain control. Similar differences may exist for the subsequent stress–strain diagram.

Cyclic Softening

An initially annealed material is observed to soften cyclically under completely reversed strain–controlled conditions such that the cycle dependent reduction in stress is equal on the tension and compression sides as shown schematically in Fig. 2. The cause of this behavior is deemed to be attributable to recovery of state if this happens at high homologous temperature. However, this evidence can be ambiguous as shearing of precipitates by dislocations alone may also cause cyclic softening, Pineau [1].

Strain Rate Path Dependence

In this case the comparison needs two specimens. The first specimen is loaded at the fast rate 1 ($\approx 10^{-4}$ 1/s) as shown in Fig. 3. This test has a short duration. A second specimen is slowly loaded at rate 2 ($\lesssim 10^{-7}$ 1/s,

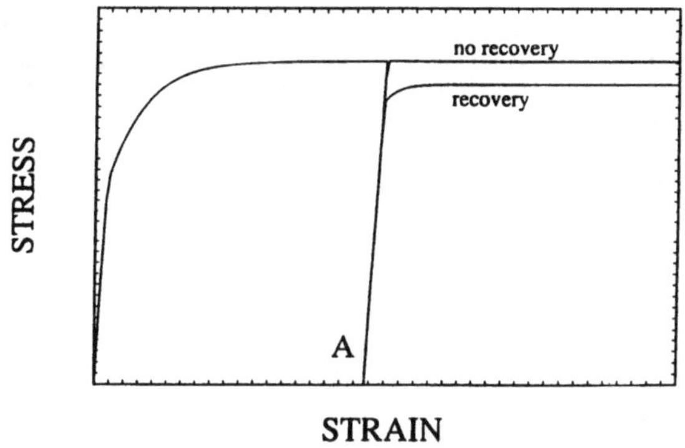

Fig.1. Schematic showing the possible influence of a rest period at A on the subsequent deformation behavior

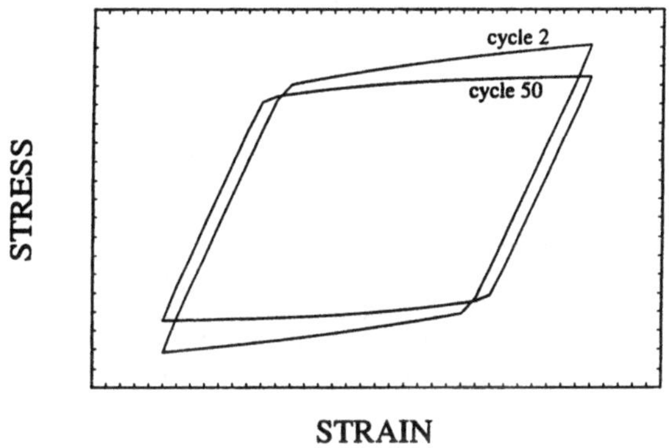

Fig.2. Schematic showing cyclic softening attributed to recovery of state under completely reversed, strain–controlled conditions

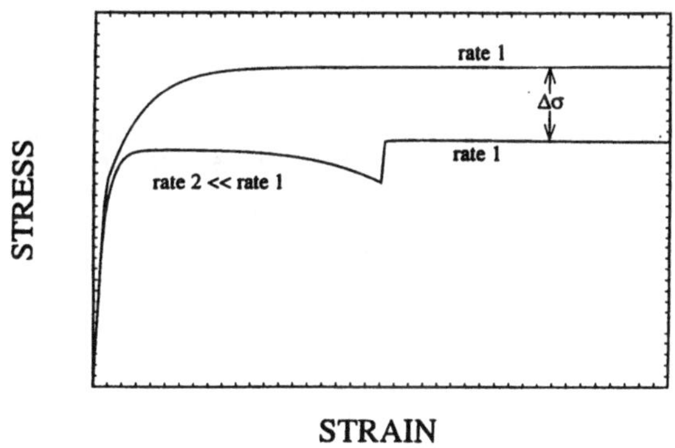

Fig.3. Schematic depicting the influence of recovery in a test with two specimens. The difference $\Delta\sigma$ is an indicator of recovery.

relaxation periods can also be used) and accumulates inelastic strains. Before the strain rate is changed to rate 1, the test time should should have been more than ten times the duration of the first test. If the two stress–strain curves with identical rate 1 merge, the effects of recovery of state are considered to be negligible. If a gap $\Delta\sigma$ remains as schematically indicated in Fig. 3 then this gap is attributed to recovery of state.

With the exception of cyclic softening none of the above effects can be reproduced by VBO in the absence of recovery of state terms. Consequently all four effects are now considered together and recovery of state terms are introduced in an attempt to model these observation with VBO.

Small Strain Viscoplasticity Theory Based on Overstress

The small strain VBO is fully described by Krempl, McMahon and Yao [2], Krempl and Yao [3] and Lee and Krempl [4] where the orthotropic case is treated. A finite strain version for isotropic behavior is given by Nishiguchi, Sham and Krempl [5]. The case of cyclic hardening is treated by Choi and Krempl [6] where other papers are also cited.

The cyclic neutral version of the theory starts with a flow law

$$\dot{\epsilon} = \dot{\epsilon}^{el} + \dot{\epsilon}^{in} = \dot{\sigma}/E + \frac{1}{E}\frac{\sigma - g}{k[\sigma - g]}. \tag{1}$$

In the above ϵ and σ denote the small strain and the stress, respectively. A superposed dot indicates differentiation with respect to time, E is the modulus of elasticity and the positive decreasing viscosity function k controls the rate dependence. It is seen that the inelastic strain rate is solely a function of the difference $\sigma - g$ which is called the overstress. The state variable g is the equilibrium stress; it is the stress which can be sustained at rest. Its growth law is given by

$$\dot{g} = \psi[\sigma - g]\dot{\epsilon} - \{(g - f)(\psi[\sigma - g] - E_t)\}|\dot{\epsilon}^{in}|/A - R[g]g \tag{2}$$

where ψ is shape function which influences the shape of the stress–strain diagram, f is the kinematic variable with the growth law

$$\dot{f} = E_t\dot{\epsilon}^{in} \tag{3}$$

which sets the tangent modulus E_t at the maximum inelastic strain of interest. The positive, increasing, even function R[g] represents the recovery of state term. The quantity A represents the rate–independent or isotropic contribution to the asymptotic stress, see Fig. 1 of [3]. If it is constant, cyclic neutral behavior is represented; for cyclic hardening a growth law is necessary, see [6]. Presently it is assumed that only softening due to recovery of state occurs which depends on time and the accumulated inelastic strain p. Therefore

$$\dot{A} = -\lambda(A - B) \tag{4}$$

where B is a threshold value and

$$\lambda = hp - \eta \tag{5}$$

with

and

$$\dot{\eta} = h|\dot{\epsilon}^{in}| - R_1[\eta]\eta \tag{6}$$

$$p = \int_0^t |\dot{\epsilon}^{in}| d\tau. \tag{7}$$

It should be noted that $\lambda \neq 0$ only in the presence of the recovery term which is controlled by the positive, increasing function R_1. The quantity h is a constant with the dimension of reciprocal time. It should also be noticed that the growth law (4) deviates from the usual format as it assumes that recovery depends on accumulated inelastic strain modified by recovery effects.

Numerical Experiments

The Mechanics of Materials Laboratory at Rensselaer Polytechnic Institute has recently completed a number of tests at 538°C on a ferritic steel, see Majors et al [7]. In the course of the modeling a set of constants were developed and the ones used here and listed in Table 1 are close to these. However, the purpose of the present tests is not to reproduce the behavior of a specific material but rather to show the influence of the recovery terms. The subsequent graphs represent the behavior of VBO when all recovery terms, the functions R[g] and $R_1[\eta]$, are nonzero.

Fig. 4 demonstrates that VBO is capable of modeling primary, secondary and tertiary creep in the quasi elastic region. Also it reproduces cyclic softening as seen in Fig. 5. The strain rate path dependence is also modeled as seen in Fig. 6. Although not explicitly shown, VBO with all recovery terms can also reproduce the influence of rest at zero stress.

It is instructive to list the capabilities of the regular VBO (eqs. (2) – (3) without recovery terms), of the recovery term in the equilibrium stress only and in the isotropic variable A only, using the constants listed in Table 1. It is believed that the results are general, i.e. they would be found with other constants. This belief must, however, be confirmed by further research. The capabilities are:

Standard VBO
- Primary creep only in the quasi elastic range
- Cyclic neutral only
- No strain rate path dependence
- No permanent effect of rest time on subsequent stress–strain diagram.

Equilibrium stress only
- Primary and secondary creep in the quasi elastic range
- Cyclic neutral
- No strain rate path dependence
- No permanent effect of rest time on subsequent stress–strain diagram.

Isotropic stress only
- Primary creep only in quasi elastic range
- Cyclic softening
- Strain rate path dependence
- Permanent softening of subsequent stress–strain diagram.

TABLE 1

MATERIAL CONSTANTS and FUNCTIONS

K_1	400000.0	1/s
K_2	63.0	MPa
K_3	14.0	1/MPa
C_1	15000.0	MPa
C_2	163100.0	MPa
C_3	0.022	1/MPa
E	165000.0	MPa
E_t	–1100.0	MPa
A	370.0	MPa
B	20.0	MPa
RA_1	0.0	MPa
RA_2	50.0	MPa
RA_3	0.01	1/s
h	0.04	1/s
RG_1	0.0	MPa
RG_2	500.0	MPa
RG_3	0.00001	1/s

VISCOSITY FUNCTION AND SHAPE FUNCTION

$$k[\sigma-g] = K_1\left[1+\frac{|\sigma-g|}{K_2}\right]^{-K_3} \qquad \psi[\sigma-g] = C_1 + (C_2-C_1)\exp[-C_3(|\sigma-g|)]$$

RECOVERY FUNCTIONS

$$R[g] = \frac{RG_3}{2}\left[\tanh(UG) + \tanh(VG)\right]$$

where

$$UG = -3.0 + 6.0\left[\frac{g-RG_1}{RG_2-RG_1}\right] \qquad VG = 3.0 + 6.0\left[\frac{g+RG_1}{RG_2-RG_1}\right]$$

$R_1[\eta]$ is same form as $R[g]$, with constants RA_1, RA_2, RA_3

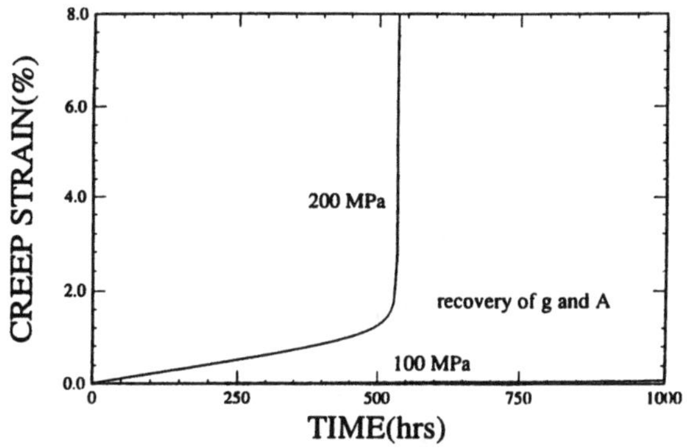

Fig.4. Numerical experiments simulating long–term creep tests. The results show that VBO can reproduce primary, secondary and tertiary creep at stress levels which are in the quasi elastic regions, see Fig. 6

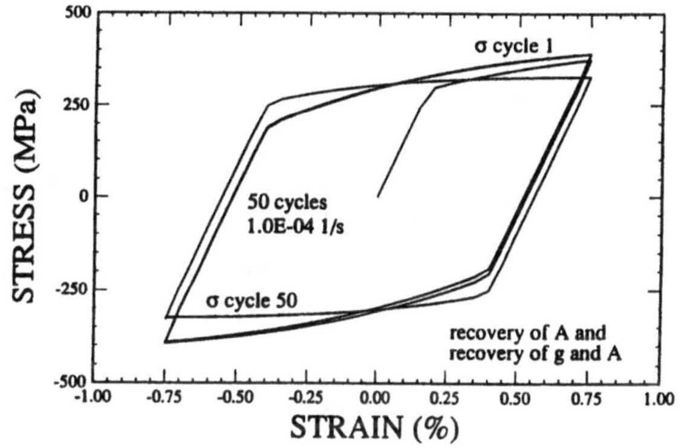

Fig.5. Numerical simulation of cyclic softening under completely reversed strain–controlled conditions.

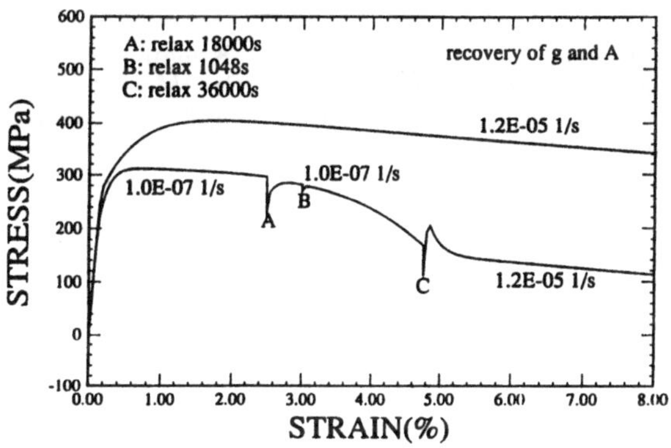

Fig.6. Numerical simulation of the schematic shown in Fig. 3. Note that relaxation periods are simulated at points A, B and C.

Conclusion

The results of Figs. 4 – 6 indicate that recovery terms for both the equilibrium stress g and the the isotropic variable A are needed to reproduce the indicators of recovery shown in Figs. 1 – 3.

The above results also aid in identifying the material constants. The recovery terms in g and A only affect secondary creep in the quasi elastic range and cyclic softening, respectively. These effects can therefore be attributed to the respective state variables.

Acknowledgement

This research was supported by the Department of Energy under Grant DE–FG02–86ER13566.

References

1. Pineau, A.: Personal communication, September 1990.

2. Krempl, E., McMahon, J.J., Yao, D.: Viscoplasticity based on overstress with a differential growth law for the equilibrium stress. Mechanics of Materials 5 (1986) 35–48.

3. Yao, D., Krempl, E.: Viscoplasticity theory based on overstress. The prediction of monotonic and cyclic proportional and nonproportional loading paths of an Aluminum alloy. Int. J. of Plasticity 1 (1985) 259–274.

4. Lee, K.D., Krempl, E.: An orthotropic theory of viscoplasticity based on overstress for thermomechanical deformations, in press Int. J. of Solids and Structures.

5. Nishiguchi, I., Sham, T.L., Krempl, E.: A finite deformation theory based on overstress. Parts I and II, in press Journal of Applied Mechanics.

6. Krempl, E., Choi, S.H.: Viscoplasticity theory based on overstress: The modeling of ratchetting and cyclic hardening of type 304 stainless steel, Rensselaer Polytechnic Institute Report MML 90–4, April 1990, submitted to Nuclear Engineering and Design.

7. Majors, P.S., Cheng, S., Ruggles, M.B., Krempl, E.: The rate(time)–dependent mechanical behavior of modified 9Cr–1Mo alloy at 538°C and its modeling using the viscoplasticity theory based on overstress, Rensselaer Polytechnic Institute Report MML 90–5, August 1990.

Some Attempts to Describe Distortional Hardening in Viscoplasticity

T. KURTYKA and M. ŻYCZKOWSKI

Institute of Mechanics and Machine Design
Cracow University of Technology, Poland

Summary

Most papers describing viscoplastic behaviour of materials allow just for isotropic or kinematic plastic hardening. However, many experiments show remarkable distortion of subsequent yield surfaces. The present paper discusses several variants of viscoplastic constitutive equations based on geometric description of distortional plastic hardening, proposed earlier by the authors for time-independent plasticity.

1 Introductory remarks

The simplest description of viscoplastic behaviour in uniaxial tension, ascribed to Manjoine, Cowper and Symonds, is given by

$$\dot{\varepsilon}^{vp} = K < \sigma - \sigma^* >^q \tag{1.1}$$

where $< >$ denote Macauley brackets, namely $< a >= a$ if $a > 0$ and $< a >= 0$ if $a \leq 0$, σ^* is the static yield-point stress, either initial, or increased due to plastic hardening, $\dot{\varepsilon}^{vp}$ is the viscoplastic strain rate, K and q are assumed as material constants depending on temperature. The difference $\sigma - \sigma^*$ is usually called "overstress" and will be denoted by σ_{over}.

Direct generalization of (1.1) to multiaxial stress states may by written as follows

$$\dot{\varepsilon}_{ij}^{vp} = K < \sigma_{over} >^q n_{ij} \tag{1.2}$$

where the overstress σ_{over} is now an invariant of stress and possibly also of internal state variables (via σ^*), turning into $\sigma - \sigma^*$ for uniaxial tension, and n_{ij} is a "directional tensor", usually derived from a viscoplastic potential G.

Effective formulation of (1.2) needs specification of quasi-static subsequent yield surfaces, evolution equations for internal state variables, definition of the overstress σ_{over} and specification of the viscoplastic potential G. Most papers make use of the effective stress based on the Huber-Mises-Hencky (HMH) hypothesis, and consider either isotropic hardening (P. Perzyna [8]) or mixed isotropic-kinematic hardening (J.L. Chaboche [1]). Some papers allow for more general hardening types, including distortion of subsequent

M. Zyczkowski (Ed.)
Creep in Structures
IUTAM Symposium Cracow/Poland 1990

yield surfaces (A. Phillips and H.C. Wu [9], M.A. Eisenberg and C.F. Yen [2], T. Inoue and S. Imatani [4]). Such distortion is observed in most experiments.

In the present paper we discuss several variants of Eq. (1.2) making use of the geometric description of distortional plastic hardening proposed by the authors in [5]. Particular attention will be paid to the respective definitions of overstress σ_{over}, whereas evolution equations will not be considered and viscoplastic potentials will just be mentioned.

2 Geometric description of distortional plastic hardening proposed by Kurtyka and Życzkowski

A geometric description of distortional plastic hardening was proposed by the authors in [5]. It is based on HMH initial yield condition, but a generalization to other conditions was given in [11], and to anisotropic bodies - in [12]; invariant formulation is given in [7], and identification of parameters - in [6]. This description makes use of Ilyushin's five-dimensional auxiliary stress space, since in that space the initial yield condition is represented by a hypersphere, and then any distortion of that hypersphere may be described in a uniform way. A.A. Ilyushin [3] introduced namely a space in which stress vector $\boldsymbol{\sigma} = \sigma_m n_m$ ($m = 1, 2, \ldots, 5$, and Einstein's summation convention holds) is defined as follows

$$\sigma_1 = \frac{3}{2} s_{xx} , \quad \sigma_2 = \frac{\sqrt{3}}{2} s_{xx} + \sqrt{3} s_{yy} , \quad \sigma_3 = s_{xy}\sqrt{3} , \quad \sigma_4 = s_{yz}\sqrt{3} , \quad \sigma_5 = s_{zx}\sqrt{3} , \tag{2.1}$$

where s_{ij} ($i, j = x, y, z$) are deviatoric stress components. The HMH initial yield condition takes then the form

$$|\boldsymbol{\sigma}| = \sqrt{\sigma_k \sigma_k} = \sigma_0 , \tag{2.2}$$

(where σ_0 denotes a uniaxial tensile yield stress), and hence is represented by a hypersphere in the five-dimensional space.

It is well known that an ellipse may be obtained from two concentric circles by a projecting procedure, if the pole of projecting radii coincides with the centres of both circles. A distorted curve may be obtained by a similar projecting procedure if we distinguish the centres of the circles and the pole. Indeed, such a curve resembles subsequent yield curves obtained from experiments for a general curvilinear trajectory. In the general five-dimensional case we introduce five hyperspheres with various radii $R_{(i)}$ and various centres $O_{(i)}$, a pole A, and a system of mutually perpendicular projecting directions.

The simplest analytical description is obtained in a moving system of coordinates $\hat{\sigma}_i$, translated and rotated with respect to the original system σ_i. The directions of $\hat{\sigma}_i$ coincide with the projecting directions, and the centre - with the pole A. In this system we define "active stresses" $\hat{\sigma}_i$ by the formula

$$\hat{\sigma}_i = Q_{ij}(\sigma_j - a_j) , \tag{2.3}$$

where Q_{ij} is the appropriate rotation matrix, and the vector a_j describes translation of the centre of moving coordinate system and may be interpreted as a vector of residual microstresses (back-stresses). The position of the centres of hyperspheres $O_{(i)}$ is defined in the moving coordinate system by five vectors $\mathbf{d}_{(i)}$. These vectors are responsible for nonelliptic (non-affine) distortion of the yield surface, hence the notation $\mathbf{d}$.

Paper [5] gives two proposals of description of distortional plastic hardening. The simpler version assumes $d_{(i)j} = 0$ for $j \neq i$ (centres of hyperspheres located on the corresponding axes), and only this version will be discussed here (Fig. 1). It contains 25

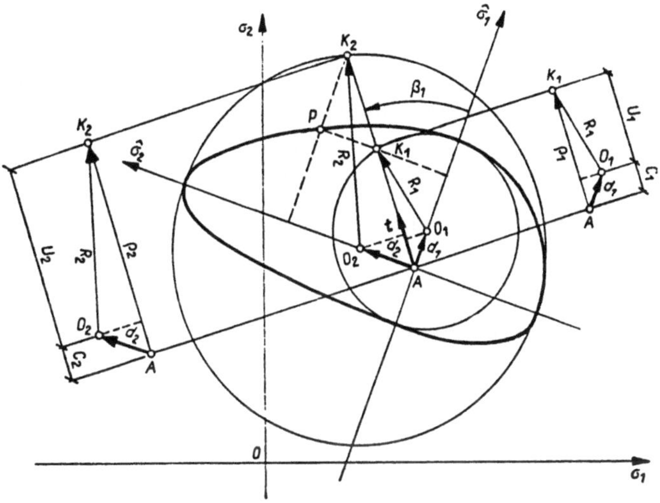

parameters (counted as scalars) versus 20 parameters describing affine hardening as proposed by Edelman-Drucker and Baltov-Sawczuk. So the increase of number of parameters is rather small. Denoting the non-zero distortional parameters $d_{(i)i}$ briefly by d_i we describe subsequent yield surfaces in Ilyushin's space by parametric equations

$$\sigma_j = Q_{ij}^{-1} \left\{ d_{\underline{i}} t_{\underline{i}} + \left[(d_{\underline{i}} t_{\underline{i}})^2 - d_{\underline{i}}^2 + R_{\underline{i}}^2 \right]^{1/2} \right\} t_i + a_j \,, \tag{2.4}$$

with summation over i and no summation over $\underline{i}$. The quantities t_i denote current parameters, namely Cartesian coordinates of the unit vector of the radius-vector in spherical coordinate system.

Each equation (2.4) contains all five parameters t_i, but if we introduce "active stresses" (2.3), then the equations become uncoupled. We may eliminate t_i, namely

$$t_i^2 = \frac{\hat{\sigma}_i^2}{R_i^2 + 2 d_{\underline{i}} \hat{\sigma}_{\underline{i}} - d_i^2} \tag{2.5}$$

and making use of the relation $t_i t_i = 1$ obtain the following implicit form

$$F = F_1 = \sum_{i=1}^{5} \frac{\hat{\sigma}_i^2}{R_i^2 + 2d_i\hat{\sigma}_i - d_i^2} = 1 , \tag{2.6}$$

(with underlining of i dropped).

Implicit equations of the type $F = 1$ are not unique. For example, we can multiply (2.6) by all denominators and then rearrange the terms so as to obtain once more equation of the type $F = 1$. Its notation in general five-dimensional case is complicated, hence we quote here just the result for a two-dimensional case

$$\begin{aligned} F &= F_2 = \frac{1}{(R_1^2 - d_1^2)(R_2^2 - d_2^2)} \left[2d_1\hat{\sigma}_1\hat{\sigma}_2^2 + 2d_2\hat{\sigma}_2\hat{\sigma}_1^2 + (R_1^2 - d_1^2)\hat{\sigma}_2^2 + \right. \\ &+ \left. (R_2^2 - d_2^2)\hat{\sigma}_1^2 - 4d_1d_2\hat{\sigma}_1\hat{\sigma}_2 - 2d_1(R_2^2 - d_2^2)\hat{\sigma}_1 - 2d_2(R_1^2 - d_1^2)\hat{\sigma}_2\right] = 1 . \end{aligned} \tag{2. 7}$$

Both equations (2.6) and (2.7) are equivalent when describing subsequent yield surfaces (in two-dimensional case), but are not equivalent beyond these surfaces, if we discuss viscoplastic behaviour, and, for example, identify F with viscoplastic potential G. Indeed, (2.6) shows singularities along the lines (hyperplanes) $\hat{\sigma}_i = -(R_i^2 - d_i^2)/2d_i$, whereas (2.7) does not involve any singularities.

3 Two classical definitions of overstress

Overstress in multiaxial case (called sometimes effective overstress) is usually defined as the distance between the point representing current stress deviator s_{ij} and a certain quasi-static stress deviator s_{ij}^* corresponding to s_{ij}. Two classical definitions find s_{ij}^* either at the same vector-radius, or as the point on the yield surface closest to s_{ij} (along the normal), [2]. We shall calculate the distance in Ilyushin's space; in general

$$\sigma_{over} = \sqrt{\sum_{i=1}^{5} (\hat{\sigma}_i - \hat{\sigma}_i^*)^2} . \tag{3.1}$$

In order to define overstress along the radius, let us construct an expanded quasi-static yield surface. Proportional expansion of (2.6) (geometric similarity with retained centre) takes place if we substitute, instead of R_i and d_i ,

$$\bar{R}_i = \psi R_i , \quad \bar{d}_i = \psi d_i , \tag{3.2}$$

with the multiplier $\psi > 1$ for expansion. So, any viscoplastic stress state $\hat{\sigma}_i$ may be considered as lying on an expanded quasi-static surface

$$F_3 = \sum_{i=1}^{5} \frac{\hat{\sigma}_i^2}{\psi^2(R_i^2 - d_i^2) + 2\psi d_i \hat{\sigma}_i} = 1 . \tag{3.3}$$

Overstress σ_{over} is defined here as the length of a part of the radius between the actual stress point $\hat{\sigma}_i$ and the point of intersection of that radius with the quasi-static surface $\hat{\sigma}_i/\psi$, namely

$$\sigma_{over} = \frac{\psi - 1}{\psi}\sqrt{\hat{\sigma}_i\hat{\sigma}_i} = \frac{\psi - 1}{\psi}\sqrt{\frac{3}{2}(s_{ij} - a_{ij})(s_{ij} - a_{ij})} \ . \tag{3.4}$$

This formula looks simple, holds for any quasi-static preloading, expanded yield surfaces are always convex and no singularities appear. On the other hand, evaluation of the parameter ψ from (3.3) requires the solution of an algebraic equation of the sixth degree (this degree equals $n+1$ for a n-dimensional case).

Overstress defined along the normal may be found by minimization of the distance. Denote $\hat{\sigma}_i^*$ any point lying on the surface (2.6) and minimize (3.1) with (2.6) as auxiliary condition. Lagrangian function $\mathcal{L}$ equals

$$\mathcal{L} = \sum_{i=1}^{5}(\hat{\sigma}_i - \hat{\sigma}_i^*)^2 + \lambda F_1(\hat{\sigma}_i^*) \ , \tag{3.5}$$

where λ stands for Lagrangian multiplier. Equating partial derivatives to zero we obtain the following system of five equations

$$(R_i^2 + 2d_i\hat{\sigma}_i^* - d_i^2)^2(\hat{\sigma}_i - \hat{\sigma}_i^*) - \lambda(R_i^2 + d_i\hat{\sigma}_i^* - d_i^2)\hat{\sigma}_i^* = 0 \ . \tag{3.6}$$

Together with (2.6) for $\hat{\sigma}_i^*$ they determine five unknown $\hat{\sigma}_i^*$ and λ. Lagrangian multiplier λ may easily be eliminated, but remaining equations are nonlinear and evaluation of $\hat{\sigma}_i^*$ and then σ_{over} from (3.1) is more difficult than in the preceding variant.

4 Overstress defined by proportional directional parameters

Both classical definitions discussed above did not results in an effective, explicit formula for overstress needed in (1.2). So, we propose also some further, more effective variants.

For any state of stress $\hat{\sigma}_i$ beyond the quasi-static subsequent yield surface, regarded as known, one can calculate the corresponding value F_1, (2.6). Denote this value by κ^2, where $\kappa^2 > 1$ (a certain limitation will be discussed below). The corresponding point $\hat{\sigma}_i^*$ on the yield surface, needed in (3.1), will now be found as follows. Calculate t_i from (2.5) and suppose that t_i^* for the point $\hat{\sigma}_i^*$ are proportional to t_i:

$$t_i^* = \frac{t_i}{\kappa} \ . \tag{4.1}$$

Solving now (2.5) with respect to $\hat{\sigma}_i^*$ one obtains

$$\hat{\sigma}_i^* = \frac{t_i}{\kappa^2}\left[d_it_i + \sqrt{(R_i^2 - d_i^2)\kappa^2 + d_i^2t_i^2}\right] \tag{4.2}$$

or, with substituted (2.5) for t_i,

$$\hat{\sigma}_i^* = \frac{\hat{\sigma}_i}{\kappa^2(R_i^2 + 2d_i\hat{\sigma}_i - d_i^2)}\left[d_i\hat{\sigma}_i + \sqrt{\kappa^2(R_i^2 - d_i^2)^2 + 2\kappa^2(R_i^2 - d_i^2)d_i\hat{\sigma}_i + d_i^2\hat{\sigma}_i^2}\right] \ . \tag{4.3}$$

Formula (4.3) may now be substituted into (3.1) and one obtains effective formula for the overstress. However, a deficiency of (4.3) lies in the singularity mentioned in Sec. 2: for $\hat{\sigma}_i = -(R_i^2 - d_i^2)/2d_i$ the overstress increases infinitely and this result cannot be agreement with experiments. So, (4.3) cannot be used within the range of large reverse stresses.

5 Two variants of overstress expressed via reduced stress

Suppose the subsequent yield surfaces to be described by the equation

$$F(\sigma_{ij}\,;\alpha_{(k)ij}\,,\kappa_m) = 0\ , \tag{5.1}$$

where $\alpha_{(k)}$, $k = 1,2,\ldots,k_n$, and κ_m , $m = 1,2,\ldots,m_n$, are tensorial and scalar internal state variables, respectively. Reduced stress σ_{red} , corresponding to a particular form of F, will be calculated by equating F for multiaxial stress to F for uniaxial tension by σ_{red} , and namely [10],

$$F(\sigma_{ij}\,;\alpha_{(k)ij}\,,\kappa_m) = F(\sigma_{red}\,,0\,,\ldots\,;\alpha_{(k)ij}\,,\kappa_m)\ , \tag{5.2}$$

where, on the right-hand side, the internal state variables α_k and κ_m should be calculated as for uniaxial tension. Eq. (5.2) should be solved with respect to σ_{red} and then the overstress σ_{over} is determined by

$$\sigma_{over} = \sigma_{red} - \sigma^*\ . \tag{5.3}$$

Making use of F_1 , (2.6), we obtain

$$F_1 = \sum_{i=1}^{5} \frac{\hat{\sigma}_i^2}{R_i^2 + 2d_i\hat{\sigma}_i - d_i^2} = \frac{(\sigma_{red} - a_1)^2}{R_1^2 + 2d_1(\sigma_{red} - a_1) - d_1^2} \tag{5.4}$$

and hence solving (5.4) as a quadratic equation,

$$\sigma_{red} = \sqrt{(R_1^2 - d_1^2)F_1 + d_1^2F_1^2} + a_1 + d_1F_1\ . \tag{5.5}$$

The corresponding yield-point stress in uniaxial tension equals $\sigma^* = R_1 + a_1 + d_1$, and finally

$$\sigma_{over} = \sqrt{(R_1^2 - d_1^2)F_1 + d_1^2F_1^2} + d_1(F_1 - 1) - R_1\ . \tag{5.6}$$

Making use of F_2 , (2.7), we arrive at

$$\sigma_{red} = \sqrt{(R_1^2 - d_1^2)F_2 + d_1^2} + a_1 + d_1\ , \tag{5.7}$$

$$\sigma_{over} = \sqrt{(R_1^2 - d_1^2)F_2 + d_1^2} - R_1\ . \tag{5.8}$$

The equations derived here show some advantages and some drawbacks. They are simplest from among all variants, in particular (5.6), where F_1 is defined by a compact formula (2.6). On the other hand, the most important drawback of (5.6) is connected with the singularities mentioned in Sec. 2; Eq. (5.8) shows no singularities. In both cases, however, the quasi-static preloading is restricted to simple loading by the stress $\hat{\sigma}_1$. Simple (proportional) loading is essential here, whereas configuration of that loading is arbitrary, since any vector σ_j may be rotated to $\hat{\sigma}_1$ via an appropriate matrix Q_{ij}, (2.3).

6 Final remarks and conclusions

1. Five variants of viscoplastic constitutive equations with distortional hardening were derived. Two classical approaches lead to nonlinear algebraic equations, but three remaining variants are more effective. Choice of an appropriate variant should be based on experimental verification and simplicity of application. In the particular case of vanishing distortion and equal radii of generating circles, the equations derived turn into the Chaboche equations.

2. Directional tensors n_{ij} were not specified, but the functions F_i proposed may serve as viscoplastic potentials. The differences in direction of the strain rate vector due to distortion may even be more important than in overstress, since the derivatives are involved here.

3. The evolution equations for the internal state variables used, a_i, d_i, R_i and Q_{ij}, are subjected to experimental verification or even evaluation.

4. Generalization to the case of initial anisotropy of material may be obtained via appropriate transformations of Ilyushin spaces described in [11] and [12].

References

[1] Chaboche, J.L.: *Viscoplastic constitutive equations for the description of cyclic and anisotropic behaviour of metals*, Bull. Acad. Polon. des Sci., Ser. Sci. Techn. **25** (1977), 1, 33-42.

[2] Eisenberg, M.A.; Yen, C.-F.: *A theory of multiaxial anisotropic viscoplasticity*, J. Appl. Mechanics **48**, (1981), 276-284.

[3] Ilyushin, A.A.: *On the relation between stresses and small strains in continuum mechanics*, Prikl. Mat. Mekh. **18** (1954), 6, 641-666 (in Russian).

[4] Inoue, T.; Imatani. S.: *A unified inelastic constitutive equation in terms of anisotropic yield function*, Trans. 10th Int. Conf. Struct. Mech. Reactor Techn., vol. L, 1-6, (1989).

[5] Kurtyka, T.; Życzkowski, M.: *A geometric description of distortional plastic hardening of deviatoric materials*, Arch. Mechaniki Stos. **37** (1985), 4-5, 383-395.

[6] Kurtyka, T.: *Parameter identification of a distortional model of subsequent yield surfaces*, Arch. Mechaniki Stos. **40** (1989), 4, 433-454.

[7] Kurtyka, T.: *Invariant formulation of a distortional model of plastic hardening*, Mech. Teor. Stos. **28** (1990), 1-2, 115-131.

[8] Perzyna, P.: *The constitutive equations for rate sensitive plastic materials*, Quart. Appl. Mathematics **20** (1963), 4, 321-332.

[9] Phillips, A.; Wu, H.C.: *A theory of viscoplasticity*, Int. J. Solids and Structures **9** (1973), 1, 15-30.

[10] Życzkowski, M.: *Combined loadings in the theory of plasticity*, Warszawa: PWN; The Hague: Nijhoff 1981

[11] Życzkowski, M.; Kurtyka, T.: *Generalized Ilyushin's spaces for a more adequate description of plastic hardening*, Acta Mechanica **52** (1984), 1-13.

[12] Życzkowski, M.; Kurtyka, T.: *A description of distortional plastic hardening of anisotropic materials*, Proc. IUTAM Symp. Yielding Anisotropic Solids, Mech. Eng. Publ., London 1990, 97-111.

A Viscoplasticity Theory of Irradiated Materials

M. V. MIĆUNOVIĆ

Svetozar Marković University, Kragujevac, Yugoslavia

Summary

In order to account for irradiation damage of reactor steels a unified theory of viscoplasticity of damaged materials is indispensable. Such a theory is presented in this paper aimed for general interaction between irradiation and viscoplasticity. Evolution equations for plastic strain rate, heat flux and neutron flux include material functions depending on pastic strain and temperature. Consideration of finite plastic and small thermoelastic strains permits significant simplifications of the theory giving rise to tensor representations of constitutive equations. The most special case of small plastic strains is also presented and discussed.

Deformation Geometry of Continuously Damaged Steels

Consider a crystalline body, $\mathcal{B}$, with continuous distribution of line and point defects as well as of voids and impurities - Fig.1. Such a body is observable in a family of deformed configurations (ψ_t) and also in an initial reference configuration (κ_0). In configurations (ψ_t) the body is subjected to surface tractions as well as to an inhomogeneous temperature field $\theta(X^K, t)$ where X^K $(K \in \{1,2,3\})$ are material coordinates and t is time. Removing the surface tractions and bringing the temperature to its constant reference value θ_0 which $\mathcal{B}$ has in (κ_0) we attain a partially unloaded configuration (κ_t). In such a configuration dislocations and voids are not removed with the consequence that an equilibrated residual stress (often named " *back-stress* " and denoted by $\mathbf{B}$) still exists. It is worth noting that, in general, $\mathbf{B}_0$ exists even in (κ_0).

Denoting by $\mathbf{F}$, $\mathbf{F}_E$ and $\mathbf{F}_P$, i.e.

$$\mathbf{F} : (\kappa_0) \rightarrow (\psi_t),\ \mathbf{F}_E : (\kappa_t) \rightarrow (\psi_t),\ \ and\ \ \mathbf{F}_P : (\kappa_0) \rightarrow (\kappa_t) \tag{1}$$

total, quasi-thermoelastic and *quasi-plastic deformation gradient tensor*, respectively, we arrive at the corresponding objective strain measures:

$$2\mathbf{E} = \mathbf{F}^T\mathbf{F} - \mathbf{1}\,, \tag{2}$$

M. Zyczkowski (Ed.)
Creep in Structures
IUTAM Symposium Cracow/Poland 1990

$$2\epsilon_P = \mathbf{F}_P^T \mathbf{F}_P - \mathbf{1}\,, \tag{3}$$

$$\epsilon_E := \mathbf{E} - \epsilon_P = \mathbf{F}_P \,(\, \mathbf{F}_E^T \mathbf{F}_E - \mathbf{1}\,)\, \mathbf{F}_P \equiv \mathbf{F}_P^T \tilde{\epsilon}_E \mathbf{F}_P \tag{4}$$

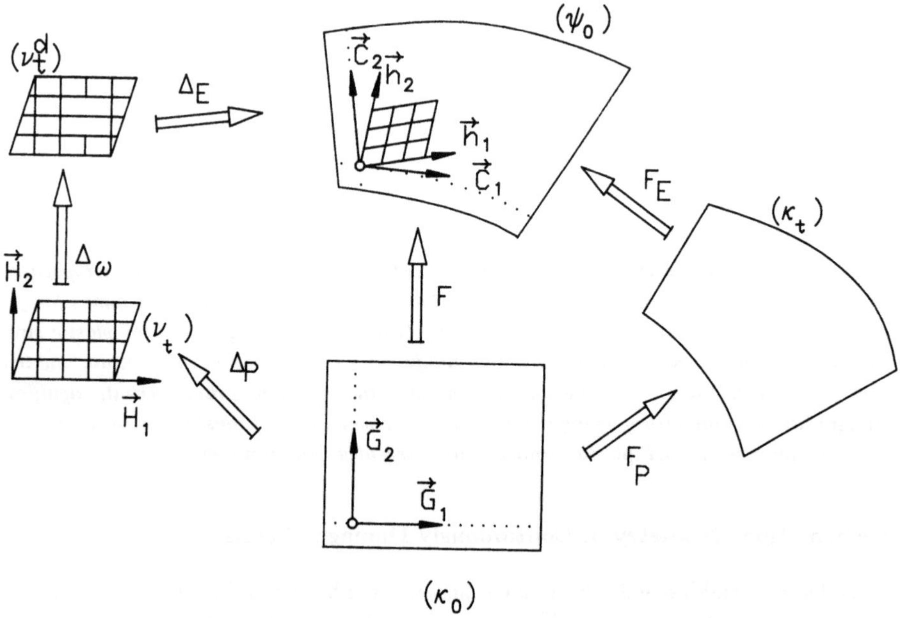

Fig. 1. Basic configurations and corresponding deformations

Another approach to deformation geometry would be take into account microstructural changes rendering lattice rearrangements. To do this we imagine that (ψ_t) is cut into small volume elements [1] (ν_t^d) relaxed from action of their neighbours with voids but without dislocations. Exhausting voids from these elements we would get the corresponding *" natural state elements "* (ν_t) (already introduced in [2–5] and elsewhere). According to this terminology (ν_t^d) - elements might be called *" damaged natural state elements "*. The corresponding mappings - second rank tensors

$$\boldsymbol{\Delta}_P : (\kappa_0) \to (\nu_t),\ \boldsymbol{\Delta}_\omega : (\nu_t) \to (\nu_t^d),\ \ and\ \ \boldsymbol{\Delta}_E : (\nu_t^d) \to (\psi_t) \tag{5}$$

are *incompatible* and for this reason are called here *plastic* [1] ,*damage* [2] and *elastic distortions* , respectively. For simplicity in the sequel it is assumed that damage distortion is spherical (somewhere called also " isotropic damage ") $\boldsymbol{\Delta}_\omega = (1+\omega)\mathbf{1}$.

[1]If instead of real crystal configuration (κ_0) an ideal crystal is chosen then the corresponding plastic distortion is not unique. This is promoted in [6] to be *replacement invariance*. Obviously (5) is replacement-invariant.

[2]Incompatibility of damage distortion $\boldsymbol{\Delta}_\omega$ is often neglected in the literature on continuum damage mechanics (cf. eg. [7] where closing of voids is assumed to give a fictitutous undamaged global configuration).

There has been recently a considerable interest in the literature concerning arbitrary rotations of natural state elements. Due to restricted space this item is not elaborated in detail here. However, a short comment is worthwhile: making polar decompositions of tensors in (5) we may either choose *isoclinicity* of (ν_t) and the corresponding global ideal crystal or eliminate rotation tensor of plastic distortion. In both cases objective microstructural deformation variables are ω, $\boldsymbol{\Delta}_P$ and $\mathbf{C}_E \equiv \boldsymbol{\Delta}_E^T \boldsymbol{\Delta}_E$ such that quasi-thermoelastic and quasi-plastic deformation gradient tensors may be expressed by the following functions:

$$\boldsymbol{\epsilon}_E = \boldsymbol{\epsilon}_E(\omega, \mathbf{C}_E, \boldsymbol{\Delta}_P) \qquad \textit{and} \qquad \boldsymbol{\epsilon}_E = \boldsymbol{\epsilon}_E(\omega, \boldsymbol{\Delta}_P). \tag{6}$$

Evolution of the Process and Constitutive Equations

In formulating a constitutive theory one of essential questions is how to choose constitutive variables and corresponding response functions. While in [9] a microstructural picture and rate-independent plasticity were chosen the primary goal of this paper is viscoplasticity and applicability to experiments. For this reason the following constitutive variables are selected here as appropriate to form the set

$$\Gamma = \left\{\boldsymbol{\epsilon}_E, \boldsymbol{\epsilon}_P, \omega, T, Grad\,T, \vec{N}, \vec{Q}, p, \delta_m, A\right\} \tag{7}$$

named *temperature-deformation configuration point*, where T - absolute temperature, $Grad\,T = \mathbf{F}^{-1} grad\,T$ - its (κ_0)-defined spatial gradient, A - a scalar dislocation density defined as length of dislocation lines in a representative volume divided by it, $\vec{Q} = J\mathbf{F}^{-1}\,\vec{q}$ and $\vec{N} = J\mathbf{F}^{-1}\,\vec{\nu}$ - are (κ_0)-defined heat and neutron flux vectors (with $J = det\mathbf{F}$) , while p is plastic deformation path length and δ_m - the maximum plastic strain distance from (κ_0) as measure of *discrete memory* [10,11] (under tacit assumption that $\delta_m(0) = 0$. The last two quantities are determined by the following expressions

$$\dot{p} \equiv \|\dot{\boldsymbol{\epsilon}}_P\| \equiv (\dot{\boldsymbol{\epsilon}}_P : \dot{\boldsymbol{\epsilon}}_P)^{\frac{1}{2}} \geq 0 \qquad \textit{and} \qquad \delta_m := \max_{\tau \in [t_0, t]} \delta(\tau) \equiv \max_{\tau \in [t_0, t]} \|\boldsymbol{\epsilon}_P(\tau)\|\,. \tag{8}$$

where the superposed dot signifies the material time derivative. To the temperature-deformation configuration point there corresponds reaction point given by the set

$$\Sigma_1 = \left\{\mathbf{S}, \mathbf{B}, u, s, \vec{\Phi}\right\} \tag{9}$$

where $\mathbf{S} = J\mathbf{F}^T\mathbf{T}\mathbf{F}$ the Piola-Kirchhoff stres tensor, u - the internal energy density, s - the entropy density and $\vec{\Phi} = J\mathbf{F}^{-1}\,\vec{\phi}$ - the (κ_0) - entropy flux vector. In this way constitutive equations are simply stated as the bijection:

$$\Sigma_1 = \hat{\Sigma}_1(\Gamma) \tag{10}$$

Also, a very important constitutive equation which delimits elastic from inelastic behaviour is intoduced by the following *dynamic yield function*

$$f = h(\mathbf{S}, T, \epsilon_P, A, \vec{N}) = \hat{f}(\Gamma) \tag{11}$$

and the corresponding *static yield function*

$$f_0 = h(\mathbf{S}^\#, T, \epsilon_P, A, \vec{N}) = \hat{f}(\Gamma) \tag{12}$$

with $\mathbf{S}^\# \equiv \mathbf{S}|_{\dot{p}=0}$ being *static stress* chosen to have the same direction as $\mathbf{S}$ and $\Delta\mathbf{S} \equiv \mathbf{S} - \mathbf{S}^\#$ is called *overstress tensor*. These functions allow the following classification:

- $f > 0, f_0 = 0, \dot{p} > 0$ -viscoplastic behaviour,
- $f = f_0 = 0, \dot{p} = 0$ -elastoplastic frontier and
- $f = f_0 < 0, \dot{p} = 0$ -elastic behaviour.

The evolution of the considered process is described by the following *evolution equations*

$$\dot{\vec{Q}} = \vec{Q}^\star(\Gamma) \tag{13}$$

$$\dot{\vec{N}} = \vec{N}^\star(\Gamma) \tag{14}$$

$$\dot{\epsilon_P} = \mathbf{E}^\star(\Gamma) \tag{15}$$

$$\dot{A} = A^\star(\Gamma) \tag{16}$$

$$\dot{\omega} = \omega^\star(\Gamma) \tag{17}$$

as well as by the balance laws. The evolution functions might be also collected into the following set $\Sigma_2 = \{\vec{Q}^\star, \vec{N}^\star, \mathbf{E}^\star, A^\star, \omega^\star, \}$. Due to the material representation (defined with regard to (κ_0)-configuration) instead of corotational time derivatives the above evolution equations contain much simpler material time derivatives.

The set of constitutive and evolution equations is too general for a practical use. However, they must obey the second law of thermodynamics in which there appear time rates and spatial gradients of the constitutive variables. These rates and gradients are not independent but connected by the balance laws and by the kinematical constraint $< f > \dot{f}_0 = 0$ (with $< x >= 1$ for $x > 0$ and $< x >= 0$ otherwise) which is a consequence of (11) and (12). Multiplying these constraint equations by Lagrange multipliers and introducing them into thus extended second law we acquire independent rates and gradients (except *Grad T* which is already included into (7)). This is the essence of Liu's theorem [12]. Therefore the extended dissipation inequality is split into the following constitutive restrictions:

$$\vec{\Phi} = \frac{1}{T}\vec{Q} \tag{18}$$

$$\mathbf{S} = \rho_0\, \partial_{\epsilon_E}\, \Psi + < f > \Lambda^f\, \partial_{\epsilon_E}\, f_0 \tag{19}$$

$$s = -\rho_0\, \partial_T\, \Psi - < f > \Lambda^f\, \partial_T\, f_0 \tag{20}$$

$$\mathbf{S} = \rho_0\, \partial_{GradT}\, \Psi + < f > \Lambda^f\, \partial_{GradT}\, f_0 \tag{21}$$

and the following residual dissipation inequality:

$$\vec{\Lambda}^q \cdot \vec{Q}^* + \vec{\Lambda}^\nu \cdot \vec{N}^* + \Lambda^A A^* + \Lambda^\omega \omega^* + \Lambda^p : \mathbf{E}^* - \frac{1}{T^2}\vec{Q} \cdot Grad\,T \geq 0, \tag{22}$$

where ρ_0 is the mass density in (κ_0)-configuration, $\Psi \equiv u - Ts$ is the free energy density and the afore mentioned Lagrange multipliers are determined as follows

$$\Lambda^\omega = -\rho_0\, \partial_\omega\, \Psi - < f > \Lambda^f\, \partial_\omega\, f_0 \tag{23}$$

$$\Lambda^A = -\rho_0\, \partial_A\, \Psi - < f > \Lambda^f\, \partial_A\, f_0 \tag{24}$$

$$\vec{\Lambda}^q = -\rho_0\, \partial_{\vec{Q}}\, \Psi - < f > \Lambda^f\, \partial_{\vec{Q}}\, f_0 \tag{25}$$

$$\vec{\Lambda}^\nu = -\rho_0\, \partial_{\vec{N}}\, \Psi - < f > \Lambda^f\, \partial_{\vec{N}}\, f_0 \tag{26}$$

$$\mathbf{M} : \Lambda^p = -\rho_0\, \partial_p\, \Psi - < f > \Lambda^f\, \partial_p\, f_0 - \mathbf{M} : [\rho_0\, \partial_{\epsilon_P}\, \Psi + < f > \Lambda^f\, \partial_{\epsilon_P}\, f_0 - \mathbf{S}]. \tag{27}$$

Here $\mathbf{M}$ is the plastic strain rate direction tensor defined by (8-1) and $\dot{\epsilon_P} = \dot{p}\mathbf{M}$.

The whole procedure of the explained thermodynamic analysis (standard for the extended thermodynamics) is presented in detail in [9].

Small Thermoelastic Strains of Isotropic Materials

In order to make the theory more explicit the next assumptions are made:

(A1) *all time rates, quasi-thermoelastic strain, temperature gradient, heat flux and neutron flux are small but plastic strain and damage are finite and*

(A2) *stress vanishes when quasi-elastic strain equals to zero,*

which are realistic for reactor steel behaviour. Due to (A1)-assumption the quasi-thermoelastic strain may be split into quasi-elastic strain and thermal strain in the simplified way:

$$\epsilon_E = \epsilon_e + \gamma\theta\mathbf{1}\,,$$

where $T_0\theta \equiv T - T_0$ and γ is the thermal expansion coefficient. Under such assumptions the stress is represented by the generalized Hooke's law (cf. also [1,9,10])

$$\begin{aligned} \mathbf{S} &= \mathcal{M} : [\epsilon_E - \gamma\theta\mathbf{1}] = \\ & c_1 i_1 \mathbf{1} + 2c_2\epsilon_E + c_3 i_3 \epsilon_P + c_5 i_5 \epsilon_P^2 + c_4(\epsilon_E\epsilon_P + \epsilon_P\epsilon_E) + c_6(\epsilon_E\epsilon_P^2 + \\ & \epsilon_P^2\epsilon_E) - (3c_1 + 2c_2)\gamma\theta\mathbf{1} - (c_3\pi_1 + 2c_4)\gamma\theta\epsilon_P - (c_5\pi_2 + 2c_6)\gamma\theta\epsilon_P^2 , \end{aligned} \quad (28)$$

where the above listed invariants are listed below

$$i_1 = \mathbf{1} : \epsilon_E,\ i_3 = \epsilon_P : \epsilon_E,\ i_5 = \epsilon_P^2 : \epsilon_E,\ \pi_1 = \mathbf{1} : \epsilon_P,\ \pi_2 = \mathbf{1} : \epsilon_P^2,\ \pi_3 = \mathbf{1} : \epsilon_P^3$$

and $\mathbf{M}$ is the fourth rank *tensor of elastic "constants"* [9]. On the other hand the back stress is assumed to depend on present state of damage and plastic strain. It is not included into evolution equations which eliminates recovery effects and allows the following representation:

$$\mathbf{B} = [b_0(p,\omega)\mathbf{1} + b_1(p,\omega)\epsilon_P + b_2(p,\omega)\epsilon_P^2][1 - b_3 exp(-p)] , \quad (29)$$

accounting (by $b_3 \neq 0$) for cyclic kinematic hardening. In the above two constitutive equations all the coefficients i.e. $c_1, \ldots, c_6$ and $b_0, \ldots, b_3$ are functions of

$$\gamma_\pi \equiv \{\pi_1, \pi_2, \pi_3, \theta, \omega, A\} .$$

As the next step, by means of tensor representation procedures (cf. [13,14]) evolution equations (11) and (12) with the assumption that the heat realxation time is negligible (cf. [9]) are simplified into

$$\vec{Q} = -[k_0\mathbf{1} + k_1\epsilon_P + k_2\epsilon_P^2] \cdot Grad\,\theta + [k_3\mathbf{1} + k_4\epsilon_P + k_5\epsilon_P^2] \cdot \vec{N} , \quad (30)$$

$$\tau_\nu\dot{\vec{N}} = [-\mathbf{1} + n_1\epsilon_P + n_2\epsilon_P^2] \cdot \vec{N} + [n_3\mathbf{1} + n_4\epsilon_P + n_5\epsilon_P^2] \cdot Grad\,\theta + [n_6\mathbf{1} + n_7\epsilon_P + n_8\epsilon_P^2] \cdot \vec{Q} \quad (31)$$

where τ_ν is the neutron relaxation time and the heat relaxation time is neglected ($\tau_q \approx 0$) which is good assumption unless very low temperatures are considered. Again, the scalar coefficients $k_0, \ldots, k_5, n_1, \ldots, n_8, \tau_\nu, d_1, ldots, d_9$ depend on the elements of the set γ_π. The equation (31) generalizes Fourier law taking into account thermal anisotropy induced by plastic strain (for k_1, k_2 different from zero) and heating by netron irradiation (by means of coefficients k_3, k_4 and k_5).

The evolution equation for plastic strain rate expressed by stress instead of elastic strain reads

$$\begin{aligned} \dot{\epsilon_P} &= \langle f \rangle \Big[d_1\mathbf{1} + d_2\mathbf{S} + d_3\mathbf{S}^2 + d_4\epsilon_P + d_5\epsilon_P^2 + \\ & d_6(\epsilon_P\mathbf{S} + \mathbf{S}\epsilon_P) + d_7(\epsilon_P^2\mathbf{S} + \mathbf{S}\epsilon_P^2) + \\ & d_8(\epsilon_P\mathbf{S}^2 + \mathbf{S}^2\epsilon_P) + d_9(\epsilon_P^2\mathbf{S}^2 + \mathbf{S}^2\epsilon_P^2)\Big] \end{aligned} \quad (32)$$

This flow rule is non-associate i.e. plastic strain rate is not nenessarily perpendicular to the yield surface. This has been confirmed recently by multiaxial experiments (cf. [15]) at small strains and small as well as medium strain rates. In [15] for such a case scalar material coefficients d_2 and d_3 being dominant were calibrated as functions of plastic strain magnitude and direction showing strong directionality effect as well as imposibillity of constructing a *"universal"* equivalent plastic strain - equivalent stress curve as essential ingredient of an associate flow rule.

Finally, let us consider the simplest case when

(A3) *plastic strain itself is small of the same order as elastic strain.*

Then the free energy may be approximated by a bilinear function (cf. also [9]):

$$\rho_0\Psi \approx \rho_0\Psi_1(\epsilon_e, \theta) + \ell_1\vec{Q}\cdot\vec{Q} + \ell_2\vec{N}\cdot\vec{N} + \ell_3\vec{Q}\cdot\vec{N}. \tag{33}$$

where ℓ_1, ℓ_2 and ℓ_3 are constants. Now, small deflections from equilibrium permit Onsager-Casimir reciprocity relations. Thus from the residual dissipation inequality there follow kinetic relations which restrict the coefficient k_3 to be

$$k_3 = -2\frac{\ell_3}{\ell_1}$$

while n_3 and n_6 may be arbitrary. Since a thermal effect on neutron flux is not observed experimentally (being small of higher order) it is reasonable to neglect terms with $Grad\,\theta$ and $\vec{Q}$ in (32). For the fully linear case considered here this would have as a consequence the next simplified evolution equation

$$\tau_\nu\dot{\vec{N}} + \vec{N} = \vec{0}.$$

Concluding Remarks

The results presented in this paper are briefly summarized as follows:

- The theory is convenient for experiments and accounts for quite general interaction between neutron irradiation creep, thermomechanical damage and viscoplasticity.
- For its application it would be necessary to perform multiaxial viscoplastic experiments during as well as after irradiation in order to calibrate the material constants.
- The flow rule is non-associate. It permits an easy explanation of mechanical as well as thermal anisotropy induced by plastic strain explicitly for the practically important case of small elastic but finite damage-plastic strain.

Acknowledgement

Discussions with Drs C. Albertini and M. Montagnani of JRC CEC - Ispra as well as Dr D. Nikezić of S. Marković University on the subject are highly appreciated.

References

1. MIĆUNOVIĆ, M.: *On low strain-rate viscoplasticity of continuously damaged materials.* In: Current advances in mechanical design and production IV, Kabil, Y. H.; Said, M. E. (eds.) 181–188. Oxford: Pergamon Press 1989

2. KROENER, E.: *Allgemeine Kontinuumstheorie der Versetzungen und Eigenspannungen.* Arch. Rational Mech. Anal. **4** (1970) 273–334.

3. STOJANOVIĆ, R.: *Reference-state problem in non-linear elasticity theory.* Physica Status Solidi **2** (1962) 566–575.

4. BILBY, B. A.: *Continuous distributions of dislocations.* Progress in Sol. Mech. **1** (1960) 329–398.

5. MIĆUNOVIĆ, M.: *A geometrical treatment of thermoelasticity of simple inhomogeneous bodies : I- Geometric and kinematic relations.* Bull. Acad. Polon. Sci.,Ser. Sci. Techn. **22** (1974) 663–642.

6. DASHNER, P. A.: *Invariance considerations in large strain elastoplasticity.* J. Appl. Mech. **53** (1986) 55–60.

7. MURAKAMI, S.: *Mechanical modelling of material damage.* ASME J. Appl. Mech. **55** (1988) 280–286.

8. MURAKAMI, S.; SANOMURA, Y.: *Creep and creep damage of copper under multiaxial state of stress.* In: Plasticity today, Sawczuk A,; Bianchi, G. (eds.) London: Elsevier 1985.

9. MARUSZEWSKI, B.; MIĆUNOVIĆ, M: *On Neutron Irradiation of an Isotropic Thermoplastic Body* Int. J. Engng. Sci. **27/8** (1989) 955–965.

10. MIĆUNOVIĆ, M. *A Thermodynamic Model of Ideal Cyclic Viscoplasticity* In: Transactions of SMIRT-9, **L** Wittmann, F. H. (ed.) 195–203. Rotterdam : Balkema 1987.

11. MROZ, Z.; SHRIVASTAVA, H. R.; DUBEY. R. N. *A Non-Linear Hardening Model and Its Application to Cyclic Loading* Acta Mechanica **17** (1973) 277–290.

12. MÜLLER, I. *The Coldness, a Universal Function in Thermoelastic Bodies.* Arch. Rational Mech. Anal. **41** (1971) 319–332.

13. TRUESDELL, C.; NOLL, W. *The Non-Linear Field Theories of Mechanics.* In: Handbuch der Physik III/3, Fluegge S. (ed.) Berlin : Springer 1965.

14. SPENCER, A. J. M. *Theory of invariants.* In: Continuum Physics, Eringen A. C. (ed.) New York : Academic Press 1971.

15. ALBERTINI, C., MONTAGNANI, M and M. Mićunović *Viscoplastic Behaviour of AISI 316 H - Multiaxial Experimental Results.* Transactions of SMIRT-10, Los Angeles 1989.

Diffusional Transformation and Creep in Steels

E. R. OBERAIGNER and F. D. FISCHER

Institute of Mechanics,
University for Mining and Metallurgy, Leoben, Austria.

Summary

The influence of creep on a material under diffusional transformation represented by a sphere is investigated. Some new results for an elasto-plastic and creeping hollow sphere are presented. It can be concluded that the transformation kinetics and, therefore, the time dependend deformation behavior is not strongly influenced by creep.

Introduction

We consider a specimen of austenite which is cooled from about 900°C to a temperature of about 650°C. At this temperature level a new phase (e.g. pearlite nodules) is built, starting to grow from the intersections of grain boundaries by diffusional transformation. Generally the new phase (or product phase) has a smaller density than the old one. This leads to an eigenstress state with significant hydrostatic pressure p exerted on the product phase, see Fig.1 and [1].

We are interested in the material behavior during and after the growing of the new phase. Of special importance is to know the macroscopic material behavior if there are many nuclei in the austhenite from where the new phase starts to grow. Questions of interest are: What is the influence of creep? How is the pressure reduced when creep occurs and how fast does this happen ? What is the effect of loading, i.e., a one dimensional global stress field additional to the internal local stress fields. To answer these questions we study first a relatively simple model system, which is planed to be refined and generalized, so that we finally end up to describe real material behavior in a rather satisfactory manner.

Mathematical Model

We consider one point in a very large or an infinitely extended specimen, where the new product phase starts to grow. The new phase itself is represented by a sphere with growing

M. Zyczkowski (Ed.)
Creep in Structures
IUTAM Symposium Cracow/Poland 1990

growing radius $r_1(t)$, see Fig.2. It is assumed that there is no external load stress and no internal stress field at the beginning. Let the density of the new emerging phase be $1/\delta$-times as large as the density of the old phase, where $\delta > 1$. Let further $v(t)$ be the growth rate of the new phase, this means the speed of the spherical border of the new phase. The radius of the sphere of the new phase then will be

$$r_1(t) = \int_{t_0}^{t} v(\tau)\, d\tau \,. \tag{1}$$

In many cases the velocity v is constant for sufficiently long time periods. See e.g. [3]. In the following it is also assumed, that there is only one such sphere of the growing new phase as inclusion in the old phase. Therefore the considered system is spherically symmetric.

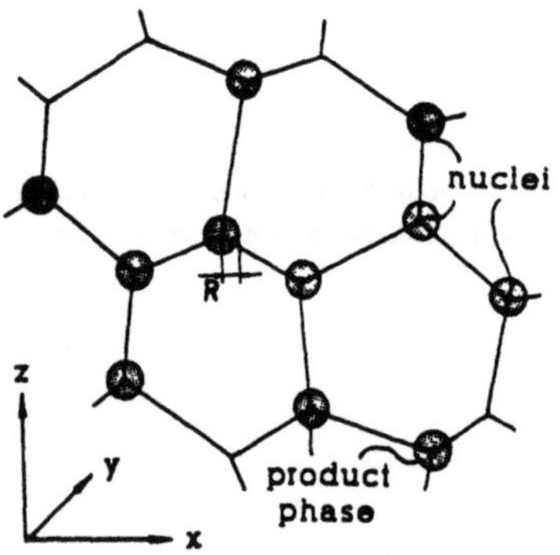

Fig.1 : Micromechanical modell — diffusional transformation

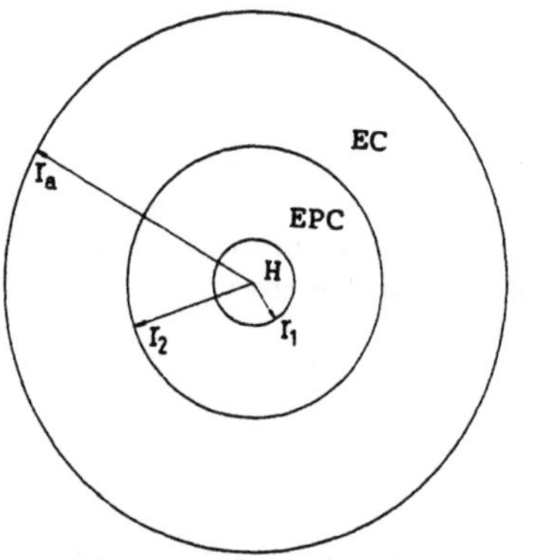

Fig.2 : Domain H : hydrostatic;
Domain EPC : elastoplastic with creep;
Domain EC : elastic with creep

If δ is large enough, the eigenstress due to phase transformation will be sufficient to cause plastification of the old phase in a certain spherical domain $(r_1 \leq r \leq r_2)$ around the inclusion. If $r_a \leq \infty$ is the outer radius of the whole considered system, which is also assumed to be spherical, then in the domain $r_2 \leq r \leq r_a \leq \infty$ the material is elastic. In both domains, the elastoplastic one and the elastic one, we will take into account creep.

The inner sphere $(0 \leq r \leq r_1)$ is under a pure hydrostatic stress state. This is a well known result in the theory of elasticity, but it can be shown, that it also holds in our case, where plasticity and creep play an essential role.

Fig.2 is to give an immediate impression of the geometrical situation discussed just before.

The mathematical model consists of three sets of equations: equilibrium conditions, kinematic equations, constitutive laws.

Due to the spherical symmetry of our system we use a spherical coordinate system.

Our variables are the radial stress σ_r, the tangent (or circumferential) stress σ_t, the radial strain ϵ_r, the tangent (or circumferential) strain ϵ_t, and the radial displacement u.

The equivalent stress is denoted by σ_e and the yield stress by σ_Y. No hardening is assumed for sake of simplicity. In the following, we will also use the superscripts EPC, EC, H, to denote the elastoplastic-creep domain, the elastic-creep, and the hydrostatic domain, respectively.

The equilibrium condition is expressed by

$$\frac{\partial \sigma_r}{\partial r} + \frac{2}{r}(\sigma_r - \sigma_t) = 0 \ . \tag{2}$$

In the hydrostatic range we have

$$\sigma_r^H = \sigma_t^H \ ,$$

hence

$$\frac{\partial \sigma_r^H}{\partial r} = 0 \quad , \qquad \sigma_r^H = const = -p \ , \tag{4,5}$$

In domain EPC we have

$$\sigma_t^{EPC} - \sigma_r^{EPC} = \sigma_Y \ , \tag{6}$$

either for the the v. Mises or for the Tresca yield criterion.

The kinematic equations are

$$\epsilon_r = \frac{\partial u}{\partial r} \quad , \qquad \epsilon_t = \frac{u}{r}. \tag{7,8}$$

It is also known from literature [2] that for hydrostatic conditions the following equation which we will need later on holds

$$u^H(r) = -p\,\frac{1-2\nu}{E}\,r \ . \tag{9}$$

Analogous analytical relations are known for the elastoplastic case and for the pure elastic case. We will derive similar equations including creep.

The constitutive equations in domain EPC are as follows: We use *Norton's* creep law to express the strain rate, i.e.,

$$\dot{\epsilon}_{ij}^c = \frac{3\,k}{2}(3\,I_2)^{\frac{n-1}{2}}\,s_{ij} = \frac{3\,k}{2}\sigma_e^{n-1}\,s_{ij} \ . \tag{10}$$

This gives in the spherical symmetric case

$$\dot{\epsilon}_r^c = -k\,(\sigma_t - \sigma_r)^n \quad , \qquad \dot{\epsilon}_t^c = \frac{k}{2}(\sigma_t - \sigma_r)^n \ . \tag{11a, b}$$

The supersript c referes to creep. k is called creep coefficient, n creep exponent. Dot means here and in the following time derivative, prime means derivative with respect to the radius r. ϵ_{ij} is the strain tensor, σ_{ij} and s_{ij} is the stress tensor and deviator, respectively. I_2 is the second stress invariant.

The elastic strain tensor is given by

$$\epsilon_{ij}^{el} = E_{ijkl}^{-1}\,\sigma_{kl} \ . \tag{12}$$

where E_{ijkl} is the elastic moduli tensor (E Young's modulus, ν Poisson's ratio). It follows

$$\epsilon_r^{el} = \frac{1}{E}(\sigma_r - 2\,\nu\,\sigma_t) \quad , \qquad \epsilon_t^{el} = \frac{1}{E}\Big((1-\nu)\,\sigma_t - \nu\,\sigma_r\Big). \tag{13a, b}$$

The superscript el refers to elastic.

The plastic strain increment can be derived from the theory of plasticity, specifically applied to an elastic-ideal plastic material, see [4]. The derivation similar to that in the literature, leads in vectorial form to

$$d\underline{\epsilon}^{pl} = \left(\frac{1}{\hat{\underline{s}}^T \cdot \underline{\underline{L}} \cdot \hat{\underline{s}}}\right) \cdot \underline{\underline{\Sigma}} \cdot \underline{\underline{L}} \cdot d\underline{\epsilon} \,. \tag{14}$$

where

$$\hat{\underline{s}} = \frac{1}{3}\begin{pmatrix} -2 \\ 1 \\ 1 \end{pmatrix} \sigma_e \,, \qquad \text{and} \quad \underline{\underline{\Sigma}} = \hat{\underline{s}} \otimes \hat{\underline{s}} \tag{15,16}$$

and finally

$$\underline{\underline{L}} = \frac{E}{(1+\nu)(1-2\nu)} \begin{pmatrix} 1-\nu & \nu & \nu \\ \nu & 1-\nu & \nu \\ \nu & \nu & 1-\nu \end{pmatrix} . \tag{17}$$

Therefore, we get

$$d\underline{\epsilon}^{pl} = \frac{1}{3} \cdot (d\epsilon_t - d\epsilon_r) \cdot \begin{pmatrix} -2 \\ 1 \\ 1 \end{pmatrix} . \tag{18}$$

Equation (18) together with equations (13) and (11) gives after some calculations

$$\dot{\sigma}_r^{EPC} = E \cdot \frac{(1-\nu)\dot{\epsilon}_r + 2\nu\dot{\epsilon}_t + (1-2\nu)\left[\frac{1}{3}(\dot{\epsilon}_t - 2\dot{\epsilon}_r) - k\left(\sigma_t^{EPC} - \sigma_r^{EPC}\right)^n\right]}{(1+\nu)(1-2\nu)} \tag{19a}$$

$$\dot{\sigma}_t^{EPC} = E \cdot \frac{\nu\dot{\epsilon}_r + \dot{\epsilon}_t + (1-2\nu)\left[\frac{1}{3}(\dot{\epsilon}_t - \dot{\epsilon}_r) - \frac{k}{2}\left(\sigma_t^{EPC} - \sigma_r^{EPC}\right)^n\right]}{(1+\nu)(1-2\nu)} \tag{19b}$$

From equation (2) and the yield criterion eqn.(6) we have

$$\frac{\partial \sigma_r^{EPC}}{\partial r} - \frac{2}{r}\sigma_Y = 0 \tag{20}$$

Eqn. (20), (19a,b), (7), and (8) are five equations for the five unknowns σ_r^{EPC}, σ_t^{EPC}, ϵ_r^{EPC}, ϵ_t^{EPC} and u^{EPC}. Later on we will present formulas for the solution.

In the next step we will establish a similar set of equations as before for the domain EC. Therefore the total strain vector is given by

$$\underline{\epsilon}^{EC} = \underline{\epsilon}^{el} + \underline{\epsilon}^{c} \tag{21}$$

with $\underline{\epsilon}^c$ given by equations (11a) and (11b) and $\underline{\epsilon}^{el}$ by equations (13a) and (13b). Insertion into eqn.(21) and elimination of ϵ_r by using

$$\epsilon_r = \frac{\partial}{\partial r}(r\,\epsilon_t) \,, \tag{22}$$

which follows from equations (7) and (8) gives two partial differential equations for the variables σ_r and ϵ_t if the stress σ_t is eliminated by eqn.(2). Elimination of ϵ_t by insertion of one of these equations into the other gives finally

$$\frac{k}{2}\left(\frac{r}{2}\frac{\partial}{\partial r}\sigma_r^{EC}\right)^n \left[(3+n) + n\frac{r\frac{\partial^2}{\partial r^2}\sigma_r^{EC}}{\frac{\partial}{\partial r}\sigma_r^{EC}}\right] + E(1-\nu)\frac{r}{2}\frac{\partial}{\partial t}\left[4\frac{\partial}{\partial r}\sigma_r^{EC} + r\frac{\partial^2}{\partial r^2}\sigma_r^{EC}\right] = 0 \,, \tag{23}$$

which is a nonlinear partial differential equation for the radial stress σ_r^{EC}. It is of first order in time and of second order in radius.

As one of the two boundary conditions of equation (23) we chose

$$\sigma_r^{EC}(r_a) = 0 \,. \tag{24}$$

The second boundary condition is

$$\sigma_r^{EC}(r_2) = \sigma_r^{EPC}(r_2) \,, \tag{25}$$

which is also the first boundary condition for the EPC case. The second boundary condition for the EPC case is

$$\sigma_r^{EPC}(r_1) = \sigma_r^{H}(r_1) = -p, \tag{26}$$

where r_1 is time dependent as given by equation (1). r_2 is time dependent, too, but is unknown yet and has to be determined from the system of equations.

The initial condition is, as we already mentioned earlier,

$$\sigma_r(r, t_0) = 0 \tag{27}$$

for any case.

It can be shown that the differential equations for the EPC case as well as for the EC case collapse to the well known simpler cases without creep discussed in literature [5], if we perform the limit $k = 0$ or $n = 1$. For $E \to \infty$ and $\sigma_Y \to \infty$ the solution agrees with [6].

Solutions of the Equations

We do not have enough space here to derive the solution in detail. Instead we will outline the way we get it and we will present the formulas.

By subtraction of eqn.(19a) from eqn.(19b) with the initial condition

$$\sigma_e(t_1) = \sigma_Y \tag{28}$$

(t_1 means here the time, when plastification first occurs) we get a relation for σ_e which is surprisingly independent on the transformation front radius $r_1(t)$

$$\sigma_e = \left[\sigma_Y^{1-n} + (n-1)\frac{3k}{2}\frac{E}{(1+\nu)}(t - t_1)\right]^{\frac{1}{1-n}} \,, \tag{29}$$

which is a monotonously decreasing function with time for $n > 1$. In the case boundary r_1 does not progress the plastified region would disappear immediately after it was established.

From eqn.(20) we get

$$\sigma_r^{EPC}(r) = 2\,\sigma_Y \ln\left(\frac{r}{r_2}\right) + \sigma_r^{EPC}(r_2) \,. \tag{30}$$

In case of EC we get by time integration from equation (23)

$$\left(\frac{\partial}{\partial r} + \frac{4}{r}\right)\sigma_r'^{EC} = \frac{E}{(1-\nu)}\frac{k}{2}\int_{t_0}^{t}\left(\frac{r}{2}\sigma_r'(r,\tau)^{EC}\right)^{n-1}\left[n\sigma_r''(r,\tau)^{EC} + \frac{n+3}{r}\sigma_r'(r,\tau)^{EC}\right]d\tau \,. \tag{31}$$

Using Green's-function methods [7] for the left hand side to get $\sigma_r'(r,t)^{EC}$ and performing a partial integration due to r afterwards yields after some calculation

$$\sigma_r'(r,t)^{EC} = \frac{1}{r^4}\frac{E}{(1-\nu)}\frac{k}{2^n}\int_{t_0}^{t}\left[r^{n+3}\sigma_r'(r,\tau)^{EC^n} - r_2(\tau)^{n+3}\sigma_r'(r_2(\tau),\tau)^{EC^n}\right]d\tau + \tag{32}$$
$$+\sigma_r'(r_2(t),t)^{EC}\left(\frac{r_2(t)}{r}\right)^4 ,$$

which is equivalent to an ordinary differential equation for the radial pressure gradient $\sigma_r'(r,t)^{EC}$. Integration of eqn.(32) due to r from r_2 to r_a leads (using eqn.(30) and (35)) to an ordinary differential equation for $\sigma_r'(r_2(t),t)^{EC}$, which can be solved analytically in the cases $n = 1,2,3$ and 4 according to [8]. For known $\sigma_r'(r_2(t),t)^{EC}$ an analytical solution of eqn.(32) is possible in the cases $n = 1,2,3$ and 4. For details see [9].

Boundary between EPC and EC domain and pressure p

The boundary radius $r_2(t)$ between the EPC domain and the EC domain can be calculated in the following way.

Let r_{free} be the radius of the freely growing interior sphere, $\bar{r}_{\text{free}}$ the inner radius of the free hollow sphere (i.e. new phase removed from the system) then the following condition holds

$$\left|u(r_1)^H\right| + \left|u(r_1)^{EPC}\right| = r_{free} - \bar{r}_{free} \tag{33}$$

where

$$\bar{r}_{\text{free}} = \frac{r_{\text{free}}}{\delta^{1/3}} . \tag{34}$$

Calculating $u^{EPC}(r_1)$ with Green's function methods from the radial stress and with the help of $u^{EC}(r_2)$ (also calculated with Green's function methods) we end up after some analysis and insertion of $u^H(r_1)$ and $u^{EPC}(r_1)$ into eqn.(33) with the formula

$$\left(\frac{r_2}{r_1}\right)^3 = 1 + \frac{3}{2}\left(1 - \frac{1}{\delta^{1/3}}\right)\frac{E}{(1-2\nu)\,\sigma_Y} \tag{35}$$

in the case $r_a \to \infty$.

The approximation $r_1 = r_{\text{free}}$ was used to get this result. Using instead of this approximation the exact formula $r_1 = r_{\text{free}} - u^{EPC}(r_1)$ leads to a nonlinear equation with a solution very close to the above formula.

Eqn.(35) gives for the data $\delta = 1.04$, $\sigma_Y = 400$ MPa, $E = 210000$ MPa and $\nu = 0.3$ the very good approximation

$$r_2 = 3\,r_1 . \tag{36}$$

Using eqn.(30) and similar calculation methods as before we get for the pressure p after some algebra (for $r_a \to \infty$)

$$p = \frac{2}{3}\sigma_Y\left\{1 + \ln\left[\frac{3}{2\sigma_Y}\frac{E}{1-2\nu}\left(1 - \frac{1}{\delta^{1/3}}\right) + 1\right]\right\} , \tag{37}$$

which gives for the above data

$$p = 1145 \text{ MPa} \quad . \tag{38}$$

The Fig.3 shows the stress distribution for increasing time with the above data set and for a constant front speed $v = 2\,\mu m\, s^{-1}$.

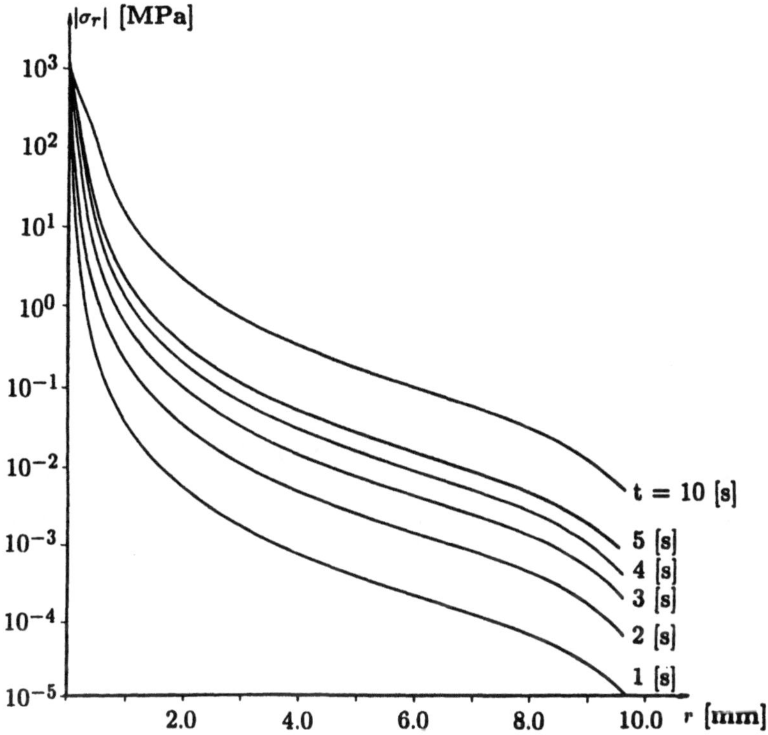

Fig.3 : Stress distribution for increasing time with the above data set and the constant front speed $v = 2\mu m\, s^{-1}$

Conclusions and outlook

The analytical and numerical computations to get the stress in dependence of radius $r_1(t)$ and time show that the only and small influence of creep exists in the elastic domain EC. The radius ratio r_1/r_2 is constant. The hydrostatic pressure of the new phase inward is independent on the creep behavior if $r_a \to \infty$. Since the stress state on the border of the product phase sphere is relevant for transformation kinetics, e.g. see [10], it can be stated that transformation kinetics is hardly influenced by creep. If r_a is finite, there are some boundary effects whose influence decreases with increasing outer radius r_a. So we have some kind of self-similarity in the case of $r_a \to \infty$.

A preliminary conclusion may be that the time dependent deformation of a transforming material is mainly influenced by the transformation itself and only to a small amount by creep.

In our forthcoming work on this subject we are investigating the boundary effects as well as the effect of a global uniaxial stress field in detail.

Acknowledgement

We thank Prof. J.A. Hult for his valuable hints and comments.

References

[1] FISCHER, F.D.: *A micromechanical model for transformation plasticity in steels.* Acta Metall & Mater. 38 (1990) 1535-1546.

[2] HAHN, H.G.: *Elastizitätstheorie.* Stuttgart: B.G.Teubner 1985.

[3] GAUTIER, E.E.; SIMON, A.: *Analysis of the kinetics of pearlitic transformation under stress or after plastic deformation.* Phase Transformations 87. London : The Institute of Metalls 1988.

[4] YAMADA, Y.; YOSHIMURA, N.; SAKURAI, T.: *Plastic stress-strain matrix and its application for the solution of elastic-plastic problems by the finite element method.* Int. J. Mech. Sci. 10 (1968) 343-354.

[5] RECKLING, K.A.: *Plastizitätstheorie und ihre Anwendungen auf Festigkeitsprobleme.* Berlin, Heidelberg, New York: Springer Verlag 1967.

[6] HULT, J.A.: *Creep in Engineering Structures.* Waltham, MA. : Blaisdell Publishing Company.

[7] BUTKOVSKIY, A.G.: *Green's Functions and Transfer Functions Handbook.* Chichester : Ellis Horwood Limited 1982.

[8] KAMKE, E.: *Differentialgleichungen — Lösungsmathoden und Lösungen I.* Stuttgart : B.G.Teubner 1977, 303.

[9] OBERAIGNER, E.R.; FISCHER, F.D.: *Interner Bericht über Kriechen von Stahl bei diffusiven Phasenünergängen.* Leoben, Institut für Mechanik, Montanuniversität Leoben, 1990.

[10] TANAKA, K.; SATO, Y.: Ingenieur-Arch. 55 (1985) 147.

Effects of Slip System and Lattice Rotation on Combined Tension-Torsion Creep of a Directionally Solidified Alloy

N. OHNO* and T. MIZUNO†
* Department of Mechanical Engineering, Nagoya University
† Nagoya Guidance and Propulsion Systems Works, Mitsubishi Heavy Industries, Ltd.

Summary

Combined tension-torsion creep tests of directionally solidified IN738LC at 850°C are first discussed to find features which become noticeable with the advance of creep deformation. Analysis based on the crystallographic slip theory of finite deformation then shows that crystalline lattice rotation explains the experimental features; it is mostly responsible for the change of the noncoaxiality between creep rate and stress in combined tension-torsion creep, and partly for the earlier appearance of tertiary creep under combined tension-torsion than under pure tension and torsion.

Introduction

Nickel-base directionally solidified and single-crystal alloys have been used recently as materials for gas turbine blades. However, multiaxial inelastic behavior of such materials has been investigated only in a few works so far, in contrast with many works for uniaxial behavior.

The present authors et al. [1, 2] therefore carried out combined tension-torsion creep tests of directionally solidified IN738LC at 850°C to examine creep anisotropy of this material in multiaxial stress states. Analyzing the experiments on the basis of the crystallographic slip theory, they found that the minimum creep rates are well simulated if the {111}<112> octahedral slip system is combined with the {100}<110> cubic system. It is noticed that while the former has been observed to be active in creep [3-9], the latter has not been reported with respect to creep [8, 9].

In the previous papers [1, 2], however, phenomena which become noticeable with the advance of creep deformation were not discussed. A possible micromechanism which brings about such phenomena is lattice rotation. Since directionally solidified alloys consist of columnar grains growing parallel in the solidification direction, their mechanical behavior is governed directly by the micromechanisms of single crystals. Therefore,

M. Zyczkowski (Ed.)
Creep in Structures
IUTAM Symposium Cracow/Poland 1990

when deformation is large, lattice rotation is a potential micromechanism which may influence markedly on the inelastic behavior of directionally solidified alloys.

In the present work, first, the experiments [1, 2] are discussed to find features which become noticeable as creep proceeds. Then, to study the effects of slip system and lattice rotation on them, the experiments are analyzed on the basis of the the crystallographic slip theory of finite deformation.

Experimental Features

In the multiaxial creep tests [1, 2], thin-walled tubular specimens of IN738LC solidified in the axial direction were subjected to tensile stress σ and torsional stress τ at 850°C. The experiments were performed under constant true stresses. The stress state in the tests was specified in terms of the equivalent stress of Mises, σ_{eq}, and the angle of stress, θ, defined as

$$\sigma_{eq} = (\sigma^2 + 3\tau^2)^{1/2}, \tag{1}$$

$$\theta = \tan^{-1}(\sigma/\sqrt{3}\tau). \tag{2}$$

The creep curves obtained under σ_{eq} = 300 MPa are shown in Fig. 1, where the equivalent strain of Mises for isotropic materials, ε_{eq}, is taken for

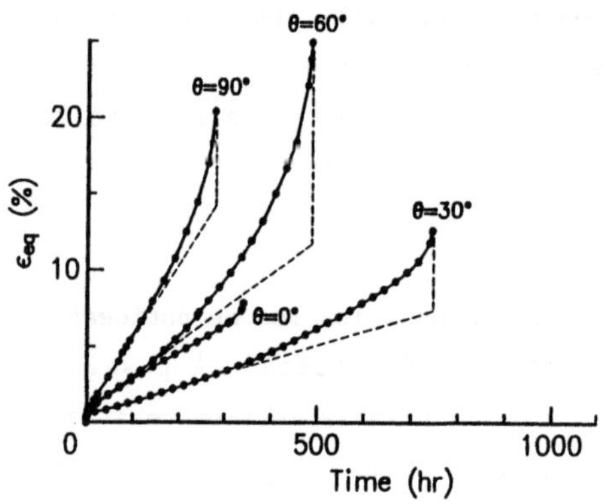

Fig. 1. Creep curves by experiments (σ_{eq} = 300 MPa).

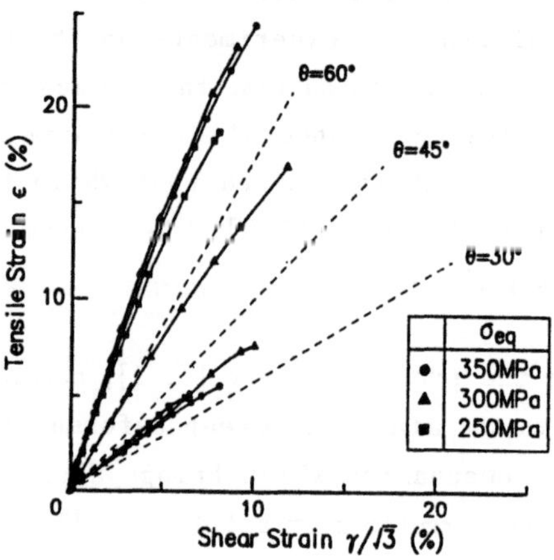

Fig. 2. Strain trajectories by experiments.

convenience in the ordinate. With tensile strain ε and shear strain γ, ε_{eq} is expressed as

$$\varepsilon_{eq} = \int(\dot{\varepsilon}^2 + \dot{\gamma}^2/3)^{1/2}dt. \tag{3}$$

We see from the figure that the change of creep rate with ε_{eq} depends on the stress state: While steady-state creep prevails up to $\varepsilon_{eq} \simeq 10$ % in the tensile creep test ($\theta = 90°$), tertiary creep becomes noticeable around $\varepsilon_{eq} \simeq 5$ % in the combined tension-torsion creep tests of $\theta = 60°$ and $\theta = 30°$. In the torsion creep test ($\theta = 0°$), on the other hand, tertiary creep is not observed except just before creep rupture.

The noncoaxiality between stress and creep rate tensors also changes, with dependence on the stress state, as creep deformation proceeds (Fig. 2). Let us define the angle of creep rate to be

$$\bar{\theta} = \tan^{-1}[\dot{\varepsilon}/(\dot{\gamma}/\sqrt{3})]. \tag{4}$$

Then, if we deal with a Mises-type isotropic material, we have $\bar{\theta} = \theta$; i.e., the dotted lines in Fig. 2 are the creep strain trajectories to be predicted. However, the trajectories observed in the experiments are not such straight lines. In the tests of $\theta = 45°$ and $\theta = 60°$, first $\bar{\theta} > \theta$, and eventually $\bar{\theta} \simeq \theta$. In the test of $\theta = 30°$, on the other hand, first $\bar{\theta} \simeq \theta$, and then slightly $\bar{\theta} > \theta$ except just before rupture. Moreover, the trajectories in the tests of $\theta = 0°$ and $\theta = 90°$ were observed to nearly coincide with the abscissa and the ordinate, respectively, though they are not shown in Fig. 2, (see Fig. 7).

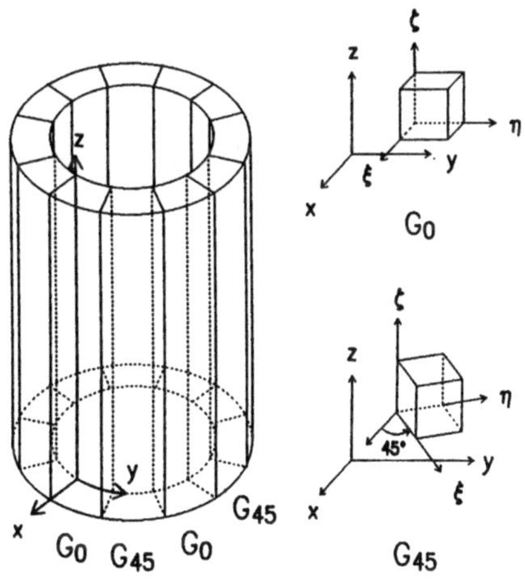

Fig.3. Model for analysis.

Model for Analysis

Directionally solidified nickel-base alloys consists of columnar grains grown in the solidification direction, for which the [001] direction is preferred. Hence the thin-walled tube was modeled simply as shown in Fig. 3, where x, y and z denote the radial, circumferential and axial directions of the tube, respectively, and ξ, η

and ζ indicate the crystalline axes. In this simple model, only two kinds of columnar grains, G_0 and G_{45}, alternate in the circumferential direction. In the grain G_0 the crystalline axes are directed in the x, y and z directions, while in G_{45} they rotate by 45° with respect to the z direction.

To specify creep rate of the grains G_0 and G_{45}, let us consider three kinds of slip systems, i.e., the {111}<110> octahedral, the {111}<112> octahedral, and the {100}<110> cubic system, reported to be active in plastic and/or creep deformations of nickel-base single crystals [8-10]. Then, on the basis of the crystallographic slip theory of finite deformation (see, e.g., [11]), creep rate and spin tensors of each grain are expressed as follows:

$$\mathbf{D} = \sum_{k=1}^{12} \dot{\gamma}^{(k)}_{oct1}\boldsymbol{\alpha}^{(k)}_{oct1} + \sum_{k=1}^{12} \dot{\gamma}^{(k)}_{oct2}\boldsymbol{\alpha}^{(k)}_{oct2} + \sum_{k=1}^{6} \dot{\gamma}^{(k)}_{cub}\boldsymbol{\alpha}^{(k)}_{cub}, \tag{5}$$

$$\boldsymbol{\Omega} = \boldsymbol{\Omega}^{s} + \boldsymbol{\Omega}^{*}, \tag{6}$$

$$\boldsymbol{\Omega}^{s} = \sum_{k=1}^{12} \dot{\gamma}^{(k)}_{oct1}\boldsymbol{\omega}^{(k)}_{oct1} + \sum_{k=1}^{12} \dot{\gamma}^{(k)}_{oct2}\boldsymbol{\omega}^{(k)}_{oct2} + \sum_{k=1}^{6} \dot{\gamma}^{(k)}_{cub}\boldsymbol{\omega}^{(k)}_{cub}, \tag{7}$$

where the subscripts oct1, oct2 and cub stand for the {111}<110> octahedral, the {111}<112> octahedral, and the {100}<110> cubic system, respectively; $\dot{\gamma}^{(k)}_{Q}$ (Q = oct1, oct2, cub) denotes the slip rate in the slip system specified by the subscripts (k) and Q; $\boldsymbol{\Omega}^{s}$ and $\boldsymbol{\Omega}^{*}$ indicate spins due to slip and rigid rotation, respectively; $\boldsymbol{\alpha}^{(k)}_{Q}$ and $\boldsymbol{\omega}^{(k)}_{Q}$ are defined in terms of the slip direction vector $\mathbf{b}^{(k)}_{Q}$ and the slip plane vector $\mathbf{n}^{(k)}_{Q}$ as

$$\boldsymbol{\alpha}^{(k)}_{Q} = \frac{1}{2}\left(\mathbf{b}^{(k)}_{Q}\otimes\mathbf{n}^{(k)}_{Q} + \mathbf{n}^{(k)}_{Q}\otimes\mathbf{b}^{(k)}_{Q}\right), \tag{8}$$

$$\boldsymbol{\omega}^{(k)}_{Q} = \frac{1}{2}\left(\mathbf{b}^{(k)}_{Q}\otimes\mathbf{n}^{(k)}_{Q} - \mathbf{n}^{(k)}_{Q}\otimes\mathbf{b}^{(k)}_{Q}\right). \tag{9}$$

The spin due to rigid rotation, $\boldsymbol{\Omega}^{*}$, rotates $\mathbf{b}^{(k)}_{Q}$ and $\mathbf{n}^{(k)}_{Q}$:

$$\dot{\mathbf{b}}^{(k)}_{Q} = \boldsymbol{\Omega}^{*}\mathbf{b}^{(k)}_{Q}, \qquad \dot{\mathbf{n}}^{(k)}_{Q} = \boldsymbol{\Omega}^{*}\mathbf{n}^{(k)}_{Q}. \tag{10}$$

For $\dot{\gamma}^{(k)}_{Q}$, we assume a simple law

$$\dot{\gamma}^{(k)}_{Q} = \mathrm{sign}[\tau^{(k)}_{Q}]\cdot K_{Q}|\tau^{(k)}_{Q}|^{n_{Q}} \qquad (Q = \text{oct1, oct2, cub}), \tag{11}$$

where $\tau^{(k)}_{Q} = \mathrm{tr}(\boldsymbol{\sigma}\boldsymbol{\alpha}^{(k)}_{Q})$, and K_Q and n_Q are material constants.

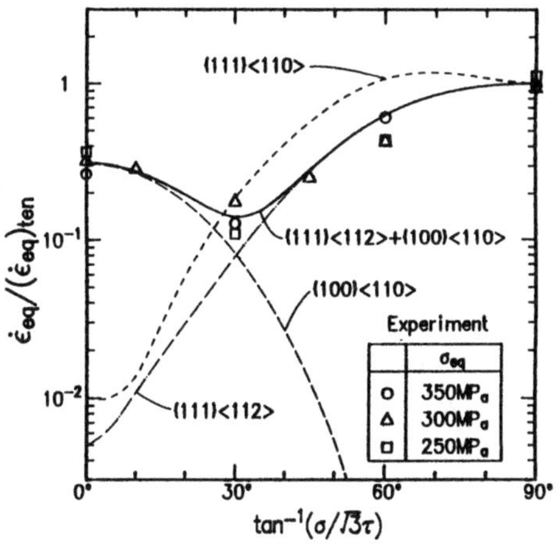

Fig.4. Dependence of creep rate on stress state.

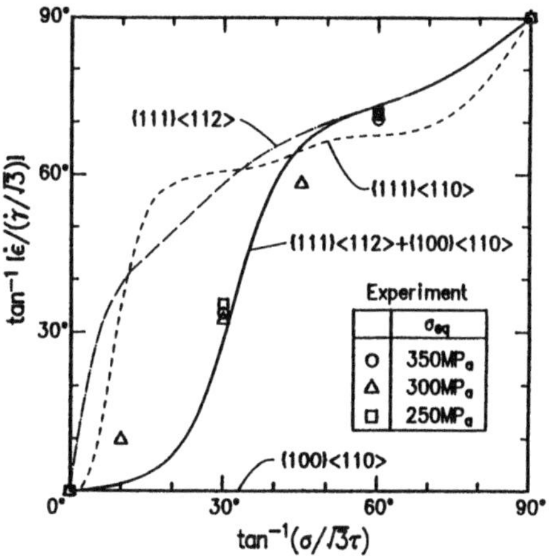

Fig.5. Noncoaxiality between creep rate and stress.

Then, combined tension-torsion creep of the directionally solidified thin-walled tube modeled as shown in Fig. 3 can be analyzed, if we take into account appropriately the interaction between the neighboring grains G_0 and G_{45}. In the present work, since we deal with considerably large deformation, it may be appropriate to assume

$$L_{zz:0} = L_{zz:45}, \quad L_{yz:0} = L_{yz:45}, \tag{12}$$

where L_{zz} and L_{yz} are components of velocity gradient tensor **L**, and the subscripts 0 and 45 denote the grains G_0 and G_{45}, respectively.

Material Constants (Simulation on Minimum Creep Rate [1, 2])

The simulation on minimum creep rate in the framework of small strain [1, 2] is reported here again for convenience in Figs. 4 and 5. The material constants used in this simulation were

$$\left.\begin{array}{l} K_{oct1} = 1.07\times10^{-23}, \quad K_{oct2} = 4.91\times10^{-24}, \quad K_{cub} = 2.37\times10^{-24} \\ n_{oct1} = n_{oct2} = n_{cub} = 9.2 \end{array}\right\}, \tag{13}$$

where units are MPa (stress), hr (time), and mm/mm (strain). As seen from Figs. 4 and 5, if we assume only one kind of slip system to be active, the experiments are not simulated accurately. However, the combination of the {111}<112> octahedral and {100}<110> cubic systems is very successful.

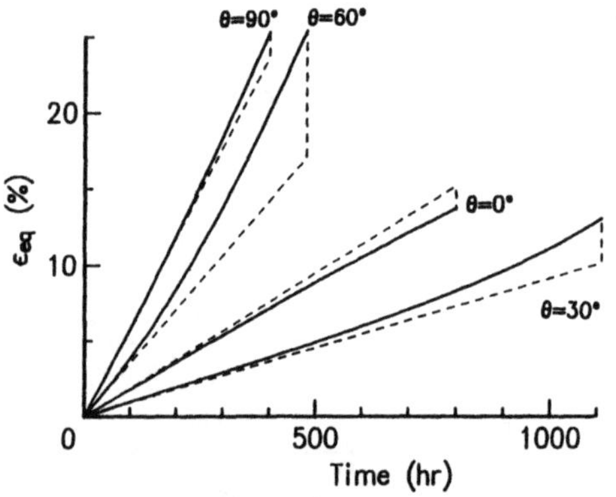

Fig.6. Creep curves by analysis based on {111}<112> and {100}<110> slip systems (σ_{eq}= 300 MPa).

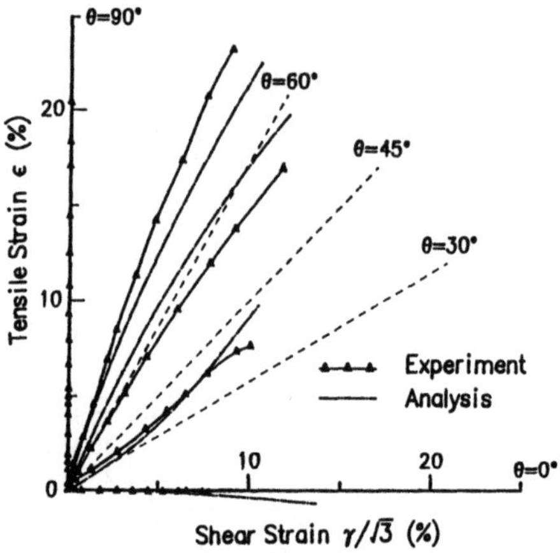

Fig.7. Strain trajectories by analysis based on {111}<112> and {100}<110> slip systems, together with experiments, (σ_{eq} = 300 MPa).

Results of Analysis and Discussion

Creep curves and creep strain trajectories resulted from the present analysis are shown in Figs. 6 and 7, respectively. The analysis was performed for σ_{eq} = 300 MPa. The combination of the {111}<112> octahedral and {100}<110> cubic systems was assumed, because it simulates accurately the minimum creep rate data (Figs. 4, 5).

As seen from Fig. 6, acceleration of creep rate occurs in the creep curves of θ = 60° and 30°, whereas rather steady-state creep prevails under θ = 90° and 0°. This trend is the same as in the experiments (Fig. 1); thus, the theory is partly successful in simulating the creep curves in Fig. 1. The strain trajectories by the analysis, on the other hand, agree well with the experiments (Fig. 7); i.e, along the trajectories of θ = 60° and 45° the noncoaxiality between creep rate and stress becomes smaller and eventually vanishes, while it becomes noticeable gradually in the case of θ = 30°.

In the present analysis, only lattice rotation is the internal change, because the creep equation employed, equation (11), does not include any internal variable. Therefore, the acceleration of creep rate and the change of the noncoaxiality in the analysis is brought about through lattice rotation.

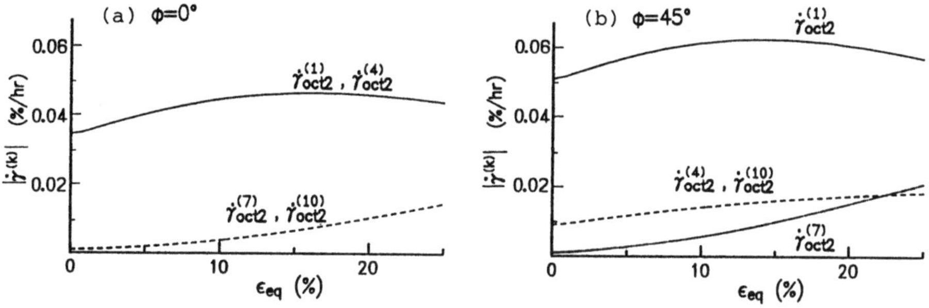

Fig.8. Change of slip rate (θ = 60°, σ_{eq}=300 MPa); (a) G_0, (b) G_{45}.

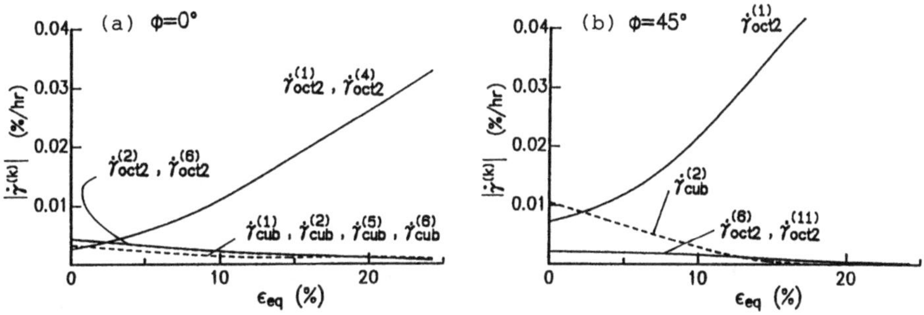

Fig.9. Change of slip rate (θ = 30°, σ_{eq}=300 MPa); (a) G_0, (b) G_{45}.

The change of the noncoaxiality has different aspects in the two cases of θ = 60° and θ = 30°, as we have seen in Fig. 7. Figures 8 and 9 show the changes of main $\dot{\gamma}^{(k)}_Q$s in these two cases. When θ = 60° (Fig. 8), only the {111}<112> octahedral system is active. The change of the noncoaxiality in this case is therefore entirely due to lattice rotation. When θ = 30° (Fig. 9), on the other hand, the {111}<112> octahedral and the {100}<110> cubic system are active; i.e., in the initial state they are equally active, resulting in $\bar{\theta} \simeq \theta$ (Fig 5), but as creep develops, the {111}<112> octahedral system becomes dominant due to lattice rotation, giving rise to $\bar{\theta} > \theta$ (Fig. 5).

Conclusions

Effects of slip system and lattice rotation on the combined tension-torsion creep of directionally solidified IN738LC at 850°C were studied. Analyzing the experiments on the basis of the crystallographic theory of finite deformation, we obtained the following results: (1) Acceleration of creep rate became noticeable around $\varepsilon_{eq} \simeq 5$ % in the combined tension-torsion creep tests, while rather steady-state creep prevailed in the pure tensile and torsional creep tests. Lattice rotation is partly responsible for this experimental feature. (2) In the combined tension-torsion creep tests, the noncoaxiality between creep rate and stress changed with the development of creep deformation. This feature is well described by lattice rotation, together with the combination of the {111}<112> octahedral and {100}<110> cubic systems.

Acknowledgment

The authors appreciate the financial support by the Ministry of Education of Japan under a Grant-in-Aid for Scientific Research C (No. 01550070) in the fiscal year of 1989. The second author, T. Mizuno, was associated with the present work while studying as a graduate student at Nogoya University.

References

1. Ohno, N.; Kawaji, H.; Mizuno, T.: Creep of a nickel-base directionally solidified alloy under combined tension and torsion. Khan, A.S.; Tokuda, M. (eds.) Advances in plasticity 1989. Oxford, New York: Pergamon Press 1989. pp. 357-360.

2. Ohno, N.; Mizuno, T.; Kawaji, H.; Okada, I.: Multiaxial creep of a nickel-base directionally solidified alloy (1st report, Experiments and analysis on minimum creep rate). Trans. JSME. Ser. A. (in print).

3. Leverant, G.R.; Kear, B.H.: The mechanism of creep in gamma prime precipitation-hardened nickel-base alloys at intermediate temperatures. Metal. Trans. 1 (1970) 491-498.

4. Paslay, P.R.; Wells, C.H.; Leverant, G.R.: An analysis of primary creep of nickel-base superalloy single crystals. ASME J. Appl. Mech. 37 (1970) 759-764.

5. Paslay, P.R.; Wells, C.H.; Leverant, G.R.; Burck, L.H.: Creep of single crystal nickel-base superalloy tubes under biaxial tension. ASME J. Appl. Mech. 38 (1971) 623-626.

6. MacKay, R.A.; Maier R.D.: The influence of orientation on the stress rupture properties of nickel-base superalloy single crystals. Metal. Trans. A. 13 (1982) 1747-1754.

7. Hopgood, A.A.; Martin, J.W.: The creep behaviour of a nickel-based single-crystal superalloy. Mat. Sci. Eng. 82 (1986) 27-36.

8. Sheh, M.Y.; Stouffer, D.C.: Anisotropic constitutive model for nickel-base single crystal superalloys. NASA report CR-182157 (1989). NASA.

9. Pope, D.P.; Ezz, S.S.: Mechanical properties of Ni_3Al and nickel-base alloys with high volume fraction of γ'. Int. Met. Rev. 29 (1984) 136-167.

10. Walker, K.P.; Jordan, E.H.: Biaxial constitutive modelling and testing of a single crystal superalloy at elevated temperatures. Brown, M.W.; Miller, K.J. (eds.) Biaxial and multiaxial fatigue. London: Mechanical Engineering Publications 1989. pp. 145-170.

11. Asaro, R.J.: Micromechanics of crystals and polycrystals. Adv. Appl. Mech. 23 (1983) 1-115.

Nonlinear Constitutive Equations for Creep of Structural Materials

K. ONARAN

Civil Engineering Faculty, İstanbul Technical University
Maslak, İstanbul, Turkey

Summary

Nonlinear constitutive equations for time dependent anisotropic materials and related kernel functions were described. Some assumptions based on structural symmetry and material restrictions suggested to simplify involved constitutive equations were overviewed. In order to verify theories thin walled tubular specimens of wood and fiber glass reinforced polyester were subjected to various combinations of tension, torsion and internal pressure. The time dependence of creep strains and the forms of kernel functions were discussed.

Introduction

It appears that various theories were developed for creep behaviour of structural materials under combined variable stresses and extensive investigations were made to verify their validity [1]-[3]. The creep phenomenon is affected by magnitudes and sequences of stress or strain in all of the past history of materials. The history dependence of nonlinear materials with memory can be expressed by a series of multiple integrals as follows [1],[3].

$$\epsilon_{ij}(t) = \sum_{k=1}^{n} \int_0^t \int_0^t \cdots \int_0^t K_{ijp_1q_1p_2q_2\cdots p_kq_k}(t-\tau_1, t-\tau_2, \ldots t-\tau_k)$$

$$\sigma_{p_1q_1}(t-\tau_1)\cdot\sigma_{p_2q_2}(t-\tau_2)\ldots\sigma_{p_kq_k}(t-\tau_k)\,d\tau_1\,d\tau_2\ldots d\tau_k \qquad (1)$$

where ϵ_{ij} and σ_{pq} are strain and stress tensor components respectively, t is the time, τ is any time between 0 and t. $K_{ijp_1q_1p_2q_2\cdots}$ is the kernel functions containing stress

M. Zyczkowski (Ed.)
Creep in Structures
IUTAM Symposium Cracow/Poland 1990

invariants, time and material constants. Eq.(1) called Cauchy type formulation may contain very large number of kernel functions for anisotropic case depending on degree of nonlinearity. However some simple and specific forms of this equation were derived by employing some assumptions based on structural symmetry and material restrictions such as orthotropy, isotropy, linear compressibility and incompressibility [4]-[6]. In this work prior developments in this field will be overviewed. The explicit form of eq.(1) for orthotropic materials will be given and the experiments performed to determine the related kernel functions will be explained.

Nonlinear Orthotropic Representation

The explicit form of eq.(1) for nonlinear orthotropic material can be obtained by employing the following orthogonal orthotropic transformation requirements: a) the strain tensor is symmetrical, $\epsilon_{ij}=\epsilon_{ji}$, b) the stress tensor is symmetrical, $\sigma_{pq}=\sigma_{qp}$, c) the integration is not affected by sequence in time arguments. The axial strain component derived from eq.(1) for third order nonlinearity in plane state case under single step loading is as follows,

$$\begin{aligned}\epsilon_{11}(t)=&M_{11}\sigma_{11}+M_{12}\sigma_{22}+M_{111}\sigma_{11}^2+M_{122}\sigma_{22}^2+2M_{112}\sigma_{11}\sigma_{22}+M_{133}\sigma_{12}^2\\&+M_{1111}\sigma_{11}^3+M_{1222}\sigma_{22}^3+3M_{1112}\sigma_{11}^2\sigma_{22}+3M_{1122}\sigma_{11}\sigma_{22}^2\\&+3M_{1133}\sigma_{11}\sigma_{12}^2+3M_{1233}\sigma_{22}\sigma_{12}^2\end{aligned} \tag{2}$$

The kernel functions in eq.(1) are replaced by the following time and material functions,

$$M_{klns}=\int\int\int K_{ijp_1q_1p_2q_2p_3q_3}(t-\tau_1,t-\tau_2,t-\tau_3)d\tau_1 d\tau_2 d\tau_3 \tag{3}$$

where k replaces ij, k=1 for ϵ_{11}, k=2 for ϵ_{22}, k=3 for ϵ_{12}. l,n,s (=1,2,3) replace $p_k q_k$; 1 for the coefficient of σ_{11}, 2 for σ_{22}, 3 for σ_{12}. For example, M_{1233} describes the contribution of $\sigma_{22}\sigma_{12}\sigma_{12}$ to ϵ_{11} . $\epsilon_{22}(t)$ and $\epsilon_{12}(t)$ can be obtained in similar way. For two-dimonsional case Cauchy type formulation contains 31 kernel functions as shown in Table 1.

Table 1 The number of kernel functions for third order plane state of stress

Type of material	1st order	2nd order	3rd order	Total
Orthotropic Cauchy	5	10	16	31
Orthotropic Green	4	6	9	19
Isotropic	2	4	6	12
Isotropic Incompressible	2	1	1	3
Isotropic linearly compressible	2	1	1	4

Further simplification may be provided by assuming that the material has a strain energy function in terms of stresses. The strain may be obtained this function and sequence of the stress does not effect the result, thus $K_{ijp_1q_1\cdots} = K_{p_1q_1ij\cdots}$. By employing this property, Green type formulation can be obtained. In this case 19 kernels are needed instead 31 as shown in Table 1. The validity of this assumption will be experimentally verified.

Nonlinear Isotropic Representation

Eq.(1) will have the following form for third order nonlinear isotropic material under single step loading [4].

$$\underline{\epsilon}(t)=I(K_1\bar{\underline{\sigma}}+K_2\bar{\underline{\sigma}}\bar{\underline{\sigma}}+K_3\overline{\underline{\sigma}\underline{\sigma}}+K_4\bar{\underline{\sigma}}\bar{\underline{\sigma}}\bar{\underline{\sigma}}+K_5\bar{\underline{\sigma}}\overline{\underline{\sigma}\underline{\sigma}}+K_6\overline{\underline{\sigma}\underline{\sigma}\underline{\sigma}})+(K_7+K_8\bar{\underline{\sigma}}+K_9\bar{\underline{\sigma}}\bar{\underline{\sigma}}+K_{10}\overline{\underline{\sigma}\underline{\sigma}})\underline{\sigma}+(K_{11}+K_{12}\bar{\underline{\sigma}})\underline{\sigma}\underline{\sigma} \quad (4)$$

Where $\underline{\epsilon}$ and $\underline{\sigma}$ are strain and stress tensor components respectively, I is unit tensor, $\bar{\underline{\sigma}}$ is the first stress invariant, $\overline{\underline{\sigma}\underline{\sigma}}$ is the second. K_i is a kernel function representing time and material dependence of strains. The number of the kernel functions are reduced from 12 to 9 for two dimensional third order nonlinearity. Further simplifications for eq.(1)

may be obtained by using some other material restrictions. By employing linearly compressibility assumption the number of kernel functions is reduced to 4, and by compressibility assumption to 3 as shown in Table 1 [6].

Time Dependence of Creep Strains

In general the shapes of creep curves under constant stress for all materials are similar. Various emprical equations such as exponential, logarithmic and power functions of time besides generalized Kelvin model are suggested to fit creep curves. They seem quite satisfactory for certain materials for rather limited time. Their predictibility for long time extrapolation may be questionable. Findley [7] observed after 26 years investigation that power function of time based on 2000 h data could predict creep strains for 260 000 h of some plastics with sufficient accuracy. The power function of time mt^n represents a particular retardation time spectrum function $\phi(\tau)$ similar to that of generalized Kelvin model as shown in the following [1].

$$\phi(\tau)=nm\,\tau^{n-1}/\Gamma(1-n) \tag{4}$$

Where m and n are material constants, $\Gamma(1-n)$ is gamma function. $n<1$ for most of the materials. This relation may be a reason for power function of time being succesfully used for large variety of materials [1] [7].

Form of Kernel Functions

A kernel function represents the contribution of certain stress term to creep strain in a creep equation. A second order creep equation for a two-step loading may be as follows,

$$\epsilon(t)=K_1(t)\sigma_1+K_2(t_1 t)\sigma_1^2+K_1(t-t_1)\Delta\sigma_2+K_2(t-t_1,t-t_1)\Delta\sigma_2^2 \\ +2K_2(t,t-t_1)\sigma_1\sigma_2 \tag{5}$$

Where σ_1 is the initial stress, $\Delta\sigma_2$ is the increase in stress at time t_1, $K_1(t)$ and $K_2(t,t)$ can be determined by a series of single step loadings. However $K_2(t,t-t_1)$ representing synergistic effects of mixed stress parameter of σ_1 σ_2 requires two-step loadings. It was observed that the time dependence of the kernel functions is similar to creep strains. Therefore they may be represented by a power function of time. For a constant t_1, $K_2(t_1,t-t_1)$ may have the following form [8],

$$K_2(t,t-t_1)=\alpha(t_1)+\beta(t_1)(t-t_1)^n \tag{6}$$

Where α and β depends on the first loading interval t_1 and material properties. A third order kernel function requires two step loading with a series of t_1 and t_2 time internals. This method used to determine the full form of kernel functions is called direct method [8]. Since it needs prothibitively large number of experiments, its practical value is limited. Several approximate methods requiring only single step loadings, such as product form of time arguments and modified superposition method [1] [9] were suggested. They seemed to represent nonlinear creep behaviour quite well. However they can not describe all of the features of creep describle by the direct multiple integral method.

Experimental Determination of Kernel Function of Orthotropic Materials

Thin walled tubular specimens of poplar wood and fiberglass reinforced polyester (FRP) were subjected to various combinations of tension, torsion and internal pressure by means of a speciaaly designed creep machine [1] [5]. The wall thickness of poplar wood was kept in an annual ring to provide circular symmetry. The plastic specimen was prepared by embedding glass fiber along two axial direction into the cylindirical polyester matrix resin. The stresses were kept below 60% of the ultimate strengths. The strains were measured by SR-4 strain gage rosets.

The creep test results showed that the time independent strains ϵ_e are considerably nonlinear for both materials while time dependent strains ϵ_c are significantly nonlinear as shown in Fig. 1.

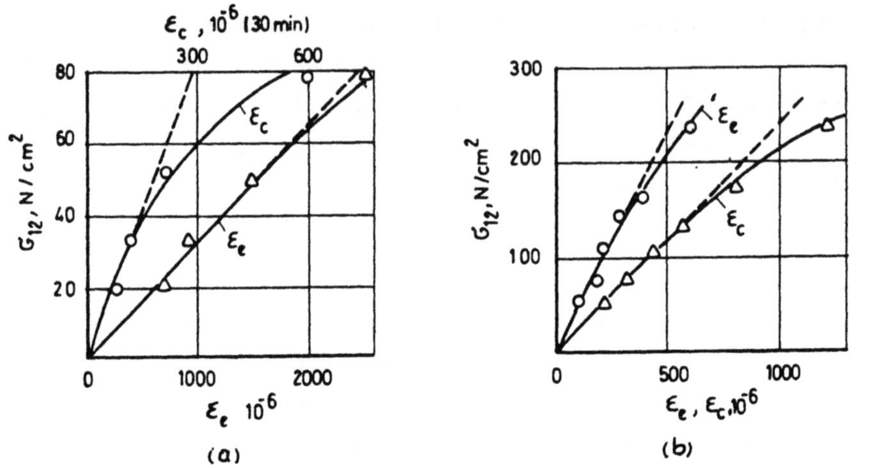

Fig.1. Isochronus shear stress-strain curves, a)poplar wood, b)FRP plastic

The strains were represented by power function of time, $\epsilon=\epsilon_e+mt^n$. ϵ_e and m depend on stress and material properties. n was independent of stress but depends on the main direction of material. Log-log plotting of creep strain-time are parallel straight lines for both material as shown Fig.2.

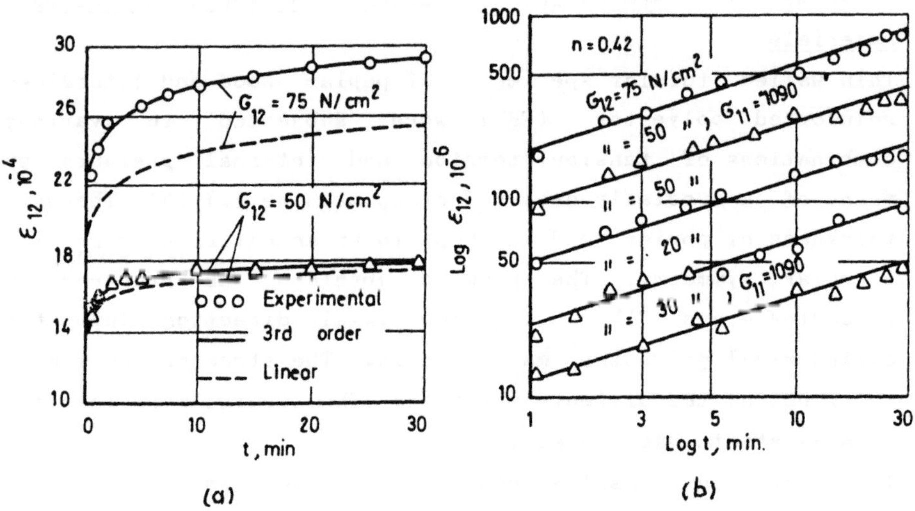

Fig.2 Shear creep strains of poplar wood, a)ϵ_{12}-t curves, logϵ_{12}-logt curves.

The creep strains were negligible in axial direction parallel to fibers, while significant creep occured in lateral direction and shear. Both Cauchy and Green type formulation were used to represent the creep strain. The related kernel functions were determined. The experimental values of first and secon order kernel functions of wood are given in Table 2. The kernel function were employed to predict the creep strains for the other loading programs.

Table 2 Kernel Functions of Wood, $M_i = M_e + M_c t^n$

Kernel	$M_e, 10^{-5}$	$M_c, 10^{-7}$	n
M_{11}	14.13	0.56	0.225
M_{12}	-23.32	-0.5	"
M_{22}	153.7	4.08	0.28
M_{21}	8.33	-0.24	"
M_{33}	290	7.30	0.42

$[M_{ij}]: 10^{-5}$ mm/mm/N/cm^2

$[M_{ijk}]: 10^{-7}$ mm/mm(N/cm^2)^2

Kernel	$M_e, 10^{-5}$	$M_c, 10^{-7}$	n
M_{111}	7.0	-0.25	0.225
M_{112}	12.48	-14.8	"
M_{122}	-347	15.1	"
M_{211}	4.15	0.11	0.280
M_{212}	108	-6.73	"
M_{222}	-83.8	672	"
M_{313}	-270	-60	0.420
M_{323}	-2245	-613	0.420

Conclusions

By taking into consideration the prior investigations and the observations on nonlinear viscoelastic representations described in this report, the following conclusions concerninşg the validity of these theories may be derived.

The creep test results obtained on the tubular specimens of poplar wood and FGP plastic were used to determine the kernel functions of those nonlinear viscoelastic orthotropic materials. It was observed that the creep strains along axial direction parallel to fibers are negligible while they are significant in laterel direction and in shear. The power function of time was employed to represent the creep strains. The power of time n has the highest value in shear, but lowest value in the axial direction. It is independent of stress in the same direction. Cauchy type formulation represents the nonlinear dependence of the orthotropic materials satisfactorily. On the other hand Green type representation yielded rather poor predictions with average discrepencies higher than 10%. Linearly compressible and incompressible assumptions provide considerable simplifications for nonlinear constitutive equation for combined loadings. The accuracy of the predictions for creep strain depend on the type of materials.

References

1. Findley, W.N.; Lai J.S,; Onaran K.; Creep and relaxtion of nonlinear viscoelastic materials. Dover Publ., (1989).

2. Pipkin, A.C.; Lectures on viscoelastic theory.Springer Verlag (1972).

3. Green, A.E.; Rivlin, R.S.; Spencer, J.M.; Mechanics of nonlinear materials with memory, part II. Arch. Rat. Mech. Anal. 3 (1959) 82-90.

4. Onaran, K.; Findley, W.N.; Combined stress creep experiments on a nonlinear viscoelastic material to determine kernel functions for a multiple integral representation of creep, Trans. of Soc. of Rheology 9/2 (1965) 299-327.

5. Erman, B.; Viscoelastic behaviour of nonlinear orthotropic materials. Ph.D. thesis, İstanbul Technical University (1974).

6. Findley, W.N.; Onaran K.; Incompressible and linearly compressible viscoelastic creep and relaxation. J. of Appl. Mechanics 41/4 (1974) 234-242.

7. Findley, W.N. ; 26 year creep and recovery of PVC and PE. Polymer Science and Engineering, 27/8 (1987) 582-585.

8. Onaran, K.; Findley, W.N.; Experimental determination of some kernel functions on multiple integral method for nonlinear creep of PVC. J. of Appl. Mechanics Series E, 38/1 (1971) 30-38.

9. Findley, W.N.; Lai, J.S.; A modified superposition principle applied to creep of nonlinear viscoelastic materials under abrupt changes in state of combined stress. Trans. Society of Rheology, 11/3 (1967) 361-376.

Intergranular and Transgranular Hardening in Viscoplasticity

P.Pilvin* , G.Cailletaud**
*Université Pierre et Marie Curie, and Centre des Matériaux de l'Ecole des Mines de Paris
**Centre des Matériaux de l'Ecole des Mines de Paris, UA CNRS 866

Summary.
This paper shows the capabilities of the micromechanical approach concerning the description of initial and subsequent yield surfaces of metals. This type of simulation produces significant information concerning the microstructural level giving rise to the hardening : it is shown that kinematic type hardening may come from intergranular or transgranular level, but that the macroscopic properties are different in the two cases. Isotropic hardening has to be modelled on a microscale, by slip system interactions.

Introduction.

The hardening of the materials is a complex phenomenon with multiple physical sources. In the metals, the microstructural levels to consider are grains and slip systems. The hardening obtained at a macroscale is the result of the combination of the effect of stress and strain heterogeneities between the grains (giving rise to intergranular hardening), and of the dislocations-precipitates interactions, dislocations-dislocations interactions,... (i.e. transgranular hardening). In this paper, we use a "micro-macro" approach in order to precisely characterize the mechanical effect of each of these components. The general framework of polycrystalline plasticity is not new, specially the initial basement (Sachs, Taylor, Lin, Kröner), but the computers were not sufficiently developped for the first investigators [1-5]. Intermediate tentative has been made ten years ago (see for instance [6,7]), but thc current use of these type of theory to represent complex loading paths is more recent [8-10]. The model used is a viscoplastic one. It takes into account a physical description of the microstructure, and explicitely defines local stresses, local viscoplastic strains in the grains, resolved shear stresses and viscoplastic shear strains on the slip systems. The constitutive equations are given on a microscale, they define the viscoplastic shear strain rate as a function of the actual state of the material (microscopic internal variables), and of the applied stresses. A more particular version of this model has been previously studied [11,12]. The formulation which is given here produces a better representation of intergranular hardening. In a same time, the grain distribution is made with a larger number of grains, so that the numerical results are no longer disturbed by the initial texture. A full account of the methodology used for grain choice and of the numerical simulations can be found in [13].

M. Zyczkowski (Ed.)
Creep in Structures
IUTAM Symposium Cracow/Poland 1990

Description of the model.

The model is based on the experimental fact that kinematic hardening (ie translation of the elastic domain in stress space) is related with internal stress development, so with heterogeneities. Two heterogeneity levels are retained for metal : grain and system. Due to strain incompatibilities from one grain to the other, or to stress redistribution inside of the grains (precipitates, cell structures,...), the local stress tensor in each grain and the effective resolved stress on each system should not be directly computed from the macroscopic stress tensor. Two sources are then registered for kinematic hardening. On the other hand, isotropic hardening is related to low range interactions; it is then simulated only at the system level. The equations coming from this very simple approach are summarized in Table 1. One can find :

- the expression of the local stress in each grain, σ^g, and of the shear stress on each system, τ^s,
- the expression of the hardening variables for the grain g, X^g, and for the system s, x^s and r^s,
- the evolution laws of the state variables β^g, α^s and v^s, corresponding respectively to intergranular kinematic hardening, transgranular kinematic hardening, and transgranular isotropic hardening. These constitutive equations are an adaptation of non linear kinematic and isotropic hardening for macroscopic approaches [14],
- the flow rules for local shear strain γ^s and for the macroscopic viscoplastic strain E_v.

In these expressions, Σ is the macroscopic stress tensor, the tensorial variables σ^g, X^g, β^g, are defined for each grain, the scalar variables τ^s, x^s, r^s, α^s, v^s, are defined for each slip system. Each grain is assumed to present the same elasticity as the aggregate, so that elastic behaviour is treated at the macroscopic level, using the classical relations for isotropic materials (Young's modulus, Poisson's ratio). In such a formulation the model can be applied to FCC metals.

$$\sigma^g = \Sigma - X^g \;\; ; \;\; \tau^s = \sigma^g : m^s \;\; ; \;\; X^g = \frac{2}{3} C \,(\beta^g - \sum_{h\in G} f^h \beta^h)$$

$$\dot{v}^s = \left\langle \frac{|\tau^s - x^s| - r^s}{k} \right\rangle^n \;\; ; \;\; \dot{\gamma}^s = \dot{v}^s \, \mathrm{sign}(\tau^s - x^s) \;\; ; \;\; \dot{\alpha}^s = \dot{\gamma}^s - d\, \alpha^s \dot{v}^s$$

$$x^s = c\, \alpha^s \;\; ; \;\; r^s = r_0 + Q \sum_{j\in g} h_{sj} \,[1 - \exp(-b\, v^j)]$$

$$\dot{\varepsilon}^g_v = \sum_{s\in g} m^s \dot{\gamma}^s \;\; ; \;\; \dot{\beta}^g = \dot{\varepsilon}^g_v - D\, \beta^g \sum_{s\in g} \dot{v}^s \;\; ; \;\; \dot{E}_v = \sum_{h\in G} f^h \dot{\varepsilon}^h_v$$

C, D , c, d	: coefficients for inter-, transgranular kinematic hardening
r_0, Q, b	: coefficients for transgranular isotropic hardening
h_{sj}	: components of the interaction matrix ($h_{sj} = 1$ in this text)
m^s	: orientation tensor of system s
f^h	: volume fraction of grain h

Table 1. Summary of the equations of the polycrystalline model.

Methodology of the numerical approach.

We first have to choose a microstructure, i.e. the number of grains, their orientation in the aggregate, the geometry of the slip planes and of the slip systems. As far as FCC materials are concerned, only octahedral slip is allowed. The set of grains is designed to be initially isotropic, and to present no coupling between tension and shear. For this purpose, primary grains are uniformly distributed on the standard triangle, each of them corresponding to 36 grains in the final distributions, after the symmetry operations. A good "quality/price" ratio is obtained with a 504 grain set, deduced from [15]. The model is implemented in the code SiDoLo [16], so that simple simulations or automatic identifications from experimental data can be performed.

Various strategies are available to get a numerical plot of the yield surfaces from a viscoplastic model. In [11], it was decided to show the viscoplastic equipotential, as being a *true* picture of the numerical response. This choice may overlook mechanical effects present in experiments. It is the reason why in this paper, the proposed yield surfaces are numerically obtained as experimental ones, by applying a star type loading in stress space, after unloading if preloading was present. The yield surface is assumed to be reached when accumulated equivalent strain reaches a given offset in the current branch of the loading. Figure 1a shows the influence of the chosen yield criterium on the initial yield surface in biaxial tension σ_{11} , σ_{22}. For a low offset value (case 1, $10^{-8}s^{-1}$), the yield criterium obeys a Tresca rule : it means that a favourably oriented slip can be found for any direction of the sollicitation. For higher value with the same criterium (case 3, $10^{-4}s^{-1}$), the simulated surface no longer corresponds to pure yield, and the curve obtained tends to a von Mises criterium. The chosen method, corresponding to a 10^{-5} offset, produces intermediate results (case 2). The same type of result can be seen in figure 1b, in σ_{11}-σ_{12} axis, the ratio σ_{11}/σ_{12} being 1.98 for case 1, 1.92 for case 2, and 1.76 for case 3.

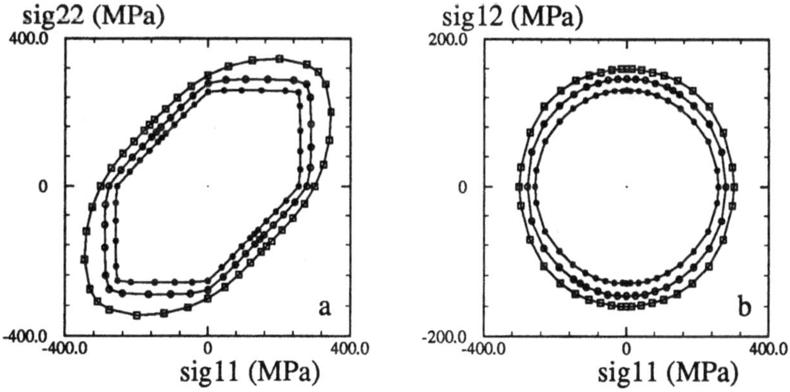

Fig.1. Influence of the numerical method on the shape of the yield surface.
a) in σ_{11}-σ_{22} plane, b) in σ_{11}-σ_{12} plane.

Simulation results.

The analysis of the subsequent yield surfaces is a good method to isolate the mechanical effect of each hardening type. So the hardening source in a material will be characterized directly from its macroscopical response. In all the simulations, the value of the coefficients are 100 MPa for r_0, 25 for n, 50 MPa $s^{1/n}$ for k. Typical responses are first shown in fig.2 : for fig.2a, only linear intergranular hardening is introduced (C=1.5μ=112500 MPa), for fig.2b, non linear intergranular hardening (C=1.5μ and D=500), then for 2c and 2d, isotropic (Q=50 MPa and b=100) and kinematic transgranular hardening (c=2500 MPa and d=50). The nonspecified coefficients remains equal to zero. The initial yield surface is the same in the four cases, and the subsequent yield surfaces are shown at maximum and minimum strain, for a cycle between +1% and -1%. For intergranular hardening, the yield surfaces present two edges, at the loading point and on the opposite side; it shrinks on σ_{12} axis (cross softening).

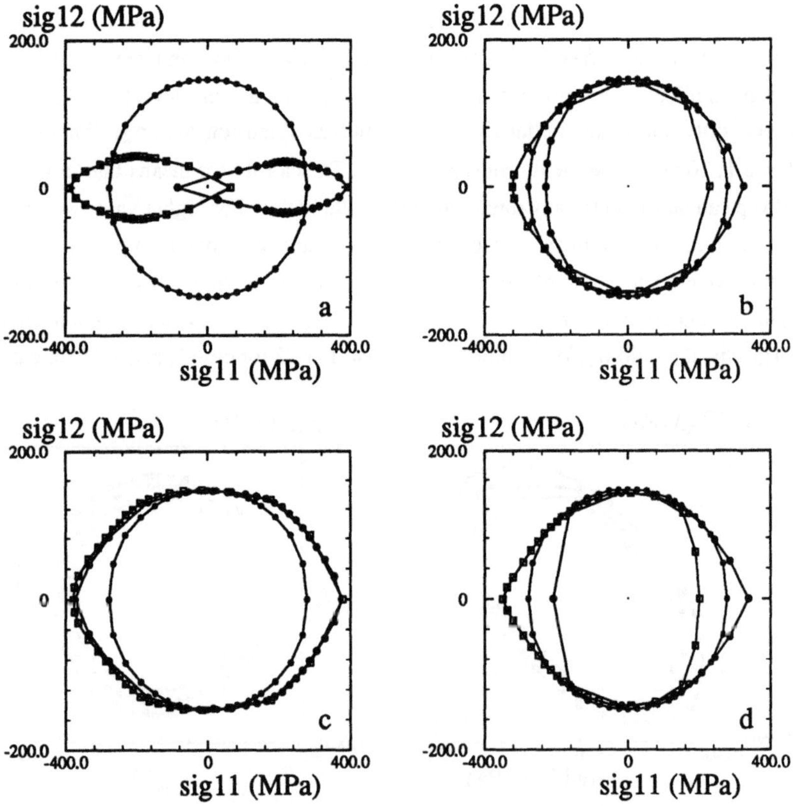

Fig.2. Subsequent yield surfaces for a tension-compression loading with each elementary hardening relation : a) linear intergranular hardening, b) non linear intergranular hardening, c) isotropic transgranular hardening, d) kinematic transgranular hardening.

This effect is more important for linear hardening than for non linear one, because the hardening modulus does decrease in this last case, so that translation and distorsion are not very high. If only pure transgranular hardening is introduced, neither cross softening nor cross hardening can be seen : the surface distorsion is centered on the origin for isotropic hardening, and is superimposed with a translation for kinematic hardening; in this last case, the side opposite to the loading point is flat, as in a number of experiments (see for instance [17,18]). A classical behaviour including all the ingredients of the model is given in fig.3 : the subsequent yield surfaces are expanded, translated and distorted versus the original one, and an edge appears at the loading point. In this case, the simultaneous presence of intergranular and isotropic transgranular hardening produces cross hardening.

The present data confirm and generalize previous results obtained with older models, for instance [19], which presented cross softening, or [20], for which the corner near the loading point is well marked. They also confirm the results obtained in [11], with an improved precision : the initial surfaces are quite smooth, and the effect of isotropic hardening no longer present an equal expansion on tension and shear axis (fig.2.c). This is due to the fact that new systems can now be activated in shear after preloading in tension; it was not previously the case, because the number of independent slip systems was too small. The individual effect of each family of coefficients being established, we will now continue the model illustration by means of microstructural data analysis.

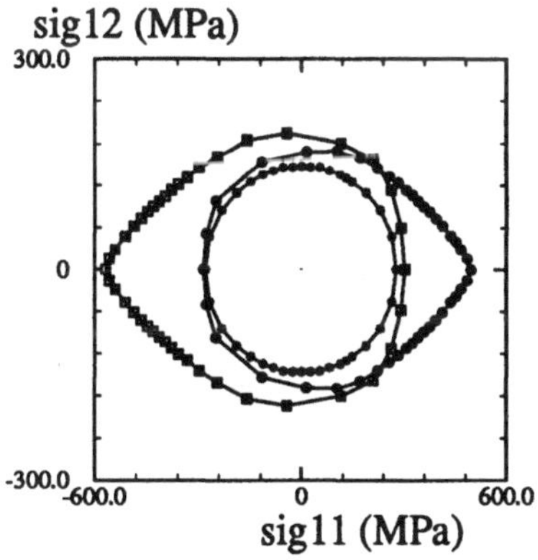

Fig.3. Subsequent yield surfaces under tension-compression loading for a full coefficient set.

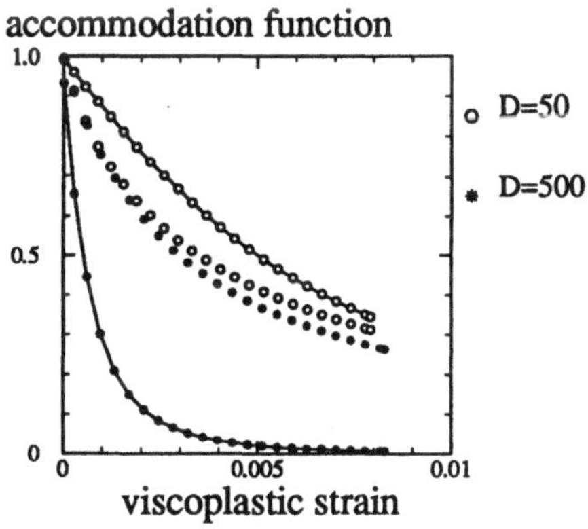

Fig.4. Evolution of the accommodation function, from [21] (symbols) and from the present model for nonlinear intergranular hardening (symbols and solid line).

Discussion.

The first point to clarify corresponds to the localization process in each grain : in the model of this paper, the accommodation can be either from elastic type (if D=0), or from plastic type, if intergranular kinematic hardening is non linear. In this last case, the approach can then be compared with other models from polycrystalline plasticity. The reference model is a simplified self-consistent scheme [21]. The accommodation for the grain g is characterized in simple tension by the ratio $A^g = \|\sigma^g - \Sigma\| / \|E_p - \varepsilon_p^g\|$; this ratio, equal to 2μ for elastic accommodation, and to $2\mu(1-\beta)$ for Kröner's model [22] (with $\beta = 2(4-5\nu)/(15(1-\nu))$), decreases in plasticity : it is written $2\mu\,(1-\beta)\,\alpha$, where $1/\alpha \cong 1 + (3\mu\,E_p / 2\,\Sigma)$. This expression of α is an integrated form, valid for simple tension. Figure 4 shows the comparison of this α function with $A^* = 1.5/C \sum_{h \in G} f^h A^g$, evaluated from the macroscopic responses, taking $C=1.5\mu$. It is made for two coefficients sets with a non zero D, D=500 taken from fig. 2b and D=50. The accommodation function evaluated from the simulated macroscopic response using E_v and Σ is in reasonable agreement with the direct computation of A^* for the second case. Let us note on the other hand that the macroscopic response, which is not shown here for the sake of brevity, is strongly dependent of the value of D : for the same loading between -1% and +1%, the stress range is 827 MPa for 2a coefficient set, and 673 MPa for 2b.

We now give a more physical view of the model behaviour through the investigation of the way the slip systems operate for each basic hardening type. This determination is made for the simulations shown in fig.2, during the first tension, the first compression and the second tension. Figures 5a, 5b, 5c and 5d respectively correspond to fig.2a, 2b, 2c, 2d. Four system categories are distinguished : (1) active slip systems, for which the instantaneous shear strain rate is non zero, (2) significantly active slip systems, for which the instantaneous strain rate is more than 0.1 times the maximum instantaneous strain rate, (3) plastified slip systems, for which the accumulated shear strain from the beginning of the simulation is non zero (they are, or have been active), (4) significantly plastified slip systems, for which the accumulated shear strain from the beginning of the simulation is more than 0.1 times the maximum accumulated shear strain. Note that the number of systems in category 3 cannot decrease. That is not the case for the other categories, due to the possible desactivations (categories 1 and 2), or to the localisation of the deformation on preferential slip systems (category 4). For linear intergranular hardening (fig.5a), the number of slip systems involved in the deformation process is very high : as 504 grains with 12 systems each are simulated, around 7.7 systems are plastified per grain for 5a. This fact should be tempered by the number of significantly plastified or significantly active systems (around 4 per grain for 5a). For non linear intergranular hardening, the number of active systems is much lower (max 3.7 per grain). A desactivation process can be seen during viscoplastic flow. As far as transgranular hardening is concerned, the number of slip systems is lower : this is partially due to the smaller level of the macroscopic stress in these simulations, but it can be explained by the fact that the material is locally harder than for the first two cases. The low amount of slip is specially obvious for isotropic hardening (fig 5c), where the number of significantly active systems is around 0.1 per grain at the end of the simulation, after a very quick activation-desactivation period during the first loading. Transgranular kinematic hardening produces around 4 plastified and 1.6 significantly plastified systems per grain, and exhibits an activation of new systems during the first compression branch (fig.5d).

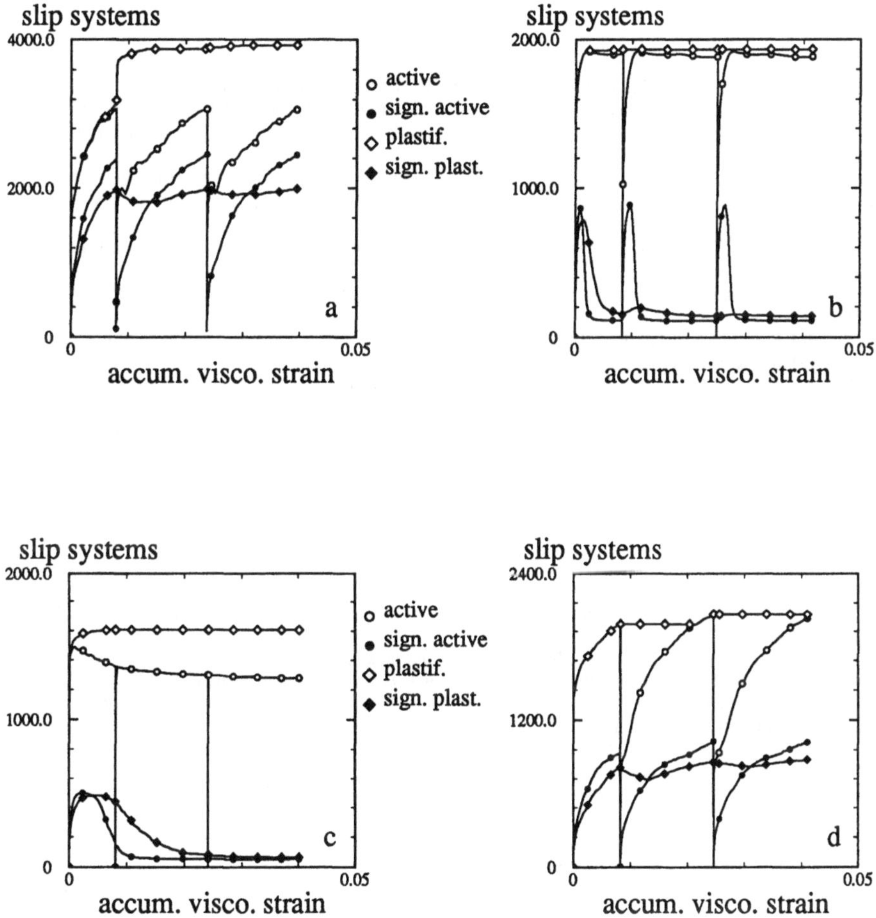

Fig.5. Count of (1) active, (2) significantly active, (3) plastified, (4) significantly plastified slip systems in a cyclic simulation : a) test 2a, b) test 2b, c) test 2c, d) test 2d.

Conclusion.

In its present formulation, the model proposed can be used as a tool for a better understanding of hardening in plasticity and viscoplasticity. It has previously been used to simulate in the same time the macroscopic response and the microscopic one (count of slip traces [23], or analysis of dislocation arrangement [24]). This study suggests also to consider the yield surfaces as a very discriminant experimental information to characterize material behaviour, as long as its shape strongly depends on the microstructural level where hardening develops. The data obtained from mechanical tests can then give information on the microstructural behaviour. For instance, the amount of grain to grain interaction can be indirectly "measured" by the fact that the surface shrinks or not.

References.

1. KOITER, W.T., "Progress in Solids Mechanics", North Holland, Amsterdam, Vol.1 (1960).
2. HILL, R., "Continuum Micro-Mechanisms of Elastoplastic Polycrystals", J.Mech.Pys.Solids 13 (1965), 89.
3. MANDEL,J., "Une généralisation de la théorie de plasticité de W.T.Koiter", Int.J.Solids Structures 1 (1965), 273.
4. BUI, H.D., "Etude de l'évolution de la frontière du domaine élastique avec l'écrouissage et relation de comportement élastoplastique des matériaux cubiques", Thesis, Univ. Paris VI, Mémorial des Sciences et Techniques de l'Armement 1 (1969), 141.
5. HUTCHINSON, J.W., "Elastic-Plastic Behaviour of Polycrystalline Metals and Composites", Proc.Roy.Soc. A319 (1970), 247.
6. WENG, G.J., "Constitutive Equations of Single Crystals and Polycrystals under cyclic loading", J.Appl.Mech. 48 (1981), 41.
7. DONADILLE, C., PENELLE, R., ENGEL, J.J., "Constitutive Equations and Structural Evolution of a 316L Stainless Steel during monotonic and cyclic loadings", ASME PVP Conf., Orlando (1982).
8. TOKUDA, M., KRATOCHVIL, J., OHNO, N., "Inelastic behaviour of Polycrystalline Metals under complex Loading Condition", Int.J.Plasticity 1 (1986), 141.
9. BERADAI, C., BERVEILLER, M., LIPINSKI, P., "Plasticity of Metallic Polycrystals under Complex Loading Paths", Int.J.Plasticity 3 (1987), 143.
10. CAILLETAUD, G., "Une approche micromécanique du comportement des polycristaux", Rev.Phys. Appl. 23 (1988), 353.
11. CAILLETAUD, G., "Numerical Simulation of Yield Surfaces by a Phenomenological Micromechanical Approach", Proc. Mecamat on "Inelastic Behaviour of Solids", Besançon, France (1988), Aug.30-Sept.1,
12. CAILLETAUD, G., "A micromechanical Approach to Inelastic Behaviour of Metals", to be published in Int.J.Plasticity.
13. PILVIN, P.,"Approches multi-échelles pour la prévision du comportement anélastique des métaux", Thesis, Univ. Paris VI (1990).
14. LEMAITRE, J., CHABOCHE, J.L., "Mécanique des matériaux solides" Dunod (1985).
15. BACROIX, B., BRUN, O., CHAUVEAU, Th., Proc. ICOTOM 9, Avignon, Sept.17-21 (1990),
16. PILVIN, P." Identification des paramètres de modèles de comportement", Proc. Mecamat on "Inelastic Behaviour of Solids", Besançon, France (1988), Aug.30-Sept.1,
17. IKEGAMI,K., "Experimental Plasticity on the Anisotropy of Metals", Int.Coll.CNRS "Comportement mécanique des solides anisotropes", Villars de Lans, France, Euromech 115 (1977).
18. PHILLIPS, A., "A Review of Quasistatic Experimental Plasticity and Viscoplasticity", Int.J.Plasticity 2 (1986), 315.
19. LIN,T.H., ITO, M., "Theoritical Plastic Strain-Stress Relationship of a Polycrystal and the Comparison with the von Mises and the Tresca Plasticity Theories", Int.J.Eng.Sci. 4 (1966), 511.
20. WENG, G.J., PHILLIPS, A., "An Investigation of Yield Surfaces Based on Dislocation Mechanics", Part 1 and 2, Int.J.Eng.Sci. 15 (1977), 45.
21. BERVEILLER, M., ZAOUI, A., "An Extension of the Self-Consistent Scheme to Plastically Flowing Polycrystals", J.Mech.Phys.Solids 26 (1979), 325.
22. KRONER, E., "Zur plastischen Verformung des Vielskristalls", Acta Met. 9 (1961), 155
23. DOQUET, V., CAILLETAUD, G., PINEAU, A., "Prediction of macroscopic multiaxial behaviour from microstructural observations", 3rd Int.Conf.on biaxial/multiaxial fatigue, Stuttgart, April 3-6 (1989).
24. HAUTEFEUILLE, L.,Thesis, Univ. Compiègne (1990).

On the State Equations for Anisotropic Elasto-Visco-Plastic Bodies

O.O. SADAKOV, V.N. MADUDIN, M.V. APAJCHEV

Institute of Mechanics,
Chelyabinsk, USSR

Summary

The paper deals with the constitutive equations of plasticity and creep of an anisotropic material under variable-repeated and cyclic loadings. Both, initial (physical) anisotropy and deformation induced one are taken into account. The Ilyushin postulate of potential symmetry is generalized to the anisotropic time-dependent materials. The approach is used to describe the rheological properties of an anisotropic materials as well as to cyclic hardening and softening and to low-cycle fatique or long-term fracture.

The objective of the report is to derive of the constitutive equations reproducing plasticity and creep of the anisotropic material under variable-repeated and cyclic loadings. The difficulty of this problem is connected with the deformation induced anisotropy in addition to the physical one with corresponding complications. Besides in a case of anisotropic media the usual scalar invariants of stress and strain tensors are not longer valid. Some more difficulties on receiving the physical correct invariant tensorial relations arise.

The first problem is resolved on the way of separate strain anisotropy description by using a structural model [1]. This allows us to deal with a perfect viscous subelements of the model. Another difficulty is overcome by using an invariant (geometrical) tensorial representation instead of the algebraic co-ordinate approach. The model is simplified by extending of the elasticity symmetry properties of material on an inelastic behaviour, that is Ilyushin's isotropy postulate adaptation.

The ideas are illustrated by methodological convenient examples of two materials with high degree of symmetry. For clarity and convenience of the customary techniques of a linear transformation employed we will use the representation of tensors $\bar{\bar{\sigma}}$, $\bar{\bar{\varepsilon}}$ as vectors $\bar{\sigma}$, $\bar{\varepsilon}$ of a suitable linear space $\mathcal{L}_6$ (chart 1). The metrics of the latter one have energetic sense hence the vectors $\bar{\sigma}$, $\bar{\varepsilon}$ stand for generalized force and displacement.

M. Zyczkowski (Ed.)
Creep in Structures
IUTAM Symposium Cracow/Poland 1990

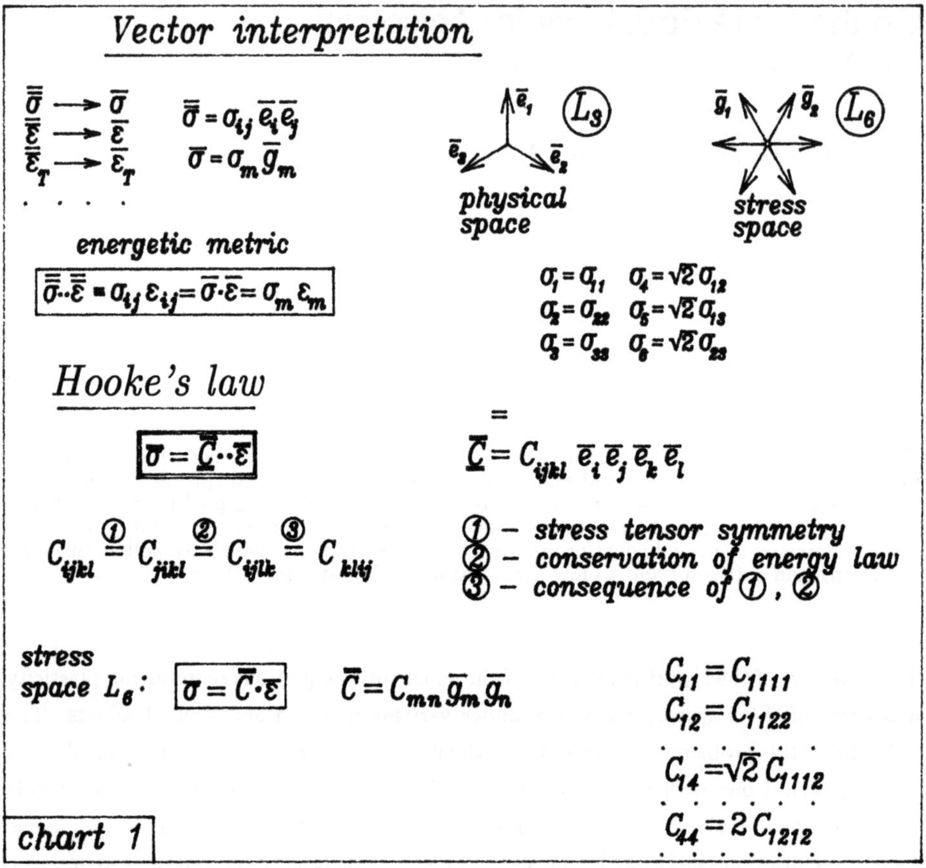

1. As in a case of the isotropic media let us assume the volume element of a material behaves as an ensemble of perfect elasto-viscous subelement with different rheologic characteristics (chart 2). It is known from isotropic media structural model practice [2] that this approach allows us to simulate automatically the strain induced anisotropy under cyclic proportional and non-proportional loading. The physical anisotropy description is reduced to the construction of a convex scalar creep potential in the space $\mathcal{L}_6$. The potentials of different subelements are setted to be similar.

Structural model

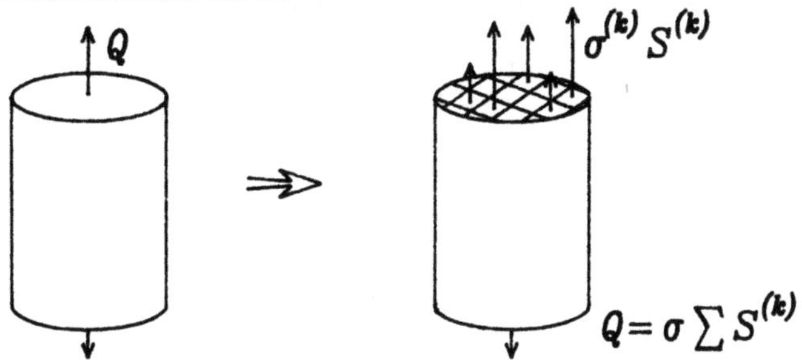

$$\bar{\varepsilon} = \bar{\varepsilon}^{(k)} \qquad \bar{\sigma} = \langle \bar{\sigma}^{(k)} \rangle = \sum \bar{\sigma}^{(k)} g_k \qquad g_k = S^{(k)} / \sum S^{(k)}$$

(k): perfect viscous

$$\bar{\varepsilon} = \bar{\varepsilon}_T + \bar{\varepsilon}_e + \bar{\varepsilon}_p$$

$$\bar{\varepsilon}_T = \bar{\alpha}(T)T - \bar{\alpha}(T_0)T_0 \approx \bar{\alpha}(T - T_0)$$

$$\bar{\varepsilon}_e = \bar{\bar{C}}^{-1}(T) \circ \bar{\sigma}$$

$$\dot{\bar{\varepsilon}}_p = \bar{\Phi}(\bar{\sigma}, T) = \nabla_{\bar{\sigma}} \Psi(\bar{\sigma}, T)$$

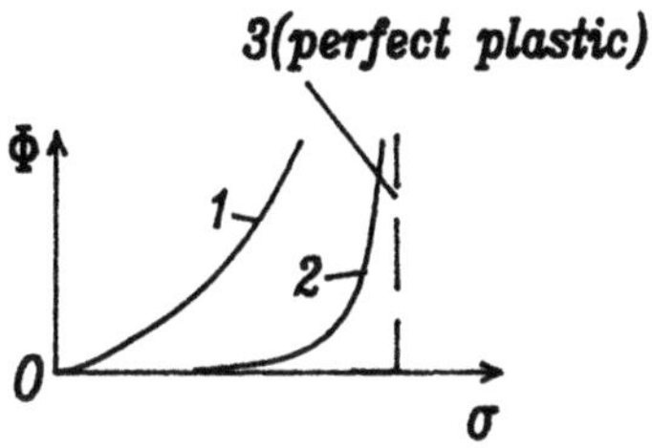

chart 2

2. In order to obtain an initial strain isotropism the creep potential must have a certain symmetry (it will be told about bellow). Proceeding from the potential convexity and taking into account the shortage of corresponding experimental data we shall adopt - with a mathematical conveniency and reducing possible error - an elliptic equipotential surfaces (chart 3).

3. The symmetry properties are defined by the material structure. They can be brought out from a Hooke's law analysis (chart 4). The elasticity tensor (in a physical space) is decomposed in an orthogonal basis of forth-rank tensors adding up to the unit (identical transformation) tensor. The basis is determined uniquely by the symmetry of material structure and by the necessary symmetry of the elasticity tensor. It decomposes tensors $\bar{\bar{\sigma}}$, $\bar{\bar{\varepsilon}}$ on orthogonal summands $\bar{\bar{\sigma}}^{(m)}$, $\bar{\bar{\varepsilon}}^{(m)}$. Then Hooke's law breaks down into a few simplest expressions with scalar elasticity coefficients.

Creep potential $\Psi(\sigma)$

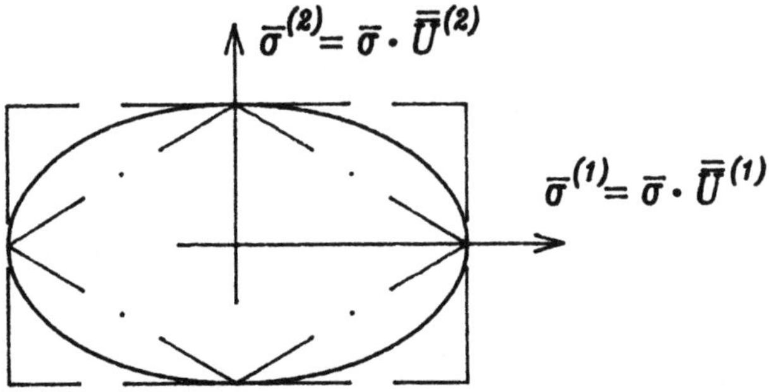

$$\Psi(\bar\sigma) \equiv \Psi(\sigma_э) \qquad \sigma_э = (\bar\sigma \cdot \bar{\bar A} \cdot \bar\sigma)^{1/2} = \sum |\bar\sigma^{(i)}| / \lambda_i^2$$

$$\bar{\bar A} = \sum \bar{\bar U}^{(i)} / \lambda_i^2 \qquad \bar\sigma^{(i)} = \bar\sigma \cdot \bar{\bar U}^{(i)}$$

chart 3

Hooke's law

A. physical spase $\boxed{\underline{\sigma} = \underline{C} \cdot\cdot\, \underline{\varepsilon}}$ $\qquad \underline{C} = \mu_m \underline{U}^{(m)}$

$$\sum_m \underline{U}^{(m)} = \underline{U} = \underset{2TS}{II}^{(s)} \quad (\underline{\sigma} \cdot\cdot\, \underline{U} = \underline{U} \cdot\cdot\, \underline{\sigma} \equiv \underline{\sigma}) \quad \underline{U}^{(m)} \cdot\cdot\, \underline{U}^{(n)} = \delta_{mn} \underline{U}^{(m)}$$

$$\underline{\sigma}^{(m)} = \mu_m \underline{\varepsilon}^{(m)} \quad \underline{\sigma}^{(m)} = \underline{\sigma} \cdot\cdot\, \underline{U}^{(m)} \quad \underline{\sigma} = \sum_m \underline{\sigma}^{(m)}$$

B. stress space $\boxed{\bar\sigma = \bar C \cdot\cdot\, \bar\varepsilon}$ $\qquad \bar C = \sum_{i=1}^{6} k_i \bar c_i \bar c_i$

$$\bar U \cdot\cdot\, \bar\sigma \equiv \bar\sigma \quad (\bar C \cdot \bar c_i = k_i \bar c_i) \quad \bar\sigma = \sum_m \bar\sigma^{(m)} \quad \bar\sigma^{(m)} = \bar\sigma \cdot\cdot\, \bar U^{(m)}$$

$$\sum_m \bar U^{(m)} = \bar U \quad \{k_i\}_6 = \{\mu_m\}_\nu \quad \bar U^{(m)} = \sum_m \bar c_i \bar c_i$$

$$\bar\sigma^{(m)} = \mu_m \bar\varepsilon^{(m)}$$

chart 1

The space $\mathcal{L}_6$ is decomposed into the corresponding vector subspaces $\mathcal{L}^{(m)}$ of terms $\bar{\sigma}^{(m)}$, $\bar{\varepsilon}^{(m)}$. These subspaces define eigenvectors of the two-rank elasticity tensor $\bar{\bar{C}}$; the eigenvalues represent the corresponding elasticity moduli (no more then 6); the eigenvector orientation characterizes additional elasticity constants of the material.

4. For the rheologic properties of the anisotropic media simulation the principal assumption adopted is the hypothesis of identity of the elasticity and creep potentials symmetry (chart 3). In a case of isotropic time-independent body this assumption leads to the von Mises yield condition and corresponds to Ilyushin's isotropy postulate in ideology (chart 8). This postulate therefore is generalized here to anisotropic time-dependent materials. Mentioned assumption may be named the postulate of a symmetry rather.

5. As examples the isotropic (chart 5) and two simplest anisotropic materials (charts 6,7) are considered. The multiple eigenvalues of elasticity tensor $\bar{\bar{C}}$ correspond to the subspaces $\mathcal{L}^{(m)}$ in which the elastic properties are isotropic (and inelastic properties are adopted isotropic, too).

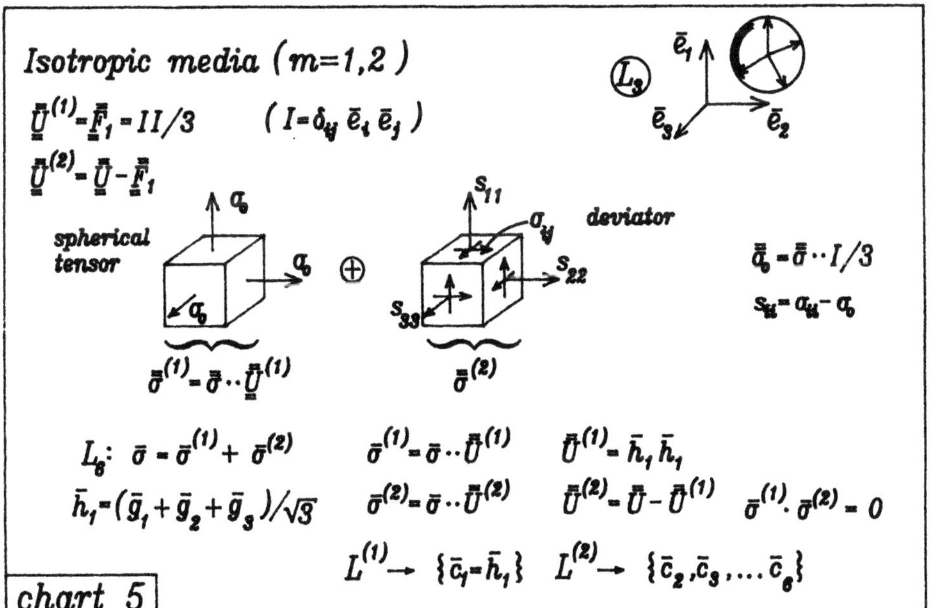

chart 5

6. So the invariant approach to the analysis of the stress and strain tensors constitution combined with a structural media model give us rather simple and effective tool to describe the anisotropic material rheological properties.

Orthoisotropic media (m=1,2,3)

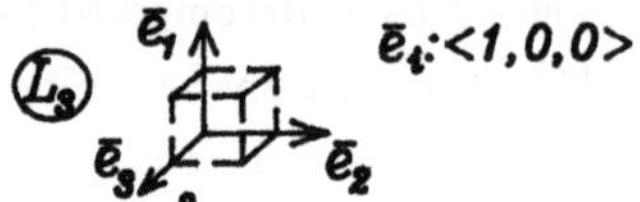

$\underline{\underline{U}}^{(1)} = \underline{\underline{F}}_1 - II/3 \quad \underline{\underline{U}}^{(2)} = \underline{\underline{F}}_2 - \underline{\underline{F}}_1 \quad \underline{\underline{U}}^{(3)} = \underline{\underline{U}} - \underline{\underline{F}}_2 \quad (\underline{\underline{F}}_2 = \sum_{i=1}^{3} \bar{e}_i \bar{e}_i \bar{e}_i \bar{e}_i)$

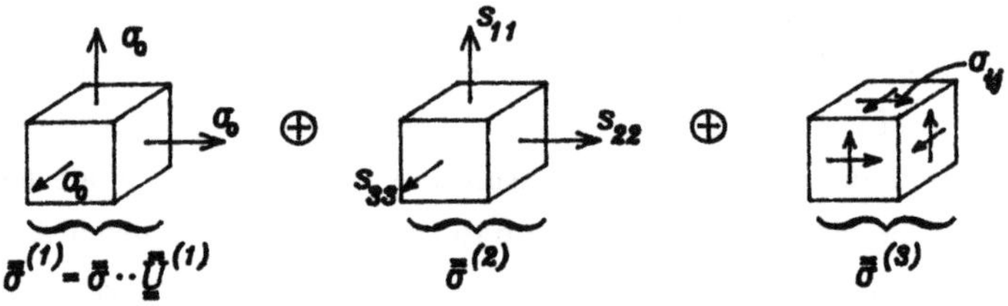

L_s: $\bar{\bar{\sigma}} = \bar{\bar{\sigma}}^{(1)} + \bar{\bar{\sigma}}^{(2)} + \bar{\bar{\sigma}}^{(3)} \qquad \bar{\bar{\sigma}}^{(1)} = \bar{\bar{\sigma}} \cdot\cdot \bar{\bar{U}}^{(1)}$

$\bar{\bar{U}}^{(1)} = \bar{h}_1 \bar{h}_1 \quad (\bar{h}_1 = (\bar{g}_1 + \bar{g}_2 + \bar{g}_3)/\sqrt{3}) \qquad L^{(1)} \rightarrow \{\bar{c}_1 = \bar{h}_1\}$

$\bar{\bar{U}}^{(2)} = \bar{h}_2 \bar{h}_2 + \bar{h}_3 \bar{h}_3 \quad (\bar{h}_1 \cdot \bar{h}_2 = \bar{h}_1 \cdot \bar{h}_3 = \bar{h}_2 \cdot \bar{h}_3 = 0) \qquad L^{(2)} \rightarrow \{\bar{c}_2 = \bar{h}_2, \bar{c}_3 = \bar{h}_3\}$

$\bar{\bar{U}}^{(3)} = \bar{g}_4 \bar{g}_4 + \bar{g}_5 \bar{g}_5 + \bar{g}_6 \bar{g}_6 \qquad L^{(3)} \rightarrow \{\bar{g}_4, \bar{g}_5, \bar{g}_6\}$

chart 6

Transversely isotropic media (m=1...4)

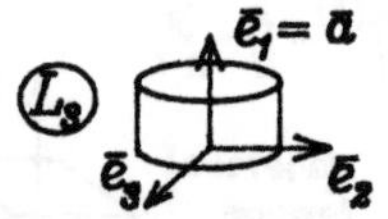

$\underline{\underline{U}}^{(1)} = 2(\underline{\underline{F}}_5 - \underline{\underline{F}}_3) \quad \underline{\underline{U}}^{(2)} = \underline{\underline{U}} + \underline{\underline{F}}_4 + \frac{\underline{\underline{F}}_3 - 3\underline{\underline{F}}_1}{2} - 2\underline{\underline{F}}_5 \qquad \underline{\underline{F}}_3 = \bar{a}\,\bar{a}\,\bar{a}\,\bar{a}$

$\underline{\underline{U}}^{(3)} = \frac{1}{1+k^2}[3\underline{\underline{F}}_1/2 + (\sqrt{2}k - 1)^2 \underline{\underline{F}}_3/2 + (\sqrt{2}k - 1)\underline{\underline{F}}_4] \qquad \underline{\underline{F}}_4 = (\bar{a}\,\bar{a}\,I + I\,\bar{a}\,\bar{a})/2$

$\underline{\underline{F}}_5 = (\bar{a}\,I\,\bar{a})^{(s)}$

$\underline{\underline{U}}^{(4)} = \frac{1}{1+k^2}[3k^2\underline{\underline{F}}_1/2 + (k+\sqrt{2})^2 \underline{\underline{F}}_3/2 - k(k+\sqrt{2})\underline{\underline{F}}_4]$

k – elastic constant

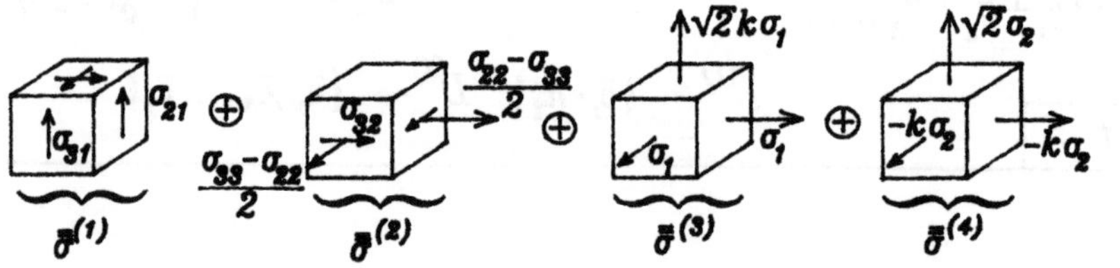

$L_s = L^{(1)} + L^{(2)} + L^{(3)} + L^{(4)} \quad L^{(1)} \rightarrow \{\bar{c}_1, \bar{c}_2\} \quad L^{(2)} \rightarrow \{\bar{c}_3, \bar{c}_4\}$

$L^{(3)} \rightarrow \{\bar{c}_5\} \quad L^{(4)} \rightarrow \{\bar{c}_6\}$

chart 7

The cyclic hardening and softening can also be reproduced within the approach by using a scalar multiplier in a creep potential depending on the loading history (on the Odquist parameter and time, for instance). Also, suggested scalar functions may be useful to describe anisotropic material low-cycle fatigue and a long-term fracture.

A.A.Ilyushin: isotropy postulate

$$L_6 = L_1 \oplus L_5 \qquad L_1 = \lambda\, \bar{h}_1 = \lambda\,(\bar{g}_1 + \bar{g}_2 + \bar{g}_3)/\sqrt{3}$$

$$L_1 \perp L_5 \qquad L_5 - \textit{deviatoric space}$$

$$\left\| \left.\begin{array}{l} \underset{2}{\bar{\sigma}}(t) = \bar{\bar{R}} \cdot \underset{1}{\bar{\sigma}}(t) \\ \underset{2}{T}(t) = \underset{1}{T}(t) \end{array}\right\} \rightleftarrows \underset{2}{\bar{\varepsilon}}(t) = \bar{\bar{R}} \cdot \underset{1}{\bar{\varepsilon}}(t) \right\| \qquad \begin{array}{l} \bar{\bar{R}} = \bar{h}_1 \bar{h}_1 + \bar{\bar{R}}_{⑤} \\ \bar{\bar{R}}_{⑤} \in L_5 \end{array}$$

Generalization of isotropy postulate

$$\bar{\bar{R}} = \sum \bar{\bar{R}}^{(m)} \qquad \bar{\bar{R}}^{(m)} \in L^{(m)} \qquad \bar{\bar{R}}^{(m)} \cdot\cdot\, \bar{\bar{R}}^{(n)} = \delta_{mn} \bar{\bar{R}}^{(m)}$$

$\bar{\bar{R}}^{(m)}$ *– mirror transform and(or) rotation in* $L^{(m)}$

chart 8

References

[1] Gokhfeld, D.A., Sadakov, O.S.: *An unified mathematical model for plasticity and creep under variable-repeated loading*, - in the present Proceedings

[2] Gokhfeld, D.A., Sadakov, O.S.: *Plasticity and creep in structural elements under repeated loading*, Moscow, Mashinostrojenije, 1984 (in Russian)

Effects of Prior Plastic Strain Path Shapes on Creep Behaviour of Incompressible Materials

M. Waniewski
Polish Academy of Sciences,
Institute of Fundamental Technological Research. Warsaw, Poland.

Summary
In this paper the discrepancies between creep processes, including proportional and non - proportional prior plastic strain deformation carried out at room temperature, are involved in constitutive relations by introducing additional tensor argument, components of which are identified with the normalized plastic prestrain tensor.
An idealization of the different types of strain trajectory, including either simple or complex sequences, such as cruciform, square or circular strain control path followed by proportional deformation, is considered in plastic prestrain space assuming tensorial generalization of the uniaxial stress-strain relation to multiaxial states of stress by using interpolation methods for tensor functions.
Results of the long - term creep tests carried out at the Nuclear Research Centre KFA Jülich, FRG, on the plastic prestrained cylindrical specimens of INCONEL 617 are compared with numerical approach taking the different shapes of prestrain trajectory into account.

Introduction

It is known from many studies, e.g. Dyson et. al. [1,2], that room - temperature straining affects the response of material to subsequent high - temperature creep conditions. Some experimental investigations by Trąmpczyński [3], Waniewski [4], Kowalewski [5] proved that the effect of prestraing under combined loading depends on its direction and magnitude causes anisotropic creep behaviour.
A theoretical model of the anisotropic creep phenomenon due to cold work may be constructed by applying non - linear constitutive equations for the secondary creep behaviour based on the representation theory of tensor functions, Betten et. al. [6]. The induced anisotropy is characterized by a prestrain tensor as an additional argument tensor of the constitutive equations. For practical applications, the simplified theories are presented by Sawczuk et. al. [7.8], involved only a linear combination of two arguments tensor generators. Generalization of the Odqvist constitutive equation is presented in [4] by employing structural tensor of rank fourth, components of which are formed by the second order plastic prestrain tensor.

Problem formulation

The prediction of the time dependent creep strains which occur when a material was subjected previously to plastic straining is a problem of current interest since some plastic

M. Zyczkowski (Ed.)
Creep in Structures
IUTAM Symposium Cracow/Poland 1990

formed structures of power producing plant operates in creep range. The question then arise as to whether it is possible, on the basis of the results of properly chosen experiments, to predict the creep strain which occurs due to any given variation in plastic prestraining. In creep mechanics, constitutive expressions of the form

$$\mathrm{d_{ij}} = \mathrm{f_{ij}}\,(\sigma_{\mathrm{pq}}, \mathrm{A_{pq}}) \tag{1}$$

represents anisotropic creep behaviour, where $\mathrm{d_{ij}}$ are the Cartesian components of the strain - rate tensor **d** as a second - order tensor - valued function of two symmetric argument tensors of rank two, the Cauchy stress tensor $\boldsymbol{\sigma}$ and a constitutive tensor **A** characterizing the anisotropy form.

The general representation of Eq.(1) is given through a linear combination

$$\mathrm{d_{ij}} = \sum_{\alpha} \varphi_{\alpha}\, {}^{\alpha}\mathrm{G_{ij}} \tag{2}$$

where the **G**'s are symmetric tensor generators of rank two involving the argument tensors $\boldsymbol{\sigma}$, **A**. The coefficients φ_{α} in Eq.(2) are scalar - valued functions of the integrity basis associated with the representation (2). Some possible ways have been discussed by Betten [6,9] in order to find the tensor generators in Eq.(2). On the other hand, by applying interpolation methods for tensor functions, it has been illustrated in detail by Betten [10] that the scalar coefficients φ_{α} in Eq.(2) can be expressed as functions of the irreducible invariants of the argument tensors and of experimental constitutive relations found in uniaxial tests.

An irreducible set of tensor generators for tensor - valued function of two argument tensors, Eq.(1), is presented in the following form [6]

$$\begin{aligned} {}^{\Theta}\mathrm{d_{ij}} = {} & \varphi_{(0,1)}\mathrm{A'_{ij}} + \varphi_{(0,2)}\mathrm{A''_{ij}} + \varphi_{(1,0)}\sigma'_{\mathrm{ij}} + \varphi_{(2,0)}\sigma''_{\mathrm{ij}} \\ & + \frac{1}{2}\varphi_{(1,1)}(\sigma'_{\mathrm{ik}}\mathrm{A'_{kj}} + \mathrm{A'_{ik}}\sigma'_{\mathrm{kj}} - \frac{2}{3}\sigma'_{\mathrm{pq}}\mathrm{A'_{qp}}\delta_{\mathrm{ij}}) \\ & + \frac{1}{2}\varphi_{(1,2)}(\sigma'_{\mathrm{ik}}\mathrm{A}'^{(2)}_{\mathrm{kj}} + \mathrm{A}'^{(2)}_{\mathrm{ik}}\sigma'_{\mathrm{kj}} - \frac{2}{3}\sigma'_{\mathrm{pq}}\mathrm{A}'^{(2)}_{\mathrm{qp}}\delta_{\mathrm{ij}}) \\ & + \frac{1}{2}\varphi_{(2,1)}(\sigma'^{(2)}_{\mathrm{ik}}\mathrm{A'_{kj}} + \mathrm{A'_{ik}}\sigma'^{(2)}_{\mathrm{kj}} - \frac{2}{3}\sigma'^{(2)}_{\mathrm{pq}}\mathrm{A'_{qp}}\delta_{\mathrm{ij}}) \\ & + \frac{1}{2}\varphi_{(2,2)}(\sigma'^{(2)}_{\mathrm{ik}}\mathrm{A}'^{(2)}_{\mathrm{kj}} + \mathrm{A}'^{(2)}_{\mathrm{ik}}\sigma'^{(2)}_{\mathrm{kj}} - \frac{2}{3}\sigma'^{(2)}_{\mathrm{pq}}\mathrm{A}'^{(2)}_{\mathrm{qp}}\delta_{\mathrm{ij}}) \end{aligned} \tag{3}$$

where $\sigma'_{\mathrm{ij}} \equiv \sigma_{\mathrm{ij}} - \sigma_{\mathrm{kk}}\delta_{\mathrm{ij}}/3$ and $\mathrm{A'_{ij}} \equiv \mathrm{A_{ij}} - \mathrm{A_{kk}}\delta_{\mathrm{ij}}/3$ are components of the stress deviator $\boldsymbol{\sigma}'$ and a constitutive deviator $\mathbf{A}'$ respectively. In Eq.(3), apart the deviators $\boldsymbol{\sigma}'$ and $\mathbf{A}'$ exist also the following deviators $\sigma''_{\mathrm{ij}} \equiv \sigma'^{(2)}_{\mathrm{ij}} - \sigma'^{(2)}_{\mathrm{kk}}\delta_{\mathrm{ij}}/3$ and $\mathrm{A''_{ij}} \equiv \mathrm{A}'^{(2)}_{\mathrm{ij}} - \mathrm{A}'^{(2)}_{\mathrm{kk}}\delta_{\mathrm{ij}}/3$.

Based on the assumption of incompressibility, the constitutive relation (3) contains eight scalar coefficients $\varphi_{(\lambda,\mu)}$, for $\lambda, \mu = 0, 1, 2$ and $\varphi_{(0,0)} \equiv 0$, the identification of which is presented in [10] by separating two tensor variables $\boldsymbol{\sigma}'$ and $\mathbf{A}'$

From the practical point of view, the simplified theory is proposed in [6] as the special case of the general representation presented in Eq.(3). Assuming incompressible and isotropic material behaviour ($d_{kk} \equiv 0$ and $A_{ij} \equiv 0_{ij}$), the constitutive Eg.(3) is reduced to the form

$$d_{ij} = \Phi_1 \sigma'_{ij} + \Phi_2 \sigma''_{ij}. \tag{4}$$

By substituting relation

$$\tau'_{ij} \equiv \frac{\sigma'_{ij} - \alpha'_{ij}}{R} \tag{5}$$

in Eq.(4), the simplified constitutive equation

$$^{\Theta}d_{ij} = \varphi_1 \tau'_{ij} + \varphi_2 \tau''_{ij} \tag{6}$$

is obtained for anisotropic creep behaviour.
In Eq.(5) two parameters have been assumed to describe the influence of room temperature plastic prestraining. The tensorial one is presented by its deviator $\alpha'_{ij} \equiv \alpha_{ij} - \alpha_{kk}\delta_{ij}/3$ while the scalar R is connected with length of the plastic strain path s^P corresponds to the accumulated plastic strain

$$s^P = \int_0^t \dot{e}^P_v dt, \tag{7}$$

where $\dot{e}^P_v$ denotes the equivalent plastic strain rate

$$\dot{e}^P_v = \int \sqrt{\frac{2}{3} d\dot{e}^P_{ij} d\dot{e}^P_{ji}}. \tag{8}$$

The scalars φ_1 and φ_2 in Eq.(6) are assumed to be functions of the creep stress only, then there can be compared with Φ_1 and Φ_2 in Eq.(4). Assuming the Norton - Bailey creep law $d = K\sigma^n$ while $d \equiv d_{11}$ and $\sigma \equiv \sigma_{11}$, it can be obtained from Eq.(4)

$$\varphi_1 + \frac{1}{3}\sigma\varphi_2 = \frac{3}{2}K\sigma^{n-1}. \tag{9}$$

Assuming a hypothesis of the equivalent dissipation rate

$$\dot{D} \equiv \sigma_{ij} d_{ij} = \sigma d, \tag{10}$$

where $\dot{D}$ denotes the rate of dissipation of creep energy, by substituting relation (10) to Eq.(9), the form of scalar functions is obtained [6]

$$\lim_{\sigma_{ij} \to \sigma} \varphi_1 = \frac{9}{2}K\sigma^{n-1}, \qquad \lim_{\sigma_{ij} \to \sigma} \varphi_2 = -9K\sigma^{n-2} \tag{11}$$

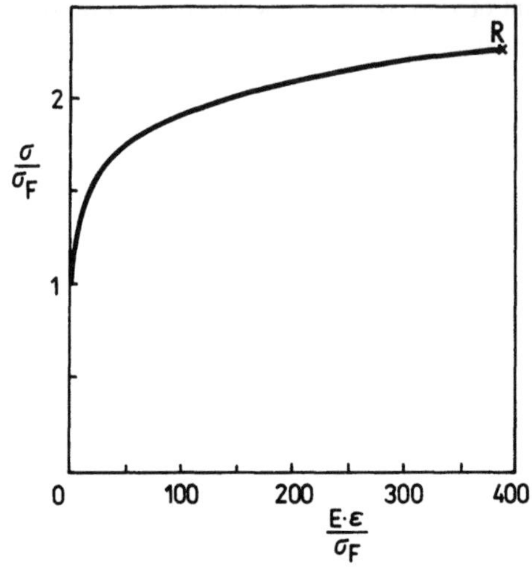

Fig 1. Idealization of stress-strain curve for Inconel 617 at 20°C, Eq.(12)

based on the uni-axial creep test.

Plastic prestraining description by tensorial generalization of the uni-axial stress-strain relation

Some materials which perform linear elastic behaviour within the range of small deformation and ductile plastic behaviour during the large deformation can be described under uni-axial stress σ^* by the following stress- strain relation, Ludwik [11]

$$\sigma^*/\sigma_F = L(Ee^*/\sigma_F)^m \tag{12}$$

where E is the modulus of elasticity, and σ_F represents the yield stress determined in an uni-axial tension test. The exponent m and constant L stimulate the elastic-plastic transition, Fig.1. For instance, a perfectly plastic behaviour is characterized by $m \to 0$.

For engineering applications it is very important to generalize the relation (12) to multi-axial states of stress, for instance, to describe plastic prestraing effect for different shapes of deformation path. This can be achieved by using tensorial generalization of Eq.(12) in the form of an isotropic tensor function [10]

$$\sigma_{ij}^{*'} = f_{ij}(\mathbf{e}^{*'}) = \sigma^*(\Psi_0\delta_{ij} + \Psi_1 e_{ij}^{*'} + \Psi_2 e_{ij}^{*'(2)}) \tag{13}$$

where the scalars Ψ_0, Ψ_1 and Ψ_2 can be determined in a similar way proposed by Betten [10] in extending the Lagrange interpolation method to a tensor - valued function, assuming the principal values of the argument tensor as interpolating points. Thus, the results are found in the form

$$\begin{aligned}\Psi_0 &= -2(1-m)/9 \\ \Psi_1 &= 2(2+m)/9e^* \\ \Psi_2 &= 4(1-m)/9e^{*2}\end{aligned} \tag{14}$$

if the uni-axial relation (12) is generalized.

The equivalent strain e^* in Eq.(14) can be determined as a function of the strain invariants while the hypothesis of the equivalent strain energy is assumed

$$\sigma^*_{ij}e^*_{ij} = \sigma^* e^*. \tag{15}$$

Thus, the result is obtained from solution of the cubic equation

$$e^{*3} + Pe^{*2} + Qe^* + S = 0 \tag{16}$$

where the abbreviations

$$P \equiv (1-m)I_1/3, \quad Q \equiv -2(2+m)I_2'/3, \quad S \equiv -2(1-m)I_3' \tag{17}$$

have been introduced. Thus, the scalar coefficients (14) are functions of the integrity basis

$$I_1 \equiv \int de^*_{kk}, \quad I_2' \equiv \int de^*_{ij}de^*_{ji}/2, \quad I_3' \equiv \int de^*_{ij}de^*_{jk}de^*_{ki}/3 \tag{18}$$

and experimental data (σ_F, E, m, L) obtained in uni-axial tension test.

The problem which now arises is an attempt to describe the different orientation of the straining path and its effect on material behaviour under creep condition. Hence, a tensorial parameter $\boldsymbol{\alpha}'$ proposed in Eq.(5) has a form similar to the relation (13)

$$\alpha'_{ij} \equiv a\sigma^{*'}_{ij} \tag{19}$$

where coefficient of proportionality is identified with material constant depended on changing in material structure due to heating up to the constant temperature in which creep process occurs. Moreover, the scalar R is assumed as follows

$$R \equiv 1 + s^P/\Delta e^* \tag{20}$$

where the path length s^P is described by Eq.(7), Δe^* denotes the maximal value of $(4I_2'/3)^{1/2}_{max}$ within plastic prestraining cycle duration, Eq.(18).

Experimental tests

Inconel 617 is a nickel-base alloy which is used for the heat exchanger elements of an HTGR plant (high temperature gass-cooled reactor). These components need to withstand temperature excursions and/or pressure transients under upset conditions due to

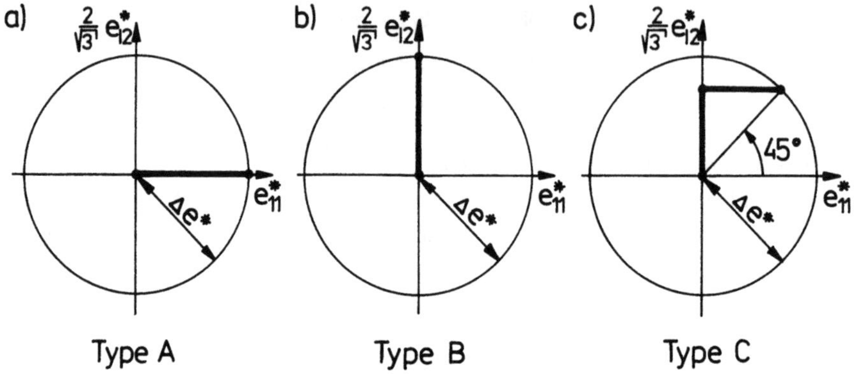

Fig 2. Plastic prestrain paths for Δe^*=0.05 or 0.10 and $\dot{s}^p = 3 * 10^{-5}/s$

the coolant PNP-helium gas operation conditions.
The experimental programme [6,12] to be carried out at the Nuclear Research Centre, Jülich, FRG, contained the long-term creep tests (up to four years duration) on cylindrical specimens plastic predeformed at room temperature in tension/torsion combined directions of type A, B and C presented in two dimensional plastic strain space (e^*_{11}, $(2/\sqrt{3})e^*_{12}$) in Fig.2. for two different values of Δe^*, Eq.(20), $\Delta e^* = 0.05$ or 0.10.

In the second part of experiment, specimens were heated up to 950°C and subjected to the uni-axial constant stress σ=15.1 MPa and σ=22.7 MPa in the PNP-helium process gas environment.

Verification of the theoretical approach

To compare the simplified theory, Eq.(6) with the general representation presented in Eq.(3) for $A'_{ij} \equiv e^{*'}_{ij}$, the following scalar coefficients are identified

$$\begin{aligned}
\varphi_{(0,1)} &= \varphi_2\Psi_2^2 I_3' a^2\sigma^{*2}/R^2 + \varphi_2\Psi_1\Psi_2 I_3' a^2\sigma^{*2}/R^2 - \varphi_1\Psi_1 a\sigma^*/R, \\
\varphi_{(0,2)} &= -\varphi_2\Psi_2^2 I_2' a^2\sigma^{*2}/4R^2 + \varphi_2\Psi_1^2 a^2\sigma^{*2}/R^2 - \varphi_1\Psi_1 a\sigma^*/R, \\
\varphi_{(1,0)} &= \varphi_1/R + \varphi_2\Psi_2 I_3' a\sigma^*/R^2, \\
\varphi_{(2,0)} &= \varphi_2/R^2,\ \varphi_{(1,1)} = -2\varphi_2\Psi_1 a\sigma^*/R^2,\ \varphi_{(1,2)} = -2\varphi_2\Psi_2 a\sigma^*/R^2,
\end{aligned} \tag{21}$$

if the assumptions (5),(7),(11),(12),(13),(15) are fulfilled and $\varphi_{(2,1)} = \varphi_{(2,2)} \equiv 0$.
For incompressible material with plastic prestraining in combined tension/torsion direction, Fig.2, the uni-axial steady-state creep rate has a simplified form [6]

Table 1:

	σ MPa	Δe^*	e^* Eq.(16)	σ^* MPa Eq.(12)	${}^{\Theta}d_{11} * 10^{-10}$ s^{-1} Eq.(22)	${}^{\Theta}d_{11} * 10^{-10}$ s^{-1} experiment
A	15.47	0.05	0.060	545.61	2.219	1.090
	16.22	0.10	0.120	597.06	3.826	2.031
	23.07	0.05	0.060	545.61	3.846	4.438
	24.04	0.10	0.120	597.06	11.900	1.141
B	14.62	0.05	0.052	535.28	0.727	0.727
	14.69	0.10	0.103	585.76	0.582	0.613
	21.97	0.05	0.052	535.28	21.326	19.367
	22.02	0.10	0.103	585.76	20.293	18.566
C	15.05	0.05	0.078	564.78	0.238	0.197
	15.52	0.10	0.156	618.03	0.544	0.043
	22.85	0.05	0.078	564.78	8.439	14.911
	23.56	0.10	0.156	618.03	7.885	16.474
isotr.	14.64	-	-	-	0.896	0.896
	22.03	-	-	-	18.060	18.060

$$
\begin{aligned}
{}^{\Theta}d_{11} &= \varphi_{(0,1)} e^*_{11} + \varphi_{(0,2)}(e^{*2}_{11} + 2e^{*2}_{12}/3)/2 + 2\varphi_{(1,0)}\sigma/3 + 2\varphi_{(2,0)}\sigma^2/9 \\
&\quad + \varphi_{(1,1)}\sigma e^*_{11}/3 + \varphi_{(1,2)}(e^{*2}_{11} + 10e^{*2}_{12}/9)\sigma/2
\end{aligned} \tag{22}
$$

with scalar functions $\varphi_{(\lambda,\mu)}$ defined in Eq.(21) and uni-axial creep stress σ.
Comparison of numerical results , Eq.(22), with experimental data obtained from tests of type A, B or C for two values of Δe^*, Fig.2, is inserted in Table 1. The room-temperature stress-strain relation in form of Eq.(12) has been derived using experimental data for Inconel 617: L=1, m=0.13, σ_F=344MPa, E=2*10^5, the ultimate stress and elongation σ_R=751MPA, A_u=66% respectively, allowing to describe the uni-axial elastic-plastic transition of tested material with good approximation.
To identify a value of material constant a, Eq.(19), equal to a=0.083 the pure torsion plastic prestrain test of type B, Fig.2. has been used with $\Delta e^* = 0.05$ and ${}^{B}d_{11} = 0.727* 10^{-10} s^{-1}$.
Moreover, the creep parameters n=7.347 and K=2.447*10^{-19} define scalar coefficients φ_1 and φ_2 in Eq.(11) as the function of creep stress σ.

Conclusions

The effect of room-temperature plastic prestrain on isochoric creep rate has been described for Inconel 617 using the tensorial constitutive equations based upon the Ludwik uni-axial stress-strain relation to be generalized to multi-axial states of stress by applying interpolation methods for tensor functions. It has been proved that the scalar coefficients in tensorial constitutive equations are functions of strain tensor invariants and experimental data.

Some examples of different plastic prestraining trajectories are compared with the numerical approach successfully describes the discrepancies between proportional and square plastic strain induced anisotropy. Further calculations are improved to specify the influence of cruciform and circular strain control paths on creep behaviour.

Two parameters, the values of which vary with straining, have been assumed in constitutive relations - the tensorial one identifies directional material hardening while the scalar one is responsible to material behaviour for isotropic hardening.

References

1. Dyson, B.F.; Rodgers, M.J.: Prestrain, cavitation and creep ductility. Metal Sci. 8(1974) 261-266.
2. Dyson, B.F.; Loveday, M.S.; Rodgers M.J.: Grain boundary cavitation under various states of applied stress. Proc. R. Soc. Lond. A. 349(1976) 245-259.
3. Trampczynski, W.A.: The influence of cold work on the creep of copper under biaxial states of stress. Acta metall. 30(1982) 1035-1041.
4. Waniewski, M.: A simple law of steady-state creep for material with anisotropy introduced by plastic prestraining. Ing. Arch. 55(1985) 368-375.
5. Kowalewski, Z.: The surface of constant dissipation energy rate under creep and its experimental determination. Arch. Mech. Stos. 39(1987) 445-459.
6. Betten, J.; Waniewski, M.: Einfluß der plastischen Anisotropie auf das sekundäre Kriechverhalten inkompressibler Werkstoffe. Rheol. Acta 25(1986) 166-174.
7. Sawczuk, A.; Anisimowicz, M.: Tensor functions approach to creep laws after prestrain. In: Creep in structures. Berlin, Heidelberg, New York: Springer Verlag: 1981.
8. Sawczuk, A.; Trampczynski, W.A.: A theory of anisotropic creep after plastic prestraining. Int. J. Mech. Sol. 24(1982) 647-653.
9. Betten, J.: Materialgleichungen zur Beschreibung des sekundären und tertiären Kriechverhaltens anisotroper Stoffe. ZAMM 64(1984) 211-220.
10. Betten, J.: Generalization of nonlinear material laws found in experiments to multiaxial states of stress. Eur. J. Mech., A/Solids 8(1989) 325-339.
11. Ludwik, P.: Elemente der technologischen Mechanik. Berlin: Springer Verlag: 1909.
12. Ennis, Ph.; Waniewski, M.: Long-term creep tests of Inconel 617 with strain-induced anisotropy. In preparation.

Some Unification of Creep Theories Based on Internal Time Concept

O. Watanabe * and S. N. Atluri †

* Institute of Engineering Mechanics, University of Tsukuba, Tsukuba 305, JAPAN
† School of Civil Engineering, Georgia Institute of Technology, Atlanta, GA 30332, U.S.A.

Summary

Among the constitutive equations for high strain rate properties of materials under dynamic loading, the 'Overstress Model' of Malvern[1] and Perzyna[2] is well known because of its simplicity and ability. This model has been widely extended for describing the quasi-static creep behavior under high temperature environment by Chaboche[3], Robinson[4], Bruhns[5], Krempl, McMahon & Yao[6] and Sahashi, Imatani & Inoue[7]. On the other hand, 'Endochronic Model' of Valanis[8] received attensions because of its ability to predict time-independent plastic deformation under complex loading histories. This model is applied for time-dependent deformations, such as dynamic loading and quasi-static creep behaviors. Wu & Yip[9] applied this model for the dynamic behavior of materials, and Chang, Sugiura & Lee[10] proposed an alternate theory of high strain rate problems. In this 'Endochronic Model' for dynamic problems, the current stress exists beyond the yield surface, and inelastic strain becomes larger as the stress goes away from the boundary of the static yield surafce. Watanabe & Atluri[11,12] applied 'Endochronic Model' for quasi-static creep behaviors, where the current stress is confined within the yield surface, and inelatic strain rate will become infinite as the stress approaches to the boundary of yield surface.

This paper gives some theoretical unification of the rate dependent constitutive modeling for dynamic and quasi-static problems developed in 'Overstress Model' and 'Endochronic Model'. By introducing a new intrinsic time derived from the fundamental stress-strain relation in the integral form, 'Endochronic Model' is shown to be general enough to relate closely to the previous 'Overstress Model'.

A New Intrinsic Time Measure

Let S be the deviator of the stress and de and $d\eta$ the deviators of the total and inelastic strain increments, respectively. The differential strain tensor $d\varepsilon$ is written as: $d\varepsilon = (de^e + d\eta) + I\, d\varepsilon_m^e$; thus the inelastic strains are purely distorsional. Further $d\eta = de - dS/2\mu_o$ where μ_o is the elastic shear modulus. In the Endochronic theory of Valanis[8], the stress S is represented in the integral form as a function of intrinsic time z based on the visco-elastic model as:

$$S = 2\mu_o \int_0^z \rho(z-z')\frac{d\eta}{dz'}dz' \;. \tag{1}$$

We assume the kernel $\rho(z)$ to be of the form:

$$\rho(z) = \rho_o\delta(z) + \rho_1(z) \;, \tag{2}$$

where $\delta(z)$ is a Dirac delta function, and $\rho_1(z)$ is a non-singular function. Use of Eq.(2) in Eq.(1) results in:

M. Zyczkowski (Ed.)
Creep in Structures
IUTAM Symposium Cracow/Poland 1990

$$S = S_y^o \frac{d\eta}{dz} + r(z) , \tag{3}$$

where

$$r(z) = 2\mu_o \int_0^z \rho_1(z-z') \frac{d\eta}{dz'} dz' , \tag{4a}$$

$$S_y^o = 2\mu_o \rho_o . \tag{4b}$$

For the time-independent plasticity problem, Valanis[8] defines the intrinsic time measure dz as follows:

$$dz = \frac{d\zeta}{f(\zeta)} , \tag{5}$$

where $d\zeta$ is defined by:

$$d\zeta = |d\eta| = (d\eta_{ij} d\eta_{ij})^{1/2} . \tag{6}$$

The function $f(\zeta)$ is non-negative and $f(0)=1$. Substituting Eq.(5) in Eq.(3) yields the following constitutive equation for plasticity:

$$S = S_y^o f(\zeta) \frac{d\eta}{d\zeta} + r(z) . \tag{7}$$

We can easily find from Eq.(7) that S_y^o is the magnitude of the initial yield surface, $f(\zeta)$ signifies isotropic hardening of yield surface expansion, and $r(z)$ denotes kinematic hardening of yield surface translation. For time-dependent creep problem, Valanis[8] suggested the following intrisic time measure dz:

$$dz^2 = \frac{d\zeta^2}{f(\zeta)^2} + \frac{dt^2}{g^2} , \tag{8}$$

where dt is a real time increment, and the function g is selected appropriately. The present authors[11,12] have proposed a simple theory of creep and creep-plasticity interaction based on the above intrinsic time dz of Eq.(8).

In the present paper, the present authors will propose an endochronic theory for creep-plasticity interaction using a new intrinsic time dz. We postulate that the fundamental stress-strain relation of Eq.(3) will hold for inelastic deformations, which leads to the following form by arranging Eq.(3):

$$d\eta = \frac{S - r(z)}{S_y^o} dz , \tag{9}$$

where $d\eta$ is the differential inelastic strain including creep and plastic strains. Fig.1 shows the current stress S and the subsequent yield surface in the deviatoric stress space. The current stress S is not always located on the subsequent yield surface f_y in creep-plasticity interaction problem, while the stress S lies just on the yield surface f_y in the time-independent plasticity problem. Now, let consider the concetric surface with the yield surface possessing the translation of surface $r(z)$ in common, on which the current stress lies accurately. We will call this surface as the loading surface f_l to distinguish from the yield surface f_y in creep-plasticity interaction problem. We introduce the following quantities N and A defined through:

$$N = \frac{S-r(z)}{|S-r(z)|}\,, \tag{10a}$$

$$A = \frac{|S-r(z)|}{S_y}\,, \tag{10b}$$

where

$$S_y = S_y^o f(\zeta)\,. \tag{10c}$$

We find that the isotropic hardening is represented through the relation for self-similar expansion of the yield surface from Eq.(10c), where S_y^o and S_y represent the magnitudes of the initial and subsequent yield surfaces, respectively. The tensor N is outward normal to the loading surface f_l, and the norm of N is a unit from Eq.(10a) as:

$$|N| = 1\,. \tag{11}$$

The scalar A represents the ratio of the magnitudes of the loading surface f_l to that of the subsequent yield surface f_y. The fundamental equation (9) can be written by the notations of Eq.(10) in the form:

$$d\eta = ANf(\zeta)dz\,. \tag{12}$$

A new intrinsic time dz can be defined by taking the norm of the fundamental equation (12) itself and using Eqs.(6) and (11) as:

$$dz = \frac{d\zeta}{Af(\zeta)}\,. \tag{13}$$

For time-independent plasticity, the loading surface coincides with the yield surface resulting in $A = 1$, which means that the intrinsic time defined by Eq.(13) is the generalized expression for Eq.(5). The relation between $d\eta$ and $d\zeta$ is calculated by substituting Eq.(13) in Eq.(12) as:

$$d\eta = N\,d\zeta\,. \tag{14}$$

Taking the norm of Eq.(14) and using Eq.(11) leads to the definition $d\zeta$ of Eq.(6).

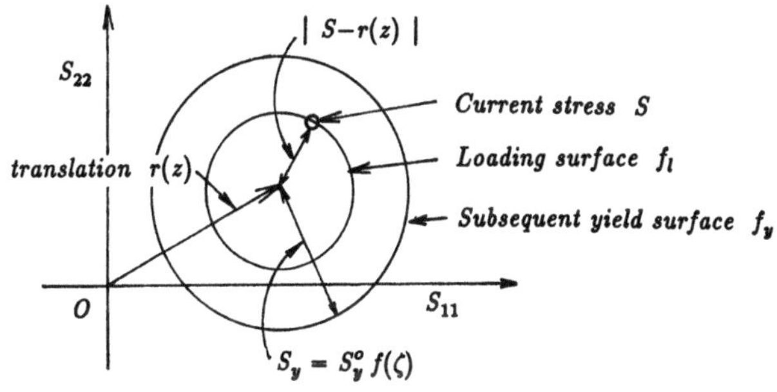

Fig.1 Loading surface and subsequent yield surface in deviatoric stress space

Understress Model

We propose a constitutive modeling for creep-plasticity interaction. During the increment of real time dt, the incremental inelastic strain $d\eta$ is assumed to consist of time-independent plastic strain $d\eta^p$ and time-dependent creep strain $d\eta^c$ as:

$$d\eta = d\eta^p + d\eta^c \,. \tag{15}$$

The constitutive equation (7) for plasticity, i.e., $d\eta^p = Nd\zeta$ should be modefied based on the fact that the current stress S does not always on the yield surface f_y, i.e., $A \neq 1$. Considering that the plasticity model approaches to Eq.(7) for $A \rightarrow 1$, we postulate the following power law with respect to A:

$$d\eta^p = A^n N d\zeta \,, \tag{16}$$

where n is a material constant. The constitutive equation for creep strain $d\eta^c$ is assumed by the following Norton type one:

$$d\eta^c = B_o A^m N dt \,, \tag{17}$$

where B_o and m are material constants. Substituting Eqs.(16) and (17) in Eq.(15) yields:

$$d\eta = (A^n d\zeta + B_o A^m dt) N \,. \tag{18}$$

Taking the norm of Eq.(18) and using Eqs.(6) and (11) gives:

$$d\zeta = \mid d\eta \mid = A^n d\zeta + B_o A^m dt \,. \tag{19}$$

Rearrangement of Eq.(19) with respect to $d\zeta$ gives:

$$d\zeta = \frac{B_o A^m}{1-A^n} dt \,. \tag{20}$$

Since the inelastic strain length $d\zeta$ is positive, we have from Eq.(20):

$$A < 1 \,. \tag{21}$$

Eq.(21) means that the loading surface f_l exists within the yield surface f_y, thus we may call this model as 'Understress Model'. Substituting Eq.(20) into Eq.(14) gives $d\eta$ as:

$$d\eta = \frac{B_o A^m}{1-A^n} N dt \,. \tag{22}$$

Comparing creep theory of Eq.(17), one may find the effects of plastic strain appear in the denominator $(1-A^n)$ in Eq.(22). This term will bring infinite inelastic strain increment as $A \rightarrow 1$, which may be regarded as the time-independent plastic strains. The creep potential Ω^c defined by $\partial\Omega^c/\partial S = \partial\eta/\partial t$ can be obtained by the following relation:

$$\frac{\partial A}{\partial S} = \frac{1}{S_y} N \,, \tag{23}$$

then, the creep potential Ω^c is written by:

$$\Omega^c = B_o S_y \int_o^A \frac{A^m}{1-A^n} dA \,. \tag{24}$$

In order to express the evolution equation of yield surface translation $r(z)$, it may be convenient to expand the kernel $\rho_1(z)$ in the exponential form:

$$\rho_1(z) = \sum_i \rho_{1i}\, exp(-\alpha_i z) \,, \tag{25}$$

where ρ_{1i} and α_i are material constants. Using Eqs.(25) and (13) in Eq.(4a), we treat each of the $r^{(i)}$ as a secondary internal variable by:

$$dr = \sum_i dr^{(i)} \,, \tag{26a}$$

where

$$dr^{(i)} = 2\mu_o\rho_{1i}d\eta - \frac{r^{(i)}\alpha_i}{Af(\zeta)}d\zeta \,. \; (no\ summation\ on\ i) \tag{26b}$$

Eq.(26b) coinsides with the non-linear hardening law of Chaboche[3]. Substituting Eq.(19) in Eq.(26b) gives an alternate expression for $dr^{(i)}$ as:

$$dr^{(i)} = 2\mu_o\rho_{1i}d\eta - \frac{r^{(i)}\alpha_i}{Af(\zeta)}\,(A^n d\zeta + B_o A^m dt) \,, \tag{26c}$$

or relation of $d\zeta$ to dt in Eq.(20) results in:

$$dr^{(i)} = 2\mu_o\rho_{1i}d\eta - \frac{r^{(i)}\alpha_i}{Af(\zeta)}\,\frac{B_o A^m}{1-A^n}dt \,. \tag{26d}$$

The evolutional equation of yield surface expansion S_y can be written for the saturated condition by:

$$dS_y = \gamma\,(S_y^\infty - S_y)\,d\zeta, \tag{27a}$$

where γ is a material constant and S_y^∞ is the magnitude of the saturated yield surface. Substituting Eq.(19) into Eq.(27a) gives:

$$dS_y = \gamma\,(S_y^\infty - S_y)(A^n d\zeta + B_o A^m dt) \,, \tag{27b}$$

or using Eq.(20) in Eq.(27a) results in:

$$dS_y = \gamma\,(S_y^\infty - S_y)\,\frac{B_o A^m}{1-A^n}\,dt \,. \tag{27c}$$

The present authors[11,12] have earlier proposed a theory for creep-plasticity interaction, which may be regarded as introducing the following $d\zeta$ instead of Eq.(19):

$$d\zeta^2 = d\eta^p{:}d\eta^p + d\eta^c{:}d\eta^c \,. \tag{28}$$

The creep strain $d\eta^c$ is given by Eq.(17), and the plastic strain is assumed to be $d\eta^p = ANd\zeta$. The derived constitutive equation using Eq.(28) is expressed by:

$$d\eta = \frac{B_o A^m}{\sqrt{1-A^2}}\,Ndt \,. \tag{29}$$

When the current stress S approaches to the boundary of the yield surface, i.e., $A \rightarrow 1$, the inelastic strain increment for a new version of Eq.(22) is approximately calculated by introducing the quantity $\Delta \equiv 1-A$ as follows:

$$d\eta \sim B_o N dt \frac{1}{n\Delta}, \tag{30}$$

while the old version of Eq.(29) gives $d\eta$ by:

$$d\eta \sim B_o N dt \frac{1}{\sqrt{2\Delta}}. \tag{31}$$

From Eqs.(30) and (31), it may be found that the singularity for $A \rightarrow 1$ using a new creep theory retains the same level regardless of the power number n of plasticity, while the singurarity in the old creep theory become greatly reduced compared to that of a new creep theory.

The recent creep theory of ONERA[13] in France based on the concept of overstress takes into account the strong nonlinearity in the thermal or time recovery term achieved in the secondary internal variable $r^{(i)}$ as:

$$dr^{(i)} = \frac{2}{3} C_i d\eta - b_i r^{(i)} d\zeta - a_i |r^{(i)}|^m \frac{r^{(i)}}{|r^{(i)}|} dt. \tag{32}$$

A similar effect can also be introduced in the isotropic hardening relation[14] as:

$$dS_y = b(Q - S_y) d\zeta - a |S_y - Q_r|^{m-1} (S_y - Q_r) dt. \tag{33}$$

Eqs.(32) and (33) of ONERA model are entirely analogous to the present expressions of Eqs.(26c) and (27b) by assuming $n=1$.

Overstress Model

Let us consider a constitutive modeling for high strain rate under impact loading. To begin with the static time-independent plastic strain $d\eta^s$ as a base to be assumed by the same form of Eq.(16) as:

$$d\eta^s = A^n N d\zeta, \tag{34}$$

since the current stress S does not lie on the yield surface. We also assume that the dynamic strain $d\eta^d$ is normal to the loading surface and proportional to the increment of time dt written by:

$$d\eta^d = KNdt, \tag{35}$$

where K is a coefficient. The total inelastic strain increment $d\eta$ may be obtained by reducing the static strain $d\eta^s$ from the dynamic effects $d\eta^d$ as:

$$d\eta = d\eta^s - d\eta^d = (A^n d\zeta - Kdt) N. \tag{36}$$

Taking the norm of Eq.(36) yields and rearranging with respect to $d\zeta$ gives:

$$d\zeta = |\, d\eta \,| = \frac{K}{A^n - 1} dt. \tag{37}$$

Since inelastic strain length $d\zeta$ is positive, from Eq.(37) we have:

$$A>1 \; . \tag{38}$$

Eq.(38) means that the current stress S, i.e., the loading surface f_l exists outside the yield surface f_y. If we assume the coefficient K as a function of A in the form:

$$K = K_o(A^n-1)^{m+1} \; , \tag{39}$$

where K_o and m are material constants. We have $d\zeta$ by introducing Eq.(39) in Eq.(37) as:

$$d\zeta = K_o(A^n-1)^m \; dt \; . \tag{40}$$

Also, substituting Eq.(40) in Eq.(14) yields:

$$d\eta = K_o(A^n-1)^m \; Ndt \; . \tag{41}$$

The viscoplastic potential Ω^v can be written by:

$$\Omega^v = K_o S_y \int_1^A (A^n-1)^m \; dA \; . \tag{42}$$

The evolutional equation for the back stress $r^{(i)}$ is expressed from Eqs.(4a) and (40) as:

$$dr^{(i)} = 2\mu_o\rho_{1i}d\eta - \frac{\alpha_i r^{(i)}}{Af(\zeta)} d\zeta \; , \tag{43a}$$

or using Eq.(40) yields:

$$dr^{(i)} = 2\mu_o\rho_{1i}d\eta - \frac{\alpha_i r^{(i)}}{Af(\zeta)} K_o(A^n-1)^m \; dt \; . \tag{43b}$$

The other recent rate-dependent constitutive equation of Chang, Sugiura & Lee[10] have defined the inelastic strain length $d\zeta$ instead of Eq.(37) by:

$$d\zeta^2 = d\eta^s{:}d\eta^s - d\eta^d{:}d\eta^d \; , \tag{44}$$

where $d\eta^s = ANd\zeta$ and $d\eta^d$ is the same as Eq.(35). It is obvious to find the difference of difinition $d\zeta$ in Eqs.(37) and (44). The obtained constitutive equation of Chang, Sugiura & Lee[10] is as follows:

$$d\eta = K_o(\sqrt{A^2-1})^m \; Ndt \; . \tag{45}$$

The above constitutive modeling corresponds to material under high strain rate as well as higher stress than the yield stress. To apply overstress model for quasi-static creep behavior at lower stress, it may be necessary to introduce the drag stress D instead of the yield stress. Assuming the quantity A defined by $A=|S-r(z)|/D$ is large enough compared to unity and power number for plasticity $n = 1$, we can obtain the following creep equation as well as the creep potential:

$$d\eta = K_o\langle A\rangle^m \; Ndt \; , \tag{46a}$$

$$\Omega^c = K_o S_y \frac{1}{m+1}\langle A\rangle^{m+1} \; , \tag{46b}$$

which coinsides with the Walker's creep model[15]. The evolutional equations for $r^{(i)}$ and D in the present theory are expressed by:

$$dD = \gamma\,(D^{\infty}-D)K_oA^m\,dt\,, \tag{47a}$$

$$dr^{(i)} = 2\mu_o\rho_{1i}d\eta - \alpha_i r^{(i)}K_1\frac{A^{m-1}}{D}dt\,. \tag{47b}$$

Conclusions

Based on the fundamental stress-strain relation in the integral form, the new intrinsic time is defined to be valid for the case that the current stress does not exist on the yield surface. Derived are the constitutive equations for quasi-static and dynamic viscoplastic deformations in a unified manner using the new internal time concept. The present internal time models are shown to be general enough to relate closely to the previous 'Overstress Model' developed in the rate-dependent problems.

References

1. Malvern, L.E.: The propagation of longitudinal waves of plastic deformation in a bar of material exhibiting a strain-rate effect. J. Appl. Mech., ASME 18 (1951) 203.
2. Perzyna, P.: The constitutive equation for rate sensitive plastic materials. Q. Appl. Math., 20 (1963) 321.
3. Chaboche, J.L.: Viscoplastic constitutive equations for the description of cyclic and anisotropic behavior of metals. Bull. L'Aca. Pol. Sci., Ser. Sci. Tech., XXV (1977) 33.
4. Robinson, D.N.: A unified creep-palsticity model for structural metals at high temperature. ORNL TM-5969, (1978)
5. Bruhns, O. ed.: Constitutive modeling in the range of inelastic deformation - a state of arts report. INTERATOM Rep. RAP-055-D (B) (1984)
6. Krempl, E.; McMahon, J.J.; Yao, D.: Viscoplasticity based on overstress with a differential growth law for the equilibrium stress. Mech. Mat. 5 (1986) 35.
7. Sahashi, T.; Imatani, S.; Inoue, T.: A constitutive model accounting for the plasticity-creep interaction condition, and the description of inelastic behavior of 2¼ $Cr-1\,M_o$ steel. Trans. JSME, A 52-476 (1986) 1126.
8. Valanis, K.C.: Fundamental consequence of a new intrinsic time measure - plasticity as a limit of the endochronic theory. Arch. Mech. 32 (1980) 171.
9. Wu, H.C.; Yip, M.C.: Strain rate and strain history effects on the dynamic behavior of metallic materials. Int. J. Solids Struct. 16 (1980) 515.
10. Chang, K.C.; Sugiura, K.; Lee, G.C.: Rate-dependent material model for structural steel. J. Eng. Mech., ASCE, 115-3 (1989) 465.
11. Watanabe, O; Atluri, S.N.: Internal time,general internal variable, and multi-yiels-surface theories of plasticity and creep: a unification of concepts. Int. J. Plast. 2 (1986) 37.
12. Watanabe, O.; Atluri, S.N.: Constitutive modeling of cyclic plasticity and creep, using an internal time concept. Int. J. Plast. 2-2 (1986) 107.
13. Chaboche, J.L.: Constitutive equations for cyclic plasticity and cyclic viscoplasticity. Int. J. Plast. 5 (1989) 247.
14. Nouailhas, D.; Policella, H.; Kaczmarek, H.: On the description or cyclic hardening under complex loading histories. Int. Conf. Constitutive Laws for Engineering Materials, Tucson, Arizona, Desai & Gallagher ed. (1983)
15. Walker, K.P.: Research and development program for non-linear structural modeling with advanced time-temperature dependent constitutive relationships. Repot PWA-5700-50, NASA CR-165533 (1981)

Characterization of Nonlinear Viscoelastic Materials for a Special Class of Loading Histories

H. WEBER

Institut für Mechanische Verfahrenstechnik und Mechanik
University of Karlsruhe, Germany

Summary

Soft polymeric materials, polymer foams and some filled polymers may be regarded as nonlinear viscoelastic materials. Structures which are made from these materials like cushions and damping elements usually are loaded by a static preload and by superimposed harmonic vibrations of small amplitudes. For such loading histories a dynamic constitutive equation is developed starting from a "simple material" approach in which the stress tensor is a functional of strain tensor. By using a single integral representation of the strain functional and after linearization with respect to the dynamic strain a special dynamic constitutive equation is derived for the stationary dynamic state. By this equation the dynamic material behaviour is characterized by 6 complex dynamic moduli. These moduli must be determined in dynamic tests under simple extension and/ or simple shear for the static preloading and the dynamic loading, respectively. The test procedure and the experimental equipment are described. For a polyethylene (PE)-foam results are reported. Problems which arise when the constitutive equation is implemented into a finite element programme are shortly discussed.

Introduction

Many soft polymeric materials like PVC, polymer foams and some rubbers may be regarded as nonlinear viscoelastic materials. They are extensively used in different fields of techniques. For instance in transportation it is usual to protect sensitive goods against vibrations and impact loading by polymer foams. Homogeneous and various kinds of filled polymers are introduced to produce damping elements and highly flexible machine parts.

In general the loading of structures made of materials like those described above consists of a static preload which mostly produces large deformations, and of superimposed vibrations with small amplitudes. For the problem of transportation-induced vibrations Ziegahn /1/ showed that in a first step we may treat the dynamic problem as one of harmonic excitation at a single frequency.

To solve structural dynamic problems for the described class of materials and of the specified loading histories for instance by a finite element programme, we have to characterize the nonlinear viscoelastic material behaviour by a constitutive equation. Here such an equation is presented. It will be shown how to determine experimentally the material parameters occuring in it. The tests which are to be performed as well as the problems of the implementation into a finite element programme are explained.

Constitutive equation

Though the materials under consideration exhibit pronounced differences in microstructure /2/ we may

M. Zyczkowski (Ed.)
Creep in Structures
IUTAM Symposium Cracow/Poland 1990

assume that within the frame of a phenomenological theory of material behaviour they can be treated as homogeneous and isotropic materials. They are regarded as "simple materials" for which Cauchy's stress Tensor **T** depends on the history of the deformation gradient **F** only, i.e. **T** is a functional of **F**. A restriction on the form of this functional arises from the principle of objectivity. It can be shown /3/ that in the following form

$$\mathbf{T}(t) = \mathbf{R}(t)\,[\,\mathbf{f}_{s=-\infty}^{t}\,\mathbf{G}(s)\,]\,\mathbf{R}(t)^{T} \tag{1}$$

the constitutive equation is objective for arbitrary functionals **f**. In eqn. (1) **G** is Green's nonlinear strain tensor and **R** represents the rotation tensor. Expressing **f** by the most general single-integral relation of the theory of nonlinear viscoelasticity we arrive at

$$\mathbf{T}(t) = \mathbf{R}(t)\,\{\int_{s=-\infty}^{t} [\,\varphi_0 \mathbf{1} + \varphi_1 \mathbf{G}(s) + \varphi_2 \mathbf{G}(s)^2\,]\,ds\,\}\,\mathbf{R}(t)^{T}\;. \tag{2}$$

Here φ_k, k = 0, 1, 2 , are relaxation functions which depend on the invariants of **G** and on t - s where s represents the time history. Development of φ_k into series with respect to the invariants of **G** to the order of $\mathbf{G}^2$ yields

$$\mathbf{T}(t) = \mathbf{R}(t)\,\{\int_{s=-\infty}^{t} [(K_1\,\mathrm{sp}\,\mathbf{G} + K_2\,\mathrm{sp}\,(\mathbf{G}^2) + K_3\,(\mathrm{sp}\mathbf{G})^2)\mathbf{1}$$

$$+K_4\,\mathbf{G} + K_5\,\mathrm{sp}\,\mathbf{G}\cdot\mathbf{G} + K_6\,\mathbf{G}^2\,]\,ds\}\;\mathbf{R}(t)^{T} \tag{3}$$

where **G** = **G** (s) and $K_i = K_i$ (t-s), i = 1, 2, ...,6, are relaxation functions. With sp ( ) we denote the trace of a tensor.

The special deformation history outlined above can be expressed by

$$\mathbf{G}(s) = \mathbf{G}_m + H(s)\,\bar{\mathbf{G}}_0\,e^{i\omega s}$$

$$\mathbf{R}(t) = \mathbf{R}_m + H(t)\,\bar{\mathbf{R}}_0\,e^{i\omega t} \tag{4}$$

$\mathbf{G}_m$ and $\mathbf{R}_m$ are the time independent strain and rotation tensors in the totally relaxed static equilibrium state due to the preloading, respectively, and $\bar{\mathbf{G}}_0$, $\bar{\mathbf{R}}_0$ are complex amplitude tensors of the superimposed harmonic vibration loading. H is the Heaviside step function and ω represents the circular frequency of excitation. It should be noted that the additive decomposition of **G** and **R** into a static and a dynamic part implies that $\bar{\mathbf{G}}_0$ and $\bar{\mathbf{R}}_0$ are not independent of the static strains and rotations.

Introducing eqn. (4) into eqn. (3) results in an additive decomposition of **T**. If we are only interested in the dynamic stresses we arrive at the following equation for the complex amplitude of the dynamic stress tensor in the stationary dynamic state, e.g. for $t \to \infty$ /4/

$$\bar{T}_o = R_m \{ [\bar{\psi}_1 \, sp \, \bar{G}_o + \bar{\psi}_2 \, sp\,(G_m \bar{G}_o) + \bar{\psi}_3 \, sp \, G_m \, sp \, \bar{G}_o] \, 1 + \bar{\psi}_4 \bar{G}_o + \bar{\psi}_5 (sp \, G_m \cdot \bar{G}_o + G_m \cdot sp \, \bar{G}_o) + \bar{\psi}_6 (G_m \cdot \bar{G}_o + \bar{G}_o G_m) \} R_m^T \qquad (5)$$

In this equation terms which are of higher order than $\bar{G}_o$ and $G_m \bar{G}_o$ and dynamic rotations are neglected. The coefficients $\bar{\psi}_k$, k = 1, 2, ..., 6, are complex dynamic moduli which depend on the circular frequency ω of the excitation.

Material testing

To determine the dynamic moduli we have to perform suitable experiments. Evaluating eqn. (5) in component form and applying it to deformation states with simple extension and/or simple shear for the static and dynamic loading, respectively, yields relations from which the $\bar{\psi}$'s may be calculated. The results are given in Table 1.

type of test	static loading	dynamic loading	stress-strain relations	involved dynamic moduli
1	simple extension	simple extension	$\bar{\tau}^o_{xx} = \bar{E}_z \bar{\varepsilon}^o_{xx}$	$\bar{E}_z = \bar{E}_z(\bar{\psi}_1, \ldots \bar{\psi}_6, \varepsilon^m_{xx}, \nu_m, \bar{\nu}_o)$
2		simple shear	$\bar{\tau}^o_{xy} = 2\bar{G}_z \bar{\varepsilon}^o_{xy}$	$\bar{G}_z = \bar{G}_z(\bar{\psi}_4, \ldots \bar{\psi}_6, \varepsilon^m_{xx} \nu_m)$
3	simple shear	simple extension	$\bar{\tau}^o_{xx} = \bar{E}_s \bar{\varepsilon}^o_{xx}$	$\bar{E}_s = \bar{E}_s(\bar{\psi}_1, \ldots \bar{\psi}_6, \varepsilon^m_{xy}, \bar{\nu}_o)$
4		simple shear	$\bar{\tau}^o_{xy} = 2\bar{G}_s \bar{\varepsilon}^o_{xy}$	$\bar{G}_s = \bar{G}_s(\bar{\psi}_4, \ldots \bar{\psi}_6, \varepsilon^m_{xy})$

Table 1: Determination of dynamic moduli

In this Table $\bar{\tau}_{xx}{}^o$ and $\bar{\tau}_{xy}{}^o$ denote the normal and shear stresses which have to be applied to the specimens to produce the strains $\bar{\varepsilon}_{xx}{}^o$ and $\bar{\varepsilon}_{xy}{}^o$, respectively, where $\bar{\varepsilon}_{xx}{}^o$ and $\bar{\varepsilon}_{xy}{}^o$ are engineering strains. They depend on the components of $\bar{G}_o$ by well known kinematical relations /5/. $\bar{E}_z$, $\bar{E}_s$, $\bar{G}_z$ and $\bar{G}_s$ constitute another set of complex moduli analogous to Young's and shear modulus in elasticity which we use for convenience. They are functions of the unknown moduli $\bar{\psi}_k$, k = 1, 2, ..., 6, as indicated in Table 1. In addition they depend on the prestrains.

Now the question arises which kinds of specimens are appropriate to realize the static and dynamic loadings according to the types 1 through 4 of Table 1.

We used

- thin tubes with axial and torsional loadings
- solid rods with circular cross sections under axial and torsional loadings and
- solid rods with square cross sections under axial and lateral loadings

Tests with solid rods with square cross sections are carried out with an electrodynamic shaker. For static preloading simple extension is used. It is performed by a loading frame shown in Figure 1

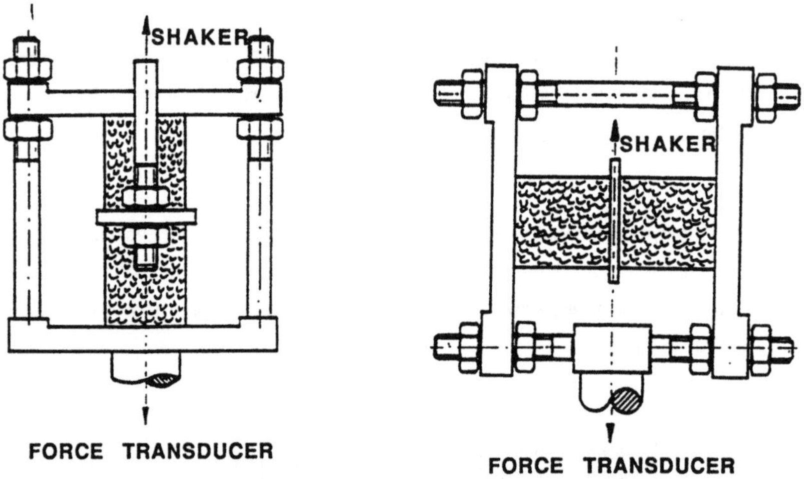

Figure 1: Loading frame

Depending on the installation of the loading frame into the shaker, the specimen may be excited dynamically by simple extension or simple shear.

Results

Results are reported here for a PE-foam. The experiments with dynamic simple extension yield the modulus $\bar{E}_Z$ whereas the experiments with dynamic simple shear result in the modulus $\bar{G}_Z$, each for different static strains ε_m and for frequencies between 10 Hz and 50 Hz. To calculate $\bar{E}_Z$ and $\bar{G}_Z$ corrections have to be introduced to take into account that the stress and deformation states in the specimens are not homogeneous and that in the shear loading bending may not be neglected /6,7/. Results are given in Figures 2 and 3

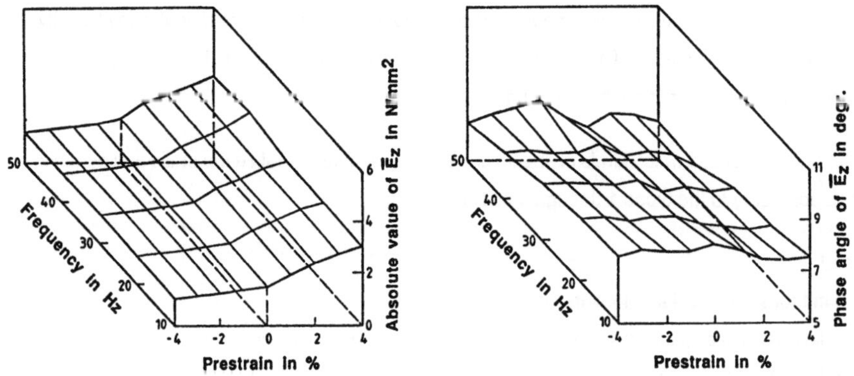

Figure 2: Dynamic moduls $\bar{E}_Z$

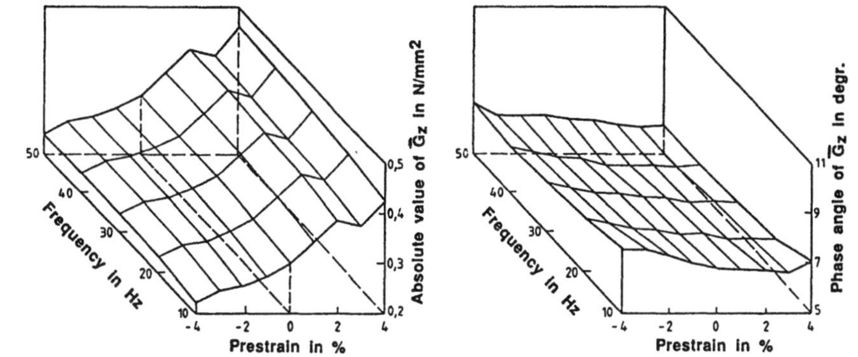

Figure 3: Dynamic modulus $\overline{G}_Z$

It turns out that for the PE-foam under investigation the absolute value of $\overline{E}_Z$ and $\overline{G}_Z$ is strongly dependent on prestrain whereas the dependence on frequency is poor. The phase angle of $\overline{E}_Z$ and $\overline{G}_Z$ is nearly independent on prestrain and frequency. With these results it is possible to evaluate the dynamic moduli $\overline{\psi}_k$, k = 1, 2, .., 6, considering the relations of Table 1 and additional relations for stress components which are zero or neglectable using a computer programme which applies least squares approximation methods /4/. As for foams it is impossible to determine dynamic lateral strains these quantities are calculated from $\overline{E}_Z$ and $\overline{G}_Z$ /4/. In Figure 4 the real and imaginary parts of the ψ_k s are plotted against frequency f

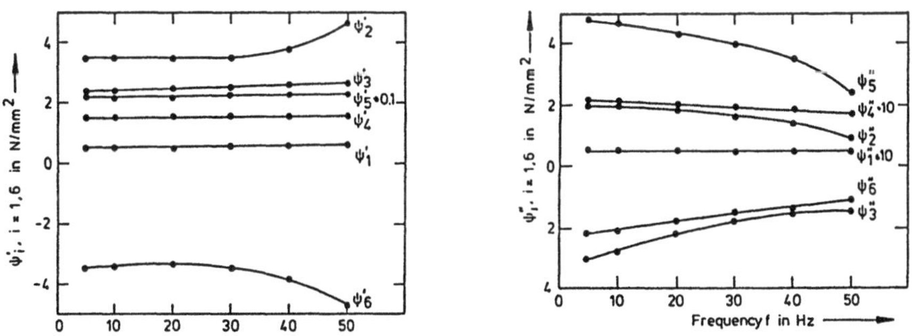

Figure 4: Real and imaginary parts of the dynamic moduli

<u>Finite element programming</u>

To develop the basic equations of a finite element programme with which the dynamic response of a structure made of the specified material may be calculated we apply the method of weighted residuals. Starting from the balance of momentum and considering the special loading history and the related dynamic constitutive equation (5) we arrive at the following global system of equations

$$(-\omega^2 [M] + [\bar{K}])\,\{\bar{u}_o\} = \{\bar{F}_o\} \tag{6}$$

Here $\{\bar{u}_o\}$ represents the column matrix of the complex dynamic nodal displacements. [M] is the symmetric mass matrix whereas $[\bar{K}]$ turns out to be a nonsymmetric complex stiffness matrix /8/. In order to use the Cholesky-algorithm which in most FE-programmes is used to solve the global system of linear equations a iterative procedure has to be applied.

With

$$[\bar{S}] = -\omega^2 [M] + [\bar{K}_{symm}] \tag{7}$$

and the initial displacement matrix

$$\{\bar{u}_o\}^{(o)} = \{0\} \tag{8}$$

the problem is described by /9/

$$\{\bar{u}_o\}^{(i)} = [\bar{S}]^{-1} (\{\bar{F}_o\} - [\bar{K}_{unsym}]\,\{\bar{u}_o\}^{(i-1)}),\ i \geq 1 \tag{9}$$

Here i denotes the iteration step.

Preliminary tests with beam structures for which analytical solutions are easily available show good agreement between FE- and analytical solutions.

Acknowledgement

The author gratefully acknowledges the financial support from Deutsche Forschungsgemeinschaft (DFG)

References

1. Ziegahn, K.-F.: Ein Beitrag zur Ermittlung und Beschreibung von mechanischen Transportbelastungen. Dissertation, Universität Karlsruhe 1987
2. Weber, H.; Geißler, A.; Kugler, H.-P.: Einsatz der experimentellen Mechanik in der statischen und dynamischen Prüfung unverstärkter Kunststoffe, VDI-Berichte 679, Düsseldorf:VDI-Verlag 1988, 289-300
3. Lockett, F. I.: Nonlinear Viscoelastic Solids. London: Academic Press 1972
4. Geißler, A.: Dynamisches Verhalten von compoundierten Kunststoffen unter Vorbelastung, dargestellt am Beispiel eines geschlossenzelligen Schaumstoffes und eines hochgefüllten Kunststoffes. Dissertation, Universität Karlsruhe, to be published
5. Becker, E.; Bürger, W.: Kontinuumsmechanik. Stuttgart: Teubner 1975
6. Del Pozo, L.; Weber, H.: Optische Untersuchung des Verformungsverhaltens von kurzen Zug/Druck-Proben aus weichem Kunststoff. Wissenschaftliche Abschlußberichte, 24. Internationales Seminar, Universität Karlsruhe 1989, 138-147
7. Wolf, H.: Berechnung komplexer dynamischer Moduln aus gemessenen Elastizitäts- und Schubmoduln. Seminararbeit, Institut für Mechanische Verfahrenstechnik und Mechanik, Universität Karlsruhe 1989
8. Mehrachsiges Verhalten nichtlinear viskoelastischer Materialien unter Schwing- und Stoßbelastung. Bericht zum DFG-Vorhaben WE 700/7-1 (Antragsteller H. Weber), Institut für Mechanische Verfahrenstechnik und Mechanik, Universität Karlsruhe 1990
9. Geißler, E.; Geißler, A.; Weber, H.: Ein Stoffgesetz für nichtlinear viskoelastische Werkstoffe unter großer statischer und überlagerter dynamischer Belastung. To be published in ZAMM 71 (1991)

Constitutive Law and Force Recovery for Large Strain Problems

Yeong-Bin Yang and Liang-Jenq Leu

Department of Civil Engineering
National Taiwan University
Taipei, Taiwan 10764, R.O.C.

Summary

For trusses with large strains, the use of identical material constants in an incremental analysis does not imply identical material behaviors for formulations with different reference states. For consistency, a transformation of the material constants from one formulation to the other based on the continuum mechanics rules is required. In calculating the member forces, it is recommended that the "total form," rather the "incremental form," be used at each step to avoid possible numerical errors.

1. Introduction

Large strain nonlinear problems are usually solved by incremental methods with iterations performed at each loading step for equilibrium. With such methods, all quantities and relations are expressed in incremental forms. In this regard, care must be taken in the definition of the material constants and in the recovery or calculation of member forces.

In describing the motion of a body, three configurations are usually required (Fig. 1): C_0—the initial undeformed state; C_1—the current (known) deformed state; and C_2—a neighboring (desired) deformed state. The total (TL) and updated (UL) formulations adopt C_0 and C_1, respectively, as the reference state. Both the TL and UL formulations can be considered as the special cases of a general Lagrangian (GL) formulation that adopts an arbitrary configuration C_m between C_0 and C_1 as the reference state [1].

Either the TL, UL, or GL formulation requires the specification of an incremental constitutive law. In terms of the 2nd Piola-Kirchhoff stress increments S and Green-Lagrange strain

M. Zyczkowski (Ed.)
Creep in Structures
IUTAM Symposium Cracow/Poland 1990

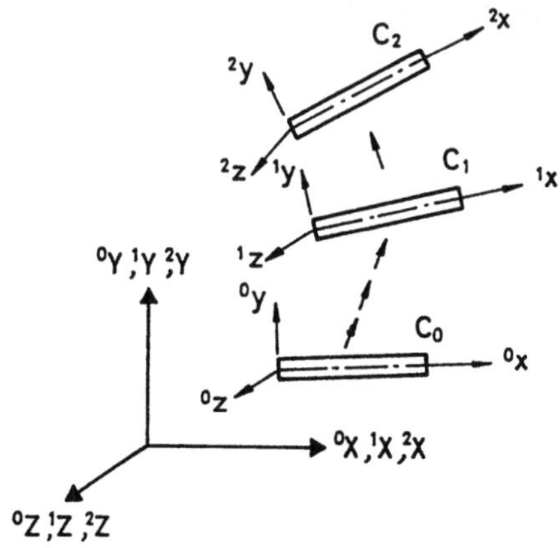

Fig. 1 Reference configurations for a moving body

increments ε, the constitutive law for truss members has been commonly assumed to possess a material constant E as follows:

$$\text{TL:} \quad {}_{\emptyset}S = {}_{\emptyset}E \; {}_{\emptyset}\varepsilon \tag{1a}$$

$$\text{UL:} \quad {}_{1}S = {}_{1}E \; {}_{1}\varepsilon \tag{1b}$$

$$\text{GL:} \quad {}_{m}S = {}_{m}E \; {}_{m}\varepsilon \tag{1c}$$

where the subscripts "Ø", "1", "m" indicate the reference states $C_{\emptyset}$, C_1, and C_m. The above stress-strain relations are incomplete at least for two aspects:

(1) The three formulations are not equivalent, even when ${}_{\emptyset}E = {}_{1}E = {}_{m}E$, due to their difference in reference states.

(2) Each of the relations in Eq. 1 implies a piecewise linear material for incremental analyses with finite step sizes.

The second problem can be solved by using a "total form" for calculating the member forces, while the first problem can be solved by using the following transformation rule [2] for transforming the material constants from one reference state to the other:

$${}_{m}C_{ijkl} = \frac{{}^{m}\rho}{{}^{n}\rho} \; {}_{n}C_{pqrs} \; \frac{{}^{m}x_{i}}{{}^{n}x_{p}} \; \frac{{}^{m}x_{j}}{{}^{n}x_{q}} \; \frac{{}^{m}x_{k}}{{}^{n}x_{r}} \; \frac{{}^{m}x_{l}}{{}^{n}x_{s}} \tag{2}$$

where x = coordinate axes; C = material constants; ρ = density of material, and the subscripts and superscripts "m" and "n" denote two arbitrary reference states C_m and C_n.

Due to the restriction of paper length, only a brief description of the force recovery procedure for truss members based on the TL, UL, and GL formulations will be given herein. The readers should refer to [3] and [4] for more details.

2. Total Form for Force Recovery

Consider a bar that is stretched with an infinite number of infinitesimal steps from the $C_\emptyset$ to C_2. Assuming the initial and final length of the bar to be ${}^{\emptyset}L$ and ${}^{2}L$, respectively, the member force at C_2 can be derived using each formulation.

(a) TL Formulation:

Let C_i denote an arbitrary configuration of the bar between $C_\emptyset$ and C_2, ${}^{i}_{\emptyset}\varepsilon$ = the Green-Lagrange strain at C_i, and ${}_{\emptyset}\varepsilon$ and ${}_{\emptyset}S$ = the strain and stress increments, respectively, from C_i to an infinitesimally close neighboring configuration. The 2nd Piola-Kirchhoff stress at C_2 can be obtained by integration,

$$ {}^{2}_{\emptyset}S = \int {}_{\emptyset}S = \int_{\emptyset}^{{}^{2}_{\emptyset}\varepsilon} {}^{i}_{\emptyset}E \; {}_{\emptyset}\varepsilon \qquad (3) $$

For conserved mass, ${}^{2}\rho / {}^{\emptyset}\rho = {}^{\emptyset}A \; {}^{\emptyset}L / ({}^{2}A \; {}^{2}L)$ and $\partial {}^{2}x / \partial {}^{\emptyset}x = {}^{2}L / {}^{\emptyset}L$, one can relate the Cauchy stress ${}^{2}\tau$ to the 2nd Piola-Kirchhoff stress ${}^{2}_{\emptyset}S$ as follows [2]:

$$ \tau^{2} = {}^{2}_{\emptyset}S \; {}^{\emptyset}A \; {}^{2}L / ({}^{2}A \; {}^{\emptyset}L) \qquad (4) $$

Consequently, the axial force ${}^{2}_{2}F$ acting at C_2 is

$$ {}^{2}_{2}F = {}^{2}\tau \; {}^{2}A \qquad (5a) $$

or

$$ {}^{2}_{2}F = {}^{2}_{\emptyset}S \; {}^{\emptyset}A \; ({}^{2}L / {}^{\emptyset}L) \qquad (5b) $$

(b) UL Formulation:

The UL formulation adopts a moving reference coordinate system. For the purpose of obtaining closed form expressions, one may convert the UL formulation into an equivalent TL formulation, which adopts a fixed reference coordinate system. The 2nd Piola-Kirchhoff stress increment for the UL formulation is ${}_{i}S = {}^{i}_{\emptyset}E \; {}_{i}\varepsilon$, which is equivalent to using ${}_{\emptyset}S = {}_{\emptyset}E^{U}_{eq} \; {}_{\emptyset}\varepsilon$ in the TL formulation based on the transformation of Eq. 2:

$$ {}_{\emptyset}E^{U}_{eq} = {}^{i}_{\emptyset}E \; ({}^{i}A/{}^{\emptyset}A) \; ({}^{\emptyset}L/{}^{i}L)^{3} \tag{6} $$

Accordingly, the equivalent stress for the TL formulation is

$$ {}^{2}_{\emptyset}S = \int_{\emptyset}^{{}^{2}_{\emptyset}\varepsilon} {}_{\emptyset}E^{U}_{eq} \; {}_{\emptyset}\varepsilon \tag{7} $$

The member force ${}^{2}_{2}F$ can be calculated from Eq. 5b as well.

(c) GL Formulation:

For the sake of comparison, the GL formulation will be transformed into an equivalent TL formulation with the tangent modulus defined as

$$ {}_{\emptyset}E^{G}_{eq} = {}^{i}_{\emptyset}E \; ({}^{m}A/{}^{\emptyset}A) \; ({}^{\emptyset}L/{}^{m}L)^{3} \tag{8} $$

It follows that the equivalent stress is

$$ {}^{2}_{\emptyset}S = \int_{\emptyset}^{{}^{2}_{\emptyset}\varepsilon} {}_{\emptyset}E^{G}_{eq} \; {}_{\emptyset}\varepsilon \tag{9} $$

and the member force ${}^{2}_{2}F$ can be obtained from Eq. 5b. It should be mentioned that the equations presented in this section are exact, once the material constant E is given.

3. Incremental Form for Force Recovery

In numerical analyses, the increment step from C_1 to C_2 is usually of finite size. Both the TL and GL formulations adopt a fixed reference system, while the UL formulation adopts a moving reference system. It will be shown numerically that the accuracy of the equations derived in this section depend on the step sizes used in analysis, except for the case when ${}^{1}_{\emptyset}E$ equals a constant in formulations with fixed reference coordinates.

The reference configuration C_m for the GL formulation is located between C_0 and C_1. Realizing that $\partial^2x/\partial^mx = {}^2L/{}^mL$, one may relate the Cauchy stress to the 2nd Piola-Kirchhoff stress as

$$ {}^2\tau = \frac{{}^2\rho}{{}^m\rho} \left(\frac{{}^2L}{{}^mL}\right)^2 ({}^1_mS + {}_mS) \tag{10} $$

With the strain increment ${}_m\varepsilon$ defined as [2]:

$$ {}_m\varepsilon = ({}^2L^2 - {}^1L^2)/(2\ {}^mL^2) \tag{11} $$

the stress increment ${}_mS$ can be obtained from Eq. 1c. Substituting Eq. 10 into Eq. 5a, and using Eqs. 1c and 11, one obtains

$$ {}^2_2F = ({}^1_mF + {}_mF)\ {}^2L/{}^mL \tag{12} $$

where for conserved mass, ${}^2\rho/{}^m\rho = {}^mA\ {}^mL/({}^2A\ {}^2L)$, ${}^1_mF$ and ${}_mF$ are defined as follows:

$$ {}^1_mF = {}^1_1F\ {}^mL/{}^1L \tag{13} $$

$$ {}_mF = {}^1_0E\ {}^mA\ ({}^2L^2 - {}^1L^2)/(2\ {}^mL^2) \tag{14} $$

Eq. 13 can be proved by letting C_2 coincide with C_1 in Eq. 12. In this case, ${}^2L = {}^1L$, ${}_mF = 0$, and ${}^2_2F = {}^1_1F = {}^1_mF\ {}^1L/{}^mL$; so the equation is proved.

By letting C_m equal C_0, the following equations can be obtained for the TL formulation:

$$ {}^2_2F = ({}^1_0F + {}_0F)\ {}^2L/{}^0L \tag{15} $$

where the initial bar force ${}^1_0F$ and bar force increment ${}_0F$ are

$$ {}^1_0F = {}^1_1F\ {}^0L/{}^1L \tag{16} $$

$$ {}_0F = {}^1_0E\ {}^0A\ ({}^2L^2 - {}^1L^2)/(2\ {}^0L^2) \tag{17} $$

Also by letting C_m equal C_1, the following equations can be obtained for the UL formulation:

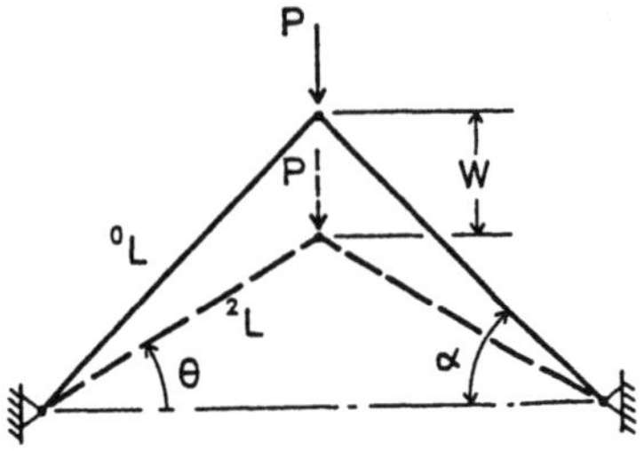

Fig. 2 Two-member truss

$$ {}^{2}_{2}F = ({}^{1}_{1}F + {}_{1}F)\ {}^{2}L/{}^{1}L \tag{18} $$

where the bar force increment ${}_{1}F$ is

$$ {}_{1}F = {}^{1}_{0}E\ {}^{1}A\ ({}^{2}L^{2} - {}^{1}L^{2})/(2\ {}^{1}L^{2}) \tag{19} $$

The equations derived in this section for the bar forces, i.e., Eqs. 12, 15, and 18, will be referred to as the "incremental form," which are piecewise linear, while those derived in Section 3 based on the integration process, i.e., Eq. 5b along with Eqs. 3, 7, and 9, will be termed the "total form," which are fully nonlinear.

4. Numerical Example

For the two-member truss shown in Fig. 2, the following are assumed for each member: ${}^{0}L = 50$ cm and ${}^{0}A = {}^{1}A = {}^{2}A = 10\ cm^{2}$. The deformed shape of the truss with a bar length of ${}^{m}L = 25\ (1+\cos\alpha)$ cm will be used as the reference C_m for the GL formulation. The material is assumed to be linear in the TL sense:

$$ {}^{i}_{0}S = E\ {}^{i}_{0}\varepsilon \tag{20} $$

Following the procedure of Section 2, the "total form" equations can be derived for the bar forces [4]:

$$ \text{TL:}\ {}^{2}_{2}F = E\ {}^{0}A\ {}^{2}L({}^{2}L^{2} - {}^{0}L^{2})/(2\ {}^{0}L^{3}) \tag{21} $$

$$ \text{UL:}\ {}^{2}_{2}F = E\ {}^{0}A\ ({}^{2}L - {}^{0}L)/{}^{0}L \tag{22} $$

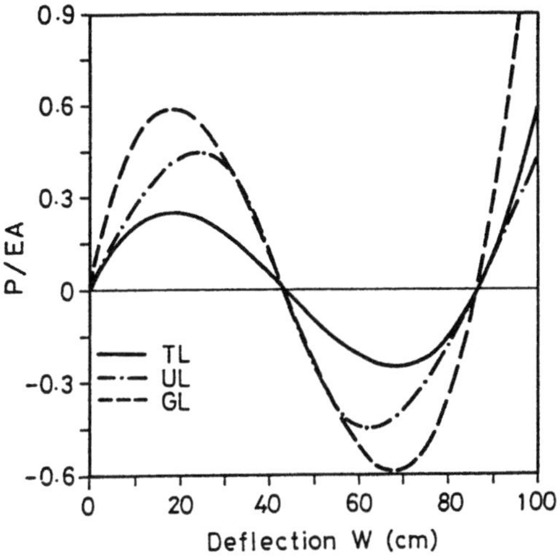

Fig. 3 Load-deflection curves obtained by various formulations

$$\text{TL:}\ {}^{2}_{2}F = E\ {}^{0}A\ {}^{2}L\ ({}^{2}L^{2} - {}^{0}L^{2})/(2\ {}^{m}L^{3}) \qquad (23)$$

With these equations, exact load-deflection curves obtained for the truss with $\alpha = 60^{o}$ have been plotted in Fig. 3. From this figure, one observes that the adoption of material laws of the type defined in Eqs. 1 will result in significantly different load-deflection responses for trusses with large strains, if formulations with different reference states are used.

In addition, the solutions obtained with the "incremental form" (Section 3) and "total form" (Section 2 or Eq. 22) for the UL formulation have been compared in Fig. 4. One observation herein is that the solutions obtained with the "incremental form" are step-size dependent, due to their piecewise linear nature in material laws. Such solutions, nevertheless, will converge to the exact one obtained by the "total form" as the step sizes are reduced.

5. Conclusions

For trusses with large strains, the use of identical material constants in the TL, UL, and GL formulations does not imply identical material properties due to their difference in

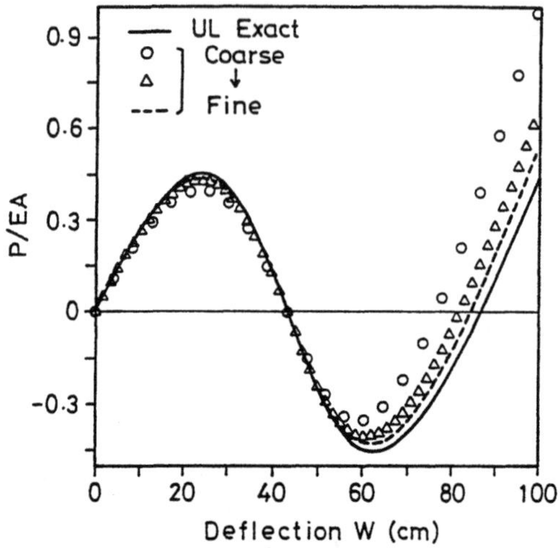

Fig. 4 Solutions obtained with incremental vs. total form

reference states. In an incremental analysis, it has further been shown that the "total form," rather the "incremental form," should always be used for calculating the bar forces, so as to avoid possible numerical errors induced by the step-wise linearization.

Acknowledgement

This research was sponsored by the National Science Council of the Republic of China under Grant Number NSC76-Ø414-PØØ2-Ø18.

References

1. Cescotto, S., Frey, F., and Fonder, G., "Total and Updated Lagrangian Descriptions in Nonlinear Structural Analysis: A Unified Approach," Energy Methods in Finite Element Analysis, edited by Glowinski, R., Rodin, E. Y., and Zienkiewicz, O. C., John Wiley & Sons, New York, N.Y., 1979, Chapter 15, pp. 283-296.

2. Fung, Y. C., Foundations of Solid Mechanics, Prentice-Hall, Englewood Cliffs, N.J. (1965).

3. Yang, Y. B., and Leu, L. J., "Postbuckling Analysis of Trusses with Various Lagrangian Formulations," AIAA Journal, American Institute of Aeronautics and Astronautics, Vol. 28, No. 5, May 199Ø, pp. 946-948.

4. Yang, Y. B., and Leu, L. J., "Constitutive Laws and Force Recovery Procedures in Nonlinear Analysis of Trusses," accepted for publication in the journal of Computer Methods in Applied Mechanics and Engineering.

Creep of Isotropic and Anisotropic Materials with Different Behaviour in Tension and Compression

A.A. ZOLOCHEVSKY

Kharkov Polytechnic Institute, USSR

Summary

A new incremental creep theory of isotropic materials based on the three invariants of stress tensor is proposed. In order to evaluate the constants in the constitutive equations the basic experiments are formulated. Special creep relations that contain less constants and invariants are derived. The approach has been applied to the problem of creep of anisotropic materials with different behaviour in tension and compression.

Introduction

Creep characteristics of different materials (such as light alloys, cast, polymers, composites etc.) mainly depend upon the loading type. This dependence as a rule is determined while comparing creep curves obtained at the same conditions of uniaxial tests on tension and compression. Traditional theory of the isotropic creep, based upon the von Mises potential, does not take into account the different behaviour in tension-compression. That is why some researchers [1-3] suggested the usage of odd invariant stresses with different refinements of classical creep dependencies.

Constitutive equations for isotropic materials

In the most theories of creep the hypothesis of the potential existence is used

$$F = \sigma_e^2 , \tag{1}$$

where the magnitude σ_e setting the equivalence of uniaxial and combined stress states is determined as the equivalent stress. Obviously expression for σ_e must contain all three invariants $I_1 = \sigma_{nn}$, $I_2 = \sigma_{kl}\sigma_{kl}$, $I_3 = \sigma_{kn}\sigma_{nl}\sigma_{lk}$ of stress tensor σ_{kl} and some constants. Let us consider the combined invariants of stress tensor and material constants A, B, C, D, K, L:

$$\sigma = BI_1 ; \quad \sigma_0^2 = AI_1^2 + CI_2 ; \quad \sigma_1^3 = DI_1^3 + KI_1I_2 + LI_3 . \tag{2}$$

Then taking into account that equivalent stress must be the homogeneous function of the stress state, let us suggest

$$\sigma_e = \alpha\sigma + \sigma_0 + \gamma\sigma_1 , \tag{3}$$

M. Zyczkowski (Ed.)
Creep in Structures
IUTAM Symposium Cracow/Poland 1990

where α, γ are the numerical coefficients. The relations (1)-(3) include as a particular cases the relations known in literature for creep potentials. For example, if we suppose in (2), (3) $\alpha = \gamma = 0$ and $C = -3A$, we shall obtain the expression

$$\sigma_e = \sqrt{\frac{2}{3}C}\sigma_i \,, \tag{4}$$

that is it coincides with accuracy to constant with expression $\sigma_e = \sigma$ in the von Mises potential on the base of the stress intensity σ_i. In the case $\gamma = 0, C = -3A$ $L = A = 1$ the well-known expression for the potential of Drucker and Prager is obtained from equivalent stress (3):

$$\sigma_e = BI_1 + \sigma_i \,. \tag{5}$$

Using the associated law

$$\dot{P}_{kl} = \lambda \frac{\partial F}{\partial \sigma_{kl}} \tag{6}$$

for the components of the tensor of creep strain rate $\dot{P}_{kl}$ and taking into account the relation (1)-(3) we obtain the equations

$$\begin{aligned} \dot{P}_{kl} &= 2\lambda\sigma_e \left[\frac{AI_1\delta_{kl} + C\sigma_{kl}}{\sigma_0} + \alpha B\delta_{kl} + \right. \\ &+ \left. \gamma \frac{DI_1^2\delta_{kl} + L\sigma_{kn}\sigma_{nl} + \frac{K}{3}(I_2\delta_{kl} + 2I_1\sigma_{kl})}{\sigma_1^2} \right] . \end{aligned} \tag{7}$$

Let us find the scalar factor λ. First of all we introduce the equivalent creep strain rate P_e for which equality

$$\sigma_e \dot{P}_e = W \tag{8}$$

is correct. Here $W = \sigma_{kl}\dot{P}_{kl}$ is power of dissipation per unit volume. Then we obtain the relation

$$W = 2\lambda\sigma_e^2 \,. \tag{9}$$

Comparing the equalities (8) and (9) we find $2\lambda\sigma_e = \dot{P}_e$. Therefore the equations (7) may be written in the following form:

$$\begin{aligned} \dot{P}_{kl} &= \dot{P}_e \left[\frac{AI_1\delta_{kl} + C\sigma_{kl}}{\sigma_0} + \alpha B\delta_{kl} + \right. \\ &+ \left. \gamma \frac{DI_1^2\delta_{kl} + L\sigma_{kn}\sigma_{nl} + \frac{K}{3}(I_2\delta_{kl} + 2I_1\sigma_{kl})}{\sigma_1^2} \right] . \end{aligned} \tag{10}$$

For the secondary steady-state creep we have $\dot{P}_e = v(\sigma_e)$. Function $v(\sigma_e)$ can be described as power dependence the law of hyperbolic sine $\sinh(\sigma_e/b)$, exponential dependence $\exp(\sigma_e/a)$ and others. In order to describe the creep damage we assume

$\dot{P}_e = v(\sigma_e)/(\varphi_* - \varphi)^m$ where $\varphi = \int w dt$, t is the time, and the magnitude φ_* corresponds to fracture.

Let us consider the foundation of six constants in constitutive equations (10). For this purpose we use the results of the basic experiments on standard specimens, such as, axial tension with the determining of longitudinal strain and transverse strain, axial compression, loading of hydrostatic pressure, pure torsion with the determining of angular strain and axial strain.

Equations (10) are written down in general form. Therefore we analyse all possible special relations, coming from the dependencies (10) and containing only three parameters. For example, if we assume $\gamma = 0$, then we have tensor-linear equations

$$\dot{P}_{kl} = \dot{P}_e \left(\frac{A I_1 \delta_{kl} + C \sigma_{kl}}{\sigma_0} + \alpha B \delta_{kl} \right) . \tag{11}$$

In the case $\alpha B = D = K = 0$ dependencies (10) are transformed into equations of creep:

$$\dot{P}_{kl} = \dot{P}_e \left(\frac{A I_1 \delta_{kl} + C \sigma_{kl}}{\sigma_0} + \gamma \frac{L \sigma_{kn} \sigma_{nl}}{\sigma_1^2} \right) . \tag{12}$$

Using the conditions $\alpha B = D = L = 0$, from dependencies (10) we have

$$\dot{P}_{kl} = \dot{P}_e \left[\frac{A I_1 \delta_{kl} + C \sigma_{kl}}{\sigma_0} + \gamma \frac{K (I_2 \delta_{kl} + 2 I_1 \sigma_{kl})}{3 \sigma_1^2} \right] . \tag{13}$$

Constitutive equations for anisotropic material

Let us obtain the connection between the components of the tensor of strain rate $\dot{\eta}_{kl}$ and the stress tensor τ_{kl} in the theory of the anisotropic creep. The anisotropic behaviour is described by the linear transformation [2] $\sigma_{mn} = M_{klmn} \tau_{kl}$; $\dot{\eta}_{kl} = M_{klmn} \dot{P}_{mn}$ on the base of fourth-rank tensor M_{klmn} and tensors σ_{mn} , P_{mn} for isotropic case. Then an invariants (2) are given by

$$\sigma = b_{ij} \tau_{ij} \ ; \quad \sigma_0^2 = a_{ijkl} \tau_{ij} \tau_{kl} \ ; \quad \sigma_1^3 = C_{ijklmn} \tau_{ij} \tau_{kl} \tau_{mn} \ ,$$

if we define

$$B_{ij} = B M_{ijpp} \ ; \quad a_{ijkl} = A M_{ijpp} M_{klqq} + C M_{ijpq} M_{klpq} \ ;$$

$$C_{ijklmn} = D M_{ijpp} M_{klqq} M_{mnrr} + K M_{ijpp} M_{klqr} M_{mnqr} + L M_{ijrs} M_{klsp} M_{mnpr} \ .$$

Therefore dependencies (10) are transformed into equations of the anisotropic creep:

$$\dot{\eta}_{ij} = \dot{P}_e \left(\frac{a_{ijkl} \tau_{kl}}{\sigma_0} + \alpha b_{ij} + \gamma \frac{C_{ijklmn} \tau_{kl} \tau_{mn}}{\sigma_1^2} \right) . \tag{14}$$

Comparison of experimental and theoretical data

Let us examine experimental control of the present theory of isotropic creep. For aluminum alloy at 473K [4] it is set that in uniaxial tension test $\dot{P}_{11} = K_+\sigma_{11}^n$, in uniaxial compression test $\dot{P}_{11} = -K_-|\sigma_{11}|^n$ and in pure torsion test $2\dot{P}_{12} = N\sigma_{12}^n$. Values of constants are: $K_+ = 5.0 \cdot 10^{-15-n}$; $K_- = 2.5 \cdot 10^{-15-n}$; $N = 11.6 \cdot 10^{-13-n}\mathrm{MPA}^{-n}\mathrm{h}^{-1}$; $n = 8$.

Then basing upon the meaning K_+, K_-, N, n we shall determine the parameters in dependencies (11), (12), (13). The results of the calculations for different combination of normal stress σ_{11} and shear stress σ_{12} are shown in the Table 1. It is seen that theoretical meanings of creep strain rates are in good agreement with experimental meanings.

Table 1:

Stresses [MPa]		Creep strain rates 10^5, h^{-1}							
σ_{11}	σ_{12}	$\dot{P}_{11}$	$2\dot{P}_{12}$	$\dot{P}_{11}$	$2\dot{P}_{12}$	$\dot{P}_{11}$	$2\dot{P}_{12}$	$\dot{P}_{11}$	$2\dot{P}_{12}$
		experiments		equations					
				(11)		(12)		(13)	
107.5	62.00	1.200	2.40	1.290	2.56	1.350	2.75	1.340	2.72
-60.6	84.40	-0.722	3.64	-0.748	4.25	-0.694	3.79	-0.754	3.89
-152.9	36.57	-1.740	1.42	-1.450	1.32	-1.430	1.27	-1.470	1.30
0.0	98.00	0.160	9.02	0.200	9.91	0.000	9.91	0.000	9.91

References

[1] Murakami, S., Sawczuk, A.: *A unified approach to constitutive equations of inelasticity based on tensor function representations*, Nucl. Eng. Dec. **65**, (1981), 33-47

[2] Betten, J.: *Creep theory of anisotropic solids*, J. Rheol., **25**, (1981), 565-581

[3] Sobotka, Z.: *Tensorial expansions in non-linear mechanics*, Praha: Akademia 1984

[4] Zolochevsky, A.A.: *Kriechen von Konstruktionselementen aus Materialien mit von der Belastung abhangigen Charakteristiken*, Tech. Mech. **9**, (1988), 177-184

Part II:
Damage, Creep Crack Growth, Creep Rupture

Localization Phenomena in Thermo-Elastoplasticity

A. BENALLAL and J. LEMAITRE

Laboratoire de Mécanique et Technologie
E.N.S. de Cachan/C.N.R.S./Université Paris 6
61,Avenue du Président Wilson,94235 Cachan,France.

Summary

The localization of deformation is understood as an instability in the macroscopic description of inelastic deformation. The effects of thermomechanical couplings in this localization are analyzed. Conditions for the onset of localization when these couplings are taken into account are exhibited and illustrated in a particular set of constitutive equations . Destabilizing effects due to couplings are underlined.

Introduction

The localization of the deformation in rate-independent (elastic-plastic,damageable,..) materials is usually treated as a bifurcation of the rate problem. A general view of the various bifurcation phenomena for possibly heteregeneous solids under general uncoupled isothermal conditions has been given by Benallal,Billardon and Geymonat [1] using modern theory of linear elliptic boundary value problems. It was in particular demonstrated that three types of localization are possible for incrementally linear solids,depending on the failure of one of the following conditions:

- The boundary complementing condition which governs instabilities at the boundary of a solid. Its failure leads to deformation modes localized at the boundary and to the appearance of stationnary surface waves (such as Rayleigh waves) in the dynamical case.
- The interfacial complementing condition (when the body is heteregeneous) governs instabilities at interfaces of a solid. Its failure leads to deformation modes localized at each side of the interface and to stationnary interfacial waves (such as Stonely waves) in the dynamical case.
- The ellipticity condition, the failure of which leads to modes involving discontinuities of the velocity gradient and which corresponds to the classical definition of localization used by Rice [5]. Its failure leads also to the appearance of stationnary acceleration waves in the dynamical context.

The same analysis and the same results can be applied to the coupled thermomechanical problem; we limit however the study in this paper to a detailed analysis of the role of thermomechanical couplings on localization phenomena of the third type ,but for the non-linear case of constitutive equations described in the next section. We remark in the sequel that, in the

M. Zyczkowski (Ed.)
Creep in Structures
IUTAM Symposium Cracow/Poland 1990

coupled problem, to the surface and interfacial modes underlined above,may correspond through the heat equation, temperature rate fields localized either at the boundaries or the interfaces.

Constitutive relations

We consider in the following a general class of rate-independent (elastic-plastic,damageable,...) materials in the small strain range. The reversible behavior of such materials is given by the knowledge of the specific free energy $\Psi(\varepsilon,\alpha_i,T)$ depending on the strain tensor ε, the absolute temperature T and a set of internal variables α_i assumed to describe various physical mechanisms. The stress tensor σ, the thermodynamical forces A_i associated to the internal variables and the entropy s are obtained by the state laws:

$$\sigma = \rho\frac{\partial\Psi}{\partial\varepsilon} \qquad A_i = -\rho\frac{\partial\Psi}{\partial\alpha_i} \qquad s = -\frac{\partial\Psi}{\partial T} \tag{1a,b,c}$$

where ρ denotes the mass density. The irreversible behavior is characterized by a reversibility domain (inside which no irreversibility occurs) in the forces space and defined by

$$f(A_j,\alpha_j,T) \leq 0 \tag{2}$$

The evolution of the internal variables is determined by the potential of irreversibility $F(A_j,\alpha_j,T)$ through the following normality rule

$$\dot{\alpha}_i = \lambda\frac{\partial F}{\partial A_i}, \qquad \lambda \geq 0, \qquad f(A_j,\alpha_j,T) \leq 0, \qquad \lambda\dot{f}(A_j,\alpha_j,T) = 0 \tag{3a,b,c,d}$$

The heat flux $\mathbf{q}$ is taken to obey a generalized Fourier's law of heat conduction

$$\mathbf{q} = -\ \mathbb{K}(A_j,\alpha_j,T).\mathbf{g} \tag{4}$$

where $\mathbf{g}$ is the temperature gradient and $\mathbb{K}$ the heat conduction matrix assumed to be positive definite in this paper.

When λ is positive, it is computed by the consistency condition f = 0 as

$$\lambda = \left\langle \frac{-\frac{\partial f}{\partial A_j}\bullet\Lambda_j : \dot{\varepsilon} + [\frac{\partial f}{\partial T} - \frac{\partial f}{\partial A_j}\bullet\Gamma_j]\dot{T}}{h} \right\rangle \tag{5}$$

where we have set:

$$h = \frac{\partial f}{\partial A_i} \bullet \Pi_{ij} \bullet \frac{\partial F}{\partial A_j} - \frac{\partial f}{\partial \alpha_k} \bullet \frac{\partial F}{\partial A_k} \tag{6}$$

$$\Lambda_j = \rho \frac{\partial^2 \Psi}{\partial\partial\ \varepsilon}, \qquad \Pi_{ij} = \rho \frac{\partial^2 \Psi}{\partial \alpha_i \partial \alpha_j}, \qquad \Gamma_j = \rho \frac{\partial^2 \Psi}{\partial T \partial \alpha_j} \tag{7a,b,c}$$

The symbol "•" is the adequate contraction depending on the nature of the internal variables α_j; ":" is the double tensor contraction and <.>, the MacCauley bracket.
With the former constitutive relations, the heat equation traducing locally conservation of energy is written as

$$\rho c \dot{T} = T\Delta : \dot{\varepsilon} + (T\,\Gamma_j + A_j) \bullet \dot{\alpha}_j + r - \mathrm{Div}\mathbf{q} \tag{8}$$

where r is the heat internal supply , c the specific heat capacity at constant strain and internal state and we have set

$$c = -T \frac{\partial^2 \Psi}{\partial T^2} \tag{9}$$

$$\mathbb{D} = \mathbb{E} + \frac{T}{\rho c} \Delta \otimes \Delta \tag{10}$$

$$\mathbb{E} = \rho \frac{\partial^2 \Psi}{\partial \varepsilon \partial \varepsilon}, \qquad \Delta = \rho \frac{\partial^2 \Psi}{\partial T \partial \varepsilon}, \qquad \mathbb{P}_j = \rho \frac{\partial^2 \Psi}{\partial \varepsilon \partial \alpha_j} \tag{11a,b,c}$$

Using the energy equation (8) in the consistency condition allows to get the plastic multiplier λ

$$\lambda = \left\langle \frac{\mathbf{v} : \mathbb{D} : \dot{\varepsilon} - \beta}{H} \right\rangle \tag{12}$$

where we have set again

$$\mathbb{D} : \mathbf{u} = [-\,\Lambda_j \bullet \frac{\partial F}{\partial A_j} - \frac{T}{\rho c}\,[\{\Gamma_j + \frac{A_j}{T}\} \bullet \frac{\partial F}{\partial A_j}].\Delta$$

$$\mathbf{v} : \mathbb{D} = -\frac{\partial f}{\partial A_j} \bullet \mathbb{P}_j + \frac{T}{\rho c}\{\frac{\partial f}{\partial T} - \frac{\partial f}{\partial A_j} \bullet \Gamma_j\}\Delta \tag{13a,b}$$

$$\beta = -\frac{1}{\rho c}[\frac{\partial f}{\partial T} - \frac{\partial f}{\partial A_j} \bullet \Gamma_j](r - \operatorname{div} q) \tag{14}$$

$$H = h - \frac{T}{\rho c}[\frac{\partial f}{\partial T} - \frac{\partial f}{\partial A_j} \bullet \Gamma_j]\,[\,(\Gamma_k + \frac{A_k}{T}) \bullet \frac{\partial F}{\partial A_k}\,] \tag{15}$$

Using the state law (1a), the energy equation (8), the evolution laws (3a) and the formula (12) ,the stress rate may be written as

$$\dot{\sigma} = \mathcal{L} : \dot{\varepsilon} + m \tag{16}$$

where $\mathcal{L}$ and m take respectively the forms

$$\mathcal{L} = \begin{cases} D = E + \frac{T}{\rho c}\Delta \otimes \Delta & \text{si } f < 0 \text{ or } f = 0 \text{ et } v{:}D{:}\dot{\varepsilon} < \beta \\ H = D - \frac{(D{:}u) \otimes (v{:}D)}{H} & \text{si } f = 0 \text{ et } v{:}D{:}\dot{\varepsilon} \geq \beta \end{cases} \tag{17}$$

$$m = \begin{cases} d = \frac{1}{\rho c}(r - \operatorname{div} q)\,\Delta & \text{si } f < 0 \text{ or } f = 0 \text{ et } v{:}D{:}\dot{\varepsilon} < \beta \\ h = \frac{1}{\rho c}(r - \operatorname{div} q)[\,\Delta + \frac{1}{H}\{\frac{\partial f}{\partial T} - \frac{\partial f}{\partial A_j} \bullet \Gamma_j\} D{:}u] & \text{si } f = 0 \text{ et } v{:}D{:}\dot{\varepsilon} \geq \beta \end{cases}$$

(18)

in the above formulae, $\mathcal{L}$ denotes the adiabatic tangent modulus, **D** the thermoelastic adiabatic modulus and **E** the isothermal elastic modulus. It is convenient for the following to define also the isothermal tangent modulus $\mathbb{L}$ (relating the stress rate to the strain rate in isothermal conditions) as:

$$\mathbb{L} - \mathbb{E} - \frac{(\mathbb{E} : a) \otimes (b : \mathbb{E})}{h} \tag{19}$$

with

$$\mathbb{E}{:}\, a = -[A_j \bullet \frac{\partial F}{\partial A_j}] \quad , \quad b{:}\mathbb{E} = [\,-\frac{\partial f}{\partial A_j} \bullet P_j\,] \tag{20a,b}$$

We assume in the following h >0 , H = 0 ,E and D positive definite.

Conditions for localization

In the classical analysis of localization , conditions for the onset of localization are obtained by seeking when bifurcation of the rate problem into modes involving discontinuities of the velocity gradient across a surface ($\mathcal{S}$) is possible. In the uncoupled problem, these conditions were given by Rice[5],Rudnicki and Rice [6] , Borré and Maier [4] for localisation inside the body and by Benallal ,Billardon and Geymonat [2] for localization at the boundary. Conditions for singular surfaces of the velocity gradient to reach interfaces can also be exhibited (see Benallal,Billardon and Geymonat [1]). As the rate constitutive equations given in (16) are bi-linear (the tangent modulus is dependent on the strain rate), two types of localization are usually recognized: continuous localization when the body is in plastic loading in each side of the singular surface and discontinuous localization when the body is in loading in one side and in unloading in the other side.
Let us suppose that a singular surface ($\mathcal{S}$) of the velocity gradient lies in the body. This surface divides the body Ω into two parts Ω_1 and Ω_2 where exist mechanical and thermal fields denoted by subscripts 1 and 2 respectively. Let us also call **n** the unit normal to ($\mathcal{S}$) directed from Ω_1 to Ω_2.
If the velocity field **v** is assumed to be continuous , then it is necessary through the Maxwell kinematical compatibility relation that the jump of the velocity gradient be in the form

$$[\, \varepsilon(\mathbf{v})] = \varepsilon(\mathbf{v}_1) - \varepsilon(\mathbf{v}_2) = \frac{1}{2}[\mathbf{a}\otimes\mathbf{n} + \mathbf{n}\otimes\mathbf{a}] \tag{21}$$

where **a** is an arbitrary vector for the time being. The balance of linear momentum imposes that the traction rate must be continuous across ($\mathcal{S}$) and therefore we have

$$[\dot{\sigma}.\mathbf{n}] = \dot{\sigma}_1.\mathbf{n} - \dot{\sigma}_2.\mathbf{n} = \mathbf{0}. \tag{22}$$

Finally,the conservation of energy leads to the continuity of the heat flux across the singular surface ($\mathcal{S}$),which will be written as

$$[\mathbf{q}.\mathbf{n}] = \mathbf{q}_1.\mathbf{n} - \mathbf{q}_2.\mathbf{n} = \mathbf{0}. \tag{23}$$

We will in the following give conditions for localization in two circumstances. We will consider the case where no restriction is imposed to the temperature field and the case where this field is assumed to remain continuous. These cases will be termed case I and case II respectively. In case I, there will be no restriction on the temperature rate;in case II however, the jump condition

(23) combined to the continuity of the temperature imposes that the temperature rate is continuous across ($\mathcal{S}$) so that

$$[\dot{T}] = \dot{T}_1 - \dot{T}_2 = 0 \tag{24}$$

The following results are given without the proofs which can be found in Benallal [3]; only the results for localization inside the body are given here, their extension to the boundary can also be found in [3].

Case I - The necessary and sufficient condition for localization inside the body is

i) $\det(\mathbf{n}.\mathbb{H}.\mathbf{n}) \leq 0$

Case II- The necessary and sufficient conditions for localization inside the body are

i) $\det(\mathbf{n}.\mathbb{H}.\mathbf{n}) \leq 0$

ii) $(\mathbf{n}.\boldsymbol{\Delta}).[\mathbf{n}.\mathbb{D}.\mathbf{n}]^{-1}.(\mathbf{n}.\mathbb{D}:\mathbf{u}) + (\Gamma + \frac{A}{T}) \bullet \frac{\partial F}{\partial A} = 0$

or equivalently

iii) $\det(\mathbf{n}.\mathbb{L}.\mathbf{n}) \leq 0$

iv) $(\mathbf{n}.\boldsymbol{\Delta}).[\mathbf{n}.\mathbb{E}.\mathbf{n}]^{-1}.(\mathbf{n}.\mathbb{E}:\mathbf{a}) + (\Gamma + \frac{A}{T}) \bullet \frac{\partial F}{\partial A} = 0$

In conditions i) and iii), the equality refers to continuous localization whereas the inequality is related to discontinuous localization.

It has to be recalled at this stage that condition iii) is exactly the necessary and sufficient condition for localization in the uncoupled problem. This will allow comparisons between coupled and uncoupled situations.

From the above conditions, it is easily seen that case I localization is more favourable than case II one.

In the general case, the localization is governed by the adiabatic tangent modulus $\mathbb{H}$. As it can be seen from relations (17) and (13) , this tangent modulus is generally unsymmetric even when the mechanical behavior is such that f = F ; from results of Rudnicki and Rice [7], it is then expected that thermomechanical couplings can lead to destabilizing effects; this will be shown in the next section through a simple example.

An example

This section is devoted to a detailed analysis of localization for a particular set of constitutive equations to which the former exhibited conditions are applied. The model considered here is based upon two internal variables (the plastic strain ε^p and the cumulated plastic strain p) describing an elastic plastic material with isotropic hardening, and corresponds in the framework described before to a free energy, an elastic domain and a plastic potential defined respectively by:

$$\rho\Psi = \frac{1}{2}\lambda[\mathrm{Tr}(\varepsilon - \varepsilon^p)]^2 + \mu(\varepsilon - \varepsilon^p) : (\varepsilon - \varepsilon^p) - [3\lambda+2\mu]\alpha(T - T_0)\mathrm{Tr}(\varepsilon - \varepsilon^p) + g(p) - bT\mathrm{Ln}T \quad (25)$$

$$f = \sigma + R - k \quad (26)$$

$$F = \sigma + R - k \quad (27)$$

λ and μ are the Lamé constants, α the thermal expansion coefficient, T_0 a reference temperature, b the heat capacity , g(p) being an arbitrary function of p ; besides is the Von Mises equivalent stress defined by $\sigma = \sqrt{\mathbf{s} : \mathbf{s}}$,R is the thermodynamical force associated to p, $\mathbf{s}$ is the stress deviator and k the yield stress.

This model is designed only to show the consequences of taking into account the thermo-mechanical couplings. All the mechanical as well as the thermal properties of the material are assumed to be temperature independant. In particular, no thermal softening is considered (this is under investigation) and the thermoelastic properies and the yield stress are kept constant. The isothermal and adiabatic tangent moduli are easily obtained by

$$L_{ijkl} = \mu\delta_{ij}\delta_{kl} + (\lambda + \mu)(\delta_{ik}\delta_{jl}+\delta_{il}\delta_{jk}) - \frac{\{\frac{2\mu}{\sigma} s_{ij}\}.\{\frac{2\mu}{\sigma} s_{kl}\}}{2\mu + g''} \quad (28)$$

$$H_{ijkl} = \mu\delta_{ij}\delta_{kl}+[\lambda+\mu+\frac{T(3\lambda+2\mu)^2\alpha^2}{b}](\delta_{ik}\delta_{jl}+\delta_{il}\delta_{jk}) - \frac{\{\frac{2\mu}{\sigma} s_{ij}+\frac{(3\lambda+2\mu)\alpha k}{b}\delta_{ij}\}.\{\frac{2\mu}{\sigma} s_{kl}\}}{2\mu + g''} \quad (29)$$

It is worthwile to notice here that this last expression is formally similar to the tangent modulus used by Rudnicki and Rice[5] in their analysis of localization for pressure-sensitive dilatant materials. The material obtained here enters this kind of material with Lamé constants given by

$$\mu^* = \mu, \qquad \lambda^* = \lambda + \frac{T(3\lambda + 2\mu)^2\alpha^2}{b} \tag{30a,b}$$

and frictional ω^* and dilatational π^* parameters (the parameters μ and β of [7]) given respectively by

$$\omega^* = 0\,, \qquad \pi^* = \frac{3k\alpha}{b + 3T(3\lambda + 2\mu)\alpha^2} \tag{31a,b}$$

This remark emphasizes to some extent the analogy between thermal effects and those of pressure and dilatancy. It allows also to transfer to our analysis some of the results obtained by Rudnicki and Rice [7]. We prefer here to use another approach which can be used to the more general case where the mechanical and thermal properties are temperature dependant. This approach is in essence geometrical.

Let us denote in the following by Σ^* the "deviatoric normal stress on the plane of localization and by S^* the shear stress on this same plane ; these stresses are respectively defined by

$$\Sigma^* = \mathbf{n.s.n}\,, \quad (S^*)^2 = (\mathbf{sn}).(\mathbf{sn}) - (\mathbf{n.s.n})^2 \tag{32a,b}$$

and by Σ and S their normalized values with respect to the Von Misès equivalent stress σ.

From (28) , (29) , localization conditions (i), (iii) (iv) given above become :

(i)
$$2\mu + g'' + \frac{4\mu^2 Z^2}{\lambda^* + 2\mu} = \frac{4\mu^2}{\lambda^* + 2\mu}(\Sigma + Z)^2 + 4\mu S^2 \tag{33}$$

(iii)
$$2\mu + g'' = 4\mu^2\{\frac{S^2}{\mu} + \frac{\Sigma^2}{\lambda + 2\mu}\} \tag{34}$$

(iv)
$$\Sigma = \frac{(\lambda + 2\mu)k}{2\mu\alpha T(3\lambda + 2\mu)} \tag{35}$$

where Z is defined by

$$Z = \frac{(3\lambda + 2\mu)k\alpha}{2b\mu} \tag{36}$$

It is then interesting to carry on the analysis in the "reduced Mohr's plane" ($\Sigma = \frac{\Sigma^*}{\sigma}$, $S = \frac{S^*}{\sigma}$) depicted in figure 1 where relation (33) and (34) represent two ellipses (EI) and (EII) whereas

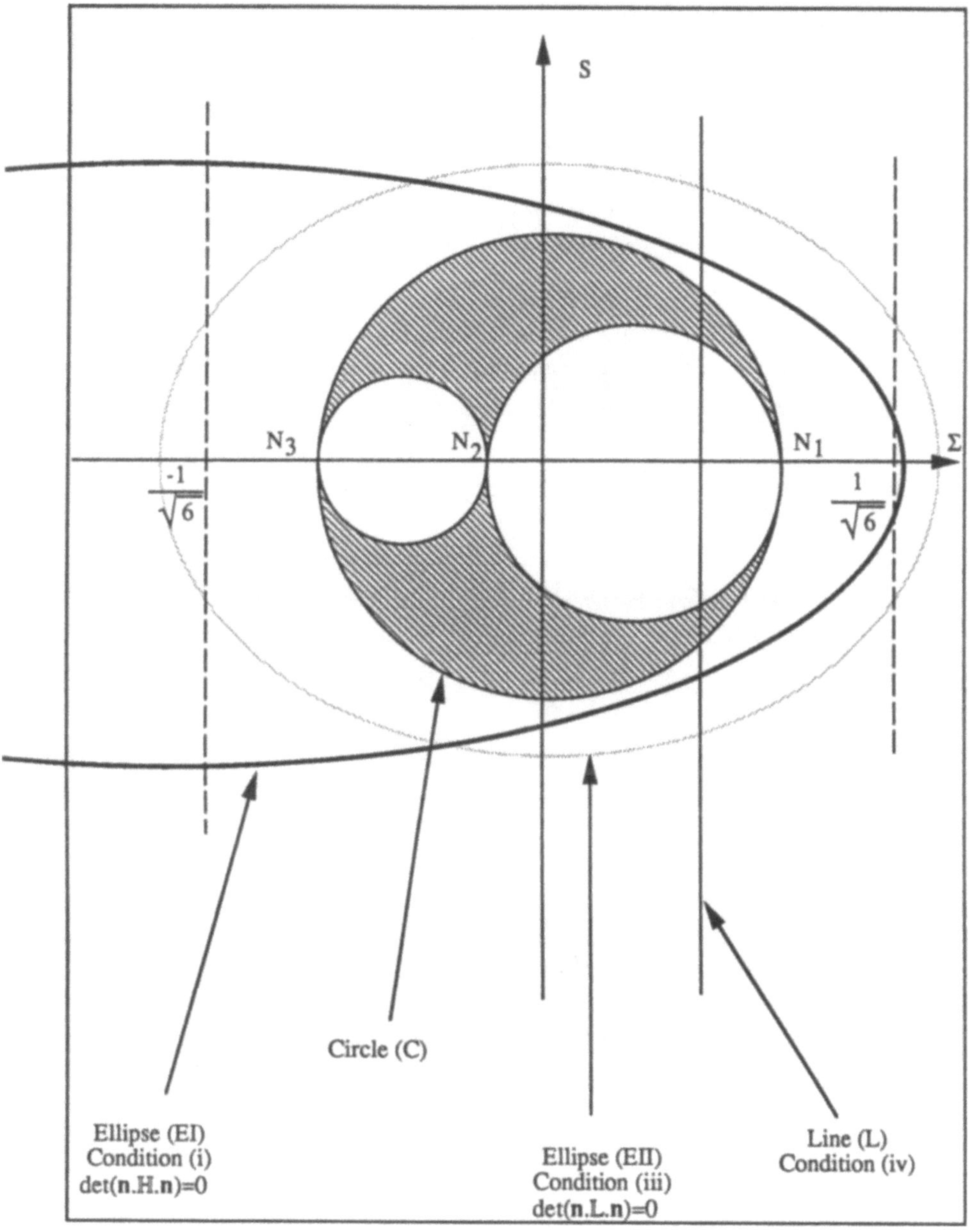

Figure 1: Interpretation of localization conditions in the reduced Mohr plane
The pricipal stresses are distinct

relation (35) is a straight line (L) parallel to the S axis. The sizes of the ellipse are linked to the "isothermal" hardening modulus $h = g''(p)$; the bigger this hardening modulus , the bigger the ellipses.

From elementary Mohr analysis, it is concluded that solutions **n** exist for the localization problem for a given state of stress and a given temperature if and only if:

- the ellipse (EI) intersects the dashed area or its boundary in case I.
- the ellipse (EII) intesects the line (L) inside or at the boundary of this dashed area in case II.

Let the state of stress σ and the temperature T fixed and assume for simplicity that the principal stresses are distinct. The former remarks allow the following conclusios:

- in case I, the biggest hardening modulus h_I leading to localization is obtained when the ellipse (EI) is tangential to the biggest Mohr circle (C) and is given by

$$\frac{h_I}{2\mu} = \frac{2\mu Z^2}{(\lambda^*+2\mu)(\lambda^*+\mu)} - \frac{2\mu Z}{\lambda^*+\mu} N_2 - \frac{(3\lambda^*+2\mu)}{2(\lambda^*+\mu)} (N_2)^2 \tag{37}$$

- in case II, the maximum hardening modulus h_{II} for localization is obtained when the line (L) and ellipse (EII) intersect on the circle (C). It is clear from figure 1 that this is possible only when

$$\frac{(\lambda+2\mu)k}{2\mu\alpha T(3\lambda+2\mu)} \leq N_1 \tag{38}$$

in which case, this modulus is given by

$$\frac{h_{II}}{2\mu} = -2[N_2 + \frac{(\lambda+2\mu)k}{2\mu(3\lambda+2\mu)\alpha T}]^2 - \frac{(\lambda+2\mu)k^2}{8\mu^2(3\lambda+2\mu)\alpha^2 T^2} \tag{39}$$

- the maximum hardening h for localization in the uncoupled isothermal problem is obtained when the ellipse (EII) is tangential to the circle (C), which gives

$$h = -\frac{(3\lambda+2\mu)}{2(\lambda+\mu)} (N_2)^2 \tag{40}$$

In the above relations, $N_1 = \frac{s_1}{\sigma}$, $N_2 = \frac{s_2}{\sigma}$ and $N_3 = \frac{s_3}{\sigma}$ are the reduced deviatoric principal stresses and s_1, s_2 and s_3 are the deviatoric principal stresses, i.e the eigenvalues of the stress deviator. These eigenvalues are arranged in such way that $s_1 \geq s_2 \geq s_3$.

When the principal stresses are distinct, it follows then that the normal to the plane of localization is perpendicular to the s_2 direction. If two of the principal stresses are equal,only the angle between the normal to the critical plane and the third of the principal axes is uniquely determined. Theses conclusions are similar to those of Rudnicki and Rice [5].

The normal **n** to the surface of localization, which lies in the extreme principal stresses plane can also be obtained by a simple Mohr analysis as shown in figure 2 where the angle θ between **n** and the biggest principal stress is depicted.

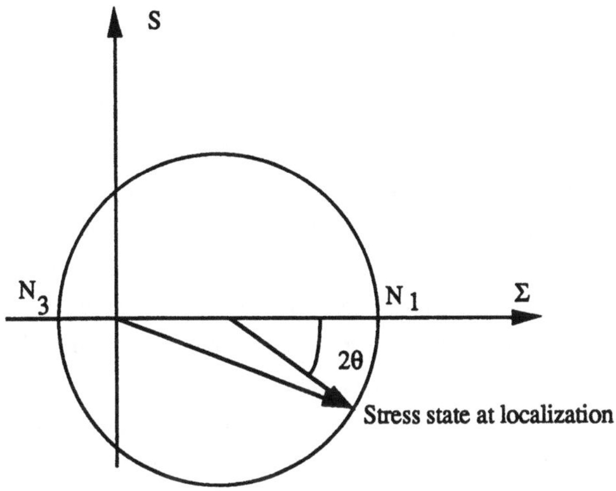

Figure 2 : Obtention of the directions of localization in the Mohr plane

This angle is given by

$$(tg2\theta)^2 = \frac{S^2}{(\Sigma - \frac{N_1+N_3}{2})^2} \tag{41}$$

which leads for the two cases

$$\text{- case I : } (tg2\theta)^2 = \frac{[\frac{N_1-N_3}{2}]^2 - [\frac{\mu}{\lambda^*+\mu}]^2[Z + (\frac{N_1+N_3}{2})]^2}{[\frac{\mu}{\lambda^*+\mu}[Z + (\frac{N_1+N_3}{2})]\,]^2} \tag{42}$$

$$\text{case II}: (\text{tg}2\theta)^2 = \frac{[\frac{N_1-N_3}{2}]^2 - [\frac{(\lambda+2\mu)k}{2\mu(3\lambda+2\mu)\alpha T} - (\frac{N_1+N_3}{2})]^2}{[\frac{(\lambda+2\mu)k}{2\mu(3\lambda+2\mu)\alpha T} - \frac{N_1+N_3}{2}]^2} \qquad (43)$$

Discussion

Although for practical reasons, case I is more important than case II, the simple example analyzed before shows that this last one is possible for given stress and temperature ranges. The conditions for localization given before show that an uncoupled analysis furnishes a conservative prediction for case II while this is not true for the case I by the example.

The example clearly shows also the destabilizing effects of thermomechanical couplings on localization phenomena . This is mainly due to the non-symmetry of the adiabatic tangent modulus which governs this localization, as already mentionned earlier and expected also from the results of Rudnicki and Rice [6]. There are cases where uncoupled analysis do not predict localization whereas the coupled approch does for the same state of stress and temperature; this is important for practical applications. It is to be emphasized here that these destabilizing effects are due to the only fact of taking into accoutnt thermomechanical couplings; no thermal softening and no temperature dependency of the thermal and mechanical properties were introduced in the example. A study under investigation suggests stong effects of this dependency.

Relation (37) examplifies the roles of each of the constitutive parameters on the critical hardening modulus for a given stress state and a given temperature. For a given temperature, this modulus has a parabolic variation with stress state through the parameter ; for a fixed state of stress, the role of the temperature which is involved only through the parameter N_2, is much more complex and must be underlined. Also of some importancc, the role of the yield stress k included in Z:the bigger the yield stress,the bigger the critical hardening modulus.

References

1. Benallal, A.; Billardon, R.; Geymonat , G. : Localization phenomena at the boundaries and interfaces of solids., Proc. 3rd Int. Conf. on Constitutive Laws for Engineering Materials,University of Arizona, Tucson, 7-12 January 1991.
2. Benallal, A.; Billardon, R.; Geymonat , G. : Phénomènes de bifurcation à la frontière d'un solide, C. R. Acad. Sci. Paris, t.308, série II, 1990, 893-898.
3. Benallal,A. ; Localization phenomena in thermo-elastoplasticity, to appear in Archives of Mechanics.

4. Borré, G.; Maier,G.: Localization in Linear versus non-linear materials, Meccanica, 24, 1989, 36-41.
5. Rice,J.R.: The localization of plastic deformation, Theoretical and Applied Mechanics, W.T.KOITER ed., North Holland, 1976, 207-220.
6. Rudnicki,J.W.; Rice,J.R;: Conditions for the localization of deformation in pressure-sensitive dilatant materials, J. Mech. Phys. Solids, 23, 1975, 371-394.
7. Rudnicki,J.W.; Rice,J.R;: A note on some features of the theory of localization of deformation, Int. J. Solids Struct., 16, 1980 ,597-605.

4. Pöch, O.; Mayer G.: Lokalisation in Linear verteilten ... Math. Mitteilungen ...

Mechanical Modeling of Irradiation Creep and its Application to the Analysis of Creep Crack Growth

S. MURAKAMI and M. MIZUNO

Department of Mechanical Engineering, Nagoya University
Chikusa-ku, Nagoya, 464-01, JAPAN

SUMMARY

After discussing the mechanisms and the physical theories of neutron irradiation reported so far, a constitutive model of irradiation creep, swelling and irradiation creep damage applicable to the inelastic analysis of nuclear reactor components was developed. It was assumed that the irradiation creep can be decomposed into irradiation-induced creep and irradiation-enhanced creep. By taking account of the physical mechanisms, the irradiation-induced creep was represented by an isotropic tensor function of order zero and one with respect to stress, as well as of neutron flux and neutron fluence. The volumetric part of the irradiation-induced creep was identified with swelling. The irradiation-enhanced creep was described by modifying Kachanov-Rabotnov creep damage theory by incorporating the effect of irradiation. The utility of the proposed constitutive equation was demonstrated by analyzing the irradiation creep and irradiation creep damage of type 316 stainless steel. As an application of the proposed equations, creep crack growth in an infinite plate under neutron irradiation was analyzed. A double cantilever beam model combined with continuum damage mechanics was employed to simplify the analysis. The effects of neutron irradiation on the rate of steady-state creep crack growth and the distributions of stress, strain and damage in front of the crack tip were discussed.

1. INTRODUCTION

Nuclear reactor components often operate at elevated temperature and under significant flux of high energy fission or fusion production particles [1]. Fuel claddings [2,3] and first walls of the fusion reactors [4], in particular, are exposed to high temperature and fast neutron flux with the energy of $E > 0.1$ MeV and $E = 14$ MeV, respectively, and are subject to severe irradiation damage and irradiation-induced inelastic deformation. Thus, the neutron irradiation in these components has crucial effects on the process of creep, creep damage and final fracture.

The present paper is concerned with the mechanical modeling of irradiation creep, swelling and irradiation creep damage of polycrystalline metals applicable to structural analyses of nuclear reactor components. A constitutive equation of irradiation creep and swelling was formulated in the framework of continuum mechanics and by taking account of physical mechanisms of neutron irradiation. Continuum damage mechanisms was employed to evaluate the irradiation damage and final fracture of materials. Then, the applicability of the proposed constitutive equation was demonstrated by analyzing the creep crack growth by means of double cantilever shear beam model. The effects of neutron irradiation on creep deformation, creep damage and creep crack growth were discussed.

M. Zyczkowski (Ed.)
Creep in Structures
IUTAM Symposium Cracow/Poland 1990

2. MECHANICAL MODELING OF IRRADIATION CREEP AND SWELLING

2.1 Mechanisms and Physical Theories of Irradiation Creep and Swelling

The irradiation of polycrystalline metals by high energy particles knocks out the constituent atoms from their lattice sites, and gives rise to the pairs of interstitial atoms and vacancies (i.e., Frenkel defects). Thus when polycrystalline metals are subjected to irradiation at elevated temperature, in addition to ordinary thermal creep, coalescence and absorption of the interstitial atoms induced by the irradiation together with aggregation of the corresponding vacancies induce the irradiation creep and swelling [5].

In order to describe the creep under irradiation, a number of physical models have been proposed so far by taking account of these mechanisms [2,6]. According to CCG (Climb–Controlled Glide) mechanisms, absorption of excess interstitials induces the climb of dislocations, and this climb facilitates the dislocation glide. By postulating that the climbed dislocations can bow out to a configuration where the line tension of the dislocation equals the externally applied stress σ, Gittus [7] expressed the creep rate $\dot{\varepsilon}$ as follows:

$$\dot{\varepsilon} = k\sigma\dot{S}\,, \tag{1}$$

where $(\dot{\ })$ is the time derivative, and k and $\dot{S}$ stand for material constant and the swelling rate. When operation of Frank–Read source is predominant, the creep rate $\dot{\varepsilon}$ is proportional to the square of stress σ^2. Hence, instead of equation (1), we have the following equation [2,6]:

$$\dot{\varepsilon} = k\sigma^2\dot{S}\,. \tag{2}$$

These mechanisms postulate the flow of excess interstitials into dislocations, and the creep rate is related to the point defect density, and hence to the swelling rate $\dot{S}$.

According to the SIPA (Stress–Induced Preferential Absorption) mechanism [8], on the other hand, the magnitude of the long range force between an edge dislocation and an interstitial atom depends on the relative orientation between the directions of applied stress and the Burgers vector, and the dislocations of different Burgers vectors have different rates of interstitial flux. Therefore, since the anisotropic rate of dislocation climb is proportional to the rate of nucleation of point defects K_i, the creep rate is given as follows [8]:

$$\dot{\varepsilon} = kK_i\sigma\,. \tag{3}$$

Since the CCG and SIPA mechanisms are not mutually exclusive but may operate concurrently, the total creep rate can be given by the combination of these two mechanisms [9]:

$$\dot{\varepsilon} = k_1\sigma\dot{S} + k_2K_i\sigma\,. \tag{4}$$

Finally the nucleation of irradiation–induced interstitial loops also depends on the direction of the applied stress; i.e., they nucleate preferentially on crystallographic plane nearly perpendicular to the applied stress. Brailsford and Bullough [10], for example, assuming this SIPN (Stress–Induced Loop Nucleation) mechanism, derived a similar equation to equation (1) for materials with low dislocation density.

The swelling rates $\dot{S}$ in equations (1), (2) and (4) have been given by the rate theory based on the diffusion rate and density of interstitials and vacancies [1,9,11], or by the empirical equations expressed in terms of stress, flux of irradiation and temperature [2,12,13].

From the view point of the constitutive equations of structural analysis, equations (1) through (4) are not only restricted to the uniaxial state of stress, do not describe the non-linear dependence of creep rates on the stress, but also have some essential limitations as regards their mechanical consistency. For example, though equations (1) through (4) give $\dot{\varepsilon} = 0$ for $\sigma = 0$ or $\dot{S} = 0$, this implies that these equations can not describe the irradiation creep and swelling under zero stress $\sigma = 0$, nor they represent ordinary thermal creep under condition of vanishing swelling rate $\dot{S} = 0$. If we identify the swelling rate $\dot{S}$ with the dilatation rate induced by the irradiation, the first terms on the right hand sides of equations (1), (2) and (4) can never include the stress σ.

2.2 Formulation of Irradiation-Induced Creep and Swelling

In nuclear reactor components operating at elevated temperature, beside the irradiation creep discussed above, the stress-controlled thermal creep under irradiation is also predominant. Therefore we will now develop a continuum theory of irradiation creep by dividing the irradiation creep into *the irradiation-enhanced creep* ε^{IEC} and *the irradiation-induced creep* ε^{IIC}; the former is the ordinary thermal creep but accelerated by irradiation, while the latter is caused exclusively by the irradiation and may occur even under vanishing stress.

If we formulate the continuum model of irradiation-induced creep ε^{IIC}_{ij} on the basis of SIPA mechanism [8] discussed in the preceding chapter, $\dot{\varepsilon}^{IIC}_{ij}$ may be given as a function of stress σ_{ij}, neutron flux ϕ and neutron fluence $\Phi = \int \phi \, dt$:

$$\dot{\varepsilon}^{IIC}_{ij} = F_{ij}(\sigma_{kl}, \phi, \Phi) \, . \tag{5a}$$

In comparison with the irradiation-enhanced creep, the stress-dependence of the irradiation-induced creep and swelling is small [2,3], and the anisotropy of material properties caused by the irradiation may be small. Then by taking account of the stress dependence of creep strain rate due to SIPA mechanism [8], the irradiation-induced creep $\dot{\varepsilon}^{IIC}_{ij}$ of equation (5a) may be expressed as an isotropic function [14,15] of stress σ_{ij} of order zero and order one:

$$\dot{\varepsilon}^{IIC}_{ij} = \eta(\phi,\Phi)\delta_{ij} + L_{ijkl}(\phi,\Phi)\sigma_{kl} \, , \tag{5b}$$

$$L_{ijkl}(\phi,\Phi) = \xi(\phi,\Phi)\delta_{ij}\delta_{kl} + \zeta(\phi,\Phi)(\delta_{ik}\delta_{jl} + \delta_{il}\delta_{jk}) \, , \tag{5c}$$

where L_{ijkl} and δ_{ij} are the fourth rank isotropic function and the Kronecker delta, respectively. The symbols η, ξ, and ζ, furthermore, are material functions of ϕ and Φ. Substitution of (5c) into (5b) furnishes

$$\dot{\varepsilon}^{IIC}_{ij} = \eta\delta_{ij} + \xi\sigma_{kk}\delta_{ij} + 2\zeta\sigma_{ij} \, . \tag{5d}$$

If we decompose the stress and strain into isotropic (or volumetric) and deviatoric parts, the volumetric part of strain rate corresponds exactly to the swelling rate $\dot{S}$. Therefore, in view of the incubation period of swelling [16] and the linear dependence of creep and swelling on neutron fluence Φ [2,8] (as observed in Fig.1 below), the constitutive equation (5) of irradiation-induced creep may be written in an alternative form:

$$\dot{\varepsilon}^{IIC}_{ij} = (1/3)\dot{S}\delta_{ij} + (3/2)P\sigma_{Dij} \, , \tag{6a}$$

$$\dot{S} = S_0(1 + Q\sigma_{kk})\phi\kappa(\Phi - \chi) \, , \tag{6b}$$

$$P = P_0\phi \,, \tag{6c}$$

where σ_{Dij} and $\kappa(\Phi - \chi)$ stand for the deviatoric part of σ_{ij} and Heaviside step function, while the coefficient $P = (4/3)\zeta(\phi,\Phi)$ in equation (6a) represents the anisotropy of the creep rate due to SIPA mechanism, and equation (6c) postulates the linear dependence of creep strain rate on neutron flux ϕ [8]. Symbols $S_0 = 3\eta(\phi,\Phi)/\phi$ and $Q = [3\xi(\phi,\Phi) + 2\zeta(\phi,\Phi)]/S_0\phi$ in equation (6b) are material functions related to swelling and express the swelling rate of zero-stress and the coefficient of stress dependence of swelling rate respectively, while χ denotes the incubation fluence for swelling. Though P_0, S_0 and Q are functions of ϕ and Φ in general, we will take them constants by assuming that their dependence on ϕ and Φ is insignificant.

Equation (6) has been derived for multiaxial states of stress by taking account of physical mechanisms of irradiation creep and swelling, and is the alternative of equations (1) through (4) proposed for uniaxial stress.

2.3 Formulation of Irradiation-Enhanced Creep

Irradiation-enhanced creep is brought about by the acceleration of the ordinary thermal creep due to irradiation effects. The thermal creep of polycrystalline metals at elevated temperature, on the other hand, is induced by the diffusion controlled dislocation motion activated by stress, and is usually accompanied by creep damage. Therefore, we will assume that the effects of irradiation on the constitutive equations and the evolution equations of the irradiation-enhanced creep ε^{IEC} and the related creep damage can be described mainly through the coefficients of the equation for unirradiated conditions.

If we assume the von Mises creep law and the Kachanov-Rabotnov creep damage theory, the constitutive equation of creep and the evolution equation of creep damage under unirradiated condition may be given as follows [17,18,19]:

$$\dot{\varepsilon}^{C}_{ij} = (3/2)A_0[\sigma_{EQ}/(1 - D)]^{n_0-1}[\sigma_{Dij}/(1 - D)] \,, \tag{7a}$$

$$\dot{D} = B_0[\sigma^{(1)}/(1 - D)]^{k_0} \,, \tag{7b}$$

where A_0, B_0, n_0 and k_0 are material constants, while $\sigma_{EQ} = [(3/2)\sigma_{Dij}\sigma_{Dij}]^{1/2}$ and $\sigma^{(1)}$ denote equivalent stress and the maximum principal stress, respectively. The symbol D in equation (7), furthermore, is an internal state variable describing damage state (damage variable), and defined as follows:

$$D = 0 \ \text{ at } \ t = 0 \qquad \text{(initial undamaged state)} \,,$$

$$D = 1 \ \text{ at } \ t = t_R \qquad \text{(final rupture state)} \,. \tag{8}$$

According to the preceding argument, equation (7) may be readily extended to the irradiation-enhanced creep as follows:

$$\dot{\varepsilon}^{IEC}_{ij} = (3/2)A(\phi,\Phi)[\sigma_{EQ}/(1 - D)]^{n(\phi,\Phi)-1}[\sigma_{Dij}/(1 - D)] \,, \tag{9a}$$

$$\dot{D} = B(\phi,\Phi)[\sigma^{(1)}/(1 - D)]^{k(\phi,\Phi)} \,, \tag{9b}$$

where $A(\phi,\Phi)$, $B(\phi,\Phi)$, $n(\phi,\Phi)$ and $k(\phi,\Phi)$ are material functions.

Up to now, any systematic experimental data applicable to determine the material functions of equation (9) are not available. However, according to Gilbert and Chin's experimental data for 20% cold worked type 316 stainless steel at 650 °C [3] (see Fig.2 below), no significant difference is observed in the stress-dependence of rupture times among unirradiated creep, creep under irradiation and post-irradiation creep; this implies

$$k(\phi,\Phi) = k_0 \,. \tag{10a}$$

As regards the creep exponent, the results observed in Reference [2] allows us to assume

$$n(\phi,\Phi) = n_0 \,. \tag{10b}$$

Since the material function A is related to the creep rate, it depends on the behavior of irradiation–induced internal defects. The effects of fluence Φ on the microstructure of the material is cumulative, and develops dislocation structures. The dislocation density and configuration will tend to an asymptotic state as Φ increase [20]. The effect of ϕ, on the other hand, is brought about by the interaction of the irradiation–induced defects with the existing defects [21]. Namely, such interaction is governed by the rate of increase of point defects induced by the irradiation together with the rate of their annihilation due to their absorption by the existing sinks, such as point defect clusters, dislocation networks, etc. Thus, sufficiently large value of ϕ again leads to a stationary defect structures [21].

Furthermore, the function B is related to the rate of creep damage. The creep damage in polycrystalline metals is brought about by the formation and the development of microcavities on the grain boundaries, which are governed mainly by the local diffusion and the matrix creep [22]. The effect of Φ on the material is embrittlement due to the cavity formation and the sessile dislocation networks. The effect of ϕ, on the other hand, enhances diffusion and matrix deformation, which in turn may be responsible to the transition of the creep rupture mode from the w–type triple point cracks to the r–type void [23]. Increase of either of them, ϕ and Φ, will lead to an asymptotic state of damage rates.

In view of these microstructural mechanisms, we will assume the following functions for A and B:

$$A(\phi,\Phi) = A_0[1 + a_1(1 - e^{-a_2\phi})][1 + a_3(1 - e^{-a_4\Phi})] \,, \tag{10c}$$

$$B(\phi,\Phi) = B_0[1 + b_1(1 - e^{-b_2\phi})][1 + b_3(1 - e^{-b_4\Phi})] \,, \tag{10d}$$

where a_1, a_2, a_3, a_4, b_1, b_2, b_3 and b_4 are constants that can be identified by comparing the creep curves of unirradiated creep, creep under irradiation and of post–irradiation creep under constant stress.

2.4 Constitutive Equations of Irradiation Creep and Swelling

Summarizing the results of the preceding discussions, the constitutive equations of irradiation creep and swelling are given by equations (6), (9) and (10) as follows:

$$\begin{aligned}\dot{\varepsilon}^{C}_{ij} &= \dot{\varepsilon}^{IEC}_{ij} + \dot{\varepsilon}^{IIC}_{ij} \\ &= (3/2)A_0[1 + a_1(1 - e^{-a_2\phi})][1 + a_3(1 - e^{-a_4\Phi})][\sigma_{EQ}/(1 - D)]^{n_0-1}[\sigma_{Dij}/(1 - D)] \\ &\quad + (1/3)\dot{S}\delta_{ij} + (3/2)P_0\phi\sigma_{Dij} \,,\end{aligned} \tag{11a}$$

$$\dot{D} = B_0[1 + b_1(1 - e^{-b_2\phi})][1 + b_3(1 - e^{-b_4\Phi})][\sigma^{(1)}/(1 - D)]^{k_0} \,, \tag{11b}$$

$$\dot{S} = S_0(1 + Q\sigma_{kk})\phi\kappa(\Phi - \chi) \,. \tag{11c}$$

In equations (11), dilatation due to swelling has been expressed as the volumetric part of the irradiation–induced creep, and the combined effects of irradiation creep and swelling under multiaxial states of stress have been formulated by satisfying the continuum mechanics consistency. By taking $\phi = \Phi = 0$, equation (11) is reduced to the thermal creep under unirradiated condition, while the case $\phi = 0$, $\Phi = \text{const.}$ accounts for the post–irradiation creep. Thus equation (11) furnishes a unified constitutive equation for unirradiated creep, creep under irradiation and of post–irradiation creep.

3. ELABORATION OF CONSTITUTIVE EQUATIONS OF IRRADIATION CREEP AND SWELLING

3.1 Elaboration of Constitutive Equation of Swelling

It has been observed that the magnitude of swelling is usually insignificant up to a specific fluence χ, and beyond this fluence, after a continuous transition from the incubation period, the swelling tends to increase almost linearly with the fluence [13,16]. Thus, in order to avoid the discontinuous increase of the swelling rate described by equation (11c) and to describe the continuous transition from the incubation to the steady state period, Bates and Korenko [13] developed empirical equations of stress–free swelling at different temperatures.

By employing the transient expression of Bates and Korenko, equation (11c) can be revised to give an elaborate constitutive equation of the stress–dependent swelling [24]:

$$\dot{S} = S_0[1 - e^{R(\chi-\Phi)}/\{1 + e^{R(\chi-\Phi)}\}](1 + Q\sigma_{kk})\phi \,, \tag{12}$$

where R denotes the curvature parameter of the swelling curve.

Fig. 1 shows the results of prediction of swelling for 20% cold worked type 316 stainless steel at 400 °C. The solid lines and symbols are the prediction of equation (12) and the corresponding experimental results of Porter et al. [16]. The results of equation (11c) are also entered by dotted lines. The calculations were performed by use of material constants

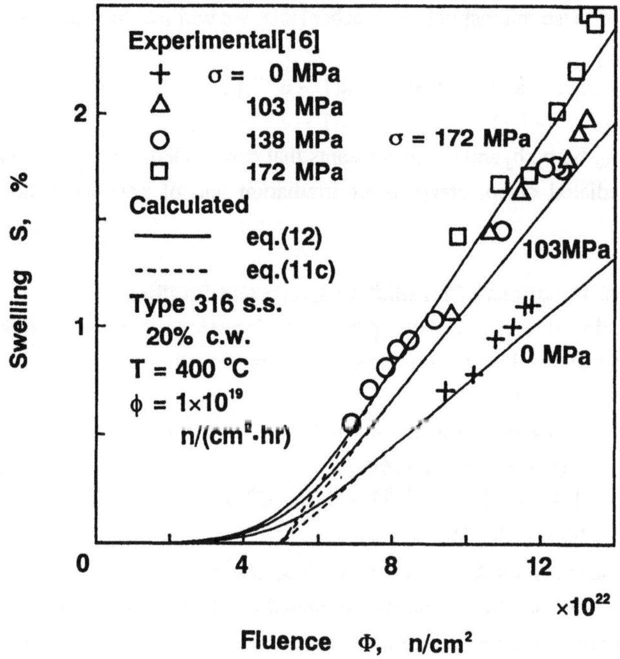

Fig.1 Swelling of 20% cold worked type 316 stainless steel at 400 °C.

determined later. As shown in Fig.1, the equation (12) represents the continuous transition from the incubation to the steady–state period without changing the characteristic of incubation and steady–state periods. Fig.1 suggests the validity of equation (12).

3.2 Constitutive Equation of Irradiation Creep Incorporating Transient Creep

Equation (9a) has been formulated by assuming the steady–state creep law of Norton [19]. However the transient creep in metals is observed not only after the application of stress, but also after the change in the magnitude or in the direction of stress. In the case of 20% cold worked type 316 stainless steel at 650 °C discussed in this paper, salient transient creep is observed also in post–irradiation creep and under–irradiation creep [3].

The transient creep in polycrystalline metals is interpreted as the un–equilibrated creep rate due to the fast movement of dislocations after the application of stress combined with the concurrent decrease of the creep strain rate due to the immobilization of these dislocations. Thus, transient creep observed under neutron irradiation is brought about mainly as a part of the stress–controlled irradiation–enhanced creep discussed above. However, since transient creep is essential only in a limited duration, we assume that neutron irradiation has no significant influence on the transient part of the stress–controlled thermal creep. Thus, if we employ McVetty's creep law which has separate expressions for transient creep and steady–state creep, the constitutive equation of irradiation–enhanced creep incorporating transient creep may be expressed as follows [24]:

$$\begin{aligned}\dot{\varepsilon}^{IEC}_{ij} &= (3/2)A^{*}a^{*}e^{-a^{*}t}\sigma_{EQ}^{\;n^{*}-1}\sigma_{Dij} \\ &\quad + (3/2)A(\phi,\Phi)[\sigma_{EQ}/(1-D)]^{n(\phi,\Phi)-1}[\sigma_{Dij}/(1-D)]\,, \qquad (13)\end{aligned}$$

where t denotes the time, and A^{*}, a^{*} and n^{*} are material constants of transient creep. The first term of the right hand side of equation (13) is based on the time–hardening theory for variable stress creep [19]. For further elaboration, we may employ more accurate constitutive equations of creep for general history of stress [19,25].

3.3 Elaborated Constitutive Equations of Irradiation Creep and Swelling

To recapitulate the above discussion, the constitutive equations of irradiation creep and swelling incorporating the continuous transition of incubation swelling and the effect of transient creep may be summarized as follows:

$$\begin{aligned}\dot{\varepsilon}^{C}_{ij} &= \dot{\varepsilon}^{IEC}_{ij} + \dot{\varepsilon}^{IIC}_{ij} \\ &= (3/2)A^{*}a^{*}e^{-a^{*}t}\sigma_{EQ}^{\;n^{*}-1}\sigma_{Dij} \\ &\quad + (3/2)A_0[1 + a_1(1 - e^{-a_2\phi})][1 + a_3(1 - e^{-a_4\Phi})][\sigma_{EQ}/(1-D)]^{n_0-1}[\sigma_{Dij}/(1-D)] \\ &\quad + (1/3)\dot{S}\delta_{ij} + (3/2)P_0\phi\sigma_{Dij}\,, \qquad (14a)\end{aligned}$$

$$\dot{D} = B_0[1 + b_1(1 - e^{-b_2\phi})][1 + b_3(1 - e^{-b_4\Phi})][\sigma^{(1)}/(1-D)]^{k_0}\,, \qquad (14b)$$

$$\dot{S} = S_0[1 - e^{R(\alpha-\Phi)}/\{1 + e^{R(\alpha-\Phi)}\}](1 + Q\sigma_{kk})\phi\,. \qquad (14c)$$

4. ANALYSIS OF IRRADIATION CREEP

4.1 Identification of Material Constants

To evaluate the validity and the applicability of equations (14), we will now perform the following analyses for 20% cold worked type 316 stainless steel at 650 °C.

Material constants in equations (14) can be identified by the procedure discussed in the preceding paper [5,24], and can be determined by use of the experimental results of References [3,16,20]. The material constants of equation (14), thus determined are as follow:

$$\chi = 5.00 \times 10^{22}\ \mathrm{n/cm^2}, \qquad S_0 = 4.00 \times 10^{-25}\ (\mathrm{n/cm^2 \cdot hr})^{-1} \cdot \mathrm{hr}^{-1},$$

$$Q = 4.75 \times 10^{-3}\ \mathrm{MPa}^{-1}, \qquad R = 1.25 \times 10^{-22}\ (\mathrm{n/cm^2})^{-1}. \tag{15a}$$

$$A^* = 1.30 \times 10^{-7}\ \mathrm{MPa}^{-n^*}, \qquad a^* = 8.50 \times 10^{-3}\ \mathrm{hr}^{-1},$$

$$n^* = 2.00. \tag{15b}$$

$$A_0 = 3.20 \times 10^{-13}\ \mathrm{MPa}^{-n_0} \cdot \mathrm{hr}^{-1}, \qquad a_1 = 5.00 \times 10^{-2},$$

$$a_2 = 2.60 \times 10^{-19}\ (\mathrm{n/cm^2 \cdot hr})^{-1}, \qquad a_3 = -9.00 \times 10^{-2}, \tag{15c}$$

$$a_4 = 2.60 \times 10^{-21}\ (\mathrm{n/cm^2})^{-1}, \qquad n_0 = 3.50.$$

$$B_0 = 6.00 \times 10^{-10}\ \mathrm{MPa}^{-k_0} \cdot \mathrm{hr}^{-1}, \qquad b_1 = -0.95,$$

$$b_2 = 4.5 \times 10^{-19}\ (\mathrm{n/cm^2 \cdot hr})^{-1}, \qquad b_3 = 1.30, \tag{15d}$$

$$b_4 = 2.60 \times 10^{-21}\ (\mathrm{n/cm^2})^{-1}, \qquad k_0 = 2.80,$$

$$P_0 = 2.50 \times 10^{-28}\ (\mathrm{MPa \cdot n/cm^2 \cdot hr})^{-1}.$$

4.2 Creep under Irradiation and Post–Irradiation Creep

Figs.2 and 3 show the comparison between the experimental results [3] for unirradiated creep, post–irradiation creep and creep under irradiation, and the corresponding predictions of equations (14). The symbols in the figures stand for the results of experiment, and the temperature, stress and irradiation condition are entered in the figures.

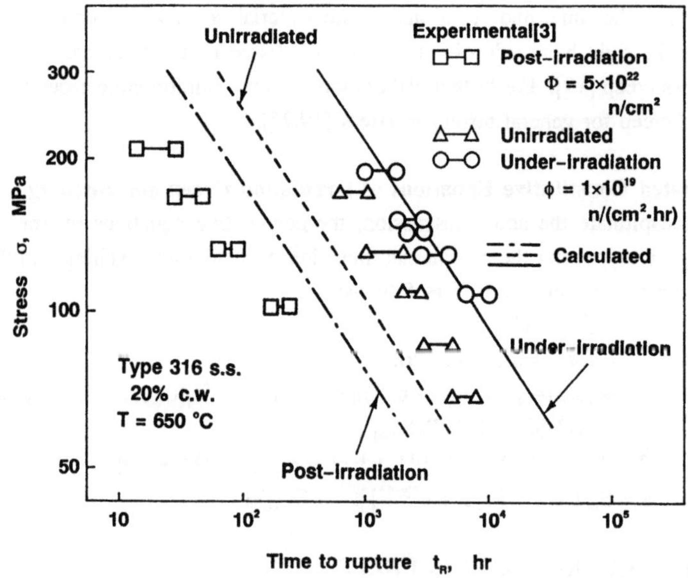

Fig.2 Stress–rupture properties of unirradiated, post–irradiation and under–irradiation creep for 20% cold–worked type 316 stainless steel.

Fig.2, to begin with, represent the stress–rupture time relations for unirradiated, post–irradiation and under–irradiation creep, predicted by equations (14). It will be observed from Fig.2 that equation (14) describes the dependence of rupture times on the irradiation condition observed in the corresponding experiments.

Fig.3, on the other hand, shows the creep curves for different conditions of irradiation. It will be observed that the constitutive equations (14) describe accurately the essential features of irradiation creep; i.e., creep curves under irradiation including transient creep stage, embrittlement due to prior irradiation, and the retardation of the start of the tertiary creep and increase of tensile ductility due to neutron flux.

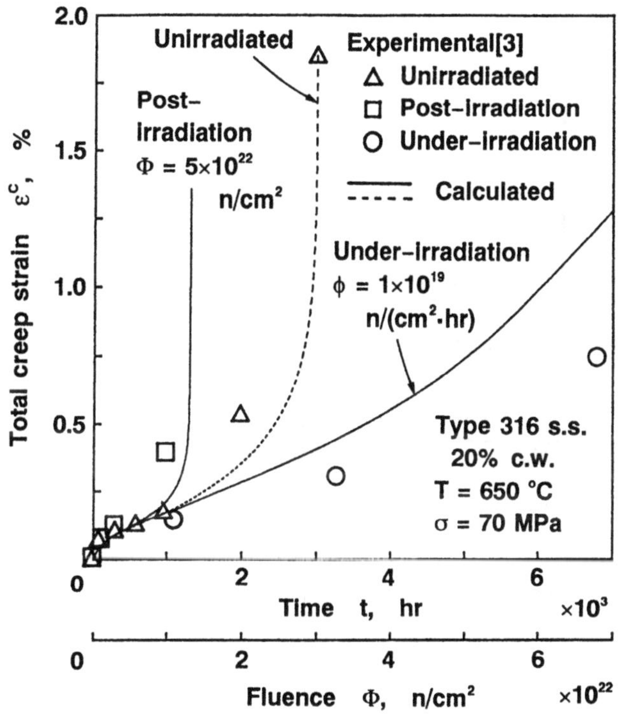

Fig.3 Creep curves of unirradiated, post–irradiation and under–irradiation creep for 20% cold worked type 316 stainless steel at 650°C.

5. SIMPLIFIED ANALYSIS OF CREEP CRACK GROWTH UNDER NEUTRON IRRADIATION

5.1 Double Cantilever Beam Model

As an application of the constitutive equations of irradiation creep and swelling developed above, we will analyze the effects of neutron irradiation on the creep crack growth in nuclear reactor components by use of the double cantilever beam model [26,27,28].

Let us now take a semi-infinite plate of height $2h$ including a crack of length a as shown in Fig.4, and model it by a double cantilever shear beam proposed by Bui et al. [28] for unirradiated creep. We assume that creep damage develops in a creep damage zone Ω_v of a strip of width 2λ and the part of the plate outside Ω_v is the shear elastic zone Ω_e. The zone Ω_D in the figure is the damaged zone including the crack.

Let us take Cartesian coordinates o–xy as show in Fig.4, and signify by $w(x)$ the distributed load acting on the beam in y–direction per unit thickness of the beam. If we assume that the state of stress in the creep damage zone Ω_v is uniaxial and uniform in y–direction, a uniform strain $\varepsilon(x)$ and a uniform stress $\sigma(x)$ are induced in Ω_v due to the displacement $u(x)$ in y–direction of the elastic zone Ω_e. The stress σ in the creep damage zone Ω_v acts on the elastic beam of Ω_e as a distributed load in negative y–direction.

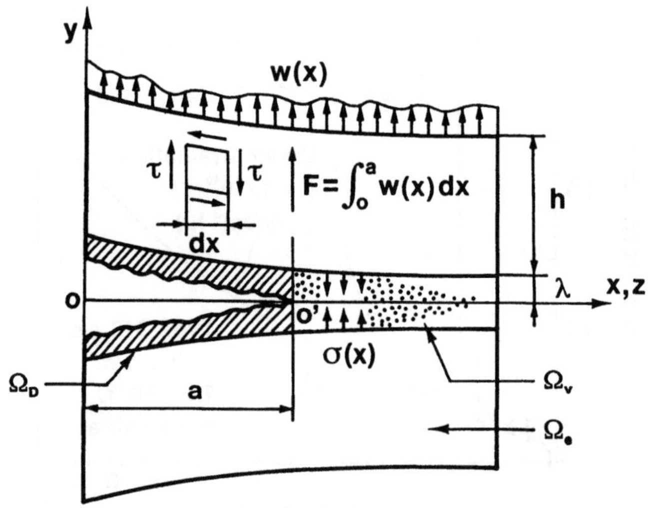

Fig.4 Double cantilever beam model.

5.2 Formulation of Steady–State Crack Growth

Let assume that the creep crack grows at a constant rate

$$\dot{a}(t) = v\,. \tag{16}$$

Then, the analysis can be simplified by taking a new Cartesian Coordinates o'–zy with origin o' at crack tip and employing the Galilean transformation

$$z = x - v\cdot t\,. \tag{17}$$

$$\partial /\partial x = (d/dz)(dz/dx) = d/dz\,, \tag{18a}$$

$$\partial /\partial t = (d/dz)(dz/dt) = -v\,(d/dz) \tag{18b}$$

which transforms the crack tip to the origin of the new coordinate system o'–zy.

If we restrict the present analysis to a cracked plate loaded by a concentrated load W at $x = 0$, the beam for $x < a(t)$ is subject only to the shear force due to W. Then, the loading condition to the beam is given by

$$F(0) = W, \quad w(x) = 0 . \tag{19}$$

In view of the above assumptions together with the relation, $\varepsilon = u/\lambda$, the governing equation in the elastic zone Ω_e is furnished from the relation between the shear stress and distributed loads and leads to [29]

$$Gh\lambda\,(\partial^2\varepsilon/\partial z^2) = \sigma . \tag{20}$$

Employing the constitutive equations of irradiation creep (9) and (10) for the uniaxial state, the governing equations in the damage zone Ω_v is expressed in following form [29]:

$$- v\,(d\varepsilon/dz) = A_0[1 + a_1(1 - e^{-a_2\phi})][1 + a_3(1 - e^{-a_4\Phi})] \times[\sigma/(1 - D)]^{m_0} + (1/3)S_0(1 + Q\sigma)\phi\kappa(\Phi - \chi) + P_0\phi\sigma , \tag{21a}$$

$$- v\,(dD/dz) = B_0[1 + b_1(1 - e^{-b_2\phi})][1 + b_3(1 - e^{-b_4\Phi})] \times[\sigma/(1 - D)]^{k_0} . \tag{21b}$$

The boundary conditions and the fracture conditions of the above equations are furnished by the condition at the crack tip $z = 0$ and that at infinity $z = \infty$ as follows [28]:

$$(d\varepsilon/dz)(0) = - W/Gh\lambda , \tag{22a}$$

$$\varepsilon(0) = \varepsilon_R , \tag{22b}$$

$$D(0) = 1 . \tag{22c}$$

$$\varepsilon(\infty) = D(\infty) = 0 . \tag{23}$$

The values of v and ε_R are unknown in advance. Thus, the differential equations (20) and (21) together with the boundary conditions (22) and (23) can be integrated numerically by specifying proper approximate values of v and ε_R, and the calculation will be iterated by modifying these values until the numerical results will satisfy the condition (23). Then, the final values of the critical strain ε_R and creep crack velocity v together with the distribution of strain, stress and damage variable in z-direction will be determined.

5.3 Results of Analysis and Discussion

Preceding equations were applied to the analysis of the steady-state creep crack growth in 20% cold worked type 316 stainless steel at 650 °C. The calculation was performed for the cases of unirradiated, post-irradiation and under-irradiation conditions. The material constants of equations (15) together with following value were employed:

$$G = 54.8 \text{ GPa} . \tag{24}$$

As an example of irradiation condition, we will take a fuel cladding of fast breeder reactors, and will specify the following values:

Unirradiated creep:

$$\phi = 0 \text{ n/cm}^2\cdot\text{hr} , \qquad \Phi = 0 \text{ n/cm}^2 , \tag{25a}$$

Post-irradiation creep:

$$\phi = 0 \text{ n/cm}^2\cdot\text{hr} , \qquad \Phi = 5.0 \times 10^{22} \text{ n/cm}^2 , \tag{25b}$$

Under-irradiation creep:

$$\phi = 1.0 \times 10^{19} \text{ n/cm}^2\cdot\text{hr} , \qquad \Phi = \infty \text{ n/cm}^2 . \tag{25c}$$

As regards the geometry of the double cantilever beam, the height $2h$ and the width of the creep damage zone 2λ are specified as

$$2h = 100 \text{ mm}$$
$$2\lambda = 1 \text{ mm}\ . \tag{26}$$

Fig.5 shows the relation between the rate of steady–state crack growth v and the average shear stress τ_0 at $z = 0$ in the elastic zone Ω_e for each irradiation condition. The relations between the reciprocal of the uniaxial creep rupture time $t_R = 1/[B_0(k_0 + 1)\sigma^{k_0}] = 1/[B_0(k_0 + 1)\tau_0^{k_0}]$ [18,19] and the corresponding stress $\sigma = \tau_0$ for each irradiation condition were also entered in the figure for reference sake. The figure indicates that v shows significant non–linear dependence on τ_0, and the non–linearity is much longer than that in $1/t_R - \tau_0$ relations. This fact implies that prediction of creep crack velocity made merely by the creep rupture properties of the material would be highly non–conservative.

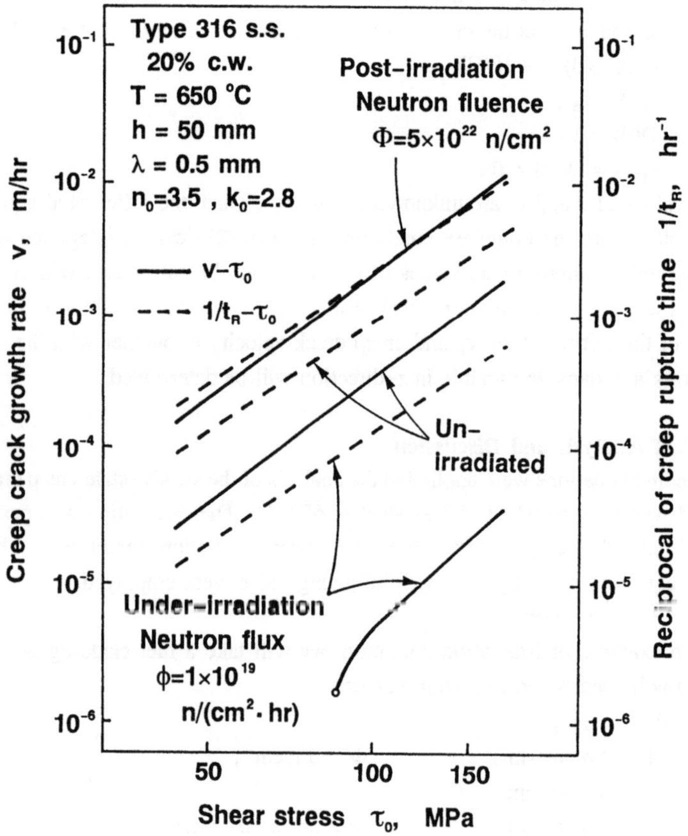

Fig.5 Creep crack growth rate under unirradiated, post–irradiation and under–irradiation condition.

While linear relations are observed in log ν – log τ_0 in the case of post–irradiation and unirradiated conditions, the differential equations (20) through (23) have the steady–state rate solution only for limited range of stress in the case of under–irradiation condition; namely, for under–irradiation creep, the differential equations (20) through (23) may have the solutions of decelerated creep crack growth rate for the stress range of $\tau_0 < 87.8$ MPa.

As observed in Fig.5, the creep crack growth rate in post–irradiation condition is about 10 times as high as that of unirradiated condition, whereas the crack growth rate under irradiation condition is decreased nearly by two digits. The increase in ν in the post–irradiation creep may be attributable to the stress concentration in the vicinity of crack tip as a result of the embrittlement and the suppression [30] of the creep rate due to the preceding irradiation. In the case of under–irradiation creep, on the other hand, the accelerated stress relaxation due to the enhanced creep rates combined with the delay in the initiation of tertiary creep as a result of irradiation [3] may account for the decrease in creep crack growth rate ν.

Fig.6, on the other hand, shows the strain distribution in front of the crack tip for the case of τ_0 = 100 MPa. The steady–state crack velocities predicted by Fig.5 are $\nu = 2.07 \times 10^{-3}$, 3.37×10^{-4} and 4.78×10^{-6} m/hr for post–irradiation, unirradiated and under–irradiation conditions. In the case of the irradiation creep, the accelerated creep rate due to irradiation and the relatively low rate of crack growth induce the creep strain in large region in front of crack tip, whereas less creep strain has developed under post–irradiation condition, as a results of the decrease of creep rate due to preceding irradiation and the related increase of creep crack growth rate.

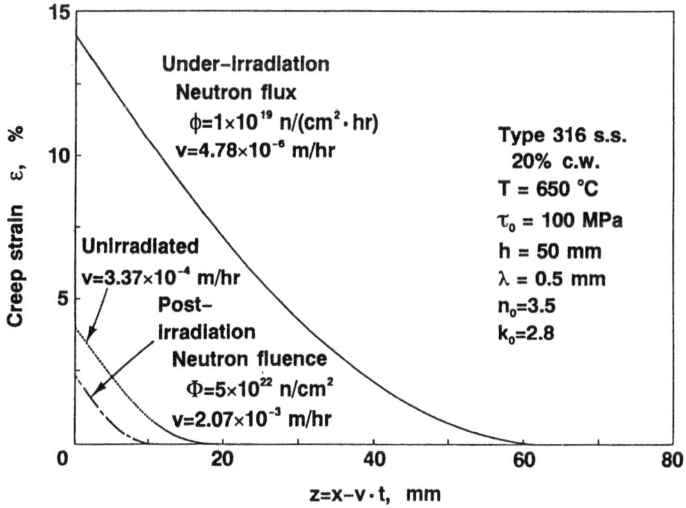

Fig.6 Distribution of creep strain in front of the crack tip under unirradiated, post–irradiation and under–irradiation condition

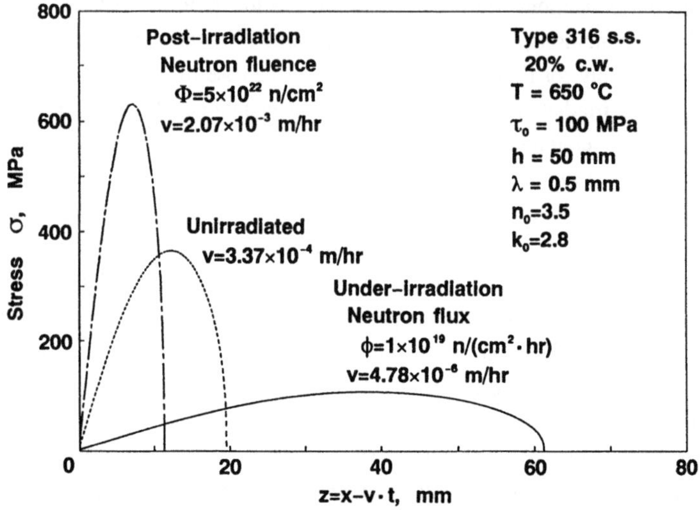

Fig.7 Distribution of stress in front of the crack tip under unirradiated, post-irradiation and under-irradiation condition

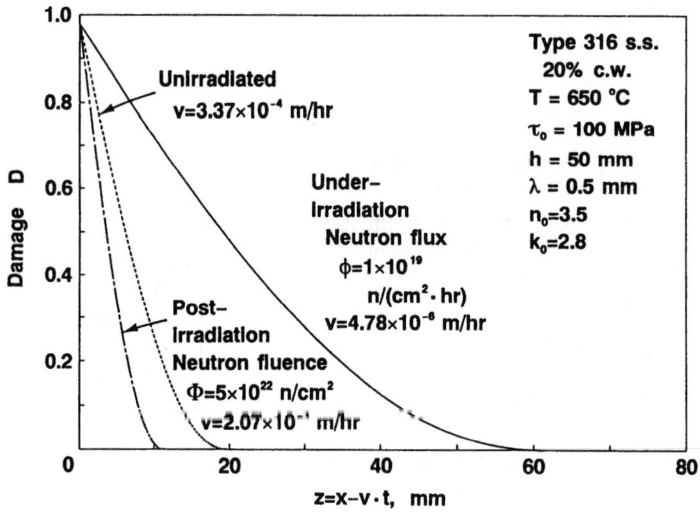

Fig.8 Distribution of damage in front of the crack tip under unirradiated, post-irradiation and under-irradiation condition.

The stress distributions in the relevant condition are shown in Fig.7. The salient difference in stress distributions for different irradiation conditions can be accounted for by the decrease of crack velocities and the enhancement of creep rate observed in Figs.5 and 6.

Finally, Fig.8 is the distribution of damage parameter for each irradiation condition. It will be observed that, while the damage zone in the case of post–irradiation creep is restricted in the vicinity of crack tip, the damage develops in large extent in the case of under–irradiation creep.

6. CONCLUSION

Modeling of constitutive equations of irradiation creep, swelling and irradiation creep damage and their elaboration were discussed first. The resulting constitutive equations were found to facilitate the unified description of unirradiated creep, post–irradiation creep and the creep under irradiation, for whole process of creep ranging from the transient creep to the tertiary creep and the final rupture. The predictions of the constitutive equations were compared with the results of the corresponding experiments.

Then the utility of the proposed constitutive equation was demonstrated by applying it to the simplified analysis of creep crack growth in nuclear reactor components. The analysis was performed by modifying the double cantilever shear beam model combined with damage mechanics. The effects of neutron irradiation on the creep crack growth rate, the distributions of strain, stress and damage parameter in front of the crack tip, together with their important features, were discussed in detail.

REFERENCES

1. Gittus, J.: Irradiation Effects in Crystalline Solids. London: Applied Science Publishers, 1978.
2. Ehrlich, K.: Irradiation creep and interrelation with swelling in austenitic stainless steels. J. Nuclear Materials **100** (1981) 149–166.
3. Gilbert, E. R.; Chin, B. A.: In–reactor creep measurements. Nuclear Technology **52** (1981) 273–283.
4. Harries, D. R.; Zolti, E.: Structural mechanics and material aspects of the next European Torus. Nuclear Engineering and Design/Fusion **3** (1986) 331–344.
5. Murakami, S.; Mizuno, M.; Okamoto, T.: Mechanical modeling of irradiation creep and swelling of polycrystalline metals. Nuclear Engineering and Design (contributed).
6. Mansur, L. K.; Reiley, T. C.: Irradiation creep by dislocation glide enabled by preferred absorption of point defects –Theory and Experiment. J. Nuclear Materials **90** (1980) 60–67.
7. Gittus, J. H.: Theory of dislocation–creep due to the Frenkel defects or interstitialcies produced by bombardment with energetic particles. Phil. Mag. **25** (1972) 345–354.
8. Heald, P. T.; Speight, M. V.: Steady–state irradiation creep. Phil. Mag. **29** (1974) 1075–1080.
9. Heald, P. T.; Harbottle, J. E.: Irradiation creep due to dislocation climb and glide. J. nuclear Materials **67** (1977) 229–233.

10. Brailsford, A. D.; Bullough, R.: Irradiation creep due to the growth of interstitial loops. Phil. Mag. **27** (1973) 49–64.
11. Weiner, R. A.; Boltax, A.: Microstructural modeling of irradiation creep and swelling. J. Nuclear Materials **66** (1977) 1–16.
12. Boltax, A.; Foster, J. P.; Weiner, R. A.; Biancheria, A.: Void swelling and irradiation creep relationships. J. Nuclear Materials **65** (1977) 174–183.
13. Bates, J. F.; Korenko, M. K.: Empirical development of irradiation–induced swelling design equations. Nuclear Technology **48** (1980) 303–314.
14. Fung, Y. C.: A First Course in Continuum Mechanics, 2nd ed. Englewood Cliffs, New Jersey: Prentice–Hall, 1977.
15. Jaunzemis, W.: Continuum Mechanics. New York: Macmillan, 1967.
16. Porter, D. L.; Takata, M. L.; Wood, E. L.: Direct evidence for stress–enhanced swelling in type 316 stainless steel. J. Nuclear Materials, **116** (1983) 272–276.
17. Kachanov, L. M.: Introduction to Continuum Damage Mechanics. Dordrecht: Martinus Nijhoff Publishers, 1986.
18. Leckie, F. A.; Hayhurst, D. R.: Constitutive equations for creep rupture. Acta Metallurgica **25** (1977) 1059–1070.
19. Boyle, J. T.; Spence, J: Stress Analysis for Creep. London: Butterworths, 1983.
20. Wolfer, W. G.: Correlation of radiation creep theory with experimental evidence. J. Nuclear Materials **90** (1980) 175–192.
21. Kiritani, M.: Radiation rate dependence of microstructure evolution. J. Nuclear Materials **169** (1989) 89–94.
22. Ashby, M. F.; Dyson, B. F.: Creep damage mechanics and micromechanisms. Proc. of the 6th International Conference on Fracture, New Dehli, 1984.
23. Gilbert, E. R.; Chin, B. A.; Duncan, D. R.: Effect of irradiation on failure mode during creep. Metall. Trans. **18A** (1987) 79–84.
24. Murakami, S.; Mizuno, M.: Constitutive equations of irradiation creep and swelling incorporating their transient behavior. Nuclear Technology, (contributed).
25. Murakami, S.; Ohno, N.: A constitutive equation of creep based on the concept of a creep–hardening surface. Int. J. Solids Struct. **18** (1982) 597–609.
26. Kanninen, M. F.: An augmented double cantilever beam model for studying crack propagation and arrest. Int. J. Fracture **9** (1973) 83–91.
27. Freund, L. B.: A simple model of the double cantilever beam crack propagation specimen. J. Mech. Phys. Solids **25** (1977) 69–79.
28. Bui, H. D.; Dang Van, K.; de Langre, E.: A simplified analysis of creep crack growth using local approach. Proc. Int. Seminar on Local Approach of Fracture pp.373–388, Moret–sur–Loing.
29. Murakami, S.; Mizuno, M.: Simplified analysis of creep crack growth under neutron irradiation. J. Mech. Phys. Solid (contributed).
30. Bloom, E. E.; Weir, J. R.: Effect of neutron irradiation on the ductility of austenitic stainless steel. Nuclear Technology **16** (1972) 45–54.

Fatigue Crack Propagation Behavior Under Creep Conditions

Kiyotsugu OHJI and Shiro KUBO

Department of Mechanical Engineering,
Faculty of Engineering, Osaka University,
2-1, Yamadaoka, Suita, Osaka 565 JAPAN

Summary

The crack propagation behavior of the SUS 304 stainless steel under creep-fatigue conditions was reviewed. Cracks propagated either in purely time-dependent mode or in purely cycle-dependent mode, depending on loading conditions. The time-dependent crack propagation rate was correlated with modified J-integral J^* and the cycle-dependent crack propagation rate was correlated with J-integral range ΔJ_f. Threshold was observed in the cycle-dependent crack propagation, and below this threshold the time-dependent crack propagation appeared. The crack propagation rates were uniquely characterized by taking the effective values of J^* and ΔJ_f, when crack closure was observed. Change in crack propagation mode occurred reversibly and was predicted by the competitive damage model. The threshold disappeared and the cycle-dependent crack propagation continued in a sub-threshold region under variable amplitude conditions, where the threshold was interposed between the maximum and minimum ΔJ_f.

Introduction

Knowledge of crack propagation under creep and creep-fatigue conditions is an essential element to determine residual service life and inspection interval of structures operating at high temperatures. A large amount of research efforts on this subject have been made for this twenty years. Reviews of these research activities on creep [1-10] and creep-fatigue [11-20] crack propagation are found in the literature.

Ten years ago, when the 3rd IUTAM Symposium on Creep in Structures and also the International Conference on Engineering Aspects of Creep were held, main interest of researchers was in characterization of creep (sustained load) crack propagation rate in terms of several candidate mechanical parameters, including stress intensity factor K_I, modified J-integral J^* (C^*-parameter), and net section stress. Hot discussion on this issue was made in these international meetings. Since then general agreement that creep crack propagation rate of ductile and environment-insensitive materials is well characterized in terms of the modified J-integral, J^* or C^*, has been established among the majority of researchers on this subject, although there still remain some different views. A considerable amount of data, which supported the validity of the modified

M. Zyczkowski (Ed.)
Creep in Structures
IUTAM Symposium Cracow/Poland 1990

J-integral characterization of creep crack propagation rates, were accumulated during this ten years.

The creep-fatigue crack propagation was not a main topic in these 1980 meetings, although a few papers were presented. However, under the circumstance mentioned above, the interest of pioneering researchers was quickly shifted from the creep crack propagation toward the creep-fatigue crack propagation. Substantial progress has been made in the creep-fatigue crack propagation research during this decade.

Creep, environmental effects and structural change of material are the three main phenomena, which affect the fatigue crack propagation rate at high temperatures. The rates of these phenomena are determined by thermally activated processes, causing time-dependent damage in material. Crack propagation may occur as a result of accumulation of damage in the vicinity of the crack tip. This means that the crack propagation rate is generally affected by the history of the values of mechanical, thermal, environmental and structural state variables, since the instantaneous values of these state variables can determine only the rate of damage at that moment.

This can easily be demonstrated, for example, in the case of creep crack propagation by a model analysis [21]. The creep crack propagation rate was correlated by the modified J-integral, J^* or C^*, which represented the current mechanical state in the vicinity of the crack tip, only when the condition was satisfied that the accumulated creep damage value ahead of the crack tip approached a steady-state value [21]. The experimental evidence is also found that the creep crack propagation rate can be correlated well with J^* at the steady-state stage (stage II) while no correlation exists between the rate and J^* at the initial stage (stage I).

Fatigue crack propagation is basically a plasticity-controlled process, and is cycle-dependent in nature. The mechanical variable (ΔK_{eff} or ΔJ_{eff}), which determines deformation at the crack tip (crack tip opening displacement), determines the crack propagation rate.

Fatigue crack propagation at high temperatures, therefore, may be determined by two opposite processes, the cycle-dependent process and the time-dependent processes. If the participation of the time-dependent processes is negligible, the cycle-dependent process may dominate the whole behavior, and vice versa. The interaction among fatigue, creep, environment and structure may or may not exist, depending on whether the micromechanisms of these four processes interact or not.

In this paper, crack propagation under creep-fatigue conditions is overviewed, mainly based on the results obtained by the authors and their colleagues. One of the most clear-cut approaches to this problem would be to exclude the effects of environment and microstructural changes as far as possible. From a survey of publications found in the literature, it was found that the 300 series austenitic stainless steels had relatively stable microstructures and were rather insensitive to air environment, while nickel-base alloys showed environment (oxygen)-sensitive crack propagation rather than creep-fatigue crack propagation. In this paper, therefore, the main interest is focused on the behavior of the type 304 stainless steel, in order to exclude the effects of air environment and microstructural changes on the creep-fatigue crack propagation.

Experiments

The SUS 304 austenitic stainless steel solution-treated at 1100°C was used. Center-cracked specimens were utilized. Precracks were introduced by fatigue at room temperature using sufficiently low values of ΔK.

Experiments were conducted mainly at 650°C in air under sinusoidal cyclic load, using servo-hydraulic machines. Unless otherwise stated, load-control mode was selected, keeping the maximum gross stress σ_g constant. Non-negative, high values of stress ratio R = (minimum stress) /(maximum stress) were selected to avoid complexity incurred by crack closure. Crack length was measured mainly using a traveling microscope.

In correlating crack propagation rates, the J-integral range ΔJ_f and the modified J-integral J^* (which is also denoted by C^*) were used. ΔJ_f was evaluated following the method proposed by Dowling [22].

$$\Delta J_f = \Delta K^2/E + S_p/B \cdot b \tag{1}$$

where ΔK was stress intensity range, E was Young's modulus, S_p was a partial area of the hysteresis loop defined by Dowling, B was the specimen thickness, and b was the ligament length. In most of the test conditions employed, the plastic component (S_p-term) of ΔJ_f was negligibly small, and ΔJ_f was approximately expressed in terms of stress intensity range ΔK and Young's modulus E.

J^* was evaluated experimentally by using the formula proposed by the authors [23].

$$J^*_{ex} = \sigma_{net}\ \dot{\delta} \tag{2}$$

where σ_{net} was the net section stress and $\dot{\delta}$ was the displacement rate. In evaluating σ_{net}, the equivalent stress for dynamic creep defined by the following equation [24] was used,

$$\sigma_e = \sigma_m \left[\frac{1}{2\pi} \int_0^{2\pi} (1+(\sigma_a/\sigma_m)\sin\theta)^n d\theta \right]^{1/n} \qquad (3)$$

$$\sigma_a = \sigma_{max}(1-R)/2, \quad \sigma_m = \sigma_{max}-\sigma_a$$

where n was the creep exponent.

Effect of Frequency [25-27]

Figure 1 shows an example of the results on the effect of frequency ν on the crack propagation rate per time *da/dt*. At high frequencies, *da/dt* increases almost proportionally with increase in ν, while at low frequencies *da/dt* approaches to a constant value for respective crack lengths. This indicates that at high frequencies the crack propagation is purely cycle-dependent, while at low frequencies it is purely time-dependent. It is also found from Fig. 1 that the transition range between the cycle-dependent crack propagation and the time-dependent crack propagation is rather narrow, less than about two logarithmic cycles of ν, and that the deviation of *da/dt* from the purely cycle-dependent or purely time-dependent crack propagation rate is small in the transition range.

Similar results were obtained by Taira et al. [28] and Ohtani [29] for type 316 stainless steel and 0.16% carbon steel in air at R = -1. The results by Solomon and Coffin [30] on iron-nickel alloy A 286 in vacuum may also be of the same nature.

Phenomenologically similar results were observed with nickel-base alloys. However, the time-dependent crack propagation in these cases may not

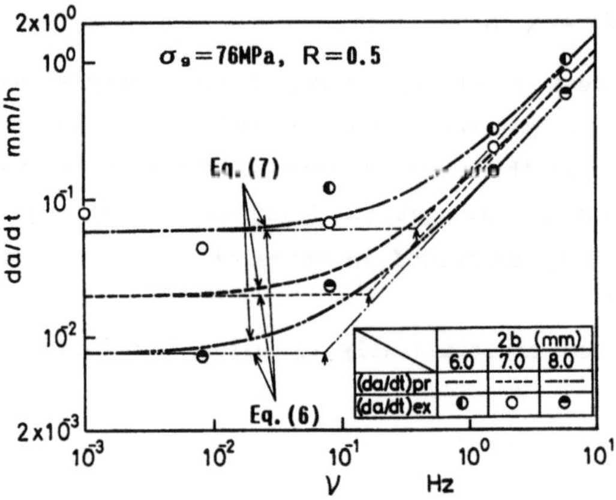

Fig.1. Effect of frequency on crack propagation rate

be a predominantly creep-induced one, but an environment-assisted one [31,32].

Correlation of Crack Propagation Rates [26,27,33]

Experimental data obtained at 650°C under the loading conditions shown in Table 1 were used.

The crack propagation rate per time *da/dt* was correlated with the experimentally determined modified J-integral J^*_{ex} as in Fig. 2. The solid line in the figure indicates the relation between *da/dt* and J^*_{ex} determined from the results obtained under pure creep conditions [34,35].

$$(da/dt)_c = 8.0\times10^{-3} J^*_{ex} \qquad (4)$$

where $(da/dt)_c$ is expressed in mm/h and J^*_{ex} in kJ/m^{2}h. The dashed lines show the scatter range of factors of 2 on $(da/dt)_c$. It is found that the *da/dt* values for low ν, high σ_g and high *R*, including the time-dependent data in Fig. 1, come within the scatter range. This indicates that *da/dt* of a crack propagating in purely time-dependent mode may be correlated with J^*_{ex}, equally for creep and creep-fatigue loading.

The same data were again correlated in the form of crack propagation rate per cycle *da/dN* with ΔJ_f as in Fig. 3. *da/dN* was obtained as $(da/dt)/\nu$. The solid line in the figure indicates the cycle-dependent $(da/dN)_f$-ΔJ_f relation determined previously under trapezoidal stresses [25] as,

$$(da/dN)_f = 1.2\times10^{-4}\Delta J_f \quad \text{for } \Delta J_f > 0.1 \qquad (5)$$

where $(da/dN)_f$ is expressed in mm/cyclc and ΔJ_f in kJ/m^2. The dashed lines show the scatter range of factors of 2 on $(da/dN)_f$. It is observed that, except for some data points below 10^{-5} mm/cycle, *da/dN* values for high ν, low σ_g and low *R*, including the cycle-dependent data in Fig. 1, fall within the scatter range. It is also found that the data within the scatter range in Fig. 2 are mostly located outside of the scatter range in Fig. 3, and vice versa. This means that *da/dN* of a crack propagating in the purely cycle-dependent mode may be correlated solely with ΔJ_f. This type of observations were first made by Taira et al. [28] and Ohtani [29] around 1980.

In Fig. 3, a threshold-like behavior is observed in the region below $\Delta J_f \simeq 0.1$ kJ/m^2. Some of these data in small ΔJ_f region, however, showed the behavior of purely time-dependent mode as in Fig. 2. In order to examine the nature of this threshold-like behavior under creep-fatigue conditions,

Table 1. Loading conditions and symbols

R	σ_g (MPa)	ν (Hz)	Symbol
0.00	61	4.0	○
		1.1	⦶
		0.30	◑
		0.080	●
0.30		1.1	◇
0.50			□
0.70			▽
0.00	76	0.080	◇ with vertical line
0.30			half-filled ◇
0.50		6.0	△
		1.6	△ with vertical line
		0.080	half-filled △
		0.0080	▲
		0.0010	△ with inner △
0.57		5.0	□
0.63		10.0	■
0.70		6.0	▷
		0.080	▶

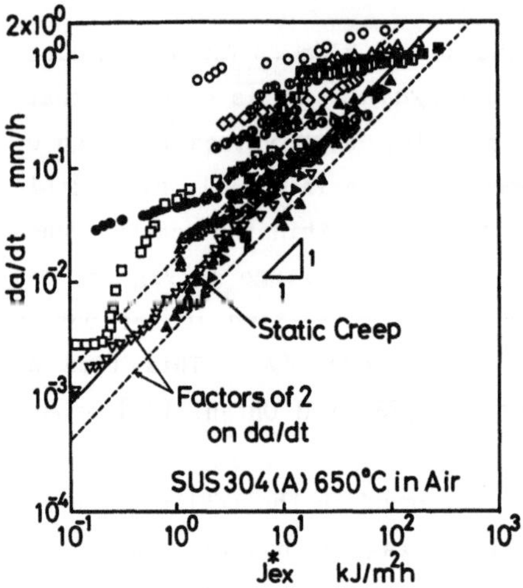

Fig.2. Correlationship between da/dt and J^*_{ex}

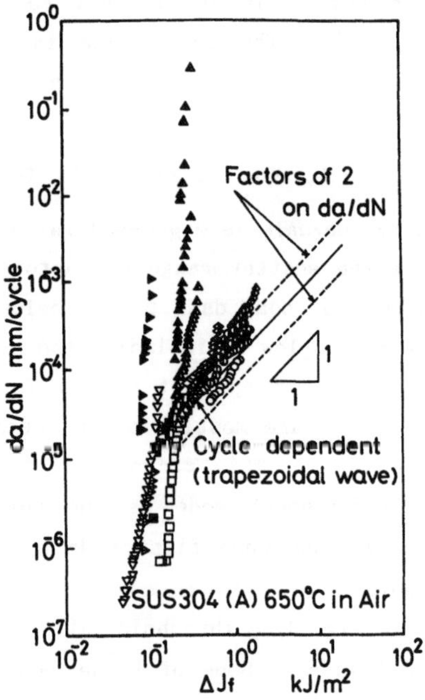

Fig.3 Correlationship between da/dN and ΔJ_f

two types of ΔJ_f-decreasing experiments were conducted [33], i.e., (1) ΔJ_f-decreasing at a constant R (decreasing σ_g), and (2) ΔJ_f-decreasing at a constant σ_g (increasing R).

Figure 4 shows the results of the ΔJ_f-decreasing tests at R = 0.5 and ν = 30 Hz. It is confirmed that a threshold $(\Delta J_f)_{th}$ of the cycle-dependent crack propagation under creep-fatigue conditions exists below and near 0.1 kJ/m^2.

When the ΔJ_f-decreasing tests were conducted at the constant σ_g conditions, the *da/dN* values initially decreased as in Fig. 5 along the cycle-dependent $(da/dN)_f$-ΔJ_f correlation line illustrated by a solid line in the figure. The *da/dN* values suddenly decreased at $\Delta J_f \simeq$ 0.1kJ/m^2 at steep slopes, leaving from the solid line, reached minimum values far below the solid line, and turned to sharp increasing. The ΔJ_f values at the minimum *da/dN* values were about the same with the $(\Delta J_f)_{th}$ value determined in Fig. 4.

In order to investigate the nature of the crack propagation below the $(\Delta J_f)_{th}$ in Fig. 5, the crack propagation rates in this portion were correlated on da/dt-J^*_{ex} diagram as in Fig. 6. In Fig. 6, (+) marks indicate the results obtained under a static creep condition. After ΔJ_f reached and

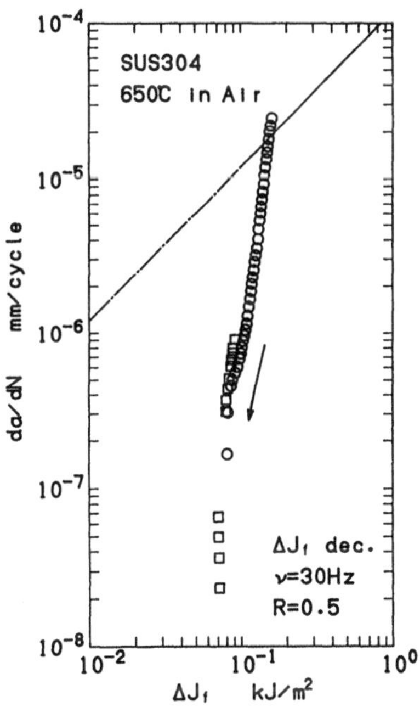

Fig.4. Results of ΔJ_f-decreasing tests at constant R

further decreased across the $(\Delta J_f)_{th}$ value determined in Fig. 5, the *da/dt* values increases with J^*_{ex} along the *da/dt*-J^*_{ex} relation under the static creep condition. This clearly shows that the crack propagation mode changed at $(\Delta J_f)_{th}$ from the cycle-dependent mode to the time-dependent mode.

From these ΔJ_f-decreasing experiments, two important conclusions were reached.

(1) A threshold value of ΔJ_f existed in the purely cycle-dependent crack propagation even at temperatures where significant creep was observed.

(2) In the region below the threshold of cycle-dependent crack propagation, the purely time-dependent crack propagation dominated.

Existence of the threshold in fatigue crack propagation at high temperatures was reported in ordinary high-temperature fatigue tests by several investigators [36,37].

In the region below the threshold, time-dependent crack propagation may appear. In the *da/dN*-ΔJ_f diagram this leads to a false threshold at arbitrary magnitudes depending on loading conditions as in Fig. 3. This false threshold may be called as "pseudo-threshold" or "creep-induced floating fatigue threshold" [33].

The experimental results described in the foregoing were obtained at 650°C. Experiments were extended to temperatures ranging from 550° to

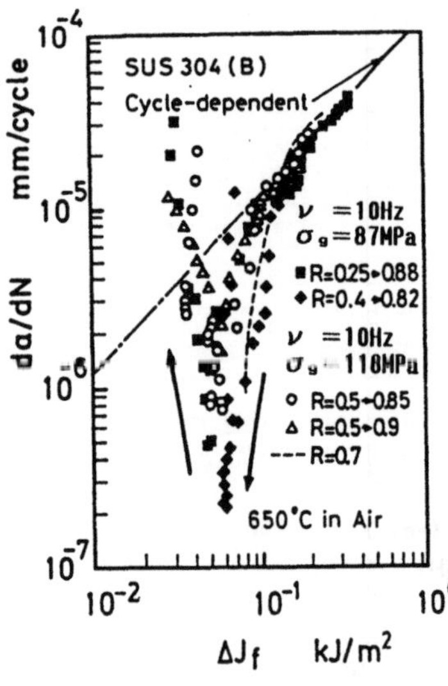

Fig.5. Results of ΔJ_f-decreasing tests at constant σ_g

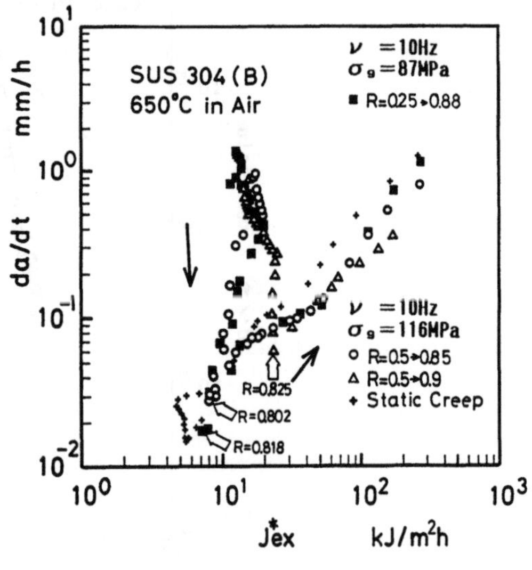

Fig.6. Relationship between *da/dt* and J^*_{ex} in ΔJ_f-decreasing tests at constant σ_g

700°C [38]. Similar results were obtained from these experiments. In general both the cycle-dependent and the time-dependent crack propagation behaviors were relatively insensitive to temperature, if the da/dN-ΔJ_f and da/dt-J^*_{ex} correlations were used.

From fractographical observations, crack propagation was transcrystalline in cycle-dependent mode and intercrystalline in time-dependent mode [39].

Effect of Crack Closure [40,41]

Crack closure is one of the most influencing factors in fatigue crack propagation at room temperature. This is also true for crack propagation under creep-fatigue conditions.

Crack closure was observed under creep-fatigue conditions at negative and low-positive values of R. The measurements of crack closure were mainly conducted at negative R values. Table 2 shows the conditions of the tests [40]. The high R-value experiments were also included in this program, in order to obtain reference data which were free from crack closure. Figure 7 summarizes the experimental results in terms of the da/dN-ΔK relations. The solid line in the figure represents the cycle-dependent da/dN-ΔK relation in the absence of crack closure. High values of threshold in ΔK are observed for negative values of R.

The closure stress was determined by the compliance method by measuring crack opening displacement along the center line for a gage length of 16mm across the crack.

Figures 8 and 9 show the da/dt-J^*_{ex} correlation and the da/dN-$(\Delta J_f)_{eff}$ correlation, respectively. For a negative value of R, J^*_{ex} was evaluated by using σ_e in Eq. (3) integrated for the positive value of σ. $(\Delta J_f)_{eff}$ was calculated by using the measured ΔK_{eff} in place of ΔK in Eq. (1). Although the scatter of the data is slightly wide in a near-threshold region, the da/dN values at negative R (with closure) agree well with those at high positive R (without closure) in Fig. 9, if the effective range of ΔJ_f is employed. By comparing Figs. 8 and 9, separation of the time-dependent and cycle-dependent crack propagation is equally good as in Figs. 2 and 3.

Similar results were also obtained for low-positive values of R.

The behavior of crack closure under creep-fatigue conditions was very much complicated both in the negative and positive R ranges, as compared with the behavior of crack closure under fatigue conditions at room temperature. The crack closure under creep-fatigue conditions may be

Table 2. Loading conditions of crack closure experiments

R	σ_g (MPa)	ν (Hz)	Symbol
-1.00	68	0.20	⊞
-0.85		0.10	◇
-0.75		0.10	◇ (with vertical line)
-0.50		0.10	◈
-0.40		1.0	▲
		0.10	◆
-0.40	76	0.10	■
-0.50	87	0.10	⊡
0.05	99	0.20	▷
0.10		0.20	▶
0.20		0.20	▶
0.50	116	10.0	○
0.60		10.0	⦶
0.70		10.0	⊙
		1.0	◬
0.80		10.0	◑
		0.010	◧
0.85		10.0	◒
0.90		10.0	●
0.50	123	10.0	▽
		0.010	△

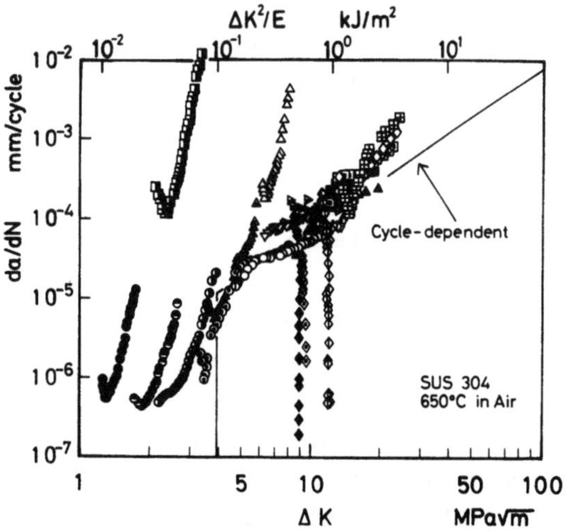

Fig.7. Correlationship between da/dN and ΔK for various values of R

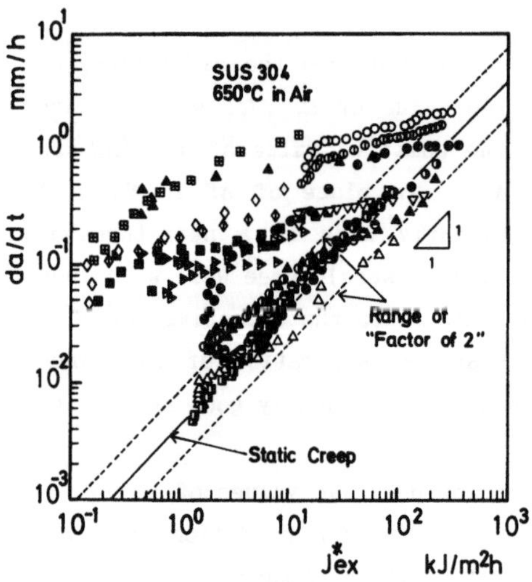

Fig.8. da/dt-J^*_{ex} correlationship of the results illustrated in Fig. 7

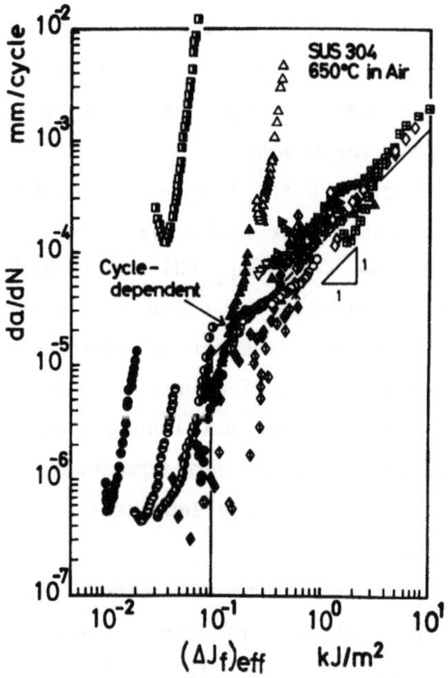

Fig.9. da/dN-$(\Delta J_f)_{eff}$ correlationship of the results illustrated in Fig. 7

affected jointly by plasticity, creep, environment (oxidation), roughness and many other factors including probably some geometrical factors. These factors cause time-dependent, cycle-dependent and mixed mode behaviors of crack closure. At the present state of the art, it is difficult to predict the general trend of crack closure under given loading conditions.

Modeling of Creep-Fatigue Crack Propagation [26,27,33]

From the preceding discussions on the behavior of crack propagation under creep-fatigue conditions, the process of crack propagation may be simplified to be controlled by one dominant process of the time-dependent (creep) mode or the cycle-dependent (fatigue) mode. The transition from one mode to the other occurs reversibly depending on the loading condition applied. This modeling belongs to the competitive (or dominant) damage hypothesis and is expressed as,

$$da/dt = \max[(da/dt)_c, \ (da/dt)_f]$$

or (6)

$$da/dN = \max[(da/dN)_c, \ (da/dN)_f]$$

$(da/dt)_f$ and $(da/dN)_c$ in Eq. (6) are the cycle-dependent crack propagation rate per time given by $(da/dN)_f \cdot \nu$ and the time-dependent crack propagation rate per cycle calculated as $(da/dt)_c/\nu$, respectively. As demonstrated in the foregoing, $(da/dt)_c$ and $(da/dN)_f$ are expressed by the sole function of J^* and $(\Delta J_f)_{eff}$, respectively, as in Eqs. (4) and (5), if the crack closure is taken into account. Figure 10 shows a map for the time-dependent and cycle-dependent modes in creep-fatigue crack propagation.

In determining $(da/dt)_c$ and $(da/dN)_f$, the scatter range of factors of 2

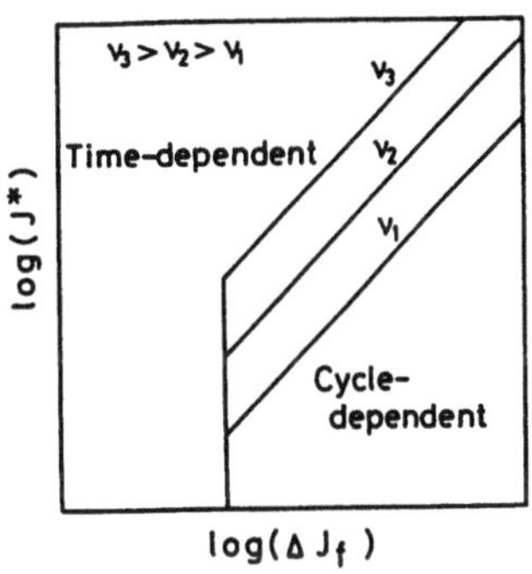

Fig.10. Map of crack propagation modes based on competitive rule

for these quantities was allowed. From experimental points of views, this means that the competitive damage hypothesis is actually equivalent to the summation (or cumulative) damage hypothesis, which is expressed as,

$$da/dt = (da/dt)_c + (da/dt)_f$$

or (7)

$$da/dN = (da/dN)_c + (da/dN)_f$$

The curves in Fig. 1 represent the results obtained by this summation damage hypothesis.

The separation of crack propagation mode into the time-dependent and cycle-dependent modes was also reported for nickel-base alloys [20,31,32]. The dominant process of the time-dependent mode in this case, however, was considered as an environment(oxygen)-assisted process, the effect of creep being negligibly small in the temperature range studied [20,32]. The dominant damage hypothesis and further the summation damage hypothesis might be applicable, on the same logical basis as in the creep-fatigue case (SUS 304 stainless steel), to nickel-base alloys as an approximation, if environment-fatigue interaction was not noticeable. The environment-fatigue interaction in the nickel-base alloys [30,32] usually might be more pronounced than the creep-fatigue interaction in the SUS 304 stainless steel.

Effect of Variable Amplitude on Threshold of Cycle-Dependent Crack Propagation [41-43]

It has been known that the fatigue threshold disappears under certain variable amplitude conditions [44], where the overload is higher and the underload is lower than the threshold level. Therefore, the effect of

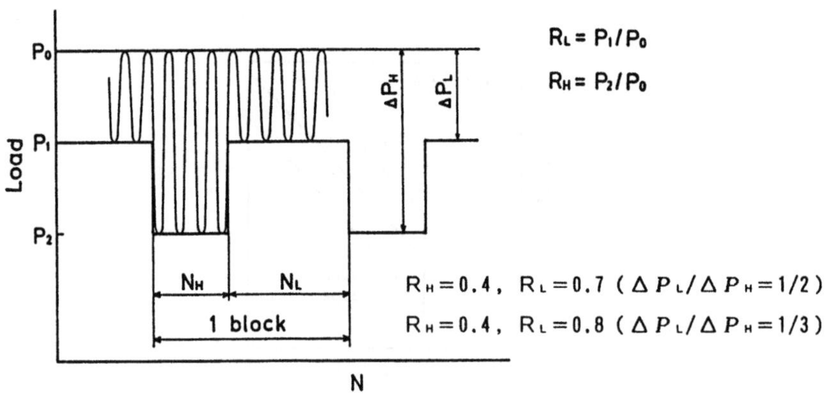

Fig.11. Loading pattern of repeated two-step tests

variable amplitude was investigated under creep-fatigue conditions [41-43].

Figure 11 shows the load pattern and the symbols used in this study. Repeated two-step loads were employed. Experiments were conducted in load-control mode by changing the minimum load P_{min} keeping the maximum load P_0 constant. Relatively high values of R were employed as indicated in Fig. 11, to avoid crack closure. In order to investigate disappearance of the threshold in cycle-dependent crack propagation, the initial lower load range ΔP_L was selected below the threshold, while the higher load range ΔP_H was larger than threshold. ν was kept at 30Hz. The crack propagation rate at ΔP_L was estimated as,

$$(da/dN)^{e*}_L = \{(da/dN)_{HL} \cdot (N_H + N_L) - (da/dN)_{\Delta JH} \cdot N_H\}/N_L \tag{8}$$

where $(da/dN)_{HL}$ was the average crack propagation rate during one load block, and $(da/dN)_{\Delta JH}$ was the rate at ΔP_H under constant amplitude condition.

Figure 12 shows the correlation between $(da/dN)^{e*}_L$ and the J-integral range $(\Delta J_f)_L$ at ΔP_L conducted at $\Delta P_L/\Delta P_H = 1/2$. The vertical solid line indicates the level of the threshold under constant amplitude conditions. It is observed that the threshold disappeared and crack propagation

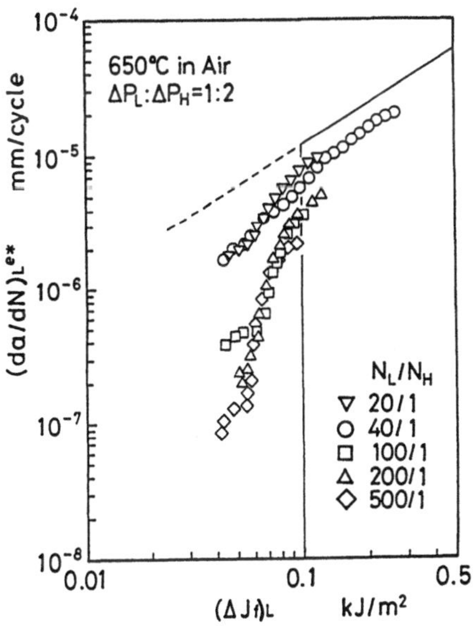

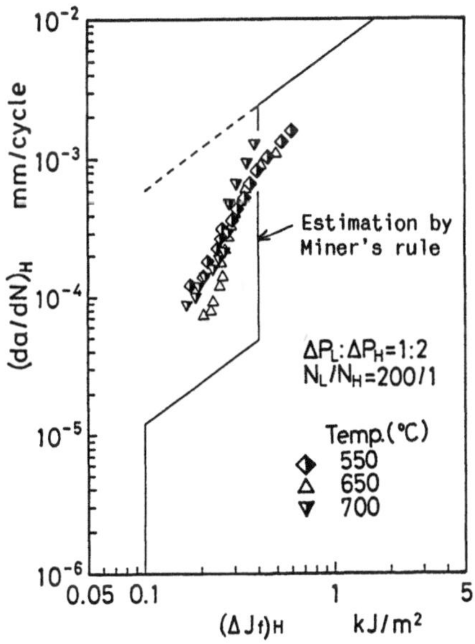

Fig.12. Relationship between $(da/dN)^{e*}_L$ and $(\Delta J_f)_L$ showing disappearance of threshold

Fig.13. Effect of temperature on $(da/dN)_H-(\Delta J_f)_H$ relationship occurred in the sub-threshold region at considerably high rates, in particular at low values of N_L/N_H. For larger values of N_L, the effect of varying amplitude decreased, implying restoration of the threshold during repeating of the lower amplitude. The same conclusions were also obtained for $\Delta P_L/\Delta P_H$ = 1/3.

Phenomenologically these results are consistent with the behavior observed in fatigue crack propagation at room temperature, although the underlying mechanisms may not be the same at room temperature and at high temperatures.

The effect of temperature was examined to obtain some information about mechanism of the threshold. Experiments were conducted at temperatures of 550°, 650° and 700°C at ν = 30Hz for N_L =200 and N_H = 1. Figure 13 shows the results in the relation between $(da/dN)_H$ and $(\Delta J_f)_H$. $(da/dN)_H$ in this figure was calculated as $(da/dN)_{HL} \cdot (N_H+N_L)/N_H$, assuming that the crack did not propagate at the sub-threshold region. $(\Delta J_f)_H$ was the ΔJ_f value at ΔP_H. No significant difference in the behavior is observed in Fig. 13 at these three temperatures. Similar results were also obtained for N_L = 40 and 20.

From fractographical observations, the crack propagation in these experiments was confirmed to be transcrystalline.

Further research should be done before the whole phenomenological structure and the physical mechanism of the threshold in creep-fatigue crack propagation are clarified.

Conclusions

Based on the experimental results of the SUS 304 stainless steel obtained by the authors, the creep-fatigue crack propagation behavior of ductile and environment-insensitive materials was overviewed.

(1) The crack propagation mode was separated into the purely time-dependent mode and the purely cycle-dependent mode, depending on the loading conditions applied.

(2) The time-dependent and cycle-dependent crack propagation rates, $(da/dt)_c$ and $(da/dN)_f$, were uniquely characterized, respectively, by modified J-integral J^* and by effective J-integral range $(\Delta J_f)_{eff}$, in which the effect of crack closure was taken into account.

(3) Threshold of the cycle-dependent crack propagation existed. Below this threshold the time-dependent crack propagation appeared, which induced in the da/dN-$(\Delta J_f)_{eff}$ diagram apparent false thresholds at arbitrary levels which depended on the loading conditions applied.

(4) The mode of crack propagation was determined by a competitive rule; the mode which gave a higher crack propagation rate dominated the whole process. The creep-fatigue interaction was very small in the case of the SUS 304 stainless steel.

Since the summation damage hypothesis is actually equivalent to the competitive damage hypothesis if the scatter of factors of 2 is allowed on damage values, the crack propagation rates may be predicted either by the competitive rule or by the summation rule.

(5) The threshold of the cycle-dependent crack propagation disappeared and the cycle-dependent crack propagation continued in the sub-threshold region under variable amplitude conditions, where the threshold was interposed between the maximum and minimum $(\Delta J_f)_{eff}$.

(6) Crack closure occurred at negative and low-positive values of *R*. Crack closure at high temperatures was induced by joint effects of plasticity, creep, environment, roughness, structural change, geometrical factors and so on, and naturally its behavior was very much complicated as compared with that at room temperature. However, both time-dependent and cycle-dependent crack propagation rates were uniquely correlated with J^* and ΔJ_f, respectively, if the effect of crack closure was taken into account in evaluating J^* and ΔJ_f.

References

1. Ohji, K.: *Theoretical and Applied Mechanics*. Univ. of Tokyo Press 27 (1977) 3-20.
2. Van Leeuwen, H.P.: *Eng. Fract. Mech.* 9 (1977) 951-974.
3. Ellison, E.G.; Harper, M.P.: *J. Strain Analysis* 13 (1978) 35-51.
4. Pilkington, R.: *Met. Sci.* 13 (1979) 555-564.
5. Fu, L.S.: *Eng. Fract. Mech.* 13 (1980) 307-330.
6. El Haddad, M.H.; Topper, T.H.; Mukherjee, B.: *J. Test. Eval.* 9 (1981) 65-81.
7. Sadananda, K.; Shahinian, P.: *Eng. Fract. Mech.* 15 (1981) 327-342.
8. Malik, S.N.: *Nucl. Eng. Design* 72 (1982) 359-371.
9. Ellison, E.G.; Musicco, G.G.: *Subcritical crack growth due to fatigue, stress corrosion and creep* (Larsson, L.H.: ed.). Elsevier (1984) Chap.14 403-448.
10. Saxena, A.; Kumar, V.: *J. Test. Eval.* 12 (1984) 189-192.
11. Sadananda, K.; Shahinian, P.: *Creep-fatigue-environment interaction* (Pelloux, R.M.; Stoloff, N.S.: eds.). Metall. Soc. AIME (1980) 86-111.
12. Ohji, K.: *Proc. the 28th Japan Nat. Symp. on Strength, Fracture and Fatigue* (1983) 31-52.
13. Lloyd, G.J.: *Fatigue at high temperature* (Skelton, R.P.: ed.). Applied Sci. Publishers (1983) Chap.5 187-258.
14. Ohtani, R.: *Bull. Japan. Inst. Mets.* 22 (1983) 190-196.
15. Ellison, E.G.: *Subcritical crack growth due to fatigue, stress corrosion and creep* (Larsson, L.H.: ed.). Elsevier (1984) Chap.17 531-561.

16. Sadananda, K.: *Nucl. Eng. Design* **83** (1984) 303-323.
17. Ohji, K.: *Role of fracture mechanics in modern technology* (Sih, G.C.; Nisitani, H.; Ishihara, T.: eds.). North-Holland (1987) 131-144.
18. Ohji, K.; Kubo, S.: *High-temperature creep-fatigue* (Ohtani, R.; Ohnami, M.; Inoue, T.: eds.). Current Japanese Materials Research **3**. Elsevier Appl. Sci. (1988) 91-113.
19. Ohtani, R.; Kitamura, T.: *ibid.* 65-90.
20. Nicholas, T.; Heil, M.L.; Haritos, G.K.: *Int. J. Fract.* **41** (1989) 157-176.
21. Kubo, S.; Ohji, K.; Ogura, K.: *Eng. Fract. Mech.* 11 (1979) 315-329.
22. Dowling, N.E.: *ASTM STP* 601 (1976) 19-32.
23. Ohji, K.; Ogura, K.; Kubo, S.: *Trans. Japan Soc. Mech. Engrs.* **44** (1978) 1831-1838.
24. Taira, S.; Tanaka, K.; Koterazawa, R.: *Proc. the 2nd Japan Congr. Test. Mats.* (1959) 55-58.
25. Ohji, K.; Kubo, S.; Yamakawa, N.; Kuri, T.: *Trans. Japan Soc. Mech. Engrs.* **50A** (1984) 1218-1227.
26. Ohji, K.; Kubo, S.; Joo, W.S.; Kuri, T.: *Trans. Japan Soc. Mech. Engrs.* **51A** (1985) 2137-2145.
27. Ohji, K.; Kubo, S., Kuri, T.; Joo, W.S.: *Int. Conf. on Creep, Tokyo.* Japan Soc. Mech. Engrs. (1986) 297-302.
28. Taira, S.; Ohtani, R.; Komatsu, T.: *J. Eng. Mat. Tech.* **101** (1979) 162-167.
29. Ohtani, R.: *Int. Conf. Engineering Aspects of Creep, Sheffield.* Inst. Mech. Engrs. (1980) **2** 17-32.
30. Solomon, H.D.; Coffin, L.F., Jr.: *ASTM STP* 520 (1973) 112-122.
31. Dimopulos, V.; Nikbin, K.M.; Webster, G.A.: *Metall. Trans.* **19A** (1988) 873-880.
32. Weerasooriya, T.: *Fracture mechanics: Nineteenth symposium* (Cruse, T.A.: ed.). ASTM STP 969 (1988) 907-923.
33. Ohji, K.; Kubo, S.; Kuri, T.; Nishizaki, K.: *Trans. Japan Soc. Mech. Engrs.* **51A** (1985) 2146-2153.
34. Ohji, K.; Ogura, K.,; Kubo, S.; Katada, Y.; Katsuhara, T.; Iwanaga, N.: *Trans. Japan Soc. Mech. Engrs.* **45A** (1979) 550-558.
35. Ohji, K.; Ogura, K.,; Kubo, S.; Katada, Y.: *Int. Conf. Engineering Aspects of Creep, Sheffield.* Inst. Mech. Engrs. (1980) **2** 9-16.
36. Matsuoka, S.; Takeuchi, E.; Nishijima, S.; McEvily, A.J.: *Metall. Trans.* **20A** (1989) 741-749.
37. Ogura, K.; Nishikawa, I.: *Fatigue 90. Proc. 4th Int. Conf. Fatigue and Fatigue Threshold, Honolulu* (Kitagawa, H.; Tanaka, T.: eds.). Materials and Component Engineering Publications Ltd. (1990) III 1413-1418.
38. Ohji, K.; Kubo, S.; Tokuhiro, T.; Iwai, S.: *Preprint of the 63rd National Meeting of Japan Soc. Mech. Engrs.* (1985) 44-45.
39. Ohji, K.; Kubo, S.; Joo, W.S.; Yamakawa, N.; Kuri, T.: *J. Soc. Mats. Sci., Japan* **34** (1985) 692-696.
40. Ohji, K.; Kubo, S.; Manabe, Y.: *J. Soc. Mats. Sci., Japan* **35** (1986) 924-929.
41. Ohji, K.; Kubo, S.; Kinugasa, H.; Gotoh, T.: *J. Soc. Mats. Sci., Japan* **37** (1988) 951-956.
42. Gotoh, T.; Ohji, K.; Kubo, S.; Nakai, Y.; Yamamoto, K.: *Preprint of the 64th Annual Meeting of Kansai Chapter,* Japan Soc. Mech. Engrs. No.894-1 (1989) 48-49.
43. Ohji, K.; Kubo, S.; Nakai, Y.; Gotoh, T.: *Preprint of the 67th National Meeting of Japan Soc. Mech. Engrs.* No.894-50 (1989) 99-100.
44. Kikukawa, M.; Jono, M.; Kondo, Y.: *Advances in fracture research (Fracture 81)* (Francois, D.: ed.). Pergamon Press (1982) **4** 1799-1806.

Creep Life Estimates for Defective Structures

R A Ainsworth
Nuclear Electric plc
Berkeley Nuclear Laboratories
Berkeley, Gloucestershire, UK

Summary

The paper describes a procedure for assessing the integrity of defective structures operating in the creep range based on reference stress techniques. Steady state behaviour and modifications to allow for stress redistribution are described. Possible approaches are given for modifying the reference stress method to predict crack growth rates in creep damaged material and to predict the remaining lifetime for continuum damage failure when crack propagation is occurring.

1. INTRODUCTION

Over the past ten years there have been significant developments in understanding the processes of incubation and crack growth which occur in defective components operating in the creep range. For components in steady state creep, the parameter C* characterises these processes. Simplified methods for estimating C* have been proposed [1-3]. At the same time, standardised experimental techniques using fracture mechanics specimens have been produced for collecting data in terms of C*. Taken together these developments have enabled procedures to be written for performing defect assessments in operating components. Section 2 outlines such a procedure which has been produced within the UK electricity supply industry. This is part of a more general high temperature assessment procedure, termed R5, which is described elsewhere in the Symposium [4].

Crack tip behaviour is not always described simply by C* but is influenced by redistribution from the stress field present on initial loading, by primary creep and by crack growth. Section 3 describes the manner in which this more complex response may be incorporated in the procedure of Section 2. An additional factor in assessing cracked components at high temperature is the development of continuum damage. Section 4 examines damage propagation rates, and possible approaches for including the effects of creep damage in life assessments of defective components.

M. Zyczkowski (Ed.)
Creep in Structures
IUTAM Symposium Cracow/Poland 1990

2. DEFECT ASSESSMENT PROCEDURES FOR STEADY LOADING

A defect assessment procedure for plant operating under steady loading has been described in [5]. It is based on the calculation of:

t_i, the incubation time during which an initially sharp crack tip blunts without any significant extension;

t_g, the subsequent growth time to a final defect size; and

t_{CD}, the time for failure by continuum damage mechanisms.

Methods for estimating t_i and t_g are given here for steady state creep. Non-steady state behaviour is discussed in Section 3 and continuum damage effects are addressed in Section 4.

The procedure [5] is based on approximate reference stress methods. The reference stress is related to the applied load, P, by

$$\sigma_{ref} = P\,\sigma_y/P_L\,(\sigma_y, a) \qquad (1)$$

where P_L is the plastic collapse load for yield stress σ_y and crack size, a. Using the reference stress, C* is estimated according to

$$C^* = \sigma_{ref}\,\dot{\varepsilon}^c_{ref}\,R \qquad (2)$$

where $\dot{\varepsilon}^c_{ref}$ is the creep strain rate from uniaxial data at the reference stress level. The length parameter R may be estimated from detailed finite-element results such as those in [1] or approximately as in [2].

Having estimated C*, the incubation time, t_i, may be obtained from analysis of a blunting crack tip by [5]

$$\varepsilon^c_{ref}(t_i) = \alpha(n)(\delta_i/R)^{n/(n+1)} \qquad (3)$$

Here ε^c_{ref} is the creep strain at the reference stress level, n is the stress exponent for creep, and δ_i is the critical crack opening displacement at the end of the incubation stage; $\alpha(n)$ is conservatively given by $\alpha(n) = 0.5$ for widespread creep conditions. This condition may be expressed

$$\varepsilon^c_{ref}(t_i) > \sigma_{ref}/E \tag{4}$$

where E is Young's modulus.

The time for propagation, t_g, is also obtained from calculations of C* and crack growth rate, $\dot{a}$, data in the form

$$\dot{a} = AC^{*q} \tag{5}$$

where A and q are material and temperature dependent constants. The growth time is determined by integration from an initial to a final defect size.

The above procedure is straightforward to use and has been applied to defects in operating plant. The basic approach has been extended to more complex situations as described in the following sections.

3. INFLUENCE OF STRESS REDISTRIBUTION

Generally crack tip behaviour is not determined simply by a steady state value of C* but is influenced by stress redistribution from a variety of sources. These are considered below along with the way in which the effects can be incorporated in the procedure of Section 2.

3.1 Redistribution from Initial Loading

On initial loading the response of a structure is elastic or elastic-plastic. As soon as creep strains develop, the crack tip field becomes similar to that in the steady state but with an amplitude, C(t), which is a function of time, t [6]. In [7, 8] it was shown that C(t) may be estimated, in terms of a non-dimensional time τ,

$$\tau = C^*t/J_o \tag{6}$$

as

$$C(\tau) = C^* \{(1 + \tau)^{n+1}/[(1 + \tau)^{n+1} - \psi]\} \tag{7}$$

Here J_o is the initial value of the J integral and ψ is a quantity in the range $0 \leq \psi \leq 1$ which is dependent on the initial plastic strains. Plasticity on initial loading ($\psi < 1$) produces crack tip fields closer to the steady state condition than those produced elastically ($\psi = 1$).

Equation (7) is similar to an estimate in [9] for elastic behaviour but is more convenient for the purposes of integration. In particular, if the estimate C(t) of eqn (7) is used to replace C* in the blunting analysis leading to eqn (3), the result becomes approximately [10] for initial elastic behaviour

$$\varepsilon^{c}_{ref}\,(t_i) + \sigma_{ref}/E = (\delta_i/R)^{n/(n+1)} \tag{8}$$

Eqn (3) with $\alpha(n) = 0.5$ is conservative when inequality (4) is satisfied.

As q in eqn (5) is usually close to n/(n+1), it is possible to include redistribution, in an approximate manner, on the crack growth rate also. This leads to the increment of growth, Δa, in time, t, being [10]

$$\Delta a \simeq A\, C^{*q}\, t\, [1 + \sigma_{ref}/E\varepsilon^{c}_{ref}] \tag{9}$$

3.2 Influence of Primary Creep

The influence of primary creep on crack tip fields is described by Riedel [11] who introduced a parameter C^*_h related to C(t). When deformation is dominated by primary creep, this reduces to eqn (2) with $\dot{\varepsilon}^{c}_{ref}$ interpreted as the instantaneous creep strain rate. Direct validation of this approach has been given in [5] by comparison with experimentally measured values of C*. Therefore, the inclusion of primary creep in the procedure of Section 2 is straightforward.

3.3 Influence of Crack Growth

The C* and C(t) parameters are applicable for stationary cracks. For moving cracks, Hui and Riedel [12] showed that a different singular field was produced in a small region at the crack tip outside which the C* and C(t) parameters describe behaviour [2, 11]. The length scale over which the C* and C(t) description breaks down depends on the rate of crack growth. Thus it is possible to use assessment procedures based on C*, provided a limit is imposed on the growth rates for which they are applied. A convenient non-dimensional crack growth rate [2] is

$$\lambda = \dot{a}\, \sigma_{ref}/E\, R\, \dot{\varepsilon}^{c}_{ref}\ . \tag{10}$$

The theoretical work suggests $\lambda << 1$ is required for C* to be the controlling parameter. Examination of experimental data [2, 5] suggests the restriction $\lambda < 0.5$ is adequate.

4. DEVELOPMENT OF CONTINUUM DAMAGE

The accumulation of widespread, significant creep damage ahead of a crack and the propagation of a dominant crack represent distinct failure mechanisms. The former is examined in sub-section 4.2 with damage propagation in non-defective components being discussed first in sub-section 4.1. The latter mechanism is discussed in sub-section 4.3 when damage is confined to a region local to the crack tip. In general, however, considerations of crack growth through more widespread creep damage are necessary and this is examined in sub-section 4.4.

4.1 Damage Propagation Rates in Non-Defective Structures

The essential features of damage propagation may be illustrated by a simple model in which the rate of damage accumulation $\dot{\omega}$ is

$$\dot{\omega} = \dot{\varepsilon}^c/\varepsilon_f \tag{11}$$

where ε_f is the ductility. Thus, 'failure' occurs at a point in a structure when a critical creep strain is accumulated and a damage front marks the boundary between 'failed' and 'unfailed' material.

Consider a damage front propagating in the z-direction and currently at $z = c$. Strain rates depend on z and on c, as $\dot{\varepsilon}^c(z,c)$ say. The condition of constant strain at the damage front defines its propagation rate as

$$\dot{c} = \dot{\varepsilon}^c(c,c)/[- d\varepsilon^c\ (z,c)/dz|_{z\ =\ c}] \tag{12}$$

It may be assumed that $\dot{\varepsilon}^c(c,c)$ is proportional to $\dot{\varepsilon}^c_{ref}$, provided the reference stress is calculated assuming that fully damaged material has no load-bearing capacity. Also, the strain gradient is likely to be proportional to the strain at the damage front, ie to ε_f. Then, eqn (12) gives

$$\dot{c} \propto \dot{\varepsilon}^c_{ref}/\varepsilon_f \tag{13}$$

Equation (13) has been checked with detailed solutions for some simple geometries. It is a reasonable approximation particularly at larger values of n. However, exact solutions for damage propagation rates are complex even for simple uncracked geometries and a simple damage law. Notwithstanding this, it transpires from numerical and experimental studies that the time, t_{CD}, for damage to propagate throughout a structure and lead to failure can be estimated reasonably accurately in terms of the reference stress of the undamaged structure:

$$t_{CD} \simeq t_r(\sigma^o_{ref}) \tag{14}$$

where $t_r(\sigma)$ is the rupture time at stress, σ, from conventional stress/time-to-rupture data [13].

4.2 Widespread Damage Without Crack Growth in Defective Structures

In defective structures, regions of high creep damage are produced early in life local to the crack tip. However, numerical studies [14] show that damage blunts the crack tip and reduces the rate of damage propagation. Consequently, defective and non-defective structures can behave in a similar manner provided the defect does not grow or follows some way behind the damage front. Thus, the lifetime of defective structures is, in some cases, described by eqn (14) with the reference stress based on the crack size at start-of-life. This has been confirmed by experiments with creep ductile materials. For the procedure of Section 2, simplified estimates of t_{CD} are, therefore, possible.

4.3 Crack Growth with Localised Damage

A number of models have been produced for predicting creep crack growth rates [11]. In [15], creep damage is assumed to be confined to a process zone at the crack tip. This leads to a predicted crack growth rate

$$\dot{a} = A' \, C^{*\,n/(n+1)} \, r_c^{\,1/(n+1)}/\varepsilon_f \tag{15}$$

where A' is a constant, r_c is the extent of the process zone and ε_f is the ductility appropriate to the stress state near the crack tip. As $1/(n+1)$ is a small fractional power, eqn (15) is insensitive to r_c and is consistent with the observed experimental data of eqn (5) if $\varphi = n/(n+1)$.

In view of eqn (2), it is apparent that eqn (15) predicts

$$\dot{a} \propto \dot{\varepsilon}^c_{ref}/\varepsilon_f \tag{16}$$

It has been noted in Section 4.1 that damage propagation rates are also proportional to creep strain rates and inversely proportional to ductility.

Therefore, crack propagation and damage front propagation will often occur simultaneously and complicate the approaches of sub-sections 4.1 and 4.2.

4.4 Interactions of Creep Damage and Crack Growth

Damage and crack growth interactions affect both the crack growth rate and the rate of damage accumulation. Possible approaches for treating these interactions are discussed below. For convenience, the discussion is separated into the effect of crack growth on continuum damage failure time and the effect of damage on crack growth.

4.4.1 The Effect of Crack Growth on Continuum Damage Failure

The reference stress of eqn (1) increases with increasing crack size. However, it should be noted from sub-section 4.1 that even in circumstances where the reference stress increases, the lifetime is still accurately predicted in terms of the initial reference stress. This is because damage occurs non-uniformly, being large in regions traversed by damage fronts but relatively small elsewhere. When a crack advances more rapidly than the damage front it is reasonable to assume that widespread damage is negligible ahead of the crack. In view of eqn (14), the remaining life at time t when the crack size is a(t) satisfies

$$\text{Remaining life} < t_r[\sigma_{ref}(a)] \tag{17}$$

When the crack is close to its initial size at substantial fractions of the life, eqn (14) is more restrictive than inequality (17). However, inequality (17) limits the lifetime when there is significant crack growth.

4.4.2 The Effect of Continuum Damage on Crack Growth

Two approaches to this problem have been described [5, 16]. With the reference stress method, the approach has been to adopt a strain hardening rule and insert instantaneous creep strain rates in eqn (2) to estimate C*. For conditions of widespread damage, this corresponds to the use of tertiary creep rates. In a similar manner to the treatment of primary creep in sub-section 3.2, for conditions in which a tertiary creep law applies throughout a structure, this is a reasonable approximation.

An alternative view is taken by Nishida et al [16] who assume that C* can still be calculated according to a secondary law but allow for damage by modifying eqn (5) in terms of the current damage, ω, to

$$\dot{a} = A\, C^{*q}/(1 - \omega) \tag{18}$$

where damage is calculated according to a life fraction rule. Alternatively, the ductility exhaustion model could be used and then eqn (18) follows from the model of [15] described in sub-section 4.3.

Because of the non-uniform nature of damage accumulation, both the above approaches are likely to be conservative: the former through the use of a strain hardening rule; the latter through the use of ω. Also, when there is widespread damage, values of C* are likely to increase: in the approach of [16] this is accounted for by using an average strain rate taken as the ductility divided by the time to rupture; in the approach of [5] instantaneous creep strain rates are used. Thus, although there are differences between the two approaches, they have similar characteristics. Indeed, Nishida et al [16] show that they give similar predictions of failure time when applied to a particular example. For practical purposes it is probably adequate to combine some of the features of both approaches by using crack growth data and strain rate data appropriate to the material in the service condition, ie allowing for effects of ageing, creep damage, etc, on the material properties.

5. CONCLUDING REMARKS

Over the past ten years there have been significant advances in understanding the behaviour of defects at high temperature. These have enabled the production of assessment procedures which are simple to apply to complex operating components and address the important phenomena of incubation, growth and continuum damage. More recently it has been possible to refine the procedures to account for the effects of stress redistribution and interactions between crack growth and continuum damage.

ACKNOWLEDGEMENT

This paper is published with permission of Nuclear Electric plc.

6. REFERENCES

1. V Kumar, M D German and C F Shih, An engineering approach for elastic plastic fracture analysis, EPRI Report EPRI No-1931, 1981.

2. R A Ainsworth, Some observations on creep crack growth, Int J Fract 20 (1982) 147-159.

3. R A Ainsworth, The assessment of defects in structures of strain hardening material, Engng Fract Mech 19 (1984) 633-642.

4. I W Goodall and R A Ainsworth, An assessment procedure for the high temperature response of structures, IV Iutam Symposium on Creep in Structures, Krakow, 1990.

5. R A Ainsworth, G G Chell, M C Coleman, I W Goodall, D J Gooch, J R Haigh, S T Kimmins and G J Neate, CEGB Assessment procedure for defects in plant operating in the creep range, Fatigue Fract Engng Mater Struct 10 (1987) 115-127.

6. K Ohji and S Kubo, Fracture mechanics evaluation of crack growth behaviour under creep and creep-fatigue conditions, in High Temperature Creep-Fatigue (eds R Ohtani, M Ohnami and T Inoue), Elsevier, London (1987) 91-113.

7. R A Ainsworth and P J Budden, Crack tip fields under non-steady creep conditions - I. Estimates of the amplitude of the fields, Fatigue Fract Engng Mater Struct 13 (1990) 263-276.

8. J Joch and R A Ainsworth, The effect of geometry on the development of creep singular fields for defects under step-load controlled loading, to appear, 1990.

9. R Ehlers and H Riedel, A finite element analysis of creep deformation in a specimen containing a macroscopic crack, ICF5 Conference, Vol 2 (1981) 691-698.

10. R A Ainsworth and P J Budden, Crack tip fields under non steady creep conditions - II. Estimates of associated crack growth, Fatigue Fract Engng Mater Struct 13 (1990) 277-285.

11. H Riedel, Fracture at High Temperatures, Springer, Berlin 1986.

12. Y Hui and H Riedel, The asymptotic stress and strain field near the lip of a growing crack under creep conditions, Int J Fract 17 (1981) 409-425.

13. I W Goodall, F A Leckie, A R S Ponter and C H A Townley, The development of high temperature design methods based on reference stresses and bounding theorems, J Engng Mater Techn 101 (1979) 349-355.

14. I W Goodall and E J Chubb, The creep ductile response of cracked structures, Int J Fract 12 (1976) 289-303.

15. Nikbin, D J Smith and G A Webster, An engineering approach to the prediction of creep crack growth, J Engng Mater Techn 108 (1986) 186-191.

16. Nishida, K M Nikbin and G A Webster, Influence of net section damage on creep crack growth, J Strain Analysis 24 (1989) 75-82.

Non-Stationary Nucleation of Cavities at Grain Boundaries

H.-A. BAHR, H. BALKE, W. POMPE and S. WERDIN[1)]

Central Institute of Solid State Physics and Material Research of the Academy of Sciences, Helmholtzstr. 20, Dresden, 8027
[1)]Technical University Dresden, Department of Mechanical Engineering, Mommsenstr. 13, Dresden, 8027

Summary

The formation of creep pores at grain boundaries as a result of stress concentration in slip processes is investigated by means of instationary nucleation theory. The development of the pore size distribution is numerically calculated. The critical stress is assessed above which persistently growing micropores are formed. It agrees better with experiment than those derived from stationary or quasi-stationary models.

1. Introduction

Polycrystalline metals and ceramics show a tendency to creep pore formation at grain boundaries [1]. The pores appear at loads much smaller than expected from stationary nucleation theory [2]. An explanation of this discrepancy would be useful for predicting critical defects under mechanical or thermal load at high temperatures. Also it would help to solve the problem of avoiding grain boundary damage in large deformation (superplasticity) [2].

As one possible reason for the deviation between conventional theory and experiment, load stress concentrations at grain boundaries have to be taken into account [2, 3]. They are expected to arise along the edges of grains, and decay by diffusion. In [4] it is emphasized that pore formation should be regarded as a non-stationary process going on under changing stress. A solution of the non-stationary problem was derived by means of a quasi-stationary approximation [3, 5] without completely removing the mentioned discrepancy. Therefore, instationary nucleation under changing stress has been calculated numerically for Al_2O_3 in [6], which led to some insight into the process.

M. Zyczkowski (Ed.)
Creep in Structures
IUTAM Symposium Cracow/Poland 1990

The present paper is aimed at a thorough foundation of the model used in [6], especially with respect to the changing stress. It provides, together with the results in [6], a reasonable lower boundary for the critical stress of pore growth.

2. Time-dependent stress concentration

The model of stress concentration after [2] is shown schematically in Fig. 1.

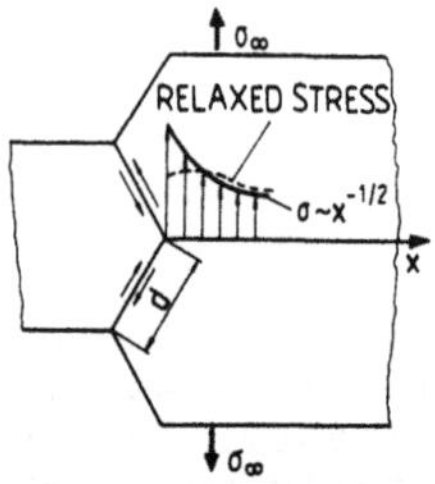

Fig. 1. Stress concentration at the grain boundary near the edge

Let the sample be subjected to external stress σ_∞ for times $t \geq 0$. The inclined grain boundaries are assumed to be free from shear stress. Then the stress concentration σ on the horizontal grain boundary near the edge can be assessed, after [2], by

$$\sigma(x,t)=\varkappa\sigma_\infty d^{1/2}\cdot f(x/(\alpha t)^{1/3})/(\alpha t)^{1/6} \quad , \quad x \geq 0$$
$$0.18 \leq \varkappa \leq 0.41 \quad , \quad \alpha = G\Omega D_b\delta_b/(2kT(1-\nu)) \tag{1}$$

d - grain size, G - shear modulus, ν - Poisson's number, k - Boltzmann's constant, Ω - atom volume, T - temperature, $D_b\delta_b$ - grain boundary diffusion coefficient.

The function f(z) is characterized by

$$f(z) = \begin{cases} 0.74 & \text{for } z = 0 \\ z^{-1/2} & \text{for } z \to \infty \end{cases}$$

and by the curve for t_0 in Fig. 2.

Since the nucleation problem is rather involved because of (1) it has to be simplified for the purpose of computation. For this reason the instant of the begin of pore formation, t_0, is chosen such that f in Fig. 2 can be approximated by a constant within a

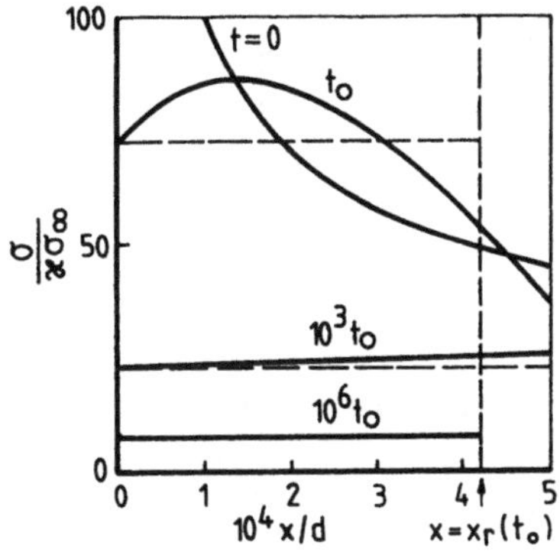

Fig. 2. Initial, relaxed and used constant stress distribution versus grain boundary coordinate

region $0 \leq x \leq x_r(t_o)$. This region must be so large that it represents a sufficiently large number of atoms. This can be achieved with

$$x_r(t_o)/(\alpha t_o)^{1/3} = k_1 \quad , \quad x_r(t_o) = k_2\Omega^{1/3} \qquad (2)$$
$$k_1 \approx 4 \quad , \quad k_2 > 5 \quad ,$$

which leads to

$$(\alpha t_o)^{1/3} = \Omega^{1/3}k_2/k_1 \quad , \quad \sigma(t)=0.74\varkappa\sigma_\infty d^{1/2}/(\alpha t)^{1/6}. \qquad (3)$$

Fig. 2 is based on $k_1=4$, $k_2=12$, $d=10^{-5}$m, $\Omega=42.5\cdot 10^{-30}m^3$ for Al_2O_3 after [7]. Obviously the constant approximate stress is below the real average stress for all $t > t_o$. Therefore, and also because of the neglection of the atoms in the range $x > x_r(t_o)$, the subsequent calculation for given load σ_∞ is expected to yield too low probabilities of above-critical pores. This remains so also with a smaller region $0 \leq x \leq x_r$, corresponding to smaller t_o, because the larger initial stress would promote the kinetic process in a stronger way than the decrease of the number of available atoms would reduce it (compare discussion of Fig. 4).

For large times the fracture-mechanical approximation (1) yields incorrect stress values $\sigma < \sigma_\infty$. Fortunately this is insignificant since nucleation ceases much earlier (compare 3.).

3. Non-stationary nucleation of cavities

The pore size distribution $Z_n(t)$ satisfies a rate equation well known from non-stationary nucleation theory [3, 5].

$$dZ_n(t)/dt = I_{n-1}(t) - I_n(t) \tag{4}$$

where

$$I_n(t) = q(Z_n(t)-Z_{n+1}(t)\cdot Z_n^e/Z_{n+1}^e) \quad , \; q=2\pi D_b\delta_b/\Omega \tag{5}$$

Z_n^e is the equilibrium distribution

$$Z_n^e = Z_o \exp(-\Delta G_n/kT) \; . \tag{6}$$

Z_o is the number of available atoms

$$Z_o = d\cdot x_r(t_o)/\Omega^{2/3} = k_2 d/\Omega^{1/3} .$$

ΔG_n is the free enthalpy of a pore of size n

$$\Delta G_n = -n\Omega\sigma(t)+(18\pi\gamma^3\zeta)^{1/3}(n\Omega)^{2/3}. \tag{7}$$

γ is the surface tension, and ζ is a pore size parameter after [2]. ΔG_n has a maximum with respect to pore size, located at

$$n_c = 16\pi\gamma^3\zeta/(3\sigma^3(t)\Omega) \tag{8}$$

and decreases monotonically beyond. Because of stress relaxation, the maximum increases and moves to larger pore sizes.

The initial and boundary conditions for (4), (5) are

$$Z_n(t_o) = 0, \quad n = 2, \quad \dots, \quad n_{max} \; ,$$

$$Z_1(t) = Z_o\exp(-\Delta G_1(t)/kT), \qquad Z_{n>n_{max}}(t) = 0 \quad .$$

$n_{max}(t)$ means the largest size at a given time t.

The computations were based on high temperature data of Al_2O_3 after [2, 7].

$$G = 1.55\cdot 10^5(1-0.35\frac{T-300K}{2320K})MPa \quad , \; \nu = 0.25 \quad , \; \varkappa = 0.41$$

$$D_b\delta_b = 10^{-9}\exp(-Q/(kT))m^3/s \quad , \quad Q = 400 \text{ kJ/mol}.$$

Pore size distributions obtained from this are shown in Fig. 3. for several values of the dimensionless time parameter qt. Also the time-dependend critical pore size n_c is indicated there.

We define a criterion for damage in such a way that damage is supposed to set in when there is at least one supercritical pore per stressed grain boundary. In order to find out whether and when this happens, we take the deterministic growth equation [5]

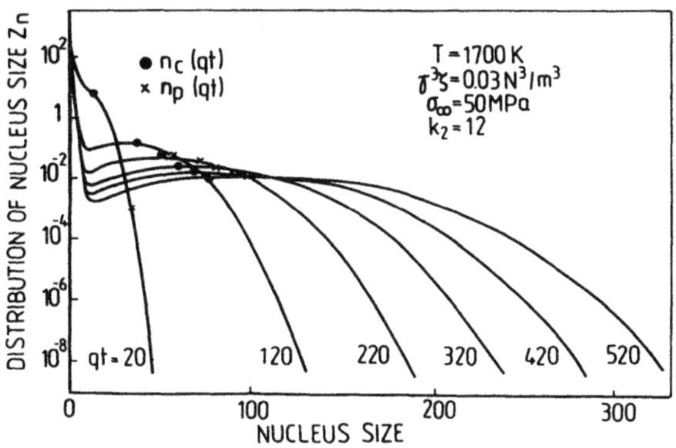

Fig. 3. Evolution of pore size distribution

$$dn_p/d(qt) = -\partial(\Delta G_n(qt)/kT)/\partial n_p$$

which can be rewritten as, by means of (7),

$$dn_p/d(qt) = (16\pi\gamma^3\zeta\Omega^2/3)^{1/3}\ (n_c^{-1/3}(qt)-n_p^{-1/3}(qt))/kT \quad (9)$$

and calculate from this the lower boundary $n_p(t')$ of the permanently growing pores of size $n \geq n_p(t) \geq n_c(t)$, making use of the condition

$$n_p \to n_c \quad \text{for} \quad qt \to \infty \quad (10)$$

The non-linear differential equation (9) is solved for large times by expansion with respect to $(n_p-n_c)/n_c$ and by numerical integration for smaller times. As expected, the calculated n_p are larger than n_c (Fig. 3).

The condition of at least one supercritical pore on average is written as

$$N \equiv \sum_{n>n_p(qt)}^{\infty} Z_n(qt,\sigma_\infty=\sigma_c) = 1 \ .$$

Since N depends on time (Fig. 4), this condition yields a time to damage under a given σ_c. In the nearly constant region, however, (11) can be regarded as an equation for σ_c. From Fig. 4 one obtains by interpolation $\sigma_c \approx 50.1$ MPa. It is seen that N depends strongly on stress. In order to make sure that N depends only weakly on the arbitrarily chosen parameter k_2, an additional curve was calculated for k = 6 (dashed line in Fig. 4).

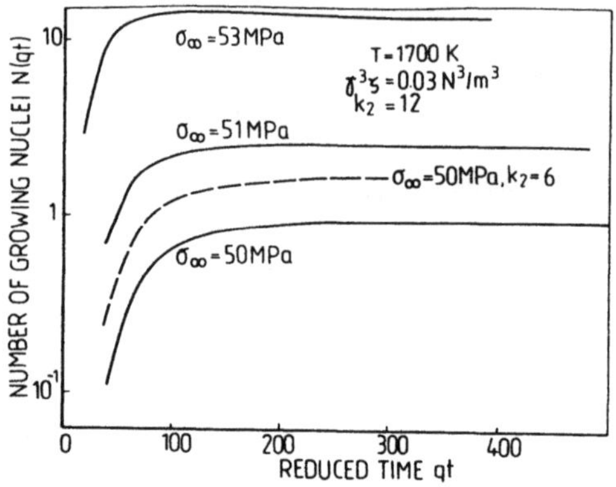

Fig. 4. Number of permanently growing pores versus reduced time

4. Discussion

The non-monotonic pore size distributions in Fig. 3 deviate essentially from the monotonically decreasing distributions known from the conventional stationary or quasi-stationary theories [3, 5]. This more realistic result leads to a reasonable lower threshold of the critical stress of pore growth. Several numerical examples were calculated in order to find out the influences of temperature and surface tension. The numerical effort was reduced by means of a transformation [6] taking into account that the stress relation (1) and the kinetics (4) - (7) depend on the same dimensionless time parameter qt. The critical stress derived

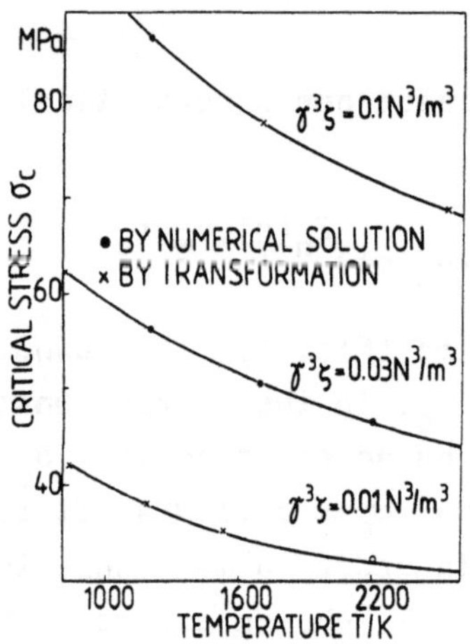

Fig. 5. Critical stress versus temperature

in this way is shown in Fig. 5. It decreases with temperature.

Note that our results have been derived with the assumption of a sudden application of load, i.e. loading rates much higher than stress decay rates. This is certainly realized in mechanical loading at lower temperatures but has to be checked if high temperatures are involved.

It must be mentioned that the damage criterion used here tentatively is to some degree arbitrary and requires additional considerations on the transition from pores to cracks.

Acknowledgement

The authors thank M. Bobeth and H.-J. Weiss for detailed discussions.

References

1. Riedel, H.: Fracture at high temperatures. Berlin, Heidelberg: Springer-Verlag 1987.

2. Evans, A.G.; Rice, J.R.; Hirth, J.P.: Suppression of cacity formation in ceramics: prospects for superplasticity. J. Amer. Ceram. Soc. 63 (1980) 368 - 375.

3. Trinkaus, H.; Yoo, M.H.: Cavity nucleation under time-dependent stress concentration. Phil. Mag. A57 (1988) 543 - 564.

4. Thouless, M.D.; Evans, A.G.: Nucleation of cavities during creep of liquid-phase-sintered materials. J. Amer. Ceram. Soc.67(1984) 721 - 727.

5. Trinkaus, H.; Yoo, M.H.: Nucleation under time-dependent supersaturation. Phil. Mag. A55 (1987) 269 - 289.

6. Bahr, H.-A.; Balke, H.; Pompe, W.; Werdin, S.: Instationäre Porenbildung an Korngrenzen. Neue Hütte 34 (1989) 201 - 207.

7. Frost, H.J.; Ashby, M.F.: Deformation-mechanism maps. Oxford: Pergamon Press 1982.

A Non-Unilateral Damage in Creeping Plates

A. BODNAR and M. CHRZANOWSKI

Institute of Structural Mechanics
Cracow University of Technology, Poland

Summary

The effect of compressive stresses upon the creep failure of structures has been analysed in the paper. A combined theory of that by D.R. Hayhurst [1] and J. Lemaitre [2] was used to describe damage evolution, whereas a non-stationary creep theory coupled with damage described the structure deformation. Isochronous limit curves in plane stress state were build, and the clamped and simply supported plates were analysed to demonstrate the influence of compressive stresses.

1 Introduction

Of the two design criteria i.e. allowance for deformation and maximum load capacity the latter in creep conditions is replaced by the time to rupture estimation, as any loading applied can lead to structure failure. Often it is a consequence of the growth of microscopic defects, and then Continuum Damage Mechanics can be used. Such an approach was used also by the authors in evaluating failure phenomenon in creeping plates [3,4], where the scalar representation of damage governed by the damage law due to D.R. Hayhurst [1] was employed. This theory takes into account compressive stresses only through effective stress. According to J. Lemaitre [2] in some circumstances the the influence of compressive stresses cannot be neglected. Consequently, the theory was proposed in [2] to include this effect. In the present paper an attempt was made to combine both approaches by introducing to damage growth law a term explicitly responsible for compressive stress induced damage. With limit curves built for such a theory, the analysis of structures became possible, and effect of compressive stresses amenable.

2 Damage evolution law

Classical theory of Kachanov-Rabotnov [5,6] uses positive principal stress σ_1 as a governing variable in damage evolution law:

$$\frac{\partial \omega}{\partial t} = A\left(\frac{\sigma_1}{1-\omega}\right)^m , \tag{2.1}$$

where ω is a damage parameter $(0 \le \omega \le 1)$, t is time, A and m are material constants.

M. Zyczkowski (Ed.)
Creep in Structures
IUTAM Symposium Cracow/Poland 1990

A material which obeys such a law can be called a *unilateral* one, as damage is being developed under positive stress only. Under constant uniaxial stress $\sigma = \sigma_1$ this equations yields time to rupture $t_R = 1/(m+1)A\sigma_1^m$.

A more general damage law in which σ_1 is replaced by an equivalent stress was proposed by D.R. Hayhurst [1]:

$$\frac{\partial \omega}{\partial t} = A\left[\alpha\frac{\sigma_1}{1-\omega} + (1-\alpha)\frac{\sigma_e}{1-\omega}\right]^m , \tag{2.2}$$

where $\sigma_e = \sqrt{3s_{ij}s_{ij}/2}$ is effective stress (s_{ij} is deviatoric stress tensor).

The parameter α reflects the sensitivity of different materials to the failure mode. For the microdefects which are formed at grain boundaries perpendicular to principal stresses direction the α parameter is assumed to be $\alpha = 1$. In the other extreme case, when slip mechanism of failure prevails $\alpha = 0$ is assumed. The intermediate values of α $(0 < \alpha < 1)$ correspond to a mixed type of failure when both mechanisms interplay. In such a formulation the negative principal stresses are taken into account implicitly through effective stress.

In the paper [2] the effect of negative stresses was taken into account by assuming that for $\sigma < 0$ in uniaxial case the damage law takes the form:

$$\frac{\partial \omega}{\partial t} = A\left(\frac{|\sigma|}{1-h\omega}\right)^m , \tag{2.3}$$

where h $(0 \leq h \leq 1)$ is a coefficient which reflects material sensitivity to compressive stress. For $h = 1$ the material is equally sensitive to positive and negative stresses (cf. Eq.(2.1)); such material behaviour will be called *bilateral* one, whereas $h = 0$ describes a material with linear time dependence of damage on negative stress. For values of h different from 1 the material will be called *non-unilateral* one. For positive stresses the Eq. (2.1) holds.

In the present paper the above theories will be combined to give the following damage evolution law:

$$\frac{\partial \omega}{\partial t} = A\left[\alpha \max\left\{\frac{<\sigma_i>}{1-\omega}, \frac{<-\sigma_j>}{1-h\omega}\right\} + (1-\alpha)\frac{\sigma_e}{1-\omega}\right]^m , \tag{2.4}$$

where $i \neq j = 1, 2$ designate principal stresses, and $<\sigma>=\sigma$ for $\sigma > 0$, $<\sigma>=0$ for $\sigma < 0$.

3 Isochronous failure curves

The Eqs. (2.1), (2.2), (2.3) and (2.4) can be integrated, and the isochronous curves constructed for fixed times to rupture. The isochronous curve corresponding to Eq. (2.1) is shown in Fig. 1 demonstrating its obvious insensitivity to compressive stresses (*unilateral*

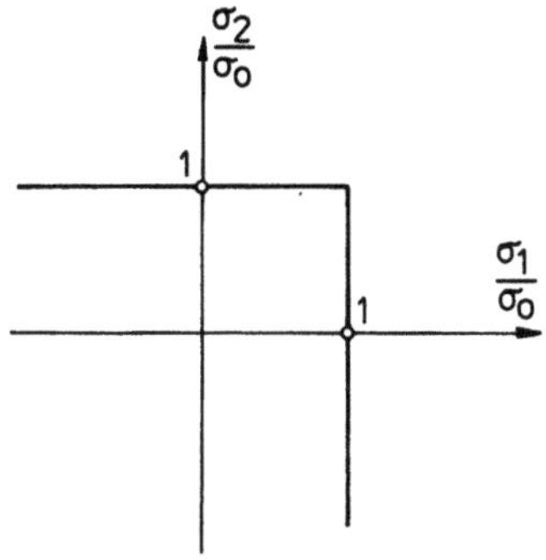

Figure 1:

material). In this figure (as well as in all others) the reference stress $\sigma_0 = [1/A(m+1)t_R]^m$ was used.

The theory represented by Eq. (2.2) gives a formal possibility of taking into account the effect of negative stresses through effective stress. This, of course, has to be justified by the observations of a real material behaviour, especially in the case when both principal stresses are negative. Two extreme cases discussed above are shown in Fig. 2 by full lines: for $\alpha = 1$ we have again the classical Kachanov-Rabotnov theory, and for $\alpha = 0$

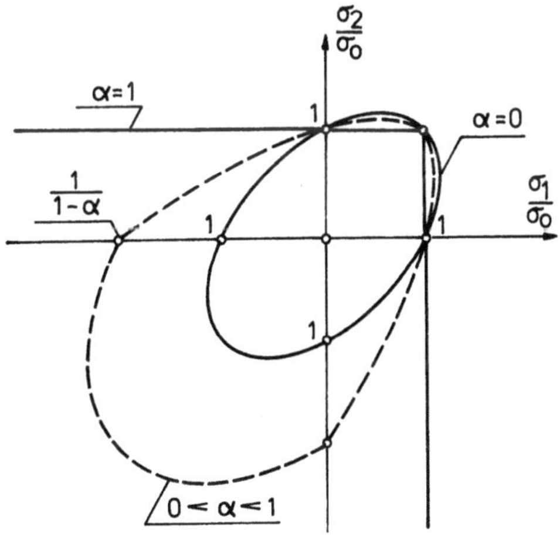

Figure 2:

the Huber-Mises ellipse is obtained.

Finally, the Lemaitre theory gives isochronous curves shown in Fig. 3. Two extreme cases shown by full lines correspond to $h = 1$ (*bilateral* material) and $h = 0$. The coordinates of intersection points on negative stress axis depend on both coefficient h and material constant m. The section $[-(1+m)^{1/m}, -1]$ of a coordinate axis represents

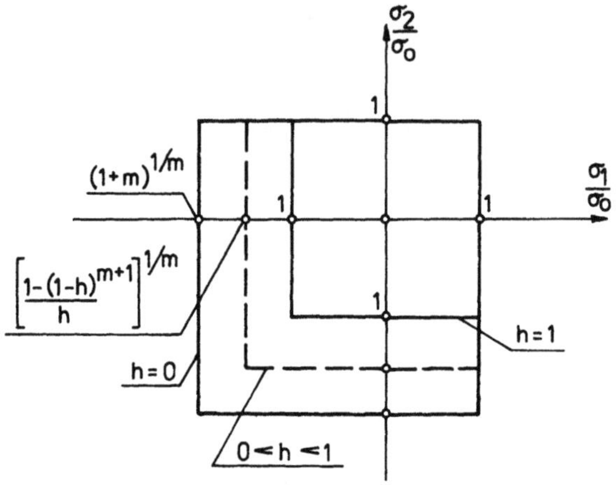

Figure 3:

material sensitivity to compressive stresses. For $m = 1$ this section becomes [-2, -1], and diminishes to 0 for $m \to \infty$.

The combined theory proposed here gives isochronous curves shown in Fig. 4 by full lines for $m = 2$ and $\alpha = 0.5$, and for three values of $h = 0, 0.5, 1$. The isochronous

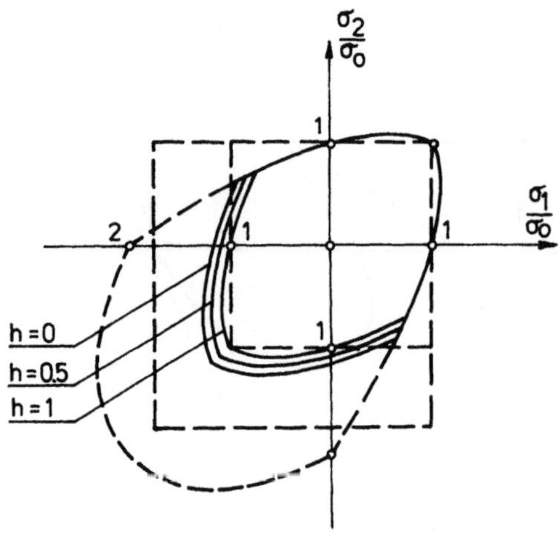

Figure 4:

curves corresponding to Eqs. (2.2) and (2.3) are shown by dashed lines for a comparison. It can be seen that the theory under consideration describes the material which possesses the properties closest to a *bilateral* material ($h = 1$). The width of "a sensitivity belt" ($h = 0, h = 1$) is relatively small for given $\alpha = 0.5$, but grows with α tending to 1.

4 Application to plates in bending

The effect of compressive stresses in structures can be different than that discussed above as a consequence of stress redistribution. This phenomenon can be described by choosing an appropriate creep law. Following the papers [3,4] we use the coupled non-stationary creep theory which is given by the set of equations:

$$\varepsilon_{ij} = \varepsilon_{ij}^{e} + \varepsilon_{ij}^{c} , \tag{4. 1}$$

$$\varepsilon_{ij}^{e} = C_{ijkl}\,\sigma_{kl} , \tag{4. 2}$$

$$\frac{\partial \varepsilon_{ij}^{c}}{\partial t} = \gamma \left(\frac{\sigma_e}{1-\omega}\right)^n \frac{\partial \sigma_e}{\partial \sigma_{ij}} , \tag{4. 3}$$

where γ and n are material constants, and C_{ijkl} is elastic constant matrix, accompanied by one of the damage evolution laws (2.2) or (2.4).

Full set of the equations governing the problem of bending of thin plates is as follows:

$$\left.\begin{aligned} &\operatorname{div}\underline{\sigma} = \underline{0} , \\ &\underline{\varepsilon} = \underline{\varepsilon}^{e} + \underline{\varepsilon}^{c} , \\ &\underline{\varepsilon}^{e} = \underline{D}^{-1}\underline{\sigma} , \\ &\dot{\underline{\varepsilon}}^{c} = \underline{\Gamma}(\underline{\sigma},\omega)\underline{\sigma} , \\ &\dot{\omega} = A\sigma_{eq}^{m}(\omega) , \end{aligned}\right\} \tag{4.4}$$

where

$$\begin{aligned} \underline{\varepsilon}^{T} &= [\varepsilon_x , \varepsilon_y , \varepsilon_{xy}] = [-z\partial^2 w/\partial x^2 , -z\partial^2 w/\partial y^2 , -2z\partial^2 w/\partial x\partial y] , \\ \underline{\sigma}^{T} &= [\sigma_x , \sigma_y , \sigma_{xy}] , \\ \underline{D} &= \frac{E}{1-\nu^2}\begin{bmatrix} 1 & \nu & 0 \\ \nu & 1 & 0 \\ 0 & 0 & 1-\nu/2 \end{bmatrix} , \quad \underline{\Gamma}(\underline{\sigma},\omega) = \gamma\frac{\sigma_e^{n-1}}{(1-\omega)^n}\begin{bmatrix} 1 & -1/2 & 0 \\ -1/2 & \cdot 1 & 0 \\ 0 & 0 & 3 \end{bmatrix} , \\ \sigma_e &= \sqrt{\sigma_x^2 + \sigma_y^2 - \sigma_x\sigma_y + 3\sigma_{xy}^2} , \quad \sigma_{1,2} = (\sigma_x + \sigma_y)/2 \pm \sqrt{(\sigma_x - \sigma_y)^2 + 4\sigma_{xy}^2}/2 , \end{aligned}$$

$w(x, y)$ is a deflection of a middle surface of a plate and E is Young's modulus, ν is Poisson's ratio, dots stand for time derivative of respective variables. The function $\sigma_{eq}(\omega)$ is given by Eqs. (2.2) or (2.4).

The above set of equations was solved by means of FE method combined with the Euler time stepping procedure. Details of the algorithm are given in [4]. The calculation are terminated when $\omega(x, y, z, t_R)$ reaches its critical value set to one.

The plates were assumed to be of square in shape (1.0m × 1.0m) and of the thickness 0.01m, uniformly loaded with p=0.2 MPa. Material properties (Ti-6Al-2Cr-2Mo at 675 K)were taken as: E=$0.102 \cdot 10^6$ MPa, $\nu = 0.33$, $n = 6.8$, $\gamma = 0.136 \cdot 10^{-23}$ $\mathrm{MPa}^{-n}\mathrm{h}^{-1}$, $m = 5.79$, $A = 1.08 \cdot 10^{-20}$ $\mathrm{MPa}^{-m}\mathrm{h}^{-1}$, α and h were free parameters.

The time to failure t_R (in hours$\cdot 10^5$) for a clamped plate and simply supported one are given in Tables 1, and 2, respectively. The corresponding graphs in Figs. 5 and 6

Table 1:

α	Unilateral	Non-unilateral		Bilateral
		h=0	h=0.5	h=1
0	5.3124	5.3124	5.3124	5.3124
0.5	4.6527	3.0277	2.8040	2.2752
1	1.9690	1.3791	1.2562	0.9093

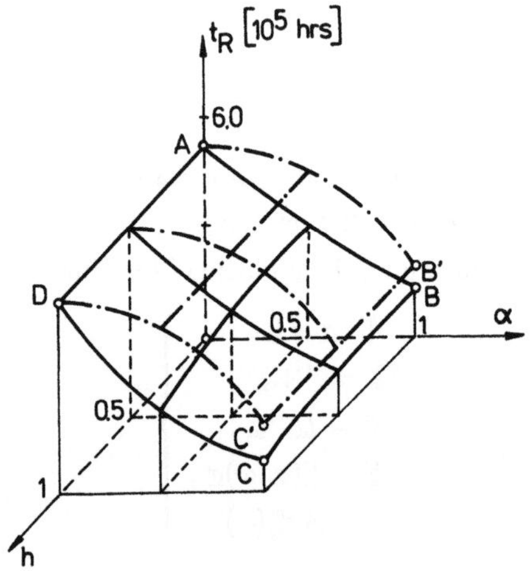

Figure 5:

illustrate the influence of compression effect upon the time to failure.

Table 2:

α	Unilateral	Non-unilateral		Bilateral
		h=0	h=0.5	h=1
0	0.2222	0.2222	0.2222	0.2222
0.5	0.3025	0.2016	0.1912	0.1723
1	0.2135	0.1556	0.1459	0.1237

5 Conclusions

1. The influence of parameter α on time to failure is structure dependent (more sensitive in the clamped plate Fig. 5).

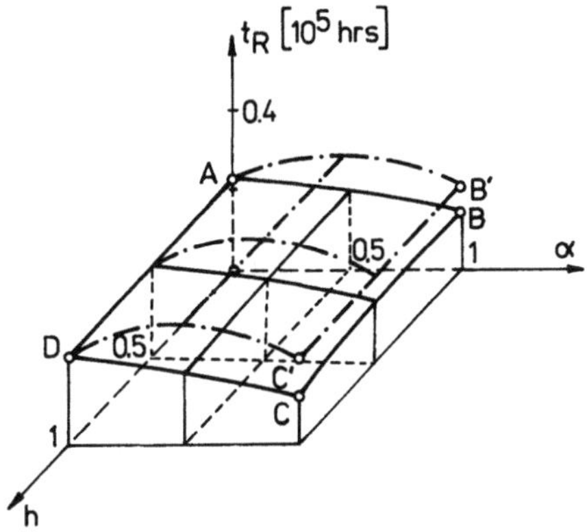

Figure 6:

2. The compressive damage reduces time to failure by an amount represented by the volume between surfaces ABCD and AB'C'D.

3. The influence of coefficient h is small, and even a rough estimation of its value affects final results very little.

According to the comments made above on the shape of isochronous curves, the influence of h becomes more pronounced for higher values of α.

References

[1] Hayhurst, D.R.: *Creep rupture under multi-axial states of stress*, J. Mech. Phys. Solids **21**, (1972), 381-390

[2] Lemaitre, J.: *Damage constitutive equations*, CISM Course on Damage Continuum Mechanics, 13-16.09.1986

[3] Bodnar, A., Chrzanowski, M.: *The effect of damage field on the transient creep in plate bending*, Trans. 9th SMIRT Conf. vol. L, (1987), 273-278

[4] Bodnar, A., Chrzanowski, M.: *Numerical analysis of damage growth in transient creep of plates subjected to bending*, Mechanika i Komputer, **9**, (1989), 101-115 (in Polish)

[5] Kachanov, L.M.: *On the time to rupture in creep conditions*, Izv. Ak. Nauk SSSR OTN **8**, (1958), 26-31 (in Russian)

[6] Rabotnov, Yu.N.: *On the mechanism of long-term failure*, Izv. Ak. Nauk SSSR Vopr. Prochn. Mat., (1959), 5-7 (in Russian)

On Microcracking due to Cavitation and Grain Boundary Sliding in Creeping Polycrystals

ERIK VAN DER GIESSEN[†] & VIGGO TVERGAARD[‡]

[†]Dep. of Engineering Mechanics, Delft University of Technology, Delft, The Netherlands

[‡]Dep. of Solid Mechanics, The Technical University of Denmark, Lyngby, Denmark

Summary

A plane strain cell model is used to analyse the development of intergranular creep fracture in polycrystalline metals at elevated temperatures. Continuous nucleation of cavities and diffusive cavity growth are accounted for in the model, and also the effect of freely sliding grain boundaries is incorporated, while the grains undergo power law creep. The numerical analyses are here mainly employed to focus on the final part of the creep rupture process, when microcracks form and start to link up.

Introduction

In polycrystalline metals at high temperatures, creep rupture occurs mainly by the nucleation and growth of microscopic cavities at the grain boundaries. Experiments have shown that cavitation occurs most rapidly on grain boundary facets normal to the maximum principal stress direction [1]. Coalescence of the cavities leads to the formation of microcracks at grain boundary facets, and several micromechanical studies of creep failure have used the time to microcrack formation as an estimate of the lifetime (see, e.g., [2,3,4,5]). However, final failure occurs somewhat later as the microcracks link up to form the macroscopic intergranular fracture surface.

The most likely mechanisms for the linking-up of microcracks are grain boundary sliding or cavitation on intermediate grain boundary facets which are significantly inclined to the direction of maximum principal stress. Such effects have been investigated by the authors [6,7] in terms of an axisymmetric model problem, which provides a reasonably realistic representation of a central grain with the associated cavitating and sliding grain boundaries. It has been found that the failure mechanisms on the adjacent inclined facets are active at an early stage of the process, and that this may accelerate failure on the central facet of the model.

M. Zyczkowski (Ed.)
Creep in Structures
IUTAM Symposium Cracow/Poland 1990

In the present paper a plane strain model is considered, analogous to the axisymmetric model considered previously [6,7]. This plane strain model gives a less realistic 3D representation of the central grain, but a more realistic description of the surrounding grains, and the model is well-suited for studies of microcrack interaction and link-up. Here, the main focus will be on the basic features of the plane strain model.

Polycrystal model

In this study we use a 2D model of a polycrystalline material consisting of a planar array of hexagonal grains (Fig. 1a). Assuming plane strain conditions, the material is subjected to remote principal stresses σ_1 and σ_2 in the x_1 and x_2 direction, respectively, with $\kappa = \sigma_2/\sigma_1 < 1$. Free sliding is assumed to take place on all grain boundaries and at least 50% of all facets normal to the x_1 direction of maximal principal stress are cavitated (emphasized in Fig. 1a). As discussed also in [3] this situation is expected to represent materials with well-separated cavitating facets. In a few cases we allow for cavitation to occur on adjacent inclined grain boundaries and finally in one case also on the remaining transverse facets. This last case of a fully cavitated material is representative of situations just before final failure (see also [5]). Due to the symmetries implied we only need to analyse the cell ABCD shown in Fig. 1b.

The cell model studied here is the plane strain analogue of the axisymmetric problems discussed in [3,6,7]. In [3] attention was confined to cavitation on transverse facets like AP in Fig. 1b, which were treated as penny-shaped facets. In [6,7] we included additional cavitation on adjacent inclined grain boundaries like PQ. As a further extension we consider here also the possibility of cavitation on intermediate transverse facets like QC.

We proceed with a very brief description of further constitutive aspects of the model and its

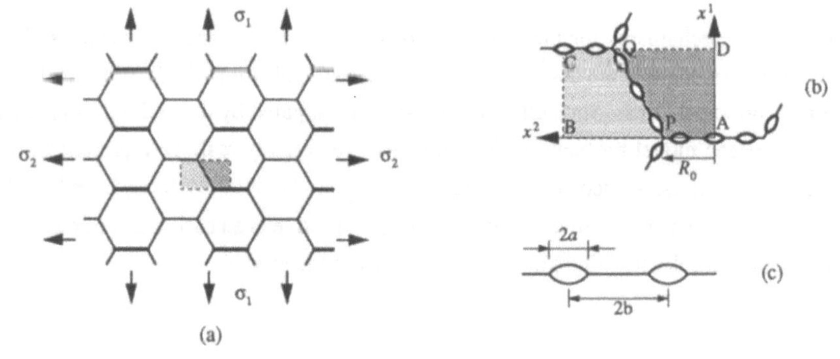

Fig. 1. The 2D polycrystal model (a). Only the hatched region is analysed (b). Spherical-caps shaped cavities on a grain boundary (c)

analysis; various details may be found in e.g. [3,6,7]. In addition to elastic deformations, the grains are taken to deform by dislocation creep modelled by a power law creep law of the form

$$\dot{\varepsilon}_e^C = \dot{\varepsilon}_0 (\sigma_e/\sigma_0)^n \tag{1}$$

for the effective creep strain-rate $\dot{\varepsilon}_e^C$ as a function of the effective Mises stress σ_e. Here, n is the creep exponent, and $\dot{\varepsilon}_0$ and σ_0 are reference strain-rate and stress parameters. Grain boundary cavities grow by diffusion as well as by creep of the surrounding material. Based on a number of detailed micromechanical analyses by various authors, a model has been proposed [8,4] for growth of a single, spherical-caps shaped cavity of radius a and half spacing b (see Fig. 1c). The diffusive path length

$$L = (\mathcal{D}\sigma_e/\dot{\varepsilon}_e^C)^{1/3} \tag{2}$$

introduced by Needleman and Rice [9] is an important stress and temperature dependent parameter in this model which determines whether coupled cavity growth occurs ($\mathcal{D}$ is a grain boundary diffusion parameter [8,9,4]). For a/L smaller than say 0.1, cavity growth is dominated by diffusion, while local creep deformations play an increasing role for larger values of a/L. Although sliding may affect the cavity shape on inclined grain boundaries like PQ, we assume that these effects are small enough so that the same growth model can be applied on these boundary facets. Cavity coalescence takes place by failure of the ligament between cavities and is taken to occur here when $a/b = 0.9$.

Grain boundary cavitation is accounted for using a "smeared out" approach in which the actual grain boundary with a distribution of discrete cavities with radius a and half spacing b is replaced by a grain boundary layer with a continuous distribution of the cavitation parameters a and b. Furthermore, this grain boundary layer is attributed a thickness δ_c — or average separation between adjacent grains — which equals the cavity volume V smeared out over a region of the grain boundary plane of radius b, i.e. $\delta_c = V/(\pi b^2)$. Thus, cavitation contributes to the macroscopic rate of deformation through the growth of δ_c. This is evidently affected by changes in the cavity spacing, which are mainly due to the nucleation of new cavities. Cavity nucleation is not clearly understood and theoretical models sofar have only been partly capable of explaining the experimental observations. Here we use the nucleation law suggested in the work of Dyson [10] and explored in some detail in [7], which involves the following evolution relation for the number of cavities N per unit undeformed grain boundary area:

$$\dot{N} = F_n (\sigma_n/\Sigma_0)^2 \dot{\varepsilon}_e^C . \tag{3}$$

where F_n is a material constant, Σ_0 is a normalization constant and σ_n is the average normal stress acting on the grain boundary. Nucleation is taken to continue until a maximum density N_{max} is reached, thus accounting for saturation of the cavity density (see e.g. [10]). This nucleation process is taken into account in an approximate way just as in previous studies [3,7].

Solutions of the model problem are obtained using a linear incremental approach based on a finite element discretization as discussed in [3,6]. Inside the grains quadrilateral elements are used built up of four linear traingular elements. All grain boundary facets are represented by way of special interface elements. These are finite elements of the grain boundary layer which account for nucleation and cavity growth as well as free sliding [6]. In the case of the transverse facets AP and QC, symmetry is enforced by using half grain boundary elements cut along their midplane. The boundary displacement rate conditions are prescribed such that the loading is consistent with the macroscopic average stress state specified by σ_1 and σ_2.

Results

We present a few results of the plane strain model for a material specified by $n = 5$ and Poisson's ratio $\nu = 0.3$. The parameter $\dot{\varepsilon}_0/\sigma_0^n$ is used to define the reference time $t_R = \sigma_e/(E\dot{\varepsilon}_e^C)$ in terms of the applied macroscopic Mises stress, which is given by $\sigma_e = \frac{1}{2}\sqrt{3}(\sigma_1 - \sigma_2)$ for pure plane strain creep, and the corresponding $\dot{\varepsilon}_e^C$ according to (1). All results to be presented are for $\kappa = \sigma_2/\sigma_1 = 0.5$ and the macroscopic stresses are prescribed such that $\sigma_e/E = 0.5\times10^{-3}$. Hence the reference time is identical to that in previous axisymmetric cell model studies [3,6,7]. Most calculations are performed using the 9×(6+6) mesh shown in the insert of Fig. 2. Whenever a grain boundary facet is taken to be cavitating, the initial cavity radius is taken as $a_I/R_0 = 0.01$, but different initial cavity densities are considered.

Figures 2 and 3 show the growth of the damage parameter a/b at various points of the grain boundaries (see insert) in the absence of cavitation on QC. In these figures the grain boundary diffusion parameter $\mathcal{D}$ is specified in terms of L relative to the initial cavity size, and is taken as $(a/L)_I = 0.025$ in Fig. 2. In Fig. 2a nucleation is neglected, thus representing a case where nucleation to a saturation value occurs at an early stage; in Fig. 2b continuous nucleation is taken into account, as specified by $F_n = 100/A_I$ with $A_I = \pi R_0^2$, starting from a very low initial cavity density. Here and in the sequel, $N_{max} = 100/A_I$ is used, but in all cases considered saturation of the cavity density takes place at lower values. The cases considered in Fig. 2 are immediate plane strain analogues of axisymmetric model studies considered in [6] and [7], respectively, but it should be noted that although the same effective Mises stress is used, the macroscopic principal stresses are higher than in the axisymmetric cases. This is one of the reasons why cavitation develops faster in the present plane strain situations than in the axisymmetric cases; the time to first coalescence in the case of Fig. 2a without cavitation on PQ e.g. is about

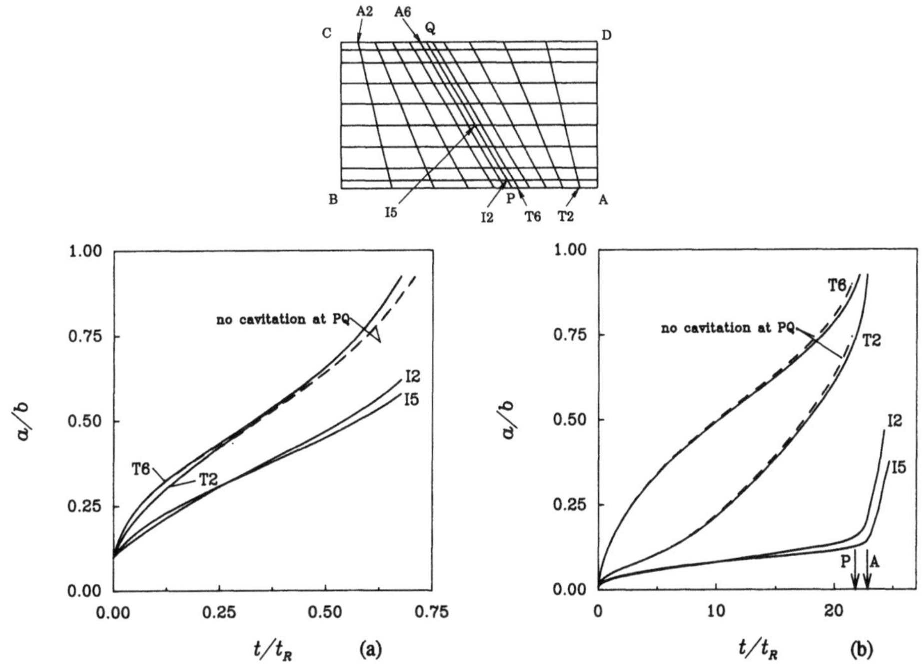

Fig. 2. Development of damage for $(a/L)_I = 0.025$ in the absence of cavitation on QC; (a) for $b_I/R_0 = 0.1$ and no nucleation; (b) for $b_I/R_0 = 1$ and $F_n = 100/A_I$. For the arrows in (b), see the text.

6 times shorter than under axisymmetric conditions. However this acceleration cannot be explained solely from the applied principal stresses since the chosen value of $\mathcal{D}$ implies that cavitation is considerably constrained by the rate of creep of the grains, which is characterized by load shedding away from the central cavitated facet AP to the surrounding grains. It turns out that this creep constraint in the planar geometry is less active than in the axisymmetric geometry. This also explains the fact that cavitation on PQ is found here to have only a slight effect, while for the same initial conditions a reduction of the time to coalescence by a factor 2 was found in the axisymmetric model [6].

For the case shown in Fig. 2b the computations are continued until after a full facet microcrack has developed on the central facet AP at $t_f/t_R \approx 23$. The arrows in that figure indicate the times at which cavity coalescence takes place first near the triple grain junction P and shortly after at the centre A of the facet. It is seen that in this case, it takes only about 5% of the time to first coalescence to develop a full facet microcrack. It is also observed that as soon as this microcrack has developed, cavitation on the inclined grain boundary near the junction P is accelerated considerably. Cavity coalescence in that region is expected to occur soon after t_f, so that the microcrack will rapidly grow on the inclined boundary; but, due to numerical instabilities, this could not be analysed in detail with the chosen mesh.

Figure 3a shows results for a lower diffusion rate as specified by $(a/L)_I = 0.1$, but otherwise with the same parameters and initial conditions as in Fig. 2b. In this case diffusion is considerably slower and the present plane strain results are virtually identical to the axisymmetric results of [7]. The development of cavitation on the transverse as well as the inclined grain boundary facets is clearly more nonuniform than in the case of Fig. 2b. Growth of a microcrack on AP now takes a relatively large period of time: the time required to develop a full facet microcrack after first cavity coalescence is about as long as the time to first coalescence. As in the previous case we see that some damage has accumulated on the inclined grain boundary prior to formation of this microcrack; but after that moment, a rapid acceleration of cavitation damage emanates from the microcrack. The growth of the microcrack on AP is shown in Fig. 3b in terms of its length $r = R_0 - x^2$ measured from the triple grain junction P to the current crack tip. It is seen that microcrack growth is very much accelerated in the later stages. This calculation has been repeated with a somewhat finer 13×(9+9) mesh which gives an improved description of the crack tip fields. The time to develop a full facet microcrack is reduced with only 2% when using the finer mesh.

Figure 4a briefly illustrates the effect when cavitation also takes place on the neighbouring transverse facet QC, for a material characterized by the same parameters as in Fig. 2. The interaction between such neighbouring facets presumably plays an important role during the linking-up process of microcracks to yield macroscopic failure. In this example, the initial cavitation states are taken to be identical on all facets, but nucleation on QC is 10 times slower than on the

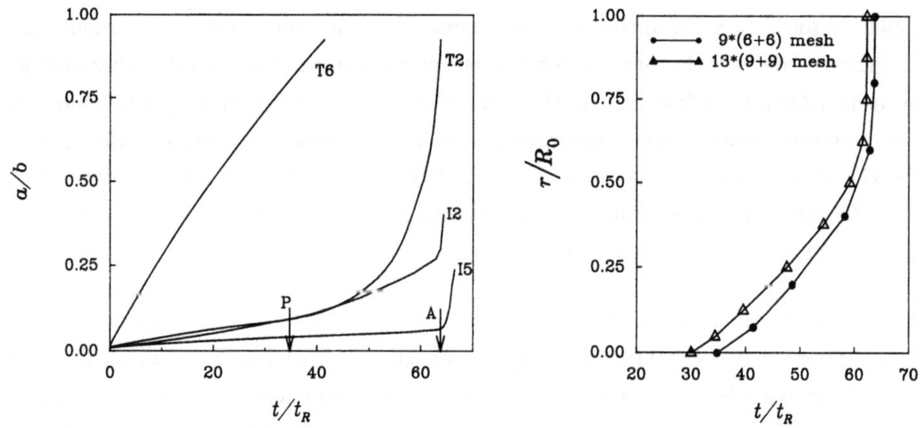

Fig. 3. Development of damage for $(a/L)_I = 0.1$, $b_I/R_0 = 1$ and $F_n = 100/A_I$ (a) and development of the microcrack extension along AP (b).

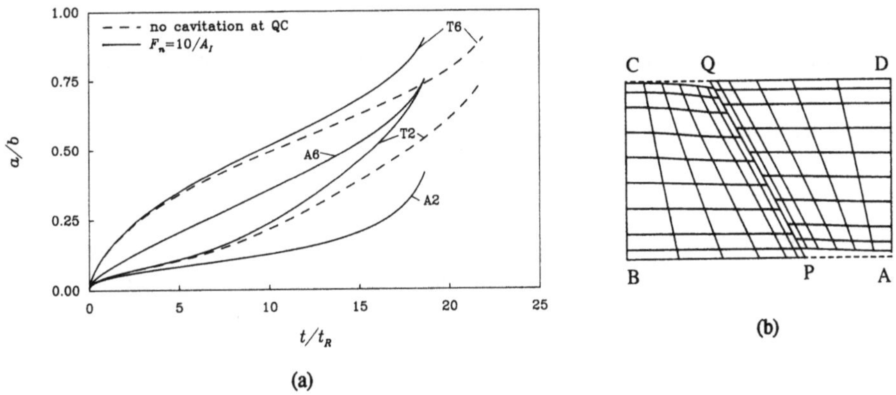

Fig. 4. Development of damage for $(a/L)_I = 0.025$ and $b_I/R_0 = 1$ with $F_n = 100/A_I$ on AP and PQ and with $F_n = 10/A_I$ on QC (a). Deformed mesh at $t/t_R = 18.5$.

other facets, i.e. $F_n = 10/A_I$ on QC. Nevertheless, we observe that cavitation on AP is enhanced considerably, in particular during later stages of the process, leading to a reduction of the time to first coalescence by 15%. Despite the fact that nucleation is 10 times slower, cavitation damage on QC is just slightly less than on AP, due to the creep constraint which involves load shedding away from QC to a much lesser extent than load shedding away from AP. Finally, the deformed mesh at the time of first coalescence at P is shown in Fig. 4b, demonstrating the presence of nonuniform cavitation and deformation states as well as sliding along PQ.

Discussion

The analyses in the present paper have shown that the predictions of the plane strain model are qualitatively in agreement with those of the corresponding axisymmetric model, although failure occurs more rapidly in the planar model for the same level of macroscopic Mises stress. One reason for this is that, when the cross-sections are identical, the central cavitating facet of the planar model represents a larger area fraction than in the axisymmetric model.

The computations in Figs. 3 and 4 are continued beyond the formation of an open microcrack by cavity coalescence on the central grain boundary facet, and it has been shown that subsequently the rate of cavity growth on the adjacent inclined facets is strongly increased, so that coalescence on these facets will occur soon after the formation of the first microcrack. Thus, the computations give an indication of the rate of damage growth during the final part of the lifetime as microcracks link up. It appears that, due to numerical instabilities, finer meshes would be needed to determine the full microcrack development; but according to Fig. 3b the prediction of the initial microcrack formation is not very mesh sensitive.

It is noted that the very high cavity growth rate on the inclined facets found in Figs. 3 and 4 after the first microcrack formation may be quite sensitive to the macroscopic stress state. In the cases considered here, the transverse stress σ_2 is rather high; for a smaller transverse stress, much less cavity growth on the inclined facets is expected and sliding may be the dominant linking-up mechanism.

The interaction between two neighbouring facets normal to the maximum principal stress is briefly illustrated in Fig. 4. However, several further analyses of this kind would be needed to obtain a more complete picture of the effect of material parameters and stress states.

Acknowledgement

The work of EvdG was made possible by a fellowship of the Royal Netherlands Academy of Arts and Sciences.

References

1. Argon, A.S.: Mechanisms and mechanics of fracture in creeping alloys. Wilshire, B.; Owen, D.R.J. (eds.) Recent Advances in Creep and Fracture of Engineering Materials and Structures. Swansea: Pineridge Press 1982, pp. 1–52.
2. Rice, J.R.: Constraints on the diffusive cavitation of isolated grain boundary facets in creeping polycrystals. Acta Metall. 29 (1981) 675–681.
3. Tvergaard, V.: Effect of grain boundary sliding on creep constrained diffusive cavitation. J. Mech. Phys. Solids 33 (1985) 447–469.
4. Tvergaard, V.: On the creep constrained diffusive cavitation of grain boundary facets. J. Mech. Phys. Solids 32 (1984) 373–393.
5. Anderson, P.M.; Rice, J.R.: Constrained creep cavitation of grain boundary facets. Acta Metall. 33 (1985) 409–422.
6. Van der Giessen, E.; Tvergaard, V.: A creep rupture model accounting for cavitation at sliding grain boundaries. Int. J. Fracture (to appear).
7. Van der Giessen, E.; Tvergaard, V.: On cavity nucleation effects at sliding grain boundaries in creeping polycrystals. Wilshire, B.; Evans, R.W. (eds.) Proceedings of the Fourth Int. Conf. on 'Creep and Fracture of Engineering Materials and Structures'. London: Elsevier Appl. Sci. Publ. 1990, pp. 169-178.
8. Sham, T.–L.; Needleman, A.: Effects of triaxial stressing on creep cavitation of grain boundaries. Acta Metall. 31 (1983) 919–926.
9. Needleman, A.; Rice, J.R.: Plastic creep flow effects in the diffusive cavitation of grain boundaries. Acta Metall. 28 (1980) 1315–1332.
10. Dyson, B.F.: Continuous cavity nucleation and creep fracture. Scripta Metall. 17 (1983) 31–37.

An Assessment Procedure for the High Temperature Response of Structures

I W Goodall+ and R A Ainsworth*
+ Nuclear Electric plc
Bedminster Down
Bridgwater Road
Bristol
England

* Nuclear Electric plc
Berkeley Nuclear Laboratories
Berkeley
Gloucestershire
England

Summary

This paper describes extensive work within Nuclear Electric, to develop a procedure which assembles knowledge in structural mechanics and materials science for assessment of components operating in the creep range. The procedure, called R5, is written in seven parts, or volumes. These volumes and the associated failure modes which are addressed are described.

The intent throughout R5 is to avoid, wherever possible, full inelastic computer solutions. The paper outlines the reference stress concept which is used in various parts of R5. Another feature of R5 is that creep damage under cyclic loading is characterised by ductility exhaustion methods. This approach is believed to describe more closely the mechanisms of creep damage. Improved understanding and the data base support this approach. Finally, the paper describes some applications of R5 to operating components and one application in some detail.

1. INTRODUCTION

The objective of R5 is to provide a comprehensive assessment document, accessible to the practitioner. It is based on 'expert knowledge' in structural mechanics and materials science and is intended to augment and replace, where necessary, the provisions of ASME Code Case N-47 and the French Code RCC-MR. There is also the need to extend these codes to enable assessments of defects and weldments.

Issue 1 of R5 [1] has been released in six volumes which are listed below along with references describing some of the contents.

Vol 1: Overview

Vol 2: Analysis/Assessment Methods for Defect-free Structures [2].

Vol 3: Creep-Fatigue Crack Initiation [3, 4].

Vol 4: Assessment of Defects under Steady Loading [5, 6].

Vol 5: Creep-Fatigue Crack Growth Assessment [3, 7].

Vol 6: Dissimilar Metal Welds.

In addition the following volume is scheduled for release in 1991:

Vol 7: Behaviour of similar welds.

M. Zyczkowski (Ed.)
Creep in Structures
IUTAM Symposium Cracow/Poland 1990

2. MODES OF FAILURE

The modes of failure explicitly covered by current issues of R5 are:

(1) Excessive plastic deformation due to single load application.

(2) Incremental collapse due to a loading sequence.

(3) Excessive creep deformation or stress rupture.

(4) The initiation of cracks in initially undefected material by creep and creep-fatigue mechanisms.

(5) The growth of flaws by creep and creep-fatigue mechanisms.

(6) Failure of dissimilar metal welds by creep-fatigue mechanisms.

Avoidance of short-term fracture is ensured by referral to the R6 procedure [8]. In using R6 for components subjected to high temperature service, the influence of creep damage and thermal ageing on the fracture and deformation properties must be considered.

The modes of failure in R5 differ from those of ASME and British Standards. No consideration is given in R5 to failure by buckling or corrosion. The interaction of the corrosion of components and the other modes of failure will be included in future issues of R5. The areas in which R5 goes considerably beyond existing codes are in the description of crack growth and the assessment of weldments.

3. THE USE OF REFERENCE STRESS TECHNIQUES AND SIMPLIFIED METHODS

3.1 Reference Stress Evaluation

The calculations in R5 are based largely on simplified methods. This is a compromise between the pessimism of elastic analysis and the cost and complexity of cyclic inelastic computation. The method adopted is the reference stress technique in which quantities of interest in a complex component are described in terms of creep data at the reference stress level. The technique was originally developed for creep deformation in steadily loaded structures but has since been extended to cover cyclic loading, creep rupture and defects [5, 9, 10]. The method generally overestimates stress levels in a component and inelastic analysis may be substituted to reduce pessimisms.

3.1.1 Calculation of the Reference Stress

In Volume 2, shakedown calculations are used to evaluate a reference stress describing the stress level at the start of a creep dwell when a steady cyclic response has been reached. If inelastic analysis is substituted, a number of computational cycles may be required to achieve this steady cyclic state. However, the application of the simplified method requires that the component is within shakedown.

Two options are provided for the calculation of reference stresses and temperatures. Option 1 uses a stress classification similar to N-47 and may be applied to structural features of thin rectangular section (plates, beams and thin shells) which remain isothermal during operation at temperatures in the creep range. The reference stress σ_{ref} is

$$\sigma_{ref} = \underset{V}{\text{Max}} \left\{\frac{P_B}{3} + \left[\left(\frac{P_B}{3}\right)^2 + {P_L}^2\right]^{\frac{1}{2}}\right\} \tag{1}$$

where V is structural volume. With Option 2 the reference stress σ_{ref} is

$$\sigma_{ref} = P\,\sigma_y/P_u \tag{2}$$

where P_u is the lower bound limit load for the yield stress σ_y and is in proportion to the working loads P.

For isothermal structures, the reference temperature is equal to the temperature during the creep dwell: a history of varying temperature is permissible. Advice is provided in R5 for the calculation of reference stresses and temperatures for non-isothermal structures, for structures of more than one material and for variable loading.

To make a creep rupture assessment the reference stress is adjusted for local strain concentrations by a factor χ which is:

$$\chi = (\bar{\sigma}_E)_{max}/\sigma_{ref} \tag{3}$$

Here, $(\bar{\sigma}_E)_{max}$ is the maximum elastically calculated equivalent stress in the structure for the same loadings used to obtain σ_{ref}. This is acceptable for $\chi \leq 4.0$; for larger values it must be considered that the stress raiser is sufficiently sharp to require treatment as a crevice or crack-like defect using Volume 4 or 5 of R5. Having evaluated χ, the rupture reference stress σ^R_{ref} is calculated from

$$\sigma^R_{ref} = \{1 + \lambda\,[\chi - 1]\}\,\sigma_{ref} \tag{4}$$

The parameter $\lambda = 0.13$ for creep ductile materials and $(1/n)$ for creep brittle materials. The reference temperature remains unaltered.

3.2 Crack Growth

The inclusion of defects in Volumes 4 and 5 is achieved by modifying the reference stress to allow for the reduced load bearing area at the defective section. From the reference stress, the crack tip parameter C* is evaluated using creep strain rate data and hence calculations of creep crack growth in the component are obtained as described more fully in [6].

The reference stress method allows the use of complex materials data often not amenable for use in finite-element codes. Particular examples are stress relaxation data incorporating internal stresses (Volume 5) and creep rupture data from cross-weld specimens (Volume 6). This direct use of data collected under relevant conditions is a significant advantage of the method. Validation for the use of complex data in the reference stress technique is not easily examined numerically and instead has been obtained by direct comparison with experimental data as reported in the various volumes. This experimental validation has been supplemented with validation obtained by comparison with numerical calculations performed with simple material laws. Although further validation is desirable in some areas it is considered that the references stress technique is sufficiently well developed for inclusion as the recommended approach in R5.

4. THE TREATMENT OF CREEP AND CREEP-FATIGUE DAMAGE

The approach used for time-dependent damage assessment, throughout R5, is based on assessing three factors:

(i) creep rupture due to loadings which are primary or which have significant elastic follow-up;

(ii) fatigue involving time independent processes which contribute to the overall damage accumulation; and

(iii) local failure characterised by cyclic stress relaxation around peak stress regions leading to the creep-fatigue interaction.

The damage factors are combined in a linear manner in R5; the details depending on the nature of the problem under consideration.

There are good reasons for separating the treatment of creep damage in items (i) and (iii). In item (i), the loading is such that the stress is known accurately and there is the possibility of large inelastic deformation leading to 'necking' instability. It is therefore appropriate to use the stress/time to rupture relationship. In (iii), consideration is being given to local stress effects where the strain accumulation is limited and 'necking' instability may be discounted. The assessment is therefore based on the 'ductility exhaustion' concept which appears to describe accurately the mechanisms of creep damage under cyclic loading. The approach is defined in more detail below.

(i) Creep Rupture

Creep damage is assessed using reference stress techniques and a life fraction rule. In this way the behaviour of the structure is related to the uniaxial stress rupture curve. The damage accumulated, D_c, is

$$D_c = \int \frac{dt}{t_f(\sigma_{ref}, T)}$$

where t_f is the failure time at temperature T and stress σ_{ref}.

(ii) Fatigue

The effect of cyclic loads on the fatigue endurance of the structure is assessed using Miner's rule. In Volume 3 the fatigue damage is

$$D_F = \Sigma \frac{N_j}{(N_o)_j}$$

Here D_F is the damage accumulated by N cycles of type j and $(N_o)_j$ is the corresponding fatigue endurance for cycle j. Volume 6, addressing transition joints, adopts the same rule but recommends factoring the strain range by a suitable fatigue strength reduction factor if cross-weld endurance data are not available.

(iii) Creep Damage in Peak Stress Analysis

The effect of combined mechanical and thermal loads on the creep damage is assessed using a ductility-exhaustion rule for 'peak stress' or 'local' features of a structure. In the terminology of Volume 3:

$$D_c = \int_o^{t_h} \frac{\dot{\varepsilon}}{\varepsilon_f(\dot{\varepsilon})} dt,$$

where D_c is the creep damage per cycle, $\dot{\varepsilon}$ is the instantaneous creep strain rate during the creep dwell, $\varepsilon_f(\dot{\varepsilon})$ is the creep ductility at that strain rate and t_h is the dwell time. If the variation of creep ductility with strain rate, for the microstructural region in which cracking takes place, is not known, the damage integral is simplified by using constant values of the local failure strain, ε_f.

Linear damage summation has been examined by comparison with laboratory data for a wide range of materials, hold times and cycle types. Results are shown in Fig. 1 using best-estimate data. Although there is considerable scatter, the approach is generally conservative. Further discussion is given in [3].

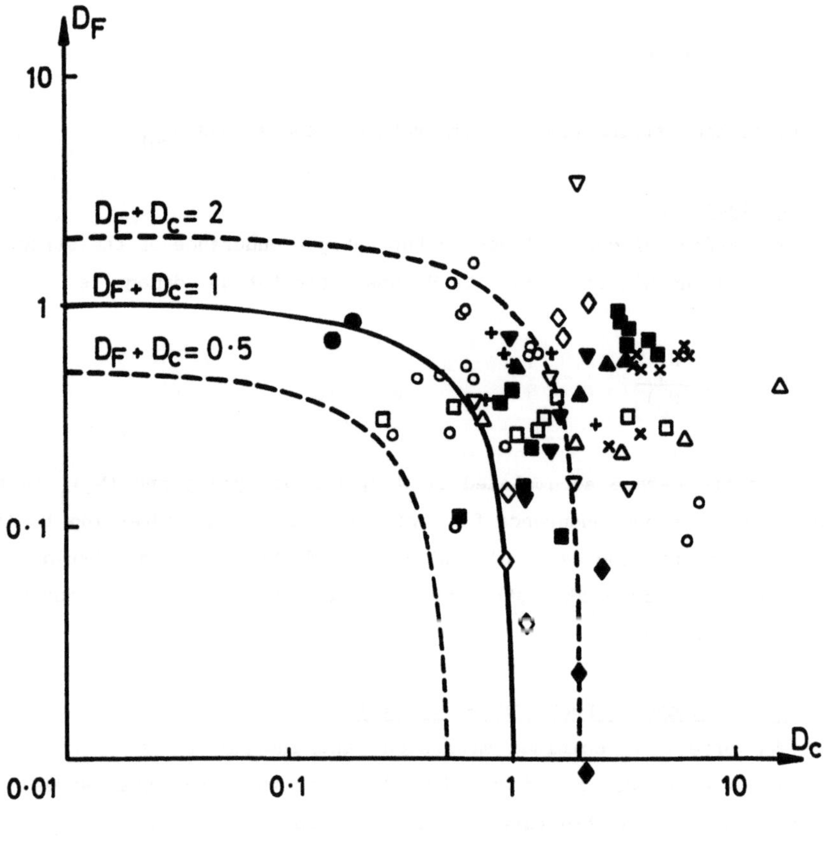

Figure 1. Creep-Fatigue Interaction Diagram

5. APPLICATIONS

R5 has been applied to a wide spectrum of plant within the electricity generation industry. In outline the applications may be summarised as

Volumes 2 and 3	Defect initiation and rupture assessment of: turbine rotors, valve chests, ligaments of steam headers, AGR insulation, AGR penetrations.
Volumes 4 and 5	Defect growth calculations for: steam headers including branch connections, AGR fuel components, AGR insulation.
Volume 6	Various assessments for conventional and nuclear plant with emphasis on joints between 2¼Cr 1Mo and type 300 steels with nickel based fillers.

An example derived from laboratory tests on a pressure vessel is indicative of the way in which the calculations may be applied. Details are contained in [11]. A schematic view of the pressure vessel is given in Fig. 2 which shows the presence of an artificial crack at section X-X.

Crack growth occurs in the region of the Heat Affected Zone which has a mixture of coarse and fine microstructure. Crack growth is either assessed by using the particular crack growth for the zone of concern or by determining an average value according to $\dot{a} = A_{HAZ}(C^*)^q$ with

$$A_{HAZ} = \frac{A_c(1+\alpha)}{A_c/A_R+\alpha} \tag{5}$$

where A_c and A_R are coefficients for coarse and refined material alone and α is the ratio of the volume of coarse to refined material.

The results of the calculations using either instantaneous crack growth rates as the crack traverses difference microstructures, or using a weighted average growth, are shown in Fig. 3. Overall, the prediction shows good agreement with experiment and shows an anticipated marked increase in growth rate as the time for continuum damage failure of the structure is approached.

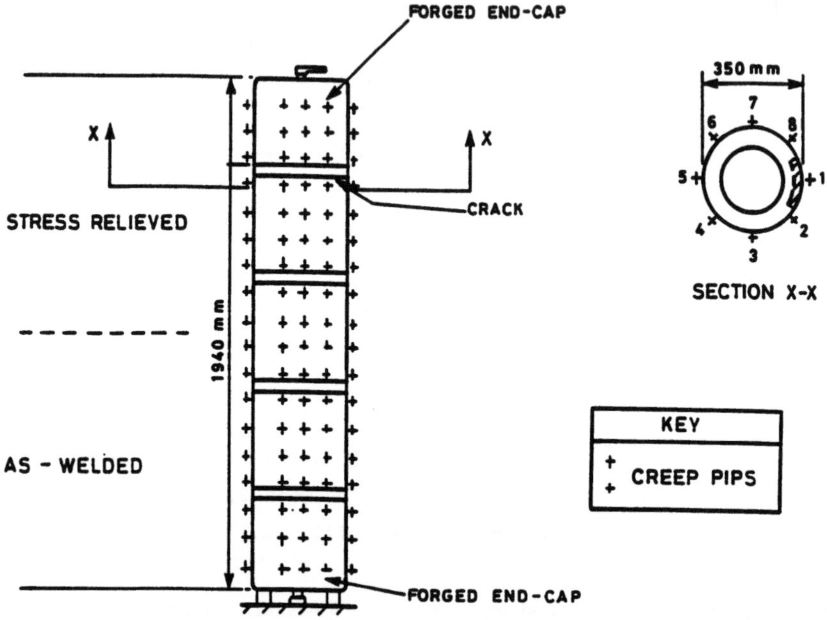

Figure 2. Schematic view of pressure vessel and defect assessed.

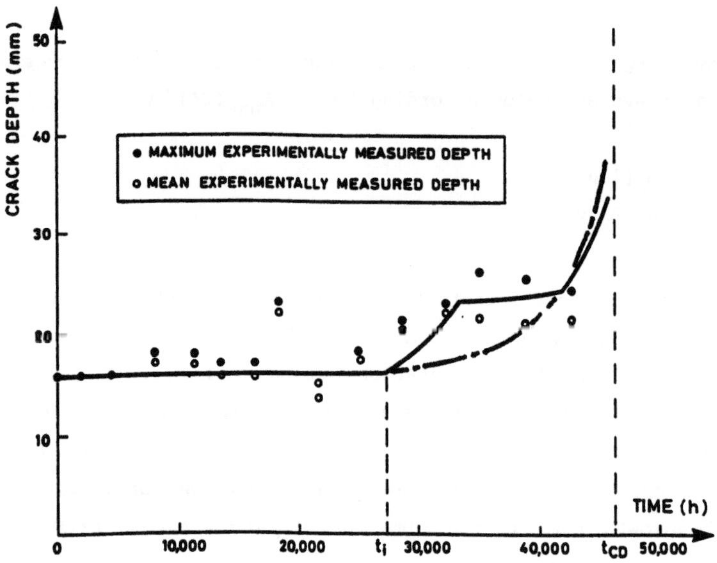

Figure 3. Comparison of experimental and calculated crack growth: solid line based on local properties; chain-dotted line based on average properties of eqn (5).

ACKNOWLEDGEMENT

This paper is published with permission of Nuclear Electric plc.

6. REFERENCES

(1) R5, An Assessment Procedure for the High Temperature Response of Structures, Berkeley Nuclear Laboratories Report, Nuclear Electric plc, UK, 1990.

(2) I W Goodall and R A Ainsworth, Structural Assessment of Metal Components, Conf. on Materials and Engineering Design, Institute of Metals, London, (1988).

(3) D A Miller and R H Priest, The Assessment of Creep-fatigue Initiation and Crack Growth, IV IUTAM Symposium on Creep in Structures, Krakow (1990).

(4) R A Ainsworth, Defect Assessment Procedures at High Temperature, SMiRT 10 Conference, Division L, Anaheim (1989).

(5) R A Ainsworth, G G Chell, M C Coleman, I W Goodall, D J Gooch, J R Haigh, S T Kimmins and G J neate, CEGB Assessment Procedure for Defects in Plant Operating in the Cree Range, Fatigue Fract. Engng. Mater. Struct. 10 (1987) 115-127.

(6) R A Ainsworth, Creep Life Estimates for Defective Structures, IV IUTAM Symposium on Creep in Structures, Krakow (1990).

(7) R H Priest, Crack Initiation and Growth During Creep-fatigue, Post-Symposium Course of the Second. Int. Symposium on Plasticity, TSU, Japan (1989).

(8) I Milne, R A Ainsworth, A R Dowling and A T Stewart, Assessment of the integrity of structures containing defects, Int. J. Pres. Ves. Piping 32, 1989, 3-104.

(9) R K Penny, and D L Marriott, Design for Creep, Mc-Graw Hill, New York (1971).

(10) I W Goodall, F A Leckie, A R S Ponter and C H A Townley, The development of high temperature design methods based on reference stresses and bounding theorems, ASME J Engng Matl Techn, 101, 349-355 (1979).

(11) M R Jones and M C Coleman, The Assessment of Creep Crack Growth in a Welded Pressure Vessel, CEGB Report RD/B/6242/R89, 1989.

Creep Continuum Damage Mechanics – Studies of Size Effects and Weldments

F.R. HALL† and D.R. HAYHURST‡

† I.R.C. in Materials for High Performance
University of Birmingham, United Kingdom
‡ Department of Mechanical and Process Engineering
University of Sheffield, United Kingdom

Summary
The paper summarizes the development of constitutive equations for creep Continuum Damage Mechanics, CDM, in metals; and solution methods are discussed for notched and cracked specimens which are identical in geometry, but different in size; and, for weldments which include three material zones. In both cases it is shown that, to achieve accurate predictions, it is essential to use constitutive equations, which accurately reflect the underlying physics of the dominant processes.

Introduction

Structures which operate at elevated temperatures over long periods of time are often subjected to multi-axial states of stress which result either from the multi-dimensional character of the loading conditions; from geometrical factors; or from the use of multi-materials within the continuum. Such structures can either fail from excessive deformation or from failure which takes place due to the softening of the microstructure of the material which might be due to dislocation kinetics or to the nucleation and growth of grain boundary defects. For component design constitutive equations are required which model the physics of the underlying processes and their dependence upon the state of multi-axial stress.

This paper described the development of creep Continuum Damage Mechanics theory which has been developed to model the deformation and damage behaviour of materials; and, to predict the time dependent development of zones of damage, and stress redistribution within structural components. This theory developed by Hayhurst [1], Hayhurst, Dimmer and Chernuka [2], Hayhurst, Leckie, Morrison [3] has been shown to be capable of accurately predicting the behaviour of a broad range of engineering structures and components from knowledge of uni-axial creep curves and of the multi-axial stress rupture criteria of materials, Hayhurst [1]. This approach has been demonstrated for plane stress plates containing stress raisers by Hayhurst [4], for axi-symmetric notch bars by Hayhurst, Leckie and Morrison [3], for plane strain edged and centre cracked tension bars by Hayhurst, Brown, and Morrison [5] and butt welds in cylindrical pressure vessels by Hall [6]. Two factors are crucial to the success of the method; firstly, the ability of the constitutive equations to model the creep strain rate process; and secondly, the accuracy with which the constitutive equations predict the multi-axial stress rupture criterion at the relevant stress and temperature levels. These criteria have been satisfied for the wide

M. Zyczkowski (Ed.)
Creep in Structures
IUTAM Symposium Cracow/Poland 1990

range of structural situations, just described, by use of equations of the form

$$\dot{v}_{ij} = f(\sigma_{ij}\,,\omega)\ ; \quad \dot{\omega} = g(\sigma_{ij}\,,\omega) \tag{1}$$

where σ_{ij}, $\dot{v}_{ij}$ are the stress and the creep strain rate tensors respectively, ω is a single state damage variable, which reflects the time-dependent material softening process, and f and g are functions to be determined from experiments. The forms of these functions and their physical significance have been discussed in detail by Hayhurst and Storakers [7]. The importance of recognising the physical significance of the single state damage variable, ω, when carrying out analysis in high temperature design has been recognised by Cox and Ashby [8], and by Leckie and Hayhurst [9], who have shown that for loading conditions which are predominantly proportional then the single state variable can be identified with a unique zone contained within the mechanisms maps of Cox and Ashby [8]. Movement from one mechanism zone to an adjacent one in these maps infers a different interpretation of ω and hence a redefinition of the functions f and g.

While the situations described above have been satisfactorily studied using CDM there are situations which have hitherto not been capable of analysis using these techniques, such as size effects in creep crack growth reported by Hayhurst, Morrison and Brown [10]; and the behaviour of multi-material components such as butt welds in cylindrical pipe work. In this paper it will be shown how the theory of creep continuum damage mechanics can be extended to cover these phenomena. In the following section the constitutive equations and numerical techniques to be used in the continuum damage mechanics analysis will be presented and reviewed. In the two sections which follow are covered the following: specimen size effects; and, multi-material components in the form of a pipe butt weldment.

Constitutive Equations and Techniques

The creep strain rate equation is given in normalised form by:

$$dV_{ij}/d\tau = 3\Sigma_e^{n-1}S_{ij}/2(1-\omega)^n \tag{2}$$

where Σ_{ij} is the normalised stress ($=\sigma_{ij}/\sigma_0$), V_{ij} is the normalised creep strain ($=v_{ij}/e_0$), σ_0 is the normalising stress ($=Ee_0$), S_{ij} is the normalised deviatoric stress ($=s_{ij}/\sigma_0$), Σ_e is the normalised effective stress, τ is the normalised time measure ($=\int(KE\sigma^{n-1}t^m/100)dt$) and K is the constant of proportionality in the un-normalised creep strain rate equation. The damage rate equation is given by:

$$d\omega/d\tau = \Delta^{\chi}(\Sigma_{ij})/V_u(1+\phi)(1-\omega)^{\phi} \tag{3}$$

where $\Delta(\Sigma_{ij})$ is the stress function $(\alpha\Sigma_1 + (1-\alpha)\Sigma_e)$, Σ_1 is the maximum principal tension stress, α is a material constant and $V_u = KE\sigma_0^{(n-\chi-1)}/100M$, and M is the constant of proportionality in the un-normalised damage rate equation. Eq. (3) can be integrated between $\omega = 0$ and $\tau = 0$ to $\omega = 1$ at failure to give $\Delta(\Sigma_{ij}) = 1$ which defines the isochronous rupture surface in stress space in terms of the parameter α [1]. Eqs. (2) and (3) are used in the creep continuum damage finite element program (termed Damage XX) to solve the combined boundary-initial value problems using classical finite element techniques, and a fourth order Runge-Kutta integration scheme with automatic time-step control. Details of the finite element analysis procedure are given elsewhere by Hayhurst, Dimmer and Morrison [11]. The finite element meshes are set up for the problems under investigation as for elastic and steady-state creep solutions. The time-variation of the stress, strain and damage fields is computed, and failure of elements is defined by $\omega > 0.9999$; when an element achieves this condition it is removed from the model and the boundary conditions are appropriately modified as discussed by Hayhurst, Dimmer and Morrison [11].

The normalising stress σ_0 is selected to be a remote boundary stress or an appropriate primary stress. In the ensuing discussion of the particular problems to be solved, the procedure as outlined above and described by Hayhurst, Dimmer and Morrison [11], will be employed with appropriate modifications.

Size Effects in Continuum Damage Mechanics

Hayhurst, Morrison and Brown [10] observed a size effect in the experimental failure times of two different sizes of both centre and edge cracked plates tested under plane strain, mode one, tensile creep conditions. The plates were made from copper and the geometry of the small, and large, externally cracked plate is given in Fig. 1. The tests were carried out at a temperature of $250 \pm 1/2°$ C. The small and large plates are geometrically similar but the large plate are twice those of the small plate. Shown in Fig. 1 are symmetrical single quadrants of the small and large test specimens. The internally cracked specimens are not discussed here since the test carried out on the specimens did not reveal a measurable size effect. For the externally cracked specimens the experimental lifetimes for a nominal stress, $\sigma_n = P/2(b-a)t = 40$ MPa, where P is the tensile load, b is the half specimen width and a is the half ligament dimension, are 1200 hr for the large specimen and 1700 hr for the small specimen, giving a difference in lifetimes of 500 hr. The lifetime predicted by Hayhurst, Morrison and Brown [10] using a local continuum damage mechanics approach is 1531 hr., which is insensitive to the specimen size. The objective of the study is to further develop the Continuum Damage Mechanics approach to take account of the size effect. As can be seen from Fig. 1 this is achieved by the introduction of a cell mesh whose dimension is controlled by the microstructural characteristics of the material. The selection of the cell size has been taken to be seven grain diameters. This figure has been arrived at from a well known empirical observation that to obtain repeatable measurements of creep curves and creep properties, a specimen thickness should be selected of at least seven grain diameters. Analyses are now discussed for the small and large externally cracked specimens which use different finite element meshes but have the same size of cell mesh. It will be assumed that every grain of material contained within the cell has its rate of damage evolution influenced by all other grains within the cell. This has been achieved by spatially averaging the damage rate parameter for each square cell and this is defined non-locally as:

$$\dot{\omega}_{cell} = \frac{1}{A_{cell}} \int_{A_{cell}} \dot{\omega}_{el}\, dA_{el} \tag{4}$$

where

$$A_{cell} = \int_{A_{cell}} dA_{el} \tag{5}$$

and where the subscripts (el) and (cell) denote parameters which refer to triangular elements and to square cells, respectively. For example ($\dot{\omega}_{el}$) is the damage rate at the centroid of the triangular finite element. If a triangular element is contained wholly within a square cell then (dA_{el}) is the total area of the element. If an element lies outside a cell then (dA_{el}) and ($\dot{\omega}_{el} dA_{el}$) for this element are not included in the numerical summation.

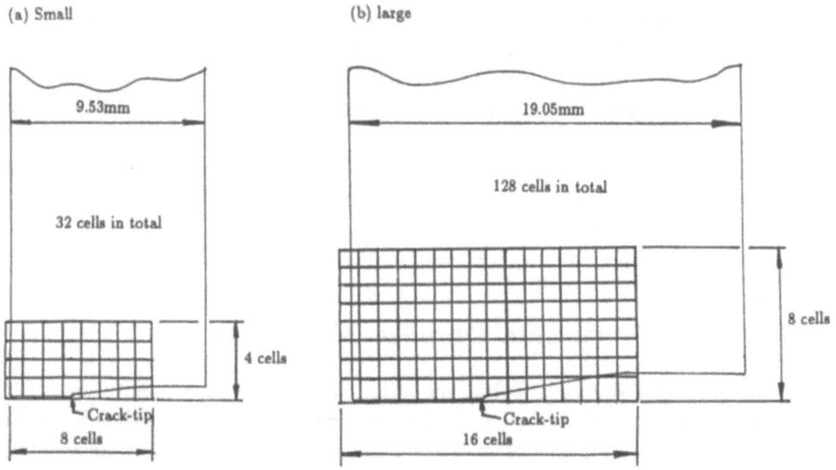

Figure 1: Positions and sizes of the cell meshes used to overlay the finite element meshes representing the small (a) and large (b) externally cracked specimens. The square cells have a side dimension of 0.9mm.

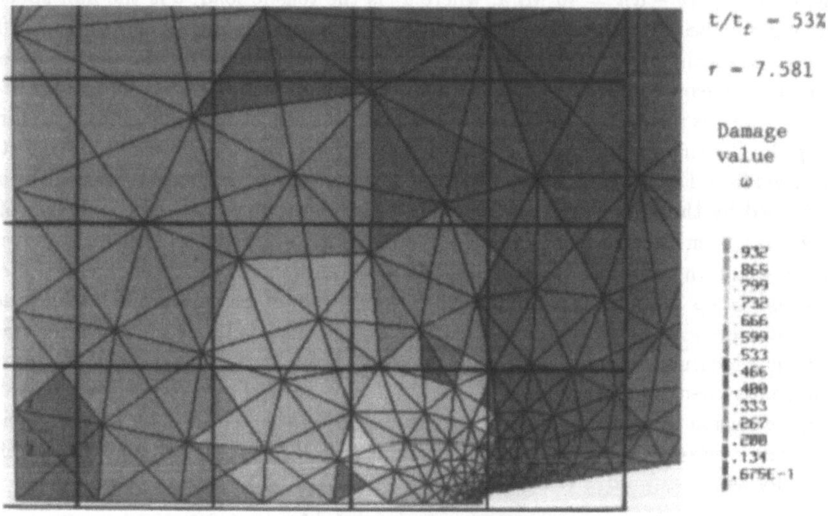

Figure 2: Contour plot showing elemental damage values, for the small externally cracked specimen, obtained using the finite domain non-local damage method at a time just before the failure of the first elements. The overlay grid shows the location of the cell mesh.

During the finite element analysis, elements which have failed ($\omega > 0.9999$) automatically have their damage rates ($\dot{\omega}_{el}$) set to zero, as the elements are removed from the global stiffness matrix.

During the analysis for each time iteration a new spatially averaged non-local damage rate ($\dot{\omega}_{cell}$) is determined using Eqs. (4) and (5) from the local damage rates ($\dot{\omega}_{el}$), and their area weights (dA_{el}). Each element contained within a cell then has the new spatially averaged damage rate ($\dot{\omega}_{cell}$). The damage rate parameter for an element crossing a cell boundary, or boundaries, will be calculated from area weighted values of ($\dot{\omega}_{cell}$) for all cells associated with the element. Therefore, smoothing the elemental damage rate step change which occurs across cell boundaries.

The results of computations carried out in this way are shown in Fig. 2 which shows a contour plot of the elemental damage values for the small externally cracked specimen at a life fraction of 53%. The figure shows how damage is maintained uniform over those elements contained entirely within a cell boundary; and, shows how damage is smoothed in those elements which lie on the cell mesh boundary, the letter can be recognised in the figure by the different intensity of shading. The lifetime predicted for the large specimen is 1390 hr. and for the small specimen is 1850 hr. yielding a difference of 460 hr. which compares with the experimental difference of 500 hr. The predicted distributions of damage at failure are shown in Figs. 3a and 3c for the small and large specimens respectively. Comparisons are made between the predicted damage distributions and those observed in the mid-thickness micrographs of Figs. 3b and 3d for the externally cracked small and large copper plates respectively. The micrographs and the respective predictions are on the same scale, and, the 2:1 size difference may by observed in Figs. 3b and 3d where the grain size is constant. It may be seen that the experimentally observed distributions of damage are reasonably well predicted by the non-local continuum damage mechanics theory. The same computations have been carried out for the internally cracked specimen and negligible size effect has been predicted. More details of this particular study can be founded in the paper by Hall and Hayhurst [12].

Creep Continuum Damage Modelling of Failure in a Pipe Weldment

It is well known that the international codes for the design of high temperature weldments in pressurised pipework systems are unsatisfactory since designs are based on the pressure stresses; on uni-axial creep rupture data for the parent pipe material; and, on factors of ignorance which bear little relationship to the physics of the material deformation and rupture processes. This work investigates the possibility to characterise the creep deformation and rupture properties of the parent pipe material; the heat affected zone region of the weldment; and the aggregated properties of the weld material by formulating creep continuum damage mechanics constitutive equations of the type given by Eqs. (2) and (3); and, to use finite element based creep CDM solutions to predict the weld behaviour. The vessel to be studied is that tested by Coleman at al [13] under constant internal pressure of 45.5 MPa at a constant temperature of 564°C. The weld material is 2.25 Cr. 1 Mo, and the pipe material is 0.5 Cr. 0.5 Mo. 0.25 V. ferritic steel. The vessel tested has sections of pipe of length 457mm with an internal diameter 230mm and a pipe thickness of 60mm; which were welded together by a 30° included angle V - sectioned circumferential weld. The width of the weld metal region, on the outer surface of the

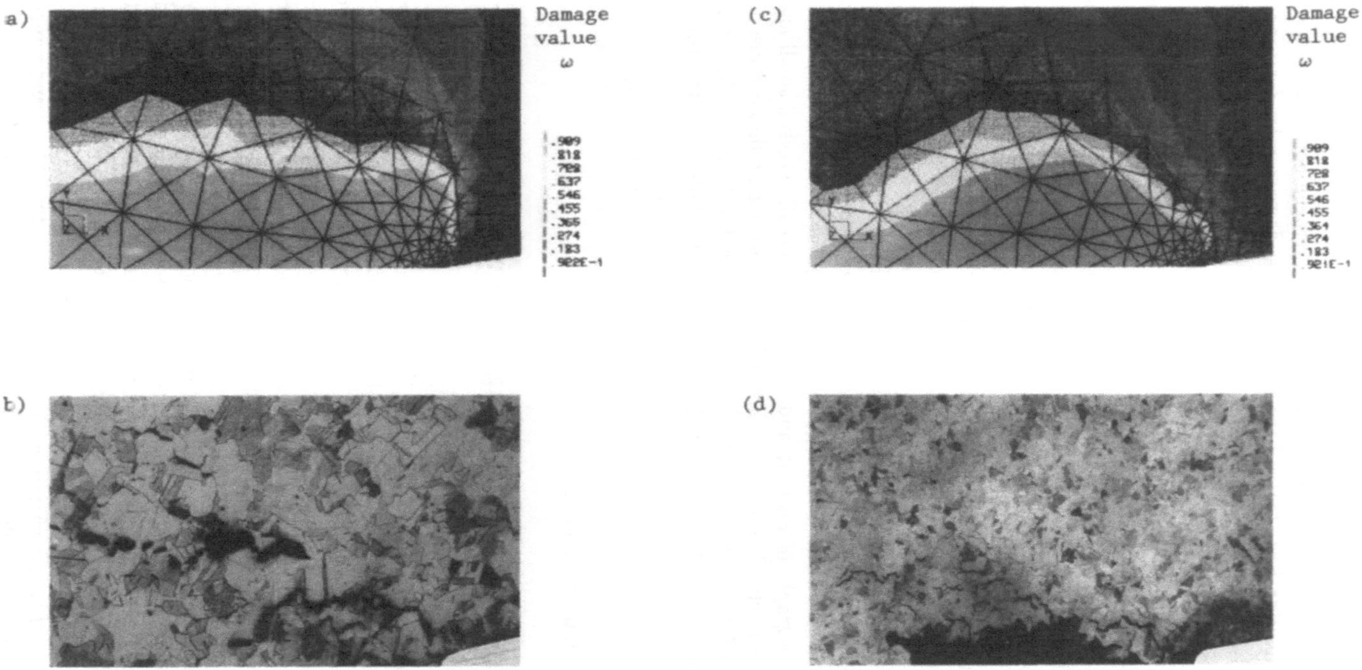

Figure 3: Comparison of the predicted damage distribution for the small (a) and large (c) externally cracked Copper plates at failure, using the finite domain non-local damage method, with a mid-thickness micrograph taken from the minimum section of a small externally cracked Copper plate specimen at a life fraction of $(t/t_f = 94\%)$, (b); and with a mid-thickness micrograph taken from the minimum section of a large externally cracked Copper plate specimen at a failure, (d).

pipe, was approximately 45mm. The ends of vessel were closed using unrestrained domed heads. The finite element idealisation chosen for the vessel is shown in Fig. 4a where the three different material zones may by clearly identified by reference to the micrograph of the same section shown in Fig. 4c. The pressure and end-loading conditions have been arranged for the axi-symmetric situation. The lower boundary of Fig. 4a shows the axis of symmetry of the weld where the displacement in the axial direction is maintained zero, and the radial components of displacements are unspecified. The creep CDM analysis has been carried out using the solver, Damage XX, discussed in the previous section.

Young's modulus for all material zones are assumed to be E =170 GPa, and the vessel was initially heat treated to be stress free. The creep deformation and failure characteristics of each material zone in the weld model have been obtained from uni-axial and multi-axial creep tests due to Cane [14]. The parameters in the constitutive Eqs. (2) and (3) have been obtained from the basic materials data using automated techniques described by Dunne et al [15]. It is evident from the test results that stresses exist in the welded structures which fall on either side of a stresses exist in the welded structures which fall on either side of a mechanism boundary as defined by Cox and Ashby [8]. This requires that for each material two sets of material constants, for Eqs. (2) and (3), had to be utilised. The separation between the mechanism zones is defined in terms of a break stress Σ_b; bellow Σ_b sufficient data is available to enable the determination of all material constants. Above Σ_b the same values of the time index m is assumed, and the values of the creep strain rate stress index, n, and of the damage rate stress index, χ, have been taken from minimum creep rate data and rupture time data respectively. All other material constants have been determined to ensure continuity of the strain rate, damage rate and the strain at the break stress Σ_b. The value of the multi-axial stress rupture criterion, α, has been determined from the uni-axial and multi-axial tests carried out by Cane [14]. The values of the multi-axial rupture parameter α are: for the parent material α =0.60; for the simulated heat affected zone, HAZ, α =0.43; and, the weld material is assumed to have the same α as the HAZ. The distribution of damage computed by Damage XX are shown in Fig. 4a at a life fraction of 53.7%. The highest levels of damage in this figure can be seen to be confined within the heat affected zone region close to the inner bore of the vessel. This situation has developed from where the damage levels were initially highest on the inner bore within the weldment region, and subsequently damage localised to the heat affected zone material, and propagated within the HAZ. This pattern of behaviour was accompanied by stress redistribution where the highest components of stress in the axial and hoop direction were initially located close to the inner bore. Subsequently damage evolution has caused these stresses to redistribute to lower levels in the regions closer to the outer bore of the pipe. Shown in Fig. 4b is the predicted damage field at failure of the vessel. It can be seen that during the last 50% of lifetime the rate of growth of damage in the HAZ on the inner bore has slowed down, and the damage has been concentrated in the fusion boundary between the HAZ and the weld material close to the outer bore. This localisation commenced on the outer bore and has moved towards the inner bore with a passage of time. The high levels of damage defined by $\omega > 0.80$ may be interpreted as the formation of a macrocrack. This same situation may be seen in Fig. 4c which shows a diametral micrograph taken from a failed vessel. The experimentally observed location of the microcrack, and that shown in Fig. 4b are extremely close. In addition, on the test vessels axial and hoop components of strain, and their variation with time, were measured throughout the test. The predicted time variation of hoop strain and axial strain were shown to be within extremely close

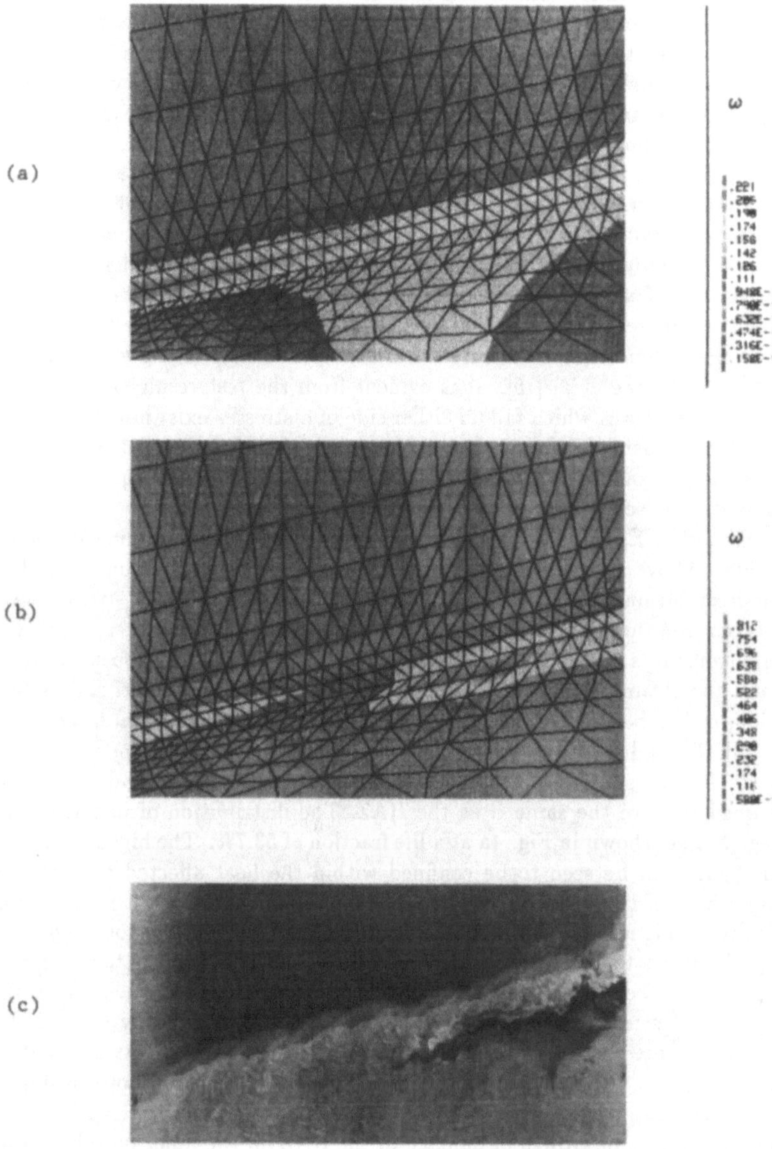

Figure 4: Damage distributions for the weld model, (a) at life fraction ($t/t_f = 53.7\%$); and, (b) at failure. The predicted regions of failure ($\omega > 0.8$) shown in (b) may be compared with the cracking shown in the micrograph of figure 4(c) taken from a diametral section of the pipe.

agreement until the latter part of the test when geometry changes became significant, so invalidating the infinitesimally small displacement assumptions of the theory. Additional information on this study may be found in the paper by Hall and Hayhurst [16]. The experimentally determined lifetime was approximately 46,000 hrs; this compares with a predicted lifetime of 43,882 hrs. Extremely close agreement being achieved, demonstrating the ability of the CDM approach to predict the microstructural response.

Discussion and Conclusions

In this paper the techniques of creep CDM have been extended to deal with two new topics. In the case of the creep crack growth in plane strain tension specimens it has been shown that creep CDM can be used provided that the rate of damage evolution predicted by Eq. (3) is treated as taking place uniformly over discrete cells of material. It is shown that if the damage rate is treated non-locally in this way, then the experimentally observed size effects can be accurately predicted. In the case of the deformation and rupture analysis of the internally pressurised butt-welded pipe it has been shown that extremely close predictions of strain histories, lifetimes, stress and damage redistribution can be predicted provided that the material characterisation be carried out correctly; and, in sympathy with the physical mechanisms which control the evolution of creep strain and damage. What is essential to the successful use of CDM is the use of Eqs. (2) and (3) with constants that pertain to particular mechanism zones in the maps due to Cox and Ashby [8]; and, that the multi-axial rupture criteria of all phases of the material be accurately represented by the parameter α. It has been shown that creep Continuum Damage Mechanics can accurately predict both size effects and the global performance of a multi-material structure as for example found in steam pipe weldments.

References

[1] Hayhurst, D.R.: *Creep Rupture Under Multi-Axial States of Stress*, J. Mech. Phys. Solids, **20**, (1972), 381-390

[2] Hayhurst, D.R., Dimmer, P.R., Chernuka, M.W.: *Estimates of the Creep Rupture Lifetime of Structures Using the Finite Element Method*, J. Mech. Phys. Solids, **23**, (1975), 335-355

[3] Hayhurst D.R., Leckie, F.A., Morrison, C.J.: *Creep Rupture of Notched Bars*, Proc. R. Soc. London., A, **360**, (1978), 243-264

[4] Hayhurst D.R.: *Stress redistribution and Rupture Due to Creep in a Uniformly Stretched Thin Plate Containing a Circular Hole*, J. Appl. Mech., **40**, (1973), 244-250

[5] Hayhurst, D.R., Brown, P.R., Morrison, C.J.: *Creep Crack Growth*, Phil. Trans. R. Soc. Lond., A, **311**, (1984), 130-158

[6] Hall, F.R.: *Development of Continuum Damage Mechanics Models to Predict the Creep Deformation and Failure of High Temperature Structures*, PhD. Thesis, University Sheffield, (1990)

[7] Hayhurst, D.R., Storakers, B.: *Creep Rupture of the Andrade Shear Disk*, Proc. R. Soc. London., A, **349**, (1976), 369-375

[8] Cocks, A.C.F., Ashby, M.F.: *On Creep Function by Void Growth*, Progress in Materials Science, **27**, (1982), No.34, 189-244

[9] Leckie, F.A., Hayhurst, D.R.: *Constitutive Equations for Creep Rupture*, Acta metall., **25**, (1977), 1059-1072

[10] Hayhurst, D.R., Morrison, C.J., Brown, P.R.: *Creep Crack Growth*, Creep in Structures, 1980 (IUTAM Symposium, Leicester, UK) (ed. A.R.S. Ponter & D.R. Hayhurst), p.564, (1981), Berlin: Springer-Verlag

[11] Hayhurst, D.R., Dimmer, P.R., Morrison, C.J.: *Development of Continuum Damage in the Creep Rupture of Notched Bars*, Phil. Trans. R. Soc. Lond., A, **311**, (1984), 103-129

[12] Hall, F.R., Hayhurst, D.R.: *Modelling of Grain Size Effects in Creep Crack Growth using Non-local Continuum Damage Approach*, (in press)

[13] Coleman, M.C., Parker, J.D., Walters, D.J.: *The Behaviour of Ferritic Weldments in Thick Section 0.5 Cr., 0.5 Mo., 0.25 V, Pipe at Elevated Temperature*, Int. J. Pres. Ves. & Piping, **18**, (1985), 277-310

[14] Cane, B.J.: *Collaborative Programme on the Correlation of Test Data for High Temperature Design of Welded Steam Pipes. Presentation and Analysis of Material Data*, CEGB Lab. Note No. RD/L/210IN81, March (1981)

[15] Dunne, F.P.E., Othman, A.M., Hall, F.R., Hayhurst, D.R.: *Representation of Uni-Axial Creep Curves using Continuum Damage Mechanics*, To Appear, I. J. Mech. Sci. (1990)

[16] Hall, F.R., Hayhurst, D.R.: *Creep Continuum Damage Mechanics. Modelling of High Temperature Deformation and Failure in a Pipe Weldment*, (in press)

Crack Growth in Linearly Viscoelastic Media

N. C. Huang

Department of Aerospace and Mechanical Engineering
University of Notre Dame
Notre Dame, Indiana 46556, USA

Summary

A computational method for the analysis of initiation and quasi-static extension of a crack in a linearly viscoelastic medium subjected to a tensile mode of deformation is presented. A fracture process zone of Dugdale type is assumed in front of the crack tip. Viscoelastic and viscoplastic deformations of the fracture process zone are analyzed, and crack initiation and crack growth are considered to follow a fracture criterion of critical energy release rate. Examples are given for demonstration of the computational method.

Introduction

Crack growth in a viscoelastic medium is recognized as a challenging problem in fracture mechanics. As the applied load is moderately small, due to material creep, the speed of crack growth is much smaller than the speed of sound and a quasi-static approach can be utilized for the study of crack growth in the medium. Under a tensile mode of deformation, a thin strip of small size is commonly assumed for the process of material failure in front of the crack tip. It is referred to as the fracture process zone [1-5]. Material particle deforms viscoelastically prior to its entry into the fracture process zone. It is then subjected to a large deformation within the fracture process zone prior to the material failure at the crack tip. This paper attempts to establish a method for analyzing the viscoelastic and viscoplastic deformations of the fracture process zone. The behavior of crack propagation is then determined by the criterion of critical energy release rate.

M. Zyczkowski (Ed.)
Creep in Structures
IUTAM Symposium Cracow/Poland 1990

Linearly Viscoelastic Model for Fracture Process Zone

In the following, we investigate the initiation of growth of a finite crack of length $2a(0) = 2a_0$ in an infinite and

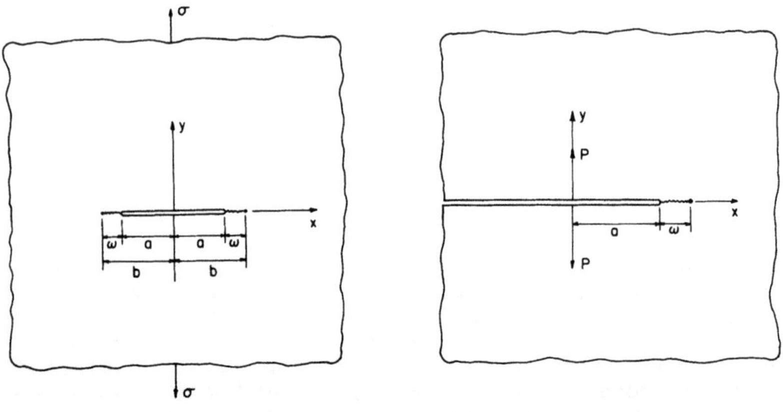

(a) (b)

Fig. 1 Goemetry of the Problem

linearly viscoelastic sheet as shown in Fig 1a. A plane stress deformation under a monotomically increasing remote tension $\sigma(t)$ for $t > 0$ is assumed. There exist fracture process zones of size $b(t)-a_0$ in front of each tip. The tensile stress within the fracture process zone is denoted by $p(x,t)$ where the origin $x = 0$ is set at the center of the crack. We have

$$\sigma(t) = \frac{2}{\pi} \int_{a_0}^{b(t)} \frac{p(x,t)}{[b^2(t)-x]^{1/2}} dx . \tag{1}$$

In the elastic case, the crack opening displacement in the fracture process zone is found to be

$$v(x) = \frac{2}{E} \left[(b^2-x^2)^{1/2} \sigma - \frac{1}{\pi} \int_{a_0}^{b} p(z) \ln \left| \frac{(b^2-z^2)^{1/2} + (b^2-x^2)^{1/2}}{(b^2-z^2)^{1/2} - (b^2-x^2)^{1/2}} \right| dz \right], \tag{2}$$

where E is the modulus of elasticity of the medium. In the viscoelastic case, the crack opening displacement $v(x,t)$ is time-dependent. For problems with time-dependent boundary conditions Graham [6] found restrictions that Radok's correspondence principle [7] can be applied to convert the elastic solution to the viscoelastic solution. It is found that Graham's conditions can be satisfied in our problem. Hence the crack opening displacement is

$$v(x,t) = \int_{t_x}^{t} C(t-\tau) \frac{\partial I(x,\tau)}{\partial \tau} d\tau , \tag{3}$$

where t_x is the time at which the tip of the fracture process zone first reaches the point x, C(t) is the viscoelastic plane stress compliance and

$$I(x,t)=2\sigma(t) \left[b^2(t)-x^2\right]^{1/2} - \frac{2}{\pi} \int_{a_0}^{b(t)} p(z,t) \ln \left| \frac{\left[b^2(t)-z^2\right]^{1/2}+\left[b^2(t)-x^2\right]^{1/2}}{\left[b^2(t)-z^2\right]^{1/2}-\left[b^2(t)-x^2\right]^{1/2}} \right| dz . \tag{4}$$

When the value of p(x,t) at the material point at x reaches p_a, the material point begins to enter the fracture process zone. If the material in the fracture process zone is linearly viscoelastic, v(x,t) and p(x,t) are related by

$$v(x,t) = \int_{t_x}^{t} F(t-\tau) \frac{\partial p(x,\tau)}{\partial \tau} d\tau, \tag{5}$$

Where F(t) is the creep compliance of the fracture process zone satisfying F(t) = 0 for t < 0. Note that F(t) is different from C(t).

If $\sigma(0) \neq 0$, there is an instantaneous elastic response due to $\sigma(0)$. For material points within the range $b(0) < x \leq b(t)$, $t_x > 0$ and equations (3) and (5) are valid. On the other hand, for material points within $a_0 \leq x \leq b(0)$, $t_x = 0$ and equations (3) and (5) must be replaced by

$$v(x,t)=C(t)I(x,0)+\int_{0}^{t} C(t-\tau) \frac{\partial I(x,\tau)}{\partial \tau} d\tau \tag{6}$$

and

$$v(x,t)= F(t)\left[p(x,0)-p_a\right]+\int_0^t F(t-\tau)\,\frac{\partial p(x,\tau)}{\partial\tau}\,d\tau \tag{7}$$

respectively. The convolution integrals involved in equations (6) and (7) can be evaluated approximately by finite series as suggested by Lee and Rogers [8].

The energy criteria for material failure proposed by Schapery [9] will be used. At the crack tip, the energy release rate is

$$G(t) = \int_0^{v(a_0,t)} p(a_0,t)\ dv(a_0,t)\ . \tag{8}$$

When G(t) reaches a critical value at $t=t_{cr}$, crack propagation initiates. Numerical results for G(t) are given in reference [5].

Viscoplastic Model for Fracture Process Zone

Let us consider Fig. 1a. The total energy release rate for an elastic-ideally plastic material in the case of plane stress is given by Cherepanov [10] as

$$G = G_e + G_p = \frac{2Y}{E}\, f(a) + \frac{\pi}{6E}\,\frac{K^3}{Y^2}\,\frac{dK}{da}\ , \tag{9}$$

where G_e is the elastic energy release rate, G_p is the rate of energy dissipation through plastic deformation,a is a measure of crack length, f(a) is a function related to v(a) by v(a) = f(a)/E, K is the mode I stress intensity factor and Y is the uniaxial tensile yield stress. It is found that Graham's conditions can be satisfied and the energy release rate at any material point x and time t can be expressed as

$$G(x,t) = \int_{t_x}^{t} C(t-\tau)\frac{\partial}{\partial\tau}\, F\left[x,a(\tau),\tau\right] d\tau\ , \tag{10}$$

where

$$F[x,a(t),t] = Y\, f[x,a(t),t] + \frac{\pi}{12}\,\frac{K^3[a(t),t]}{Y^2}\,\frac{d}{da}\,K[a(t),t] \quad (11)$$

and C(t) is again the plane stress compliance of the medium.

The load is first applied to the body at $t = 0$. The crack initiation time t_{cr} is obtained by solving the equation

$$G_{cr} = \int_0^{t_{cr}} C(t_{cr}-\tau)\,\frac{\partial}{\partial\tau}\,f(a_0,a_0,\tau)\,d\tau \,. \quad (12)$$

The growth of the crack can be analyzed under the assumption that $\dot{a}$ is constant over a sufficiently small time interval $t_1 \le t \le t_2$. Hence

$$\dot{a}(t) = \frac{\omega\left[a(t_1),t_1\right]}{t_2-t_1} \quad (13)$$

for $t_1 \le t \le t_2$. Thus the entire deformation history of the material point at $x = a(t_2)$ can be considered in the analysis. In the numerical computation, the convolution integral of equation (13) is evaluated by Lee and Rogers technique [8].

Examples

Let us consider the problem as shown in Fig 1a. From Dugdale solution [2], the expressions of $\omega(t)$ and $v(x,t)$ can be found and the analysis of crack initiation and propagation follows through the use of equations (10), (12), and (13). The results are presented in terms of the dimensionless quantities $\bar{a} = a/a_0$, $\bar{\sigma} = \sigma/Y$, $\bar{G}_{cr} = \pi G_{cr}/(Ya_0)$ and $\bar{C}(t) = YC(t)$. Fig.2 shows the variation of crack initiation time t_{cr} as function of $\bar{\sigma}$ for $\bar{G}_{cr} = 0.005$, where an empirical relation for E(t) of PMMA at 80 °C given by Muki and Sternberg [10] is used. There are two asymptotes at $\bar{\sigma} = \bar{\sigma}_L$ and $\bar{\sigma} = \bar{\sigma}_H$. For $\bar{\sigma} \le \bar{\sigma}_L$, no crack initiation will occur. On the other hand, for $\bar{\sigma} \ge \bar{\sigma}_H$, crack initiation occurs immediately after application of the load. It is followed

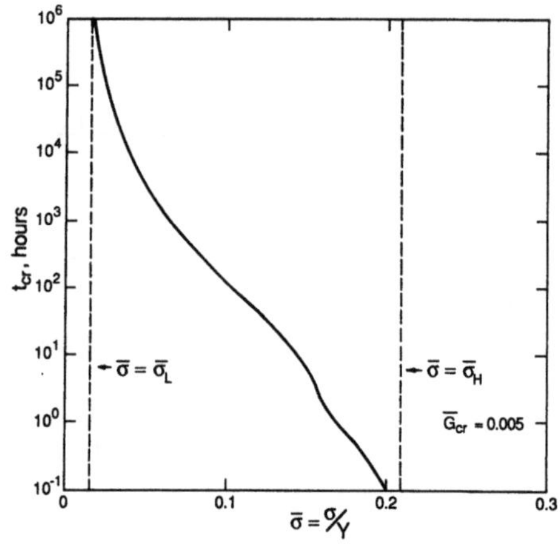

Fig.2 Crack Initiation Time versus Applied Stress for Example 1.

by an unstable crack propagation. For $\bar{\sigma}_L < \bar{\sigma} < \bar{\sigma}_H$, crack initiation occurs at a finite time and is followed by a crack growth with increasing speed as shown in Fig. 3.

Next, let us consider a second example of a semi-infinite crack in an infinite sheet as shown in Fig. 1b. The crack is subjected to a pair of equal and opposite concentrated forces of

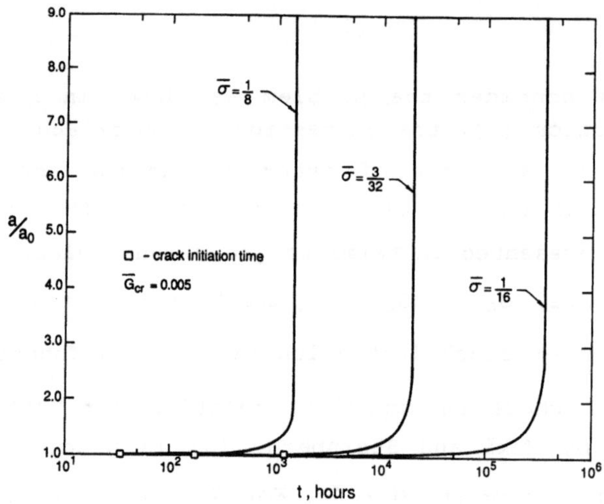

Fig. 3 Crack Half-length versus Time for Example 1.

magnitude P per unit thickness. Setting the origin x=0 at the point of application of the loads, the length of the fracture process zone $\omega(t)$ is determined by the requirement that the stress singularity due to the applied load is cancelled by a yielding in the fracture process zone. With the known $\omega(t)$ and the crack opening displacement in the fracture process zone, we are able to analyze the crack initiation and crack propagation by equations (10), (12), and (13). The results are presented in terms of the dimensionless quantities defined previously and $\overline{P} = P/(Ya_0)$.

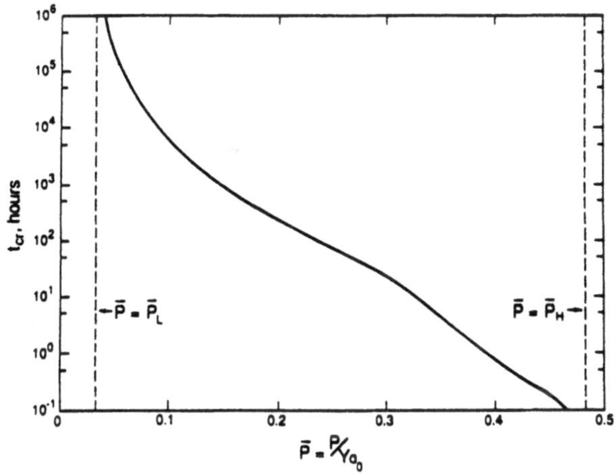

Fig. 4 Crack Initiation Time versus Applied Load for Example 2.

In contrast to the previous example, it is found that the present problem is characterized by a stable crack growth. Figure 4 shows the crack initiation time as a function of the loading. There is a load $\overline{P}_L$ such that the crack initiation will not occur for $\overline{P} \leq \overline{P}_L$ and a load $\overline{P}_H$ such that for $\overline{P} \geq \overline{P}_H$, the initial crack extension will resemble that in an elastic material. For $\overline{P}_L < \overline{P} < \overline{P}_H$, the critical time for crack initiation t_{cr} decreases with increasing load. Since the loading produces a stable crack propagation in the elastic material, the crack decelerates under constant loading after its initiation.

Eventually, the crack speed will decrease to a level where the creep deformation becomes appreciable, resulting in viscoelastic crack growth at determinate speed. Finally, crack arrest occurs as time approaches infinity.

References

1. Barenblatt, G.I., "The Mathematical Theory of Equilibrium Cracks in Brittle Fracture", Advances in Applied Mathematics, Vol. 7, 1962, pp. 55-129.

2. Dugdale, D.S., "Yielding of Steel Sheets Containing Slits", Journal of Mechanics and Physics of Solids, Vol. 8, 1960, pp. 100-108.

3. Schapery, R.A.,"A Theory of Crack Initiation and Growth in Viscoelastic Media-I. Theoretical Development", International Journal of Fracture, Vol. 11, 1975, pp. 141-159.

4. Russell, S.G. and Huang, N.C., "A Numerical Method for Crack Growth Analysis in Linearly Viscoelastic Materials", Theoretical and Applied Fracture Mechanism, Vol. 9, 1988, pp. 97-108.

5. Zhang, L.H., and Huang, N.C., "Initiation of Crack Growth in Linearly Viscoelastic Materials", Technical Report, 1989, University of Notre Dame, Notre Dame, Indiana, to appear in Computer and Structures, An International Journal.

6. Graham, G.A.C., "The Correspondence Principle of Linear Viscoelasticity Theory for Mixed Boundary Value Problems Involving Time-dependent Boundary Regions", Quarterly of Applied Mathematics, Vol. 26, 1968, pp.167-174.

7. Radok, J.R.M., "Viscoelastic Stress Analysis", Quarterly of Applied Mathematics, Vol. 15, 1957, pp.198-202.

8. Lee, E.H. and Rogers, T.G., "Solution of Viscoelastic Stress Analysis Problems Using Measured Creep on Relaxation Functions", Journal of Applied Mechanics, Vol. 30, 1963, pp.127-133.

9. Cherepanov, G.P., Mechanics of Brittle Fracture, McGraw Hill, New York, 1979, pp. 343-351.

10. Muki, R. and Sternberg, E., "On Transient Thermal Stress in Viscoelastic Materials With Temperature-dependent Properties", Journal of Applied Mathematics, Vol. 28, 1961, pp. 193-207.

Mixed-Mode Creep and Creep/Fatigue Crack Growth

T H HYDE AND A C CHAMBERS*
Department of Mechanical Engineering
University of Nottingham
Nottingham
NG7 2RD

* Now at Rolls-Royce plc
PO Box 31
DERBY
DE2 8BJ

SUMMARY

Pure mode-I, pure mode-II and mixed-mode fatigue, creep and creep/fatigue crack growth tests have been carried out on Jethete M152, an aeroengine casing material, at 550°C. A reasonably good correlation of the creep crack growth rates was obtained using the C* parameter. Simple superposition of fatigue and creep crack growth results underestimates the creep/fatigue crack growth rates.

INTRODUCTION

A practical compact-mixed-mode (CMM) specimen, Fig. 1, has been developed (1) to investigate the fatigue (eg. 2), creep (eg. 3) and creep/fatigue (eg. 4) behaviour of materials. Results have been obtained for Jethete M152 at 550°C, 316 stainless steel at 600° and a lead alloy at 20°. In this paper, a summary of the pure mode-II and mixed-mode (with initial $K_I/K_{II} = 1.6$) results for Jethete are presented; these are compared with pure mode-I results obtained from compact-tension (CT) tests. The typical chemical composition of Jethete M152 is 0.12%C, 0.8% Mn, 11.5% Cr, 2.5% Ni, 1.75% Mo, 0.3% V, Fe making up the balance.

From uniaxial creep tests (3), a Norton (power law) fit to the secondary creep rate data was obtained, ie.

$$\dot{\epsilon}^c_{min} = 2.933 \times 10^{-27} \sigma^{9.5} \qquad (1)$$

and a Norton-Bailey fit to the primary creep data was found to be

$$\epsilon^c = 6.9 \times 10^{-16} \sigma^{5.2} t^{0.37} \qquad (2)$$

where σ and t have units of MPa and h, respectively. The creep rupture data is given in Figs. 2(a) and (b).

Relatively little scatter, compared with most engineering materials, was found to exist in the creep and creep rupture behaviour of the Jethete M152 material.

M. Zyczkowski (Ed.)
Creep in Structures
IUTAM Symposium Cracow/Poland 1990

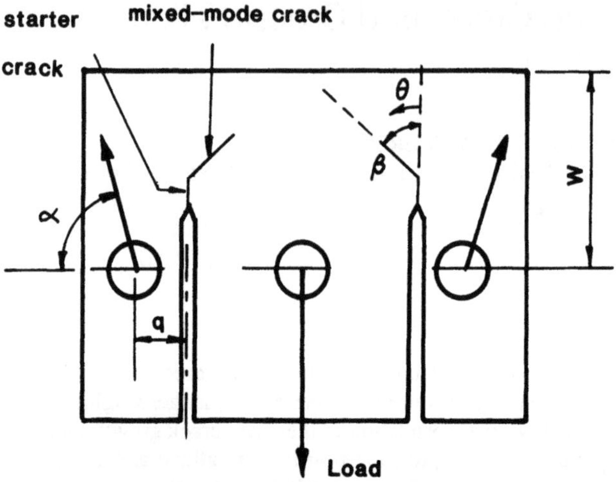

Fig. 1 The compact-mixed-mode specimens; details are given in reference 1. (W = 33 mm and thickness, B = 3.18 mm).

CREEP CRACK GROWTH

Initial cracks were produced in the CT and CMM specimens by fatigue loading using an Amsler Vibrophore at 80Hz. The method of producing straight cracks in the CMM specimens is described in reference (1). The CT and CMM tests were carried out in 5 ton Denison static load, creep testing machines and a Mayes servo-electric machine, respectively. Load-line displacements were obtained from high temperatures extensometers for the CT specimens and from cross-head displacements for the CMM specimens. Crack lengths were obtained using DC potential drop equipment.

Using a similar approach to that of Harper and Ellison (5), a general method for determining C*, for mode-I and mixed-mode loading, from load-line displacement rates, $\dot{\Delta}$, has been obtained (6); the resulting formulation is

$$C^* = \frac{n}{n+1} \cdot \frac{F\dot{\Delta}}{BW} \frac{1}{\sigma_R^L} \cdot \frac{d\sigma_R^L}{d(a/W)} \tag{3}$$

where F is the applied load, a is the crack length and σ_R^L is the limit load reference stress

For the CT tests, the resulting C* parameter is obtained as

$$C^* = \frac{n}{n+1} \frac{F\dot{\Delta}}{BW} \left[\frac{1}{(1 + a/W)} + \frac{2}{(1 - a/W)} \right] \tag{4}$$

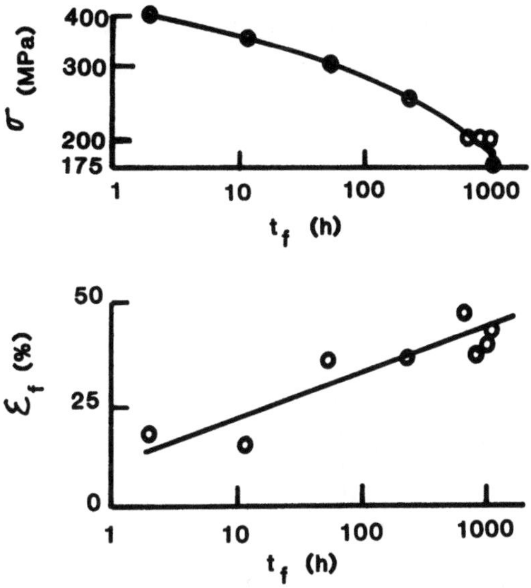

Fig. 2 Creep rupture data for Jethete M152 at 550°C; log (stress, σ) versus log (failure time, t_f) and failure strain, ϵ_f versus log (failure time, t_f)

When α = 106° (see Fig. 1), the crack is subjected to mode-II conditions and equation (3) can be used to obtain C*, ie.

$$C^* = \frac{n}{n+1} \cdot \frac{F\dot{\Delta}}{BW} \quad \frac{1}{(1 - a/W)} \tag{5}$$

When α = 90°, the crack is subjected to mixed- mode conditions (K_I/K_{II} = 1.6) and C* is given by

$$C^* = \frac{n}{n+1} \frac{F\dot{\Delta}}{BW} \left[\frac{2}{(1 - a/W)} - \frac{(1-a/W)}{(2q/W)^2 + (1 + a/W)^2} \right] \tag{6}$$

The results of four CT tests, three mode-II tests and three mixed-mode tests are given in Fig. 3. It can be seen that the mode-II results and mixed-mode results fall below the mode-I results. However, the results indicate that reasonable predictions of mixed-mode creep crack growth rates can be obtained using the CT data and a C* approach.

Under mode-II conditions, creep crack growth was at θ = 0° (see Fig. 1). However under the mixed-mode conditions, two cracks occurred; one at θ = 0° and one at θ = 90°. The crack at 90° was significantly larger than that at 0° and this was used

$\theta = 90°$. The crack at 90° was significantly larger than that at 0° and this was used to produce the results in Fig. 3. Photographs of typical surface creep crack growth are shown in Fig. 4.

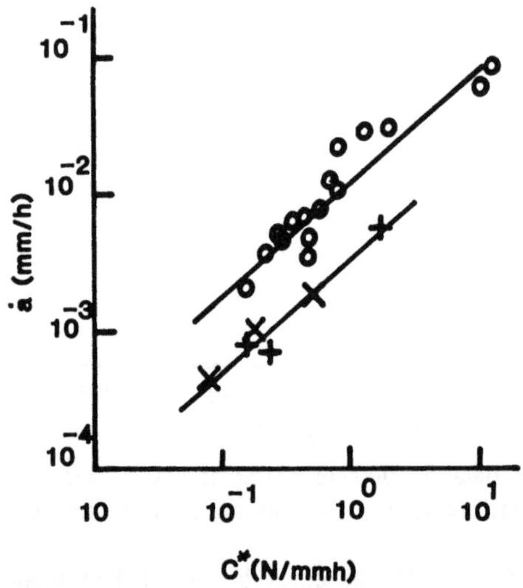

Fig. 3 Variation of creep crack growth rate, å, with C* (0 = mode-I; + = mode-II and X = mixed-mode (K_I/K_{II} = 1.6))

CREEP/FATIGUE CRACK GROWTH

The loading cycles applied to the CT and CMM specimens are shown in Fig. 5. Tensile dwell periods of between 0 and 1800 seconds were included in the test programme. Crack growth rates per cycle, da/dN, were obtained using DC potential drop equipment. Nickbin and Webster (7) have shown that creep/fatigue data can be conveniently represented, for mode-I results, on a plot of log (da/dN) versus log (f), where f is the loading cycle frequency; this is illustrated in Fig. 6. Creep dominated crack growth predictions and fatigue dominated crack growth predictions can be obtained from static load creep and rapid cycling fatigue data (7). In the present investigation, frequency f_1 and the rapid cycling fatigue crack growth rate $(da/dN)_{fatigue}$, shown in Fig. 6, are used as convenient normalising parameters. The approach (7) has also been extended from its use on mode-I data and the results of a

CT, three pure mode-II and four mixed-mode creep/fatigue tests, normalised with respect to f_1 and $(da/dN)_{fatigue}$, are presented in Fig. 7 Results from two CT and seven CMM fatigue tests (ie. rapid cycling) are also included in Fig. 7.

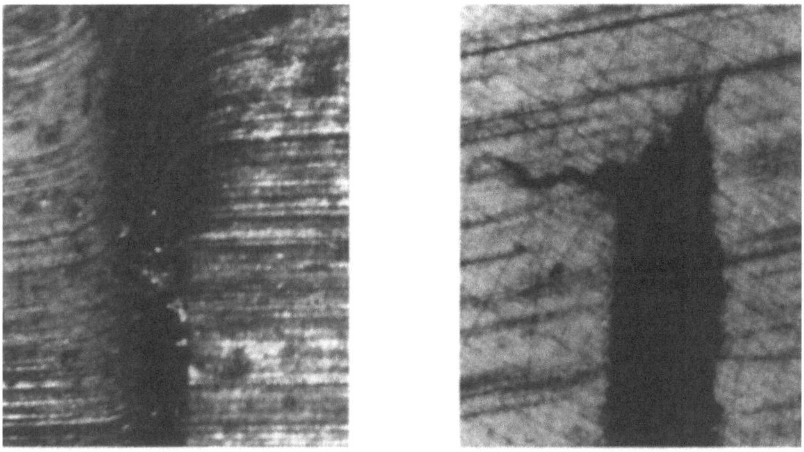

(a) (b)

Fig. 4 Photographs of typical surface creep crack growth; (a) mode-II, (b) mixed-mode (K_I/K_{II} = 1.6)

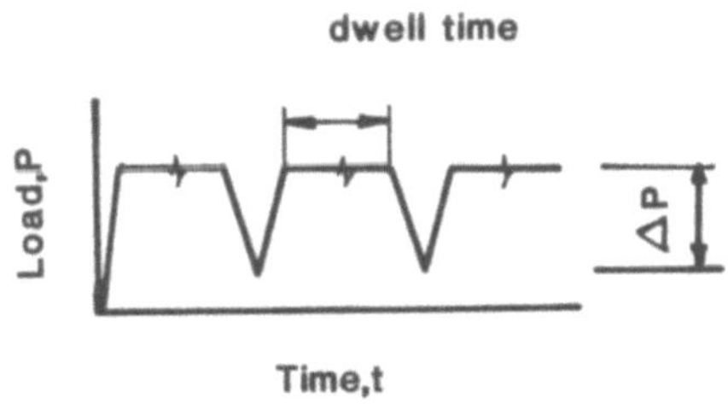

Fig. 5 Loading cycle used in the creep/fatigue tests.

From Fig. 7 it can be seen that the creep/fatigue data are above the creep dominated and fatigue dominated predictions. Linear and vector (because the directions of fatigue and creep crack growth for the same mixed-mode conditions are different) summations of the creep and fatigue dominated predictions also underestimate crack growth rates.

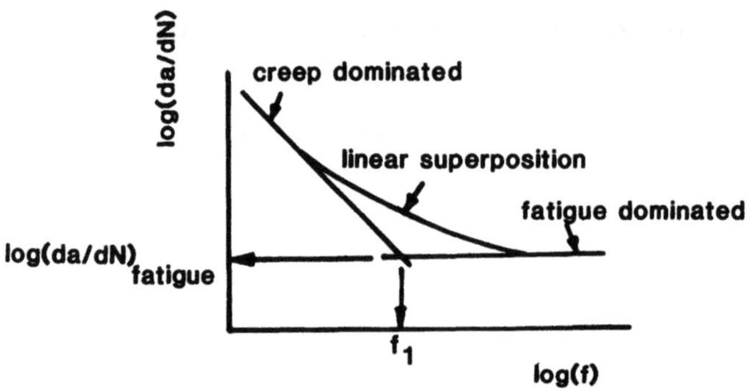

Fig. 6 Schematic diagram of log (da/dN) versus log (f). The frequency, f_1, and crack growth rate, $(da/dN)_{fatigue}$, used to normalised the creep/fatigue results are indicated.

The directions of crack growth under mode-II and mixed-mode (initial K_I/K_{II} = 1.6) are shown in Fig. 8(a) and (b), respectively. It can be seen that the directions of crack growth fall between the corresponding fatigue and creep crack growth directions.

CONCLUSIONS

A mixed-mode specimen suitable for elevated temperature creep, fatigue and creep/fatigue tests has been developed and successfully used to obtain mixed-mode data.

The use of da/dt - C* data obtained from CT tests overestimates the creep crack growth rates under pure mode-II and mixed-mode conditions (Fig. 3)

Linear and vector summations of fatigue and creep crack growth rate predictions underestimate crack growth rates under creep/fatigue conditions (Fig. 7).

The direction of crack growth in mode-I, mode-II and mixed-mode are different. The directions of crack growth in each case is also different for creep and fatigue

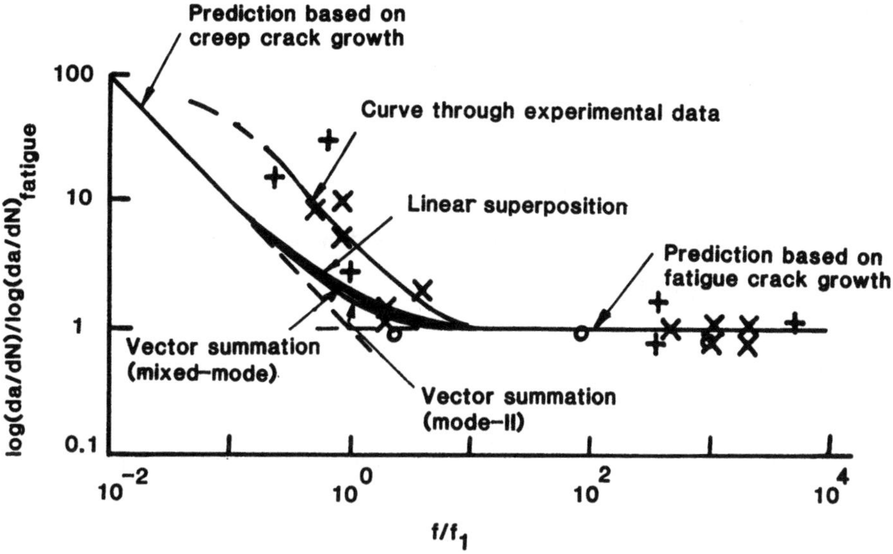

Fig. 7 Variation of normalised crack growth rate, log (da/dN)/log $(da/dN)_{fatigue}$, with normalised frequency, log (f/f_1) (0 = mode-I, + = mode-II and X = mixed-mode (K_I/K_{II} = 1.6)).

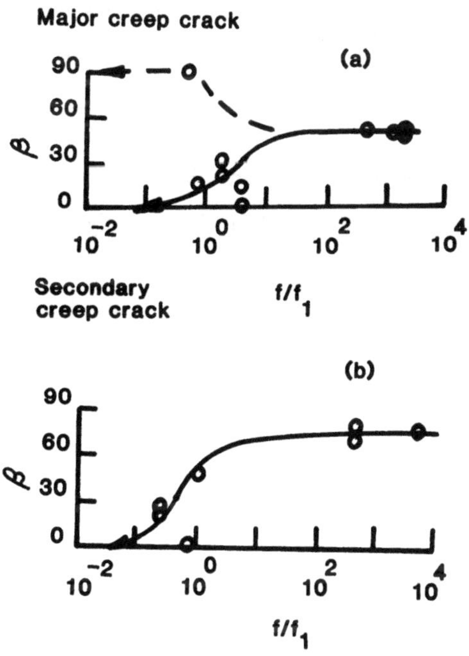

Fig. 8 Direction of crack growth, ß, versus normalised frequency, log (f/f_1);
(a) mode-II,
(b) mixed-mode (K_I /K_{II} = 1.6).

ACKNOWLEDGEMENTS

This work was carried out at the University of Nottingham under a SERC/Rolls-Royce cooperative programme; Professor T H Hyde and Dr J J Webster were joint holders of the grant. The authors gratefully acknowledge the financial support provided. Dr J J Webster was continuously involved in the work and Mr B Smith assisted with the experimental testing.

REFERENCES

1. Hyde, T H and Chambers, A C: A compact mixed-mode (CMM) fracture specimen. J. Strain Analysis 23 (1988) 61-66.

2. Chambers, A C, Webster, J J and Hyde, T H: Mixed-mode fatigue crack growth at 550°C under plane stress conditions in Jethete M152. Submitted to Engineering Fracture Mechanics.

3. Hyde T H and Chambers, A C: An experimental investigation of mixed-mode creep crack growth in Jethete M152 at 550°C. Submitted to High Temperature Technology.

4. Hyde, T H and Chambers, A C: Creep/fatigue crack growth in Jethete M152 at 550°C under mixed-mode conditions. Submitted to J. Strain Analysis.

5. Harper, M P and Ellison, E G: The use of the C* parameter in predicting creep crack propagation rates. J. Strain Analysis 12 (1977) 167-179.

6. Hyde, T H: Estimates of the creep parameter C* in terms of reference stress. J Strain Analysis 21 (1986) 117-119.

7. Nickbin, K M and Webster, G A: Prediction of crack growth under creep/fatigue conditions. ASTM STP 942 (1988) 281-292.

Theoretical, Experimental and Numerical Investigations of Creep Crack Growth

R. Kienzler and T. Hollstein

Fraunhofer-Institut für Werkstoffmechanik, Wöhlerstr. 11,
D-7800 Freiburg

Abstract

The paper shows the derivation of integral fracture concept from conservation laws of general continuum mechanics. Its applicability to cracks under steady-state creep conditions is demonstrated experimentally. The numerical procedure necessary for a reliable evaluation of the crack driving parameter from experimental data is discussed.

Introduction

For the assessment of mechanically loaded components under high-temperature conditions, inelastic time-dependent deformations must be considered in metallic materials. Creep takes place in a homogeneous stress state by diffusion processes and cavity nucleation and growth and may be described by damage mechanics concepts [1,2]. Stress concentrations in real structures, however, cause a highly localized accumulation of damage. Macroscopic cracks develop, and crack growth occurs with damage almost being absent elsewhere in the structure. Therefore, fracture mechanics concepts are expected to be applicable.

The application of integral criteria is of common use in fracture mechanics. Crack initiation and crack growth are described by Rice's J integral [3] in elastic-plastic materials, whereas crack growth under steady-state creep conditions is correlated with the so-called C* integral [4]. The common basis of both, J and C*, is the translational invariance of the general energy balance of continuum mechanics [5].

Whether those concepts really describe crack initiation and crack growth, must be determined, of course, experimentally. This re-

M. Zyczkowski (Ed.)
Creep in Structures
IUTAM Symposium Cracow/Poland 1990

quires, that - independent of size and geometry of the cracked component - these events occur at the same applied magnitude of the respective parameter. To this end, several specimens, different in size and geometry are tested under different loading and temperature conditions in different high-temperature materials.

The experimental evaluation of C* relies on an approximation formula which involves a geometry-dependent η factor, which is shown numerically to deliver quite accurate results. For specimen and crack geometries with factors not tabulated in the standards, η is calculated. Only in this way is it possible at all to perform a reliable evaluation of C* from experimental data for those specimens.

Integral criteria of fracture mechanics

Conservation laws in terms of surface integrals have been derived by several authors [6-9]. Buggisch et al. [10] showed that J-type integrals follow from the translational invariance of the energy balance equation of general continuum mechanics. Let S_o be the boundary of a simply connected domain in the reference state at time $t = t_o$ with unit outward normal vector $\vec{n}$. The conservation law described in Lagrangian (or material) co-ordinates reads

$$I_k = \int_{S_o} B_{kj} \, n_j \, dA = 0 \qquad (1)$$

(Indices range from 1 to 3 for three-dimensional and from 1 to 2 for two-dimensional problems. The summation convention is used for repeated indices).

From the general expression (1), the components of B_{kj} may be specified by introducing constitutive relations for a particular material behaviour. If steady-state creep conditions are assumed to prevail, the C* integral of creep fracture mechanics (CFM) emerges from (1)

$$\text{CFM:} \; I_k \rightarrow C_k^* = \int_{S_o} (W^* \, \delta_{kj} - \sigma_{ij} \, \dot{u}_{i,k}) \, n_j \, dA \qquad (2)$$

(creep potential W*, δ_{kj} Kroneker's tensor of unity, Piola-Kirchhoff stress tensor σ_{ij}, displacement vector u_i, dots denote time derivation, indices following a comma denote spatial derivation).

If the material law is rate-independent and dynamic as well as thermal effects are neglected, the J integral of linear-elastic and elastic-plastic fracture mechanics (LEFM, EPFM) is re-derived from (1)

$$\left.\begin{matrix}\text{LEFM}\\ \text{EPFM}\end{matrix}\right\} : I_k \rightarrow J_k = \int_{S_o} (W\, \delta_{kj} - \sigma_{ij}\, u_{i,k})\, n_j\, ds \tag{3}$$

(stress work density W). Similar expressions may be given for thermo-elastic problems [5].

Whereas the J integral ($J \equiv J_1$) has been proven as a reliable fracture mechanics parameter controlling crack initiation and growth in EPFM (e.g. [11, 12]), the C* integral ($C^* \equiv C_1^*$) seems to have the widest range of applicability in CFM. It describes creep crack-growth rates independent of geometry under steady-state creep conditions [4,13,14].

<u>Experimental evaluation of C*</u>

For tests under constant force, a formula has been developed [15,16] to calculate C* from measured quantities during a creep crack-growth test

$$C^* = \eta \frac{F\, \dot{V}}{B(W-a)} \tag{4}$$

(specimen thickness B, and width W). Force F, crack-opening displacement V in the load line, and crack length a are to be continuously measured during the test [17]. The η factor depends mainly on the specimen geometry and only very weakly on the specimen size, the crack depth and the material creep law involved . For standard test specimen, η values are tabulated [15] or may be calculated numerically [14].

<u>Numerical evaluation of C*</u>

In the course of a Finite Element (FE) calculation, C* may be obtained directly from the line- (or surface-) integral definition (2) or by the virtual crack extension method [18,19]. The appropriate singularity is modelled by applying the collapsed element technique [20,21]. In order to include numerical simulation of crack propagation, the node shifting and releasing technique [22] is employed. The crack propagation law is a measured material resistance curve $C^*(\dot{a})$. An implicit time-integration scheme secures unconditionally stable solutions [23]. The numerical results presented in the following were obtained with the finite-element programme ADINA [24] supplemented with subroutines to calculate J and C* [25].

Results

For the commercial high-temperature alloy Incoloy 800 H, different specimen geometries were tested at a temperature of T = 800°C [26]. Figure 1 shows C* values of a test with a compact tension specimen determined experimentally (eq. 4, η = 2) and numerically (plane strain).

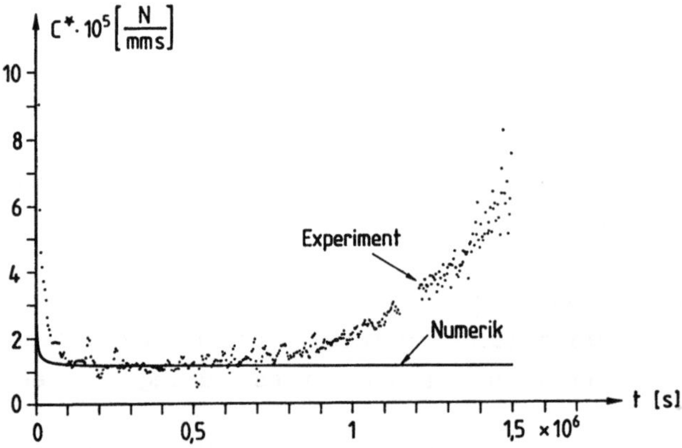

Fig. 1: Values of C* determined experimentally and numerically, CT specimen, Incoloy 800 H

For a secondary region, $0.1 \cdot 10^6$ s $\leq t \leq 0.8 \cdot 10^6$ s, the agreement is very good. The increase of C* in the experiment is due to creep crack growth, which was not modelled in this numerical calculation.

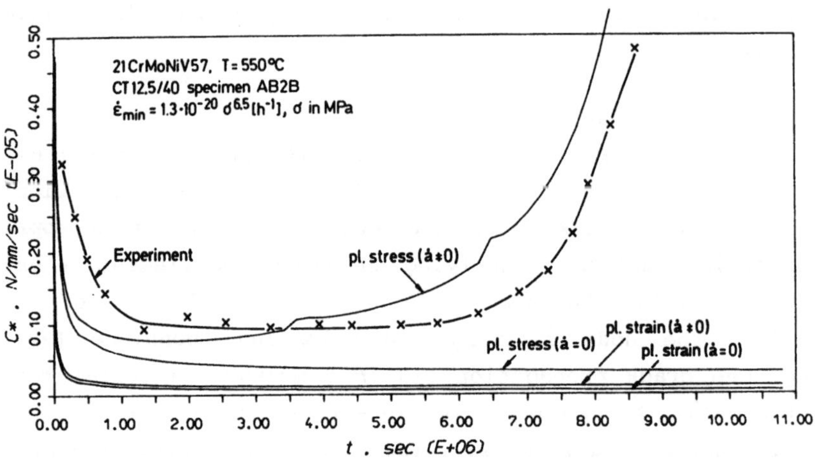

Fig. 2: CT specimen, 1% Cr steel, experimental and numerical evaluations of C* vs. time

Creep crack growth experiments with the 1% Cr steel 21 CrMoNiV 5 7 were carried out at 550°C with side-grooved compact tension (CT) specimens [27]. Figure 2 compares measured and calculated C* vs. time curves from one typical experiment and from the finite element simulation.

The calculations without consideration of crack growth ($\dot{a}$ = o) significantly underestimate C* for all times. Even the plane-strain solution with consideration of crack growth fails to meet the experimental curve by about one order of magnitude. The plane-stress solution including crack propagation ($\dot{a}$ ≠ o) compares fairly well with the experimental curve.

Creep crack growth rates, $\dot{a}$, are plotted in Fig. 3 as a function of the C* integral for various specimen geometries under different loading conditions [14]. C* and $\dot{a}$ are correlated by a power law within a narrow scatter band independent of size and geometry of specimen and crack, and loading conditions. In this way assessment of structural components with flaws and remaining life time predictions under high-temperature conditions become possible.

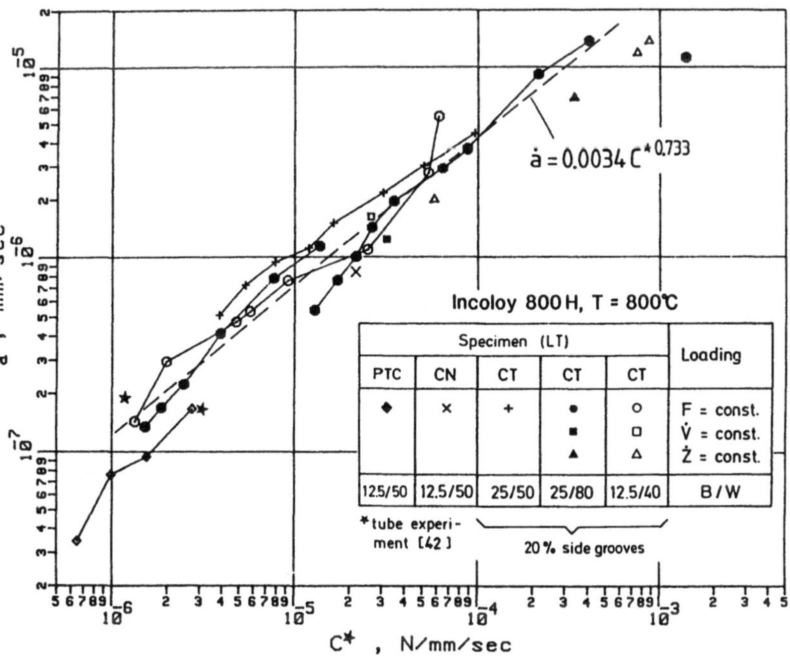

Fig. 3: Incoloy 800 H, crack growth rates $\dot{a}$ vs. C*

Conclusions

In the paper, it is sketched, how integral concepts for non-linear fracture mechanics problems are derived from energy (or thermo-dynamic) balance equations, leading to physically meaningful fracture criteria. The applicability of these concepts is shown experimentally: Creep crack-growth rates can be correlated with the loading parameter C* by a power law within a relatively narrow scatterband.

Reliable finite element methods are available to simulate creep fracture experiments numerically. The agreement between experimental and numerical results is good.

A more detailed presentation of theory and results with further references may be found in [5,14,28].

Acknowledgement

The financial support of Bundesministerium für Forschung und Technologie, of Stiftung Volkswagenwerk and of Deutsche Forschungsgemeinschaft is gratefully acknowledged.

References

[1] H. Riedel: Fracture at High Temperatures. Springer-Verlag Berlin, Heidelberg, New York, London, Paris, Tokyo 1987

[2] J. Lemaitre and J.-L. Chaboche: Mechanics of Solid Materials. Cambridge University Press, Cambridge 1990

[3] J.R. Rice: A Path Independent Integral and the Approximate Analysis of Strain Concentrations by Notches and Cracks. Journal of Applied Mechanics 35 (1968) 379-386

[4] H. Riedel and J.R. Rice: Tensile Cracks in Creeping Solids. In: Fracture Mechanics, ASTM STP 700, American Society for Testing and Materials (1980) 112-130

[5] R. Kienzler: On Integral Criteria of Fracture Mechanics and Local Parameters Based on the Energy-Momentum Tensor. In: Proceedings of the Symposium on Elastic-Plastic Fracture Mechanics: Elements of Defect Assessment, October 9-11, 1989, Freiburg, FRG, to be published.

[6] J.D. Eshelby: The force on an elastic singularity. Philosophical Transactions of the Royal Society London A 244 (1951) 87-112

[7] W. Günther: Über einige Randintegrale der Elastomechanik. Abhandlungen der Braunschweigischen Wissenschaftlichen Gesellschaft 14 (1962) 53-72

[8] J.K. Knowles and E. Sternberg: On a Class of Conservation Laws in Linearized and Finite Elasto-Statics. Archive for Rational Mechanics and Analysis 44 (1972) 187-211

[9] A.E. Green: On Some General Formula in Finite Elastostatics. Archive for Rational Mechanics and Analysis 50 (1973) 73-80

[10] H. Buggisch, D. Gross und K.-H. Krüger: Einige Erhaltungssätze der Kontinuumsmechanik vom J-Integral-Typ. Ingenieur-Archiv 50 (1981) 103-111

[11] J.A. Begley and J.D. Landes: The J-Integral as a Fracture Criterion. In: Fracture Toughness, ASTM STP 514, American Society of Testing and Materials (1972) 1-23

[12] J.W. Hutchinson and P.C. Paris: Stability Analysis of J-Controlled Crack Growth. In: Elastic-Plastic Fracture, ASTM STP 668, American Society of Testing and Materials (1979) 37-64

[13] K.M. Nibkin, D.J. Smith and G.A. Webster: Prediction of Creep Crack Growth from Uniaxial Creep Data. Proceedings of the Royal Society London A 396 (1984) 183-197

[14] T. Hollstein and R. Kienzler: Fracture Mechanics Characterization of Crack Growth under Creep and Fatigue Conditions. Journal of Strain Analysis 23 (1988) 87-96

[15] V. Kumar, M.D. German and C.F. Shih: An Engineering Approach for Elastic-Plastic Fracture Analysis. Report to Electric Power Research Institute, EPRI NP-1931, Palo Alto 1981

[16] T.Hollstein, F. Djavanroodi, G.A. Webster and S.R. Holdsworth: High Temperature Crack Growth in Alloy 800H and a 1% Cr Mo V Steel -The Results of an EGF Round Robin. In: E. Czoboly (ed.) Failure Analysis, Theory and Practice, ECF 7 II (1988) 656-668

[17] T. Hollstein: Experimental Investigation of Creep Crack Growth for Two High Temperature Alloys. In: H.C. van Elst an A. Bakker (eds.) ECF 6; Fracture Control of Engineering Structures 1 (1986) 451-461

[18] D.M. Parks: A Stiffness Derivative Finite Element Technique for Determination of Elastic Crack Tip Stress Intensity Factors. International Journal of Fracture 10 (1974) 487-502

[19] H.G. DeLorenzi: Energy Release Rate Calculations by the Finite Element Method. General Electric Technical Information Series, Report No. 82 CRD 205 (1982)

[20] R.D. Henshell and K.G. Shaw: Crack Tip Finite Elements are Unnecessary. International Journal for Numerical Methods in Engineering 9 (1975) 496-507

[21] R.S. Barsoum: Triangular Quarter-Point Elements as Elastic and Perfectly-Plastic Crack Tip Elements. International Journal for Numerical Methods in Engineering 11 (1977) 85-98

[22] H.G. DeLorenzi: J-Integral and Crack Growth Calculations with the Finite Element Program ADINA. Methodology for Plastic Fracture. Report to Electric Power Research Institute, EPRI SRD-78-124 (1978)

[23] K,-J. Bathe: Finite Element Procedures in Engineering Analysis. Prentice-Hall, Inc. Englewood Cliffs, New Jersey 1982

[24] K.-J Bathe: ADINA, a Finite Element Program for Automatic Dynamic Incremental Nonlinear Analysis. Report AE 84-1, Massachussetts Institute of Technology, Cambridge, Massachussetts, U.S.A. (1984)

[25] IWM-CRACK: Subroutine-Package for Crack Problems. Fraunhofer-Institut für Werkstoffmechanik, Wöhlerstr. 11, D-7800 Freiburg

[26] R. Kienzler and T. Hollstein: Theoretisch-numerische Untersuchungen zu Experimenten mit Kriechrißwachstum. Zeitschrift für Werkstofftechnik 17 (1986) 393-397

[27] T.Hollstein and R. Kienzler: Investigation of Creep Crack Growth: A Comparison between Experimental and Numerical Results. In: P.Bensussan et al. (eds.) High Temperature Fracture Mechanisms and Mechanics. Proceeding of a Seminar held in Dourdan, France, 1987. Proceeding of Mechamat (1987) III/79-III/91

[28] R. Kienzler and T. Hollstein: Numerical Treatment of Creep Crack Growth. IWM-Bericht W9/90, Fraunhofer-Institut für Werkstoffmechanik, Wöhlerstr. 11, D-7800 Freiburg

Characterization of Creep-Fatigue Failure Process Based on Stochastic Behavior of Small Crack Initiation and Growth

Takayuki KITAMURA, Naoya TADA, and Ryuichi OHTANI

Dept. of Engineering Science, Kyoto University,
Yoshida-honmachi Sakyo-ku Kyoto, Japan

Summary

In creep-fatigue conditions of usual ductile steels and alloys, multiple small cracks initiated on the specimen surface and their growth and coalescence brought about the failure. The process of crack initiation and growth was highly dependent on the microstructure of materials and the test conditions. A stochastic model was developed in order to simulate the random behavior of small cracks. Two variables, F and K, were used to characterize and simulate the initiation and growth behavior, respectively. A normalization and characterization method of failure process was proposed on the basis of the ratio of K to F. The failure process was dominated by the initiation and coalescence of multiple small cracks for a small value of K/F, while it was governed by the growth of restricted number of cracks for a large one.

Introduction

The failure process in creep-fatigue has been experimentally investigated by the authors[1-4] using smooth specimens of several heat resisting alloys. As a result, multiple intergranular cracks initiated on the specimen surface by a grain boundary length between adjacent triple points and their growth and coalescence brought about the failure. The small cracks exhibited irregular behavior due to the microstructural inhomogeneity of materials. Moreover, the behavior was remarkably sensitive to the test conditions such as strain range, strain waveform, etc. A stochastic model was developed for the random initiation and growth of small cracks and the failure process was numerically simulated by means of the Monte Carlo method[5,6].
In this study, a characterization of the creep-fatigue failure process is proposed on the basis of the experimental observation and the simulation in terms of stochastic behavior of small cracks.

Experimental Observation

Smooth specimens of Type 304 stainless steel were subjected to

M. Zyczkowski (Ed.)
Creep in Structures
IUTAM Symposium Cracow/Poland 1990

strain-controlled creep-fatigue conditions (Fig.1 and Table 1) at 923K(650℃) and the initiation and growth of small cracks on the surface was carefully observed by means of an optical microscope [1,3,4]. Fig.2 shows the morphology of small cracks in the cp- and cs-fatigue with a strain range of 1% in air. Here, the compressive strain rate is particularly different between them. Numerous cracks initiated in the cp-fatigue and the main crack was formed by not only their growth but also the coalescence. On the other hand, only a few cracks initiated in the cs-fatigue and they grew without coalescence. It implies that the compressive creep strain remarkably repressed the crack initiation though it affected little on the crack growth rate. The sensitivity of the test condition to this kind of behavior brings complexity to the failure analyses in the creep-fatigue. The effects of various factors on the crack initiation and growth were discussed in detail elsewhere[4] based on the experimental results.

<u>Simulation Model</u>

The simulation model[5,6] is summarized as follows; (1) Grain boundaries on the specimen surface are projected and linked on a straight line perpendicular to the applied strain. The length is a random variable with a normal distribution (average =20μm and standard deviation =20μm for the 304 stainless steel). (2) The

Table 1 Test conditions and the values of F and K.

Test	$\Delta\varepsilon$	$\dot{\varepsilon}_{ten}$	$\dot{\varepsilon}_{com}$	Env.	N_f	K	F	K/F
cp	1.5	10^{-4}	2×10^{-2}	air	100	10.7	3.0	3.6
	1.0	10^{-3}	1	air	374	2.5	0.4	6.3
	0.7	10^{-4}	2×10^{-2}	air	449	2.25	0.08	28
cc	1.0	2×10^{-3}	2×10^{-3}	air	822	1.75	0.04	44
cs	1.0	2×10^{-3}	3×10^{-4}	air	643	1.75	0.02	88
cp	1.0	10^{-3}	1	vac.	388	4.3	0.63	6.8
cc	1.0	2×10^{-3}	2×10^{-3}	vac.	734	4.3	0.13	33

$\Delta\varepsilon$: Strain range (%)
$\dot{\varepsilon}_{ten}$: Tension strain rate (%/s)
$\dot{\varepsilon}_{com}$: Compression strain rate (%/s)
Env.; Environment
vac.: Vacuum
N_f: Number of cycles to failure

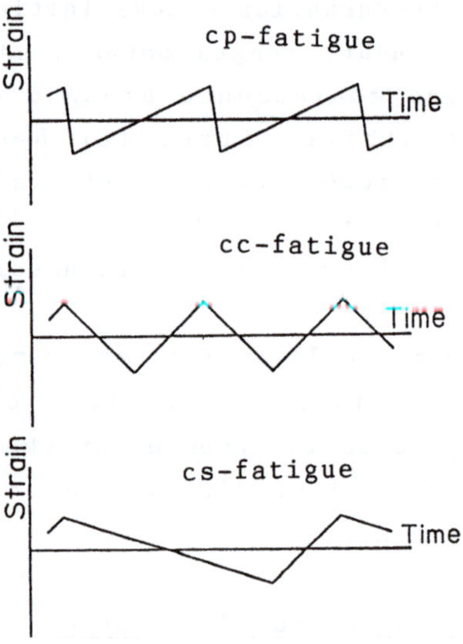

Fig. 1 Strain waveforms.

grain boundaries have random fracture resistance of uniform distribution between 0 and 1000. (3) Fracture driving force is given to each grain boundary by $D=F+K(c/\ell)$ where F and K are deterministic variables and c and ℓ are a crack length and mean grain boundary length. (4) The fracture resistance decreases by the driving force in every strain cycle and a crack is formed at the boundary where the residual resistance reaches zero. The crack initiation is defined as the cracking at a grain boundary isolated from the other cracks, and the crack growth at the grain boundary being adjacent to a crack. The variable, F, governs the crack initiation in the model, while K dominates the crack growth [5]. Considering that the creep J-integral is proportional to the crack length in an infinite body, the value of K is obtained by extrapolation of the fracture mechanics law for a long crack (relation between crack propagation rate and creep J-integral range[7]). On the other hand, since a determination method of value F is not explicitly developed yet, it is determined, here, by adjusting the simulation to the experimental data on crack initiation at the very early stage of life through trial and error. The values of F and K in various test conditions are

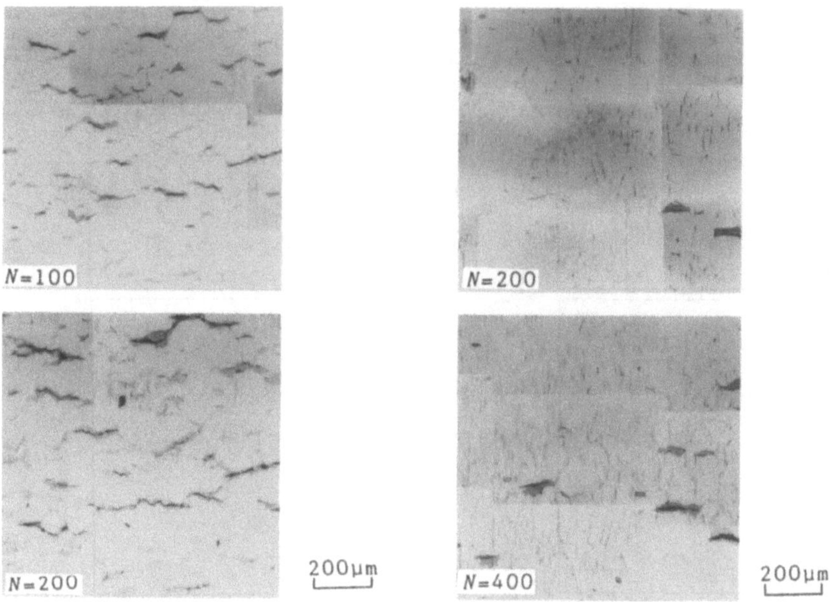

(a) cp-fatigue (b) cs-fatigue

Fig. 2 Photographs of small surface cracks in creep-fatigue.

listed in Table 1. Fig.3 illustrates the simulated small cracks in the cp- and cs-fatigue by means of a computer using the Monte Carlo method[5,6]. The image of the crack distribution patterns is analogous to the actual morphology in Fig.2. The crack density and probabilistic distribution of crack length were quantitatively compared to those in the experimental observation and the validity of the simulation was confirmed by the coincidence[5,6].

In order to characterize the failure process, a comparison is made on the following two cases where the ratio of K to F is the same; (a)F=a, K=b and (b)F=xa, K=xb. Here a,b and x are arbitrary constants. Because the reduction of fracture resistance at each grain boundary in the case (b) is accelerated by x times as fast as that in the case (a), the number of strain cycles to cracking is shortened exactly by x times in the case (b). In other words, the probabilities of crack initiation and growth at N strain cycle in the case (a) are the same as those at N/x cycle in the case (b). It means that both cases go through the similar failure process in terms of the image of multiple cracks except the changing rate against the strain cycle. Thus, the failure process

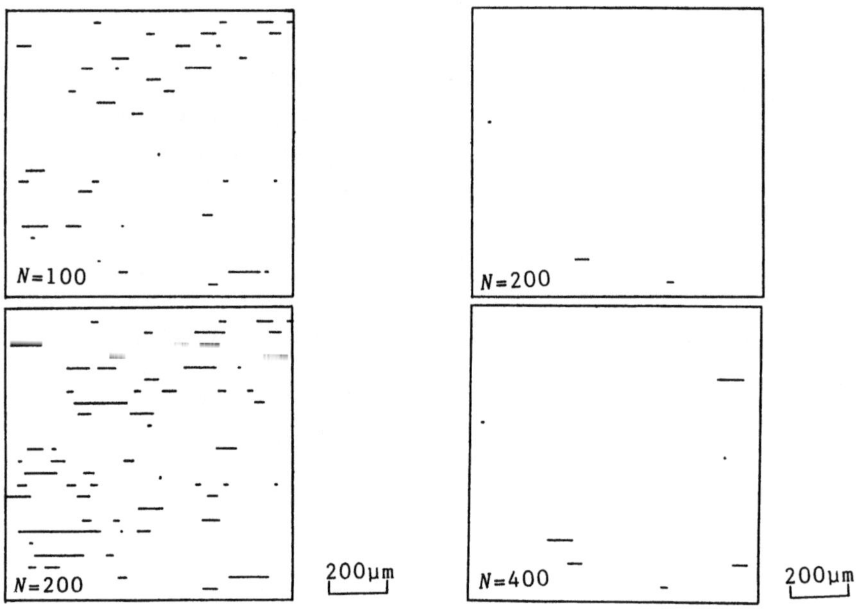

(a) cp-fatigue (b) cs-fatigue

Fig. 3 Distribution patterns of small cracks obtained by the simulation for the cp- and cs-fatigue.

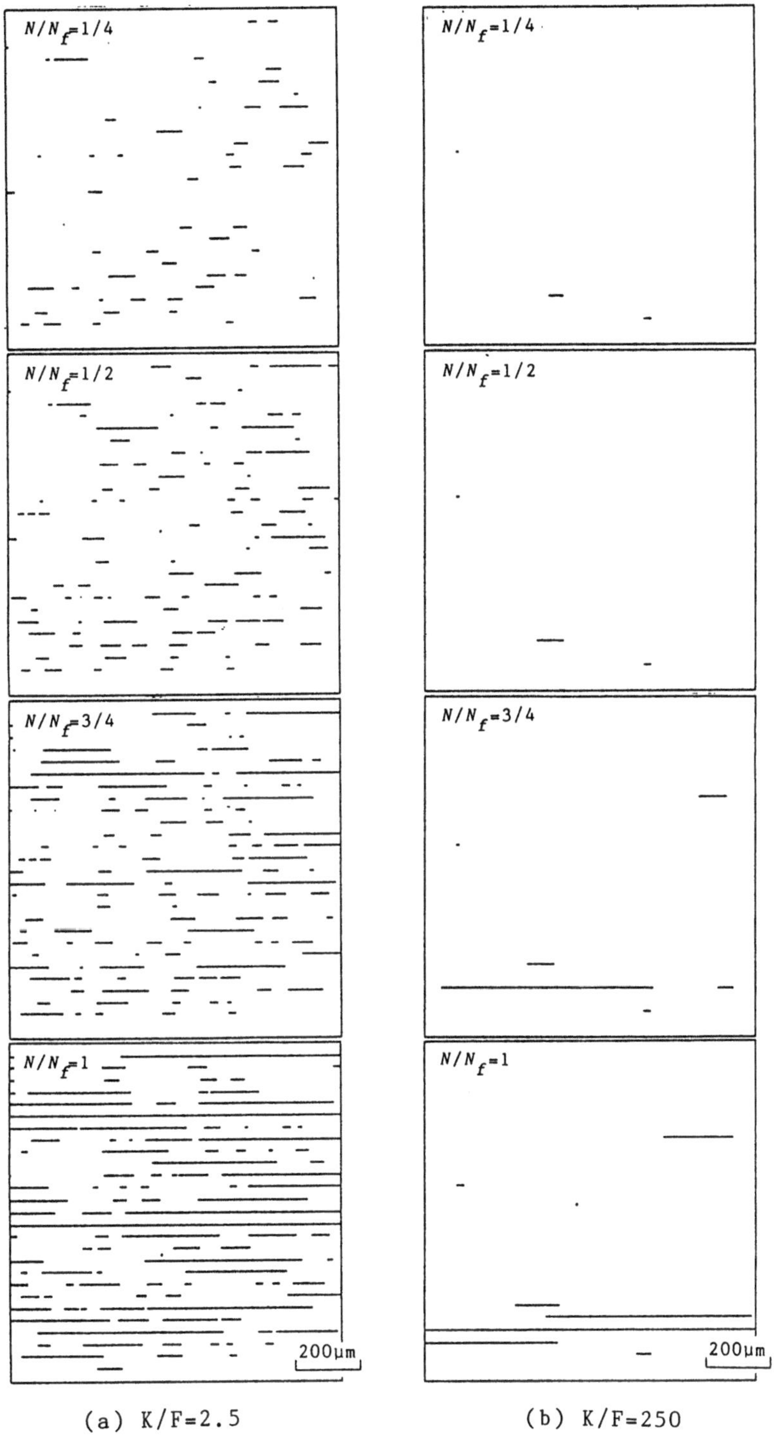

Fig. 4 Distribution patterns of small cracks.

is represented by the value of K/F as the cycle can be normalized by a standard cycle (i.e. the failure life of specimen) eliminating the factor, x.

Characterization Of Failure Process

The Monte Carlo simulation was carried out in a wide range of K/F value on 1000 lined grain boundaries which correspond to the area of about $1mm^2$ for the 304 stainless steel.

Fig.4 shows the images of small cracks in K/F=2.5 and 250. Here, the line of grain-boundaries is cut and realigned by every 40 boundaries for a help to visual comprehension. The number of strain cycles, N, is normalized by the failure life, N_f, which is defined as the number of cycles when the longest crack reaches 1mm. It is clear from the figure that the failure process is highly dependent on the K/F value. Enormous cracks initiate and their coalescence forms the several long cracks for K/F=2.5, whereas a few cracks which initiate at the early stage of life grow individually and bring about the failure for K/F=250. Fig.5 shows the change in the crack density which is defined as the number of cracks per unit surface area. There is a clear distinction in the density. The decrease at the late stage for K/F=0.25 and 2.5 is caused by the crack coalescence. Fig.6 shows the distribution of crack length. It is found that the cumulative

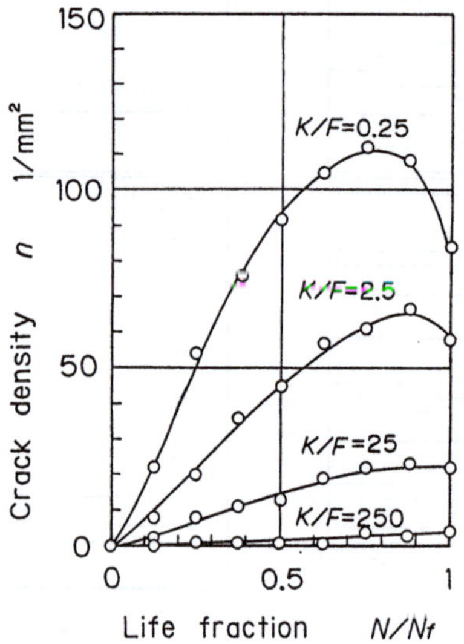

Fig. 5 Change in crack density for various values of K/F.

probability curve tends to change the shape at the side of larger crack length for K/F=0.25 in Fig.6(a), and that it keeps a similar shape with increasing number of cycles for 25, the higher value of K/F, in Fig.6(b). This means that the former case (small K/F) increases the crack length not by their own growth but by their coalescence, whereas the latter (large K/F) does by the individual growth of propagating cracks. The difference is illustrated clearly in Fig.7 which shows the growth curves of a main crack.

Referring the value of K/F in Table 1, it becomes clear that the faster compressive strain rate is and the larger strain range is, the more frequent crack coalescence contributes to the failure.

The fracture mechanics approach could be applied to the crack growth in the creep-fatigue with a high value of K/F by taking

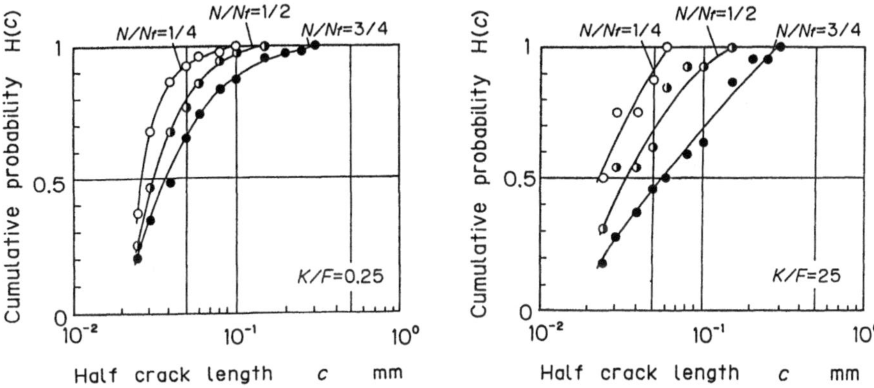

(a) K/F=0.25 (b) K/F=25

Fig. 6 Cumulative probability of half crack length.

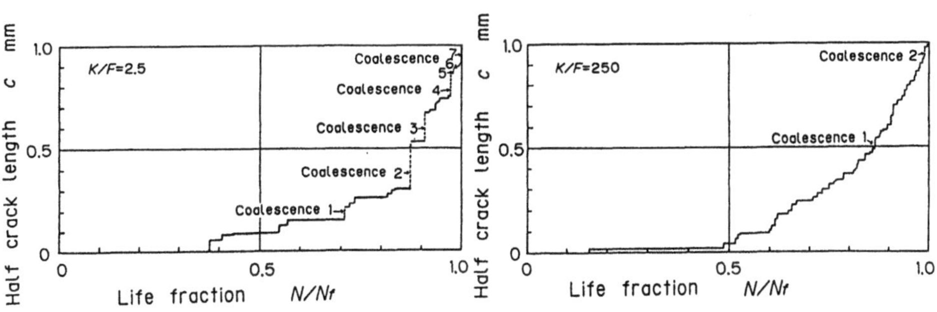

(a) K/F=2.5 (b) K/F=250

Fig. 7 Growth curve of main crack.

into consideration the effect of microstructural randomness[5,6]. The concept of damage mechanics approach might be effective to the analyses in the forming process of long cracks due to the coalescence of small cracks for a low value of K/F. The random behavior, however, should be carefully included in the analyses [8]. Moreover, it should be noted that the damage (small cracks) are formed only on specimen surface in the creep-fatigue condition of the present investigation, although it was originally assumed that the damage distributed uniformly in the entire body.

References

1. Ohtani,R.; Kinami,T.; Sakamoto,H.; Small crack propagation in high temperature creep-fatigue of 304 stainless steel, Proc. Int. Conf. on Role of Fracture Mechanics in Modern Technology", Elsevier Science Pub. (1987) 381-392.

2. Ohtani,R.; Kinami,T.; Murayama,H.; Watabe,T.; Observation of initiation and early growth of small cracks during high-temperature low-cycle fatigue in 304 stainless steel, J. of Society of Mat. Sci., Japan, 36 (1987) 974-979(in Japanese).

3. Ohtani,R.; Kitamura,T.; Behavior of initiation and growth of creep and creep-fatigue small cracks in austenitic stainless steels, Proc. Int. Conf. on Fracture 7, Vol.1, Pergamon Press (1989) 213-221.

4. Ohtani,R.; Kitamura,T.; Initiation and propagation of microstructurally small cracks under creep-fatigue condition, Proc. 4th Int. Conf. on Creep and Fracture of Engineering Materials & Structures, Univ. Swansea (1990) 791-802.

5. Ohtani,R.; Kitamura,T.; Tada,N.; Numerical simulation of initiation and early propagation of creep-fatigue small cracks based on a model of random fracture resistance of grain boundaries, ASME PVP, Vol.163, Book No.H00478, ASME (1989) 123-127.

6. Ohtani,R.; Kitamura,T.; Tada,N.; Stochastic simulation of initiation and growth of small surface cracks in creep-fatigue condition, Proc. 4th Int Conf. on Fatigue and Fatigue Thresholds, Vol.IV, Materials and Component Engineering Pub. LTD. (1990) 2143-2148.

7. Ohtani,R.; Kitamura,T.; Nitta,A.; Kuwabara,K.; High-temperature low-cycle fatigue crack propagation and life laws of smooth specimens derived from the crack propagation laws, ASTM STP 942, ASTM (1988) 1163-1180.

8. Ohtani,R.; Kitamura,T.; Stochastic damage variable based on randomly distributed micro-cavities during creep, Proc. 2nd. Int Conf. on Plasticity and Its Applications, Pergamon Press (1989) 687-689.

A New Damage Approach for Creep Crack Growth Prediction

S. KRUCH, J.L. CHABOCHE, P.M. LESNE

Office National d'Etudes et de Recherches Aerospatiales (ONERA)
Division OR/CE
B.P. n 72 - 92322 CHATILLON CEDEX (France)

Summary

The incapacity of global fracture criteria (like K or G) to describe correctly a crack evolution for highly dissipative materials implies the use of local fracture criteria. One of these may be a classical damage law, but a new difficulty appears named "strain localization" characteristic of strain softening materials. To avoid this problem, it will be necessary to develop a non local damage law introducing a new material intrinsic parameter in the modelization. All these aspects are developed in this paper and illustrated with some significant examples.

I. INTRODUCTION

From long time, damage mechanics is widely used to describe microcracks evolution inside strongly loaded materials. Classically, damage is represented by a scalar D varying from 0 to 1. These values represent respectively the virgin state and the break state of the volume element. If the purpose of the modelization consists in simulating only the crack initiation then this approach is sufficient. In other case, if the simulation consists also to simulate the crack propagation, the use of a damage law carries theoretical and numerical problems. In fact, as soon as D reaches its critical value D_{cr} (set by the user, in general $D_{cr} = 1$ representing the crack initiation), equations of the problem lose their ellipticity and the solution becomes nonunique. Numerically, using the Finite Element Method, this solution will be strongly mesh dependent and unsuitable. This problem, named "strain localization", prevents to carry on with calculation. It could be possible to simulate the crack initiation (1st element of the mesh reaching $D = D_{cr}$) but impossible to simulate the crack growth in structures. Some possibilities exist, limiting strain localization. Two solutions are described below.

1) A local fracture criterion (Fc-Rc) where Fc represents the critical value of a mechanical parameter (like stress or strain) and Rc represents a critical intrinsic material length. The crack grows as soon as F reaches Fc at the critical distance Rc from the crack tip or from the notch. This approach has been successfully used with brittle materials like ceramics or with metals in plasticity, but, for elastoviscoplastic materials the expression of F could be too complicated carrying other numerical difficulties.

M. Zyczkowski (Ed.)
Creep in Structures
IUTAM Symposium Cracow/Poland 1990

2) A non local damage law. It's necessary to introduce a new parameter in the formulation. This parameter represents again an intrinsic material length, characterizing the size of the greatest micro-structural defect. This solution, adopted in this work, consists of a slight modification of classical viscoplastic formulation coupled with damage. Only a non local variable is introduced, all the others internal variables remain unchanged [1],[2].

All these aspects, described above, will be developed in this paper and applied to the prediction of creep crack growth in a structure constituted by a viscoplastic material.

II INTERACTION BETWEEN CREEP AND DAMAGE

Global criteria used in Fracture Mechanics are not able to predict correctly the crack growth in strongly non linear materials (i.e. with a large plastic zone at the crack tip, large damage or delamination). Then, it's necessary to develop a criterion able to take into account local phenomena. Damage laws are suitable to simulate the crack initiation and growth if this crack, at the macroscopic level (i.e. structure scale), stem from the process of nucleation, growth and coalescence of micro-defaults at microscopic level. (In fact, this microscopic level could be modelized by micromechanics but this is not our purpose here). We consider an isotropic creep damage characterized by a scalar parameter D acting on stresses. The effective stress tensor is related to the Gauchy stress tensor by the following relation (Kachanov [3]) :

$$\tilde{\sigma} = \sigma \ / \ (1 - D)$$

The coupling between viscoplastic formulation and damage may be done at various levels :

a) No coupling between the elastoviscoplastic behaviour and damage. At the end of each step in the calculation, the value of σ is used to compute the value of D, but D doesn't affect σ.

b) The coupling is applied exclusively on the viscoplastic behaviour. The elastic stiffness of the structure is not affected by D.

c) Like b) but when D reaches its critical value D_{cr}, the Young's modulus of the considered element drops to zero. This is the method Hayhurst [4],[5] has been using for a long time.

d) Total coupling between the elastoviscoplastic behaviour and damage. The elastic stiffness of the structure decreases continuously to zero as damage increases.

Only the calculation type d) must describe objectively the crack process studied here [6]. The results presented in this paper are exclusively related with this type of calculation.

III CONSTITUTIVE EQUATIONS AND STRAIN LOCALIZATION

The equations governing the problem can be written as follow :

a) Elastic behaviour :

$$\varepsilon^e = \frac{1+\nu}{E(1-D)}\,\sigma - \frac{\nu}{E(1-D)}\,tr(\sigma)\;1$$

b) Viscoplastic behaviour :

$$\left\{\begin{array}{l} \dot{\varepsilon}_p = \dfrac{3}{2}\,\dot{p}\,\dfrac{s-X}{J(s-X)} \\ \dot{p} = \left\langle \dfrac{J(\tilde{s}-\tilde{X})}{K} \right\rangle^n \dfrac{1}{1-D} \\ \dot{X} = C\,[\,a\,\dot{\varepsilon}_p - X\,\dot{p}\,] \\ \dot{K} = K_v\,(\,K_s - K\,)\,\dot{D} \end{array}\right.$$

c) Damage law :

$$\left\{\begin{array}{l} \dot{D} = \left\langle \dfrac{\Delta(\sigma)}{A} \right\rangle^r (1-D)^{-k(\Delta(\sigma))} \\ \Delta(\sigma) = \alpha\,\sigma_{P_{max}} + 3\,\beta\,\sigma_H + (\,1-\alpha-\beta\,)\,J(\sigma) \\ k(\Delta) = k_0 + k_1\left(\dfrac{\Delta}{100}\right) + k_2\left(\dfrac{\Delta}{100}\right)^2 \end{array}\right.$$

Where :

- X is the kinematic hardening tensor
- s is the stress deviatoric tensor
- $K, n, K_v, K_s, A, r, \alpha, \beta, k_0, k_1, k_2$ are material parameters
- $J(\;)$ is the Von-Mises equivalent stress
- $\sigma_{P_{max}}$ is the maximum principal stress
- σ_H is the hydrostatic stress

As mentioned previously in the introduction, this formulation with a total coupling between the elastoviscoplastic behaviour and damage is only able to predict correctly the macroscopic crack initiation.

The simulation of the crack growth by the same way is not possible because when one element of the mesh breaks, the equations lose their ellipticity (i.e. the solution is non unique). Numerically, using FEM, the solution is strongly mesh dependent. Damage localize in a band of elements. The more small mesh, the more small band, to lead to the aberrant result of a crack growing without dissipation of energy.

To show this, a numerical simulation of a cylindrical specimen loaded in traction has been made (figure 1). Three different meshes were used (figure 2). Figure 3 shows the damage zone in the middle of the specimen for two meshes. As can be seen, the crack length is more important for the refined mesh (local level), at the same time, the global response, plotted in figure 4, doesn't present any difference for both meshes at this state of load.

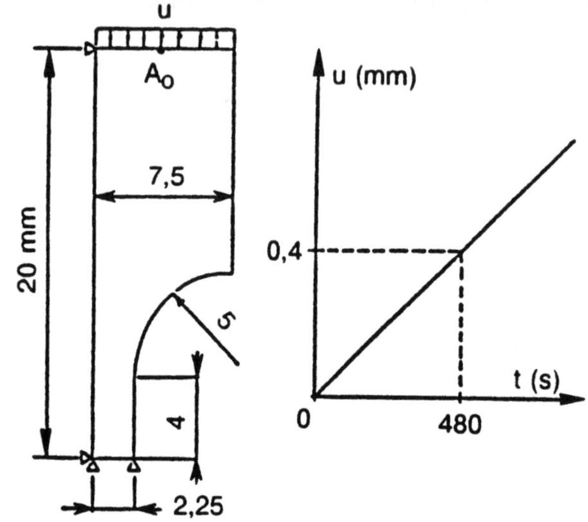

Fig. 1. Idealization of the tensile round bar

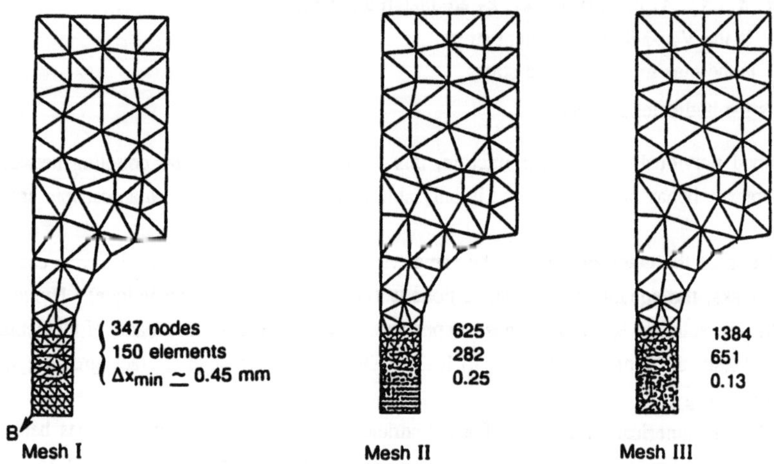

Fig. 2. Different finite element idealizations of the tensile round bar

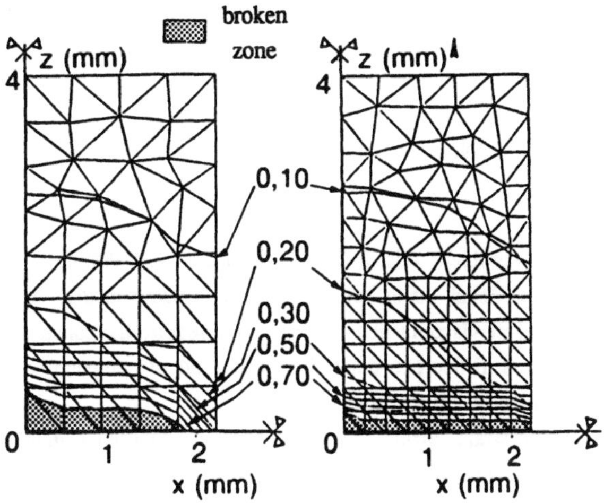

Fig. 3. Damage zones obtained with mesh I and II (local damage)

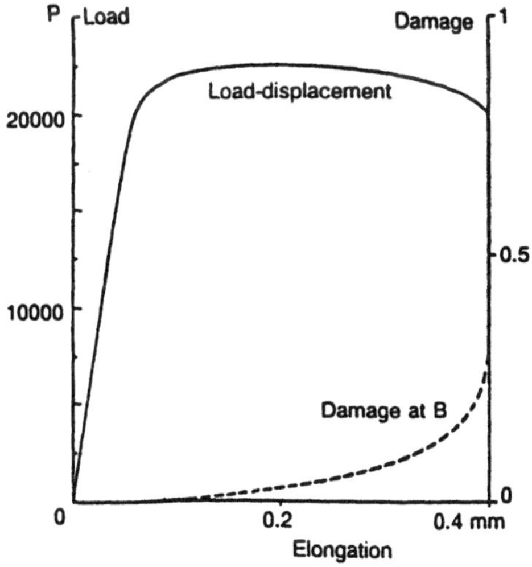

Fig. 4. Load-elongation curve and damage evolution

To avoid this strain localization problem, a new formulation is developed using a non local damage law.

IV NON LOCAL DAMAGE LAW

An other way to solve the problem raised before, consists in developing a material constitutive equation at the microscopic level using micromechanics models instead continuum mechanics. Micromechanics, in the field of viscoplasticity and damage, require the construction of models with a large number of parameters difficult to determine. This kind of approach is not compatible with our models nor with our numerical algorithms.
The easier solution to solve strain localization, in the field of continuum damage mechanics, consists in introducing a new parameter in the formulation. This parameter is an intrinsic material characteristic length.
Experimentally is possible to determine a length (a surface or a volume) in which take place the process of nucleation, growth and coalescence of microdefects. This volume is called the "process zone". For example, the process zone diameter of concrete is on 30 mm, and, 0.2 mm for steel 17-12 SPH at 600°C. Generally this diameter represents the size of the greatest material default (i.e. the aggregate for concrete or the grain for steel). If the size of the smallest mesh element used in the numerical simulation is greater than or equal to the process zone size, then the results will be good and any modification will be necessary in the formulation. In the other case, i.e. the smallest mesh element is less than the characteristic size, is necessary to transform the formulation acting on the parameter D. The non local damage formulation consists in averaging D over the characteristic volume with a Gaussian function. The new damage parameter $\tilde{D}$ must be written as follow :

$$\tilde{D}(x) = \frac{\int_v D(x,X)\, g(X)\, dv}{\int_v g(X)\, dv}$$

where :
$X \in v$ (characteristic volume)
$x \in \Omega$ (structure)
$g(X) = \exp\left[- \left(\frac{X}{d_{ch}} \right)^2 \right]$ (Gaussian function)
d_{ch} : characteristic volume diameter

This new value of damage will be introduced in the elastoviscoplastic equations.

V RESULTS

Numerical simulation with this new formulation, on the same specimen presented above shows a perfectly convergence of the damaged zone with the mesh. Figure 5 represents the damaged zone

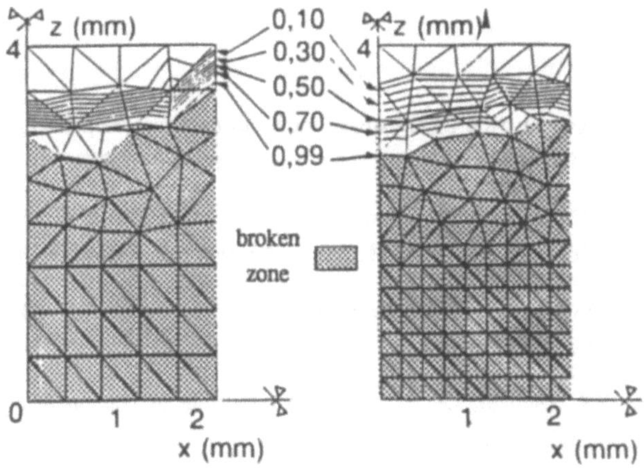

Fig. 5. Damage zones obtained with mesh I and II (nonlocal damage)

for 2 meshes. The size of the mesh doesn't have any influence on the damage repartition, the crack growth is then well described with this formulation.

Other materials and other geometries were tested, particularly notched specimen (structures with an initial concentration of stresses, CT specimen widely used in fracture mechanics). Comparison between these numerical results and experience shows a good agreement for local parameters (stresses, strains, damage ...) as well as for global parameters (strength, displacements ...) [7], [8].

VI CONCLUSION

Using local fracture criteria is possible to describe creep crack growth better than using a global approach for strongly dissipative materials with locally important non linearities. A fully coupled damage law is able to simulate the crack process, particularly the crack initiation but, as is true for any coupled local approach, the essential problem is the localization of the mechanical fields. Mathematical and numerical problems (i.e. non uniqueness of the solution and mesh dependency) prevent the crack growth simulation. In order to solve this problem it's sufficient to introduce a new intrinsic material parameter in the damage equation. The local approach, with a non local damage, can provide an excellent prediction of the structure lifetime with initial stress concentrations. However, is necessary to carry on with the study of the mathematical problem concerning the bifurcation point and the stability analysis of equations at this critical point.

REFERENCES

1. P.M. Lesne, K. Saanouni; Non local damage model to describe creep fracture in the framework of damage mechanics. La Recherche Aerospatiale, vol. 1, pp. 23-26, 1990.

2. K. Saanouni, J.L. Chaboche, P.M. Lesne; On the creep crack growth prediction by a non local damage formulation. European Journal of Mechanics A/solids, vol. 8 n° 6, pp. 437-459, 1989.

3. L.M. Kachanov; Time of the rupture process under creep conditions. Isv. Akad. Nauk. SSR, Otd Teckh. Nauk., n° 8, pp. 26-31, 1958.

4. D.R. Hayhurst, P.R. Dimmer, C.I. Morrison; Development of continuum damage in the creep rupture of notched bar. Phil. Trans. R. Soc. Lond. A311.

5. F.A. Leckie, D.R. Hayhurst; Creep rupture of structures. Phil. Trans. R. Soc. Lond. A240, 1974.

6. K. Saanouni, J.L. Chaboche, C. Bathias; On the creep growth prediction by a local approach. Eng. Fract. Mech., vol. 25, n° 6, 1986.

7. K. Saanouni; Sur l'analyse de la fissuration des milieux elastoviscoplastiques par la theorie de l'endommagement continu. These d'Etat es Sciences. Universite de Compiegne, 1988.

8. A. Diboine; Etude de la propagation de fissures en fatigue-fluage oxydation. These ENSM, 1988.

Small-Scale and Large-Scale Creep Predictions of Notch Stress and Strain Under Creep Conditions

Shiro KUBO and Kiyotsugu OHJI
Department of Mechanical Engineering,
Faculty of Engineering, Osaka University,
2-1, Yamadaoka, Suita, Osaka 565 JAPAN

Summary
Simple expressions are proposed for predicting the notch stress and strain, by applying the concept of the state of small-scale and large-scale creep conditions with the path-independence of the J-integral. A simple equation is obtained, which predicts a stress concentration factor varying with time under the small-scale creep conditions from the elastic stress concentration factor K^e_t. An ordinary differential equation is derived for estimating the stress concentration factor under the small-scale to large-scale creep conditions. Expressions for a strain concentration factor are also proposed. The applicability of the proposed predictions is demonstrated using finite element calculations.

1. Introduction

Evaluation of notch stress and strain under creep conditions is important in designing structures and their components for high temperature use, since cracks usually initiate at notches. Finite element method or other numerical methods may be used for the stress and strain evaluation, but a simple and convenient prediction scheme is preferable for that purpose. Many schemes [1-9] have been proposed for estimating notch stress under elastic-plastic conditions, while no scheme is available for predicting the notch stress and strain varying with time under creep conditions.

The present authors applied the concept of small-scale creep, which was introduced originally for a creep crack [10,11], for obtaining approximate expressions of the notch stress and strain under creep conditions [12-15]. The expressions have been shown to be valid under the small-scale creep conditions, where load duration is short and a creep-dominated region developing near the crack tip is well confined by an elasticity-dominated region.

In this paper expressions applicable under the small-scale creep and large-scale creep conditions are sought by extending the previous prediction methods. The applicability of the proposed expressions is examined by comparing the predictions with finite element calculations.

M. Zyczkowski (Ed.)
Creep in Structures
IUTAM Symposium Cracow/Poland 1990

2. Small-Scale and Large-Scale Creep Predictions

2.1 Hypothesis of Path-Independence of the J-Integral

Consider an elastic-nonlinear-viscous material, whose total strain is given by a sum of elastic and creep components:

$$\varepsilon = \varepsilon_{el} + \varepsilon_{cr} \tag{1}$$

The creep strain rate $d\varepsilon_{cr}/dt$ is assumed to be expressed by the Norton law,

$$d\varepsilon_{cr}/dt = A \cdot \sigma^n , \tag{2}$$

where n denotes creep exponent and A is a constant.

The J-integral [16] is defined for a two-dimensional notch having flat surfaces parallel to the x_1-axis (Fig. 1) as,

$$J = \int_{\Gamma} (W \nu_1 - \sigma_{ij} \nu_j \partial u_i / \partial x_1) d\Gamma , \tag{3}$$

where Γ denotes a path enclosing the notch root, σ_{ij} the stress tensor, ν_j the outward unit normal to Γ, u_i the displacement vector, and $d\Gamma$ the line increment of Γ. The strain energy density denoted by W may be defined by an integral along a stress-strain trajectory as,

$$W = \int_{history} \sigma \, d\varepsilon . \tag{4}$$

Numerical results have shown that the path-independence of the J-integral holds well for creep cracks under the small-scale creep conditions [10,17,18]. The path-independence of the modified J-integral J^* (C^*

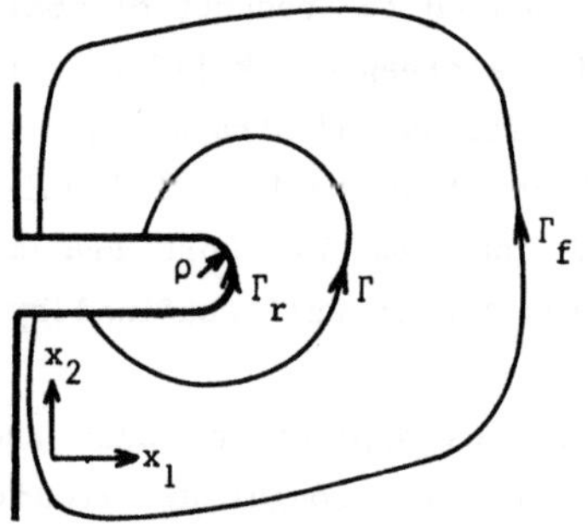

Fig.1. A notch and integration paths of the J-integral

parameter) [19-21] implies that the path-independence of the J-integral holds under the large-scale creep conditions, where the creep-dominated region develops and the stress state approaches to that under the steady-state creep conditions. The path-independence of the J-integral for a notch is then expected to hold well for the whole time range from the small-scale creep to the large-scale creep. The hypothesis of the path-independence of the J-integral can be expressed as,

$$J(\Gamma_r, t) = J(\Gamma_f, t), \tag{5}$$

where $J(\Gamma, t)$ denotes the value of the J-integral evaluated along a path Γ at time t, and Γ_r and Γ_f are paths taken along the notch surface and in the region far from the notch root, respectively.

2.2 Evaluation of the J-Integral along Γ_r and Γ_f

Strain energy density W in the first term of the integrand of the J-integral may be expressed as a sum of elastic and creep components as,

$$W = W_{el} + W_{cr} = H\sigma^{*2}/(2E) + \sigma^* \varepsilon^*_{cr}[n/(n+1)], \tag{6}$$

where σ^* and ε^*_{cr} represent the Mises type equivalent stress and equivalent creep strain, respectively, and H is defined in terms of Poisson's ratio ν as,

$$H = 1 \quad \text{(for plane stress)} \tag{7}$$

$$H = (1-\nu^2)/(1-\nu+\nu^2) \quad \text{(for plane strain)} \tag{8}$$

Since the distribution of equivalent stress σ^* and equivalent creep strain ε^*_{cr} on Γ_r can be expressed as their maximum values $\sigma_r^*(t)$ and $\varepsilon_r{}^*_{cr}(t)$ at the notch root multiplied by some distribution functions, substitution of Eq. (6) and the condition that the traction is free ($\sigma_{ij}\nu_j = 0$) into Eq. (3) leads to the following expression for $J(\Gamma_r, t)$ [14,15].

$$J(\Gamma_r, t) = \Phi_1 \rho H \sigma_r{}^{*2}/(2E) + \Phi_2 \rho\, \sigma_r{}^* \varepsilon_r{}^*_{cr}[n/(n+1)] \tag{9}$$

where Φ_1 and Φ_2 are constants dependent on the distribution functions of σ^* and ε^*_{cr} along the notch surface, and ρ being the radius of notch root curvature.

The J-value $J(\Gamma_f, t)$ evaluated for path Γ_f far from the notch root may be composed of elastic and creep terms, and the latter can be given as the modified J-integral J^*_{st} under the steady-state conditions multiplied by t:

$$J(\Gamma_f, t) = J_{el} + J_{pl} = J_{el} + J^*_{st} \cdot t \tag{10}$$

By substituting Eqs. (9) and (10) into Eq. (5) one has

$$J_{el} + J^*_{st} \cdot t = \Phi_1 \rho H \sigma_r{}^{*2}/(2E) + \Phi_2 \rho \sigma_r{}^* \varepsilon_r{}^*{}_{cr} [n/(n+1)]. \tag{11}$$

Since the second terms in the left and right hand sides of Eq. (11) vanish at $t = 0$, J_{el} is given as,

$$J_{el} = \Phi_1 \rho H \{\sigma_r{}^*(0)\}^2/(2E). \tag{12}$$

For very large t, where the first terms in Eq. (11) are small compared with other terms, Eq. (11) with Eq. (2) leads to

$$J^*_{st} \cdot t = \Phi_2 \rho A\{\sigma_r{}^*(\infty)\}^{n+1} \cdot t[n/(n+1)], \tag{13}$$

where $\sigma_r{}^*(\infty)$ denotes the steady-state value of the notch stress. This equation can be used to estimate $\sigma_r{}^*(\infty)$ from the value of J^*_{st}:

$$\sigma_r{}^*(\infty) = [J^*_{st}(n+1)/(n \cdot \Phi_2 \rho A)]^{1/(n+1)}$$

By substituting Eqs. (12), (13) and (2) into Eq. (11), the following relation is obtained.

$$\{\sigma_r{}^*(0)\}^2 + \alpha EA\{\sigma_r{}^*(\infty)\}^{n+1} \cdot t$$
$$= \{\sigma_r{}^*(t)\}^2 + \alpha EA \sigma_r{}^*(t) \int_0^t \{\sigma_r{}^*(\xi)\}^n d\xi \tag{14}$$

where α is given by

$$\alpha = \alpha'[2n/(n+1)]/H$$
$$\alpha' = \Phi_2/\Phi_1 \tag{15}$$

By examining the distributions of stress and creep strain under plane stress condition, the following values of α' have been proposed [13,15].

$$\alpha' = 5/4$$
$$\alpha' = 9\pi/32 \tag{16}$$

2.3 Small-Scale Creep (SSC) Prediction

A non-dimensional time τ is defined in terms of t, nominal stress σ_{nom} and material constants as,

$$\tau = EA \sigma_{nom}{}^{n-1} t \tag{17}$$

Stress concentration factor K^*_σ at non-dimensional time τ is defined as,

$$K^*_\sigma = \sigma_r{}^*/\sigma_{nom}. \tag{18}$$

The initial value of K^*_σ is equal to the theoretical elastic stress concentration factor K^*_t, and the steady-state value of K^*_σ is denoted by K^*_{st}:

$$\begin{aligned} K^*_t &= \sigma_r{}^*(0)/\sigma_{nom} \\ K^*_{st} &= \sigma_r{}^*(\infty)/\sigma_{nom} \end{aligned} \tag{19}$$

By using stress concentration factors K^*_σ, K^*_t, K^*_{st} and the non-dimensional time τ, Eq. (14) is rewritten as,

$$K^*_t{}^2 + \alpha\ K^*_{st}{}^{n+1}\cdot\tau = K^*_\sigma{}^2 + \alpha\ K^*_\sigma \int_0^\tau K^*_\sigma{}^n\ d\tau'. \tag{20}$$

Equation (20) means that the behavior of stress concentration factor K^*_σ is predicted using the values of K^*_t and K^*_{st}.

Under the small-scale creep conditions, τ is small and the second term on the left hand side of Eq. (20) can be neglected. Then differentiation of Eq. (20) and some rearrangements lead to

$$[K^*_t{}^2/K^*_\sigma{}^{n+1} - K^*_t{}^{1-n}]/(n+1) + [K^*_\sigma{}^{1-n} - K^*_t{}^{1-n}]/(n-1) = \alpha\tau. \tag{21}$$

When the values of K^*_t and τ are given, this equation can be solved for the stress concentration factor K^*_σ.

If the strain rate concentration factor $K^*_{\dot{\varepsilon}}$ is defined using the rate of equivalent strain $\dot{\varepsilon}_r{}^*$ as

$$K^*_{\dot{\varepsilon}} = \dot{\varepsilon}_r{}^*/(A\cdot\sigma_{nom}{}^n), \tag{22}$$

$K^*_{\dot{\varepsilon}}$ can be given by the following equation.

$$K^*_{\dot{\varepsilon}} = K^*_\sigma{}^n \tag{23}$$

The strain concentration factor K^*_ε may be defined as,

$$K^*_\varepsilon = \varepsilon_r{}^*/(\sigma_{nom}/E), \tag{24}$$

where $\varepsilon_r{}^*$ denotes the equivalent strain at the notch root. It may be assumed that the total equivalent strain $\varepsilon_r{}^*$ at the notch root consists of elastic and creep components:

$$\varepsilon_r{}^* = (2/3)(1+\nu)\sigma_r{}^*/E + \int_0^t A\{\sigma_r{}^*(\xi)\}^n\, d\xi \tag{25}$$

Then, K^*_ε can be expressed as,

$$K^*_\varepsilon = (2/3)(1+\nu)K^*_\sigma + \int_0^\tau K^*_\sigma{}^n\ d\tau' \tag{26}$$

Since the second term of K^*_ε in Eq. (26) is expressed in terms of K^*_t and K^*_σ using Eq. (20), K^*_ε can be written as,

$$K^*_\varepsilon = (2/3)(1+\nu)K^*_\sigma + [K^*_t{}^2/K^*_\sigma - K^*_\sigma]/\alpha. \tag{27}$$

With the value of K^*_σ predicted by Eq. (21), Eq. (27) predicts the behavior of K^*_ε.

The prediction of K^*_σ and K^*_ε by Eqs. (21) and (27) is called small-scale creep (SSC) prediction.

2.4 Small-Scale to Large-Scale Creep (SLSC) Prediction

Under the large-scale creep conditions, where creep-dominated region is not confined, and the transient condition between the small-scale creep and large-scale creep conditions, the second term on the left hand side of Eq. (20) can not be neglected. Differentiation of Eq. (20) leads to

$$dK^*_\sigma/d\tau=\{K^*_{st}{}^{n+1}/K^*_\sigma - K^*_\sigma{}^n\}/[(1+K^*_t{}^2/K^*_\sigma{}^2)/\alpha + K^*_{st}{}^{n+1}\cdot\tau/K^*_\sigma{}^2]. \tag{28}$$

By solving numerically this differential equation with the initial condition $K^*_\sigma = K^*_t$ at $\tau = 0$, the K^*_σ value varying with time can be predicted from the values of K^*_t, K^*_{st} and τ. The strain concentration factor is estimated by combining Eq. (26) with Eq. (20) as,

$$K^*_\varepsilon = (2/3)(1+\nu)K^*_\sigma + [(K^*_t{}^2 + \alpha\ K^*_{st}{}^{n+1}\cdot\tau)/K^*_\sigma - K^*_\sigma]/\alpha \tag{29}$$

The prediction of K^*_σ and K^*_ε by Eqs. (28) and (29) is called small-scale to large-scale creep (SLSC) prediction.

3. Finite Element Analyses for Assessing the SSC and SLSC Predictions

To assess the applicability of the proposed prediction method, finite element stress and strain calculations based on the initial strain method were made for center notched plate specimens shown in Fig. 2 under plane stress condition. A quarter of the specimen was modeled using about 1500 triangular elements.

As an example, Fig. 3 illustrates variation of K^*_σ with τ for n = 5, a/w = 0.3, and ρ/w = 0.02, where a is the half length of the notch and w is the half plate width. The ordinate on the right of the figure makes it possible to read $K^*_{\dot\varepsilon}-\tau$ relation from $K^*_\sigma-\tau$ relation, using the power relation between $K^*_{\dot\varepsilon}$ and K^*_σ given by Eq. (23). In the figure finite element results are shown by open symbols. The SSC (small-scale creep) predictions with α' = 5/4 and $9\pi/32$ are shown by chained and dashed lines, respectively. The dotted line is obtained using the SLSC (small-scale to

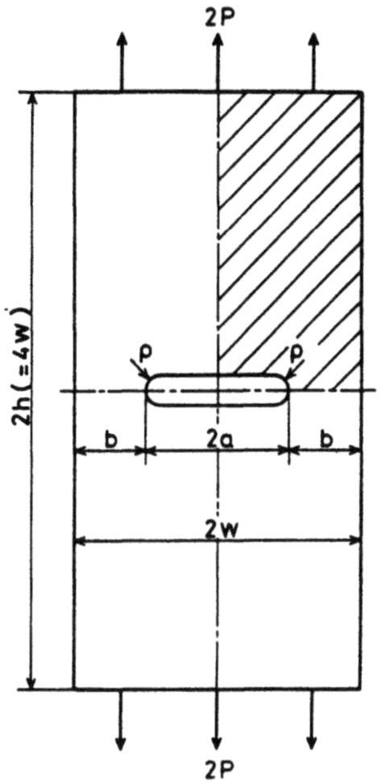

Fig.2. Specimen geometries used for finite element calculations

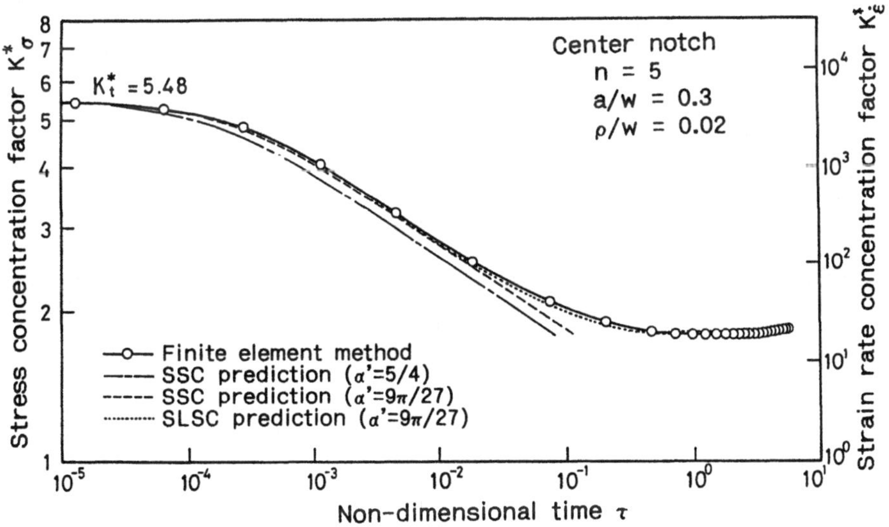

Fig.3. Comparison of SSC (small-scale creep) and SLSC (small-scale to large-scale creep) predictions with finite element results

large-scale creep) prediction with $\alpha' = 9\pi/32$. The values of K^*_σ for $\tau = 0$ and for a large τ value obtained by the finite element analyses are used as K^*_t and K^*_{st} in these predictions.

The figure shows that for short load duration where the small-scale creep condition is satisfied, the SSC prediction with $\alpha' = 9\pi/32$ agrees well with the results of the finite element analyses, while the prediction deviates from the finite element results for τ values larger than 0.02. The SLSC prediction gives a good approximation of K^*_σ for the entire time range.

Calculations made for other configurations also demonstrate the applicability of the SSC and SLSC predictions.

References

1. Neuber, H.: *J. Appl. Mech.* **28** (1961) 544-550.
2. Conle, A.; Nowack, H.: *Exp. Mech.* **17** (1977) 57-63.
3. Seeger, T; Heuler, P.: *J. Test. Eval.* **8** (1980) 199-204.
4. Ogura, K.; Miki, N.; Ohji, K.: *Trans. Japan. Soc. Mech. Engrs., Ser. A* **47** (1981) 55-62.
5. Gemma, A.E.: *Eng. Fract. Mech.* **21** (1985) 495-501.
6. Taira, S.; Ohtani, R.; Ito, T.: *Proc. 14th Japan Congr. on Mats. Res.*, Soc. Mats. Sci., Japan (1971) 158-164.
7. Leis, B.N; Gowda, C.V.B.; Topper, T.H.: *J. Test. Eval.* 1 (1973) 341-348.
8. Molski, K.; Glinka, G.: *Mats. Sci. Eng.* **50** (1981) 93-100.
9. Glinka, G.: *Eng. Fract. Mech.* **22** (1985) 839-854.
10. Ohji, K.; Ogura, K.; Kubo, S.: *J. Soc. Mats. Sci., Japan* **29** (1980) 465-471.
11. Riedel, H.; Rice, J.R.: *ASTM STP* 700 (1980) 112-130.
12. Kubo, S.; Ohji, K.: *Preprint of the 31st Annual Meet. of Soc. Mats. Sci., Japan* (1982) 261-263.
13. Ohji, K.; Kubo, S.; Hirahara, H.: *Trans. Japan. Soc. Mech. Engrs., Ser. A* **50** (1984) 1583-1590.
14. Kubo, S.; Ohji, K.: *Int. Conf. on Creep*, Japan. Soc. Mech. Engrs. (1986) 417-422.
15. Kubo, S.; Ohji, K.; Kumamoto, H.: *Proc. ASME 1989 PVP Conf., Structural design for elevated temperature environments- Creep, ratchet, fatigue, and fracture.* PVP **163** (1989) 95-100.
16. Rice, J.R.: *J. Appl. Mech.* **35** (1968) 379-386.
17. Bassani, J.L.; McClintock, F.A.: *Int. J. Solids Struct.* 17 (1981) 479-492.
18. Biner, S.B.; Wilkinson, D.S.; Watt, D.: *Eng. Fract. Mech.* **21** (1985) 315-328.
19. Ohji, K; Ogura, K.; Kubo, S.: *Preprint of Japan. Soc. Mech. Engrs.* 740-11 (1974) 207-209
20. Ohji, K.; Ogura, K.; Kubo, S.: *Trans. Japan. Soc. Mech. Engrs.* **42** (1976) 350-357.
21. Landes, J.D.; Begley, J.A.: *ASTM STP* 590 (1976) 128-148.

Creep Rupture of Metals Under Tri-Axial State of Stress

A.LITEWKA and Z.LIS

Technical University of Poznań,
Institute of Building Structures,
Poznań, Poland

Summary

The aim of this paper is to present a theoretical model describing the creep damage for a multi-axial state of stress. The creep rupture criterion proposed consists of the damage evolution equation and the failure criterion for material with deteriorated structure. The emphasis is given to the solution of the damage evolution equation formulated for the tri-axial state of stress in the form of a differential equation. The possible application of the equations derived is shown by using the results of the tests performed on the tubular specimens subjected to an internal pressure and axial load where a pronounced tri-axial state of stress is produced.

Introduction

The actual knowledge of the creep damage and creep rupture process for metals which operate at elevated temperature depends mainly on the quality of the experimental data available. The great number of the experimental results for an uniaxial tension [1,2] made it possible to formulate the appropriate theories used to predict the material behaviour. These theories proposed by Hayhurst [3] and Murakami and Ohno [4] are also used for multi-axial state of stress and, in fact, these theories are the generalization of the classical Kachanov's theory [5] formulated for creep rupture prediction in uniaxial tension.

The aim of this paper is to show how the creep rupture criterion proposed by one of the authors [6,7] can be applied in description of the experimental results obtained for the specimens subjected to tri-axial state of stress. The similar problem was discussed by Hayhurst and Leckie [8], Dyson and Loveday [9] but their considerations were confined to the tri-

M. Zyczkowski (Ed.)
Creep in Structures
IUTAM Symposium Cracow/Poland 1990

axial state of stress obtained in the notched bars designed by Bridgman [10]. Creep rupture process under tri-axial loading was also analysed by Hayhurst and Felce in [11] where special experimental technique was developed. Some preliminary results concerning the Bridgman notched specimens are also presented in this paper but the main attention is paid to the tri-axial state of stress produced in tubular specimens subjected to an internal pressure and axial load. The creep rupture experimental data collected for steels by Browne, Lonsdale and Flewitt [12] are compared with the theoretical predictions obtained from the theoretical model presented in [6,7] and from formulae derived in this paper.

Theoretical model

The detailed description of the theoretical model used in this paper together with its physical background is presented in [6] that is why only final form of the basic equations will be shown here. It is assumed that the creep damage growth is governed by the damage evolution equation

$$\dot{\underset{\sim}{\Omega}} = C\Phi_e^2 \underset{\sim}{\sigma}^* \tag{1}$$

where $\dot{\underset{\sim}{\Omega}}$ is a time derivative of the Murakami-Ohno tensorial damage measure [4], C stands for material constant and $\underset{\sim}{\sigma}^*$ denotes modified stress tensor whose compressive principal stresses are replaced by zeros, whereas tensile principal stress components are left unchanged. The scalar multiplier Φ_e is the elastic strain energy density of the damaged solid expressed in the following form

$$\Phi_e = \frac{1-2\nu}{6E}\, \mathrm{tr}^2\underset{\sim}{\sigma} + \frac{1+\nu}{2E}\, \mathrm{tr}\underset{\sim}{S}^2 + \frac{D_1}{2(1+D_1)E}\, \mathrm{tr}\underset{\sim}{\sigma}^2\underset{\sim}{D} \tag{2}$$

where E and ν are Young's modulus and Poisson's ratio of the undamaged material, and $\underset{\sim}{\sigma}$ and $\underset{\sim}{S}$ are the stress tensor and stress deviator. The second rank symmetric damage effect tensor $\underset{\sim}{D}$ is associated with damage tensor $\underset{\sim}{\Omega}$ by means of their principal values

$$D_i = \frac{\Omega_i}{1-\Omega_i} \qquad ; \qquad i = 1,2,3$$

According to the theoretical model used in this paper creep rupture of the material occurs when damage amount, represented by $\underset{\sim}{\Omega}$ or $\underset{\sim}{D}$ growing in the material subjected to the given state of stress $\underset{\sim}{\sigma}$, satisfies the failure criterion

$$C_1 \mathrm{tr}^2 \underset{\sim}{\sigma} + C_2 \mathrm{tr} \underset{\sim}{S}^2 + C_3 \mathrm{tr} \underset{\sim}{\sigma}^2 \underset{\sim}{D} - \sigma_u^2 = 0 \tag{3}$$

where C_1, C_2 and C_3 are temperature and damage dependent material constants to be determined from the simple tests described in [6] and σ_u is the temperature dependent uniaxial ultimate strength of the undamaged material.

Application of the creep rupture criterion represented by Eqs (1) and (3) in the case of uniaxial and biaxial state of stress can be found in [6,7,13]. An integrated form of the damage evolution equation for tri-axial state of stress depends on the state of stress. The following three solutions can be distinguished:

$$-\frac{1}{2\,b} \ln\left| \frac{b}{a}\,\Omega_1^2 - \Omega_1 + 1 \right| + \frac{2b-a}{b\sqrt{\Delta}} \left(\arctan\frac{2b\Omega_1-a}{\sqrt{\Delta}} - \right.$$

$$\left. - \arctan\frac{-a}{\sqrt{\Delta}} \right) = k\sigma_1^5 t \tag{4}$$

for $\Delta = 4ab - a^2 > 0$,

$$-\frac{1}{2\,b} \ln\left| \frac{b}{a}\,\Omega_1^2 - \Omega_1 + 1 \right| + \frac{a-2b}{b(2b\Omega_1-a)} + \frac{a-2b}{ab} = k\sigma_1^5 t \tag{5}$$

for $\Delta = 0$, and

$$-\frac{1}{2\,b} \ln\left| \frac{b}{a}\,\Omega_1^2 - \Omega_1 + 1 \right| + \frac{a-2b}{b\sqrt{-\Delta}} \left(\mathrm{arth}\frac{2b\Omega_1-a}{\sqrt{-\Delta}} - \right.$$

$$\left. - \mathrm{arth}\frac{-a}{\sqrt{-\Delta}} \right) = k\sigma_1^5 t \tag{6}$$

for $\Delta < 0$

where: $k = C/4E^2$

$$\sqrt{a} = M = \frac{2(1+\nu)}{3}\left[(1 + m_1^2 + m_2^2) - (m_1 + m_1 m_2 + m_2)\right] +$$

$$+ \frac{1-2\nu}{3}(1 + m_1 + m_2)^2$$

$$b = 2M(1 + n_1 m_1^2 + n_2 m_2^2)$$

$$m_1 = \sigma_2/\sigma_1 \quad , \quad m_2 = \sigma_3/\sigma_1$$

$$n_1 = \Omega_2/\Omega_1 = \begin{cases} m_1 & \text{for } 0 \le m_1 < 1 \\ 0 & \text{for } m_1 < 0 \end{cases}$$

$$n_2 = \Omega_3/\Omega_1 = \begin{cases} m_2 & \text{for } 0 \le m_2 < 1 \\ 0 & \text{for } m_2 < 0 \end{cases}$$

The failure criterion (3) and Eqs (4), (5), (6) make it possible to calculate the creep rupture time for an arbitrary tri-axial state of stress.

Experimental verification

Creep rupture data presented by Browne, Lonsdale and Flewitt [12] were obtained for tubular specimens with the ratio of the outer radius to the wall thickness equal to 4.22. It means that internal pressure and axial load produce in such specimens a pronounced tri-axial state of stress with radial and circumferential stresses varying through the wall thickness. It was proposed by Hayhurst [14] that in the case of nonhomogeneous state of stress the creep rupture is governed by a mean stress in the specimen. Following this line the theoretical creep rupture time was calculated from Eqs (3) and (4) for the tri-axial state of stress in these points of the wall thickness where the mean value of the circumferential stress is observed. The comparison of the theoretical results with experimental data for 1%Cr$\frac{1}{2}$%Mo and 12%Cr1%MoVW steels is shown in Figs 1 and 2. Broken lines added in these figures represent theoretically determined curves for two specific states of stress: pure shear ($m_1 = -1$, $m_2 = 0$) and equal biaxial tension ($m_1 = 1$, $m_2 = 0$).

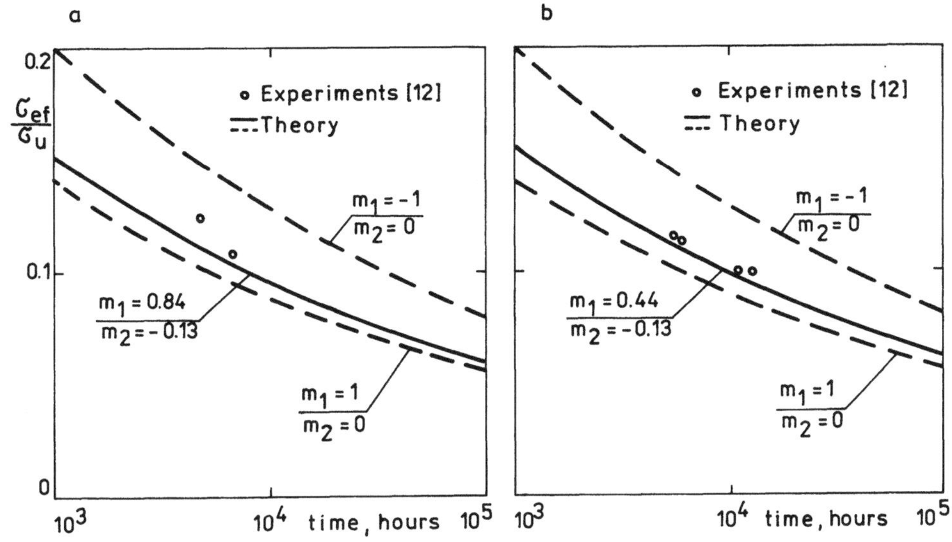

Fig.1. Creep rupture time versus Huber-Mises effective stress for 1%Cr$\frac{1}{2}$%Mo steel at temperature of 848K under tri-axial state of stress: a) $m_1 = 0.84$, $m_2 = -0.13$, b) $m_1 = 0.44$, $m_2 = -0.13$

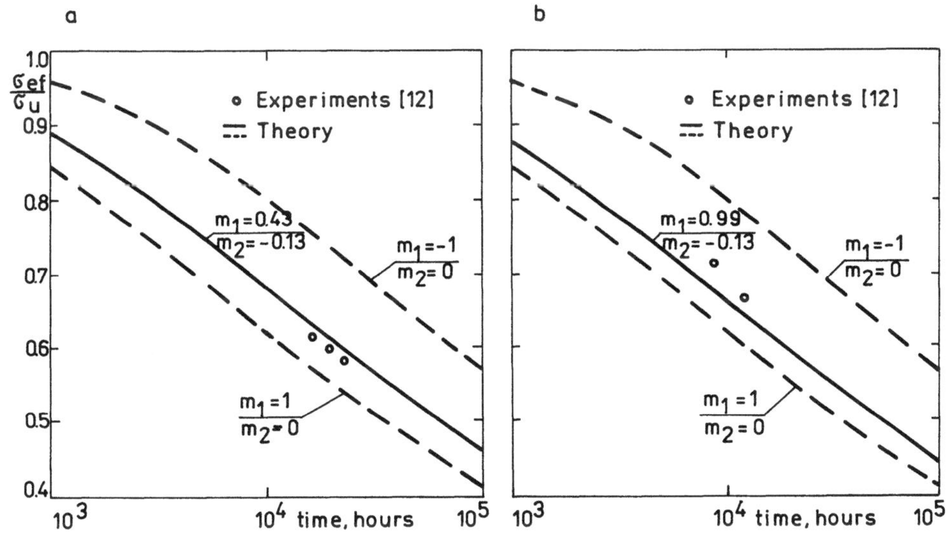

Fig.2. Creep rupture time versus Huber-Mises effective stress for 12%Cr1%MoVW steel at temperature of 848K under tri-axial state of stress: a) $m_1 = 0.43$, $m_2 = -0.13$, b) $m_1 = 0.99$, $m_2 = -0.13$

Creep rupture experimental data for the Bridgman notched bar made of Nimonic 80A and obtained at temperature of 1023K by Dyson and Loveday [9] are shown in Fig.3. Solid line shown in this figure is obtained from Eqs (1) and (3) for the Bridgman effective stress $\sigma_{ef} = 0.72\sigma_a$ where σ_a is the net section stress at notch throat.

The theoretical curves shown in Figs 1, 2 and 3 were obtained for the material constants presented in Table 1. The uniaxial

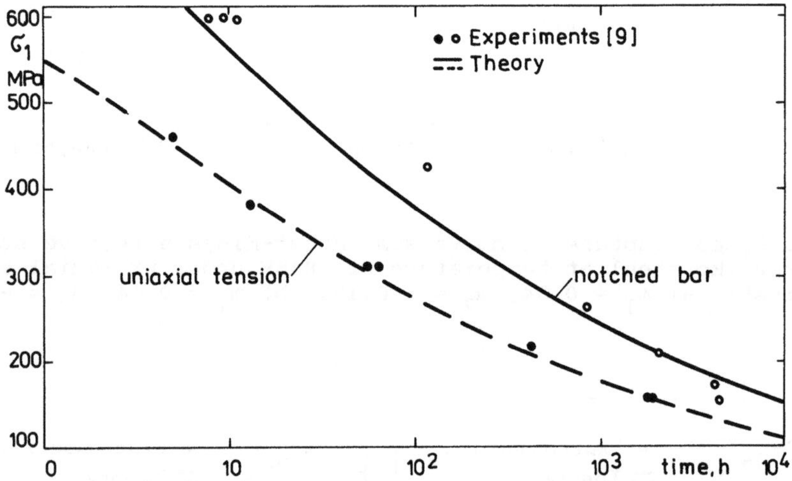

Fig.3. Creep rupture time versus maximum principal stress for Nimonic 80A at temperature of 1023K

Table 1. Material constants

Material	Temperature K	σ_u MPa	k $1/MPa^5h$	ν
12%Cr1%MoVW steel	848	196	$6.16\cdot10^{-16}$	0.3
1%Cr$\frac{1}{2}$Mo steel	848	600	$5.56\cdot10^{-14}$	0.3
Nimonic 80A	1023	650	$3.03\cdot10^{-15}$	0.3

experimental creep rupture results for these materials presented in [9,12] were used to determine the constant k according to rules described in [6]. An example of such uniaxial results together with the relevant theoretical curve is shown for Nimonic 80A in Fig.3.

Conclusions

Satisfactory agreement of the theoretical and experimental result observed in Figs 1 and 2 corroborates the validity of the theoretical model proposed. The preliminary theoretical results for Bridgman notched bar presented in Fig.3 show that Eqs (1) and (3) could be used to describe the creep rupture process under tri-axial tension. However, this requires further studies on the stress distribution at a notch throat as well as further analysis of the failure criterion (3) for material subjected to tri-axial tension.

References

1. Conway, J.B.: Stress-rupture parameters: origin, calculation and use. New York: Gordon and Breach 1969.

2. Goldhoff, R.M.: Towards the standardization of time-temperature parameter usage in elevated temperature data analysis. J. Test. Eval. 2 (1979) 387-424.

3. Hayhurst, D.R.: Creep rupture under multi-axial state of stress. J. Mech. Phys. Solids. 20 (1972) 381-390.

4. Murakami, S.; Ohno, N.: A continuum theory of creep and creep damage. In: Creep in Structures. A.R.S. Ponter; D.R. Hayhurst (eds.), Berlin: Springer Verlag 1981, 422-444.

5. Kachanov, L.M.: On time to rupture in creep conditions. Izv. Akad. Nauk S.S.S.R., O.T.N., 8 (1958) 26-31, (in Russian).

6. Litewka, A.: Creep rupture of metals under multi-axial state of stress. Arch. Mech. 41 (1989) 3-23.

7. Litewka, A.: Analytical and experimental study of fracture of damaging solids. In: Yielding, Damage and Failure of Anisotropic Solids. J.P. Boehler (ed.), London: Mechanical Engineering Publ. 1989, 653-663.

8. Hayhurst, D.R.; Leckie, F.A.: Design of notched bars for creep-rupture testing under tri-axial stress. Int. J. Mech. Sci. 19 (1977) 147-159.

9. Dyson, B.F.; Loveday, M.S.: Creep fracture in Nimonic 80A under triaxial tensile stressing. In: Creep in Structures. A.R.S. Ponter; D.R. Hayhurst (eds.), Berlin: Springer Verlag 1981, 406-421.

10. Bridgman, P.W.: Studies in large plastic flow and fracture. New York: McGraw Hill 1952.

11. Hayhurst, D.R.; Felce, I.D.: Creep rupture under tri-axial tension. Eng. Fract. Mech. 25 (1986) 645-664.

12. Browne, R.J.; Lonsdale, D.; Flewitt, P.E.J.: The role of stress state on the creep rupture of 1%Cr$\frac{1}{2}$%Mo and 12%Cr1%MoVW tube steels. In: Creep and Fracture of Engineering Materials and Structures. B. Wilshire, D.R.J. Owen (eds.), Swansea: Pineridge Press 1981, 545-558.

13. Litewka, A.; Hult, J.: One parameter CDM model for creep rupture prediction. Eur. J. Mech., A/Solids. 8 (1989) 185-200.

14. Hayhurst, D.R.: Stress redistribution and rupture due to creep in a uniformly stretched thin plate containing a circular hole. Trans. ASME, J. Appl. Mech. 40 (1973) 244-250.

The Investigation of the Metal Damage at the Creep by the Method of Electrical Resistance Measuring

A.M. LOKOSHTCHENKO

Institute of Mechanics,
University of Moscow, USSR

Summary

In this paper the method of damage measuring of metal structure for the creep is proposed. In contrast to other methods, it allows to carry the measuring of this damage in the process of test for creep without unloading and cooling of specimens. Experimental damage curves for creep are obtained as a result of calculating processing of test data by use the suggested method. The analysis of these curves allows to obtain a conclusion that the material damage for rupture is monotonically decreasing function of applied stress. This conclusion is experimental confirmation of the theoretical results, obtained earlier.

Introduction

One of the basic problems of estimation of strength of constructions exploited at high temperature, is the determination of rupture time for these constructions. The study of creep of metals up to rupture by use of phenomenological method of two mechanisms are usually considered: ductile and brittle. The ductile rupture is appeared for the process of creep which is characterized by constant structure of material the the large deformations. The hypothesis of brittle fracture is applied for analysis of creep rupture which is appeared on account of the structure change of tested materials. In this paper the physical parameter which characterizes the structure change of metals for creep, and the method of the measuring of this parameter are suggested.

The method of measuring of metal damage at the process of creep test

The certain experimental methods of damage ω study do not give a possibility to carry out measuring of structure damage for the process of creep test. At the measuring of value ω by these methods it is necessary to stop a test; for the use of the metallographical methods it is necessary besides to cut specimen.

In this paper the method of the measuring of the damage of metal structure for the creep is proposed; this method allows to carry this measuring in the process of test for high temperature without unloading and cooling of specimens. In order to receive the function $\omega(t)$ it should be measured the electrical resistance $R(t)$ of cylindrical specimens

M. Zyczkowski (Ed.)
Creep in Structures
IUTAM Symposium Cracow/Poland 1990

for pure tension and the length $l(t)$ of tested specimens at the same time t. The electrical resistance of tensional cylindrical specimens for the tests process increases on account of two reasons. The first reason consists in the increase of specimen length and in the corresponding decrease of its cross-section. The second reason of increase of $R(t)$ is explained by damage of material structure which is characterized by appearance of voids and microcracks, by their accumulation and by confluence right up to formation of the main crack. Let us think that these two factors act mutually independently. The electrical resistance $R(t)$ of the loaded specimen is possible to represent as the sum of 5 terms:

$$R(t) = R_0 + \Delta R_1 + \Delta R_2 + \Delta R_3(t) + \Delta R_4(t) \ , \tag{1}$$

R_0 - the electrical resistance of unloaded specimen at the tested temperature T; ΔR_1 and ΔR_2 are the increments of electrical resistance for the momentary loading obtained correspondingly in consequence of solid specimen lengthening and of appearance of structure defects; $\Delta R_3(t)$ and $\Delta R_4(t)$ - the increments of electrical resistance for the creep, also obtained in consequence of this factors.

At first we consider the case of absence of microdefects in specimen. In this case we take into account the following conditions: the constancy of the specimen volume for homogeneous deformation (till appearance of the neck), the constancy of the specific resistance of material ϱ for the process of test, the direct proportionality of the electrical resistance R of specimen to his length and the inverse proportionality of to the area of specimen cross-section. In this case we obtain from (1):

$$R(t) = R_0 + \Delta R_1 + \Delta R_3(t) = \frac{\varrho l(t)}{F(t)} = \frac{\varrho l_0}{F_0}\left(\frac{l}{l_0}\right)^2 = R_0\left(1 + \frac{\Delta l_0}{l_0} + \frac{\Delta l_t}{l_0}\right)^2 \ , \tag{2}$$

$$l_0 = l_{00}(1 + \alpha T) \ , \tag{3}$$

where l_{00} and l_0 are the lengths of the unloaded specimen at room temperature and at tested temperature T, F_0 and $F(t)$ are the areas of the cross-section of unloading rod at temperature T for initial time ($t = 0$) and for arbitrary time respectively; α is the coefficient of the linear temperature expansion, Δl_0 is the lengthening of specimen in consequence of momentary loading, Δl_t is the lengthening accumulated during the creep process. We obtain from (2):

$$\Delta R_1 = \frac{2R_0\Delta l_0}{l_0}\left(1 + \frac{\Delta l_0}{2l_0}\right) \ , \quad \Delta R_3(t) = \frac{2R_0}{l_0}\Delta l_t\left(1 + \frac{\Delta l_0}{l_0} + \frac{\Delta l_t}{2l_0}\right) \ . \tag{4}$$

We obtain from (1) for the case

$$R_0 + \Delta R_1 + \Delta R_2 = \tilde{R}_0 \ , \tag{5}$$

where $\tilde{R}_0$ is the value of the electrical resistance of specimen just after the loading, this value is measured in tests. The next equation can be obtained for $\Delta R_4(t)$, if substitute (4) and (5) into (1):

$$\Delta R_4(t) = R(t) - \tilde{R}_0 - \Delta R_3(t) \ . \tag{6}$$

The value $\Delta R_4(t)$ determines the increment of the electrical resistance, characterizing the structure damage of specimen during the creep process. We call the ratio

$$\omega(t) = \Delta R_4(t)/R_0 \ , \tag{7}$$

as the measure of the structure damage.

The utilization of the suggested method for analysis of test results

The results of tests of the cylindrical copper specimens [1] are utilized as an example of application of suggested method. Test in [1] were carried out at temperature 400°C under constant tensional force; initial stresses σ_0 were equal 40, 50, 60 and 70 MPa (longitudinal strength of material of these specimens at temperature 400°C was equal 120 MPa). Two measures of strain were considered for the analysis of experimental data in [1]: the usual strain measure

$$\varepsilon(t) = \ln \frac{l(t)}{l_0} \ , \tag{8}$$

utilizing the data about the lengthening of specimens for creep process, and

$$\varepsilon_R = 0.5 \ln \frac{R(t)}{R_0} \ , \tag{9}$$

using the results of measuring of electrical resistance of these specimens. It was supposed in (9) that microdefects are absent in specimens. Values l_0 and R_0 in equations (8) and (9) are the length and the electrical resistance of specimen at tested temperature (400°C) before the loading respectively. The creep curves $p(t)$ and $p_R(t)$, showed in [1], are obtained by use of equations (8) and (9) correspondingly. It is showed, that at first the creep curves, obtained by the two different methods, practically coincide. Beginning with a certain time, the creep strain $p_R(t)$, measured by the electrical resistance method, increases greater than the strain $p(t)$ found from the geometric elongation of the specimen. The analysis of experimental data showed that creep for high stresses (from considered diapason) is realized mainly by shear deformations across the body of the grains, and therefore the difference of $p_R(t)$ and $p(t)$ is relatively small. The difference of these curves increases for the low stresses; the creep process for low stresses is characterized by the early development of voids and cracks along grain boundaries, by embrittlement and by subsequent intergranular rupture.

The experimental curves $l(t)$ and $R(t)$ were obtained in tests for every specimen, and then they were used for calculating of the dependence of the damage parameter $\omega(t)$ from time t for various stresses. The results of utilization of the proposed method for the analysis of experimental data, corresponding to rupture time T^*, for all 8 tested specimens are given in the Table 1. Values σ_0 in the Table 1 are measured in MPa, t^* - in hours, Δl_t^* in mm, all values of electrical resistance - in mikroohm. Values of all parameters,

Table 1: The results of analysis of experimental data [1] by use of the proposed method

σ_0	t^*	R_0	$\tilde{R}_0$	R^*	Δl_t^*	ΔR_3^*	ΔR_4^*	ω^*	ω_0^*
40	68.0	96.8	101.0	127.0	3.96	15.8	10.2	0.101	0.089
	66.5	99.4	102.5	131.2	4.97	20.6	8.1	0.077	
50	40.0	123.4	126.5	162.6	5.41	27.9	8.2	0.059	0.071
	37.0	121.9	125.0	167.7	6.14	31.5	11.2	0.083	
60	24.0	148.5	154.5	209.5	7.40	46.2	8.8	0.046	0.054
	27.8	148.5	158.4	220.0	7.90	50.3	11.3	0.062	
70	11.8	170.0	182.4	272.4	10.80	80.7	9.3	0.027	0.030
	11.5	171.5	187.6	275.0	10.30	77.3	10.1	0.032	

corresponding to rupture time are marked by star. We consider ω_0^* as the average values among sizes ω^* for specimens, which were tested for the same stress σ_0. Table 1 shows that there is the evident monotonically decreasing dependence of the limit values ω_0^* on stress σ_0.

The Fig. 1 shows the creep curve $p(\bar{t})$ and the damage curve $\omega(\bar{t})$ for stress $\sigma = 60$ MPa. A value $\bar{t}$ is a dimensionless time equal to the ratio between the real time t and the rupture time for considered stress. The Fig. 1 shows that curve $\omega(\bar{t})$ leads to more exact determination of the transition from secondary creep to tertiary creep than curve $p(\bar{t})$.

The results of the metallographical investigation of structure of these specimens are given in [2]; this investigation was performed on microsections, which were cut from the specimens midsections along the axial direction. In the paper [2] the quantitical measure of the damage is considered which uses the results of metallographical analysis of specimens structure. Two methods of measuring of structure damage, based on use of the different principles (the metallographical analysis in [2] and the utilization of the electrical resistance data in this paper) give the same results relatively of the decreasing kind of function $\omega_0^*(\sigma_0)$.

For the analytical description of this result we may apply the kinetic equations of H. Broberg [3], which take into account the material damage at the momentary loading. It is showed in [4], that these equations lead to creep rupture condition

$$\left.\frac{d\omega}{dt}\right|_{t=t^*} \to +\infty ,$$

value $\omega^* = \omega(t^*)$ is limit quantity. In [2,3] is shown, that the equations of Broberg lead to the monotonic decreasing of limit damage value for increasing of stress σ_0. The same result is obtained for the analysis of experimental data [1] using the proposed method.

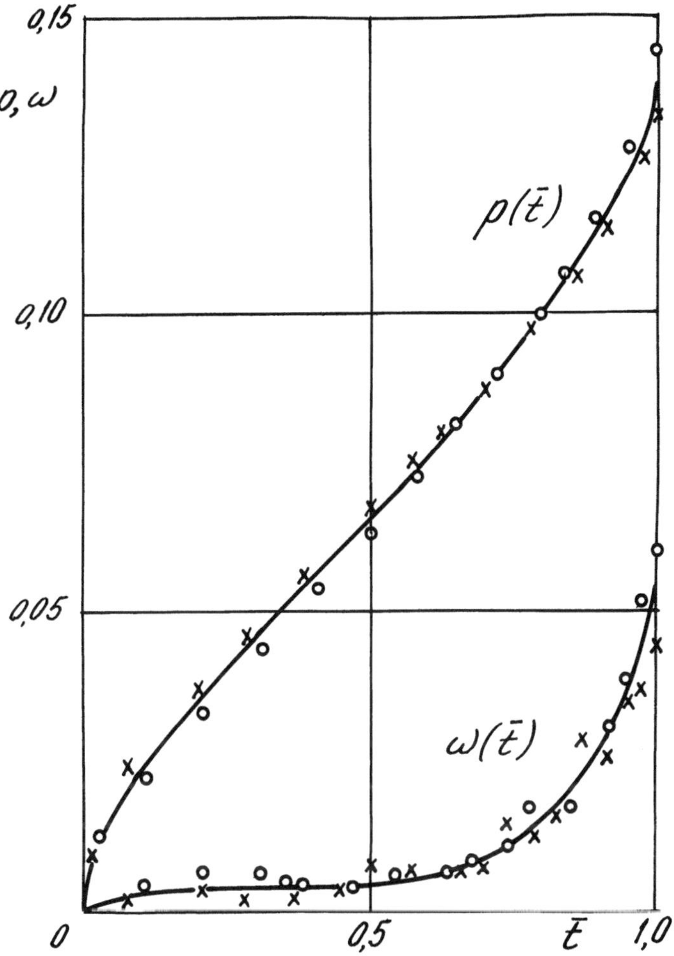

Figure 1: The creep curve $p(\bar{t})$ and the damage curve $\omega(\bar{t})$ obtained for $\sigma_0 = 60$ MPa

References

[1] Shesterikov, S.A., Lokoshtchenko, A.M., Myakotin, E.A.: *Use of the method of measuring electrical resistance in studing the strength and creep of metals*, Probl. prochn., AN Ukr. SSR, **10**, (1984), 32-35 (in Russian)

[2] Lokoshtchenko, A.M.: *Investigation of material damage under creep and creep strength*, Prikl. mekhan. i tekhn. fiz., **6**, (1982), 129-133 (in Russian)

[3] Broberg, H.: *A new criterion for brittle rupture*, J. Appl. Mech., **41**, (1974), 3, 809-811

[4] Boström, P.O., Broberg, H., Bräthe, L., Chrzanowski, M.: *On failure conditions in viscoelastic media and structures*, Int. Symp. on Mechanics of Viscoelastic Media and Bodies, Gothenburg, 2-6 Sept. 1974, pp 302-311, Springer-Verlag, Berlin 1975

[5] Dacheva, M.D., Lokoshtchenko, A.M., Shesterikov S.A.: *Model representation of the ultimate strain during creep*, Prikl. mekhan. i tekhn. fiz., Sib. Otdel. AN SSR, **4**, (1984), 139-142 (in Russian)

Creep of Concrete During Load-Controlled Experiments

J. Luijerink

Fac. of Civil Engineering,
Delft University of Technology
2628 CN, The Netherlands

Summary
The present work deals with a method based on the theories of deformation kinetics and has been developed to describe the total compressive time dependent deformation of materials during load controlled experiments, in particular for concrete.
The anelastic deformation of concrete is here considered as a total of the elastic deformation and the time dependent creep deformation (the contributed external work $V \cdot W = V \cdot \int \sigma \cdot \delta \varepsilon$). The anelastic deformation during loading, depends on microstructural displacements of material elements, the so-called "particles".
A statistical mechanical approach of the energy states of these elements, leads to the theory of the time dependent deformation rate processes, with particle-parameters as the apparent free enthalpy of activation G and the activation volume V.
Description of the total stress-strain diagram of materials was possible with this theory.
The mean values of the $\Delta G = \Delta F - V \cdot W$ associated with the forward jumps of material elements versus the load σ, show a maximum at a point in stress-strain diagrams (about 65% of the strength), when the poisson's ratio looses its linearity. This means that, untill this point, the material is able to build up higher energy barriers against the work contributed by the external effective stress.
This method includes a dammage function, which means that it can predict the rupture time and/or the ultimate load of a material.

Introduction

During the anelastic deformation of a material under an applied load, a energy-flow from one state into another takes place and the deformation depends on microstructural displacements of elements of the material. Elements, the so-called "particles", are entities. They themselves may be identical assemblies of particles, in which case one has an ensemble [1]. In case of this a material element can be treated as a continuum [2].

M. Zyczkowski (Ed.)
Creep in Structures
IUTAM Symposium Cracow/Poland 1990

If the distribution among the energy states of the particals in a system is known, the macroscopic properties of the system can be determined. The physical properties of these microscopic material elements contribute to the macroscopic measurable properties of the solid. The Maxwell-Boltzmann energy-distribution function is often used in the theory of deformation kinetics and relates the deformation rate with the energy states in a solid. Particles try to get into energetical favorable positions. Some particles contribute to the external measurable deformation by only a displacement, while others jump, by thermal activation, an energy barrier and a bond will break. Other elements fall backward in an energy pitch and show a bond mending [2] and [3]. In the equilibrium state bond breaking and mending are in balance; figure 1.

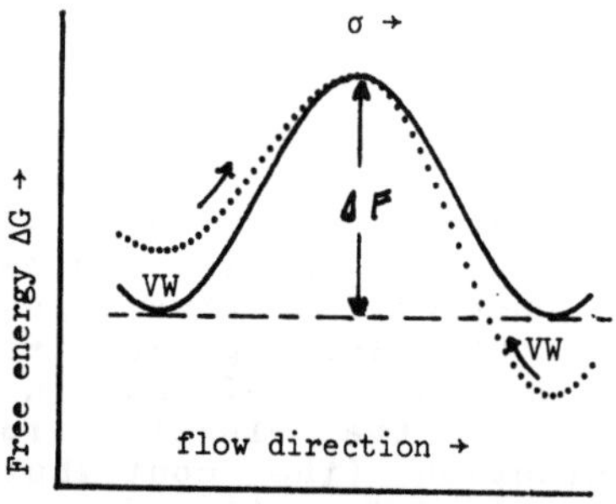

fig 1 Energy barrier

Theory

The deformation rate $\delta\varepsilon/\delta t$ is proportional to the frequency of forward jumps minus backward jumps of particles over the energy barriers and is according the established equations:

$$\delta\varepsilon/\delta t = C_f \cdot \exp(-G_f/k \cdot T) - C_b \cdot \exp(-G_b/k \cdot T) = \quad (1a)$$

$$\delta\varepsilon/\delta t = C_f \cdot e^{-\frac{F-V \cdot W}{k \cdot T}} - C_b \cdot e^{-\frac{F+V \cdot W}{k \cdot T}} = \quad (1b)$$

$$\frac{\delta\varepsilon}{\delta t} = K \cdot \frac{\delta\varepsilon}{\delta\sigma} = \{A \cdot \sinh(\frac{V \cdot W}{k \cdot T}) + B \cdot \cosh(\frac{V \cdot W}{k \cdot T})\} \cdot e^{-\frac{F}{k \cdot T}} \quad (1c)$$

For a load controlled experiment, where $\delta\sigma/\delta t = K$ (constant), is the strain rate in the origin ($\sigma=0$ and $W=0$) given by: $K/E_0 = B \cdot \exp(-F_0/k \cdot T)$. Equation (1c) can now be rewritten as:

$$K\cdot\frac{\delta\varepsilon}{\delta\sigma} = \{A\cdot e^{-\frac{F_0}{kT}}\cdot\sinh(\frac{VW}{kT}) + \frac{K}{E_0}\cdot\cosh(\frac{VW}{kT})\}\cdot e^{-\frac{F-F_0}{kT}} \qquad (2a)$$

Rearranging with coth(C)=A/B leads to:

$$E_0\cdot\frac{\delta\varepsilon}{\delta\sigma} = e^{-\frac{F-F_0}{kT}}\cdot\frac{\sinh(VW/kT + C)}{\sinh(C)} \qquad (2b)$$

The subscript $_0$ indicates the values at the time $t = \sigma/K = 0$.

To solve (2) for load-controlled experiments, both sides of (2b) have to be multiplied by $V\sigma\cdot\delta\sigma/kTE_0$ and after rearranging, one gets:

$$\frac{V.\delta W/kT}{\sinh(VW/kT + C)} = \frac{V\cdot e^{-\frac{F-F_0}{kT}}\cdot\sigma\cdot\delta\sigma}{kT\cdot E_0\cdot\sinh(C)} \qquad (3)$$

Integrating (3) gives:

$$\sinh(\frac{VW}{kT} + C) = \frac{1}{\sinh[D - \frac{V\cdot\sinh(D)}{kT\cdot E_0}\cdot\int_0^{\sigma}\sigma\cdot e^{-\frac{F-F_0}{kT}}\delta\sigma]} \qquad (4)$$

Here is sinh(D)=1/sinh(C).
The dimensionless rate now becomes, see (2b):

$$E_0\cdot\frac{\delta\varepsilon}{\delta\sigma} = \frac{e^{-\frac{F-F_0}{kT}}\cdot\sinh(D)}{\sinh[D - \frac{V\cdot\sinh(D)}{kT\cdot E_0}\cdot\int_0^{\sigma}\sigma\cdot e^{-\frac{F-F_0}{kT}}\delta\sigma]} \qquad (5)$$

In these equations are:
- F (Nm) the mean true activation free energy of the particles
- V (m^3) the volume of a particle
- A, B, C, D and K the constants
- E_0 (N/m^2) is the modulus of elasticity.
- k (J/K) the Boltzmann constant = $1.38066\cdot10^{-23}$
- T (K) temperature

- $W = \int_0^{\sigma} \sigma \cdot \delta\varepsilon$ (N/m^2) the external work per volume.

It is easy to understand that, when $\delta\varepsilon/\delta\sigma = \infty$ and $W = \infty$, a solid will break; this is when the denominator on the right hand side of (4) and (5) becomes zero for a rupture load σ_r, see figure 2.

Untill now the activation free energy $\Delta F = F - F_0$ is not expressed in a time dependent function. The function $\Delta F = m \cdot kT \cdot \ln(t/d + 1)$ was used in [4] and [5] and was suitable to be used for the whole creep-curve. For a load controlled experiment, with $t = \sigma/K$, one can now use succesfully $\Delta F = F - F_0 = m \cdot kT \cdot \ln(\sigma/d \cdot K + 1)$.

The apparent free energy of activation $\Delta G = \Delta F - V \cdot W$ associated with the forward jumps of material elements versus the load σ, shows a maximum at a point for σ–ε diagrams see figure 2c, when the poisson's ratio looses its linearity (about 65% of the strength). So $\delta(\Delta G)/\delta\sigma = \delta(\Delta F)/\delta\sigma - V \cdot \sigma \cdot \delta\varepsilon/\delta\sigma = 0$ is used to calculate this critical point and together with (5) this gives the equation:

$$\sinh\left[D - M \cdot \int_0^{\sigma_c} \left\{\frac{d \cdot K}{\sigma + d \cdot K}\right\}^m \cdot \sigma \cdot \delta\sigma\right] = \frac{M \cdot \sigma_c d \cdot K}{m} \cdot \left\{\frac{\sigma_c + d \cdot K}{d \cdot K}\right\}^{1-m} \quad (6)$$

With $M = \dfrac{V \cdot \sinh(D)}{k \cdot T \cdot E_0}$

This maximum means that untill this point, the material was able to build up higher energy barriers against the work contributed by the external effective stress σ. Untill this stage bond breaking and mending are in balance. The internal energy F will then only increase due the internal displacements.

After this critical point, however, the material does not have this capacity; micro-crack propagation occurs and the solid starts to break. During this part of the deformation in the material concrete micro-crack propagation in the particles takes place. The anelastic behaviour is also due to internal crack propagation.

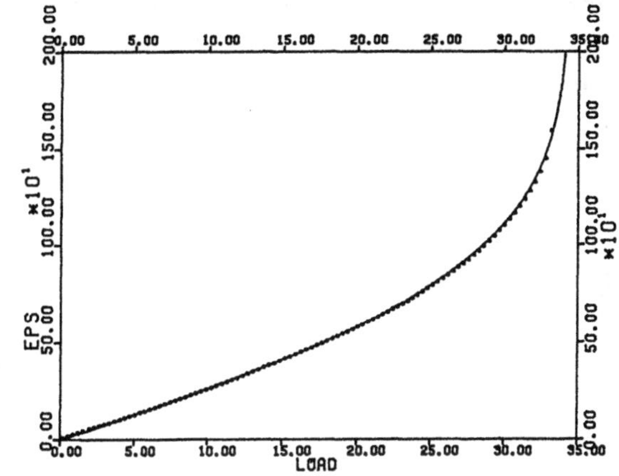

fig 2a

EPS. IN MICROSTRAIN
ε

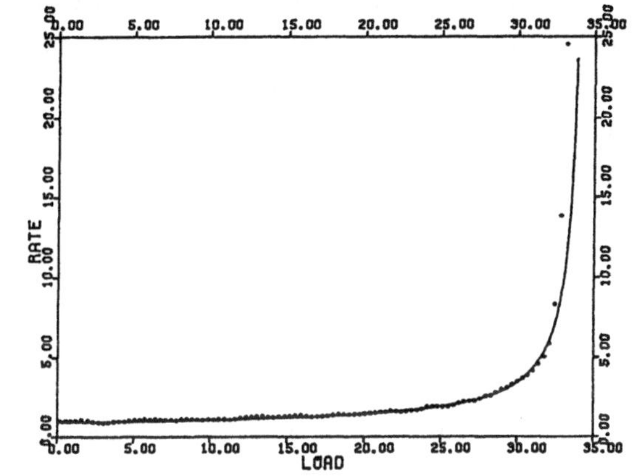

fig 2b

RATE $(E_0 \times \Delta EPS/\Delta SIGM)$
$\delta\varepsilon/\delta\sigma$

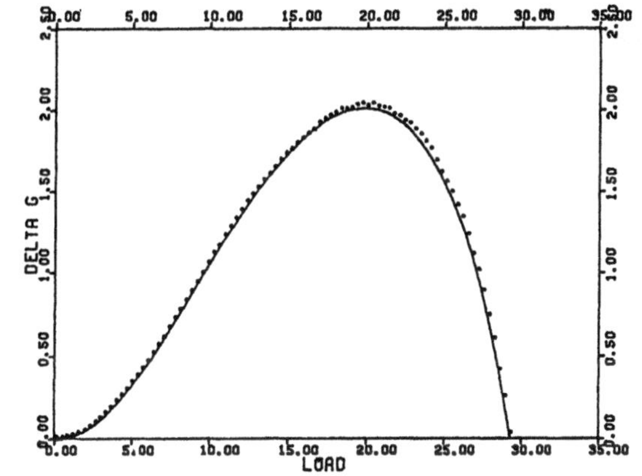

fig 2c

$\Delta G \times 10^{21}$ IN J

LOAD IN N/MM2

The internal free energy F of the material includes the potential energy of the particles and/or also for example the surface energy $2\cdot\gamma\cdot S$ of the cracks in the system.

To calculate the ultimate load σ_r, one can use the denominator:

$$\int_0^{\sigma_r}\{\frac{d\cdot K}{\sigma + d\cdot K}\}^m\cdot\sigma\cdot\delta\sigma] = D\cdot\frac{k\cdot T\cdot E_0}{V\cdot\sinh(D)} = \frac{D}{M} \qquad (7)$$

The experiments were carried out with the material under a constant loading rate $\sigma = K\cdot time$.

It was possible to obtain correlation coefficients between the theoretical function and the experimental data better than 0.99 by fitting stress-strain curves. The fitting was done according equation (4) with m=1, see (6) and (7). The running parameters where the activation volume V and the constant C.
In this case $\exp\{-(F-F_0)/kT\} = dK/(\sigma + dK)$.

The parameters such as the activation energy and the activation volume are in the same range as the corresponding values obtained from creep curves of concrete.
The value of the activation volume V of a micro-structural particle was $45\cdot10^{-25}$ (m^3) (cubic edge is 16.5 nm) and the value of the activation energy F_0 that this particle have to overcome for irreversible deformation was $56\cdot10^{-21}$ (N.m).

In the cementious material concrete the time dependent behaviour takes place in the cementpaste, because this material is already porous and initiated by cracks.

References

1. Sears, F.W.; Salinger, G.L.: Thermodynamics, Kinetic Theory, and Statistical Thermodynamics. Massachusetts: Addisn-Wesley Publishing Company 1982.

2. Kröner, E.; Teodosiu, C.; Lattice defect approach to plasticity and viscoplasticity. Problems of Plasticity, A. Sawczuk, Leyden: Noordhoff Int. Publishing (1974) 45-88.

3. Krausz, A.S.; Eyring, H.; Deformation Kinetics. New York: John Wiley&Sons,Inc. 1975.

4. Luijerink, J.; Creep Deformation Kinetics of Concrete. Creep in Structures, A.R.S. Ponter and D.R. Hayhurst, Berlin: Springer-Verlag (1981) 362-367.

5. Luijerink, J.; Deformation Kinetics of Concrete. Fundamental Research on Creep and Shrinkage of Concrete, F.H. Wittmann. The Hague Martinus Nijhoff Publisher (1982) 27-34.

Evaluation of Creep Damage in Some Major Components of Steam Turbine

J. MATOUŠEK

National Research Institute of Machine Design, 190 11 Prague

Summary

The paper presents a finite element analysis of some physically non-linear problems of material creep in application for machine parts operating under creep conditions. The numerical solution is based on a statistically stable analytical relationship for plastic deformation which expresses all three stages of creep process. For the variable load and/or temperature modelling at multiaxial state of stress the equivalence of local damage according to the hypothesis of linear cumulation of damage is used. The suggested method is suitable for creep calculations of geometrically complicated bodies with respect of local damage at the points of deformation and stress concentration. Physically real boundary conditions were included in the FEM numerical technique. Dispersion of the creep tests is accounted for and the numerical results are compared with measurements.

1. Introduction

The proposed method facilitates the application of the computational technique founded on the knowledge of more detailed modelling of both analytical expressions of creep curves for all three stages and the regression dependency (Larson-Miller form) for long-time creep strength of heat-resistant steels. It is not sufficient to base the mathematical description of the creep behaviour of material exclusively on the results of the experiments, but also the trends underlying the data at long term creep conditions of heat-resistant steels must be taken account of. This concerns especially the long term material properties - creep strength for 10^5 to 2×10^5 hours. The s.-c. degrees of creep damage of material (given in percents) represent the principal results of these procedures, together with complementary evaluations of the permanent deformations in selected critical locations. The following paragraph is aimed to demonstrate a mathematical model of the process of creep.

M. Zyczkowski (Ed.)
Creep in Structures
IUTAM Symposium Cracow/Poland 1990

2. Mathematical Model of Properties of Heat-resistant Steels

The process of establishment of a strain dependence model has been based upon the principle of dislocation mechanism for which stochastic behaviour of material during the strain mechanism has been assumed. Creep deformation tests (by which plastic elongation ε^c is measured) provide the fundamental experimental data. The constitutive equation describing the development of deformation can be derived [3] for the mean value of ε^c as

$$\varepsilon^c = \varepsilon_o \left(\frac{\varepsilon_l}{\varepsilon_o}\right)^{g(D)} , \qquad (1)$$

where the damage parameter D replaces the parameter of time, t. Relationship

$$\varepsilon_o = \frac{\sigma}{E_T} \qquad (2)$$

is valid for the initial deformation ε_o in the region of elastic stress/strain relationship in which E_T stands for the modulus of elasticity at temperature T.

The deformation limit ε_l can be determined by extrapolation of the non-accelerating creep curve up to fracture. Thus only the collapse phase of the creep process is excluded from the present mathematical model, this stage, however, representing only approximately 1 p.c. of the overall time to fracture. The expression

$$\varepsilon_l = \exp\left[A_1 + A_2 \tanh\left(\frac{\ln t_r - A_3 - A_4\, T}{A_5}\right)\right] + \varepsilon_o \qquad (3)$$

holds for ε_l. The time to rupture t_r for forged steels can be expressed by the equation

$$\log t_r = B_1 + B_2 \log\left|\frac{1}{T} - \frac{1}{B_5}\right| + B_3 \log\left|\frac{1}{T} - \frac{1}{B_5}\right| \times$$
$$\times \log \sinh(B_6 \sigma T) + B_4 \log \sinh(B_6 \sigma T) . \qquad (4)$$

The hardening function g(D) can be expressed

$$g(D) = [D(t)]^N \, \frac{(1 + \exp[-2(D(t))^K])^M}{1 + \exp(-2)^{\,M}} , \qquad (5)$$

where $D(t)=t/t_r$ is the material damage at time t. Numerical values of the regression parameters A_i, B_i, K, M, N, have been obtained for Cr1.25Mo0.65V0.25, Cr1.0Ni0.7Mo0.9V0.3, and Cr1.25Mo0.65V0.75W0.55 grade steels (steam turbine rotors) and also for Cr1.25Mo1.0V0.25 steel (the internal high-pressure casing of the 200 MW turbine) by experimental identification of the creep process.

3. Linearization of Creep Damage Process

This non-linear creep problem can be numerically reduced to a step by step linear one by dividing the plastic flow process of material into time intervals Δt, where the theoretical time to fracture

$$t_f = \sum_i (\Delta t)_i \tag{6}$$

corresponds to the resulting damage

$$D = \sum_i \left(\frac{\Delta t}{t_r}\right)_i = 1 \tag{7}$$

and plastic strain

$$\varepsilon^c = \sum_i (\Delta \varepsilon^c)_i = \varepsilon_l \ . \tag{8}$$

The damage parameter is according to (7) defined by the s.-c. "time fracture rule".

For modelling the creep process, the equivalence of local damage at corresponding stress levels expressed according to the von Mises criterion by the second invariant of the stress tensor and the linear hypothesis of cumulative damage in each element are used instead of either the strain hardening or the time hardening theories [2]. The above mentioned transition from individual levels of uniaxial creep tests has been verified by both experiments and practical observations, and illustrated in Fig.1.

For evaluation of individual plastic strain components and thus stress relaxation process the model uses the Prandtl-Reuss relationship, which has form

$$\varepsilon^c_{kk} = 0 \quad , \tag{9}$$

$$d\varepsilon^c_{ij} = \frac{\Delta\varepsilon^c}{2\sigma_{ef}}\; 3\, s_{ij} \quad , \tag{10}$$

where s_{ij} denotes the components of stress deviator, σ_{ef} effective stress of von Mises hypothesis, ε^c_{ij} tensor of creep deformation, and $d\varepsilon^c_{ij}$ the components of increment of creep strain. The effective plastic deformation $\Delta\varepsilon^c$ in each element of the structure and each time step is effected by assuming the linear hypothesis of cumulative damage on the premise of the known real operational temperature and mechanical loading. Resulting creep damage value D accumulated after i-th solved interval is

$$D = \sum_i \frac{{}^2t_i - {}^1t_i}{t_{r_i}} \quad . \tag{11}$$

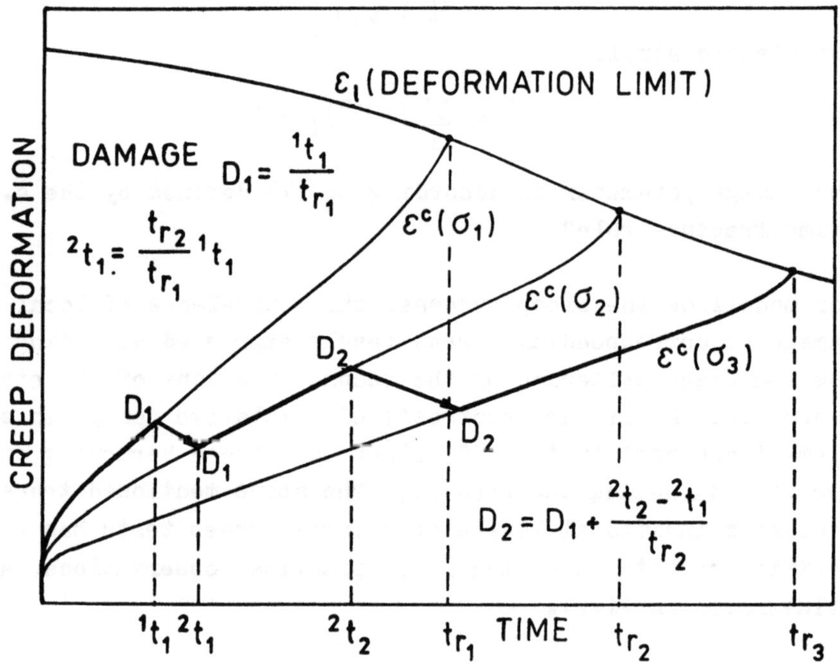

Fig.1. Transition between creep curves according to the achieved value of creep damage

4. Analysis of Creep Damage by the FE Method

The meshes with triangular elements with linear shape functions (plane stress/strain, rotationally symmetrical and spatial problems) has been used for establishment of stationary temperature fields and the creep process. The basic equation for modelling creep by FEM has the form

$$[K]\{\delta\} = \{R\} + \{F^c\} \quad , \qquad (12)$$

where F^c is additional fictional force which will increase in each time interval Δt by ΔF^c:

$$\{\Delta F^c\} = [\alpha]^T \int_{(V)} [\beta]^T [E] \{d\varepsilon^c\} \, dV \quad . \qquad (13)$$

In the above equations denotes:
$[K]$ the overall stifness matrix; $\{\delta\}$, $\{R\}$ the column vector of displacements and external forces in noidal points; $\{\alpha\}$, $\{\beta\}$ a matrix of polynominal coefficients of element and the matrix of differential operators, $[E]$ the matrix of elasticity, V the volume of an element and the exponent T denotes the transposition of a matrix.

The calculation of the time intervals, which is the most important factor of any numerical procedure of creep evaluation, is performed automatically by the present algoritm according to the maximum value of the stress intensity or the set of intervals can be prescribed. From 3 to 5 reccurent solutions are sufficient in each time interval. The optimisation of the numerical solution is controlled by the commonly accepted Wilson's iterative procedure [4]. This procedure permits very large time steps using the average equivalent stress for each time step instead of assuming that the stress remains constant during Δt.

This method has been used for assessment of the residual service-life of the 1st medium-pressure rotor disc of 110 and 200 MW steam turbines manufactored by Škoda Plzen and for estimation of the damage by material creep of the control stage disc of the 200 MW high-pressure rotor during stationary operation [1].

In another application of this numerical process, the procedure was used to assess creep damage of the rotationally symetrical model of the internal steam turbine casing under non-proportional load conditions following the s.-c. start-up [5].

For prediction of the 200 MW medium-pressure disc life-time, the maximum creep rate at the limit of the scatter band of creep curves was assumed for the material near the rotor boring characterized by worsened creep properties. The model was modified using a correction coefficient of 1.25 [3] for the effective value of stress in relationships (1) to (4). This coefficient linearly decreases to its mean value of 1.0 at the outer radius of the disc.

To evaluate the creep process in material of the steam turbine rotors, it was necessary to incorporate in the FEM analyses the plane boundary conditions [1] at both sides of the disc solved separately, and to ensure a free axial displacement at zero end rotation of the grooved disc circumference during creep.

The pattern of creep damage levels over the solved disc is shown in Fig.2. The intensity of creep damage is the greatest in the corners of the blade grooving where the pressure resulting from the centrifugal forces of the blades bears. The damage values in the other dangerous locations after 130 000 hours of creep at the disc operational temperature of 532 °C, i.e. in the transitional curve from the disc to the rotor body and at the rotor boring surface, are smaller by an order.

The same relations hold for the effective plastic deformation in these critical locations of the structure. In the transitional curve from the disc to the rotor body, a creep strain of 0.05 p.c. accumulates during 130 000 h of operation, the extreme permanent strain at the axial boring being about the same. In the corner of the blade fixation grooving, local plastic strain of cca 0.5 p.c. after 130 000 h of creep was determined by the FEM computation, although the mean plastic strain values were assumed in this region of the 200 MW disc.

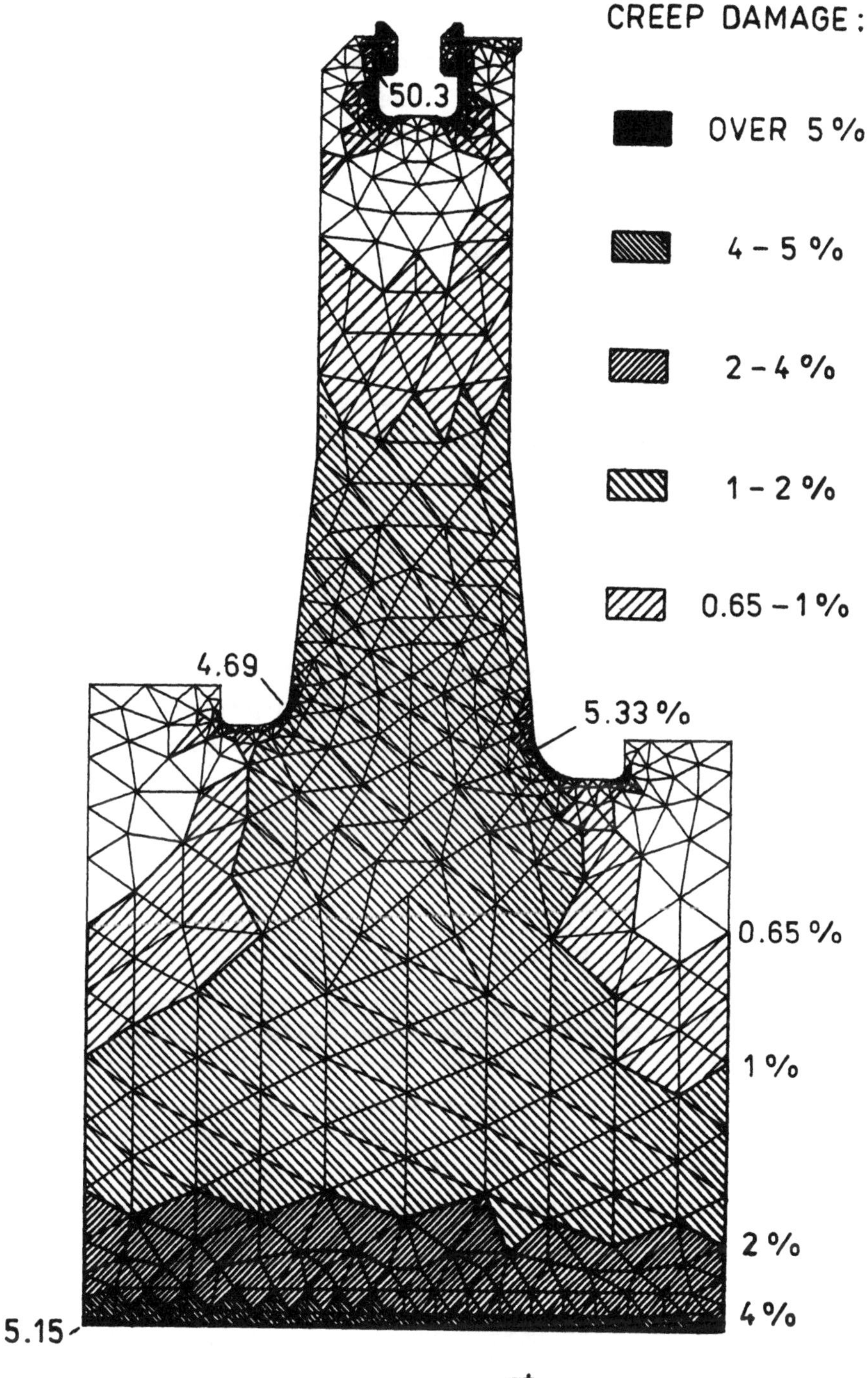

Fig.2. The pattern of damage in the 1st disc of 200 MW medium-pressure rotor after 130 000 h of creep

5. Conclusions

A detailed analysis of rotors and casings of steam turbines in creep regime was performed based on extensive theoretical and experimental studies using advanced computer programmes. Levels of material damage by creep in critical locations have been assessed by the mathematical model of the creep process and its statistical properties.

Creep damage computations of a single rotor disc of a Škoda steam turbine were performed in order to evaluate the residual life-time of rotors which had already been in operation. The computations of permanent strain of the rotor boring under 1st disc are in agreement with two measurements which were performed on two separate rotors of turbines of identical output, the proposed method of power plant components thus being verified. The permanent strain of a 110 MW rotor boring radius after 126 000 hours of creep has been computed as 0.037 p.c. The on-site measurement of the permanent strain of boring under 1st disc of 110 MW medium-pressure rotor has yielded the value of 0.04 p.c. after 112 000 operational hours.

References

1. Matoušek, J.; Drahý, Z. et al.: Creep Damage in Rotors of Škoda Steam Turbines. Monographs and Memoranda No.40, Nat.Res.Inst.for Machine Design, Praha 1990 (in print).

2. Bushnell, D.: A Strategy for the Solution of Problems Involving Large Deflections, Plasticity and Creep. Int.Journ.for Num.Method in Enging 11 (1977) 683-708.

3. Bína, V.; Pooh, R.: General Mathematical Model of Creep Process in CrMoV Steels. Machinery Eng. (Strojírenství) 30 (1980) 227-236 (in Czech).

4. Wilson, E.A.: A Design Oriented Approach to Creep and Plasticity in Finite Elements Programs. Journ.of Enging for Industry 93 (1971) 793-798.

5. Matoušek, J.; Švarc, M.; Valenta, J.; Löbl, K.; Bína, V.: Evaluation of Creep Damage in Some Major Components of Power Generator. ISIJ International 30 (1990) 451-456, Tokyo.

Uniaxial Viscoplastic Experiments of Predamaged AISI 316H

M. MONTAGNANI , C. ALBERTINI [1] and M. MIĆUNOVIĆ[2]

Summary

Two groups of specimens, being previously subjected either to creep or low cycle fatigue, are examined by tension at 550°*C in the strain-rate range* $(10^{-3}, 10^{3})\, s^{-1}$. *Specimens previously treated by creep have suffered diverse levels of creep strains while specimens predamaged by LCF had various LCF-strains critical cycle number (as a measure of initial life time). Evolution equations are calibrated for as-received, creep-damaged as well as LCF-damaged specimens covering stress-strain-strain rate relations and accounting for strain and strain rate hardening. Material constants entering the equations are damage dependent.*

Experimental Evidence and Statement of the Problem

The test material AISI 316 H stainless steel had been received as a mill-annealed 50 mm thick hot-rolled plate with the following chemical composition [wt %] : $Cr-16.9$, $Ni-12.4$, $Mn-1.65$, $Mo-2.45$, $Nb-0.006$, $C-0.053$, $Si-0.35$ and $S-0.008$.The observed grain size range of the as-received material was $[70, 90]\,\mu m$.

The predamage treatment had been as follows:

- LCF testing had been carried out by P.N.C. Tokyo at the University of Tokyo with the hourglass-shaped specimens (depicted on Figure 1A) by making use of an electro-hidraulic closed loop fatigue testing machine under the conditions of triangular wave shape and strain rate of approximately $0.001\, sec^{-1}$ (details are given in [1]). The specimens had been heated by an induction heater and the temperature at the test section maintained in the range $550°C \pm 2°C$. Cycle strain range had been either 0.6 % or 1 % with cycle ratios n/N_F equal to 0.2, 0.4 and 0.6 .

- Creep tests performed at JRS-Ispra had been accomplished also at $550°C$ with a constant tensile stress of $300 MPa$ and a creep life time of $T_c = 2000h$ with the specimen shown on Figure 1B. These tests had been interrupted at $423h$, $823h$ and $1220h$ (i.e. approximately $t/T_c = 0.2$, 0.4 and 0.6 of the life time). As a preliminary action for actual testing thin outer layers of the specimens , previously subjected to LCF or creep , were removed in order to exclude surface microcracks. Subsequently the bulk specimens were split longitudinally by electric discharge machining so that

[1]Joint Research Centre CEC Ispra, Italy
[2]Svetozar Marković University, Kragujevac, Yugoslavia

M. Zyczkowski (Ed.)
Creep in Structures
IUTAM Symposium Cracow/Poland 1990

the final test small specimens presented on Figure 1C were machined out from each singular piece.

Actual tensile testing was carried out in the following three regimes :

- at the lowest strain rate of $3.5 \times 10^{-3} s^{-1}$ by an Instron machine ;
- at medium strain rates in the range $[10^{-1}, 10^{2}] s^{-1}$ by a hydropneumatic machine installed in a hot cell of the Medium Activity Laboratory of the JRC in Ispra and
- at high strain rates in the range $[10^{2}, 10^{3}] s^{-1}$ by a modified Hopkinson bar with a prestressed loading device installed in a hot cell of the High Activity Laboratory of the JRC in Ispra.

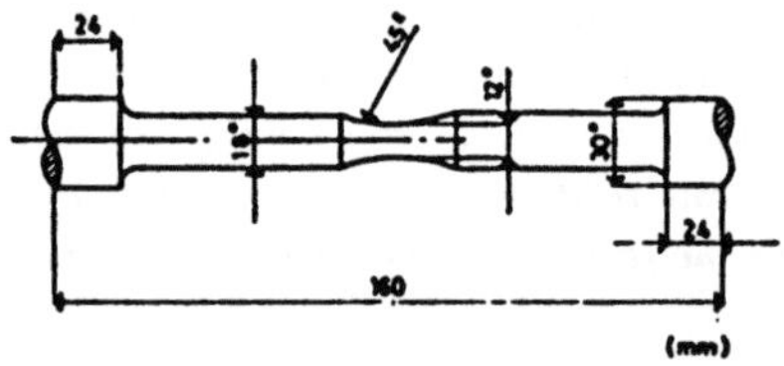

HOURGLASS TYPE SPECIMEN
Fig. 1A

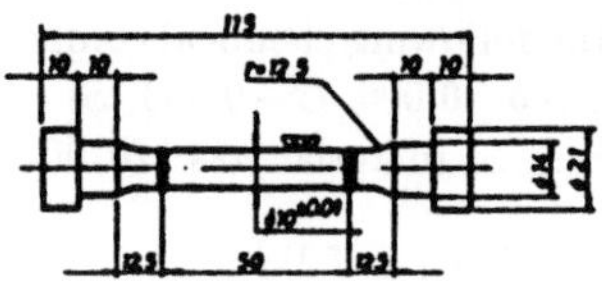

CREEP SPECIMEN
Fig. 1B

SPECIMEN

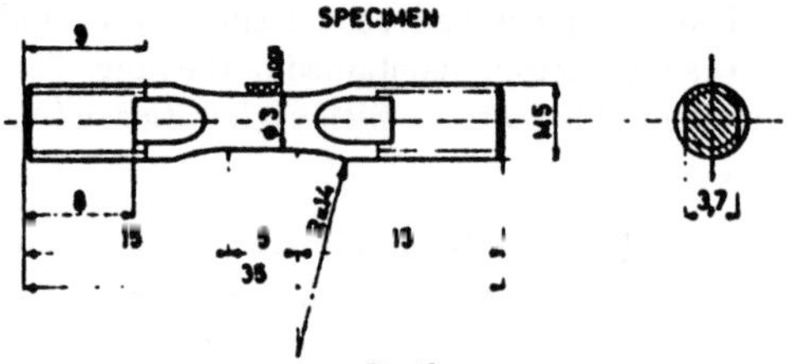

Fig. 1C

Fig. 1. LCF, creep and test specimens

The last two apparatuses had been developed in Ispra and were described in [2]. Concerning the test results the typical (strain rate dependent) strain versus true stress curves are depicted in Figure 2 for one case of previous creep and in Figure 3 for one case of previous LCF. As a general tendency both damage processes provoke for a fixed strain rate :

An increase in the initial yield stress and (not so marked) ultimate tensile stress ; a reduction of ductility represented by the fracture strain and (more pronounced) uniform strain.

The first phenomenon might be called conditionally a *damage hardening* while the second , leading to higher sensitivity of damaged materials caused by void accumulation during a damage process , is also important and must not be forgotten.

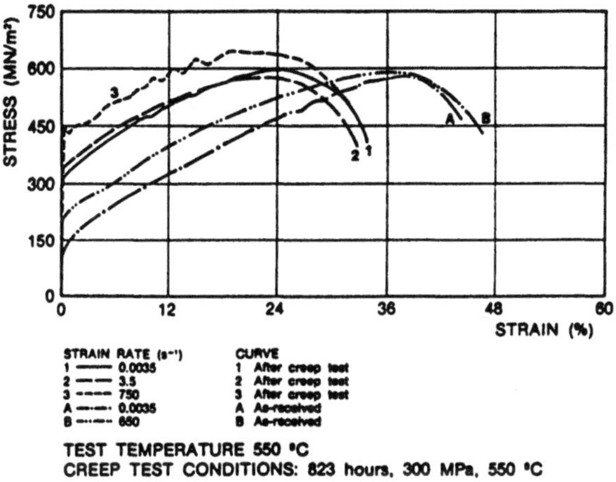

Fig. 2. Strain versus true stress curves for AISI 316 H as-received and predamaged by creep

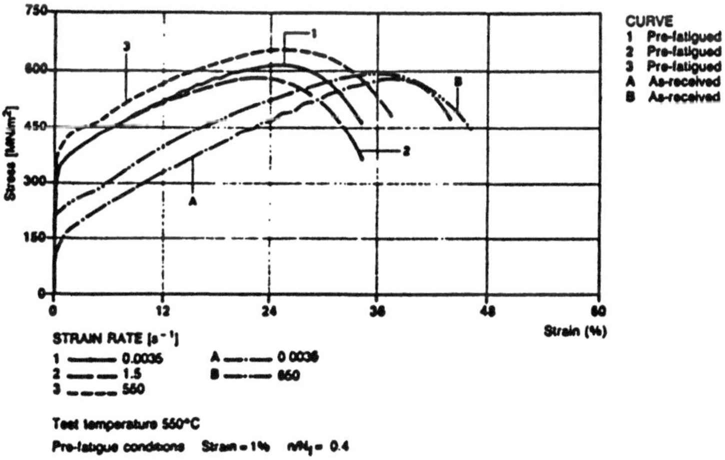

Fig. 3. Strain versus true stress curves for AISI 316 H as-received and predamaged by LCF

From the same figures we observe also that up to large strain values for a fixed strain there exits strain rate hardening i.e. an increase of flow (yield) stress with strain rate increase. Finally, the observed dynamic strain ageing serrations are analysed elsewhere (cf. [3]) and are not considered in this paper.

If we want to give some analytical expression accounting for the above discussed phenomena,then the following simple yet general formula seems to be appropriate :

$$\begin{aligned} uniaxial\,flow\,stress \;=\; & f\,(\,strain,\ strain\,rate,\ damage \\ & history,\ plastic\,strain\ history, \\ & plastic\,strain\,rate\,history) \end{aligned} \tag{1}$$

Its calibration obtained from a specified model is the subject of the next section.

Model and Its Calibration

The principal question to be answered here is the following :

Is it possible to propose a unified evolution equation among stress , strain and strain rate with a minimal number of (temperature and history dependent) material " constants " in such a way that a maximal collection of experimental stress-strain diagrams is described within a beforehand prescribed accuracy ?

Such a question is a natural consequence of the law (1) specified to applicability in FEM-codes which do require a unified equation for an improvement of safety analysis of reactor structures subjected to complex thermomechanical processes.

The common notion in viscoplasticity is the *overstress* (cf. [4],[5],[6]) i.e. the difference between the actual ("dynamic") stress and its static counterpart (at negligible plastic strain rate). According to this concept the following uniaxial equation is most natural to be accepted :

$$\sigma = \sigma^{\star}(\epsilon_P, \theta, h)\,[\,1 + \Delta(\dot{\epsilon}_P, \epsilon_P, \theta, h)\,], \tag{2}$$

where σ is the engineering dynamic stress, $\sigma^{\star}$ - its static counterpart, ϵ_P and $\dot{\epsilon}_P$ Lagrangian uniaxial plastic strain and its material time derivative , h stands for history whereas

$$\Delta \equiv \frac{\sigma - \sigma^{\star}}{\sigma} \tag{3}$$

is a relative overstress coeficient.

For the static stress we propose here the following slight (but essentially nonlinear) modification of Ludwik-Hollomon-Swift empirical equation (cf. also [7]) :

$$\sigma^{\star} = Y_0(\theta, h)\,[\,1 + m_1(\theta, h)\epsilon_P\,]^{m_2(\theta,h)} \tag{4}$$

where Y_0 is the initial yield stress. The alternative model by Johnson and Cook [8] in the form

$$\sigma^* = (A + B\epsilon_P^C)\,[1 - \theta^D] \tag{5}$$

is easier for a (linear) best fit analysis but has the important disadvantage that its slope for small plastic strains is too steep which is not realistic as easily seen from figures 2. and 3.

For the relative overstress coeficient we have at the disposal either the Ludwig-Prandtl equation

$$\Delta = a(\theta)\, ln[\,\dot{\epsilon}_P/b(\theta)\,] \tag{6}$$

or the Zener-Hollomon equation :

$$\Delta = \left[1 + \frac{\dot{\epsilon}_P}{b(\theta)}\right]^{c(\theta)} - 1 \tag{7}$$

where the later behaves smoothly also for small strain rates. In [8] a modification of (6) is given in the following form :

$$\Delta = C ln[\dot{\epsilon}_P/\dot{\epsilon}_{Pmin}] \tag{8}$$

where $\dot{\epsilon}_{Pmin}$ is the "static" strain rate (i.e. the smallest strain rate of the considered strain-rate range) and C is a constant giving rise to a constant strain-rate hardening.

In this paper we propose the following two generalizations of (6) and (7) :

$$\Delta = m_5(\theta, h)\,[m_6(\theta, h) + \epsilon_P]^{m_3(\theta,h)}\,\,[\,ln(\dot{\epsilon}_P/\dot{\epsilon}_{Pmin})\,]^{m_4(\theta,h)} \tag{9}$$

for the Ludwig-Prandtl equation and

$$\Delta = m_5(\theta, h)\,[m_6(\theta, h) + \epsilon_P]^{m_3(\theta,h)}\,\,\dot{\epsilon}_P{}^{m_4(\theta,h)} \tag{10}$$

for the Zener-Hollomon equation.

Once the model is chosen its calibration is far from trivial due to essential nonlinearity (i.e. imposibility of its transformation into a linear model) of (4) and (9) or, equvivalently, of (4) and (10). The complexity of the encountered problem requires a multi-step procedure. The general approach (explained in detail in [9] for multi-damage case) is reduced to minimization of the *chi-square function*. In our case this function is the sum of squares of differencies of calculated (by the proposed model) and experimental stress values over all discrete values of plastic strains and strain rates for a fixed particular predamaged history case.The identification strategy is split into the following two procedures :

P1) For a fixed strain rate the equation (4) is adopted not only for static but for dynamic strain processes as well. Due to nonlinearity of (4) a numerical procedure is very sensitive to initial guesses for Y_0, m_1 and m_2. These guesses are first located by a polynomial representation (linear best fit) and then by three-point threading through the experimental data. The algebraic nonlinear equations obtained in such a way are solved by application of Banach's theorem applied to its fixed point. The subsequent minimization of the chi-square function of the so reduced problem follows then closely Levenberg-Marquardt method for nonlinear best fit either by the steepest descent method or by the variable multidimensional metric method [10]. In such a way Y_0 , m_1 and m_2. are determined not only for σ^* but for all dynamic stresses.

P2) For the unified set of equations (4) and (9) (or alternatively (4) and (10)) the above procedure served for a linear best fit of the function $ln\,\Delta$ with m_6 -constant held fixed in the interval $[0, 1]$. Repeating such a procedure for a set of discrete values of m_6 we choose its value which corresponds to maximum of the correlation coefficient. Thus with Y_0 , m_1 and m_2. determined for $\sigma^* \equiv \sigma(\dot{\epsilon}_P min)$ we have also $m_3, \ldots, m_6$ from the described linear best fits.The acquired material constants serve now as initial guesses for the above mentioned Levenberg-Marquardt method for final nonlinear best fit simultaneously in all seven material constants.

The results of the identification are grouped into the table 1 for the case of generalized

Table 1. Material Constants at 550°C for as-received and predamaged AISI 316 H according to the unified set (4) and (10)

Group (damage)	$Y_0[MPa]$	m_1	m_2	m_3	m_4	m_5	m_6	Correlation coefficient
A (virgin)	127.71	52.973	0.3872	−0.416	0.1117	0.0570	0.0255	0.9946
B_1 (creep)	318.92	29.609	0.2161	−0.469	0.5842	0.0012	0.0328	0.9974
B_2 (creep)	308.05	63.885	0.1672	−0.243	0.3303	0.0116	0.0039	0.9975
B_3 (creep)	324.08	25.133	0.2391	−3.280	0.2767	0.0026	0.4266	0.9931
C_1 (LCF)	262.32	36.984	0.2403	−4.720	0.2296	0.0278	0.9461	0.9715
C_2 (LCF)	272.06	39.828	0.2309	1.3658	0.3060	0.0159	0.4877	0.9907
C_3 (LCF)	273.72	42.512	0.2206	−0.611	0.3188	0.0124	0.8731	0.9922
D_1 (LCF)	337.25	83.612	0.1242	−1.253	0.3185	0.0098	0.8311	0.9955
D_2 (LCF)	337.49	51.629	0.1566	−3.288	0.8198	0.0002	0.6465	0.9957
D_3 (LCF)	320.2	93.459	0.1460	0.430	0.515	0.003	0.900	0.9864

Zener-Hollomon equation i.e. using of the unified set of equations (4) and (10). The alternative set represented by (4) and generalized Ludwig-Prandtl equation (9) had much smaller correlation coefficient. Uncertaintines of the constants in the above table which follow from Monte Carlo simulation [10] were also determined having satisfactory small values in [1, 2]%.

Concerning items in the above table B_1 , B_2 and B_3 correspond to 423, 823 and 1220 h of creep time ; C_1 , C_2 and C_3 - cycle number ratios $n/N_F = 0.2$, 0.4 and 0.6 for the LCF strain amplitude of 0.6% while D_1 , D_2 and D_3 -LCF predamage with $n/N_F = 0.2$, 0.4 and 0.6 for $\epsilon_{LCF} = 1\%$.

Concluding Remarks

General remarks which hold for as-received and predamaged by creep as well as by LCF specimens are :

- Generalized Zener-Hollomon equation with very high correlation coefficients in the interval $[0.9715, 0.9975]$ is much better than generalized Ludwik-Prandtl equation with rather low coefficients in the interval $[0.3201, 0.8182]$.
- Strain rate hardening is observed in all cases given by $m_5 > 0$ and $m_6 > 0$.
- Further identification of analitical expressions for damage dependence of the obtained material constants at this stage of research is not advisable due to small number of tests in isolated damage groups – for creep, LCF at 0.6% and LCF at 1%.

Concerning overall material constants for AISI 316H predamaged by creep it can be said that :

- The almost universally applied assumption on additivity of plastic and creep strains has led to extremely low correlation coefficients ; thus previous creep strain has inevitably a nonlinear influence on the present plastic strain.
- The initial yield stress is higher than for as-received AISI 316H with a nonlinear dependence on creep time with a curious minimum at the medium creep time to be checked by subsequent tests.

For damage history realized by LCF the following remark is interesting :

- The increase in LCF strain amplitude causes an increase in the initial yield stress suggesting that $Y_0(\epsilon_{LCF}$ might be established with as-received case being treated as the special case with $\epsilon_{LCF} = 0$.

Acknowledgements

Thanks are due to Prof. K. Iida of the Tokyo University for the LCF tests, to Dr. R. Matera for creep tests, to Messrs. M. Forlani and A. Pachera for the dynamic tests, Mr. M. Radosavljević of ZCZ-Kragujevac for computer programs and to Prof. G. Milovanović of the University of Niš for discussions concerning numerical modeling of data.

References

1. IIDA, K.; YOSHIE, S.; ALBERTINI, C.; MONTAGNANI, M.: *Impact properties of AISI 316 H fatigued in L.C.F.* In: 2nd Int. Conf. on LCF. Munich: VDM 1973.

2. ALBERTINI, C.; MONTAGNANI, M.: *Dynamic material properties of several steels for fast breeder safety analysis.* EUR 5787 EN-Report (1977)

3. ALBERTINI, C.; ELEICHE, A. M.; MONTAGNANI, M.: *Instabilities at high strain rate of AISI 316 H damaged by creep, fatigue and irradiation. Discussion of adiabatic temperature oscilations.* Inst. Phys. Conf. Ser. **102/10** (1989) 503–510.

4. PERZYNA, P.: *Temperature and rate dependent theory of plasticity of crystalline solids.* Rev. Phys. Appliqu'e. **23** (1988) 445–459.

5. CERNOCKY, E. P.; KREMPL, E.: *A theory of thermoviscoplasticity based on infinitesimal total strain.* Int. J. Solids Structures **16** (1980) 723–741.

6. MIĆUNOVIĆ, M. *A Thermodynamic Model of Ideal Cyclic Viscoplasticity.* Transactions of SMIRT-9, L, Wittmann, F. H. (ed.) 195–203. Rotterdam : Balkema 1987.

7. EL-MAGD, E.; HELLWIG, K.J.; HOCK, H. G.; HOMAYUN, M.: *Mechanical behaviour of austenitic steel and aluminium with respect to elastic-plastic wave propagation.* Mat. Techn.-Steel Res. **58/1** (1987) 33–39.

8. JOHNSON, G. R.; COOK, W. H.: *A constitutive model and data for metals subjected to large strains, high strain rates and high temperatures.* 7th Int. Symp. on Ballistics, **P4.5**, 541–547. The Hague 1983.

9. ALBERTINI, C.; MIĆUNOVIĆ, M. MONTAGNANI, M.; RADOSAVLJEVIĆ. M.: *Material constants for viscoplastic behaviour of damaged AISI 316 H from uniaxial dynamic experiments* – in press.

10. PRESS, W. H.; FLANNERY, B. P.; TEUKOLSKY, S. A.; VETTERLING, W. T.. *Numerical recipes - the art of scientific computing.* Cambridge : University Press 1986.

Creep Rupture in Fibre Bundles with Hardening-Softening Fibre Material

JONAS M NEUMEISTER

Division of Solid Mechanics, Chalmers University of Technology
S-412 96 Göteborg, Sweden

Summary
Creep rupture of a fibre bundle consisting of fibres with hardening-softening creep behaviour is studied for a constant tensile load. Locally varying fibre properties such as fibre modulus and fibre strength are considered. Different modes of final rupture are identified. The total time to rupture is divided into an incubation time, with all fibres intact, and a failure propagation time, where fibres deteriorate and fail progressively. Governing equations are given for both time intervals. Two damage parameters are introduced, one for the fraction of fibres having passed the strain rate corresponding to the peak stress level, and one for fraction of completely failed fibres. It is shown that the failure propagation time interval and the critical state of damage are specific for every bundle and thus independent of the load.

1 INTRODUCTION

Significant for fibrous and other non homogeneous materials is a larger variation, both locally and globally, of properties due to the diversification of its constituents. Effects of scatter in fibre properties have been studied by several authors. An important theorem about fibre bundles was found by Pierce[1] who studied the influence of stocastically distributed flaws. Daniels[2], in a major work on the subject, introduced the concept of the ideal (or classical) fibre bundle, which disregards the interaction among fibres due to a surrounding matrix. Extensive work about fibre composites has also been undertaken by Phoenix *et al.*[3-7] who have studied more general fibre bundles under different conditions.
Creep rupture due to the reduction of the fibre cross section was analyzed by Hoff[8]. Creep and subsequent failure in structures of parallell elements has been treated by Martin and Leckie[9]. Brittle fibre rupture is not necessarily a relevant description of failure under creep conditions. Necking of a fibre at a section along its length will show an ascending-descending stress curve for a monotonically increasing rate of strain. A coated fibre may show the same behaviour. Hult and Travnicek[10] examined the static carrying capacity of bundles consisting of fibres with a bi-linear stress-strain curve. Creep rupture of fibre bundles with varying properties was studied by Chrzanowski and Hult[11] for a linear creep law and subsequently by Neumeister[12] considering more general distributions and constitutive relations. A more comprihensive version of the present paper with an example is available in [13].

M. Zyczkowski (Ed.)
Creep in Structures
IUTAM Symposium Cracow/Poland 1990

2 DESCRIPTION OF MODEL AND PROBLEM

In the present paper creep rupture of bundles consisting of fibres with a hardening-softening constitutive behaviour, i.e. fibres wich still can take load after reaching its ultimate strength, is considered under a constant tensile load. It is assumed here that these bundles consist of fibres of equal length and area without any mutual interaction, often referred to as ideal fibre bundles[2]. Such a model would yield a longer time to final failure than one with brittle fibres, also final rupture would generally occur with a greater number of fibres having reached their ultimate strength.The study investigates the effects of such a material behaviour (compared to brittle behaviour) rather than explaining the physical process of material deterioration. As in [11] and [12], the total time to rupture is divided into an incubation time, with all fibres intact, and a failure propagation time where fibres deteriorate and fail progressively. But here two damage parameters are necessary to describe the state of the bundle, one for the fraction of completely failed fibres and one for the part of fibres on the descending part of the constitutive law. It has been shown[10] that the analysis of such bundles is considerably simplified by using the model of a fibre continuum for bundles a with large number of fibres. A fictitious arrangement sideways of all fibres allows a visualization of distributions ofproperties, stress etc, through the bundle, cf. [10-12]. In this context the ascending part of the stress-strain rate curve is modelled as a power law behaviour, the ordinary Norton creep law, whereas the descending part is described by another power law relation with different modulus and exponent. The model degenerates to a bi-linearly viscous material by setting both exponents equal to unity, see Fig. 1. The Norton creep law (the ascending part) is formulated as

$$\dot{\varepsilon} = \frac{1}{\tau_0}\left(\frac{\sigma}{N}\right)^n \qquad 2.1$$

where N is the creep modulus (with dimension stress) and n is the creep exponent, τ_0 is a time constant. The descending part is modelled as

$$\dot{\varepsilon} - \dot{\varepsilon}_m = \frac{1}{\tau_0}\left(\frac{\sigma_m - \sigma}{S}\right)^s \qquad 2.2$$

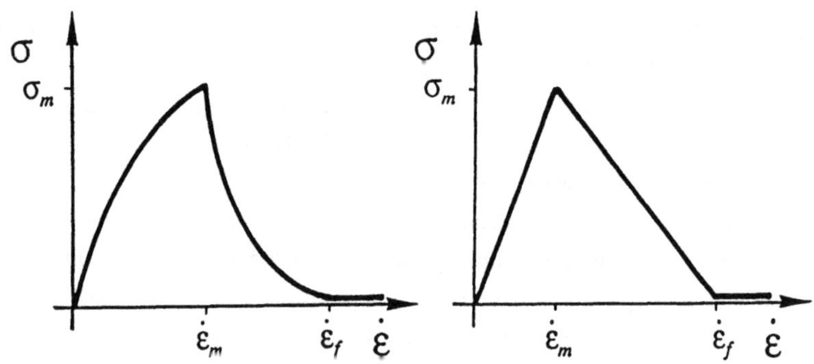

Figure 1. Stress versus strain rate for fibre materials with a hardening-softening creep relation

Here S and s are softening creep modulus and corresponding creep exponent respectively, σ_m is the maximum stress which can be sustained by the material under uniaxial loading, and $\dot{\varepsilon}_m$ is the resulting strain rate at this stress. At the strain rate $\dot{\varepsilon}_f$, which is determined by the properties defined above, the fibre is said to have failed finally.

As in [12], it is here assumed that the strength of the fibres varies among them and is distributed throughout the bundle in some arbitrary way. The arrangement of the fibres allows the variation of strength σ_m within the bundle to be expressed as

$$\sigma_m(x) = \overline{\sigma}_m f(x) \qquad 2.3$$

where $\overline{\sigma}_m$ is the arithmetic mean value and $f(x)$ is a distribution function with unity mean value; it depends on the sorting of the fibres. In the same manner the creep moduli vary along another shape function

$$N_m(x) = \overline{N} g(x) \qquad 2.4$$

and

$$S_m(x) = \overline{S} g(x) \qquad 2.5$$

while both creep exponents n and s are taken to be constant in the bundle.

The variation of constitutive parameters as moduli and exponents would generally be of a pronouncedly lower order than the scatter of severity of flaws and defects and consequently the distribution of ultimate stress σ_m. However, the ideal fibre bundle model states fibres with equal cross section, which is not necessarily granted for an arbitrary bundle. A scatter of fibre geometry can here be accounted for by changing both definitions of stress, ultimate stress, and fibre modulus to correspond to a uniform fibre area. This also justifies the assumptions of letting both hardening and softening vary along the same distribution function $g(x)$ while the creep exponents n and s are held constant in the formulations above. Under these conditions of varying properties and of material behaviour the object is now to establish equations to determine the time to rupture and the mode of rupture for such a bundle.

3 INCUBATION OF DAMAGE

The rearrangement of the fibres mentioned previously is done in a certain manner, viz. the fibres are sorted from left to right in order of beginning failure, i.e. the order in which they reach their ultimate stress σ_m. In that way they will start to fail, to pass their peak stress, from left to right. Then the damage suffered to the bundle can be interpreted as the width of a damaged zone. As mentioned previously two parameters are needed to describe the state of damage. Both damage processes will start to occur at x=0 and grow monotonically into the bundle. This sorting of the fibres settles the distribution functions f and g; it implies the condition

$$\frac{\mathrm{d}}{\mathrm{d}x}\left(\frac{f(x)}{g(x)}\right) \geq 0 \qquad 3.1$$

Equality relates to domains with identical fibres or domains of fibres which will rupture at the same elongation. It is expedient to assume that both f and g are smooth functions to facilitate the mathematical formulation. Generally this is not the case but the validity of the results presented still applies.

The stress in the bundle can be separated into a time dependent part and a geometric part. Adopting the logarithmic strain definition it reads

$$\sigma(t,x) = \sigma_0 e^{2\mu\varepsilon} g(x) \tag{3.2}$$

Here σ_0 is the initial load (at t=0) and μ is Poisson's ratio, usually assumed to be 0.5 (constant volume). Eqs. (3.2) and (2.3) are sufficient to determine the strain ε_i at the end of the incubation time t_i

$$\varepsilon_i = \frac{1}{2\mu} \log \frac{\overline{\sigma}_m f(0)}{\sigma_0 g(0)} \tag{3.3}$$

Insertion of expression (3.2) into the creep law (2.1) together with eq. (2.4) yields a separable differential equation in ε with integration limits 0 and ε_i. The evaluated integral yields the incubation time

$$t_i = \frac{\tau_0}{2\mu n} \left(\frac{\overline{N}}{\sigma_0}\right)^n \left[1 - \left(\frac{\sigma_0 g(0)}{\overline{\sigma}_m f(0)}\right)^n\right] \tag{3.4}$$

A thorough derivation of eqs. (3.1)-(3.4) is presented in [12].

4 GROWTH OF DAMAGE

In a bundle with scatter of properties the initiation of damage growth does not have to lead to final rupture, especially not when the individual fibres are able to carry a load also after having passed their peak stress level. This depends on whether the load can be redistributed among the remaining fibres without loss of stability.

When the first fibre reaches its peak stress level (at x=0) a damage front will start to move into the bundle. The coordinate of this front will be denoted c_m. In all fibres to the right of that front, the stress is still increasing, they are governed by the ascending part of the creep law, while on the left the stress in each fibre is decreasing again. In each of these domains the stress profile can be expressed in terms of the strain rate using the appropriate part of the constitutive relation.

The strain rate corresponding to the peak stress is distributed through the bundle in a manner determined by the scatter of properties and the (ascending part of the) creep law

$$\dot{\varepsilon}_m = \frac{1}{\tau_o} \left(\frac{\overline{\sigma}_m}{\overline{N}}\right)^n \left(\frac{f(x)}{g(x)}\right)^n \tag{3.5}$$

Eq. (3.5) can also be used to express the actual strain rate $\dot{\varepsilon}$ for a partly damaged bundle in terms of the damage front coordinate c_m and bundle properties by inserting x=c_m.

Solving now eqs. (2.1) and (2.2) respectively, for the stress and substituting expression (3.5) for $\dot{\varepsilon}_m$ and $\dot{\varepsilon}$ respectively, gives the stress profile in the two domains. In the undamaged part it is

$$\sigma(x) = \bar{\sigma}_m \frac{f(c_m)}{g(c_m)} g(x) \qquad 3.6$$

whereas in the interval $[0,c_\mathrm{m}]$ the stress has the form

$$\sigma(x) = \bar{\sigma}_m f(x) - \bar{S} g(x) \left[\left(\frac{\bar{\sigma}_m}{\bar{\bar{N}}} \right)^n \left(\frac{f(c_m)}{g(c_m)} \right)^n - \left(\frac{\bar{\sigma}_m}{\bar{\bar{N}}} \right)^n \left(\frac{f(x)}{g(x)} \right)^n \right]^{\frac{1}{s}} \qquad 3.7$$

Eq. (3.7) is valid as long as no fibre fails completely, i.e. as long as $\dot{\varepsilon}_f$ is not exceeded in any fibre. The variation of $\dot{\varepsilon}_f$ can be derived from the descending part of the creep law (2.2), setting $\sigma = 0$ and substituting eq. (3.5) for $\dot{\varepsilon}_m$

As the first fibre fails completely at x=0 a second damage front c_f will move into the bundle. Its location depends on c_m in a manner given by eq. (3.8). For x=c_f the stress vanishes and eq. (3.7) is sufficient to obtain the correlation between c_m and c_f

$$\left(\frac{\bar{\sigma}_m}{\bar{S}} \right)^s \left(\frac{\bar{\bar{N}}}{\bar{\sigma}_m} \right)^n \left(\frac{f(c_f)}{g(c_f)} \right)^s + \left(\frac{f(c_f)}{g(c_f)} \right)^n = \left(\frac{f(c_m)}{g(c_m)} \right)^n \qquad 3.8$$

Eq (3.8) defines $c_\mathrm{f}(c_\mathrm{m})$, or vice versa, everywhere but for domains where the ratio f/g is constant. There may be values of c_m for which no unique corresponding value c_f exists. It will later be shown that no particular difficulties arise from this. Setting c_f=0 in (3.8) determines the value c_m^0 of the first damage parameter at which a second damage front, c_f, starts to develop. Fig. 2 shows possible stress profiles in partly damaged bundles.

If c_m exceeds c_m^0 a third region with $\sigma \equiv 0$ will occur within the interval $[0,c_\mathrm{f}]$, beside eqs. (3.6) and (3.7). Integrating the stress over the whole interval [0,1] yields the mean stress which also is stated by (3.2). This relation between damage c_m and strain reads

$$\sigma_o e^{2\mu\varepsilon} = \bar{\sigma}_m \left\{ [F(x)]_0^{c_m} + \frac{f(c_m)}{g(c_m)} [G(x)]_{c_m}^1 \right\} - \bar{S} \left(\frac{\bar{\sigma}_m}{\bar{\bar{N}}} \right)^{\frac{n}{s}} \int_0^{c_m} g(x) \left[\left(\frac{f(c_m)}{g(c_m)} \right)^n - \left(\frac{f(x)}{g(x)} \right)^n \right]^{\frac{1}{s}} dx \qquad 3.9$$

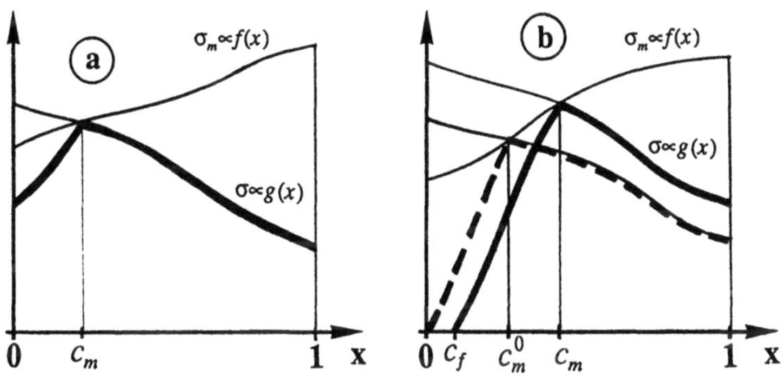

Figure 2. Stress profiles in a bundle with only one damage front c_m (a) and one with the second damage front c_f developed (b)

for $c_m < c_m^0$ and

$$\sigma_o e^{2\mu\varepsilon} = \bar{\sigma}_m \left\{ [F(x)]_{c_f}^{c_m} + \frac{f(c_m)}{g(c_m)} [G(x)]_{c_m}^{1} \right\} - \bar{S} \left(\frac{\bar{\sigma}_m}{\bar{N}} \right)^{\frac{n}{s}} \int_{c_f}^{c_m} g(x) \left[\left(\frac{f(c_m)}{g(c_m)} \right)^n - \left(\frac{f(x)}{g(x)} \right)^n \right]^{\frac{1}{s}} dx \qquad 3.10$$

for $c_m > c_m^0$. Here F and G are primitive functions of f and g respectively. If the lower integration limit c_f in eq. (3.10) is not unambigiously determined by eq. (3.8) any value c_f fulfilling (3.8) can be used since $\sigma = 0$ for all choices and $\sigma = 0$ by definition for $x < c_f$.
Note that the right hand members in the two expressions (3.9) and (3.10) are functions of c_m only and contain all information about the bundle's material properties and their scatter. Also c_m^0, and thus the transition between eqs. (3.9) and (3.10) depends only on bundle properties. The two equations are rewritten in the compact form

$$\sigma_o e^{2\mu\varepsilon} = Q(c_m) \qquad 3.11$$

where $Q(c_m)$ is one of the right hand members of eqs (3.9) or (3.10) depending on whether or not $c < c_m^0$. The function $Q(c_m)$ is the equivalent of the scatter function $R(c)$ introduced in [12]. But here, depending on the more complex model of material behaviour, it contains information not only about the scatter of properties but also the values of the properties themselves.

5 FAILURE CONDITIONS AND MODES

Solving eq. (3.11) for ε yields

$$\varepsilon = \frac{1}{2\mu} \log Q(c_m) - \frac{1}{2\mu} \log \sigma_0 \qquad 3.12$$

Apparently the initial load σ_0 affects the height but not the shape of the ε-c_m curve. The intersection with the ε-axis gives the incubation strain ε_i.
The relation (3.12) between strain and damage c_m is derived solely from equilibrium, without consideration of stability. Since the strain cannot decrease under constant load, all domains where $Q(c_m)$ has a negative slope must be invalid. Differentiation of eq (3.12) with respect to time gives

$$2\mu\sigma_o e^{2\mu\varepsilon}\dot{\varepsilon} = Q'(c_m)\dot{c}_m \qquad 3.13$$

Since the left hand member of (3.13) is strictly positive, the damage front velocity $\dot{c}_m$ increases without limit when its coefficient $Q'(c_m)$ vanishes. That occurs at certain damage levels denoted c_m^* where $Q(c_m)$ has a maximum. Also these damage levels depend on bundle properties only. As a bundle failure criterion may be taken the damage level at which the damage rate approaches infinity. Several such levels may exist. Final bundle failure will occur at the damage level c_m^* where $Q(c_m^*)$ has a maximium corresponding to the highest strain (note that $Q(c_m)$ may have a maximum for c_m=0). Different shapes of the relations (3.19), which lead to different modes of final rupture, are shown in Fig. 4. The two mainly different modes are instantaneous (a) and gradual failure (b), but also combinations of these two exist where a limited part of the fibres fails instantaneously without causing complete failure (c).

After having passed a maximum of $Q(c_m)$, the damage grows instantaneously. As only viscous deformation is considered, the strain ε remains unchanged during such a damage leap and the path in the ε-c_m plane is a straight horisontal line. If this line again intersects the original curve ε-c_m, this damage leap stops and equilibrium is once more established, and damage will grow progressively until the next maximum is reached.

The domains where c_m grows instantaneously and where it grows progressively depend only on the shape of $Q(c_m)$.

Equation (3.13) can also be used for an alternative expression of the strain rate. The left hand member still contains $Q(c_m)$, and thus

$$\dot{\varepsilon} = \frac{Q'(c_m)}{2\mu Q(c_m)} \dot{c}_m \qquad 3.14$$

The strain rate is also stated by the constitutive relation and can be expressed in terms of the first damage front location, c.f. eq (3.5). Combining these two now yields a separable differential equation in c_m which governs the damage evolution in time.

$$\frac{Q'(c_m)}{2\mu Q(c_m)} \dot{c}_m = \frac{1}{\tau_0} \left(\frac{\overline{\sigma}_m}{\overline{\overline{N}}} \right)^n \left(\frac{f(c_m)}{g(c_m)} \right)^n \qquad 3.15$$

The failure propagation time interval Δt_p can be calculated directly from this by integrating over the first damage front from c_m=0 up to its critical value $c^*_{m,crit}$. The domains where c_m grows in leaps have, of course, to be excluded from the integration.

$$\Delta t_p = \tau_0 \int_{\{c_m\}} \frac{Q'(c_m)}{2\mu Q(c_m)} \left(\frac{\overline{N}}{\overline{\overline{\sigma}}_m} \right)^n \left(\frac{g(c_m)}{f(c_m)} \right)^n dc_m \qquad 3.16$$

It is seen from eq. (3.16) that the propagation of damage and thus subsequent failure are independent of the initial load σ_0. This result was already hinted at in [11] and later proved for more general conditions[12]. It is here shown to hold also for fibre materials with post critical strength.

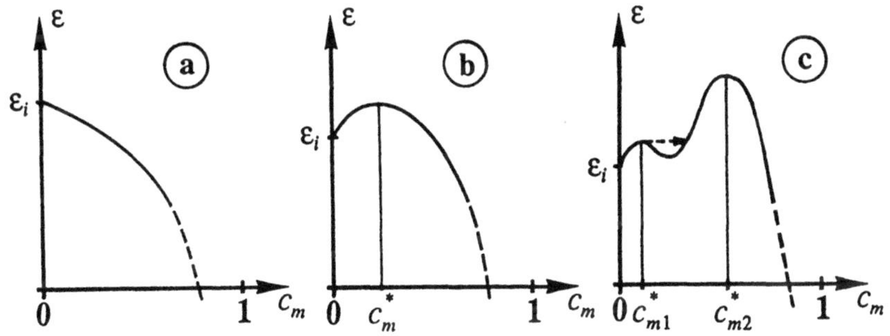

Figure 3. Different distributions af material properties lead to different modes of final rupture

Acknowledgment - This project was partly financed by the Swedish Board for Technical Development (STU). Valuable discussions with Professor J. Hult are gratefully acknowledged.

REFERENCES

[1] F. T. Pierce, Tensile tests for cotton yarns vs "the weakest link". *J.Textile Inst.* **17**, T355-T368 (1926)

[2] H. E. Daniels, The statistical theory of the strength of bundles of threads-I. *Proc. Roy. Soc. London* **A183**, 405-435 (1945)

[3] S. L. Phoenix and H. M. Taylor, The asymptotic strength distribution of a general fibre bundle. *Adv. appl. Prob.* **5**, 200-216 (1973)

[4] S. L. Phoenix, Stochastic strength and fatigue of fibre bundles. *Int. J. Fracture* **14**, 327-344 (1978)

[5] D. G. Harlow and S. L. Phoenix, The chain-of-bundles probability model for the strength of fibrous materials I: Analysis and conjectures. *J. Compos. Mater.* **12**, 195-214 (1978)

[6] S. L. Phoenix and R. L. Smith, A comparison of probabilistic techniques for the strength of fibres under local load-sharing among fibres. *Int. J. Solids Structures* **19**, 479-496 (1983)

[7] D. C. Lagoudas, C.Y. Hui and S. L. Phoenix, Time evolution of overstress profiles near broken fibres in a composite with a viscoelastic matrix. *Int. J. Solids Structures* **25**, 45-66 (1989)

[8] N. J. Hoff, Necking and rupture of rods under tensile loads. *J. Appl. Mech.* **20**, 105-108 (1953)

[9] J. B. Martin and F. A. Leckie, On creep rupture of structures. *J. Mech. Solids* **20**, 223-238 (1972)

[10] J. Hult and L. Travnicek, Carrying capacity of fibre bundles with varying strength and stiffness. *J. Méc. Théor. Appl.* **2**, 643-657 (1983)

[11] M. Chrzanowski and J. Hult, Ductile creep rupture of fibre bundles. *Eng. Fracture Mech.* **28**, 681-688 (1987)

[12] J. Neumeister, Creep and rupture behaviour of fibre bundles. *Eng. Fracture Mech.* **35**, 1081-1092 (1990)

[13] J. Neumeister, Creep rupture in fibre bundles with hardening-softening fibre materials. *Chalmers University of Technology, Division of Solid Mechanics*, Publication F136 (1990)

Creep-Fatigue Life and Cyclic Constitutive Relation of MAR-M247 Directionally Solidified Superalloy at High Temperature

M.OHNAMI, M.SAKANE and T.HASEBE

Faculty of Science and Engineering, Ritsumeikan University, Kyoto, Japan

Summary

This paper describes the effect of solidification direction on creep-fatigue life of Mar-M247 directionally solidified (DS) superalloy. The specimen of which axis agrees to the solidification direction exhibited the maximum life, while the specimen which has 45° against the solidification direction exhibits the minimum life. Transversely directed specimen had the intermediate life. The difference in creep-fatigue life due to the loading direction was resulted from the anisotropy of Young's modulus. A life prediction method for DS superalloy, considering the anisotropy of the Young's modulus, was proposed and successfully applied to the creep-fatigue data.

Introduction

This paper analytically estimates the cyclic constitutive relation of Mar-M247 directionally solidified (DS) superalloy under creep-fatigue condition considering the anisotropy of the Young's modulus and predicts creep-fatigue lives based on the derived constitutive relation. Creep-fatigue damage of the DS plate, which is composed of five columnar grains, was calculated using three dimensional finite element analysis. This paper also discusses the effect of crystal rotation about DS axis as well as deviation of solidification direction from [001] axis on the damage and the creep-fatigue life prediction.

Inelastic Constitutive Relation

For the creep-fatigue life prediction, the estimate of stress amplitude is essential, since the damage of superalloy is controlled by stress amplitude rather than inelastic strain amplitude. This paper develops an inelastic constitutive relation taking account of the anisotropy of the Young's modulus.

Total strain increment, $d\varepsilon_t$, is assumed here to be a linear summation of elastic, $d\varepsilon_e$, plastic, $d\varepsilon_p$, and creep, $d\varepsilon_c$, strain increments:

$$d\varepsilon_t=d\varepsilon_e+d\varepsilon_p+d\varepsilon_c$$

No interaction between plastic and creep strains is considered for the

M. Zyczkowski (Ed.)
Creep in Structures
IUTAM Symposium Cracow/Poland 1990

simplicity of the theory. For the Mar-M247 DS superalloy, Young's modulus has significant anisotropy due to the crystallographic structure. It is expressed as a function of two angles, θ and ϕ, and compliances, s_{ij}, as follows[1].

$$E(\theta,\phi)=[s_{11}-2\{s_{11}-s_{12}-(1/2)s_{44}\} \times \sin^2\theta\{\cos^2\theta+(1/4)\sin^2\theta\cdot\sin^2 2\phi\}]^{-1} \quad (1)$$

Two angles, θ and ϕ, respectively denote angles between specimen axis and [001] direction, and rotation angle about the specimen axis as shown in Figure 1. In the figure the specimen axis is represented by a heavy vector. Figure 2[2] shows the Young's modulus as a function of θ and ϕ. The figure clearly shows that the Young's modulus significantly depends on θ and ϕ.

On the contrary to the Young's modulus, there is no anisotropic relation between plastic strain, creep strain and stress amplitude, which has been confirmed in experiments[2]. Thus the plastic strain can be written as $\varepsilon_p=(\sigma/K)^{1/n}$ and creep strain as $\varepsilon_c=a\sigma^{\alpha}t^m$. K and n are the material constants related to the plastic strain and have the values of 2016MPa and 0.24 respectively[2]. The material constants, a, α and m, appeared in the creep equation have the values of 3.11×10^{-42}%, 14.7 and 2.04 in the large stress range and they have 13.11×10^{-46}%, 18.2 and 2.09 in the small stress range, where σ is taken in MPa and t in hour. Material constants in the creep equation was determined from the cyclic relaxation tests. The inelastic constitutive equation is finally expressed as:

$$d\varepsilon_t=[(1/E)+(1/n)\cdot(\sigma/K)^{(1/n)-1}]d\sigma+ma\sigma^{\alpha}t^{m-1}dt \quad (2)$$

Experimental Procedure

Test material used in this paper is Mar-M247LC Ni-base directionally solidified (DS) superalloy. Table 1 lists the chemical composition in weight percent ratio and heat treatment of the material. Specimen used was a small cylinder with 5 mm diameter and 15 mm gage length. Three kinds of specimens were employed as illustrated in Figure 3. They were cut out from DS plates to provide three kinds of specimen axes of 0°, 45° and 90°. Cutout angles of 0°, 45° and 90°, correspond to the angle of θ defined in Figure 1. The width of columnar grain is between 5 and 10 mm. There exists a deviation between solidification direction and [001] axis. The maximum deviation is about 15°, which was confirmed by X-ray Laue Method.

An electric hydraulic servo-type uniaxial fatigue machine conducted total strain range controlled low cycle fatigue tests with a trapezoidal strain wave at 1173 K in air. Hold times introduced at the tensile peak strain were 10 and 60 minutes. Strain rate of the test was 0.1%/sec. A high-frequency induction furnace heated the specimen. The number of cycles to failure, N_f, was defined as the cycle at which the tensile stress amplitude decreases to 75 % from saturation.

Experimental Results and Discussion

(1) Cyclic Stress Response

Figures 4(a), (b) and (c) compare the hysteresis loops obtained in experiment with those calculated from Equation (2) for 10 min. hold-time test. On the experimental results represented by solid lines, the $\theta=45^o$ specimen exhibits the largest stress amplitude, the $\theta=90^o$ specimen the intermediate and the $\theta=0^o$ specimen the smallest. As is usually seen in hysteresis loops of superalloys, the plastic strain has a small percentage in the total strain but elastic and creep strains are predominant in the hysteresis loops.

The theoretical constitutive equation developed in this paper and is expressed by Equation (2) quite well follows the actual cyclic stress-strain relation with 10 min. hold-time. Thus, we can conclude that the difference in the cyclic stress amplitude due to the cutout direction is resulted from the anisotropy of the Young's modulus.

(2) Creep-Fatigue Life Prediction

Figure 5 shows the creep-fatigue lives in hold-time test and no hold-time tests for the three kinds of specimens. The former is plotted with open marks and the latter with closed marks. The open circle, square and triangle respectively indicate the creep-fatigue lives of the 0°, 45° and 90° specimens in the tests with 10 min. hold-time. The angle θ has a significant effect on the creep-fatigue lives. The failure life of the 0° specimen is the largest, while that of the 45° specimen the smallest and the 90° specimen exhibits the intermediate. The life reduction due to the hold-time is about 55% for the 45° specimen and about 30% for the 90° specimen. The 0° specimen, however, exhibits no life reduction due to 10 min. hold-time.

Now we will make a creep-fatigue life prediction using the constitutive relation expressed in Equation (2). In predicting the creep-fatigue life, cyclic stress amplitude was calculated using Equation (2) and then the creep-fatigue damage was estimated based on the following equation.

$$N_f(D_f+D_c)=D \qquad (3)$$

D_f is the fatigue damage per cycle calculated from the stress-failure cycle relation. D_c is the creep damage per cycle and is written as $D_c=\int_0^{t_H}(dt/t_{r,cyclic})$, where the cyclic creep rupture time, $t_{r,cyclic}$, is assumed to be ten times of the monotonic creep rupture time.

Figure 6 compares the predicted creep-fatigue life with the experimental life. All the data fall into a factor of two scatter band so that creep-fatigue life prediction method developed above is valid for DS superalloys.

(3) Creep-Fatigue Life of DS Plate

In the previous section, we have discussed the creep-fatigue life of the DS superalloy but the specimen size is small so that the discussion is on the single crystal model. Actual turbine blades, however, are composed of several columnar grains, and that the treatment of the DS plate as an aggregate of several grains is necessary. In this section, we will calculate the creep-fatigue damage of the DS plates which are composed of five columnar grains.

Figure 7 shows DS plates with different solidification directions of 0°, 26.6^o, 45°, 63.4^o and 90° to the loading direction z[3]. These DS plates have the dimensions of 50×50×8 in millimeter. Two types of DS plates are considered; the coaxis and non-coaxis models[3]. The coaxis model means that the solidification direction of all the five columnar grains agrees to [001] axis but four of five columnar grains have the different initial rotation angles about [001] axis. The grains ② and ④ have $\phi_0=45^o$, ① and ⑤ $\phi_0=25^o$ and only the grain ③ has no rotation. The non-coaxis model is similar to the coaxis model but additionally has an initial deviation in the solidification direction. The initial deviation angle between solidification direction and [001] axis, θ_o, is assumed to be $\pm15^o$ for the grains ①, ③ and ⑤. Five DS plates were stretched in z-direction and creep-fatigue damage was estimated.

Figures 8(a) and (b) respectively show the creep damage, D_c, and fatigue damage, D_f, per cycle for the coaxis model at the total strain ranges of 0.7, 1.0, 1.2 and 1.5% with 10 min. hold-time. The figures show that D_c and D_f significantly depend on the orientation angle, θ. D_f yields the maximum value at around $\theta=45^o$ at all the total strain ranges. D_c calculated, on the other hand, has the maximum value at $\theta=90^o$ at the large total strain ranges.

Figure 9 shows the predicted creep-fatigue lives for the coaxis and non-coaxis models, where (a) is for the plate of $\theta=26.6^o$ and (b) for that of

63.4°. The solid line denotes the predicted results of the coaxis model while the chained and broken lines respectively denote those of the non-coaxis models with θ_0=15° and -15° deviations. The predicted life apparently depends on initial deviation angle, θ_0. The creep-fatigue life for θ_0=-15° is shorter than that for θ_0=15° for the θ=63.4° model and the reversed relation appears for the θ=26.6° model.

Conclusion

Inelastic plastic creep constitutive equation taking account of the anisotropy of the Young's modulus was developed. The stress response calculated from the constitutive relation in combination with the linear damage law was successfully applied to the life prediction of the small DS specimen. The creep-fatigue life prediction method for DS plates composed of five columnar grains was proposed.

Acknowledgments

The authors express their gratitude to Mr. S. Ohama, General Manager, Aero-Engine & Space Operation, Ishikawajima-Harima Heavy Industries Co. Ltd., for supplying the DS superalloy and Dr. F. Masuyama, Assistant Chief Research Engineer, Materials & Welding Laboratory, Nagasaki Research & Development Center, Mitsubishi Heavy Industries, Ltd., for machining the specimens. This work is partly supported by Grant-in-Aid for Scientific Research, B 63460074, the Ministry of Education, Science and Culture, Japan.

References

1. Nye,J.F.: Physical properties of crystals, Oxford University Press, 1957, pp.143-145.

2. Hasebe,T.; Sakane,M.; Ohnami,M: High temperature low cycle fatigue and cyclic constitutive relation of Mar-M247 directionally solidified superalloy, Jour. Engng. Mater. Techno., Trans. ASME (to appear).

3. Hasebe,T; Sakane,M.; Ohnami,M.: Elastic Anisotropy of Directionally Solidified Superalloy. Jour. Engng. Mater. Techno. Trans. ASME (to appear).

Table 1. Chemical composition and heat treatment of the material tested

(a)Chemical composition (wt%)

C	Si	Mn	Al	B	Nb	Co	Cr
0.071	0.007	<0.001	5.55	0.015	<0.05	9.2	8.0

Cu	Fe	Hf	Mo	P	Ta	Ti	W
<0.001	0.034	1.4	0.50	<0.005	3.2	0.70	9.4

Zr	V	Ni
0.022	<0.002	Bal

(b)Heat treatments

Solution heat treatment
(vacuum furnace)
: 1221℃×2hr Ar GFC

Two-step aging heat treatment
: 982℃×5hr Ar GFC
871℃×20hr Ar GFC

(Ar GFC: Ar gas forced cooling)

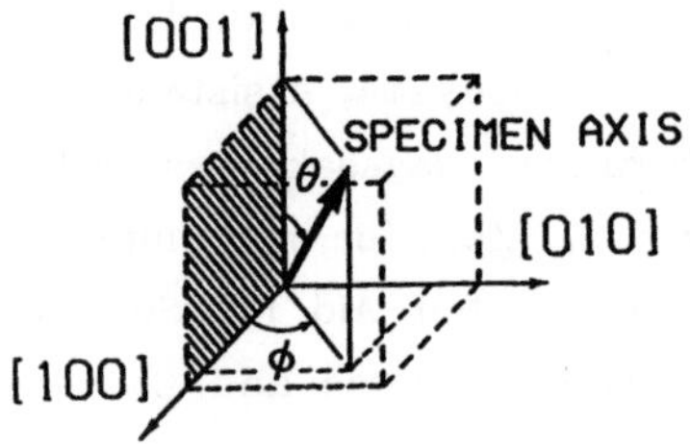

Fig.1. Definition of the orientation angle θ and φ.

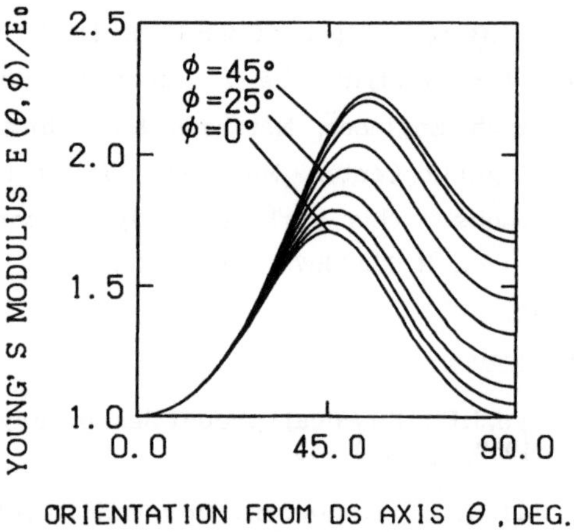

Fig.2. Variation of the Young's modulus with two angles, θ and φ[2].

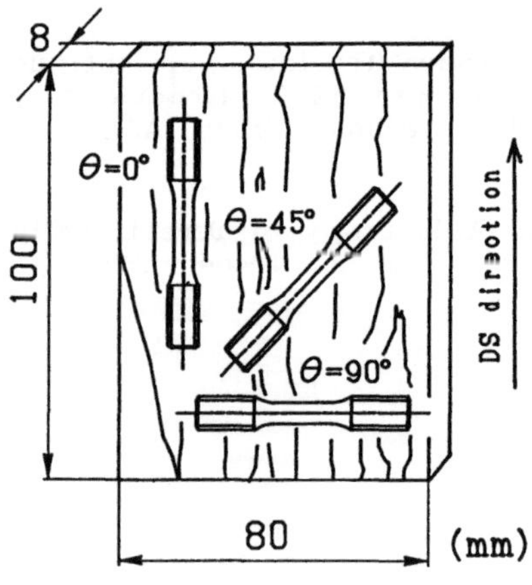

Fig.3. Cutout direction of the small cylindrical specimens.

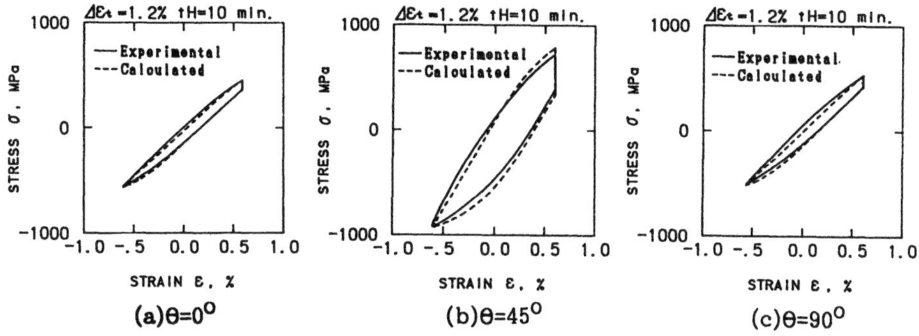

Fig.4. Comparison of the calculated hysteresis loops for small specimen with the experimental results in the 10 min. hold-time test.

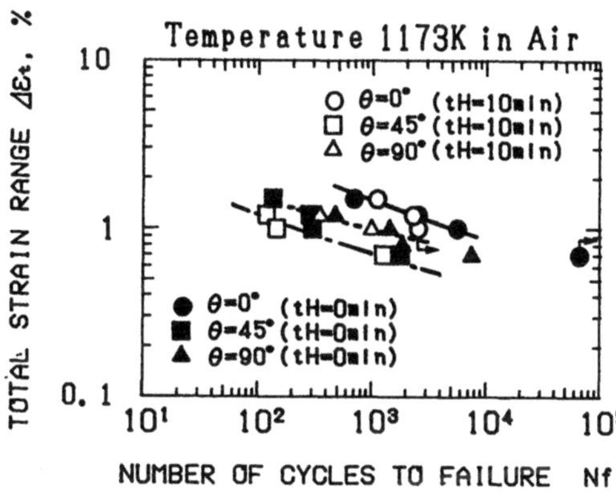

Fig.5. Creep-fatigue life of DS small specimens in 10 min. hold-time and no hold-time tests. Open symbols denote the hold-time test while solid symbols no hold-time test.

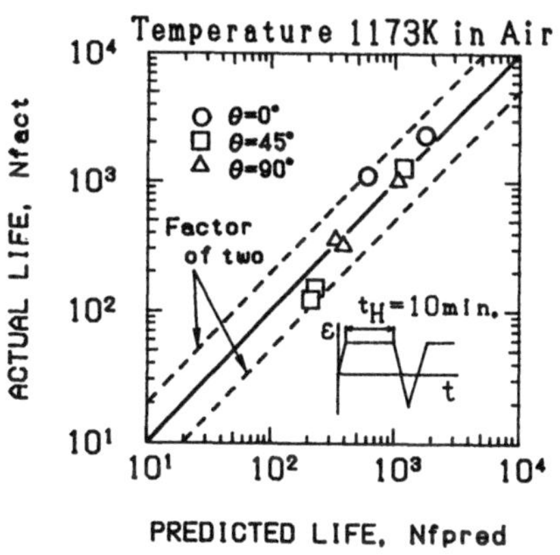

Fig.6. Comparison between the predicted and the exprimental creep-fatigue lives for DS small specimen.

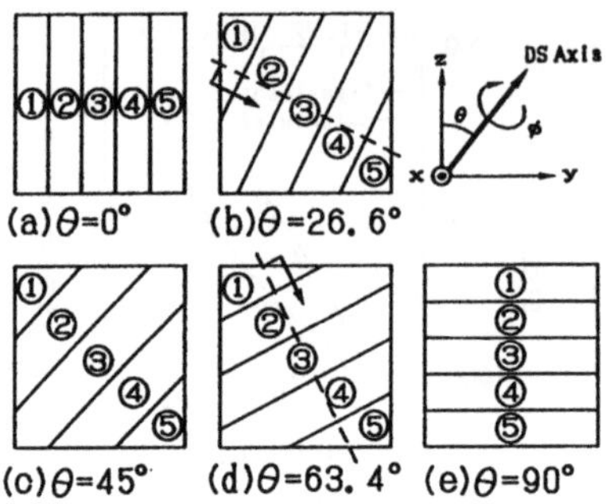

Fig.7. Five DS plates analized.

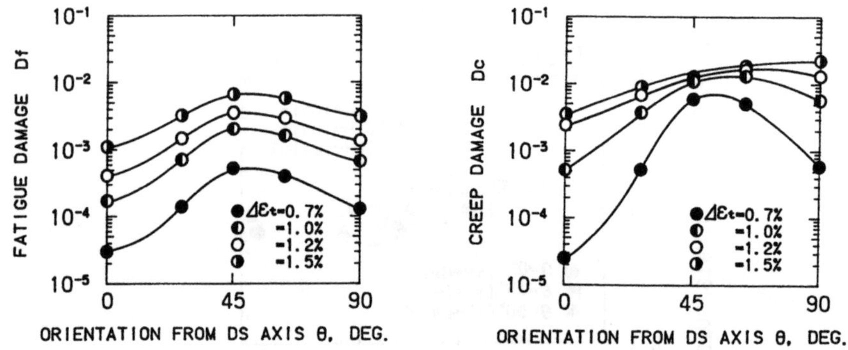

(a)Fatigue damage

(b)Creep damage

Fig.8. Variations of fatigue and creep damage with orientation angle for the coaxis model.

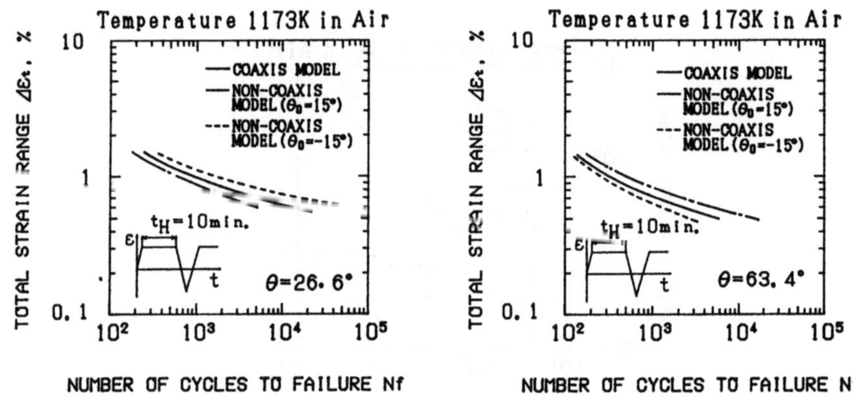

(a)θ=26.6^{O}

(b)θ=63.4^{O}

Fig.9. Predicted creep-fatigue lives of coaxis and non-coaxis plates for (a)θ=26.6^{O} and (b)θ=63.4^{O}.

Percolation Models as a Basis of Material Failure

Martin Ostoja-Starzewski
Department of Metallurgy, Mechanics and Materials Science
Michigan State University
East Lansing, MI 48824-1226, U.S.A.

ABSTRACT

We use a paradigm of damage formation in a polycrystalline layer as a basis for development of percolation models of material failure. The model is set in the framework of a graph-representable microstructure with damage being confined to the honeycomb network of grain boundaries. Strengths of grain boundaries are assumed to be described by space-homogeneous statistics. A hierarchy of Markov random fields is introduced to grasp the attractive effects in the spread of damage between the first, second, third ... nearest neighbors. The simplest case in the hierarchy corresponds to a Bernoulli percolation, which can easily be solved. The higher cases of nontrivial Markov fields may be analyzed via a reduction to a problem of phase transition on a graph. While the focus here is on equilibrium reversible phenomena, various extensions to more realistic and richer models are possible.

Introduction

Damage is a phenomenon of failure of a solid's microstructure involving at least two spatial scales. Considering that many materials may conveniently be represented by graphs [1,2] and that their microscale physical and geometric characteristics are random, stochastic processes (fields) on graphs offer an avenue for analysis of damage phenomena [3]. In this paper we outline recent advances of this research program and specifically, we focus on a mapping of problems in damage formation onto a class of stochastic processes that naturally grasp the cooperative spatio-temporal effects: Bernoulli percolation and Markov random fields. Such a mapping has a great advantage - of reducing the number of computations - over the existing computer-intensive methods which rely on simulations of failures of moderately large lattices (see e.g. [4,5]). We outline this mapping using an example problem of damage formation in a thin polycrystalline layer. We discuss basic facts about Markov random fields and phase transitions that may occur on them, with a phase transition being indicative of a change in global behavior from partial- to macro-damage. We conclude the paper with a discussion of possible extensions of this model as well as its relation to more conventional methods of solid mechanics.

M. Zyczkowski (Ed.)
Creep in Structures
IUTAM Symposium Cracow/Poland 1990

The Percolation Problem

As mentioned in the Introduction we consider a problem of damage formation in a very thin layer on a substrate such as that of aluminum oxide Al_2O_3 on an aluminum panel, see Fig. 1. The layer which is made of crystals with spatially random crystallographic axes orientations, lends itself to a planar *graph representation* of a Voronoi-type, see Fig. 1b) in [3]. Here $G'(V',E')$ is a graph of the polycrystalline tessellation, in which E' is the set of all grain boundaries (i.e. edges) and V' is the set of all end vertices of these edges.

While the qualification of any single edge $e \in E'$ to be in only one of two states - intact or damaged - seems to be a very simplistic one, it affords a possibility of grasping the spatial cooperative phenomena on the set E'. Accordingly, we introduce a *binary random variable* Z describing the state of any single edge $e \in E'$:

$$Z(e,\omega) = \begin{cases} z_1 & \text{if } e \text{ is intact} \\ z_2 & \text{if } e \text{ is damaged} \end{cases} \tag{1}$$

ω denotes a particular realization of geometric (e.g. orientations of Al_2O_3 crystals separated by e) and physical (e.g. grain boundary strengths) randomness. The damage phenomenon is assumed reversible so that Z may switch back and forth between z_1 and z_2 according to the strength condition at e being reached or not. The external loads are given in terms of either a macroscopic strain $\bar{\varepsilon}$ or a macroscopic stress $\bar{\sigma}$ applied quasi-statically to the substrate. Our goal is to find $\bar{\varepsilon}$, or $\bar{\sigma}$, at the onset of macro-damage when the type of randomness - i.e. description of all $\omega \in \Omega$ - is specified.

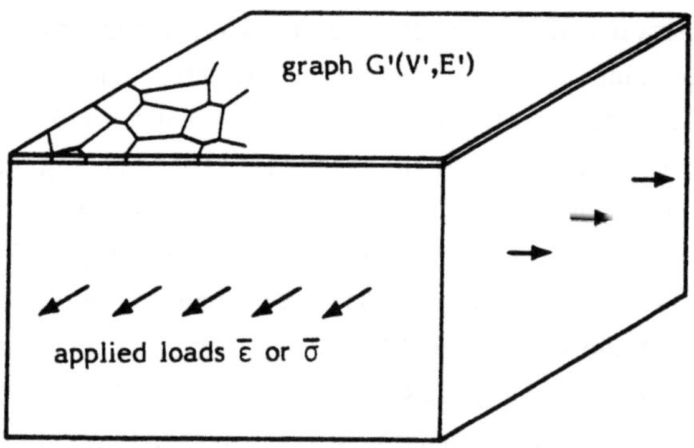

Fig. 1. Thin layer of Al_2O_3 with a graph microstructure (planar tessellation) on a thick substrate of Al.

The states of damage z of the layer are realizations of a *binary random field*

$$Z:\ E' \times \Omega \rightarrow \{z_1, z_2\} \tag{2}$$

Thus $Z = \{z_1, z_2\}^{E'}$ is the *damage state space*, having, in principle, $2^{|E'|}$ different elements; $|E'|$ denotes the number of elements in E′. We want to characterize Z in terms of a probability measure P on (Z, F), where F is a σ-algebra of Z. More specifically, P has to be parameterized by the applied loading, say $\bar{\varepsilon}$, so that $P_{\bar{\varepsilon}}\{Z(\omega) = z\}$ is a probability of a specific state z under a given $\bar{\varepsilon}$. The assumption of reversibility of Z at all $e \in E'$ qualifies $P_{\bar{\varepsilon}}$ as an equilibrium distribution.

While the layer has *clusters* $C_1, C_2, \cdots$ of connected damaged edges, we are interested in a cluster C_B spanning the entire body B of the layer, see Fig. 1. We say that

$$\text{if } C_B \begin{matrix} \text{does not} \\ \text{does} \end{matrix} \text{ exist, then B is in a state of } \begin{matrix} \text{partial–damage} \\ \text{macro–damage} \end{matrix} \tag{3}$$

Since the phenomenon is stochastic we employ a *percolation probability*

$$P_{\bar{\varepsilon}}(C_B) = \begin{cases} 0 & \text{then B is in a state of partial–damage} \\ x>0 & \text{then B is in a state of macro–damage} \end{cases} \tag{4}$$

Thus our original goal may be restated as follows: find $\bar{\varepsilon}$ such that $P_{\bar{\varepsilon}}(C_B)$ makes a transition from 0 to positive.

The global description of the random field Z in terms of $P_{\bar{\varepsilon}}$ depends on the local description given in terms of conditional probabilities $P_{\bar{\varepsilon}}\{z(e) \mid z(\bar{e})\}$ for all possible states z(e) and $z(\bar{e})$ ($\bar{e} = E' - \{e\}$)

$$P_{\bar{\varepsilon}}\{z(e) \mid z(\bar{e})\} = P_{\bar{\varepsilon}}\{z(e) \mid z(\bar{e}),\ {}^{e}F,\ {}^{e}n,\ {}^{\bar{e}}\{n\}\} \tag{5}$$

in which ${}^{e}F$ is a random failure condition of the edge e, ${}^{e}n$ is the orientation of e, and ${}^{\bar{e}}\{n\}$ is the set of orientations of all edges of $\bar{e}$. The failure condition ${}^{e}F$ may be stated in terms of the actual surface traction t at e, or the relative displacement at e, etc. For example, letting ${}^{e}F_R(r)$ denote a probability distribution function of random strength R of edge e (r is any value in the range of R), we may calculate the reliability (probability of no failure) of e

$$P_{\bar{\varepsilon}}\{Z(e) = z_1 \mid z(\bar{e})\} = P_{\bar{\varepsilon}}\{t(\bar{\varepsilon}) < R\} = 1 - {}^{e}F_R(t(\bar{\varepsilon})) \tag{6}$$

Clearly, in order to use the above we need a relation $t(\bar{\varepsilon})$. This relation should reflect the effects of presence (state z_1) and absence (state z_2) of damage at nearest and further neighbors surrounding the edge e.

Markov Random Fields

We shall say that the triplet $(Z,F,P_{\bar{\varepsilon}})$ describes a Markov random field if these three conditions are satisfied:

i) a *positivity condition*

$$P_{\bar{\varepsilon}}\{z\} > 0 \qquad \forall z \in Z \tag{7}$$

ii) a *neighborhood condition*

$$P_{\bar{\varepsilon}}\{z(e) \mid z(\bar{e})\} = P_{\bar{\varepsilon}}\{z(e) \mid Z(N_s^e)\} \qquad \forall e \in E' \tag{8}$$

in which N_s^e - called a *neighborhood* - is the set of all edges other than e that can be reached in a maximum of s steps from edge e on the graph G′(V′,E′), see Fig. 2. The successive cases of s = 1, 2, 3, ... are *s-Markov (random) fields* with nearest and longer range (or nonlocal) dependencies, although the 1-Markov field is often simply termed a *Markov (random) field.* We also consider the case of s = 0, without conditioning, which corresponds to a strictly white noise random field that may also be called a *0-Markov (random) field.*

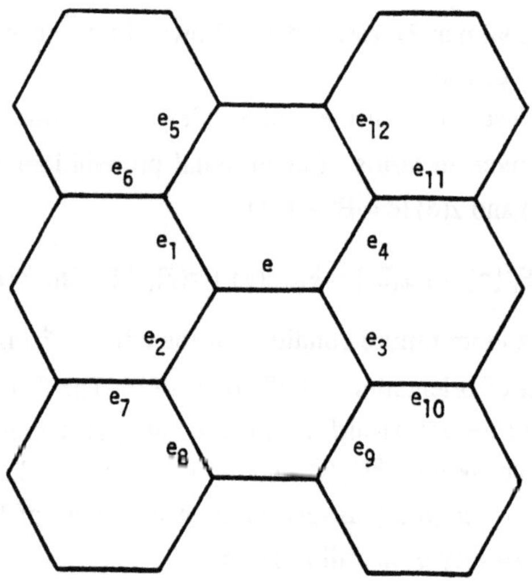

Fig. 2. Edge e and neighborhoods $N_1^e = \{e_1, e_2, e_3, e_4\}$ and $N_2^e = \{e_1, \cdots, e_{12}\}$.

iii) a *consistency condition* expressed generally as

$$P_{\bar{\varepsilon}}\,(A \cap B \mid C) = P_{\bar{\varepsilon}}(A \mid B \cap C)\, P_{\bar{\varepsilon}}(B \mid C) \tag{9}$$

where A = interior, B = neighborhood, C = exterior, and which for, say, a 1-Markov field results in a following relation

$$p_1\,(1 - P_{\bar{\varepsilon}}\{z_1(e) \mid z(N_1^e)\}) = p_2\, P_{\bar{\varepsilon}}\{z_1(e) \mid z(N_1^e)\} \tag{10}$$

In (10) $p_i = P_{\bar{\varepsilon}}\,\{z_i(e),\, z(N_1^e) \mid z(C)\}$, $i = 1, 2$, and $C = E' - N_1^e - \{e\}$.

It is seen from point ii) above that the simplest type of a random field (i.e. 0-Markov field) is one for which the neighborhood condition (8) simplifies to

$$P_{\bar{\varepsilon}}\{z(e) \mid z(\bar{e})\} = P_{\bar{\varepsilon}}\{z(e)\} \qquad \forall e \in E' \tag{11}$$

Here $\bar{\varepsilon}$, being prescribed via a substrate, is a uniform strain field of the layer B, so that $t(\bar{\varepsilon})$ is no obstacle. It follows from (11) that Z is a random field with independence everywhere, so that behavior of every edge is statistically the same, and the problem of determination of a relation between $P_{\bar{\varepsilon}}\{z(e)\}$ and $P_{\bar{\varepsilon}}\{C_B\}$ becomes a *classical (Bernoulli) percolation* problem.

Employing a solution of an isotropic bond-percolation problem [6] on a given regular graph (p_c = critical probability)

$$\left.\begin{array}{ll} \text{honeycomb:} & 0.653 \\ \text{square:} & 0.5 \\ \text{triangular:} & 0.347 \end{array}\right\} p_c = P_{\bar{\varepsilon}}\{z(e)\} \tag{12}$$

we can find $\bar{\varepsilon}$ at the point of percolation of damage. If $\bar{\varepsilon}$ is not a hydrostatic strain field (i.e. $\bar{\varepsilon}_I \neq \bar{\varepsilon}_{II}$), then we use anisotropic percolation results.

We turn now to s-Markov fields, which model ever longer attractive effects in the spread of damage. In the following we assume:

a) the graph $G'(V',E')$ to be hexagonal (honeycomb),

b) same ${}^eF_R(r)$ strength distribution for all $e \in E'$,

c) same elastic constants of all grains (no crystal anisotropy).

The fascinating aspect of Markov random fields on graphs is that they exhibit *phase transitions* when the interactions between neighbors (i.e. edges in our model) become sufficiently strong. These interactions are reflected here by the right hand side in (8). Strictly speaking, this right-hand side implies that a so-called *specification* of the s-Markov field Z is given

$$\Pi_{\bar{\varepsilon}} = \{P\{z_i(v) \mid z_j(N_s^e)\};\ \ \forall e \in E',\ i=1,2,\ \ \forall j \in J\} \tag{13}$$

J in the above is a set of all possible configurations on the set of neighbors N_s^e; for example, in case of a binary 1-Markov field on a honeycomb graph J consists of 2^4 elements. In view of the assumptions a)-c) above, Z is a spatially homogeneous field, so that a description of conditional probabilities for a single vertex will suffice to give the specification (13).

Now, if we recall that a random field is described by the probability measure $P_{\bar{\varepsilon}}$, we are faced with the following question: is the description in terms of $P_{\bar{\varepsilon}}$ the same as in terms of $\Pi_{\bar{\varepsilon}}$? The answer is: sometimes yes and sometimes no. Specifically, for nearest neighbor interactions being sufficiently weak there is a unique correspondence of $P_{\bar{\varepsilon}}$ and $\Pi_{\bar{\varepsilon}}$, while for strong interactions this is not true (see e.g. [7]). The very point of change from one to the other behavior is termed a *phase transition*, or a critical point in the language of statistical physics. It is important to note that a phase transition may occur only for the range of interactions $s \geq 1$ (i.e. not for 0-Markov fields) and on lattices (or graphs) in two or more dimensions and of infinite extent.

It is known that there is a strong disorder in realizations z of the random field Z in case of uniqueness of $P_{\bar{\varepsilon}}$ and $\Pi_{\bar{\varepsilon}}$. On the other hand, there is a high and long range order in z's in case of nonuniqueness of $P_{\bar{\varepsilon}}$ and $\Pi_{\bar{\varepsilon}}$ and two different global behaviors are possible - one of z_1-type and another of z_2-type. The point of phase transition is characterized by percolation and fractal geometry of z's on E′. A condition for absence of a phase transition is expressed by an inequality [8] to be satisfied by $\Pi_{\bar{\varepsilon}}$. Thus, the objective is to use this inequality to find a maximum $\bar{\varepsilon}$ for which the phase transition does occur. In a research program currently underway we calculate $t(\bar{\varepsilon})$ under all possible conditions $z(N_1^e)$, which in turn permits calculation of $\Pi_{\bar{\varepsilon}}$ according to (6) and a determination of $\bar{\varepsilon}$ at the onset of macro-damage.

3. Discussion and Comments

i) The successively larger neighborhoods N_1^e, N_2^e, N_3^e, ... represent the increasing ranges of interactions between edges of E′ in the hierarchy of s-Markov fields. As pointed out before, for each s all possible conditions $z(N_s^e)$ have to be considered separately; the numbers of these conditions are: $|N_1^e| = 2^4$, $|N_2^e| = 2^{12}$, $|N_3^e| = 2^{26}$, ... Thus the applicability of the Markov field approach is in those phenomena where a restriction to the first and second nearest neighbors is justified, otherwise, the number of computations that need to be carried out grows tremendously. The key idea is this: damage is most strongly felt at e - due to the stress concentration around the crack tip - if it occurs at N_1^e. Therefore, a restriction to a 1-Markov field, with a very small number of computations, may give a very useful approximation. At this point, it is interesting to note that a similar problem of irreversible damage formation in layers has been studied in [9] using computer-intensive simulations of large lattices.

ii) The Markov field model may be extended to time-dependent irreversible phenomena as well as those situations where more than two states at the microscale need to be considered (e.g. elastic, plastic, broken).

iii) It was already shown in [3] that the percolation theory approach justifies the functional form of Weibull-type statistics (except for the volume dependence) and brings out the fractal character of damage phenomena.

iv) It was also shown in [3] that the one-to-one relation between probabilities on Markov fields and Gibbs probability measures provides a link between an information-theoretic picture of damage geometry and the thermomechanics of damage. On this basis the deterministic formulation of thermodynamic orthogonality due to Ziegler (see e.g. [10]) may be employed to determine constitutive damage laws of random microstructures in a continuum approximation. To that end, the free energy and the dissipation function - which now are random functionals - may be assessed (e.g. averaged) with respect to a probability measure that is obtained from a micromechanical model; this procedure should rely on results of [11].

Acknowledgments

We thank Dr. Keith J. Bowman of Purdue University for many stimulating discussions. This research is supported by the Air Force Office of Scientific Research under Grant No. AFOSR-89-0423; Lt. Col. G.K. Haritos is the program manager.

References

1. Ostoja-Starzewski, M.: Graph approach to the constitutive modelling of heterogeneous solids. Mech. Res. Comm., 14(4) (1987) 255-262.

2. Ostoja-Starzewski, M.; Wang, C.: Linear elasticity of planar Delaunay networks: random field characterization of effective moduli. Acta Mech., 80 (1989) 61-80.

3. Ostoja-Starzewski, M.: Damage in a random microstructure: size effects, fractals, and entropy maximization. Appl. Mech. Rev., 42(11, Part 2) (1989)

202-212; also in Steele, C.R.; Leissa, A.W.; Crespo da Silva, M.R.M. (eds.) Mechanics Pan-America, New York: ASME Press 1989.

4. Hinrichsen, E.L.; Hansen, A.; Roux, S: A fracture growth model. Europhys. Lett. 8(1) (1989) 1-7.

5. Hansen, A.; Roux, S.; Herrmann, H.J.: Rupture of central-force lattices. J. Phys. France 50 (1989) 733-744.

6. Ziman, J.M.: Models of disorder, Cambridge: Cambridge University Press 1979.

7. Preston, C.: Gibbs states on countable sets, Cambridge: Cambridge University Press 1974.

8. Kindermann, R.; Snell, J.L.: Markov random fields and their applications, Contemporary Mathematics 1, Providence, RI: Am. Math. Soc. 1980.

9. Skjeltorp, A.T.; Meakin, P.: Fracture in microsphere monolayers studied by experiment and computer simulation. Nature 335 (1988) 424-426.

10. Ziegler, H.; Wehrli, Ch.: The derivation of constitutive relations from the free energy and the dissipation function. Adv. Appl. Mech. 25 (1987) 183-238.

11. Ostoja-Starzewski, M.: A generalization of thermodynamic orthogonality to random media. ZAMP (1990) in press.

Global and Local Approaches to Creep Crack Initiation and Creep Crack Growth

R. PIQUES and A. PINEAU
Centre des Matériaux, Ecole des Mines
B.P. 87 - 91003 EVRY - Cédex (FRANCE)
URA CNRS N°866

ABSTRACT :

Creep crack initiation and creep crack growth behaviours were studied in two creep ductile materials, an austenitic stainless steel and a low alloy ferritic steel. A number of limitations of the global approach to creep cracking, in which unique correlations between time to initiation or creep crack growth rate and the load-geometry parameter, C*, are shown. A local approach to creep cracking is developed in the austenitic steel. In this material a damage function based on the measurement of intergranular damage in notched specimens is established and then applied ahead of the crack tip to simulate numerically crack growth. A good agreement between observed and calculated crack growth curves is obtained using crack initiation data.

INTRODUCTION

Creep cracking has been extensively studied over the last past 10 years in a number of structural materials, (for a review see, for example, Ref 1 to 3). Most of the published work has been devoted to creep crack growth (CCG) only and not to creep crack initiation (CCI), defined here as the time, Ti, necessary for an existing defect to grow by a small predetermined increment, X_c, under creep conditions, although this latter stage may represent a large part of the rupture life of a cracked component. Most studies have been devoted to the identification of unique correlations between CCG rates, da/dt, and load-geometry parameters. This is the global approach to creep cracking. In this approach two kinds of materials are distinguished : creep brittle and creep ductile materials, which fail by creep cracking under, respectively, small and large-scale creep conditions. Based on the Riedel and Rice theoretical work on the fracture mechanics of creeping solids (FMCS) [4,5], the elastic stress-intensity factor, K, and the creep C*-integral, which characterize the time-dependent crack tip stress-strain field for short and long times, have often been proposed as correlating parameters with da/dt in creep brittle and creep ductile materials, respectively.

The present paper deals with CCI and CCG in two creep ductile materials, 316 L stainless steel and a low alloy Cr steel taken from a turbine casing which operated for about 150 000 hours. Specimens were taken from the hot part of the component (~ 540°C) and from the cold part (~ 200°C). They are referred to as aged and unaged material, respectively. These materials can be considered to be creep ductile since the transition time t_{tr} calculated from the Riedel and Rice expression, ie :

$$t_{tr} = K^2 / E\,(n_2+1)\,C^* \qquad (1)$$

is much smaller than the observed times to failure [6-9].

M. Zyczkowski (Ed.)
Creep in Structures
IUTAM Symposium Cracow/Poland 1990

In the first part of the paper, a number of limitations of the global approach to creep cracking are shown. Full details about the experimental procedures used to investigate both CCI and CCG behaviour are given elsewhere, ie in Ref [6–8] for 316 steel and in Ref [9] for ferritic steel. Here it is enough to say that the load-geometry parameter C* was defined from the measurement of the load–line displacement rate, $\dot{\Delta}_{exp}$, as:

$$C^*_{exp} = An_2 / (n_2+1) \,.\, P \,.\, \dot{\Delta}_{exp} / B\,(W-a) \tag{2}$$

where n_2 is the exponent of the stationary creep law ($\varepsilon^\circ = B_2\sigma^{n_2}$), A is a coefficient (A ≃ 2 for CT specimens), P is the applied load, B the specimen thickness and (W–a) the ligament size. It should also be added that C* parameter can be calculated using either the results of finite element computations [10] or a simplified approach based on reference stress and reference length concept [11–12], without introducing direct measurements of $\dot{\Delta}_{exp}$. These limitations have partly motivated the development of the local approach to creep cracking which is briefly presented in the second part of this paper. This approach is largely based on the investigation of creep rupture micromechanisms. A quantitative damage function relating intergranular damage with the maximum principal stress and the equivalent Von–Mises strain is established from tests on notched bars. This damage function is then used to model both CCI and CCG behaviour. To this end, the crack tip stress–strain field is calculated either numerically by finite element method (FEM) or analytically using the theoretical solutions given by FMCS. It can therefore be said that the global and the local approaches to creep cracking are more complementary than contradictory.

GLOBAL APPROACH TO CREEP CRACKING

Creep Crack Initiation Times

CCI time was defined conventionally as the time T_i necessary for initial crack to propagate over a critical distance, X_c = 50μm [2, 6, 7]. Test results on 316 steel carried out on CT, DEN and axisymmetrically (AX) cracked specimens at 575°C, 600°C and 650°C are reported in Fig 1a. These results show a unique correlation between T_i and C^*_{exp}, independent of specimen geometry, temperature and initial test conditions, given by :

$$T_i \, C^{*\alpha}_{exp} = \text{constant} \tag{3}$$

Similar results were obtained on a Cr steel, tested at 550°C as shown in Fig 1b. In this material the correlation between T_i and C^*_{exp} is independent on the initial condition of the material (aged or unaged). The slope α was found to be equal to 0.65 in 316 steel and 0.85 in the ferritic material.

Creep Crack Growth Rates

CCG rates measured on AX specimens at 600°C are presented versus C^*_{exp} in Fig 2a. Similar results were obtained on other specimen geometries, such as CT and DEN specimens. CCG rates measured on 1 Cr–1 Mo–0,25 V steel tested in the unaged condition at 550°C with CT specimens are given in Fig 2b. Similar results were obtained on the aged material. In both steels the CCG rates show a typical two-stage behaviour, as indicated in Fig 2b for a specific test (specimen 8–14) carried out on the ferritic material. During the initial Stage I regime, $\dot{\Delta}_{exp}$ and C^*_{exp} remain essentially constant, while da/dt increases rapidly. Thus Stage I depends strongly on initial test conditions. No unique correlation can therefore be established in Stage I between da/dt and C^*_{exp}. Conversely, a unique quasi-linear correlation appears to exist in both materials between C^*_{exp} and Stage II CCG rates, that is, for da/dt $\geq$ 5µm/h. Such a quasi-proportionality betweeen da/dt and C^*_{exp} in the high crack growth rates Stage II regime has often been reported in the literature (see, eg ref [13–16]), especially where C* is estimated on the basis of load-line displacement rates (Eq 2). It is very important to remember that, during a crack growth test, the variation of load-line displacement is due not only to creep deformation, but also to crack growth [1–3, 7–9], such as;

$$\Delta_{exp} = f(a, t) \text{ and } \dot{\Delta}_{exp} = \partial f(a,t) / \partial t + (\partial f / \partial a) . (da / dt) \quad (4)$$

Only the first term in Eq 4 is controlled by creep deformation. It was shown using the reference stress and reference length concept that, for high CCG rates, the second term in Eq 4 may become predominant [7–9]. The quasi-proportionality observed experimentally between Stage II da/dt and C^*_{exp} reduces, under these conditions, essentially to a simple correlation between da/dt and da/dt. This may bring into doubt some published data where only high CCG rates were studied. It appears therefore necessary to limit the correlation between da/dt and C^*_{exp} at times lower than a critical time, T_2, such as the first term in Eq 4 is larger than the second term. This time which corresponds to the transition between both stages of CCG is reported in Fig 1 for both materials. It can easily be calculated, as shown previously [7–9]. Fig 1 shows that in both materials, especially in the ferritic steel, T_2 is of the same order of magnitude than T_i. Moreover it can be shown that, in materials such as 316 steel and low alloy Cr–Mo–V steels, times T_i or T_2 represent a large part of the time to creep failure, T_F, such as the time corresponding to Stage II crack growth, T_G is not predominant [17].

LOCAL APPROACH TO CREEP CRACKING IN 316 STEEL

Intergranular damage function

Metallographic examinations showed that CCI and CCG occurred along heavily damaged grain boundaries. A study was specifically devoted to the quantitative measurement of intergranular creep damage in 316 L stainless steel tested with notched bars of various geometries to investigate the influence of stress state [18–19]. Creep damage, D, was defined as the ratio of the length of cracked or cavitated grain boundaries observed on a unit section over the length of initial grain boundaries on the same surface. Typical results obtained on two geometries, ie a 5 mm notch and a 1 mm notch, tested at 600°C under an applied stress of 330 MPa and 404 MPa, respectively, are shown in Fig 3. These specimen geometries proved to be quite interesting since they lead to widely

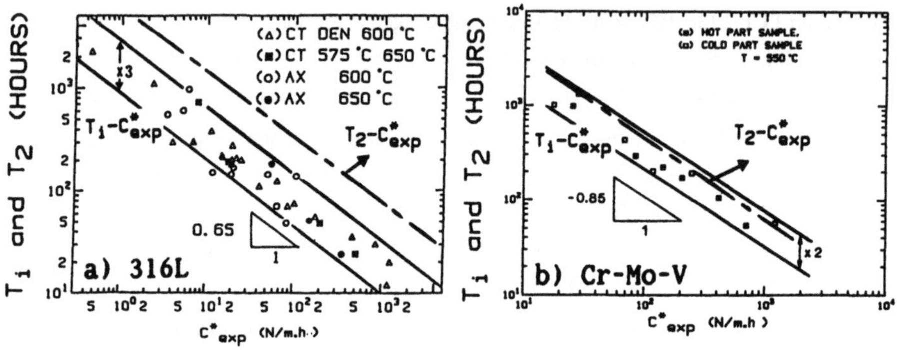

Figure 1 : Correlations T_i - C^*_{exp} and T_2 - C^*_{exp} :
a) 316L stainless steel - CT and AX specimens
b) Cr-Mo-V steel - CT specimens.

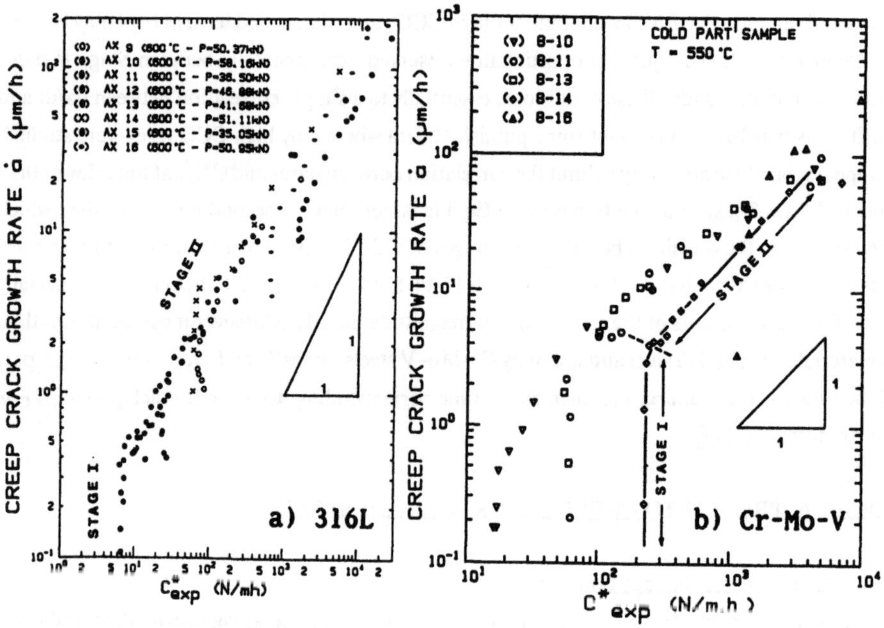

Figure 2 : Creep crack growth rates versus C^*_{exp} :
a) 316L stainless steel - AX specimens
b) Cr-Mo-V unaged steel - CT specimens.

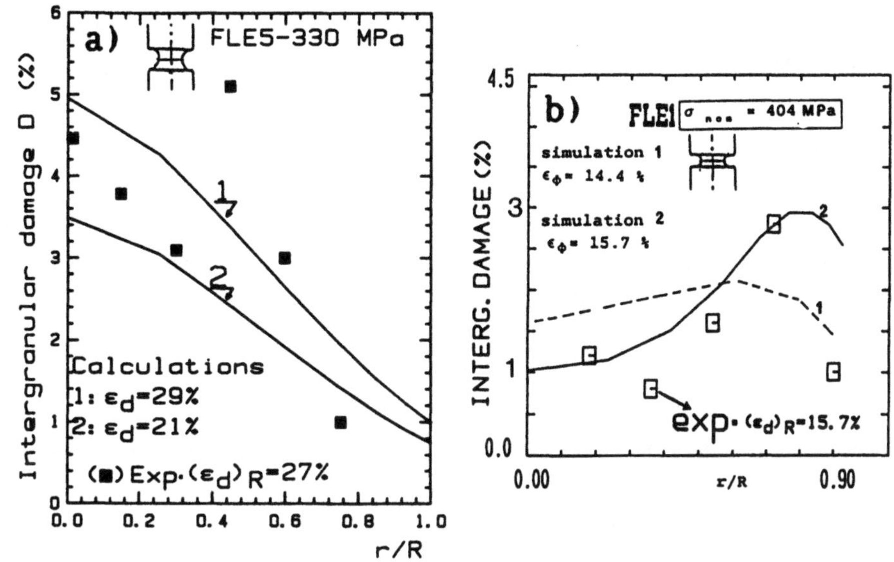

Figure 3 : Metallographically measured and computed intergranular creep damage in the minimum section of 316L notched specimens tested at 600°C : a) 5mm notch, σ = 330MPa; b) 1mm notch, σ = 404MPa.

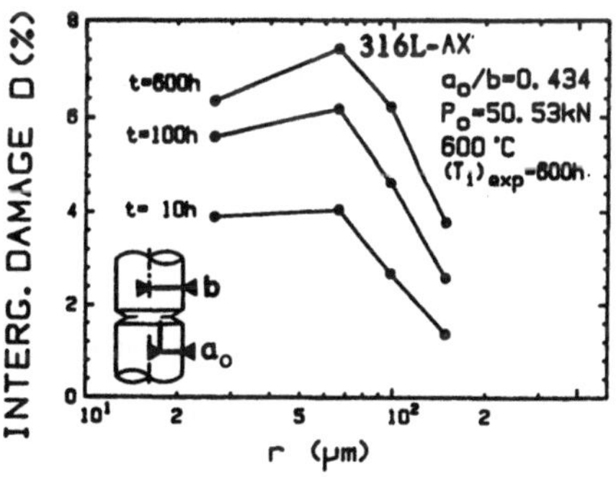

Figure 4 : Computed intergranular creep damage at the crack tip of 316L-AX cracked specimen at 600°C.

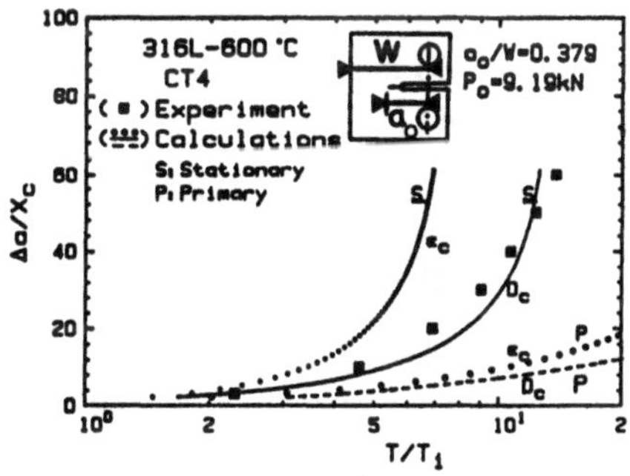

Figure 5 : Comparisons between experimental and calculated creep crack growth curves in 316L-CT4 specimen tested at 600°C.

different distributions of intergranular damage. FEM was used to calculate these specimens. Creep damage D could be expressed, in an incremental form, as :

$$dD = B|\Sigma|^2(\varepsilon_{eq})^{-0.5}d\varepsilon_{eq} \tag{5}$$

where Σ is the maximum principal stress, ϵ_{eq} is the Von-Mises equivalent creep strain, while A is a constant. Fig 3 shows that this expression integrated along the creep stress-strain history up to creep fracture, which is represented by the diametral deformation of the specimens, $(\epsilon_d)_R$, gives a satisfactory representation of the creep damage distribution measured across the minimum section of both geometries. It is also observed that, at failure, intergranular damage reaches a critical value D_C of the order of few percent (3 to 5 %).

Application to crack tip situation

Eq 5 was used to model both CCI and CCG behaviour in 316 steel. A X specimens were calculated by FEM using fine meshes to account for crack blunting effect. The results of intergranular damage calculations ahead of crack tip are shown in Fig 4 as a function of time. In this specific case, the CCI time was measured (T_i = 600 h). It is observed that, at a critical distance X_C of 50μm, which is a distance consistent with the definition of crack initiation, CCI takes place at 600 h for a critical damage of 7 %, which is of the same order of magnitude than the critical damage measured on notched bars, yet slightly larger.

FEM calculations on CT specimens showed that the theoretical asymptotic stress field, derived from FMCS, was relatively well verified at distances from the crack tip larger than 50μm. Furthermore it was observed that, close to the crack tip, the stress distribution corresponded rapidely to the primary or stationary creep stabilized stage. These results therefore suggested that a simplified CCG model could be developed by using, for instance, the stationary analytical stress-strain field given by :

$$\sigma_{ij}(\theta,r) = [C^* / B_2 I_{n_2} r]^{1/n_2+1}\tilde{\sigma}(\theta) \tag{6a}$$

$$\varepsilon_{ij}(\theta,r) = B_2[C^* / B_2 I_{n_2} r]^{n_2/n_2+1}\tilde{\varepsilon}(\theta)t \tag{6b}$$

where In_2 is a constant ($\simeq \Pi$), $\tilde{\sigma}(\theta=0) \simeq 1.10$ and $\tilde{\varepsilon}(\theta=0) = 1$ for plane stress conditions. Crack initiation can be modeled by integrating Eq 5 with the stress-strain field given by Eq 6. This leads to a simple form written as Eq 3, where $\alpha = (1+n_2)/(n_2+1) \simeq 1.34$ with $n_2 = 7.69$. Thus the calculated value of the α exponent is much larger than the experimental value. This might be partly related to the fact that the C^*_{exp} used in Eq 3 does not necessarily correspond only to stationary creep. To model crack propagation, a similar procedure can be used, except that it is necessary to take into account damage accumulation ahead of the growing crack [6]. Typical results obtained on a CT specimen are shown in Fig 5, where the calculated and the observed crack growth curves are plotted in adimensionnal coordinates, $\Delta a/X_C$ versus T/T_i. In these numerical simulations, the experimental correlation between T_i and C^* (Eq 3) was used to initialize the calculations. In Fig 5 we have also reported the results obtained with C^* calculated from primary creep. Two damage laws were introduced : the intergranular damage, D_c and the ductility exhaustion concept, ϵ_c. As a rule it was noticed that, as in this specific case, a better agreement was obtained with stationary creep law and intergranular damage as a failure criterion. It is also worth adding that, provided that the T_i–C^* experimental correlation is accepted, this local approach methodology to model creep crack propagation is fully predictive, since it contains no adjustable parameter.

CONCLUSIONS AND PERSPECTIVES

1- In 316 L steel and in 1 Cr–1 Mo–0.25 V steel, which are creep ductile materials, there exists a unique correlation between the creep crack initiation time, T_i and the global load–geometry parameter, C^*. More extensive work is necessary to extend this correlation to other materials and to test its applicability to other loading conditions, including stress relaxation under constant displacement and non–isothermal conditions.

2- The apparent correlations between stage II da/dt and C^* which have often been proposed in the literature for creep ductile materials seem to be very doubtful and of little use for life predictions purposes. Much more work should be devoted to the study of CCI instead of CCG.

3- A local approach based on intergranular damage accumulation is proposed for creep cracking of 316 steel. This model, based on critical intergranular damage, D_C, measured on notched specimens and a critical distance, X_C, ahead of a growing crack, is shown to provide results in good agreement with creep crack growth data. The development of this approach requires also much more extensive work in at least four directions.

(i) Constitutive equations which should be able to account for interactions between creep and plastic deformation at least in materials such as austenitic stainless steels.

(ii) Quantitative measurements of physical damage, such as creep intergranular damage in 316 steel. These measurements must be performed at a scale consistent with the mechanical tools used to model creep cracking.

(iii) Modeling of the crack tip situation. This is certainly one of the most difficult facets of the problem. Numerical techniques in which a coupling effect between damage and constitutive equations is introduced may be useful to solve, at least partly, the difficulties associated with the process zone concept (see eg [20]).

(iv) Applications to more complex situations, including non–isothermal conditions and cyclic deformation.

4- Simplified methods must be developed for practical application purposes. For creep ductile materials, the reference stress and reference length concepts, based on limit load analysis, are useful tools to estimate the value of the load–geometry parameter, C^*. In addition to fracture mechanics of creeping solids, they can be used to develop methods of defect assessment at elevated temperature which could account for transient effects, for example, the effect of hold times on fatigue crack growth rates.

REFERENCES

1- Bensussan P,. An experimental and theoretical study of creep crack growth in 2219 Aluminium alloy. PhD Thesis, Departement of Materials Science and Engineering, Massachusetts Institute of Technology, Cambridge, MA, Feb. 1984.
2- Maas E., Propagation des fissures par fluage dans l'acier inoxydable austenitique Z3CND 17-13. Docteur Ingénieur Thesis, Ecole Nationale Supérieure des Mines Paris, April 1984.
3- Bensussan P., Approches mécaniques globale et locale de l'amorçage et de la propagation de fissures par fluage dans l'alliage léger Al-Cu 2219. Doctorat es-Science Thesis, Université de Paris Sud - Orsay - France, Feb 1986.
4- Riedel H., Journal of the Mechanics and Physics of Solids, Vol 29, (1981), pp. 35-49.
5- Riedel H., and Rice J.R., Fracture Mechanics : Twelfth Conference, ASTM STP 700, American Society for Testing and Materials, Philadelphia, (1980), pp. 112-130.
6-Mass E. and Pineau A., Engineering Fracture Mechanics, Vol. 22, (1985), pp 307-325.
7- Bensussan P., Maas E., Pelloux R. and Pineau A., Creep Crack Initiation and Propagation : Fracture Mechanics and Local Approach. Journal Pressure Vessel Technology, Vol 110, (1988), pp. 42-50.
8- Bensussan P., Piques R. and Pineau A., A critical assessment of global mechanical approaches to creep crack intitiation and creep crack growth in 316 L steel. Non-linear Fracture Mechanics : Volume I - Time- Dependent Fracture, ASTM STP 995, (1989), pp.27-54.
9- Molinié E., Piques R. and Pineau A., Behaviour of a 1Cr-1 Mo-0,25 V steel after long term exposure. Part II : Creep crack initiation and creep crack growth. To appear in Fatigue and Fracture of Engineering Materials and Structures.
10- Kumar V. and Shih C.F., Fully plastic crack solutions with applications to creep crack growth. Proceedings of the International Conference on Engineering Aspects of Creep. I Mech. E. Conf. Publ., Vol 2, Edmunds, England, Paper N°C228/80, pp. 211-214.
11- Ainsworth R.A., Chell G.G, Coleman M.C, Goodall I.W., Gooch D.J., Haigh J.R., Kimmins S.I. and Neate G.J., CEGB assessment procedure for defects in plant operating in the creep range. Fatigue Fract. Eng. Mater. Struct., Vol 10, (1987), pp. 115-127.
12- Piques R., Molinié E. and Pineau A., Comparison between CEGB and EMP assessement for defects in the creep range, submitted to Fatigue Fract. Eng. Mater. Struct., (1990).
13- Ohji K., Ogura K., Kubo S. and Katada Y., Paper N°C 240/80. Proceedings of the International Conference on Engineering Aspects of Creep, Vol 2, Institution of Mechanical Engineers Conference Publication, Edmunds, England (1980).
14- Taira S., Ohtani R. and Kitamura T., Journal of Engineering Materials and Technology, Vol 101, (1979), pp 154-161.
15- Koterazawa R. and Mori T., Journal of Engineering Materials and Technology, Vol 99, (1977), pp. 298-305.
16- Nibkin K. and Webster G., in Micro and Macro Mechanics of Crack Growth, K. Sadananda, B.B. Rath and D.J. Michel, Eds., American Institute of Mechanical Engineers, New-York, (1982), pp.137-148.
17- Pineau A., Assessment Procedures for Defects in the Creep Range, Int. Conf. High Temperature Structural Design, Venice 24-26 october 1990 (Italy).
18- Yoshida M., Levaillant C., Piques R. and Pineau A., Quantitative study of intergranular damage in an autenitic stainless steel on smooth and notched bars. Int. Seminar on High Temperature Fracture Mechanisms and Mechanics, October 13-15, (1987), Dourdan-France.
19- Yoshida M., Levaillant C. and Pineau A., Metallographic measurement of creep intergranular damage and creep strain. Influence of stress state on critical damage at failure in an austenitic stainless steel. Int. Conf. on Creep, Tokyo, (1986), pp. 327-332.
20- Saanouni K., Chaboche J.L. and Lesne P.M., On the creep crack growth prediction by a non local damage formulation. Eur. J. Mech. A/Solids, Vol 8, (1989), pp. 437-459.

The Assessment of Creep-Fatigue Initiation and Crack Growth

R H PRIEST and D A MILLER

Nuclear Electric plc
Bedminster Down, Bridgwater Road
Bristol, BS13 8AN, UK

Summary

An outline of Nuclear Electric's Assessment Procedure for the High Temperature Response of Structures ('R5') for creep-fatigue initiation and crack growth is given. A unified approach is adopted for both regimes. For initiation, total damage is described in terms of separate creep and fatigue components. Ductility exhaustion is used for estimating creep damage whilst continuous cycling endurance data are used to evaluate the fatigue damage term. Evidence supporting this approach is given through the successful prediction of creep-fatigue endurances for a range of materials, cycle types, dwell period times, etc. Creep-fatigue crack growth is similarly described in terms of separated creep and fatigue components. Crack growth rates for each component are characterised in terms of fracture mechanics parameters. It is shown that creep crack growth rates can be rationalised on a ductility basis. Creep-fatigue interactions are accommodated in the cyclic growth component through the use of materials coefficients which depend on dwell time.

Introduction

It is now well established that components in power generating equipment can be subject to thermally-induced cyclic strain deformations interspersed with periods of relatively constant loading, ie. simultaneous creep-fatigue cycles. The consequent formation of structurally significant defects broadly comprised two stages; crack initiation and crack growth. Over recent years much work has been performed in the examination of the creep-fatigue response of a range of materials on a crack initiation basis; less information is available on the assessment of crack growth under simultaneous creep-fatigue conditions.

In the current paper, life prediction pertaining to creep-fatigue initiation is described in terms of the ductility exhaustion approach. Its application to integrity assessments through Nuclear Electric's Assessment Procedure for the High Temperature Response of Structures (R5) is outlined. Data obtained on creep-fatigue crack growth are also reviewed. The success of combining creep and fatigue crack growth contributions in

M. Zyczkowski (Ed.)
Creep in Structures
IUTAM Symposium Cracow/Poland 1990

producing rate equations for the simultaneous creep-fatigue situation is indicated and its use in the R5 procedure described.

Creep-Fatigue Initiation

Much work has been performed on studying the creep-fatigue crack initiation characteristics of a range of materials, and there are a number of detailed reviews of these data in the literature [1-3]. The question is,how do we use these data to predict the creep-fatigue response of a range of materials?

The study of life prediction for components subject to simultaneous cyclic and creep deformation has led to the development of several possible techniques [3]. The most widely used approach is based on the linear super-position of fatigue and creep damage. Indeed, the mainstay of present design procedures that account for creep and fatigue is the linear life fraction rule which forms the basis of the ASME Boiler and Pressure Vessel Code Case N47 [4]. This combines separated time independent and time dependent damage contributions in a linear fashion as described by;

$$\Sigma N/N_p + \Sigma \tau/T_d = D \qquad (1)$$

where N/N_p is the ratio of the number of cycles at a given strain range to the fatigue endurance at that strain range, τ is the time at a given stress during the creep hold period and T_d the time to rupture at that stress. In Code Case N47, the parameter D is unity for assessments which include elastic stress analyses. For inelastic analyses, D represents a total damage envelope which must not be transgressed if failure is to be avoided.

More recent work has shown the benefit to be gained by assessing creep damage through a strain fraction rule rather than the traditional time fraction rule [3]. This development is incorporated in R5, [5] where the life fraction rule is written;

$$\Phi_F + \Phi_C = 1 \text{ at failure} \qquad (2)$$

In equation (2), the creep damage, Φ_C, is described by the ductility term;

$$\Phi_C = N\int_0^{t_h} \dot{\epsilon}/\epsilon_f(\dot{\epsilon})dt \qquad (3)$$

where N is the number of operational cycles, t_h is the dwell time in the cycle and ϵ_f is the creep ductility which is a function of strain rate $\dot{\epsilon}$. Thus, a full integration of exhausted creep ductility over the range of strain rates encountered in any given hold period is used. Elastic follow-up effects are characterised by the parameter Z, Figure 1, and can be used in conjunction with equation (3) by defining $\epsilon = {}^{-Z}/_{E}\ {}^{d\sigma}/dt$ where σ is the stress during relaxation.

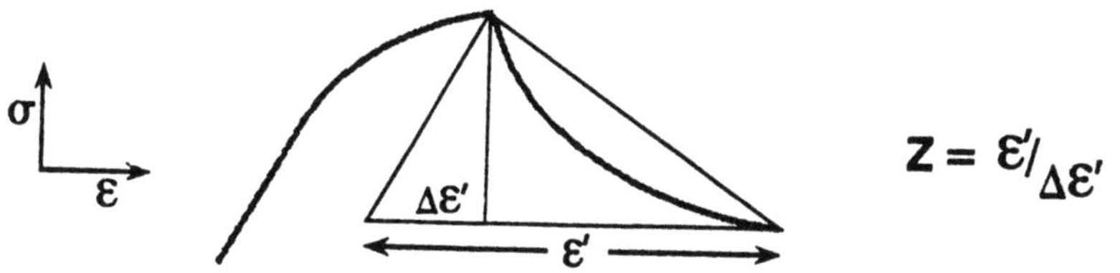

Fig. 1. Definition of Z

The fatigue damage, Φ_F, is deduced from the number of cycles, N_0, to grow a defect to size a_0 under continuous cycling conditions at the surface strain range, strain rate and temperature such that;

$$\Phi_F = {}^{N}/N_0 \quad (4)$$

Unlike previous procedures, specimen size effects are explicitly taken into account with the R5 procedure [5]. Where the size a_0 is larger than the crack size, a_l, corresponding to failure, N_l, in a laboratory specimen, it is conservative to set $N_0 = N_l$. However, particularly for thin section applications, a_0 may be smaller than a_l and it is necessary to deduce N_0 from N_l. It is assumed that N_0 may be obtained from the number of cycles, N_i, to produce a crack of size a_i and the subsequent number of cycles to grow the crack to size a_0 [6, 7]. The number of cycles, N_0, to initiate and grow a crack to a depth a_0 is given by;

$$N_0 = MN_l + (1-M)N_i \quad (5)$$

where $N_i = 0.0366\ N_l^{1.306}$ for $15<N_l<50{,}000$ cycles [8] and M is a function of a_0, a_i and a_l and depends on the short crack growth rate equation used to describe growth from a_i to a_0. In R5, a_i is taken to be 20 μm [6], the crack depth in austenitic steels beyond which well defined fatigue striations have been identified.

It can be seen from equation (2) that a crack size a_0 is assumed to be formed at the surface when the total damage equals unity. The assumption of linear damage has been examined by calculating Φ_F and Φ_C at failure in laboratory tests covering a wide range of materials, hold times and cycle types. Best estimte fatigue and stress relaxation data were used in the calculations and the results are as shown in Figure 2. Although there is considerable scatter, the R5 approach [5] can be seen to have resulted in generally conservative predictions. The use of lower bound data when calculating damage fractions (as would be used in component assessments) significantly increases the degree of conservatism.

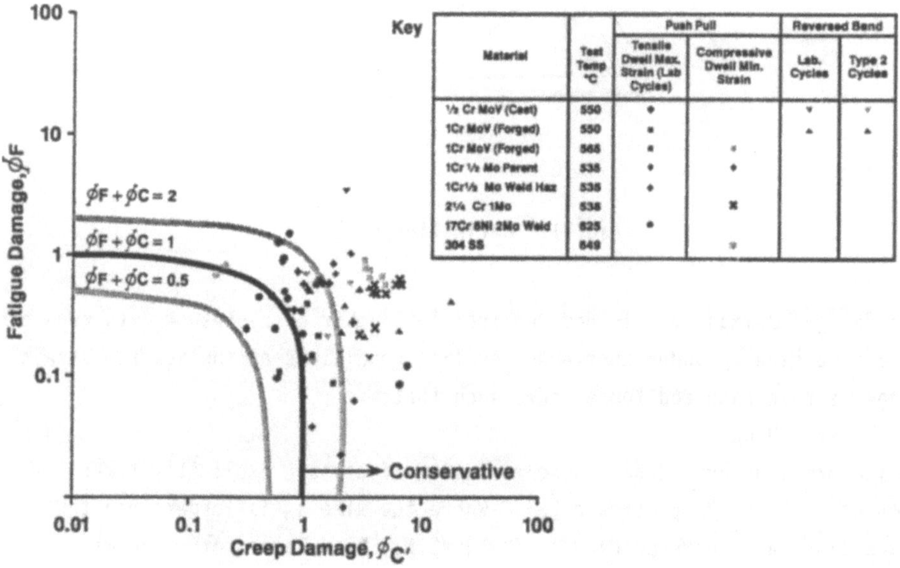

Fig. 2. Damage Component Map using R5 Procedure with Best Estimate Data.

Creep-Fatigue Crack Growth

Having dealt with the number of creep-fatigue cycles to initiate a significant crack, we now examine methods of quantifying crack growth. Previous work on assessing creep-fatigue crack growth has been performed under both load-controlled and displacement-controlled conditions [9-17]. These data have exhibited a range of behaviour from crack growth occurring predominantly during the tensile cyclic ramp [12] to crack growth occurring mainly in the hold periods [13]. The fracture mechanics parameters ΔK and C^*, respectively, have been successfully used to correlate crack growth rates in these tests.

More recent studies carried out on a range of ferritic and austenitic steels [14-16] have indicated crack growth increments during both the cyclic excursions and the hold periods. On the basis of these data, the R5 procedure for assessing crack growth during creep-fatigue [18] is addressed in a similar manner to the linear life fraction summation of creep-fatigue initiation damage, equation (2). Crack growth is taken to be the sum of fatigue and creep components;

$$da/dN = (da/dN)_f + (da/dN)_c \tag{6}$$

The fatigue crack growth per cycle is obtained from a Paris law,

$$(da/dN)_f = C(\Delta K_{eff})^l \tag{7}$$

where C and l are obtained from laboratory test specimen data and may be functions of time to reflect any enhancement in growth rates due to creep-fatigue interactions. In equation (7), ΔK_{eff} represents the stress intensity factor range for which the cracks remain open [19]. Values of C and l for a range of engineering materials are given in R5 [18]. The need for time dependent values of C and l results from the observation that, for brittle materials, increasing hold periods result in increasing cyclic crack contributions [14], Figure 3. This observation is interpreted as accelerated fatigue crack growth occurring through creep cavitated material ahead of the main crack, ie. a creep-fatigue interaction. This behaviour has been correlated with damage zone sizes ahead of the main crack tip as measured by quantitative metallography [20].

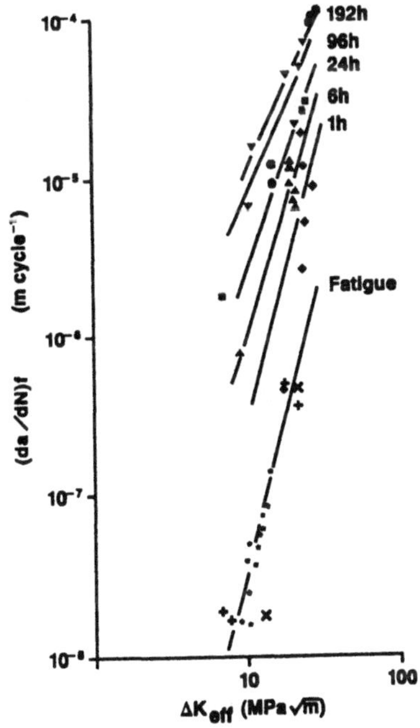

Fig. 3. Cyclic Contributions to Growth in Type 347 Weld at 650°C [14]

The creep crack growth contribution is described by:

$$(da/dN)_c = \int_0^{t_h} AC^{*q}\, dt \tag{8}$$

where the constants A and q are obtained from test specimen data (such as static creep crack growth); values for a range of engineering materials are given in R5 [18]. This result has been shown to hold for a range of constant displacement creep-fatigue crack growth results [14-16] as well as for constant load and stress relaxation creep-fatigue data [17]. Typical data obtained on ferritic steels are shown in Figure 4, [16, 21, 22].

In order to use an equation such as (8) during integrity assessments, simplifications are requried. R5 adopts the reference stress approximation for C* [23];

$$C^*_{ref} = \sigma_{ref}\, \epsilon_{ref}\, R' \tag{9}$$

where ϵ_{ref} is the creep strain rate corresponding to a reference stress, σ_{ref} and R′ is a characteristic dimension given by $R' = K^2/\sigma_{ref}^2$. It should be noted that for application to creep-fatigue crack growth rate data, ϵ_{ref} values are obtained from stress relaxation tests rather than steady state forward creep. Equation (9) can be further simplified for long hold time to [14];

$$C^*_{ref} = \frac{BK^2}{Et} \tag{10}$$

where K is evaluated at the beginning of the hold period.

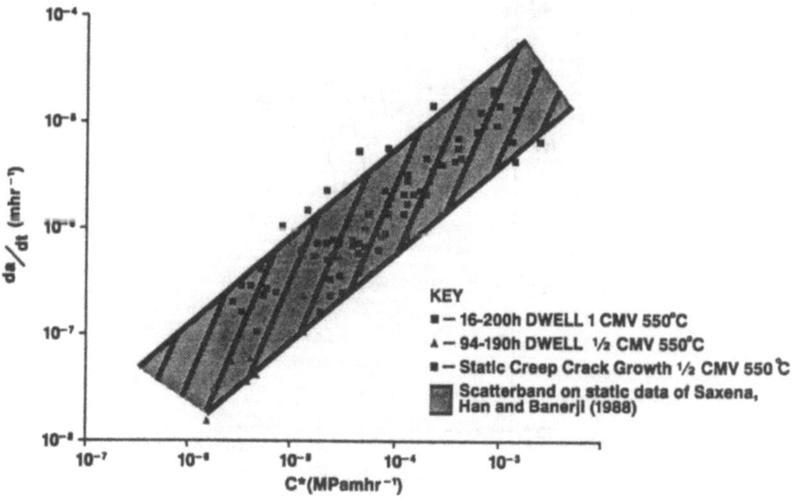

Fig. 4. Crack Growth rates During Constant Displacement Hold Periods Correlated with C*.

Data from a range of materials [14-16, 22] have been used to show good correlation between C* values measured during creep-fatigue hold periods and equation (10), Figure 5. Furthermore validation of this approximation in conjunction with equation (6) has been demonstrated for a feature weld geometry [24].

Finally, it is recalled that ductility arguments have been invoked to derive a reliable method of predicting the number of creep-fatigue cycles to crack initiation. For crack growth assessments, similar ductility considerations can be applied. For example, Nikbin et al [25] have correlated static creep crack growth in a range of materials in terms of C^*/ϵ_f, where ϵ_f is a measure of the materials ductility. These workers indicated that, for plane stress, static creep crack growth could be described by equations of the form:

$$da/dt = \frac{0.03\ C^{*0.85}}{\epsilon_f} \tag{11}$$

An attempt has been made to extend this approach to creep-fatigue crack growth results, as indicated in Figure 6. A good correspondence is evident from Figure 6, equation (11) yielding generally conservative predictions when used with lower bound ductility values.

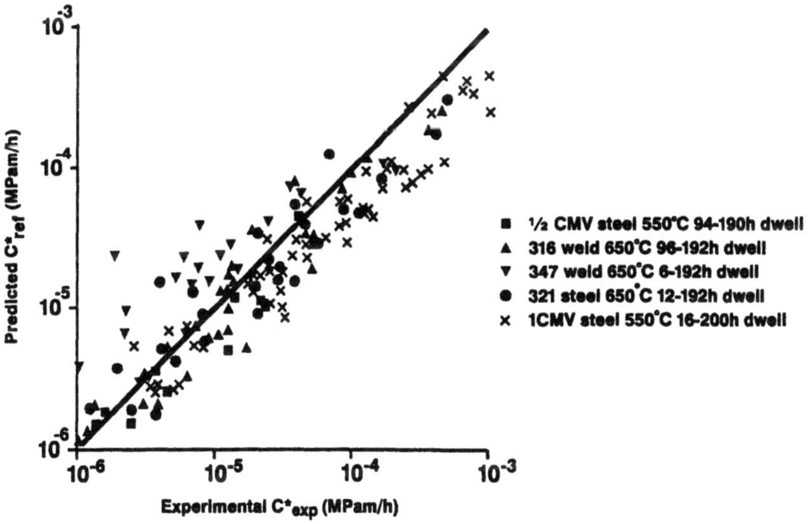

Fig. 5. Correlation of Experimental C* with Reference Stress Approximation from Equation (11)

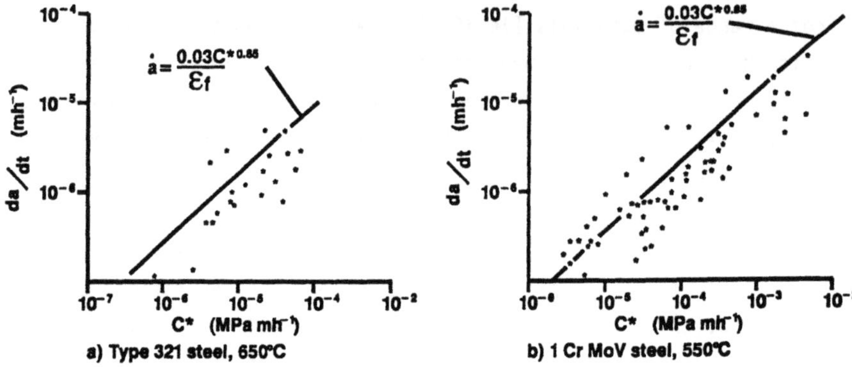

Fig. 6. Creep Crack Growth Data from Creep-Fatigue Tests as a Function of Ductility

Conclusions

The assessment of creep-fatigue initiation and growth via the R5 approach has been described. It is concluded that:

1. Linear summation of cyclic and creep components adequately describes both initiation and growth.
2. Ductility can be used to rationalise creep damage.
3. For creep-fatigue initiation, account is taken of size effects and elastic follow-up.
4. For crack growth, creep-fatigue interactions are accommodated through time dependent terms in the cyclic growth component. Creep crack growth contributions are described by a simplified C* equation.
5. Validation has been obtained on a wide range of materials subjected to varying test types, temperature, etc., including simple structural features test pieces.

References

1. Glenny, E.: Met. Reviews, 6 24 (1961)

2. Coffin, L.F.: Proc. I. Mech. E., 188, 19/74 (1974).

3. Miller, D.A.; Priest, R.H.; Ellison, E.G.: High Temp. Materials and Processes, 6, 3 & 3 (1984).

4. ASME Boiler and Press. Vess. Code, Section III, CODE CASE N47 (1974).

5. Chell, G.G.; Raynor,D.; Skelton, R.P.; Hales, R.; Miller, D.A.; Priest, R.H.: CEGB Report RD/L/3403/R88.

6. Pineau, A.: 'Fatigue at High Temperature'. (ed. R.P. Skelton), 305 (1983).

7. Skelton, R.P.: ASTM STP770, 337 (1983).

8. Ainsworth, R.A.: SMIRT 10, Anaheim, Vol. L, 79-90 (1989).

9. Lloyd, G.J.: Met. Sci., 13, 39 (1979).

10. Ohji, K.; Kubo, S.; Kiri, T.; Joo, W.S.: Proc. Int. Conf. Creep, JSME, 197 (1986).

11. Jaske, C.E.: ASME Conf. 'Advances in Life Prediction Methods', 93 (1983).

12. Skelton, R.P.: CEGB Report TPRD/L/2927/N85 (1986).

13. Neate, G.J.: CEGB Report MID/SSD/86/0018/R/MT (1986).

14. Gladwin, D.N.; Miller, D.A.; Priest, R.H.: Mat. Sci. Tech., 5, 40 (1989).

15. Gladwin, D.N.; Miller, D.A.; Neate, G.J.; Priest, R.H.: Fat. Frac. Engng. Mat. Stuct., 11, 355 (1988).

16. Priest, R.H.; Miller, D.A.; Gladwin, D.N.; Maguire, J.: Proc. ASM Conf. Power Plant Rehabilitation, 31 (1989).

17. Ohtani, R.; Kitamura, T.; Nitta, A.; Kuwabara, K.: ASTM STP 943, 1163 (1988).

18. Goodall, I.W.; Ainsworth, R.A.; Hales, R.; Miller, A.G.; Chell, G.G.; Skelton, R.P.; Miller, D.A.; Muscati, A.; Neate, G.J.; Priest, R.H.; CEGB Report TPRD/B/1007/R87.

19. Skelton, R.P.; Beech, S.M.; Holdsworth, S.R.; Neate, G.J.; Miller, D.A.; Priest, R.H.: To be Published (1990).

20. Senior, B.A.; Maguire, J.; Evans, C.A.: CEGB Report ESTD/L/0041/R89 (1989).

21. Saxena, A.; Han, J.; Banerji, K.: J. Press. Ves. Tech. (1989).

22. Gladwin, D.N.; Miller, D.A.; Priest, R.H.; Proc 4th Int. Conf. Creep and Fracture of Engng. Mats. and Structs., 765-778 (1990).

23. Ainsworth, R.A.: Int. J. Fract., 20, 147 (1982).

24. Gladwin, D.N.; Miller, D.A.; Priest, R.H.; Fat. Fract. Engng. Mater. Struct., 12, 3, 187-200 (1989).

25. Nikbin, K.M.; Smith, D.J.; Webster, G.A.: J. Eng. Mats. Tech. 108, 186 (1986).

Acknowledgement

This paper is published with permission of Nuclear Electric plc.

A Constitutive Model for Creep and Damage in Composite Materials

R. PYRZ

Institute of Mechanical Engineering
University of Aalborg
Pontoppidanstræde 101, 9220 Aalborg, Denmark.

Summary

The concepts underlying slip theories of polycrystalline materials are implemented for the determination of constitutive equation for composite materials. The creep deformation and an influence of damages are described on the local planes. The area length density of cracks serves as an indicator for the extent of deterioration of the material and may be estimated by the use of stereological methods.

Introduction

The nonlinear response of composites is a macroscopic manifestation of the irreversible changes in its microstructure. The most common modes of the irreversible microstructural re–arrangements are viscoelastic or viscoplastic flow and loss of cohesion along the constituents boundaries and within each constituent fase. The degree to which the two classes of micro–structural rearrangements influence the macroscopic response of a composite and the complexity of evolution explain the apparent separation of approaches employed in formulating constitutive theories for these materials. The common formalism used in the prediction of particular damage mode is that of fracture mechanics and micromechanical modelling. Although adding a great deal to our understanding of various microdamage mechanisms the complexity of such modelling has made its application to practical problems of predicting the composite performance rather limited. The question still remains whether the fracture mechanics can be applied to composites weakened by a multitude of microcracks distributed over a large portion of the volume. Since the number of microdamages is extremely large and their locations are mostly random it is inevitable to treat the densely cracked composite as a continuum. The continuum damage mechanics has emerged as a viable framework for continuum theories that introduce damage variables reflecting the influence of density, distribution, shape and orientation of microdefects on the macroscopic elastic response. The major weakness of the continuum damage models resides in the arbitrariness

M. Zyczkowski (Ed.)
Creep in Structures
IUTAM Symposium Cracow/Poland 1990

of the choice of evolution equations for damage descriptors. On the other hand the description of a nonlinear response of composites coming from inelastic properties of the matrix material usually overlook a simultaneous nucleation of microdefects.

The objective of this study is to develop a model that is capable to take into account both the damage and creep response in unidirectional composite material.

Basic concepts

A fundamental hypothesis introduced here assumes that the constitutive properties of a composite may be characterized separately on planes of various orientations within the material and then in some way superimposing the effects from all the planes. This is the original idea of slip theories of polycrystalline materials developed in [1]. In order to simplify the analysis the plane stress conditions are assumed. The fixed off–axis coordinate system $\alpha-\beta$ determines the position of the material on the plane, and is rotated from the principal material axes 1–2 by an off–axis angle Θ, Fig. 1. The mobile coordinate system y–z introduced on the unit circle characterizes the position of the local plane with the strain components ε_{zz} and ε_{zy}. The total strain in fixed coordinate system is obtained by superimposing contributions from all planes as mobile coordinate systems moves along the circumference of the unit circle i.e.

$$\epsilon_{\alpha\beta} = \frac{2}{\pi} \int_0^{\pi} \epsilon_{zz}\, n_{\alpha z}\, n_{\beta z}\, d\phi + \frac{1}{\pi} \int_0^{\pi} \epsilon_{zy} \left(n_{\alpha z}\, n_{\beta y} + n_{\alpha y}\, n_{\beta z} \right) d\phi . \quad (1)$$

where n represents direction cosines between axes. The constitutive equation on the plane of arbitrary orientation must reflect the anisotropy of the material. The stress–strain relation in an arbitrary y–z–axis coordinate system is

$$[C]_{yz} = [T]^T\, [A]\, [T]. \quad (2)$$

where [T] is the transformation matrix and [A] is the compliance matrix expressed in terms of engineering constants in principal directions. Having the compliance matrix [C] the strain components on an arbitrary plane may be obtained

$$\begin{aligned} \varepsilon_{zz} &= C_{12}\, \sigma_{yy} + C_{22}\, \sigma_{zz} + C_{26}\, \sigma_{zy} \\ 2\, \varepsilon_{zy} &= C_{16}\, \sigma_{yy} + C_{26}\, \sigma_{zz} + C_{66}\, \sigma_{zy}. \end{aligned} \quad (3)$$

The integer indexes denote the position of the matrix elements according to composite mechanics notation and are not related to principal material coordinate system. The stresses in the Eq. 3 are defined as

$$\sigma_{yy} = n_{ry}\, n_{sy}\, \sigma_{rs}, \quad \sigma_{zz} = n_{rz}\, n_{sz}\, \sigma_{rs}, \quad \sigma_{zy} = n_{rz}\, n_{sy}\, \sigma_{rs}. \quad r,s = \alpha,\beta \tag{4}$$

with σ_{rs} being the components of the macroscopic applied stress.

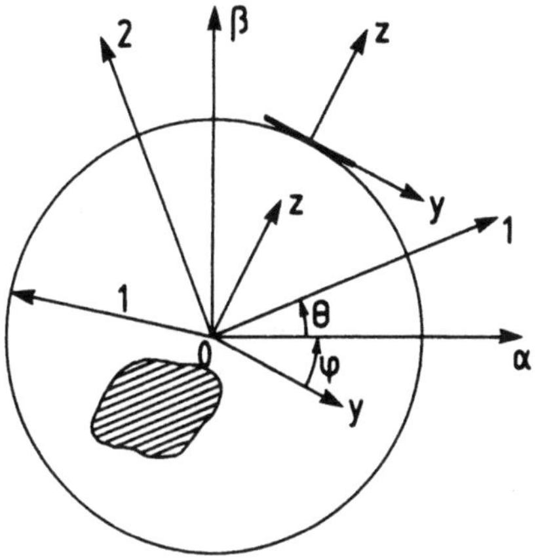

Fig. 1. Fixed and mobile coordinate system.

In deriving the constitutive equation the microstructure of the composite has been constrained statically, i.e. the stresses on planes of various orientations are the resolved components of the macroscopic stress (4). An alternative formulation with the kinematic constraint has been used in [2].

<u>Constitutive model with creep</u>

The nonlinear creep deformation is usually attributed to viscoelastic properties of the matrix material. Experimental investigations concerning viscoelastic behaviour suggest that a dominant contribution to creep deformation results from the application of shear stresses. Figure 2 shows typical creep curves for the unidirectional carbon fibre – PEEK composite loaded uniaxially in different off–axis configurations. It is seen that the more compliate configurations are shifted towards fiber direction which would suggest that the tensile stress along the fibres accelerates the shear creep. However, in the first approximation the creep

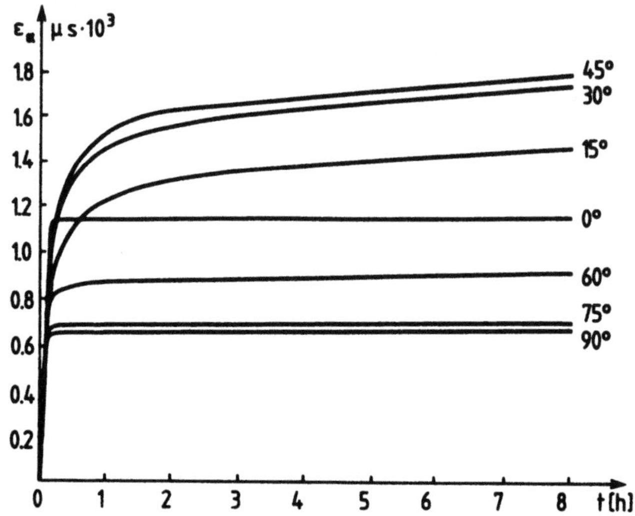

Fig. 2. Creep curves for unidirectional composite at stresses equal 80% of ultimate strength in corresponding off–axis configurations.

contribution on planes may be related exclusively to shear strains and stresses on the plane. Thus instead of Eq. $(3)_2$ we write

$$2\,\epsilon_{zy} = C_{16}\,\sigma_{yy} + C_{26}\,\sigma_{zz} + C_{66}\,\sigma_{zy} + g(\phi)\ b\sinh\frac{\sigma_{zy}}{\sigma_o}\cdot t^n. \qquad (5)$$

where a nonlinear power law for creep is used [3] and b, σ_o, n are constants. The function $g(\phi)$ describes the orientation of planes which contribute to overall creep deformation. A privilege shearing plane is always directed along the fibres independently of the off–axis angle Θ. The planes declining from the privilege direction contribute less and less as plane approaches the direction perpendicular to fibres. The orientation distribution function $g(\phi)$ should have a bell–like character with its maximum located at an angle corresponding to privilege direction. Among many orientation distribution functions investigated in the circular statistics two seems to be specially suited for the present purposes. The von Mises distribution is defined as follows [4]

$$g(\phi) = \frac{1}{2\,\pi\,I_o(k)}\ \exp\,[k\,\cos\,2\,(\phi-\mu_o)]. \qquad (6)$$

where $I_o(k)$ is the modified Bessel function of the first kind and order zero. The parameter μ_o is the privilege direction while the parameter k is the concentration parameter; for k = 0, $g(\phi) = 1/\pi$ and all planes contribute equally, while if $k \to \infty$ only plane along the privilege direction contributes, Fig. 3a.

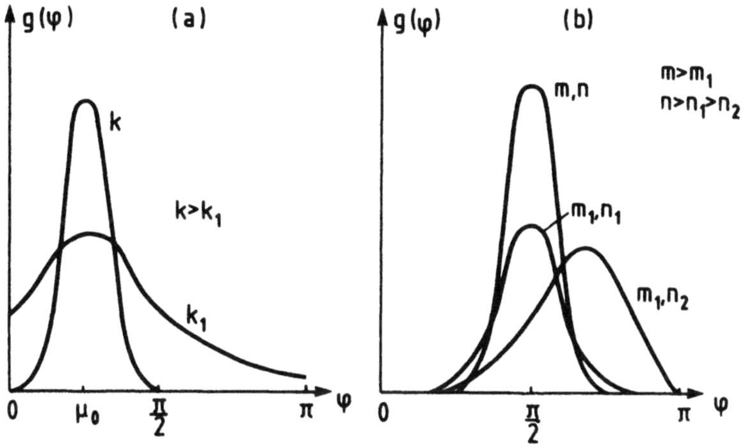

Fig. 3. Planes orientation distribution functions.

The other candidate for the orientation distribution of shearing planes is given by the expression [5]

$$g(\phi) = A\left(\int_0^{\pi} A d\phi\right)^{-1}, \quad A = (\sin\phi/2)^{2m-1}(\cos\phi/2)^{2n-1}. \qquad (7)$$

with m, n ≥ 1/2. The degree of peakedness increases as the values of parameters increase and the degree of skewness increases with an increase in the difference between the two parameters, Fig. 3b. Parameters of the orientation distribution and the power law should be determined from the micromechanical analysis which is left to further studies. To make the expression for $g(\phi)$ more suitable for calculations it can be convenient to introduce the

$$G(\phi) = \frac{\pi}{W}\sum_{j=1}^{N} w_j\,\delta(\phi-\phi_j), \quad W = \sum_{j=1}^{N} w_j. \qquad (8)$$

quantity where δ denotes the Dirac delta function, ϕ_j determines the discrete privilege directions of shearing planes and w_j weights the contributions from each plane. Equation (8) gives the discrete approximation for continuous orientation distribution functions (6) and (7). For composite materials with high fiber volume fraction the orientation distribution should be closely clustered along a privilege direction and it may suffice to take only one term in the expression (8). In cases where the influence of normal stresses on the shear creep is significant and the normal stresses themselves contribute to the total creep as in the case of creeping fibres, an appropriate term in the Eq. (3) should be added in the way demonstrated only for shear components on the plane of various orientation.

Constitutive model with damage

Non–destructive testing and microscopic observation indicate that the creep deformation is accompanied with nucleation and growth of microcracks which in unidirectional composite are alligned along the fiber direction in an off–axis configuration. An appearance of microcracks accelerates the creep deformation and consequently contributes to nonlinear response of the composite.

If microcracks tend to decline from the fibre direction an estimation of the microcracks orientation may be obtained from microphotographs taken during the deformation. For the unidirectional composite the distribution function (8) seems to be sufficient. Under the simplifying assumption that the cracks affect only the normal components of stresses and strains on each cracking plane the expression $(3)_1$ may be rewritten as follows

$$\varepsilon_{zz} = C_{12}\sigma_{yy} + C_{22}\, F(L)\, g(\phi)\sigma_{zz} + C_{26}\sigma_{zy}. \tag{9}$$

where the factor F(L) characterizes the crack length per unit area of the specimen. Let us identify the cracks with geometrical quantities i.e. consider them as curves distributed over the plane of the specimen. The area length density L may be calculated from microphotographs of area A designated in Fig. 4 as a planar domain S which contains a set of curves B showing some preferred orientation determined by an angle α. The area length density is calculated indirectly by placing the grid of testing lines T perpendicular to preferred orientation and counting the number of intersections N_T per unit length of testing lines [6],

$$L = 2N_T \left\{ \int_0^{2\pi} |\cos\psi| \, f(\psi) \, d\psi \right\}^{-1} . \qquad (10)$$

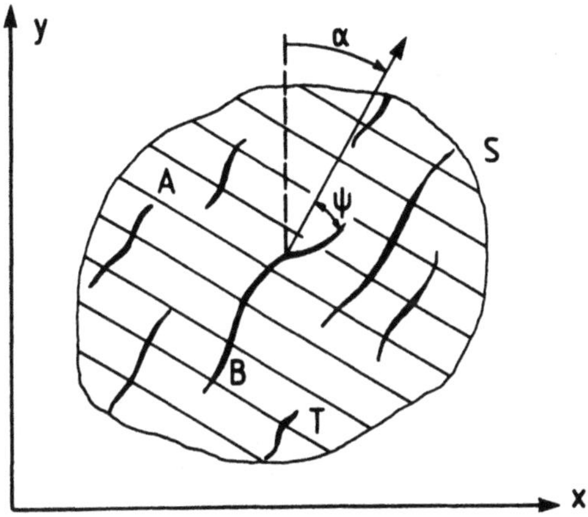

Fig. 4. Estimation of length density.

where the orientation function $f(\psi)$ describes the distribution of deviations ψ of lines from preferred orientation. The function $f(\psi)$ may be taken in the form (6), (7) or (8) if the cracks are perfectly straight. The evolution of the parameter L as the function of stresses and time follows from consecutive analysis of microphotographs as deformation progresses. If the structure of composite does not allow for optical monitoring of microcracks the calculation of area length density may be carried out by sectioning the specimen [7]. Having found the area length density the factor F(L) is taken as a nondimensional quantity

$$aF(L) = \frac{L_u - L}{L_u} . \qquad (11)$$

where a is adjustment parameter and L_u is the area length density at failure or at saturation point for the given type of cracks. In view of the relation between only normal components ε_{zz} and σ_{zz} in (9) the normal stress on the cracking plane may be interpreted as the compilation of the elastic stress and of the additional relaxation stress caused by all cracks of the same orientation as the cracking plane.

Conclusion

The adventage of the model is that the stress–strain relations are primarily defined on the microplane level thus it is possible to use more elaborated equations which would have to reflect micromechanical aspects in more detailed form. The model does not reflect spatial interactions between the planes. It takes into account the micro–interactions among various orientations but not those at distance. The latter ones are perhaps best described by nonlocal concepts. The proposed phenomenological model of composite materials conveys the influence of micromechanisms in an appropriately smoothed sense retaining the fundamental aspects of phenomena and neglecting the less important details.

References

1. Batdorf, S.B.; Budiansky, B.: A mathematical theory of plasticity based on the concept of slip. Technical Note No. 1871, Nat. Advisory Committee for Aeronautics Washington, D.C., 1949.

2. Pyrz, R.: Inelastic constitutive modelling of composites – rate and damage effects. In Durability of Polymer Based Composite Systems for Structural Applications. A.H. Cardon, G. Verchery (eds.), London: Elsevier 1991.

3. Dillard, D.A.: Creep and creep rupture of laminated graphite–epoxy composites. Ph.D. Dissertation, VPI & SU, Blacksburg, VA, 1981.

4. Mardia, K.V.: Statistics of directional data. London: Academic Press 1972.

5. Maekawa, Z.; Hamada, H.; Yokoyama, A.: Lamination theory of composite material with complex fiber orientation distribution. In Proc. ICCS 5, I.H. Marshall (ed.), London: Elsevier 1989 701–714.

6. Pyrz, R.: A hybrid model for nonlinear characterization of composite materials. In Inelastic Deformation of Composite Materials, G.J. Dvorak (ed.), New York: Springer 1990.

7. Weibel, E.R.: Stereological methods - theoretical formulation, vol. 2. New York: Academic Press 1980.

Part III:
Structures, Analytical and Numerical Methods, Optimal Design

Advances in Finite Element Methods for Elasto-Plastic and Creep Analysis

Klaus-Jürgen Bathe
Massachusetts Institute of Technology
Cambridge, MA 02139 U.S.A.

and

Miloš Kojić and Jan Walczak
ADINA R & D, Inc.
Watertown, MA 02172 U.S.A.

Summary

Practical inelastic analysis of structures requires effective kinematic and finite element formulations, the use of physically appropriate constitutive relations and efficient computational procedures.

The objective in this lecture is to survey some recent advances in finite element methods for elasto-plastic and creep analysis: effective kinematic formulations, efficient finite element discretizations and the effective-stress-function algorithm for the integration of the highly nonlinear stress relations. Experiences with these computational procedures are presented and discussed to point out their strengths and further desirable developments.

This is an extended abstract of the lecture.

Introduction

The solution of complex elasto-plastic and creep problems has received increasing attention during the last two decades. On the one side, there has been a competitive incentive and a need to design structures that are less expensive to build, that operate more efficiently and over longer time periods, and that possibly use new materials, while on the other side, the use of finite element methods has opened up new possibilities to analyze complex structural designs. The use of finite element methods for inelastic analysis is now widely accepted in design -- although, of course, only for advanced

M. Zyczkowski (Ed.)
Creep in Structures
IUTAM Symposium Cracow/Poland 1990

analysis -- and we can certainly expect a continued significant increase in this usage [1].

However, although the first more general finite element program capabilities for nonlinear analysis were released already in the early 1970's (the programs Marc and NONSAP), there has been, during the last decade, most significant further progress in the development of more comprehensive and reliable general procedures for nonlinear analysis. The objective in this survey lecture is to review some of these developments that have considerably strengthened elasto-plastic and creep analysis capabilities. Since the field of inelastic finite element analysis is indeed very large, we concentrate, due to the limited amount of time allocated to our talk, on the research and development results that we have accomplished in our Finite Element Research Group at M.I.T. and in our company ADINA R & D. We aim to emphasize in this presentation the very significant new and further developments that have been achieved during the recent years.

These developments fall into the following categories, (i) the continuum mechanics kinematic formulations of finite element elasto-plastic and creep analysis, (ii) the finite element discretization of these governing equations, and (iii) the numerical solution of the finite element equations.

The text given here is an extended abstract of our lecture. For the details presented in the lecture, we refer to the references cited on the subject areas.

<u>Continuum Mechanics Formulations</u>

In general, inelastic analysis problems can be categorized into three problem areas; (1) problems that only exhibit material nonlinearities, i.e. in which the displacements are infinitesimally small, (2) problems with material nonlinearities, large displacements and large rotations but in which the strains are small, and (3) problems with material nonlinearities and in which the displacements and strains are large [2].

In addition, of course, contact conditions can be encountered in each of these problem areas [3].

The formulation of the materially-nonlinear-only problems is a straight-forward extension of linear elastic infinitesimal displacement analysis [2]. The difficulties in these analyses lie in the accurate iterative solution of the finite element equations.

The formulation of large displacement and large rotation but small strain analysis has been presented in refs. [2,4], and the total Lagrangian formulation given there is, in general, probably the most effective solution approach.

The formulation of large displacement and large strain analysis, has, on the other hand, presented considerable research questions. Here rate-type formulations have been, and are still commonly used (with the Jaumann or related stress rate measures)[5], and such solution approach has only been significantly improved upon during the recent years by use of a total formulation [6-10]. Such total formulation can be arrived at by a total Lagrangian [8] or an updated Lagrangian approach [10]. These formulations admit larger load steps (as does the total Lagrangian formulation in large displacement, small strain analysis [2]) and are conservative in a closed elastic strain path [11].

Finite Element Discretization

The finite element discretization of the continuum mechanics incremental formulations is effectively performed using isoparametric elements that, however, are formulated using a mixed (u/p) interpolation (u = displacement, p =

pressure) [12]. The u/p interpolation is necessary in plane strain, axisymmetric and three-dimensional analyses in order to prevent locking in almost incompressible conditions [13].

The elements that can be recommended for general two-dimensional and three-dimensional analyses are the two-dimensional 9/3 element (9 nodes for the displacement interpolations and three variables for the pressure interpolation) and three-dimensional 27/4 element. Namely, these elements satisfy the inf-sup condition for stability and optimal convergence and show good behavior in practice [12]. It should be noted that elements which do not satisfy the inf-sup condition may perform satisfactorily in many inelastic analyses, but under general conditions, it is very desirable to use elements that are mathematically well-founded. Specifically, in analyses with large strains that are modeled as (almost) incompressible, the above elements show much better predictive capability than those that do not satisfy the inf-sup condition.

Of course, the better performance of elements can only be seen and measured when stresses and strains are displayed that are not "smoothed" to always, in any case, obtain continuous stress contours [14].

Stress Integration

The accuracy of an inelastic solution depends entirely on how accurate the stress integration is performed given a total strain increment. We have developed the effective-stress-function (ESF) algorithm for the integration, which is a generalization of the radial return method [15,16]. The ESF algorithm can be used for elasto-plastic and creep analysis with thermal effects. The accuracy of the procedure lies in that a governing response equation in the effective stress is solved with high precision and all other inelastic variables are then evaluated.

We have developed this solution technique for the analysis of two- and three-dimensional solids, including for plane stress conditions, and for plate, shell, and piping analyses. The attractive feature of the technique is that it is applicable to elasto-plasticity and creep with general creep laws and that it shows excellent behavior to give stable and accurate solutions.

Sample Solutions

Figures 1 and 2 show the results of two analyses performed with the techniques briefly summarized above. Additional problem solutions are given in the cited references.

References

[1] Nonlinear Finite Element Analysis and ADINA, Conference Proceedings, Volumes 1 to 7; Vol. 7 in J. Computers & Structures, Vol. 32, No.3/4, 1989.

[2] K.J. Bathe. Finite Element Procedures in Engineering Analysis. Prentice-Hall, Englewood Cliffs, N.J., 1982.

[3] K.J. Bathe and A. Chaudhary. A solution method for planar and axisymmetric contact problems. Int. J. Numer. Methods in Eng., 21, 65-88, 1985.

[4] K.J. Bathe, E. Ramm, and E.L. Wilson. Finite element formulations for large deformation dynamic analysis. Int. J. Numer. Methods in Eng., 9, 353-386, 1975.

[5] D.R.J. Owen, E. Hinton, and E. Oñate, editors. Computational Plasticity: Models, Software and Applications. Pineridge Press, April 1987.

[6] W.D. Rolph III and K.J. Bathe. On a large strain finite element formulation for elasto-plastic analysis. In K.J. Willam, editor, Constitutive Equations: Macro and Computational Aspects, pages 131-147. ASME, AMD, December 1984.

[7] K.J. Bathe, M. Kojić, and J. Walczak. Some developments in methods for large strain elasto-plastic analysis. In D.R.J. Owen, E. Hinton, and E. Oñate, editors, Computational Plasticity: Models, Software and Applications, pages 263-279. Pineridge Press, April 1987.

[8] J.C. Simo. A framework for finite strain elastoplasticity based on maximum plastic dissipation and the multiplicative decomposition: Part I. Continuum formulation, Part II. Computational aspects. J. Computer Methods in Applied Mech. and Eng., 66,199-219, 1988, 68,1-31, 1988.

[9] A. Eterovic and K.J. Bathe. A note on the use of the additive decomposition of the stress tensor in finite deformation inelasticity. J. Computer Methods in Applied Mech. and Eng., 1990.

[10] A. Eterovic and K.J. Bathe. A hyperelastic-based large strain elasto-plastic constitutive formulation with combined isotropic-kinematic hardening using the logarithmic stress and strain measures. Int. J. Numer. Methods in Eng., 1990.

[11] M. Kojić and K.J. Bathe. Studies of finite element procedures -- stress solution of a closed elastic strain path with stretching and shearing using the Updated Lagrangian Jaumann formulation. J. Computers & Structures, 26, No. 1/2, 175-179, 1987.

[12] T. Sussman and K.J. Bathe. A finite element formulation for nonlinear incompressible elastic and inelastic analysis. J. Computers & Structures, 26, No. 1/2, 357-409, 1987.

[13] F. Brezzi and K.J. Bathe. A discourse on the stability conditions for mixed finite element formulations. J. Computer Methods in Applied Mech. and Eng., 1990.

[14] T. Sussman and K.J. Bathe. Studies of finite element procedures -- stress band plots and the evaluation of finite element meshes. J. Engineering Computations, 3, 178-191, 1986.

[15] M. Kojić and K.J. Bathe. The 'effective-stress-function' algorithm for thermo-elasto-plasticity and creep. Int. J. Numer. Methods in Eng., 24, 1509-1532, 1987.

[16] M. Kojić and K.J. Bathe. Thermo-elastic-plastic and creep analysis of shell structures. J. Computers & Structures, Vol. 26 No. 1/2, 135-143, 1987.

[17] J.K. Lee, R.H. Wagoner, E. Nakamachi. A benchmark test for sheet forming analysis. Report no. ERC/NSM-S-90-22. The Ohio State University, Columbus, Ohio, July 1990.

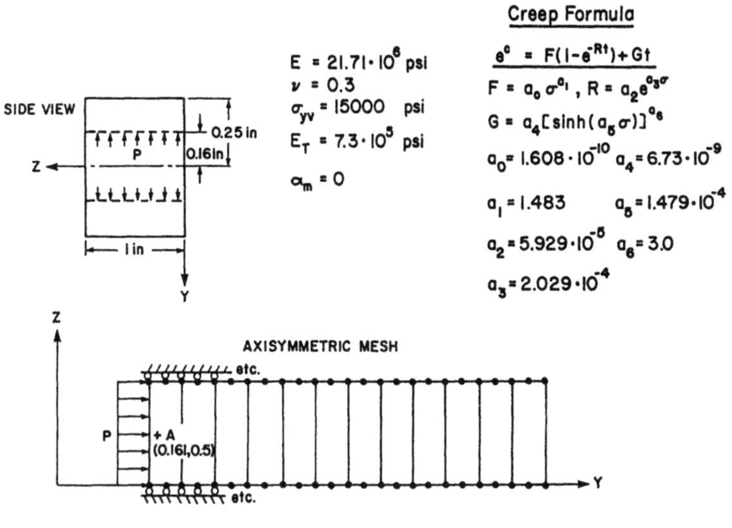

Finite Element Mesh and Material Data

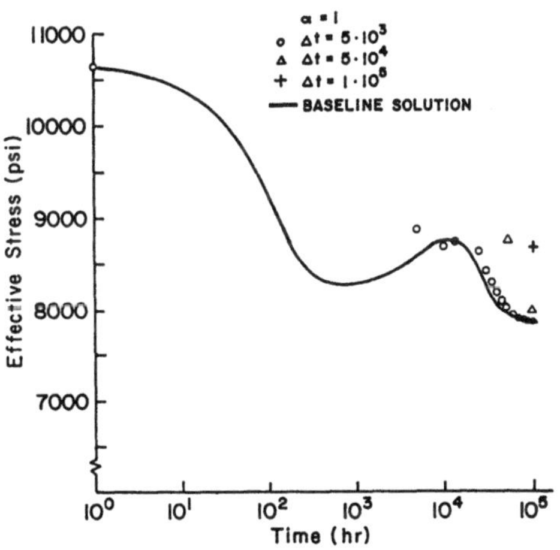

Effective Stress at Location A

iteration	
i = 1	0.82×10^{-1}
2	0.90×10^{-4}
3	0.82×10^{-8}
4	0.13×10^{-15}

$\Delta t = 10^5$, unbalanced energy

$({}^{t+\Delta t}\underline{R} - {}^{t+\Delta t}\underline{F}^{(i)})^T\, \Delta\underline{U}^{(i)}$

Figure 1 Creep analysis of a cylinder from ref. [15]. The solution shows the efficiency of the ESF algorithm.

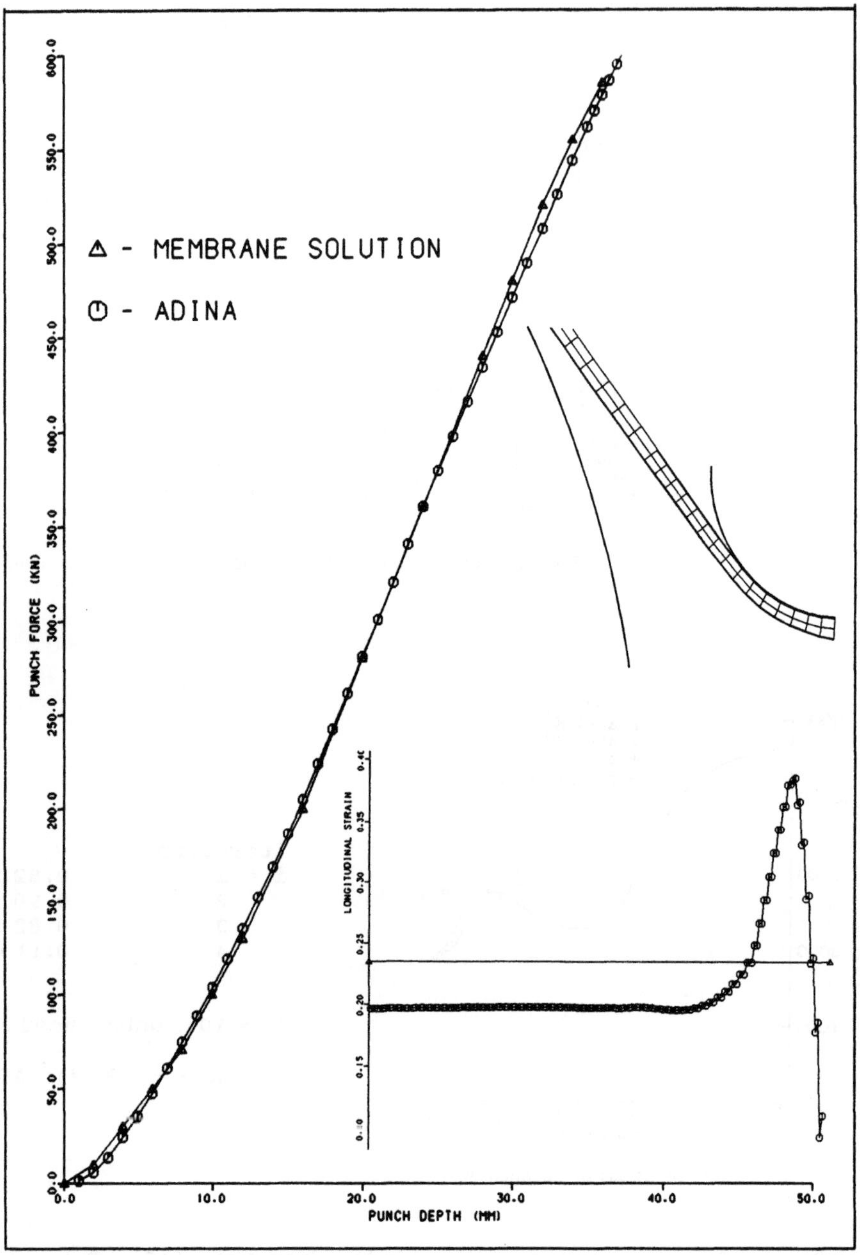

Figure 2 Analysis of a sheet metal forming problem defined in ref. [17]. The analysis is performed with ADINA using 130 4-node axisymmetric elements.

Large Displacements and Rotations for Shells of Revolution due to Axisymmetric Elastic, Creep and Plastic Deformation

J.F. BESSELING

Laboratory for Engineering Mechanics
Delft University of Technology, The Netherlands

Summary

With the radial and axial displacements and the angle of rotation of the normal to the midsurface as the kinematic variables in the nodal circles, the properties of a finite element are derived on the basis of Kirchhoff's hypothesis. The deformation of the element is determined by seven generalized strains. For the dual generalized stresses the constitutive equations are derived by application of the local constitutive equations in sampling points. For the elastic behaviour they are taken to be two Gaussian integration points, while for the creep and plasticity properties a number of points along the normal in the middle section are used. Since in the case of creep and plasticity the variation of the state of stress over the thickness is history dependent, only one section is considered, thus limiting the number local stresses, that have to be stored in the course of a loading program. A strategy of solution is discussed that is robust and that will give qualitative insight, without undue emphasis on the representation of the local creep and plasticity model, which generally is also of a qualitative nature.

Introduction

Though axisymmetric loading of shells of revolution, including circular plates, has extensively been discussed in the literature [1,2,3], the way how to handle creep and plastic behaviour has received less attention. In particular the combination of inelastic behaviour and arbitrarily large displacements and rotations is of interest in connection with pressforming of axisymmetric objects. However, this paper is only concerned with the derivation of the finite element

M. Zyczkowski (Ed.)
Creep in Structures
IUTAM Symposium Cracow/Poland 1990

properties for the case that the local strains remain small, while the solution strategy does not consider the problem of a growing contact area as in pressforming.
In the finite element model care is taken that the elastic properties are accurately represented since the elastic constants are usually known with a fair amount of accuracy. In the representation of the creep and plastic properties it is made sure that the element as a whole shows an inelastic behaviour, consistent with the underlying material model, without undue emphasis on the accuracy of the derivation of the global element properties from the local creep and plastic properties in the inhomogeneous state of deformation of the element.
Similarly the solution strategy aims at an accurate representation of the elastic behaviour of the shell as a whole, while giving qualitative insight into the effect of creep and plasticity without undue computing effort.

Principal of virtual work and deformations

The statics of an axisymmetric shell may be formulated in terms of the principal of virtual work: "The work of the external loading shall be equal to zero for all virtual displacements and rotations that produce no material deformations".
For the shell of revolution of Fig. 1 we consider as external loading an internal pressure p and endloads and -moments F_r, F_z and T. Since we shall only consider cases in which local strains remain small, the radial displacement u may be neglected with respect to the radial coordinate r ($\varepsilon_t = {}^u/_r \ll 1$). The condition of no deformation in the principle of virtual power is taken into account by the introduction of the dual stresses σ_s and σ_t as multipliers. Furthermore Kirchhoff's hypothesis is added as a variational expression with multipliers $\sigma_s{}^*$ and $\sigma_t{}^*$, while the strains of the midplane and the curvatures are expressed in terms of the displacements and rotation with the aid of a variational expression with multi-

pliers N_s, N_t, M_s, M_t. Finally with a multiplier λ a variational expression should be added to express the relation between the displacements and angle of rotation φ.

$$
\begin{aligned}
&\int_0^{\ell} p\left\{\delta u\left(\sin\alpha + \frac{dw}{ds}\right) - \delta w\left(\cos\alpha + \frac{du}{ds}\right)\right\} 2\pi r ds \\
&\quad + 2\pi r^0 (F_r{}^0 \delta u^0 + F_z{}^0 \delta w^0 + T^0 \delta\varphi^0) + 2\pi r^1 (F_r{}^1 \delta u^1 + F_z{}^1 + T^1 \delta\varphi^1) \\
&\quad = \int_0^{\ell} \int_{-\frac{h}{2}}^{\frac{h}{2}} (\sigma_s \delta\varepsilon_s + \sigma_t \delta\varepsilon_t)\ 2\pi r dz ds \\
&\quad + \delta \int_0^{\ell} \int_{-\frac{h}{2}}^{\frac{h}{2}} \{\sigma_s{}^* (z k_s + \varepsilon_s{}^0 - \varepsilon_s) + \sigma_t{}^* (z k_t + \varepsilon_t{}^0 - \varepsilon_t)\}\ 2\pi r dz ds \\
&\quad + \delta \int_0^{\ell} [N_s \left\{\frac{du}{ds}\left(\cos\alpha + \frac{1}{2}\frac{du}{ds}\right) + \frac{dw}{ds}\left(\sin\alpha + \frac{1}{2}\frac{dw}{ds}\right) - \varepsilon_s{}^0\right\} + \\
&\qquad + N_t \left({}^{u}/{}_{r} - \varepsilon_t{}^0\right) + M_s \left(\frac{d\varphi}{ds} - k_s\right) + M_t \left(\frac{\sin(\varphi+\alpha) - \sin\alpha}{r} - k_t\right) + \\
&\qquad + \lambda \left(\varphi - \tan^{-1} \frac{\left(\sin\alpha \dfrac{du}{ds} + \cos\alpha \dfrac{dw}{ds}\right)}{\left(1 + \cos\alpha \dfrac{du}{ds} + \sin\alpha \dfrac{dw}{ds}\right)}\right)]\ 2\pi r ds
\end{aligned}
\tag{1}
$$

By considering the variations of ε_s and ε_t we may eliminate immediately the multipliers $\sigma_s{}^*$ and $\sigma_t{}^*$, as well as ε_s and ε_t ($\sigma_s{}^* = \sigma_s$, $\sigma_t{}^* = \sigma_t$).

Though the curvatures k_s and k_t are given by simple expressions in φ, for the angle φ itself we have a complicated expression in u and w.

The statics of our finite element model will be based upon the variational condition of the principle of virtual work in the form (1). However we shall see that the use of the complicated expression for φ can be avoided.

Each finite element will be represented by a conical element with a linearly varying thickness. In the case of discontinuities of thickness the adjoining elements will be given a common nodal circle as interface, thus simplifying the static interaction.

Though by Kirchhoff's hypothesis the axial and radial displacements would completely determine the deformations of the shell, we shall consider in each nodal circle in addition to these displacements the angle of rotation φ of the normal to the midsurface, thereby eliminating the need for more than C^0 continuity of the displacements. Thus we have six nodal displacements and rotations for each element. These 6 kinematic variables determine 6-3=3 deformation parameters in a meridional plane, but furthermore each kinematic variable describing a radial movement of material will imply an extra deformation. This is true both for the radial displacements u as well as for the rotations φ because of the finite thickness. Hence we may choose 3+4=7 independent deformation parameters for one finite element.

We select the deformation parameters such, that they are valid for arbitrarily large rotations φ and displacements w and convenient for the case of small strains, to which our treatment is restricted.

The integral of the meridional strain $\varepsilon_s{}^0$ and the tangential endstrains, u_1/r_1 and u_2/r_2, are a suitable choice for midplane deformations. As illustrated in Fig. 2 the meridional

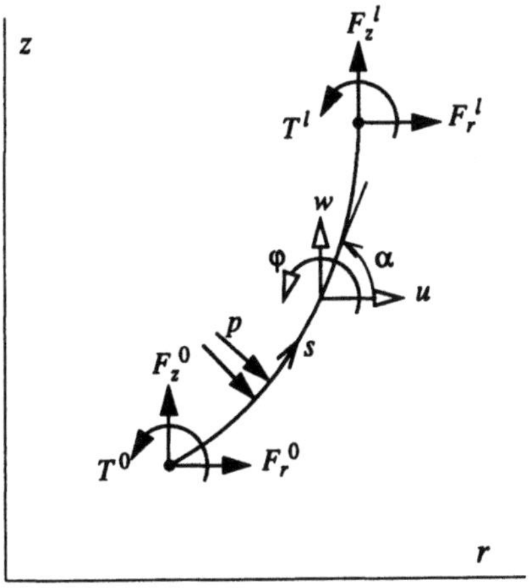

Fig. 1. Geometry and loading.

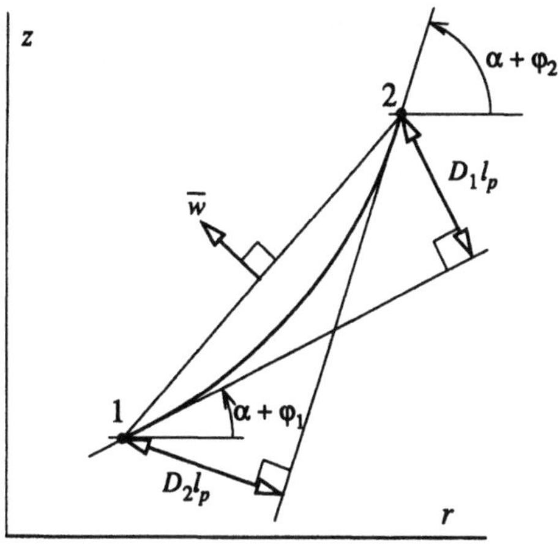

Fig. 2. Bending deformation.

bending can be represented by the differences in slope at the ends between the deformed midplane and the connecting line between nodal points. The tangential bending is finally expressed by the tangential curvatures at the ends.

It can be observed in Fig. 2 that the meridional bending can be described as an additional displacement $\bar{w}$, perpendicular to the connecting line of the displaced nodal points. Since at the ends $\bar{w}$ must be equal to zero and the slopes can be expressed in the meridional bending parameters D_1 and D_2, we can write for $\bar{w}$ a cubic interpolation, provided $D_1 \ll 1$ and $D_2 \ll 1$.

$$D_1 = -\left(\cos\alpha + \frac{u_2 - u_1}{\ell_p}\right)\sin(\alpha+\varphi_1) + \left(\sin\alpha + \frac{w_2 - w_1}{\ell_p}\right)\cos(\alpha+\varphi_1) \qquad (2a)$$

$$D_2 = \left(\cos\alpha + \frac{u_2 - u_1}{\ell_p}\right)\sin(\alpha+\varphi_2) - \left(\sin\alpha + \frac{w_2 - w_1}{\ell_p}\right)\cos(\alpha+\varphi_2) \qquad (2b)$$

$$\bar{w} = -\ell_p D_1(\xi^3 - 2\xi^2 + \xi) - \ell_p D_2(-\xi^3 + \xi^2), \quad 0 \le \xi \le 1. \qquad (2c)$$

The relation between $\bar{w}$ and φ is given by

$$\frac{d\varphi}{ds} = \frac{1}{\ell_p{}^2}\frac{d^2\bar{w}}{d\xi^2} = \frac{1}{\ell_p}\,[D_1(4-6\xi)+D_2(-2+6\xi)], \quad (3)$$

which is consistent with the following interpolation for φ

$$\varphi = \varphi_1+D_1(4\xi-3\xi^2) + D_2(-2\xi+3\xi^2). \quad (4)$$

Since

$$\varphi_2 = \varphi_1+D_1+D_2 \quad \text{or} \quad \varphi_2-\varphi_1 = D_1+D_2 << 1, \quad (5)$$

we have a linear interpolation of the meridional curvature over the length of the finite element,

$$\frac{d\varphi}{ds} = \frac{1}{\ell_p}\,[D_1(4-6\xi) + D_2(-2+6\xi)], \quad (6)$$

that is consistent with Kirchhoff's hypothesis without the need of employing the complicated relationship between the angle φ and the displacements.

The tangential curvatures will be based upon a linear variation of $\{\sin(\varphi+\alpha)-\sin\alpha\}$ between the endvalues. Numerical difficulties for r=0 are avoided by introducing u_1/r_1 and u_2/r_2 as nodal kinematic variables instead of the radial displacements u_1 and u_2 themselves. A linear variation between endvalues then approximates the tangential strain of the midplane.

In addition to D_1 and D_2 of (2a and 2b) we now have the following deformation parameters or generalized strains, that completely determine the state of deformation of the finite element.

$$D_3 = \sin(\varphi_1+\alpha)-\sin\alpha, \tag{7a}$$

$$D_4 = \sin(\varphi_2+\alpha)-\sin\alpha, \tag{7b}$$

$$D_5 = \int_0^{\ell_p} \left\{\frac{du}{ds}\left(\cos\alpha + \frac{1}{2}\frac{du}{ds}\right) + \frac{dw}{ds}\left(\sin\alpha + \frac{1}{2}\frac{dw}{ds}\right)\right\} 2\pi r ds, \tag{7c}$$

$$D_6 = u_1/r_1, \tag{7d}$$

$$D_7 = u_2/r_2. \tag{7e}$$

By putting the righthand side of equation (1) equal to $\sigma_i \delta D_i$ (i=1...7), we obtain by numerical integration from the variational condition the following finite element equations, where a circumflex above the symbol indicates the value of the variable in the integration point i.

$$\begin{vmatrix} \delta\tilde{\varepsilon}_{si}{}^{0} \\ \delta\tilde{\varepsilon}_{ti}{}^{0} \end{vmatrix} \Longrightarrow \begin{vmatrix} \tilde{N}_{si} \\ \tilde{N}_{ti} \end{vmatrix} = \int_{-\frac{\tilde{h}_i}{2}}^{\frac{\tilde{h}_i}{2}} \begin{vmatrix} \tilde{\sigma}_{si} \\ \tilde{\sigma}_{ti} \end{vmatrix} dz, \tag{8a}$$

$$\begin{vmatrix} \delta\tilde{k}_{si} \\ \delta\tilde{k}_{ti} \end{vmatrix} \Longrightarrow \begin{vmatrix} \tilde{M}_{si} \\ \tilde{M}_{ti} \end{vmatrix} = \int_{-\frac{\tilde{h}_i}{2}}^{\frac{\tilde{h}_i}{2}} \begin{vmatrix} \tilde{\sigma}_{si} \\ \tilde{\sigma}_{ti} \end{vmatrix} zdz, \tag{8b}$$

$$\begin{vmatrix} \delta\tilde{N}_{si} \\ \delta\tilde{N}_{ti} \end{vmatrix} \Longrightarrow \begin{vmatrix} \tilde{\varepsilon}_{si}{}^{0} \\ \tilde{\varepsilon}_{ti}{}^{0} \end{vmatrix} = \begin{vmatrix} D_5/(2\pi\ell_p\tilde{r}_i) \\ \alpha_{i1}D_6+\alpha_{i2}D_7 \end{vmatrix}, \tag{8c}$$

$$\begin{vmatrix} \delta\tilde{M}_{si} \\ \delta\tilde{M}_{ti} \end{vmatrix} \Longrightarrow \begin{vmatrix} \tilde{k}_{si} \\ \tilde{k}_{ti} \end{vmatrix} = \begin{vmatrix} \beta_{i1}D_1 + \beta_{i2}D_2 \\ \gamma_{i1}D_3 + \gamma_{i2}D_4 \end{vmatrix}, \tag{8d}$$

$$\begin{vmatrix} \delta D_1 \\ \delta D_2 \\ \delta D_3 \\ \delta D_4 \\ \delta D_5 \\ \delta D_6 \\ \delta D_7 \end{vmatrix} \Longrightarrow \begin{vmatrix} \sigma_1 \\ \sigma_2 \\ \sigma_3 \\ \sigma_4 \\ \sigma_5 \\ \sigma_6 \\ \sigma_7 \end{vmatrix} = 2\pi\ell_p \begin{vmatrix} \sum_i w_i \; \tilde{r}_i \; \beta_{i1} \; \tilde{M}_{si} \\ \sum_i w_i \; \tilde{r}_i \; \beta_{i2} \; \tilde{M}_{si} \\ \sum_i w_i \; \tilde{r}_i \; \gamma_{i1} \; \tilde{M}_{ti} \\ \sum_i w_i \; \tilde{r}_i \; \gamma_{i2} \; \tilde{M}_{ti} \\ N_s/(2\pi\ell_p) \\ \sum_i w_i \; \tilde{r}_i \; \alpha_{i1} \; \tilde{N}_{ti} \\ \sum_i w_i \; \tilde{r}_i \; \alpha_{i2} \; \tilde{N}_{ti} \end{vmatrix} . \qquad (8e)$$

Two integration points with a weight $w_i = \frac{1}{2}$ seems to be an optimum choice for this type of element, in which case we have

$$\begin{bmatrix} \alpha_{11} & \alpha_{12} \\ \alpha_{21} & \alpha_{22} \end{bmatrix} = \begin{bmatrix} \left(\frac{1}{2} + \frac{1}{6}\sqrt{3}\right) & \left(\frac{1}{2} - \frac{1}{6}\sqrt{3}\right) \\ \left(\frac{1}{2} - \frac{1}{6}\sqrt{3}\right) & \left(\frac{1}{2} + \frac{1}{6}\sqrt{3}\right) \end{bmatrix} , \qquad (9a)$$

$$\begin{bmatrix} \beta_{11} & \beta_{12} \\ \beta_{21} & \beta_{22} \end{bmatrix} = \frac{1}{\ell_p} \begin{bmatrix} (1+\sqrt{3}) & (1-\sqrt{3}) \\ (1-\sqrt{3}) & (1+\sqrt{3}) \end{bmatrix} , \quad \begin{vmatrix} \tilde{r}_1 \\ \tilde{r}_2 \end{vmatrix} = \begin{bmatrix} \alpha_{11} & \alpha_{12} \\ \alpha_{21} & \alpha_{22} \end{bmatrix} \begin{vmatrix} r_1 \\ r_2 \end{vmatrix} , \qquad (9b)$$

$$\begin{bmatrix} \gamma_{11} & \gamma_{12} \\ \gamma_{21} & \gamma_{22} \end{bmatrix} = \begin{bmatrix} \alpha_{11}/\tilde{r}_1 & \alpha_{12}/\tilde{r}_1 \\ \alpha_{21}/\tilde{r}_2 & \alpha_{22}/\tilde{r}_2 \end{bmatrix} , \quad \begin{vmatrix} \tilde{h}_1 \\ \tilde{h}_2 \end{vmatrix} = \begin{bmatrix} \alpha_{11} & \alpha_{12} \\ \alpha_{21} & \alpha_{22} \end{bmatrix} \begin{vmatrix} h_1 \\ h_2 \end{vmatrix} . \qquad (9c)$$

Indicating the kinematic nodal variables by a common symbol v, and the dual force and moment variables by a common symbol f, we obtain from the principle of virtual work the following equilibrium equations

$$\delta v_k \Longrightarrow f_k = \sigma_i \, D_{i,k} \; , \quad D_{i,k} \equiv \frac{\delta D_i}{\delta v_k} \; , \qquad (10)$$

where the summation convention applies to the index i.
The constitutive equations for the generalized stresses σ_i are to be obtained from the local constitutive equations with the aid of (8a and b).

<u>Elastic properties</u>

The elastic material properties will be described by the constitutive equations for generalized plane stress of an isotropic material. Anisotropy consistent with the presupposed axial symmetry could be a useful extension of the treatment given here.
Substitution of the relations

$$\begin{vmatrix} \tilde{\sigma}_{si} \\ \tilde{\sigma}_{ti} \end{vmatrix} = \frac{E}{1-\nu^2} \begin{bmatrix} 1 & \nu \\ \nu & 1 \end{bmatrix} \begin{vmatrix} z\,\tilde{k}_{si} + \tilde{\varepsilon}_{si}{}^{0} \\ z\,\tilde{k}_{ti} + \tilde{\varepsilon}_{ti} \end{vmatrix} \tag{11}$$

into (8a) and (8b) and evaluation of the expressions (8e) for σ_i leads to a symmetrical elasticity matrix, in which the bending and membrane action are uncoupled.

$$\begin{vmatrix} \sigma_i{}^{b} \\ \sigma_j{}^{m} \end{vmatrix} = \begin{bmatrix} S_{ik}{}^{b} & 0 \\ 0 & S_{jl}^{m} \end{bmatrix} \begin{vmatrix} D_k{}^{b} \\ D_l{}^{m} \end{vmatrix}, \quad \begin{matrix} k,i = 1,..4, \\ l,j = 1,..3. \end{matrix} \tag{12}$$

Because of the geometric nonlinearity in the expressions for $D_1..D_5$ we shall have to analyse the elastic deformation problem by a step by step procedure, combined with Newton-Raphson iterations, if a test by a suitable criterion reveals the significance of this nonlinearity for the problem at hand. The incremental equations have the form (employing summation convention)

$$(D_{i,k}S_{ij}D_{j,l} + \sigma_i D_{i,kl})\,\Delta v_l = \Delta f_k,\quad D_{i,kl} = \frac{\partial^2 D_i}{\partial v_k \partial v_l}\,. \tag{13}$$

Especially for thin, initially flat circular plates under pressure loading the meridional stress resultant N_s will very soon (i.e. for relatively low stresses) lead to geometrically nonlinear effects, the membrane action taking over from the bending action.

Inelastic properties

For time independent plasticity we define a yield limit by the condition

$$\varphi(\sigma_s, \sigma_t) = \sigma_s^2 - \sigma_s\sigma_t + \sigma_t^2 - \sigma_y^2 = 0, \tag{14}$$

the yield stress in tension being given by σ_y.
Creep is defined in terms of a dissipation function π, depending on the temperature T and the stresses σ_s and σ_t.

$$\pi = \gamma(T) \left(\frac{\sigma_s^2 - \sigma_s\sigma_t + \sigma_t^2 - (1-\beta)\sigma_y^2}{\beta\sigma_y^2} \right)^n, \tag{15}$$

where

$$0 \leq \beta \leq 1, \quad n \geq 1 ,$$

provided

$$(\sigma_s^2 - \sigma_s\sigma_t + \sigma_t^2) \geq (1-\beta)\ \sigma_y^2,$$

else $\pi=0$.
In many cases significant creep will occur only at the higher stresses. Neglecting creep at the lower stresses can be expressed by putting $\beta<1$.
The constitutive equations can now be written as follows

$$\dot{\sigma}_i = (S_{ij} - (1-h)Y_{ij})(\dot{\varepsilon}_j - c_j), \tag{16}$$

with

$$S = \frac{E}{1-\nu^2}\begin{vmatrix} 1 & \nu \\ \nu & 1 \end{vmatrix}, \quad Y_{ij} = \frac{S_{im}\dfrac{\partial\varphi}{\partial\sigma_m}\dfrac{\partial\varphi}{\partial\sigma_n}S_{nj}}{\dfrac{\partial\varphi}{\partial\sigma_k}S_{kl}\dfrac{\partial\varphi}{\partial\sigma_l}} \quad \begin{array}{l}\text{if } \varphi=0,\ \dot{\varphi}=0, \\ \text{else } Y_{ij}=0,\end{array}$$

$$c_j = \frac{\partial\pi}{\partial\sigma_j}\pi\Big/\left(\sigma_i\frac{\partial\pi}{\partial\sigma_i}\right) \tag{17}$$

A (small) isotropic hardening or softening may be included by giving the quantity h in (16) a small value different from zero, in which case the condition $\dot{\varphi}=0$ has to be changed accordingly.

For the irreversible process under consideration the constitutive equations could only be determined as stress rate equations, in which the present stresses act as state variables. Since in the inhomogeneous deformation of the finite element the stress distribution will change drastically if inelastic deformation occurs, the necessity arises to keep a record of these state variables in a sufficient number of sampling points from the moment an element undergoes inelastic deformations. By means of this record of state variables the history dependence of the creep and plastic properties of a finite element is described. Again, by putting β in the creep dissipation function π (15) smaller than 1, creep at the lower stresses is being neglected, now with the advantage that no sampling point stresses have to be stored to keep track of the history of the process.

The sampling point stresses may be looked upon as hidden, or internal state variables by which a better description of the dissipative process is obtained without an increase of the number of independent kinematic variables. In other words the effect of the dissipative processes in the different material points, taken as sampling points, overshadows the effect of local disturbances of mechanical equilibrium, the number of equilibrium equations being equal to the unchanged number of kinematic variables.

In choosing the sampling points it is essential that the variation of state over the thickness can be represented. In contrast to this inhomogenity of state, the variation over the length of the finite element can always be reduced by taking a smaller element length, which because of the small local bandwidth of the system equations does not unacceptably increase the computational effort. Therefore we choose for the simplest state representation for the derivation of the element constitutive equations: we determine the yield matrix y and the creepvector c in (16) for a number of m sampling points, evenly distributed over the thickness of the mid-section of the element. Next the rate equations are substituted in the expressions (8a and b), after these have been differentiated with respect to time.

$$\begin{vmatrix} \dot{\tilde{N}}_{si} \\ \dot{\tilde{N}}_{ti} \end{vmatrix} = \frac{E}{1-\nu^2} \begin{vmatrix} 1 & \nu \\ \nu & 1 \end{vmatrix} \begin{vmatrix} \tilde{h}_i \dot{\varepsilon}_{si}{}^o - \sum_m c_s{}^m \Delta\tilde{z}_i \\ \tilde{h}_i \dot{\varepsilon}_{ti}{}^o - \sum_m c_t{}^m \Delta\tilde{z}_i \end{vmatrix}$$

$$- (1-h) \sum_m \begin{bmatrix} Y_{11}{}^m & Y_{12}{}^m \\ Y_{21}{}^m & Y_{22}{}^m \end{bmatrix} \begin{vmatrix} (\tilde{z}_i{}^m \dot{\tilde{k}}_{si} + \dot{\tilde{\varepsilon}}_{si}{}^o - c_s{}^m)\Delta\tilde{z}_i \\ (\tilde{z}_i{}^m \dot{\tilde{k}}_{ti} + \dot{\tilde{\varepsilon}}_{ti}{}^o - c_t{}^m)\Delta\tilde{z}_i \end{vmatrix},$$

$$\begin{vmatrix} \dot{\tilde{M}}_{si} \\ \dot{\tilde{M}}_{ti} \end{vmatrix} = \frac{E}{1-\nu^2} \begin{bmatrix} 1 & \nu \\ \nu & 1 \end{bmatrix} \begin{vmatrix} \tilde{h}_i{}^3/12\ \dot{\tilde{k}}_{si} - \sum_m \tilde{z}_i{}^m c_s{}^m \Delta\tilde{z}_i \\ \tilde{h}_i{}^3/12\ \dot{\tilde{k}}_{ti} - \sum_m \tilde{z}_i{}^m c_t{}^m \Delta\tilde{z}_i \end{vmatrix}$$

$$- (1-h) \sum_m \begin{bmatrix} Y_{11}{}^m & Y_{12}{}^m \\ Y_{21}{}^m & Y_{22}{}^m \end{bmatrix} \begin{vmatrix} (\tilde{z}_i^{m\,2} \dot{\tilde{k}}_{si} + \tilde{z}_i{}^m \dot{\tilde{\varepsilon}}_{si}{}^o - \tilde{z}_i{}^m c_s{}^m)\Delta\tilde{z}_i \\ (\tilde{z}_i^{m\,2} \dot{\tilde{k}}_{ti} + \tilde{z}_i{}^m \dot{\tilde{\varepsilon}}_{ti}{}^o - \tilde{z}_i{}^m c_t{}^m)\Delta\tilde{z}_i \end{vmatrix} \tag{18}$$

Evaluation of the constitutive equations for the element as a whole gives equations of the type

$$\dot{\sigma}_i = (S_{ij} - (1-h)\ Y_{ij})\ (\dot{D}_j - c_j). \tag{19}$$

Unlike the elasticity matrix S_{ij} the yield matrix Y_{ij} gener-

ally produces a coupling between the bending and the membrane action.
It can be shown that for the yield function φ of (14) and the dissipation function π of (15) holds

$$\dot{\sigma}_i = (S_{ij}-(1-h)Y_{ij})\ \dot{D}_j\ -h\ S_{ij}\ c_j \text{ if } Y_{ij} \neq 0. \tag{20}$$

Thus for the element as a whole a model holds, that is consistent with the local material model, the layer stresses playing a similar role as hidden state variables as the fraction stresses in the fraction model of material behaviour [4], useful for the representation of anisotropic effects like the Bauschingereffect.

Solution strategy

If in a loading program creep and time independent plastic deformation occurs, the following set of highly nonlinear equations has to be solved

$$[D_{i,k}(S_{ij}-(1-h)Y_{ij})D_{j,l}+\sigma_i D_{i,kl}]\ \dot{v}_l = \dot{f}_k+D_{i,k}S_{ij}c_j. \tag{21}$$

Simultaneously the layer stresses have to be updated by the local constitutive equations.
For the numerical simulation the equations (21) are replaced by incremental equations. A possible deviation from equilibrium, accrued from preceding steps,

$$f_k - \sigma_i D_{i,k} \neq 0,$$

is added to the righthand side.

$$[D_{i,k}(S_{ij}-(1-h)Y_{ij})D_{j,l}+\sigma_i D_{i,kl}]\Delta v_l = \dot{f}_k\Delta t+D_{i,k}S_{ij}c_j\Delta t + f_k-\sigma_i D_{i,k}. \tag{22}$$

By putting $\Delta t=0$ an iteration procedure based upon (22) is essentially a Newton Raphson iterative procedure, restoring equilibrium to a degree of accuracy, limited only by the computer error if convergence occurs. In the case of elastic behaviour this also ensures the accuracy of the solution, but in the history dependent inelastic case the accuracy of the solution strongly depends on the stepsize in the incremental procedure.

Time independent plastic deformation ($c_j=0$) is the most difficult case because of the strong nonlinearity in the form of the discontinuity at the yield limit. Even with small load steps the layer stresses will gradually move outside the yield limit, either because this yield limit will be reached before the end of a step, or because the change of stress along the tangent to the yield limit, permitted by the yield matrix, may build up to an appreciable distance to the yield limit by a number of finite steps. In the literature a number of remedies are found for this growing error in the state of stress. It should be realized, however, that just to return the state of stress to the yield limit does not necessarily improve the accuracy of the solution, because the history dependence of the process in principle does not permit a meddling with the stress path.

The problem of exceeding the yield limit is in author's opinion best met by updating the state of stress (the layer stresses in the present case) by a subincremental procedure, the number of subincrements being for instance a tenfold of the total calculated increment Δv.

Because of plasticity and geometrical nonlinearities the loads will have ultimate values. The system matrix becomes singular and the equations (22) with $c_j=0$ do not permit a computation beyond the ultimate loads.

We may consider the following equations

$$\begin{bmatrix} D_{i,k}(S_{ij}-(1-h)Y_{ij})D_{j,l}+\sigma_i D_{i,kl} & -\dfrac{df_k}{d\lambda} \\ -\dfrac{df_l}{d\lambda} & 0 \end{bmatrix} \begin{vmatrix} \Delta v_l \\ \Delta\lambda \end{vmatrix} = \begin{vmatrix} f_k-\sigma_i D_{i,k} \\ -W \end{vmatrix} \qquad (23)$$

Time t has been replaced by a loading parameter λ. The equation

$$\frac{df_l}{d\lambda} \Delta v_l = W \ (> 0), \tag{24}$$

added to the system to obtain a symmetric, nonsingular matrix, or

$$\frac{df_l}{d\lambda} \Delta\lambda\Delta v_l = \Delta f_l \Delta v_l = \Delta\lambda W,$$

shows that after reaching the ultimate load at a point of instability with $\Delta\lambda=0$, the computation may be continued, giving in general negative values for $\Delta\lambda$.

If the time dependent inelastic deformation of creep is added to the equations we have to return to the time increments. If the solution of the equations with prescribed values of $\dot{f}_k$,

$$\begin{bmatrix} D_{i,k}(S_{ij}-(1-h)Y_{ij})D_{j,l}+\sigma_i D_{i,kl} & -(\dot{f}_k+D_{i,k}S_{ij}c_j) \\ -(\dot{f}_l+D_{i,l}S_{ij}c_j) & 0 \end{bmatrix} \begin{vmatrix} \Delta v_l \\ \Delta t \end{vmatrix} = \begin{vmatrix} f_k-\sigma_i D_{i,k} \\ -W \end{vmatrix} \tag{25}$$

gives values for Δt approaching zero, it is clear from the equations that the structure will collapse, since $\Delta v/\Delta t$ grows beyond any bounds. At a finite constant value of the load, producing stresses beyond the treshold value for creep ($\beta<1$), collapse will ultimately occur due to the geometrical non-linearity of the problem. However it should be kept in mind that in the present treatment the restriction to small local strains has been introduced. Because of the limiting effect on Δv in the last equation of (25), the process of collapse can be computed to some extent.

It is clear that a program based upon the equations (23) or (25) is more robust than a program based directly upon the equations (22), where the values of Δv may increase easily beyond values, for which an incremental updating is allowable. A scaling back procedure may of course remedy this situation, without however improving the ultimate load analysis very much.

Concluding remarks

A Turbo-Pascal computer program, including the source, implementing the approach discussed in this paper, is available for interested readers at the costs of the disk and the postage. Rather than producing results for some particular cases, tests carried out by the reader were thought to be more illuminating. Furthermore in author's opinion computer results in scientific papers should by now be accompanied by full accessibility of the program, including the source.

References

1. Fried, I., Nonlinear finite element analysis of the thin elastic shell of revolution, Comp. Methods Appl. Mech. Eng., 48, 283-299 (1985).
2. Wagner, W., A finite element model for nonlinear shells of revolution with finite rotations, Int. J. Numer. Methods Eng., 29, 1455-1471 (1990).
3. Makowski, J., Stumpf, A., Finite axisymmetric deformation of shells of revolution with application to flexural buckling of circular plates, Ingenieur-Archiv 59, 456-472 (1989).
4. Besseling, J.F., Models of metal plasticity: theory and experiment. In: Plasticity Today (ed. by A. Sawczuk and G. Bianchi), Elsevier Appl. Sci. Publ., London, 97-113 (1985).

Finite Element Methods in Stress Analysis for Creep

JOAO NISAN C. GUERREIRO, ABIMAEL F.D. LOULA & JAMES T. BOYLE[¶]

Laboratorio Nacional de Computacao Cientifica
Rio de Janeiro, Brasil

[¶]University of Strathclyde,
Glasgow, Scotland

Summary

This paper provides an overview of classical and novel solution methods for finite element analysis of creep in structures. The reader is asked to question more thoroughly the accuracy of routine finite element analysis of components subject to creep and to ask whether the resulting solution is indeed meaningful.

Introduction

The past decade has seen an increase in computational power, coupled with the availability of advanced commercial finite element systems such as ABAQUS or ANSYS, which has allowed the engineer to attempt quite complex problems in elevated temperature design in a time scale unforeseen ten years ago. Indeed in 1983 [1] it was stated that "... the numerical solution techniques ... have general applicability, yet for large structures and complex loading histories they can be expensive, if not beyond available computing capacity ..."; today, on an advanced engineering workstation, this observation is no longer valid.

Yet, with the availability of such an advanced capability in creep stress analysis, the question of reliability of solution has been hardly addressed. Typical benchmarks for creep analysis generally relate to rather simple problems and are not exhaustive - typically only requiring to demonstrate that a particular algorithm has been correctly coded. In any case it is arguable whether or not the use of benchmarks in nonlinear finite element analysis is appropriate. A better approach to the assessment of the the reliability of finite element approximations would be a numerical analysis including stability and convergence studies. Only recently have engineers become aware of these questions, in particular concerning constrained problems such as creep [2].

M. Zyczkowski (Ed.)
Creep in Structures
IUTAM Symposium Cracow/Poland 1990

This paper examines classical and novel formulations and solution algorithms for stress analysis of structures for creep. It is worth remarking that finite element creep analysis has never been discussed in any detail in this series of symposia. In 1970 it was remarked by Finnie [3] that although computer solutions using finite element techniques should be of interest, there were no papers on this topic. Further in 1980 Leckie [4] stated that "... the procedure for computing the time varying stress ... is well established ...", while in that symposium only two papers made any reference to finite element analysis. The implication to be drawn from this could be that such analysis is straightforward; indeed in the writers' opinion much finite element creep analysis is carried out on a routine basis using commercial software with little questioning of the results. Research on adequate solution methods has reduced to the search for efficiency, with little apparent treatment of the problem of accuracy. It is the main aim of this paper to question this faith.

Background

The procedure for finite element analysis of time dependent structural creep problems is simple and well known being based on a method described by Mendelson et al [5], among others, over thirty years ago. This method was based on similar procedures for plasticity [6] and is generally known as the *initial strain* method wherein the increments (or rates of change) of the basic field variables of displacement and stress corresponding to an increment in inelastic strain over a specified time increment were obtained from the solution of an elastic problem [1]. Variations on this, such as the initial stress method, were also suggested. The use of this initial strain procedure specifically for finite element analysis was first described in the general literature by Gallagher et al [7] and later in more detail by Greenbaum & Rubenstein [8]. Most commercial finite element analysis software use this basic procedure for creep, and more general inelastic analysis. In the form described originally the incrementation procedure is known as an *explicit single step* procedure. The basic field variables are estimated at the end of a time step from their values and the inelastic strain rate at the beginning of the step. Information from previous time steps (a multi-step method) are not required; for reasons of storage efficiency most finite element solution strategies are single step. This simple incrementation scheme can be identified as the basic first order Euler solution algorithm for ordinary differential equations forming an initial value problem [1,9]. This method is known to be only *conditionally stable*. That is a bounded solution can only be obtained if the time increment is less than a specific (but usually unknown) value. A large number of small time steps are usually required for an accurate solution. To overcome this problem higher order multistep methods could be used, but these are inefficient for finite element analysis. Rather a class of *implicit single step* algorithms may be used: in these the increment of the field variables over a time step depend on their values at the beginning and at the end of the time step. Thus some iteration over the time step is necessary. The commercial software ABAQUS can use this type of procedure. Such methods can be shown to be unconditionally stable, but naturally at the expense of increased solution time through time step iteration. Nevertheless finite element creep analysis is fairly routine on state of the art work-stations and personal computers which can be dedicated to a specific analysis; it is then only a matter of what locally constitutes an acceptable solution time.

Although commercial finite element creep analysis is largely based on the simplest (explicit or implicit) single step algorithms, research on solution procedures has continued, mostly related to efficiency and automatic procedures for time step control. Commercial software usually also include simple methods of adjusting time step size automatically based on the increments to the field variables compared to their current values if the analyst requests this. Such methods are typically only included to avoid the use of an unnecessarily small time increment, and are not intended to provide control over possible stability problems. A pioneering paper by Cormeau [10] summarised methods for time step control to ensure numerical stability with a simple single step incremental procedure. Much work on this topic over the past decade has origins in the papers by Hughes & Taylor [11] and Argyris et al [12] which described a class of unconditionally stable implicit solution algorithms for inelastic analysis. These methods were examined in more detail and improved by Snyder & Bathe [13], whose work in turn was improved by Heaps & Mansfield [14]. The latter demonstrated a particularly efficient implicit procedure which required iterated solution of a set of nonlinear equations for stress and displacement rate at node level. This technique was identified as being related to a well established method in optimisation known as Uzawa's algorithm; this algorithm has been extensively studied mathematically, in particular its convergence properties were assessed in the creep problem [14] using similar techniques used when applied to the Navier Stokes equation [15]. Kojic & Bathe [16] also later modified the original Synder & Bathe procedure implemented in ADINA to improve overall efficiency in addition to improved accuracy in stress evaluation. This was achieved by replacing the iteration over each time step to the solution of a single equation (the 'effective stress function').

Discussion over the relative merits of explicit or implicit solution procedures continue in the literature. Explicit single step methods are only conditionally stable, requiring a large number of time steps. Time step control is usually determined by the most highly stressed elements where the rate of change is significant, even though much of the structure is less highly stressed and could allow the use of larger time increments. On the other hand implicit methods can be unconditionally stable, but require iteration over each time step. Analysis codes such as ABAQUS allow the use of both procedures depending on the load history and the requirements of the analyst. Another possibility is the use of combined explicit-implicit algorithms where for efficiency some of the elements are incremented using explicit methods, others with implicit as indicated by the analysis. The techniques have been used successfully in dynamics [17] and have recently been applied with some success to creep problems [18]. These combined techniques do seem promising in particular in problems which include creep damage.

It should be clear to the reader that much routine finite element creep analysis using commercial software is based on well established but inefficient, solution methods. Further most recent research has emphasised the pursuit of increased efficiency and stability of the solution algorithm. Very little work has been done on the accuracy of solution; this is understandable given the complexity of the underlying equations. Studies of the accuracy of solution, and its relation to the apparent need for efficiency, have largely been inconclusive [19, 20]. Further most studies of accuracy and stability relate to the finite element discretisation, and do not address the problem of approximation to the basic problem formulation in mechanics (the 'exact' solution). In linear elastic finite element analysis the analyst now has access to error estimators (such as the energy norms in ANSYS); for creep, and indeed most inelastic analysis, the analyst has no choice but to carry out the analysis to the best of his ability, assuming the basic software has been adequately benchmarked. He has no means of quantifying the accuracy of his solution, other than his judgment and experience and the knowledge that the solution method has been successful on a range of simple problems which can be easily checked, and has no recourse but to believe the solution. It will be shown in this paper that this belief could be misplaced, due to inadequacies in the conventional formulations themselves when applied to some simple problems. The use of benchmarks, coupled with the availability of error estimators, has proven crucial in conventional linear elastic finite element analysis, bringing to light numerous inadequacies both in software and user [21,22]. In the writers' opinion benchmarks are inappropriate for inelastic analysis, only serving to assess correct coding of specific algorithms, and that the only reliable way forward is a mathematical analysis of the numerical approximations. The main purpose of this paper is to outline a possible basis for this analysis which has already generated some remarkable results [2].

Problem formulation

One aim of this paper is to demonstrate some difficulties with the classical formulation of finite element creep analysis which do not appear to have been fully appreciated before, thus highlighting the need for novel formulations which will be described here. It has been discussed in the preceding that the writers believe that this may only be fully recognised using a mathematical basis and modern concepts from numerical analysis. In order to facilitate this study, the transient elasto-creep problem will be limited to cases where the loads are constant in time (and further, but without restriction, that the boundaries are simply clamped - the homogeneous Dirichlet boundary condition). Under such conditions it will be expected that the transient stress field should eventually become constant in time, reaching the so-called steady state, accompanied by constant displacement and strain rates [1].

The basic field equations for the elasto-creep problem relate to an inelastic body which occupies the domain Ω of one, two or three dimensional space R^n, $1 \le n \le 3$ with smooth boundary Γ. The body is subject to a volume force f which is constant in time; if the material is assumed to exhibit time dependent creep then the *transient elasto-creep problem* consists in finding, for each time t in an interval $[t_0, t_1]$, the Cauchy stress tensor $\sigma:\Omega \to R^n \times R^n$ and the displacement rate (velocity) field $v:\Omega \to R^n$, with $\dot{\varepsilon}$ the related strain rate, satisfying the field equations of equilibrium,

$$div\sigma + f = 0 \tag{1}$$

and strain displacement,

$$\dot{\varepsilon} = \frac{1}{2}(\nabla v + \nabla v^T) = Bv \tag{2}$$

which both must be satisfied in $\Omega \times [t_0, t_1]$ with the initial condition

$$\sigma(x, t_0) = \sigma^e(x) \tag{3}$$

in Ω where σ^e is the stress field which satisfies the associated elastic problem, together with the (for simplicity but without loss in generality) homogeneous Dirichlet boundary conditions

$$v(x) = 0 \tag{4}$$

which should be satisified on $\Gamma \times [t_0, t_1]$.

These equations are completed by the constitutive relation, which here is assumed to take the simple time hardening form which decouples elastic and creep strains (and without loss in generality the time function may be removed from the creep strain assuming similarity); thus

$$\dot{\varepsilon} = C\dot{\sigma} + A(\sigma) \tag{5}$$

where C is the inverse of the elasticity tensor and $A(\sigma)$ is the nonlinear function of the deviatoric part of the stress tensor

$$S = \sigma - pI \tag{6}$$

with p being the hydrostatic pressure. In general the function A is subject to the internal constraint, $trA = A:I = 0$, which implies that the creep strain rate must satisfy the *incompressibility condition*, $div\,v = 0$. The presence of this condition is the cause of the main difficulties in constructing finite element approximations. This is particularly evident in the associated *steady state creep problem* where it is assumed that stress and velocity are constant in time, with the simplified constitutive relation,

$$Bv = A(\sigma) \tag{7}$$

The solution of the steady state creep problem will be denoted by σ^{∞} and v^{∞}.

In the classic form given above it is difficult to construct a meaningful mathematical or numerical analysis. However if the equations are recast in the so-called *variational form* and the methods of functional analysis are adopted, as in modern theories of numerical approximation, then it is possible to attempt such an analysis. Unfortunately the notation and terminology of modern functional analysis are obscure for the uninitiated, although the writers are witness to the fact that it is not impenetrable. The reader is referred to some simple introductions [23-25] for a more complete description of the underlying principles. The most difficult concept here is the need for *function spaces* and the associated concepts of norm and inner product. In essence these concepts are required to ensure certain conditions of continuity, differentiability, intergrability and boundedness of the stress and velocity fields which will be examined. It is only under such conditions that basic results in approximation theory can be used to ensure such considerations as stability and convergence and to develop meaningful error estimates. For clarity a complete definition of the basic function spaces will not be given as the reader is referred to more detailed work presented elsewhere [2,26-29]. Two principal spaces are defined: firstly the space of symmetric stress tensors of order n

$$U = \{\tau - [\tau_{ij}], \tau_{ij} = \tau_{ji} \in L^p(\Omega); i,j = 1,\ldots\ldots n\}$$

with norm

$$\|\tau\|_U = \left(\int_\Omega |\tau|^p d\Omega\right)^{\frac{1}{p}} \quad \forall \tau \in U$$

where $L^p(\Omega)$ is the Banach space (of functions whose absolute values are p-integrable) with inner product (more strictly duality pairing)

$$(f,g)=\int_{\Omega} fg\,d\Omega \quad \forall f\in L^p(\Omega) \quad \forall g\in L^q(\Omega) \quad \text{with} \quad \frac{1}{p}+\frac{1}{q}=1$$

Secondly the space of velocity vectors of order n

$$V=\{v=\{v_i\}, v_i\in W_o^{1,q}(\Omega), i=1,\ldots,n\}$$

with norm

$$\|v\|_V=\left(\int_{\Omega}(|v|^q+|\nabla v|^q)d\Omega\right)^{\frac{1}{q}} \quad \forall v\in V$$

where

$$W_0^{1,q}(\Omega)=\{f\in W^{1,q}(\Omega), f=0 \quad \text{on} \quad \Gamma\}$$

is a subset of the Sobolev space $W^{1,q}(\Omega)$ of functions whose first derivative is also in $L^q(\Omega)$ and which satisfy the homogeneous boundary condition.

Additional subspaces are also necessary: the subspace of zero average pressure fields

$$L_0^p(\Omega)=\{g\in L^p(\Omega);(g,1)=0\}$$

the subspace of stress tensors with zero average trace

$$U_0=\{\tau\in U;(tr\tau,1)=0\}$$

and the subspace of deviatoric stress tensors

$$U_T=\{T\in U;trT=0\}$$

With these definitions the *variational formulation of the elasto-creep problem* is derived from the basic field equations Eqns.(1), (2) & (5) by multiplying eqn.(1) by an arbitrary vector $v \in V$ and eqn.(2) by an arbitrary tensor $\tau \in U$ and integrating over the domain Ω. Then the problem takes the so-called variational (or weak) form:

Find $(\sigma(t), u(t)) \in U \times V$ satisfying for each time $t \in [t_0, t_1]$,

$$c(\dot{\sigma}, \tau) + (A(\sigma), \tau) = b(\tau, v) \quad \forall \tau \in U \tag{8}$$

$$b(\sigma, v) = f(v) \quad \forall v \in V \tag{9}$$

with initial condition

$$(\sigma(t_0), \tau) = (\sigma^e, \tau) \quad \forall \tau \in U \tag{10}$$

The following functional pairings have been defined:

$$(A(\sigma), \tau) = \int_\Omega A(S){:}\tau \, d\Omega \quad \forall \sigma, \tau \in U$$

$$b(\tau, v) = \int_\Omega \tau{:}Bv \, d\Omega \quad \forall \tau \in U \quad \forall v \in V$$

$$c(\dot{\sigma}, \tau) = \int_\Omega C\dot{\sigma}{:}\tau \, d\Omega \quad \forall \dot{\sigma}, \tau \in U$$

$$f(v) = \int_\Omega f.v \, d\Omega \quad \forall v \in V$$

It may be assumed that the elastic functional is bounded below, that is there exists some constant $\gamma > 0$ such that

$$c(\tau, \tau) \geq \gamma \| \tau \|_U^2 \quad \forall \tau \in U$$

Further, a complete characterisation of the elasto-creep problem will depend on the particular creep constitutive relation defined by the operator *A(.)*. In general this is a nonlinear monotone operator which must satisfy certain conditions of ellipticity and continuity, specifically there exists a constant α such that the condition

$$(A(T_2)-A(T_1), T_2-T_1) \geq \alpha \| T_2-T_1 \|_U^p \quad \forall T_1, T_2 \in U_T \tag{11}$$

is satisfied with an additional condition on boundedness. Clearly it is possible to verify the inequality expressed in Eqn.(11) for the familiar Norton-Odquist power law of creep,

$$A(\sigma) = \mu |S|^{p-2} S \tag{12}$$

in which μ and p are material constants. It is important to note that the condition given in eqn.(11) is verifiable only on the subspace of deviatoric stress. A different ellipticity hypothesis was used by Loula & Guerreiro [29] (where the exponent is changed from p to 2 in eqn.(11)). In the writers view this deficiency is not crucial for the development of the behaviour of the finite element approximation studies, although eqn.(12) will be used in the numerical studies to be presented later.

A similar *variational statement of the steady state creep problem* may be given; in this case the problem is formulated in the reduced space U_0. It has been demonstrated by Loula & Guerreiro [29] that a unique solution $(\sigma^\infty, v^\infty) \in U_0 \times V$ exists to the steady state problem, subject to the conditions given above (eqn.(11) and Lipschitz continuity after Scheurer [30]). In addition it has been shown by Guerreiro [2] that at large times the solution of the elasto-creep problem tends to the steady state such that

$$\lim_{t \to \infty} \sigma(t) = \sigma^\infty + cI$$

$$\lim_{t \to \infty} v(t) = v^\infty$$

where c represents a constant hydrostatic pressure. For constant loads, as discussed here, it is usual to determine the steady state solution from the solution of the transient problem for sufficiently large times. Some aspects of this methodology are problematic when applied to discrete finite element approximations: this observation will form the main issue to be discussed in this paper.

Finite element approximation

For simplicity this part of the presentation will be restricted to two dimensional creep problems, that is $\Omega \in R^2$ is a polygonal domain discretised by the *uniform finite element mesh* of N elements such that

$$\overline{\Omega} = \bigcup_{e=1}^{N_e} \overline{\Omega}^e \quad \text{and} \quad \Omega^e \cap \Omega^f = \emptyset, \quad e \neq f$$

where Ω^e denotes the interior of the eth element and $\overline{\Omega}^e$ represents its closure (that is including the boundary). The *mesh parameter*, h is defined as $h = \max h_e, e = 1,2,\ldots,N_e$ with h_e being the diameter of the element e. It is then necessary to define finite element approximations for the function spaces U and V as the subspaces $U_h^l \subset U$ of C^{-1} piecewise polynomial finite element interpolations of stress of degree l, and $V_h^k \subset V$ of C^0 (continuous) piecewise polynomial finite element interpolations of displacement of degree k with zero value on the boundary. Similar additional subspaces of U_0 and U_T must also be defined.

The classical *Galerkin approximation for the elasto-creep problem* is then given as:

For each $t \in [t_0, t_1]$ find $(\sigma_h, v_h) \in U_h^l \times V_h^k$ such that

$$c(\dot{\sigma}_h, \tau_h) + (A(\sigma_h), \tau_h) = b(\tau_h, v_h) \quad \forall \tau_h \in U_h^l \tag{13}$$

$$b(\sigma_h, v_h) = f(v_h) \quad \forall v_h \in V_h^k \tag{14}$$

with the initial condition

$$(\sigma_h(t_0), \tau_h) = (\sigma^e, \tau_h) \quad \forall \tau_h \in U_h^l \tag{15}$$

A similar *Galerkin approximation for the steady state problem* may also be formulated (see [27,29]); the finite element approximation in this case will be denoted $(\sigma_h^\infty, v_h^\infty) \in U_{0h}^l \times V_h^k$ in similar notation to the above. However problems occur here; a detailed analysis of the Galerkin approximation for steady creep has been given by Loula & Guerreiro [29] following Scheurer [30]; the details are not given here, but the main observations are described:

The main difficulty relates to the incompressibility condition; as presented here the Galerkin approximation for the steady state problem cannot be shown to be numerically stable and also gives rise to possible nonuniqueness of solution. Firstly, to construct stable Galerkin finite element approximations for this type of constrained problem, it is necessary to segregate the pressure field from the deviatoric part of the stress tensor and to then interpolate the pressure, the deviatoric stress and the velocity fields independently using different interpolation orders. This of course causes severe limitations in the approximation of this type of problem using the classical approach. Secondly loss of uniqueness may be evident as *spurious (nontrivial) pressure modes*. For equal order approximations, $l = k$, discontinuous pressure and continuous velocity interpolations there may be many nontrivial solutions, especially with lower order interpolations. These spurious modes cause the instability in Galerkin approximations to the steady state problem (these spurious pressure modes naturally also appear in finite element approximations for constrained elasticity [31]).

For the elasto-creep problem, even for equal order interpolation, the finite element problem has a unique solution [2] *but it does not approximate well the real solution of the continuous problem*: In the limit at large times it would be expected that the pressure solution to the finite element elasto-creep problem would tend to the pressure solution to the finite element steady state problem, with spurious pressure modes filtered out, but including the projection of the spurious modes on the pressure initial condition. In addition any other misbehaviour in the finite element solution such as locking [32] and spurious pressure oscillations typical of unstable approximations to constrained problems would be masked. Examples will be given later. To overcome these difficulties a mixed Petrov Galerkin method has been suggested by Loula & Guerreiro [26,27]:

This novel *Petrov Galerkin approximation for the elasto-creep problem* is based on a similar treatment of the Timoshenko beam problem by Loula et al [33] and is given as:

For each $t \in [t_0, t_1]$ find $(\sigma_h, v_h) \in U_h^l \times V_h^k$ such that

$$c(\dot{\sigma}_h, \tau_h) + (A_h(\sigma_h), \tau_h) = b(\tau_h, v_h) - g_h(\tau_h) \quad \forall \tau_h \in U_h^l \tag{16}$$

$$b(\sigma_h, v_h) = f(v_h) \quad \forall v_h \in V_h^k \tag{17}$$

with the initial condition

$$(\sigma_h(t_0), \tau_h) = (\sigma^e, \tau_h) \quad \forall \tau_h \in U_h^l \tag{18}$$

Modified functional pairings are given as

$$(A_h(\sigma_h),\tau_h) = (A(\sigma_h),\tau_h) + \frac{\delta h^2}{\vartheta}(div\,\sigma_h, div\,\tau_h)_h \quad \forall \sigma_h, \tau_h \in U_h \tag{19}$$

$$g_h(\tau_h) = \frac{\delta h^2}{\vartheta}(f, div\,\tau_h)_h \quad \forall \tau_h \in U_h^l \tag{20}$$

with δ being a positive scalar, ϑ a viscosity parameter and the product $(.,.)_h$ defined as

$$(\mu_h, \phi_h)_h = \sum_{e=1}^{N_e} \int_{\Omega^e} \mu_e . \phi_e \, d\Omega$$

where μ_e and ϕ_e are the restrictions of μ_h and ϕ_h to the element e.

The δ term in the above corresponds to the least squares residual of the equilibrium equations which is added to the Galerkin approximation to form the Petrov Galerkin approximation as in [33]. It is crucial to the performance of the method:

A *Petrov Galerkin formulation for the steady creep problem* may also be given; a detailed analysis is presented in [29]. In particular it is shown that a unique solution to the steady creep problem can be obtained, and a simple error estimator can be developed from finite element interpolation theory [34]. The δ term is critical in establishing the stability of the method. Existence and uniqueness of the finite element approximation can be established for $\delta > 0$ and $k \geq 2$. As in the Galerkin approximation nonuniqueness of the pressure field can result for $k = 1$, but in this case a 'checker-board' type mode results, which can be easily filtered out in post-processing.

With this ensured stability of the associated steady state problem, the asymptotic behaviour of the finite element solution of the transient elasto-creep problem can be demonstrated, [2]. For $\delta > 0$ and $k \geq 2$ the finite element Petrov Galerkin approximation of the elasto-creep problem tends to the associated approximation to the steady state problem such that

$$\lim_{t \to \infty} \sigma_h(t) = \sigma_h^{\infty} + c_h I \tag{21}$$

$$\lim_{t \to \infty} v_h(t) = v_h^{\infty} \tag{22}$$

as in the above, where c_h represents a constant hydrostatic pressure. Thus with the Petrov Galerkin approximation the solution of the steady state problem can be taken as the limit of the corresponding transient solution with confidence.

Examples

The detailed numerical procedures required to implement the Petrov Galerkin aprroximation will not be discussed in any detail here; the reader is referred to previous publications [2, 26-29]. Indeed as discussed in the background section above similar problem areas relating to efficiency of solution can be introduced. The transient problem can be resolved using first order single step implicit procedures as previously described. At each time step (or for the solution of the steady creep problem) a nonlinear set of equations must be solved. In the examples presented here a Uzawa's type algorithm is adopted, but this is different from other iterative methods [13,14] which completely decouple the iteration for stress and velocity. Here there is coupling but it is possible to demonstrate convergence [29], which cannot be established for the Heaps & Mansfield algorithm for example. At each iteration the nonlinear equations can be solved at the element level using iterative methods such as Newton's or modified Newton's Methods. However numerical convergence problems can sometimes be encountered [29]; recent additional work has shown that a more robust algorithm for iteration, inspired by those proposed by Glowinski & Marroco [35], can overcome these problems.

Three simple problems have been examined in considerable detail: the classical pressurised cylinder, the driven cavity problem (where a unit square is fixed on all three sides with a specified displacement on the remaining side) and a plane block (fixed on three sides with a pressure on part of the remainder). Selected results are presented here:

The *driven cavity problem* is modelled with a 8x8 mesh of 9 noded (biquadratic) elements with a creep exponent, $p = 4$. The result of the Petrov Galerkin solution (with $\delta = 1$) and classical Galerkin solution for the steady state and converged limiting transient solution are compared in Figure 1 as hydrostatic pressure surfaces over the domain. As can be seen, the Petrov Galerkin solutions compare well with each other (as expected), but for the Galerkin solution the pressure fields are highly oscillatory and the transient solution does not agree with the erroneous steady state solution!

The *plane rectangular block* is shown in Figure 2 with finite element mesh superimposed. The problem is modelled as plane strain completely clamped on the boundaries $x = 0, x = 2$ and $y = 0$ and subjected to a partially distributed load on the boundary $y = 1$; the material creep exponent is taken as $p = 4$. Again equal order biquadratic interpolations were adopted for both stress and velocity fields. Post processed pressure distributions at various depths are shown in Figure 3

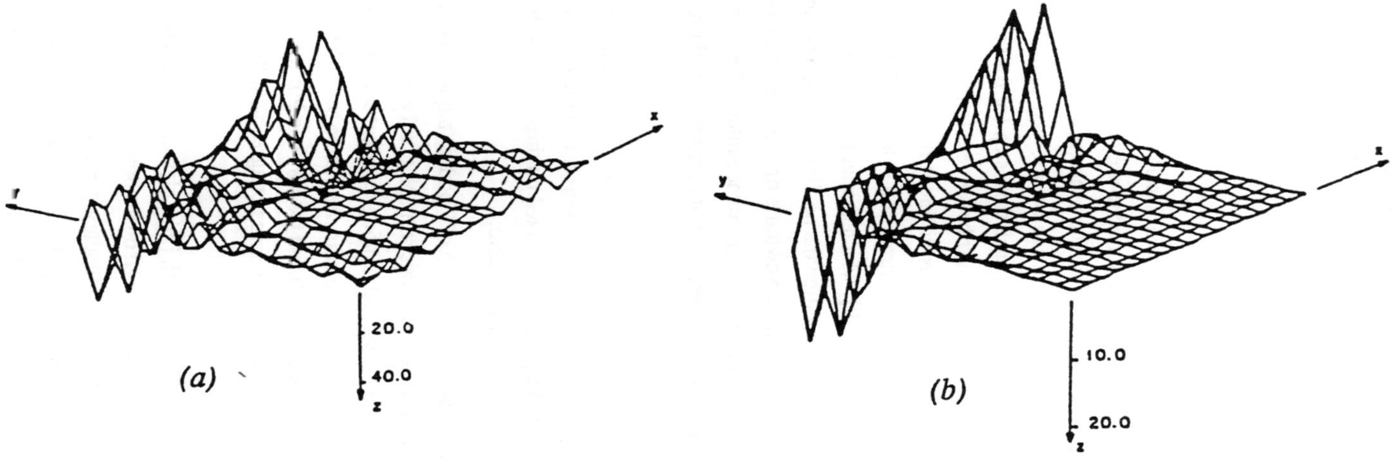

Figure 1a: Pressure field for Galerkin method
(a) Steady state, (b) Transient

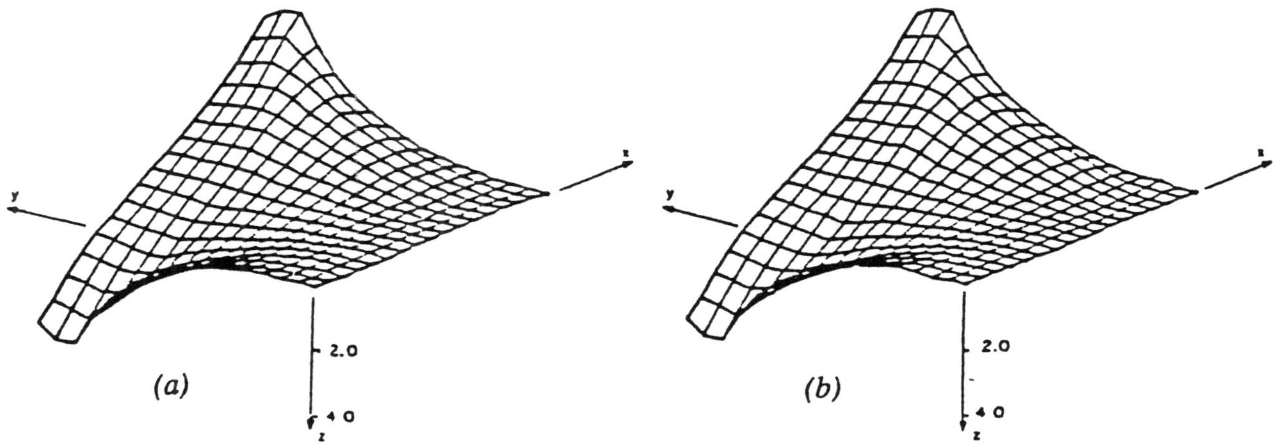

Figure 1b: Pressure field for Petrov Galerkin method
(a) Steady state, (b) Transient

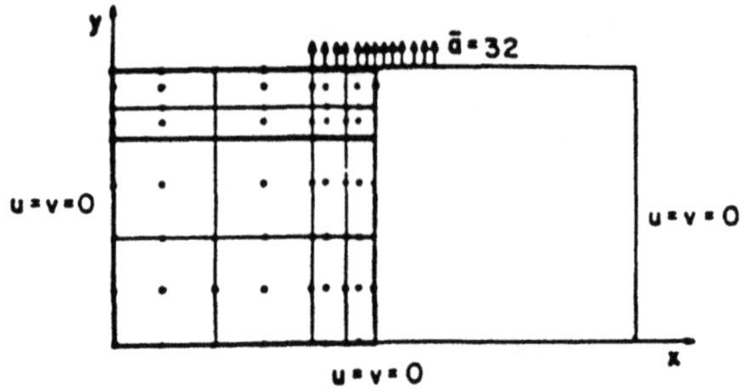

Figure 2: Geometry of plane rectangualr block

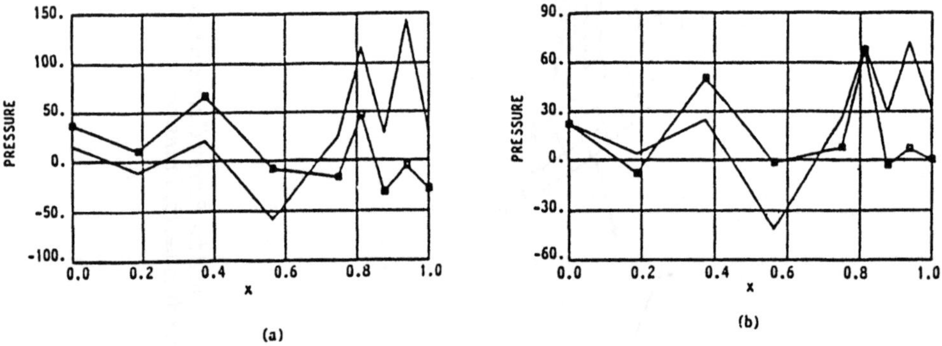

Figure 3a: Post processed pressures for Galerkin
(a) Steady state, (b) Transient limit
—— y = 0.5625, y = 0.9325 —■—

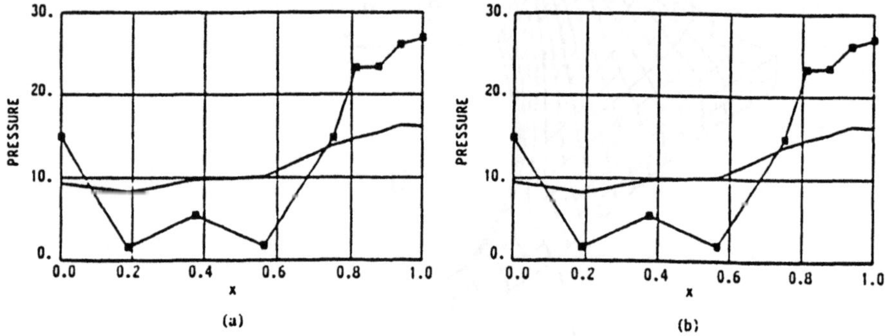

Figure 3b: Post processed pressures for Petrov Galerkin
(a) Steady state, (b) Transient limit
—— y = 0.5625, y = 0.9325 —■—

for Petrov Galerkin and Galerkin solutions for a steady state and transient limit solution. Once again the Galerkin solutions are not in agreement with each other and are different from the correct Petrov Galerkin solution.

Conclusions

It has been demonstrated here that erroneous solutions to fairly simple problems can be obtained using classical Galerkin finite element methods. Why has this not been noted before? One possible reason is that many elementary benchmark problems address plane stress problems; the instability and lack of convergence is not manifest in this type of problem and can be overcome in other problems using reduced integration elements (through experience). The other possible reason of course is that the analyst has not been expecting such results. Indeed the writers would suggest the problems described as possible benchmarks for creep analysis.

Of course the material behaviour described here is fairly simple and the question arises as to the implications for more general inelastic (elasto visco plastic) materials. Studies of the numerical solution of such problems using finite element procedures appear regularly in the literature [11,36-39]. Since these material models also usually relate to inherent incompressibility conditions, and the proposed finite element approximations are wholly based on Galerkin methods, and since constant load problems are often used as benchmarks, it would be interesting to extend the the present Petrov Galerkin formulations to these also. For peace of mind at least ...

References

1. JT Boyle & J Spence: *Stress Analysis for Creep* Butterworths, 1983
2. JNC Guerreiro: *New Mixed Finite Element Methods for Creep* DSc dissertation, COPPE/UFRJ, 1988
3. I Finnie: Reflections on the past and future of creep. Proceedings Conf on *Creep in Structures*, Gothenburg, 1970 Springer, 1972
4. F Leckie: Advances in creep mechanics. In *Creep in Structures*, Leicester, 1980 Springer, 1981
5. A Mendelson et al: A general approach to the practical solution of creep problems. Trans ASME Journ Basic Engineering Vol.82D, 689-694, 1960
6. AA Il'yushin: *Plasticite*. Ed Eyrolles, 1956
7. RH Gallagher et al: Stress analysis of heated complex shapes. ARS Journ Vol.32, 700-707, 1962
8. GA Greenbaum & MF Rubenstein: Creep analysis of axisymmetric bodies using finite elements. Nucl Eng & Design, Vol.7, 379-397, 1968
9. CW Gear: *Numerical Initial Value Problems in Ordinary Differential Equations* Prentice Hall, 1971
10. I Cormeau: Numerical stability in quasistatic elasto-visco-plasticity. Int Journ Num Methods in Eng., Vol.9, 109-127, 1975
11. TJR Hughes & RL Taylor: Unconditionally stable algorithms for quasi static elasto-visco-plastic finite element analysis. Comp Struct, Vol.8, 169-172, 1978
12. JH Argyris et al: Improved solution methods for inelastic rate problems. Comp Meth Appl Mech Eng, Vol.16, 231-277, 1978

13. MD Snyder & KJ Bathe: A solution procedure for thermo elastic plastic and creep problems. Nucl Eng & Design, Vol.64, 49-80, 1981

14. CW Heaps & L Mansfield: An improved solution procedure for creep problems. Int Journ Num Meth Eng, Vol.23, 525-532, 1986

15 R Teman: *Navier Stokes Equations*. North Holland, 1977

16. M Kojic & KJ Bathe: The effective stress function algorithm for thermo elasto plasticity and creep. Int Journ Num Meth Eng, Vol.24. 1509-1532, 1987

17 TJR Hughes & WK Liu: Implicit explicit elements in transient analysis. Trans ASME Journ Appl Mech, Vol.45, 371-378, 1978

18. GG Chen & TR Hsu: A mixed explicit implicit algorithm for creep stress analysis. Int Journ Num Meth Eng, Vol.26, 511-524, 1988

19. M Ortiz & EP Popov: Accuracy and stability of integration algorithms for elasto plastic constitutive equations. Int Journ Num Meth Eng, Vol.21, 1561-1576, 1985

20. R Mahnken & E Stein: Adaptive time step control in creep analysis. Int Journ Num Meth Eng, Vol.28, 1619-1633, 1989

21. GAO Davies: Results for selected benchmarks. Benchmark, April 1989

22. MJ Wheeler & SM Yunus: An efficient error approximation technique for use with adaptive meshing. 2nd Int Conf on *Quality Assurance and Standards in Finite Element Methods*, NAFEMS, Stratford, 1989

23. JN Reddy: *Applied Functional Analysis and Variational Methods in Engineering* McGraw-Hill, 1986

24. F Hartmann: *The Mathematical Foundation of Structural Mechanics* Springer, 1985

25. C Johnson: *Numerical Solutions of Partial Differential Equations by the Finite Element Method* Cambridge University Press, 1987

26. AFD Loula & JNC Guerreiro: A new mixed finite element formulation for inelastic analysis. IX COBEM, Florianopolis, Brasil, 1987

27. AFD Loula & JNC Guerreiro: A new finite element method for nonlinear creeping flows. Proc Pan American Congress of Applied Mechanics, Rio de Janeiro, 198. Published in Appl Mech Rev

28. JNC Guerreiro & AFD Loula: Applications of mixed Galekin and Petrov Galerkin finite element methods in creep analysis. Proc 10th Int Conf on Structural mechanics in Reactor Technology, Los Angeles, 1989

29. AFD Loula & JNC Guerreiro: Finite element analysis of nonlinear creeping flows. Comp Meth in Appl Mech & Eng, Vol.79, 87-109, 1990

30. B Scheurer: Existence et approximation de point-selle pour certains problemes nonlinearies. RAIRO Vol.11, 369-400, 1977

31. JT Oden & K Kikuchi: Finite element methods for constrained problems in elasticity. In Journ Num Meth in Eng, Vol.18, 701-725, 1982

32. TJR Hughes: *The Finite Element Method* Prentice Hall, 1987

33. AFD Loula, TJ Hughes, LP Franca & I Miranda: Mixed Petrov Galerkin methods for the Timoshenko beam. Comp Meth in Appl Mech & Eng, Vol.63, 133-154, 1987

34. PG Ciarlet: *The Finite Element Method for Elliptic Problems* North Holland, 1978

35. R Glowinski & A Marroco: Sur lapproximation par elements finis d'ordre un et la resolution par penalization dualite d'une classe de problem de Dirichlet nonlineaires. RAIRO Vol.2, 41-76, 1975

36. V Kumar et al: Numerical integration of some stiff constitutive models of inelastic deformation. Trans ASME Journ Eng Mat Tech, Vol.102, 92 96 (1980)

37. M Ortiz & JC Simo: An analysis of a new class of integration algorithms for elastoplastic constitutive equations. Int Journ Num Meth Eng, Vol.23, 353-366, 1986

38. D Cordts & FG Kollmann: An implicit time integration scheme for inelastic constitutive equations with internal state variables. Int Journ Num Meth Eng, Vol.23, 533-554, 1986

39. K Hornberger & H Stamm: An implicit integration algorithm with a projection method for viscoplastic constitutive equations. Int Journ Num Meth in Eng, Vol.28, 2397-2421, 1989

Thermal, Mechanical & Creep Behavior of Metal Matrix Composites

Stefan Jansson

Frederick A. Leckie

Department of Mechanical and Environmental Engineering

University of California

Santa Barbara, CA 93106

1. INTRODUCTION

The high strength/weight ratio and thermal conductivities of metal matrix composites are properties which make them attractive for application in modern components for which low weight and good thermal efficiency are essential. An intrinsic difficulty in establishing the mechanics of these materials is that they are at a very early stage of their development when changes in their manufacture are frequent and many. As a consequence the amount of material available for comprehensive testing is limited. Furthermore, the test specimens tend to be small and have high strength so that problems associated with gripping are considerable. Consequently, the mechanics describing the behavior of metal matrix composites must be achieved with only very little information. However, the recent advances in computational techniques combined with established mechanics principles are helpful in establishing the material properties and their effect on component performance.

This paper shall describe some of the advances with special reference to the properties of a composite consisting of an aluminum-lithium matrix reinforced with continuous alumina fibers. This material combination is distinguished by the strong bond which exists between matrix and fiber so that stress and displacement continuity conditions may be assumed. Attempts shall be made to describe the general approach which is being developed. It should be emphasized that the procedures are not fully developed but sufficient progress has been made to believe that the prospects of success are high.

M. Zyczkowski (Ed.)
Creep in Structures
IUTAM Symposium Cracow/Poland 1990

diminished or even lost if multi-laminate systems are adopted. Consequently, if full advantage is to be taken of the dominant strength characteristics then the fibers should be aligned in the direction of maximum stress transmission. In this circumstance the transverse properties of the material are also important since there must be sufficient strength in the matrix to support the secondary stresses applied in the transverse direction. In addition to the strong bond existing between the fiber and the matrix another important feature of the composite is the large mismatch in the coefficients of thermal expansion so that fluctuations in operating temperature induce substantial thermal stresses. It is the goal of this study to establish the behavior of the composite when subjected to transverse stress and cycles of temperature. This study shall be used to illustrate how the shakedown concept can be used to describe the material behavior when subjected to the described loading conditions and when creep effects are also important.

2. EXPERIMENTAL PROGRAM

The composite studies is Du Pont's FP/AL with continuous fibers in a unidirectional lay-up. The fiber volume fraction was determined tto be 55%. The FP fiber consists of 99% pure crystalline α-alumina (Al_2O_3) coated with silica that improves the strength of the fiber and aids the wettting by the molten metal. The fibers have a diameter of approximately 12 μm, a modulus of 345 to 380 GPa, a tensile strenth of 1.9 to 2.1 GPa for 6.4 mm gauge length, and a fracture sttrain of 0.3-0.4%. The matrix material is a 2 wt% Li-Al binary alloy. The modulus is 68.9 GPa and Poisson's ratio is 0.32 The lithium promotes the wetting of the alumina fibers that forms a strong matrix-fiber interface. The composite is fabricated by preparing the FP fibers into tapes by using a fugitive binder and the tapes subseqnelty laid up in a metal mold in the desired orientation. The binder is burned away and the mold is vacuum-infiltrated with the molten matrix. The composite was available in the form of a plate 150 x 150 x 12.5 mm thick.

The specimen used for transverse tests is shown in Fig. 1. It has a relatively large radius at the transition from the gripping section to the reduced gauge section to provide a low stress concentration and a short specimen to prevent specimen buckling during

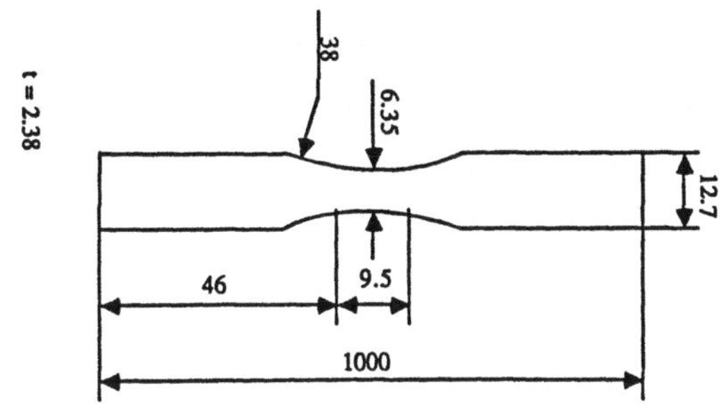
38
6.35
t = 2.38
12.7
46
9.5
1000

Fig. 1

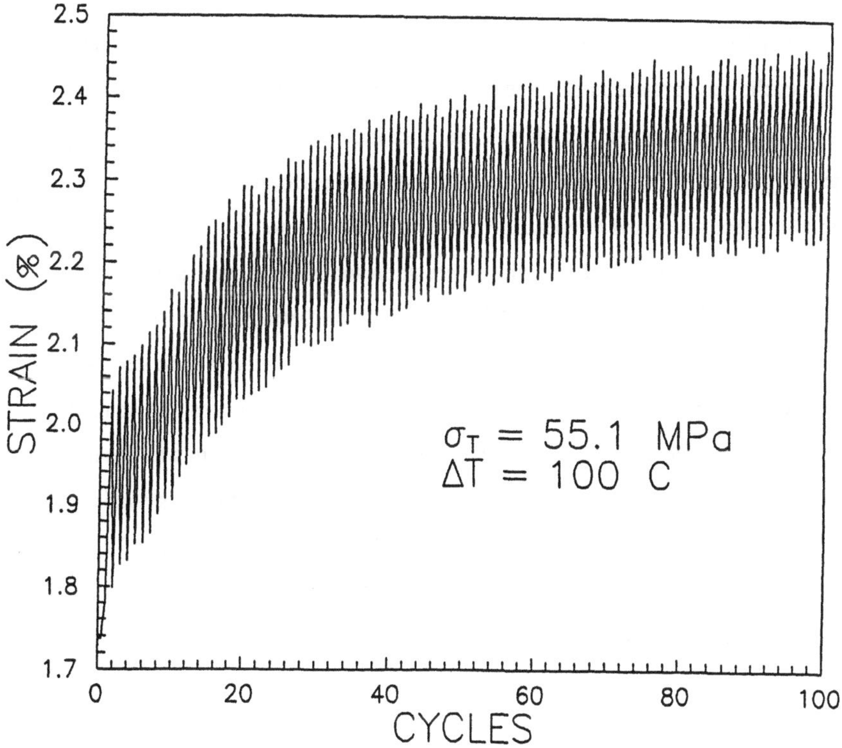
STRAIN (%)
CYCLES
σT = 55.1 MPa
ΔT = 100 C
2.5
2.4
2.3
2.2
2.1
2.0
1.9
1.8
1.7
0
20
40
60
80
100

Fig. 2

compressive loading. The gauge section is 10 mm long and 6 mm wide and 2.4 mm thick. The strains were measured with 3.2 mm strain gauges. Specimen was loaded by an M. T. S. hydraulic machine and it was heated by means of induction coils. The control of the temperature was achieved by using thermocouple measurements..

The tests reported in this study involved a constant transverse stress in combination with cycles of temperature with cycle time 150 s. The temperature and strain variations were continuosly recorded.and a typical example of the accumulation of strain is illustrated in Fig. 2.

The transverse stress strain curve at room temperature is shown in Fig. 3, from which a deviation from linearity is observed to occur at 75 MPa. The tultimate strength is 200 MPa and the strain to fracture is 0.8%. The ultimate strength is about 50% higher than the ultimate matrix strength while the failure strain of 0.8% is only 4% of the failure strain of the matrix [Sakui and Tamura, 1969] A representative test for the constant transverse stress and cyclic temperature is given in Fig. 2 which indicates that transient behavior is followed by a cyclic response for which there is an increment of strain after each cycle, i.e., ratcheting occurs in both examples. Similar tests were performed at different values of constant transverse stress and temperature cycle. In Fig. 4 contours of constant values of $d\varepsilon_p/dN$ are plotted, where ε_p is the plastic ratchet strain and N is the cycle number.

There are combinations of σ_T and ΔT for which no ratcheting occurs when the initial transient reponse is followed by elastic cycle reponse. This condition is indicated in Fig. 4 as the shakedown condition. When the operating conditions exceed the shakedown condition ratcheting occurs at rates indicated in Fig. 4. The failure strain was observed to vary according to the operating condition as indicated in Fig. 5. The failure strain is 0.08% for high stress and low thermal load whereas for low transverse stress of 30 MPa the failure strain reaches 12%. Microscopic examination of the failure surface indicates the presence of voids in the matrix, with no evidence of damage in either the fiber or fiber-matrix interface. This observation suggests that the failure mechanism in the matrix is associated with void growth.

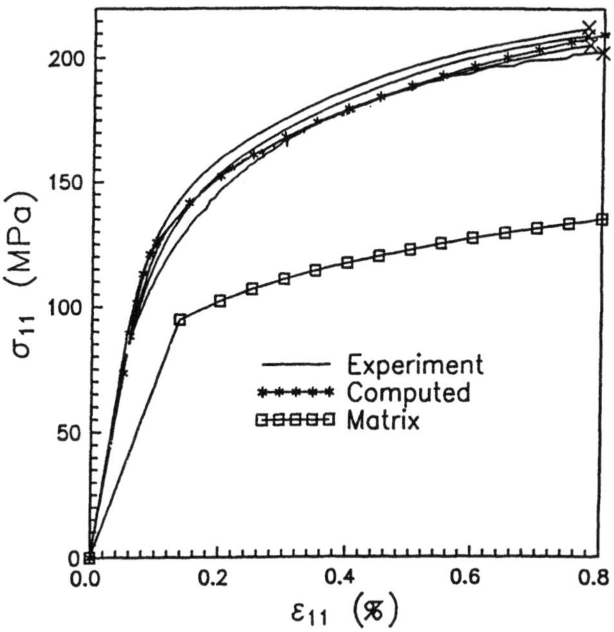
200
150
100
50
0
0.0
0.2
0.4
0.6
0.8
σ11 (MPa)
ε11 (%)
Experiment
Computed
Matrix

Fig. 3

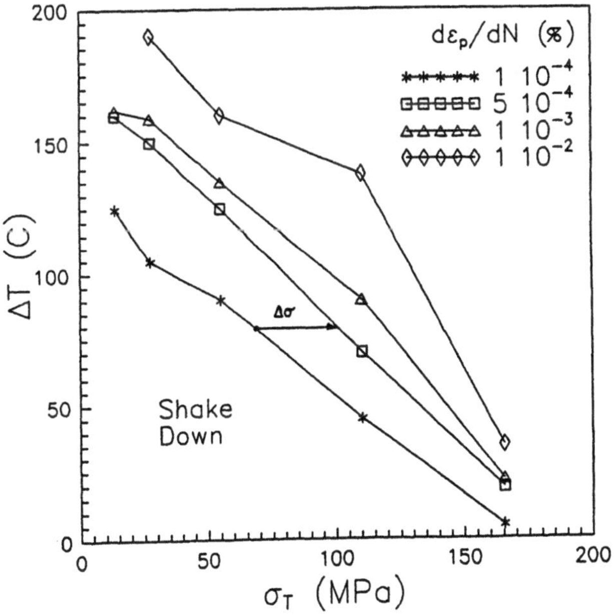
200
150
100
50
0
0
50
100
150
200
ΔT (C)
σT (MPa)
dεp/dN (%)
1 10^-4
5 10^-4
1 10^-3
1 10^-2
Δσ
Shake
Down

Fig. 4

The contours of constant ratchet strain rate plotted in Fig. 4 are generally parallel to the shakedown surface. This observation suggests that the driving force which drives the ratchet strain rate is given by

$$\alpha = \left[\frac{T}{T_s} + \frac{\sigma_T}{\sigma_S} - 1\right]$$

where T_S and σ_S are the ordinates defining the shakedown condition. The relationship of $d\varepsilon_p/dN$ has the form

$$\frac{d\varepsilon_p}{dN} = f(\alpha)$$

where $f(\alpha)$ is linear for small values of α and increases exponentially as α approaches the value 1 - Fig.6.

3. COMPUTATIONAL STUDIES AND THEIR IMPLICATIONS

The homogenization technique has been widely developed and can be used to predict the composite stress-strain response. The procedure can then be used to determine the response to the different loading which form the basis of developing constitutive equations Jansson [1990a]. A difficulty with the approach is the lack of properties of the constitutents of the composite when they are subject to a complex history of time and temperature during the manufacturing process.

The elastic properties of fiber and matrix are not greatly affected by the history of processing and heat treatment of the composite and this makes it possible to use data from the literature. However, the flow properties of the Al Li matrix alloy are strongly dependent on heat treatment and cold-working [Stark et al., 1981 and Sikui and Tamura, 1969]. Details of the processing and post heat treatments of the composite are not availble. Hence, the exact state of the matrix is not known and the information is insufficient to extract the flow properties of the matrix from the literature. This leaves as the only option

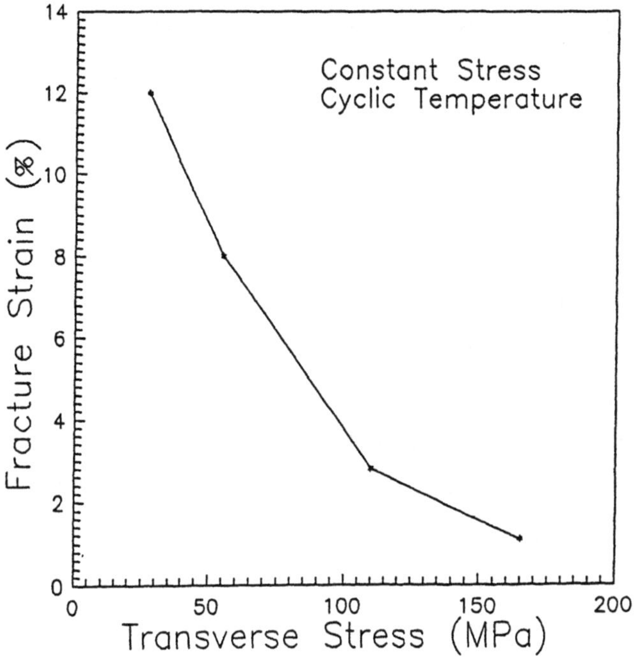
Constant Stress
Cyclic Temperature
Fracture Strain (%)
Transverse Stress (MPa)
0
2
4
6
8
10
12
14
0
50
100
150
200

Fig. 5

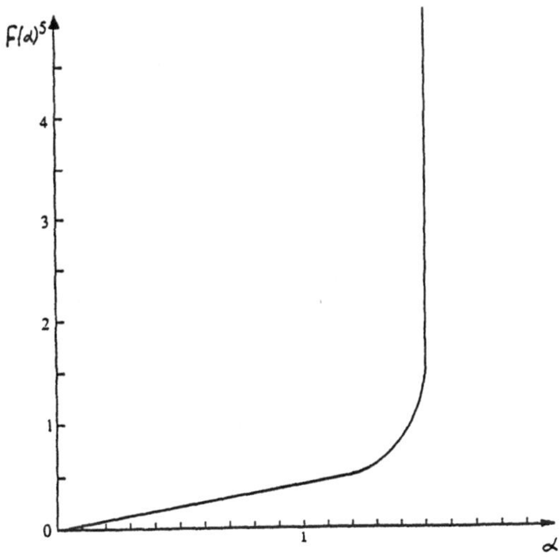
F(α)
0
1
2
3
4
5
1
α

Fig. 6

to determine the flow properties of the matrix in the composite is to fit a calculated response, by varying the flow propoerties of the matrix, to an experimentally determined matrix dominated stress strain curve for the composite. This procedure has been performed by Jansson [1990b] who determined that the initial yield stress of the matrix is 94 MPa with a hardening exponent n = 5 (Fig. 3). The matrix properties were then used to predict the performance of the compsite when subjected to shear and multiaxil stress loadings. The predictions of the computations agreed well with the experimental observations and provided confidence that predictions of the computations using the material data from one experiment could be applied to different loadings. In practice this could lead to significant savings since complex testing programs may be avoided.

The present composite consists of long fibers in a unidirectional lay up that are randomly distributed in the transverse plane. In the model to be analyzed the fibers are assumed to be long parallel cylinders arranged in a hexagonal array, Fig. 7. This periodical array has the mechanical properties with the closest symmetries to randomly distributed fibers that are transversely isotropic. Both systems are transversely isotropic when the constituents are linear elastic but the hexagonal array has a weak deviation from transverse isotropy when the matrix exhibits a nonlinear stress strain relation Jansson [1990b]. The deviation is most pronounced for a perfectly plastic matrix. However, reasonable results can be expected if effective properties are calculated for loadings that do not permit slip planes unconstrained by the fibers. This resembles the closest condition to a composite with randomly distributed fibers.

In the present calculations the fibers are assumed to be linear elastic and the matrix behavior is modelled with a small strain J_2 perfectly plastic theory using the properties given in Fig. 7.

The governing boundary value problem for the effective properties of the unit cell is two dimensional and has been solved with the Finite Element method. The displacement field is interpolated with nine nodes isoparametric elements and reduced integration is used to avoid locking, 2 x 2 for the hydrostatic component and 3 x 3 for the deviatoric component of the stress tensor. The nonlinear system of equations is solved with a

FIBROUS COMPOSITE

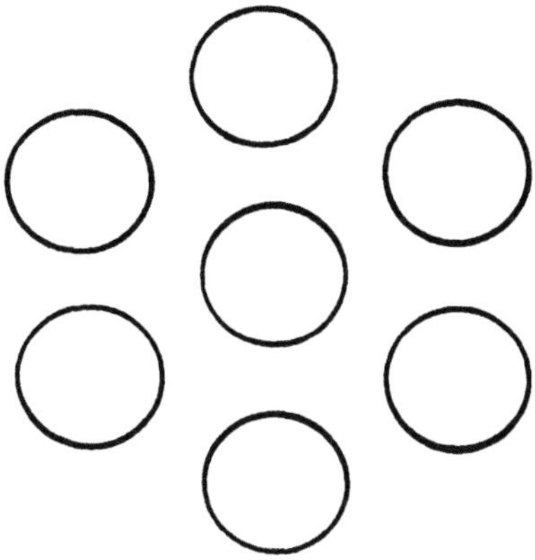

HEXAGONAL ARRAY

PROPERTIES OF MATRIX (aluminum-lithium)

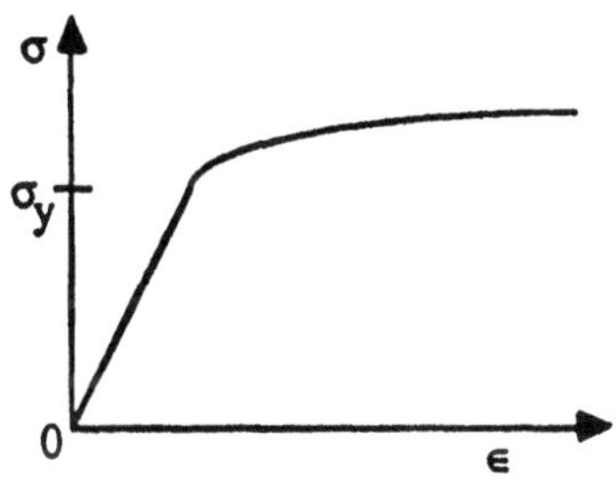

$E = 69.0$ GPa

$\sigma_y = 95$ MPa

$\nu = 0.32$

$\alpha = 23.6 \times 10^{-6}/°C$

PROPERTIES OF FIBER (alumina)

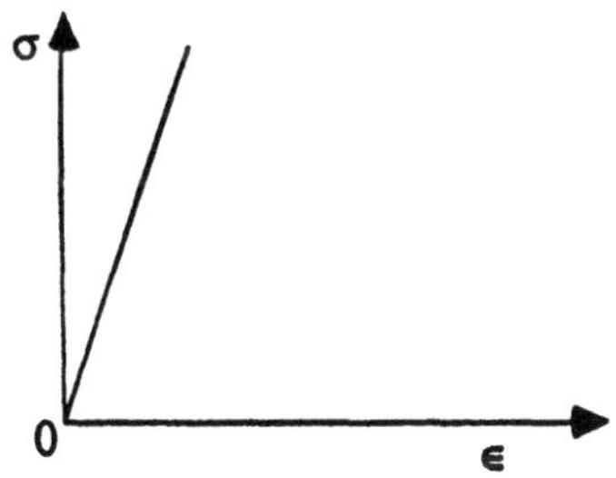

$E = 345$ MPa

$\nu = 0.26$

$\alpha = 8.6 \times 10^{-6}/°C$

$V_f = 55\%$

Fig. 7

Newton-Raphson scheme. The loading of the unit cell, Fig. 7, is symmetric with respect to the y_1 and y_2 axis. This implies that only an eighth of the indicated unit cell need be analyzed. A detailed description of the finite element mesh of the implementation of the method and deviations of the boundary conditions for different loadings are given in Jansson [1990a].

In performing the elastic-plastic calculation it was possible to determine the shakedown boundary defined in Fig. 4. This has been expressed in terms of the dimensionless loadings $E\Delta\alpha\Delta T/\sigma_y$ and σ_T/σ_y. In terms of the material parameters for the present system the experimental determination of the shakedown surface is also shown in Fig. 4. The agreement is reasonably good but it is noted that the experimental results indicate a straight line relationship exists between ΔT and σ_T whereas the computed boundary is somewhat convex. The differences in the experimental and computational predictions is likely to be related to deficiencies in the constitutive model for the matrix. The computational procedures involved very heavy computation since more than 100 cycles are often required before a steady state condition was reached. For this reason no attempt was made to determine the behavior for operating points outside the shakedown conditon, and this is left to a bound method developed by Ponter and Cox [1983].

The upper bound applies for loading conditions which exceed the shakedown condition by a modest amount. The upper bound is given by

$$\frac{\Delta\varepsilon}{\Delta N} \geq 4\Delta e$$

where $\Delta\varepsilon$ is the increment of elastic strain given by

$$\Delta e = \frac{\Delta\sigma_1}{E_T}$$

$\Delta\sigma_1$ being the stress increment in excess of the shakedown condition and E_T is the elastic modulus in the transverse direction.

For the optional point shown in Fig. 4 $\Delta\sigma = 35$ MPa and with $E_T = 150$ GPa [Jansson, 1990] the upper bound as ratchet strain is

$$\Delta\varepsilon = \frac{4x35}{150} = 0.93x10^{-3} / \text{cycle}$$

which compares with the experimental value of 0.5×10^{-3}/ cycle.

The failure strain can differ by over an order of magnitude depending on the operating condition. It is known for Al-Li [Pilling and Rindly, 1986] that ductile failure is the result of void nucleation and growth from small particles. In the studies of Hancock and Mackenzie [1976] it is suggested that when failure is the result of void nucleation and growth the failure strain ε_f for multiaxial states of stress has the form

$$\varepsilon_f = 1.65\varepsilon_0 \exp\left\{-\frac{3}{2}\frac{\sigma_M}{\bar{\sigma}}\right\} \tag{3}$$

where ε_0 is the uniaxial failure strain, σ_M is the mean stress and $\bar{\sigma}$ the effective stress. The failure strain for the present matrix in uniaxial tension is reported to be approximately 0.2 [Sakui and Tamura, 1969]. Jansson [1990b] has reported in tests on the composite under consideration that the strain for in-plane shear parallel to the fibers is also 0.2. From the computer studies it has been determined that transverse loading alone introduces a multiaxial stress state for which $\frac{\sigma_M}{\bar{\sigma}} = 2.5$ over a large region of the matrix.

Applying the formula for the failure strain gives the failure strain for transverse loading

$$\varepsilon_{11}^{f} = 0.008$$

which compares well with the observed failure. As the transverse stress decreases the degree of triaxiality decreases and the failure strain is expected to incease.

4. EXTENSION TO CREEP CONDITIONS

Creep experiments have not yet been performed on the composite but these are now in hand. However, it is possible to gain some insight into the creep behavior using bounding
theorems for components subjected to cyclic loading conditions. Based on the Bailey Orowan theory, constitutive equations which describe the creep strains for variable stress histories can be expressed by the potential developed by Ponter and Leckie [1976]. The potential form is given by

$$\Omega = F(\phi - s) + G(s)$$

from which the strain and internal variable rates can be determined to be,

$$\dot{\varepsilon}_{ij}^{c} = f(\phi - p)\frac{\partial \phi}{\partial \sigma_{ij}} \qquad [F' = f]$$

$$\dot{s} = h(s)f(\phi - p) - r(p) \qquad \left[G' = \frac{r(p)}{h(p)}\right]$$

where $\phi^2 = \frac{3}{2} s_{ij} s_{ij}$ and p the internal variable is a scalar

The function $f(\phi\text{-}p)$ is defined by

$$f(\phi - p) > 0 \qquad \text{when } \phi = p$$
$$= 0 \qquad \text{when } \phi \leq p$$

Following the expression for $f(\phi - p)$ strains only occur when $\phi = p$.

For fast loading $\phi = p$ and

$$\dot{\varepsilon}^{p}_{ij} = \frac{\dot{p}}{h(p)} \frac{\partial \phi}{\partial \sigma_{ij}}$$

so that h(p) can be determined from the short time stress-strain curve.

At steady state conditions $\dot{p} = 0$ and

$$\dot{\varepsilon}_{ij} = \frac{r(p)}{h(p)} \frac{\partial \phi}{\partial \sigma_{ij}} = \dot{\varepsilon}_0 \left(\frac{\sigma}{\sigma_0} \right)^n \frac{\partial \phi}{\partial \sigma_{ij}}$$

where σ_0 is the reference stress and $\dot{\varepsilon}_0$ is the corresponding strain rate. The above relation can then be used to establish the function r(p). Following the procedure developed by Ponter [1976], Cocks and Leckie [1988] found an optimized bound for the creep deformations when the component is subjected to cyclic loading. The bound applied to the current problem when the composite is subjected to transverse stress σ_1 and cyclic thermal loading gives the result

$$\dot{\varepsilon}_1 \leq \frac{\sigma_y}{\sigma_u} \dot{\varepsilon}_0$$

where σ_u is the maximum transverse stress corresponding to a matrix yield stress σ_y matrix. Reference to Fig. 3 shows the value

$$\frac{\sigma_u}{\sigma_y} = 1.5$$

Consequently

$$\dot{\varepsilon}_1 \leq \frac{\dot{\varepsilon}_0}{1.5}$$

The result only applies when the loading condition does not exceed the shakedown condition for a yield stress corresponding to the reference stress σ_0. However the shakedown condition is known for σ_y and a simple scaling produces the shakedown for the reference stress σ_0. A proviso for the application of the result is that the continuity conditions between fiber and matrix be maintained. Diffusion mechanisms have been reported which reduce the shear stress which can be supported at the interface. In these circumstances the result developed above could be seriously non-conservative.

CONCLUSIONS

It has been demonstrated that the shakedown concept provides useful insight into the behavior of a metal matrix composite when loaded with transverse stress and cyclic thermal loading. It is also shown that a slight modification in which the reference stress is used in conjunction with the shakedown condition is sufficient to give estimates on the creep behavior.

ACKNOWLEDGEMENTS

The authors acknowledge the support of a grant from NASA-Lewis.

REFERENCES

Campion, A.R., Krueger, W.H., Hartman, H.S. and Dhingra, A. K., [1978], "Fiber F.P. Reinforced Metal Matrix Composites," Proc. 2nd International Conference on Composite Materials, Toronto, 1978, Metallurgical Soc. AIME, pp. 883-904.

Cocks, A.C.F. and Leckie, F.A., [1988] "Deformation Bounds for Cyclically Loaded Shell Structures Operating Under Creep Conditions," J.Appl. Mech., p. 56, 1988.

Hancock, J.W., and Mackenzie, A.C., [1976] "On the Mechanisms of Ductile Fracture in High Strength Steels Subject to Multiaxial States of Stress," J. Mech Physics of Solids, 24, pp. 141-169, 1976.

Jansson, S., [1990a] "Homogenized Nonlinear Constitutive Properties and Local Stress Concentrations for Composites with Periodical Internal Structure," Department of Mechanical Engineering, UC Santa Barbara, 1990.

Jansson, S., [1990b] "Mechanical Characterization and Numerical Modeling of Non-Linear Deformation and Fracture of a Continuous Fiber Reinforced Metal Matrix Composite," Department of Mechanical Engineering, UC Santa Barbara, 1990.

Pilling, J. and Ridley, N., [1986] "Role of Hydrostatic Pressure on Cavitation During Superplastic Flow d Al-Li Alloy," Aluminum-Lithium Alloys III, The Institute of Metal, London, pp. 184-190, 1986

Ponter, A.R.S. and Leckie, F.A., [1976] "Constitutive Relationships for Time-Dependent Deformation of Metals," J. Eng. Mat. Tec., ASME, 47, 1976.

Ponter, A.R.S., [1976] "Deformation Bounds for the Bailey-Orowan Theory of Creep," J. Appl. Mech., 42, p. 619, 1976.

Ponter, A.R.S. and Cocks, A.C.F., [1983] "The Incremental Strain Growth of an Elastic-Plastic Body Loaded in Excess of the Shakedown Limit," Department of Engineering, Leicester University, 1983.

Sakui, S. and Tamura, M., [1969] "Yielding Phenomena of an Al-3% Li Alloy," Transactions of the Japan Institute of Metals, Vol. 10, pp. 343-350,1969.

Starke, E.A., Sanders, T.H. and Palmer, J.G., [1981], "New Approaches to Alloy Development in Al-Li Systems," Journal of Metals, Vol. 33, pp. 24-32, 1981.

Problems of Structural Optimization Under Creep Conditions

M. ŻYCZKOWSKI

Institute of Mechanics and Machine Design
Cracow University of Technology, Poland

Summary

Optimal structural design under creep conditions, initiated over 20 years ago, brings a fairly large variety of problems, interesting both from theoretical and engineering point of view. Existence of the time factor leads to optimization for a prescribed lifetime and the number of independent variables in governing equations increases by one. Practical importance is connected with design of metal structures working at elevated temperatures, structures made of plastics, concrete etc. Here we discuss optimization problems with constraints imposed on brittle and ductile creep rupture, stiffness, creep buckling, dynamic behaviour. Some further possible directions of development are mentioned as well.

1 Introductory remarks

Structural optimization under creep conditions was initiated by four papers published in the years 1967-68 by M.I. Reitman [14], W. Prager [13], Yu.V. Nemirovsky [10] and M. Życzkowski [24]. An extensive survey published by the author in 1988, [25], discussed over 100 papers published since that time. In the present paper we are going just to show and classify typical problems of structural optimization under creep conditions and quote some results obtained. Recent achievements, exceeding [25], will also be reviewed.

Typical design objective in structural optimization is the volume of a structure; this volume should be minimized. Indeed, minimal volume corresponds to minimal material costs (if the material is preselected), and also is important in exploitation: so, this design objective remains also under creep conditions without any change. Similar remark pertains to design variables. Conversely, essential changes are seen in constraints and equation of state: they contain a new independent variable, namely time t. On the other hand, sometimes it is convenient to interchange the roles of the design objective and of a constraint (mutual or dual formulation), and then one can also consider new design objectives or criteria.

Optimization constraints under creep conditions are usually connected with strength (rupture), stiffness and stability - as in the elastic case - but also some quite new constraints may be imposed: for example, limited stress relaxation, limited residual displacements after impact and so on. However, even in the case of classical constraints the factor of time brings some new elements: structures under creep conditions should be designed

M. Zyczkowski (Ed.)
Creep in Structures
IUTAM Symposium Cracow/Poland 1990

for a given lifetime under given loadings. So, the problem of minimal volume may be replaced here not only by a problem of maximal loading parameter for prescribed volume and lifetime, but also by a problem of maximal lifetime for prescribed volume and loading.

2 The problem of dependence of optimal shape on the constitutive equations of creep

Considering elastic or perfectly plastic structures one usually does not discuss the dependence of optimal shapes on the constitutive equations adopted (except the effect of possible anisotropy). Conversely, in the case of structures subject to creep, this problem becomes very essential in view of large variety of constitutive equations in rheology, equations of damage, rupture etc. One may expect that the optimal shapes depend on such equations. However, A. Gajewski [4,5] - who paid much attention to this problem - separated many problems of optimal structural design in which final shape does not depend on constitutive equations adopted. We mention here homogeneous statically determinate beams (provided axial force is absent), homogeneous circular plates designed for minimal power of external loadings, and homogeneous rotating disks designed for constant stress intensity, but optimal shapes are independent of constitutive equations in many other cases as well.

3 Constraints imposed on brittle creep rupture

We are now going to discuss the problems and examples of structural optimization connected with typical constraints under creep conditions.

The simplest theory of brittle creep rupture proposed by Kachanov, is based on continuum damage mechanics: in the first period one considers evolution of continuity function ψ or damage function $D = 1 - \psi$ and the condition $\psi = 0$ or $D = 1$ at a point of the body means formation of a macrocrack at that point. Further, in the second period, Kachanov discussed propagation of the front of damaged zone up to total rupture of the element. However, the second period is usually much shorter then the first one and hence the formation of the first macrocrack - much easier to be evaluated - may be assumed as determining lifetime up to brittle rupture.

Suppose evolution law for continuity function in Kachanov's form

$$\frac{d\psi}{dt} = -R\left(\frac{\sigma_{red}}{\psi}\right)^{\nu} , \tag{3.1}$$

where R and ν are material constants depending on temperature, and the reduced stress is a certain stress invariant, for example, as proposed by Sdobyrev and Rabotnov,

$$\sigma_{red} = \delta\sigma_I + (1 - \delta)\sigma_e , \tag{3.2}$$

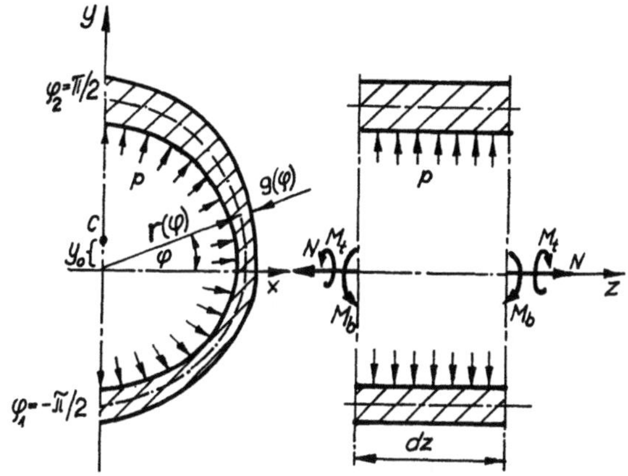

Figure 1:

with σ_I denoting maximal normal stress, σ_e - the Huber-Mises-Hencky effective stress, δ - material constant. Then, integrating (3.1) up to formation of the first macrocrack, we obtain

$$\sup_{X \in V} R(\nu+1) \int_0^{t_R} [\sigma_{red}(X, t)]^\nu \, dt = 1 . \tag{3.3}$$

A relatively broad class of structures optimal with respect to brittle creep rupture may be found from among the structures of uniform creep strength. We use the term "structure of uniform creep strength (for brittle rupture)" if (3.3) is satisfied simultaneously either for any $X \in V$, or along certain characteristic lines or surfaces. In the first case the second period is reduced to zero (macrocracks appear simultaneously), and this is typical for optimality; in the second case, e.g. in structural elements subject to bending, the integral (3.3) may be constant just in outer fibres of all cross-sections, but even here the name "uniform strength" is justified.

If we consider just stationary creep (both physically and geometrically), then $\sigma_{red}(X, t) = \text{const}(t)$ and instead of (3.3) we can write the condition of uniform creep strength in a much simpler form

$$R(\nu+1)\sigma_{red}^\nu t_R = 1 , \tag{3.4}$$

it means also $\sigma_{red} = \text{const}(X)$ (at least along certain lines or surfaces).

As the first example, consider a circular cylindrical shell of radius r and thickness $g = g(\varphi)$, under internal pressure p, bending moment M_b, torque M_t and normal force N, Fig. 1, (M. Życzkowski, M. Rysz [27]). Assume the Norton-Odqvist creep law; then, besides the shearing stresses $\tau_{z\varphi}$ and circumferential stresses σ_φ determined directly from equilibrium equations

$$\tau_{z\varphi} = \frac{M_t}{2\pi r^2 g} , \qquad \sigma_\varphi = \frac{pr}{g} , \tag{3.5}$$

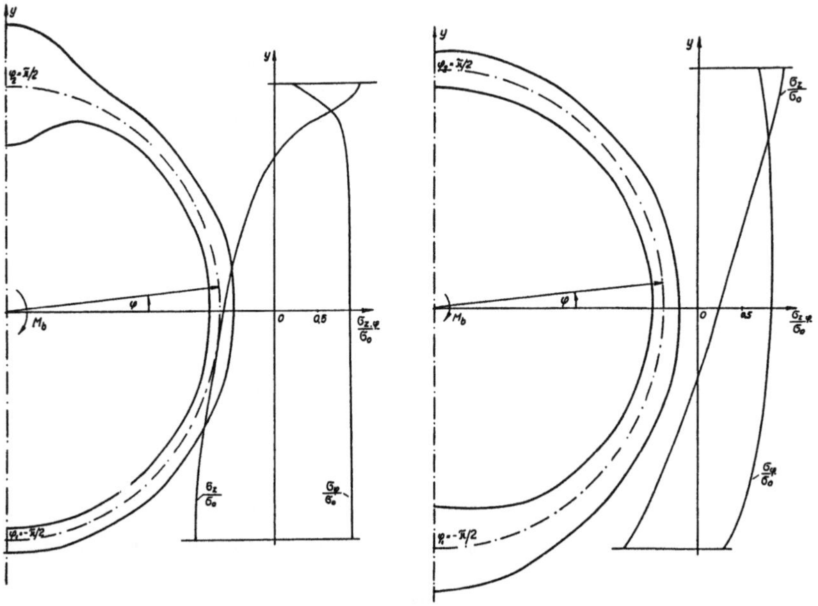

Figure 2:

one obtains from the constitutive equation, eliminating $g(\varphi)$ and $\tau(\varphi)$,

$$\dot{\kappa} r \sin\varphi + \dot{\varepsilon}_0 = \frac{K}{2}\left(\sigma_z^2 + \sigma_\varphi^2 - \sigma_z\sigma_\varphi + 3m^2\sigma_\varphi^2\right)^{\frac{n-1}{2}}(2\sigma_z - \sigma_\varphi) . \tag{3.6}$$

In this equation $m = M_t/2\pi p r^3$ joins the loadings and the design variable r; $\dot{\kappa}$ and $\dot{\varepsilon}_0$ denote rate of curvature and rate of unit elongation, respectively, K and n - constants in Norton-Odqvist law. With geometry changes neglected, one can look for the optimal shell from among the shells of uniform creep strength, (3.4). This approach may be justified if creep buckling constraints are not employed. Here, we neglected such constraints presuming that the resulting thickness will be sufficiently large; optimization for creep buckling will be discussed in Sec. 6. So, assuming reduced stress in the form (3.2) we write the condition of uniform creep strength in the form

$$\frac{\delta}{2}\left[\sigma_z + \sigma_\varphi + \sqrt{(\sigma_z+\sigma_\varphi)^2 + 4m^2\sigma_\varphi^2}\right] + (1-\delta)\sqrt{\sigma_z^2 + \sigma_\varphi^2 - \sigma_z\sigma_\varphi + 3m^2\sigma_\varphi^2} = \sigma_0 . \tag{3.7}$$

Equations (3.5), (3.6) and (3.7) determine all four functions $\tau_{z\varphi}$, σ_φ, σ_z and g; just the radius r hidden in m is subject to further optimization under the constraints $M_b = \text{const.}$ and $N = \text{const.}$, determining $\dot{\kappa}$ and $\dot{\varepsilon}_0$. Minimizing the cross-sectional area A we find optimal r, and then also optimal $g(\varphi)$. Resulting shapes and the relevant stress distributions are shown in Fig. 2 for $\delta = 0$ (Kachanov-Huber-Mises) and $\delta = 1$ (Kachanov-Galileo). Essential effects of the parameter δ are seen. Indeed, the Huber-Mises hypothesis considers circumferential tension with axial compression as particularly dangerous, whereas the Galileo hypothesis joins maximal danger with both normal stresses being positive.

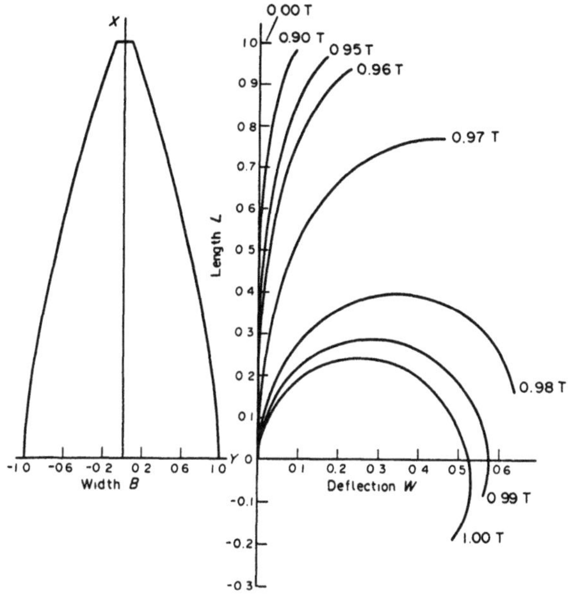

Figure 3:

Optimization of shells with double curvature under the creep rupture constraint was discussed by V.M. Dulman [3].

The situation changes essentially if geometry changes are taken into account. Then the stresses change in time and (3.4) should be replaced by (3.3) holding for any X, but, moreover, a structure of uniform creep strength designed by (3.3) is, in general, not optimal. It is just useful as a first step towards further optimization.

Consider, for example, an eccentrically compressed slender column with I-section (W. Świsterski, A. Wróblewski, M. Życzkowski [20]). Finite deflections and compressibility of axis were allowed for and material was described by Norton's creep law. Width of the column β was assumed as the design variable. A system of six equation of state was derived and integrated numerically with the additional condition of uniform creep strength (3.3). The resulting shape is shown in Fig. 3 with deflections lines corresponding to some fractions of the lifetime $t_R = T$. However, this column is not optimal. Further optimization was carried out by imposing corrections of shape

$$\Delta\beta = A\left(\cos\alpha\xi - \frac{\sin\alpha}{\alpha}\right) , \quad \xi = x/l . \tag{3.8}$$

These corrections do not change the volume, but may result in an essential increase of the lifetime in Kachanov's sense. Optimal values of the parameters A and α were found as follows: $A = -0.37$, $\alpha = 205°$, and the column optimal within the class (3.8) is shown in Fig. 4, compared with the column of uniform creep strength. Dimensionless lifetime t_R equals here $t_R = 1.903T$, and hence the gain is visible. A further optimization may be

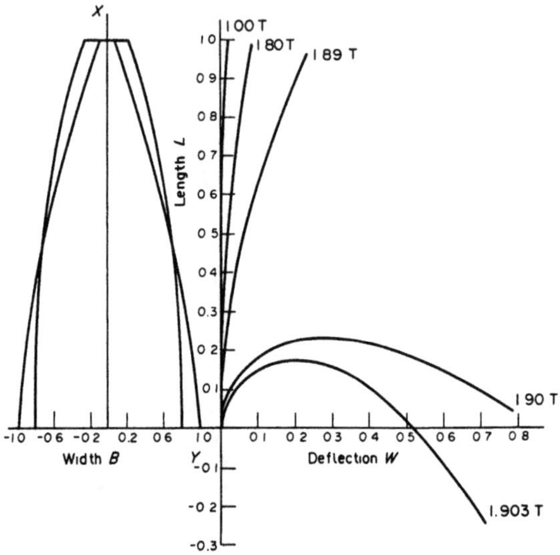

Figure 4:

performed within a class broader then (3.8).

Prestressed structures with stress relaxation taken into account give another example of possible discrepancy between the shape of uniform creep strength and optimal shape. The first shape is described here by (3.3), since stresses change in time. This shape is either optimal, or gives a good first step towards the optimal one. Optimization of prestressed rotating disks with respect to brittle creep rupture was performed by A. Ganczarski and J. Skrzypek [6,7].

Problems of optimality of structures of uniform strength under creep conditions was also discussed by Yu.V. Nemirowsky, at the present symposium, [11].

4 Constraints imposed on ductile creep rupture

Ductile creep rupture, as described by Hencky and Hoff, requires an analysis of finite strains up to infinity; lifetime is expressed here by the condition of vanishing transverse dimensions. Such a constraint is very difficult to be employed in optimal design, and hence the solutions obtained are limited just to constant principal directions.

Problems of optimal design with respect to ductile creep rupture, mentioned in [24], are discussed in detail by K. Szuwalski, in his habilitation thesis [19], and also in the papers [17] and [18] (this Volume). One can introduce the concepts of structures of uniform strength with respect to ductile creep rupture and of structures of uniform deformability; such structures may be optimal, or may serve a first step towards optimality. Some

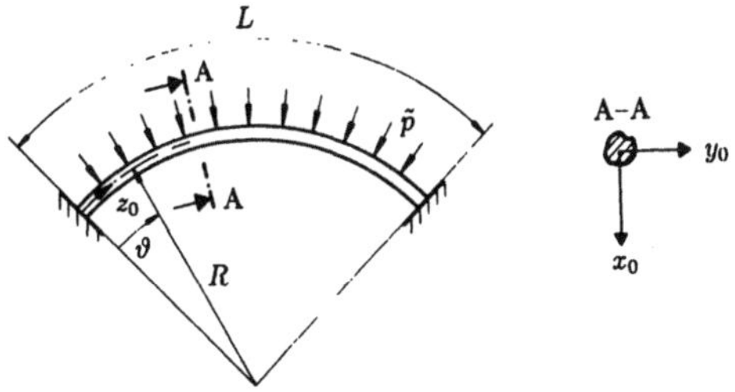

Figure 5:

fundamental theorems regarding uniform strength, uniform deformability and optimal design were proved in [19].

5 Constraints imposed on creep stiffness or compliance

There is a large variety of criteria of stiffness or compliance under creep conditions; they were discussed in detail by M. Życzkowski in [24]. The first paper by W. Prager [13] used the power of external loadings as a criteria of compliance. Other criterion may be expressed by some norms of velocities or strain rates; certain examples and comparison of results are given in [24].

6 Constraints imposed on creep buckling

Many approaches to creep buckling problems were proposed and all of them may serve as a basis to formulate optimization constraints. We mention here, for example, the theories of creep buckling of straight columns (Shanley, Gerard, Rabotnov-Shesterikov) and of imperfect columns (Rzhanitsyn, Kempner-Hoff, Fraeijs de Veubeke). Most of these theories were generalized to plates and shells. The first optimal shapes of a column subject to creep buckling were determined by M. Życzkowski and R. Wojdanowska-Zając and presented at the second symposium Creep in Structures in Göteborg [28].

We quote here two examples of optimization, using two different creep buckling theories. First, consider a circular clamped-clamped arch under uniform external pressure (Fig. 5), subject to creep as described by strain-hardening law

$$\Phi = \dot{\varepsilon}^C (\varepsilon^C)^\mu - \Gamma \sigma^n = 0 \tag{6.1}$$

where $\varepsilon^C = \varepsilon - \sigma / E$. We are looking for the optimal distribution of cross-sectional area along the axis of arch as to minimize the volume under given loading and critical time according to the Rabotnov-Shesterikov theory (A. Wróblewski, M. Życzkowski [23]). Multimodal optimization will be applied, two in-plane and two out-of-plane buckling modes will be considered, both symmetric and antisymmetric.

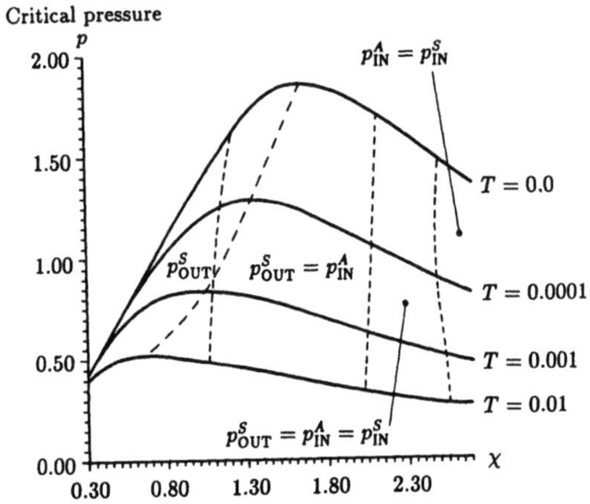

Figure 6:

The Rabotnov-Shesterikov theory in the first step considers creep in fundamental state, namely pure compression described directly by (6.1) (funicular arch). Then it assumes that buckling is described by variation of (6.1) generalized for multiaxial stress state,

$$\delta s_{ij} = \frac{s_{ij}}{\sigma_e^2}\left(\frac{\sigma_e}{\varepsilon_e} - \frac{\delta\sigma_e}{\delta\varepsilon_e}\right) s_{kl}\delta\varepsilon_{kl} + \frac{2}{3}\frac{\sigma_e}{\varepsilon_e}\delta\varepsilon_{ij}\ , \tag{6.2}$$

where s_{ij} and ε_{ij} denote deviatoric stresses and strains, respectively (incompressibility of material is assumed). These equations are written separately for in-plane buckling, and separately for out-of-plane buckling, with torsion included. Various types of loading behaviour during buckling are analyzed. Optimal design requires minimization of the volume of the arch for given loading and critical time:

$$V = \int_0^L A(z)\, dz \to \min, \quad p\,, t_{cr} = \text{const.} \tag{6.3}$$

In fact, it was more convenient to maximize critical pressure (all four critical pressures for individual buckling modes) at constant volume and constant time. Rectangular cross-section with a constant ratio $B/H = \chi$ was assumed. Critical loadings in terms of the parameter χ are shown in Fig. 6 for the direction of loading fixed in space, and for arch steepness $L/R = \pi/2$. The ranges shown by dashed lines correspond in turn to uni-, bi-, tri- and bimodal optimization. Optimal ratio χ depends on time, it is shown by long dashed line. Optimal shapes are shown in Fig. 7.

As the second example consider a non-circular cylindrical shell under overall bending by concentrated moment M_b and tension-compression due to normal force N, M. Rysz and M. Życzkowski [16]. Both senses of M_b and N are admitted, and hence the cross-section of the shell is assumed bisymmetric. We look for minimal volume, equivalent here to minimal cross-sectional area

$$A = 2\int_{-H}^{H} h\sqrt{1 + x'^2}\, dy + 2A_0 \to \min \tag{6.4}$$

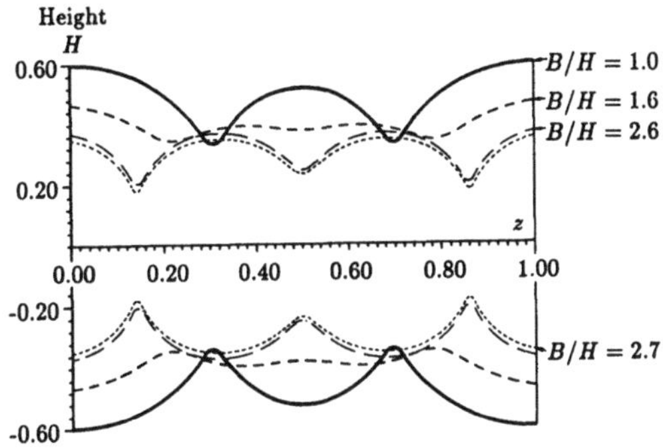

Figure 7:

under the constraints of given bending moment and normal force:

$$\begin{aligned} M_b &= 2\int_{-H}^{H} \sigma_z y h\sqrt{1+x'^2}\,dy + A_0 H(\sigma_{r1}-\sigma_{r2}) \ , \\ N &= 2\int_{-H}^{H} \sigma_z h\sqrt{1+x'^2}\,dy + A_0(\sigma_{r1}+\sigma_{r2}) \ , \end{aligned} \tag{6. 5}$$

where A_0 denotes the area of a rib at outer fibers (such ribs are necessary to carry bending in an optimal manner), σ_{r1} and σ_{r2} are stressed carried by ribs, H is overall height (depth) of the cross-section, $h = h(y)$ denotes wall thickness. So, in the problem under consideration we have two functions $x(y)$ and $h(y)$ and two parameters H and A_0 as design variables.

The stability constraint will now be formulated as follows. We introduce the concept of a shell of uniform creep stability in a manner anologous to that proposed by M. Życzkowski and J. Krużelecki [26,8] for elastic shells. For such a shell local stability condition is satisfied at any point of the contour. Here, this concept will be combined with the simplest theory of creep buckling, namely that proposed by Gerard, defining critical state by critical strain. It is supposed that at any point

$$\varepsilon = \varepsilon_{cr} = \beta\frac{h}{\varrho} \ , \tag{6.6}$$

where β is a numerical coefficient (e.g. $\beta = 0.6$ according to linear theory), and ϱ is the radius of curvature of the central line of the profile. Moreover, for outer fibres (and ribs) Kachanov's constraint of brittle creep rupture is imposed. Eq. (6.6) eliminates one function, e.g. $h = h(y)$, and the second function $x = x(y)$ is found from the Euler-Lagrange equation. The resulting optimal shape and stress distribution is shown in Fig. 8.

A similar solution based on the Rabotnov-Shesterikov creep buckling theory, better justified than that of Gerard, is presented by M. Rysz at the same Symposium [15]. Optimal design of circular plates under the constraint of creep buckling is discussed by A. Wróblewski in this Volume [22].

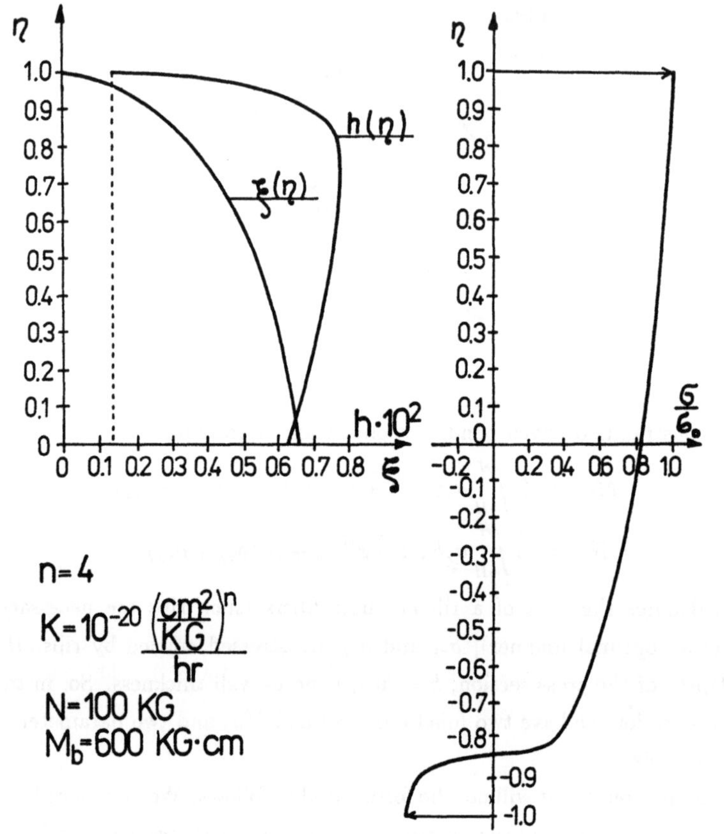
η
1.0
0.9
0.8
0.7
0.6
0.5
0.4
0.3
0.2
0.1
0
h(η)
ξ(η)
h·10²
0 0.1 0.2 0.3 0.4 0.5 0.6 0.7 0.8
ξ
n=4
K=10⁻²⁰ (cm²/KG)ⁿ / hr
N=100 KG
M_b=600 KG·cm
σ/σ₀
-0.2 0.2 0.4 0.6 0.8 1.0
-0.1
-0.2
-0.3
-0.4
-0.5
-0.6
-0.7
-0.8
-0.9
-1.0

Figure 8:

7 Dynamic problems

Optimal design of structures subject to dynamic loadings under creep conditions,though very scarce, goes in two directions: optimal design for vibrations and for impact loadings. We mention here just some recent papers, following those reviewed in [25].

R.A. Meric [9] applied sensitivity analysis to optimization of viscoelastic structures under transient response. C.P. Chen and R.S. Lakes [2] looked for optimal material properties for viscoelastic impact absorbes; this problem resembles optimization of material for safety belts, discussed by A. Trojnacki and M. Życzkowski [21]. E. Cegielski [1] found optimal shapes of nonprismatic viscoplastic circular bars subject to dynamic twisting loading for various shapes of the impulse.

8 Final remarks

A fairly large variety of optimization problems under creep conditions has been shown. It is connected with the time factor appearing, but also with variety of creep rupture and creep buckling theories, stiffness criteria and so on. Many important phenomena, like creep crack growth, creep-fatigue interaction, creep anisotropy, have not been used as optimization constraints at all. Optimal design of composite structures subject to creep is also of great practical importance; besides the papers reviewed in [25] we mention here a recent paper by P.U. Post [12]. Further progress will probably be achieved by broader application of optimal control methods, sensitivity analysis, and development of suitable methods of numerical optimization.

References

[1] Cegielski, E.: *Parametric optimization of viscoplastic bars under dynamic twisting loadings*, Mech. Teor. Stos. **28** (1990), 1-2,35-44.

[2] Chen, C.P.; Lakes, R.S.: *Design of viscoelastic impact absorbers: optimal material properties*, Int. J. Solids Struct. **26** (1990), 12, 1313-1328.

[3] Dulman, V.M.: *Optimal structural design of shells with double curvature under creep conditions*, Lit. Mekh. Sbornik **29** (1987), 24-32. (in Russian)

[4] Gajewski A.: *Optymalne kształtowanie wytrzymałościowe w przypadku materiału o nieliniowości fizycznej*, Zeszyty Nauk. Polit. Krak. **5**, Kraków 1975.

[5] Gajewski A.: *Effects of physical nonlinearities on optimal design of structures*, Proc. IUTAM Symp. on Physical Nonlinearities, Senlis 1980, Springer 1981, 81-84.

[6] Ganczarski, A.; Skrzypek, J.: *Optimal prestressing and design of rotating disk against brittle rupture under unsteady creep conditions*, Rozpr. Inż (Eng. Trans.) **37** (1989), 4, 49-71.

[7] Ganczarski, A.; Skrzypek, J.: *Optimal shape of prestressed disks against brittle rupture under unsteady creep conditions*, Struct. Optimiz. (submitted for publication).

[8] Krużelecki, J.; Życzkowski, M.: *Optimal design of an elastic cylindrical shell under overall bending with torsion*, Solid Mech. Arch. **9** (1984), 3, 269-306.

[9] Meric, R.A.: *Shape variations for viscoelastic solid under transient response*, Z. angew. Math. Mechanik **70** (1990), 11, 509-516.

[10] Nemirovsky, Yu.V.: *Optimal design of creeping structures*, III Vsesoyuzny Syezd po Mekh., Moskva 1968. (in Russian)

[11] Nemirovsky, Yu.V.: *Problems of creep rupture and optimization of structures under creep conditions*, IUTAM Symp. Creep in Structures IV, Kraków 1990 (not submitted for publication).

[12] Post, P.U.: *Optimale Auslegung viskoelastischer Faserverbundstrukturen unter hygrothermischen Einflüssen*, Beiträge zur Maschinentechnik, Universität Siegen 1990, 103-127.

[13] Prager, W.: *Optimal structural design for given stiffness in stationary creep*, Z. angew. Math. Physik **19** (1968), 2, 252-256.

[14] Reitman, M.I.:*On the theory of optimal design of structures made of plastics with the time factor taken into account*, Mekh. Polimerov (1967), 2, 357-360. (in Russian)

[15] Rysz, M.: *Optimal shaping of a cross section of a cylindrical shell under combined loadings with respect to creep stability*, IUTAM Symp. Creep in Structures IV, Kraków 1990 (not submitted for publication).

[16] Rysz, M.; Życzkowski, M.: *Optimal design of a cylindrical shell under overall bending and axial force with respect to creep stability*, Structural Optimization **1** (1989), 1, 29-36.

[17] Szuwalski, K.: *Optimal design of bars under nonuniform tension with respect to ductile creep rupture*, Mechanics of Struct. Machines **17** (1989), 3, 303-319.

[18] Szuwalski, K.: *Bars of uniform strength vs. optimal with respect to ductile rupture time*, Proc. IUTAM Symp. Creep in Structures IV (this Volume).

[19] Szuwalski, K.: *Optymalne kształtowanie elementów konstrukcji z uwagi na czas zniszczenia ciągliwego w warunkach pełzania*, Zeszyty Nauk. Polit. Krak., Kraków 1991.

[20] Świsterski, W.; Wróblewski, A.; Życzkowski, M.: *Geometrically non-linear eccentrically compressed columns of uniform creep strength vs. optimal columns*, Int. J. Non-Linear Mech. **18** (1983), 4, 287-296.

[21] Trojnacki, A.; Życzkowski, M.: *Effect of plastic crushing of the car body on optimization of rheological properties of safety belts*, Rozpr. Inż (Eng. Trans.) **29** (1981), 1, 115-130.

[22] Wróblewski, A.: *Optimal design of annular plates against creep buckling*, Proc. IUTAM Symp. Creep in Structures IV (this Volume).

[23] Wróblewski, A.; Życzkowski, M.: *On multimodal optimization of circular arches against plane and spatial creep buckling*, Structural Optimization **1** (1989), 4, 227-234.

[24] Życzkowski, M.: *Optimal structural design in rheology*, 12th Int. Congr. Theor. Appl. Mech., Stanford 1968; J. Appl. Mech. **38** (1971), 1, 39-46.

[25] Życzkowski, M.: *Optimal structural design under creep conditions*, Appl. Mech. Reviews **41** (1988), 12, 453-461.

[26] Życzkowski, M.; Krużelecki, J.: *Optimal design of shells with respect to their stability*, Proc. IUTAM Symp. Optimization in Structural Design, Warszawa 1973; Springer 1975, 229-247.

[27] Życzkowski, M.; Rysz, M.: *Optimization of cylindrical shells under combined loading against brittle creep rupture*, Proc. IUTAM Symp. Inelastic Behaviour of Plates and Shells, Rio de Janeiro 1985; Springer 1986, 385-401.

[28] Życzkowski, M.; Wojdanowska-Zając, R.: *Optimal structural design with respect to creep buckling*, Proc. IUTAM Symp. Creep in Structures II, Göteborg 1970; Springer 1972 371-387.

Modelling of Viscoelastic Behaviour of Plates

H. ALTENBACH

Technische Universität "Otto von Guericke"
Magdeburg, Germany

Summary

The behaviour of viscoelastic plates can be described by several engineering theories, based on different assumptions and theoretical conceptions. One of them - the direct approach - is used in this paper to modelize the viscoelastic behaviour of plates. This theory is physically and mathematically correct, but there are other problems, especially the identification of the effective properties in the constitutive equations. The paper deals with a Timoshenko-Reissner-Mindlin type theory, so the main problem is the determination of the classical-plate-properties for stretching, plane shear, bending and torsion and the properties for transverse shear. It is shown by a simple analytical method, that all properties can be found out for viscoelastic plates without any restrictions for the distribution of the material properties in the thickness (symmetric, unsymmetric) and the linear-viscoelastic behaviour (differential or integral relations, complex moduli). The proposed theory is suitable for homogeneous and inhomogeneous (e.g. laminated) plates. For some special cases there are given examples.

1 Introduction

A plate is a plane deformable structure with one geometrical dimension (thickness) smaller then other two dimensions. There are different approaches for constructing plate theories. The two-dimensional treatment is connected with the formulation of the system of partial differential equations in two independent spatial variables and a set of boundary/initial conditions. General approaches to develop this system are based on the method of hypotheses, method of expansion, asymptotic integration technique, iterative methods, methods of successive correction or the direct approach [1] - [4]. On Fig. 1 is given a classification of several approaches.

The same situation we can obtain for laminated plates [5] - [7]. In this contribution we consider a theory of anisotropic, linear-viscoelastic plates, based on the direct approach. The theory allows the analysis of plates with variable over the thickness material properties, so the investigation of multilayered plates is possible.

2 Assumptions and basic two-dimensional equations

The real plate (three-dimensional) is replaced by a generalized two-dimensional deformable continuum (deformable surface). Each point of this continuum is an infinitesimal

M. Zyczkowski (Ed.)
Creep in Structures
IUTAM Symposium Cracow/Poland 1990

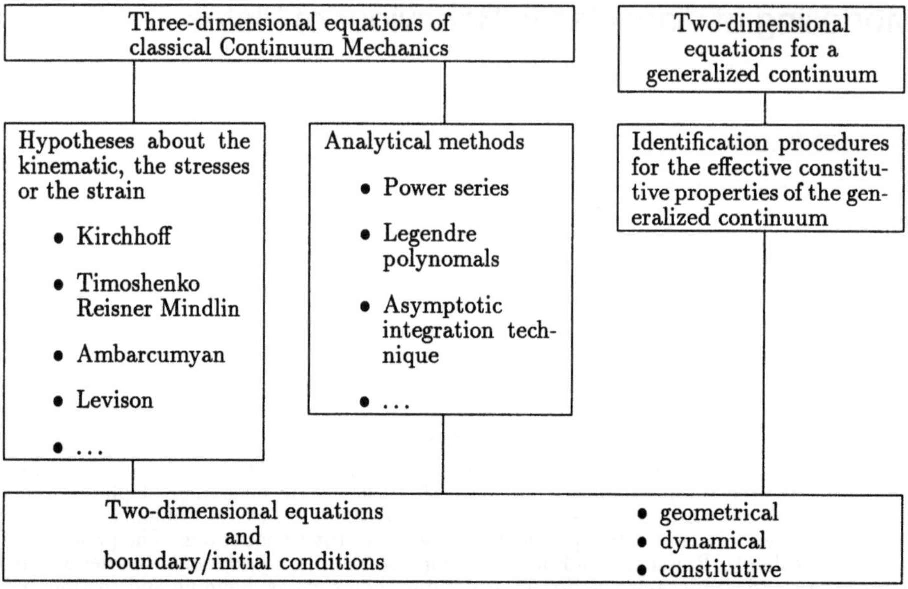

Figure 1: Classification of approaches for constructing plate theories

small rigid body with 5 degrees of freedom (3 translations and 2 rotations). The basic linear equations (geometrical relations and equations of motion) we get from [8]-[11] in the following form

$$\nabla\mathbf{T} + \mathbf{q} = \varrho(\mathbf{u} + \mathbf{J}_1^T \cdot \boldsymbol{\varphi})^{\cdot\cdot} \tag{1}$$

$$\nabla\mathbf{M} + \mathbf{T}_\times + \mathbf{m} = \varrho(\mathbf{J}_1 \cdot \mathbf{u} + \mathbf{J}_2 \cdot \boldsymbol{\varphi})^{\cdot\cdot} \tag{2}$$

$$\boldsymbol{\mu} = (\nabla\mathbf{u} \cdot \mathbf{a})^{sym}\,; \quad \boldsymbol{\gamma} = \nabla\mathbf{u} \cdot \mathbf{n} + \mathbf{c} \cdot \boldsymbol{\varphi}\,; \quad \boldsymbol{\kappa} = \nabla\boldsymbol{\varphi} \tag{3}$$

$$\mathbf{u} = u_1\mathbf{e}_1 + u_2\mathbf{e}_2 + w\mathbf{n}\,; \quad \boldsymbol{\varphi} = -\varphi_2\mathbf{e}_1 + \varphi_1\mathbf{e}_2$$

$$\mathbf{a} = \mathbf{e}_1\mathbf{e}_1 + \mathbf{e}_2\mathbf{e}_2\,; \quad \mathbf{c} = -\mathbf{a} \times \mathbf{n}$$

$\mathbf{T}, \mathbf{M}$ force tensor, couple tensor; $\mathbf{q}, \mathbf{m}$ surface force, surface couple; $\mathbf{u}$, $\boldsymbol{\varphi}$ translation vector (displacements), rotation vector; ϱ surface mass (density); $\varrho\mathbf{J}_1$, $\varrho\mathbf{J}_2$ surface tensor of inertia; $\mathbf{T}_\times$ vector's invariant of $\mathbf{T}$; $\boldsymbol{\mu}$, $\boldsymbol{\gamma}$, $\boldsymbol{\kappa}$ plane deformation tensor, transverse shear deformation vector, flexural and torsional deformation tensor; ∇ Nabla operator (space derivatives); $(\ldots)^{\cdot}$ time derivatives; $\mathbf{e}_1$, $\mathbf{e}_2$ unit base vectors of the surface; $\mathbf{n}$ unit normal vector to the surface; $(\ldots)^T$ transposed, $(\ldots)^{sym}$ symmetric part; $\cdot$ scalar product; $\times$ vector product.

In analogy with the three-dimensional theory of viscoelasticity [12] we derive the following constitutive equations [9]

$$\mathbf{T} \cdot \mathbf{a} = \int_{-\infty}^{t} \left[{}^{(4)}\mathbf{A}(t-\tau) \cdot\cdot \dot{\boldsymbol{\mu}}(\tau) + {}^{(4)}\mathbf{B}(t-\tau) \cdot\cdot \dot{\boldsymbol{\kappa}}(\tau) + \boldsymbol{\gamma}(\tau) \cdot {}^{(3)}\boldsymbol{\Gamma}_1(t-\tau) \right] d\tau \tag{4}$$

$$\mathbf{T}\cdot\mathbf{n} = \int_{-\infty}^{t} \left[\mathbf{\Gamma}(t-\tau)\cdot\boldsymbol{\gamma}(\tau) +^{(3)}\mathbf{\Gamma}_1(t-\tau)\cdot\cdot\dot{\boldsymbol{\mu}}(\tau) +^{(3)}\mathbf{\Gamma}_2(t-\tau)\cdot\cdot\dot{\boldsymbol{\kappa}}(\tau)\right] d\tau \qquad (5)$$

$$\mathbf{M}^T = \int_{-\infty}^{t} \left[\dot{\boldsymbol{\mu}}(\tau)\cdot\cdot^{(4)}\mathbf{B}(t-\tau) +^{(4)}\mathbf{C}(t-\tau)\cdot\cdot\dot{\boldsymbol{\kappa}}(\tau) + \boldsymbol{\gamma}\cdot^{(3)}\mathbf{\Gamma}(t-\tau)\right] d\tau \qquad (6)$$

${}^{(4)}\mathbf{A},\ldots,\mathbf{\Gamma}$ are the tensors of the effective properties of the two-dimensional, linear-viscoelastic continuum. They depend on the physical properties of the plate material and the geometrical parameters of the thickness. In general these tensor contain 68 components. The number may be reduced analyzing the material symmetry (kind of anisotropy), the thickness symmetry etc. Some examples are given in tabl. 1.

	${}^{(4)}\mathbf{A}$	${}^{(4)}\mathbf{B}$	${}^{(4)}\mathbf{C}$	${}^{(3)}\mathbf{\Gamma}_1$	${}^{(3)}\mathbf{\Gamma}_2$	$\mathbf{\Gamma}$	Σ
general	16	16	16	8	8	4	68
anisotropic	6	9	6	6	6	3	36
orthotropic	4	5	4	0	0	2	15
orthotropic/symmetric	4	0	4	0	0	2	10
isotropic	2	2	2	0	0	1	7
isotropic/symmetric	2	0	2	0	0	1	5

Table 1: Number of tensor components in dependence of the kind of anisotropy and symmetry in the thickness directions

3 Identification of the effective properties tensor components

The two-dimensional basic equations are dealing with some properties connected with physical values of real materials. E.g. density and the tensors of inertia can be given in the following form [8], [11]

$$\varrho = <\varrho_*>, \quad \varrho\mathbf{J}_1 = -<\varrho_* z>\mathbf{c}, \quad \varrho\mathbf{J}_2 = <\varrho_* z^2>\mathbf{a} \qquad (7)$$

ϱ_* density of the real material, z thickness variable, $<(\ldots)> = \int_{-h/2}^{h/2}(\ldots)dz$, h thickness.

The relation between the tensor $\mathbf{T}$, $\mathbf{M}$ and the stress tensor σ ($\sigma = \sigma^T$) can be given by [8], [11]

$$\mathbf{T} = <\mathbf{a}\cdot\sigma>, \quad \mathbf{M} = <\mathbf{a}\cdot\sigma\cdot\mathbf{c}z> \qquad (8)$$

The identification procedure for the effective properties tensor is shown on Fig. 2.

This procedure is proved in the case of elastic homogenous plates and shells and for elastic laminates [11]. An expansion of this method for viscoelastic material is possible. It is founded on elastic solutions and Boltzmann's analogy principle of elastic and similar viscoelastic solutions [12].

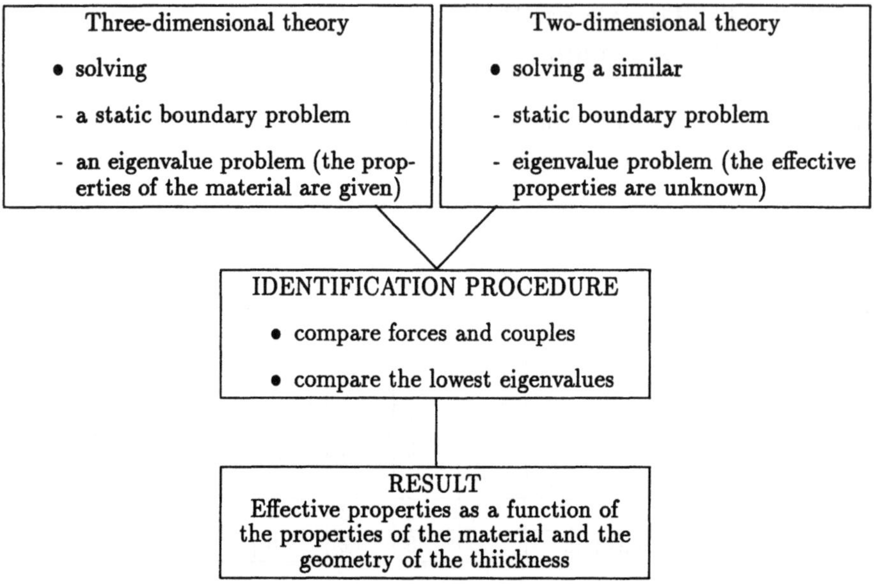

Figure 2: Identification methods for effective properties

4 Orthotropic viscoelastic behaviour

In this case we should identificate following 15 tensor components

$$
\begin{aligned}
{}^{(4)}\mathbf{A} &= A_{11}aa + A_{12}(aa^* + a^*a) + A_{22}a^*a^* + A_{44}c^*c^* & (9)\\
{}^{(4)}\mathbf{B} &= B_{13}ac + B_{14}ac^* + B_{23}a^*c + B_{24}a^*c^* + B_{42}c^*a^* & (10)\\
{}^{(4)}\mathbf{C} &= C_{33}cc + C_{34}(cc^* + c^*c) + C_{44}c^*c^* + C_{22}a^*a^* & (11)\\
\mathbf{\Gamma} &= \mathbf{\Gamma}_1\mathbf{e}_1\mathbf{e}_1 + \mathbf{\Gamma}_2\mathbf{e}_2\mathbf{e}_2\,; & (12)\\
a^* &= \mathbf{e}_1\mathbf{e}_1 - \mathbf{e}_2\mathbf{e}_2\,; \quad c^* = \mathbf{e}_1\mathbf{e}_2 + \mathbf{e}_2\mathbf{e}_1
\end{aligned}
$$

Following [12] in writing the three-dimensional, linear-viscoelastic constitutive equations for orthotropic material as

$$\sigma_{ij} = \int_{-\infty}^{t} R_{ijkl}(z\,, t-\tau)\varepsilon_{kl}(x_1\,, x_2\,, \tau)d\tau \ , \tag{13}$$

where R_{ijkl} are components of the relaxation functions tensor (9 linear independent values - $R_{1111}\,, R_{1122}\,, R_{1133}\,, R_{2222}\,, R_{2233}\,, R_{3333}\,, R_{1313}\,, R_{1212}\,, R_{2323}$); ε_{kl} are the deformation tensor components and x_1 , x_2 are the spatial coordinate variable. Using

$$d/dt \int_{0}^{1} K_{ijkl}(z\,, t-\tau)R_{ijkl}(z\,, t-\tau)d\tau = 1 \ , \tag{14}$$

we get the inverse constitutive equations. K_{ijkl} are the components of the creep functions tensor.

PROBLEM 1 (flexure and stretching):

In this case we have to solve the following problem

$$\begin{aligned} \mathbf{u} &= D_1(t)x_1\mathbf{e}_1 + D_2(t)x_2\mathbf{e}_2 - 0.5\left[R_1(t)x_1^2 + R_2(t)x_2^2\right]\mathbf{n} \\ \boldsymbol{\varphi} &= -R_2(t)x_2\mathbf{e}_1 + R_1(t)x_1\mathbf{e}_2 \end{aligned}$$

and the similar three-dimensional boundary value problem. After the comparison of the forces and couples we get

$$\begin{aligned} \left[\tilde{A}_{11}(s)\,;\tilde{B}_{13}(s)\,;\tilde{C}_{33}(s)\right] &= \langle F_1(z\,,s)\left(1\,;-z\,;z^2\right)\rangle \\ \left[\tilde{A}_{12}(s)\,;\tilde{B}_{23}(s)\,;\tilde{B}_{14}(s)\,;\tilde{C}_{34}(s)\right] &= \langle F_2(z\,,s)\left(1\,;-z\,;z\,;z^2\right)\rangle \\ \left[\tilde{A}_{22}(s)\,;\tilde{B}_{24}(s)\,;\tilde{C}_{44}(s)\right] &= \langle F_3(z\,,s)\left(1\,;z\,;z^2\right)\rangle \end{aligned}$$

with

$$\begin{aligned} F_1(z\,,s) &= \left[\tilde{K}_{1111}(z\,,s) + \tilde{K}_{2222}(z\,,s) - 2\tilde{K}_{1122}(z\,,s)\right]/(4s^2\Delta) \\ F_2(z\,,s) &= \left[\tilde{K}_{2222}(z\,,s) - \tilde{K}_{1111}(z\,,s)\right]/(4s^2\Delta) \\ F_3(z\,,s) &= \left[\tilde{K}_{1111}(z\,,s) + \tilde{K}_{2222}(z\,,s) + 2\tilde{K}_{1122}(z\,,s)\right]/(4s^2\Delta) \\ \Delta &= \tilde{K}_{1111}(z\,,s)\tilde{K}_{2222}(z\,,s) - \tilde{K}_{1122}(z\,,s)^2 \\ \tilde{f}(s) &= \int_0^\infty f(t)\exp(-st)dt \quad \text{(Laplace-transformation)} \end{aligned}$$

PROBLEM 2 (plane shear):

In this case we have to solve the following problem

$$\begin{aligned} \mathbf{u} &= B_1(t)(x_2\mathbf{e}_1 + x_1\mathbf{e}_2) - B_2(t)x_1x_2\mathbf{n} \\ \boldsymbol{\varphi} &= -B_2(t)(x_1\mathbf{e}_1 - x_2\mathbf{e}_2) \end{aligned}$$

and the similar three-dimensional boundary value problem. The solution is

$$\left[\tilde{A}_{44}(s)\,;\tilde{B}_{42}(s)\,;\tilde{C}_{22}(s)\right] = \langle(1\,;-z\,;z^2)/(s^2\tilde{K}_{1212})\rangle = \langle(1\,;-z\,;z^2)\tilde{R}_{1212}$$

PROBLEM 3 (torsion):

In this case we have to solve the following problem

$$\begin{aligned} \mathbf{u} &= u_2(x_1\,,t)\mathbf{e}_2\,; \\ \boldsymbol{\varphi} &= -\varphi_2(x_1\,,t)\mathbf{e}_1 \end{aligned}$$

and the similar three-dimensional boundary value problem. After non-trivial calculations we get

$$\tilde{\boldsymbol{\Gamma}}_{22}(s) = \lambda^2[\tilde{A}_{44}(s)\,\tilde{C}_{22}(s) - \tilde{B}_{42}(s)^2]/\tilde{A}_{44}(s)$$

$\lambda(s)$ is the solution of the Sturm-Liouville problem

$$\left[\tilde{R}_{2323}(z\,,s)\tilde{Z}(z\,,s)_{,z}\right]_{,z} + \lambda(s)^2\tilde{R}_{1212}(z\,,s)\tilde{Z}(z\,,s) = 0$$
$$z = \pm h/2\,:\;\tilde{Z}(z\,,s)_{,z}$$

Solving an analogous problem we get the last component

$$\tilde{\Gamma}_{11}(s) = \eta^2\left[\tilde{A}_{44}(s)\,\tilde{C}_{22}(s) - \tilde{B}_{42}(s)^2\right]/\tilde{A}_{44}(s) = 0$$
$$\left[\tilde{R}_{1313}(z\,,s)\,\tilde{Z}_1(z\,,s)_{,z}\right]_{,z} + \eta(s)^2\tilde{R}_{1212}(z\,,s)\,\tilde{Z}_1(z\,,s)$$
$$z = \pm h/2\,:\;\tilde{Z}_1(z\,,s)_{,z}$$

The problems 1 - 3 are the foundation of the determination of all orthotropic effective properties. The main problem now is the retransformation of these values into the real time space (Laplace- retransformation). More details about the identification procedures in the case of orthotropic elastic or isotropic viscoelastic (integral and differential relations, complex moduli) properties are given in [13], [14].

5 Examples

Here we present a simple viscoelastic bending problem for a quadratic homogeneous plate with sinusoidally distributed load. The plate is simply supported and the material has a transverse-isotropic behaviour ($\mathbf{n}$ is the axis of symmetry). In this case the elastic solution is

$$w = w_0\left\{1 + G(3K+G)/[G_n(3K+4G)](h/a)^2\right\} = w_0(1+\sigma)$$

w_0 Kirchhoff solution; G, K shear and bulk modulus in x_1, x_2 directions; G_n transverse shear modulus; a side length of the plate.

The viscoelastic behaviour should be investigated by the following assumptions: in the x_1, x_2 directions we propose elastic behaviour, in the transverse direction - the Maxwell-body behaviour

$$R_{G_n} = G_n \exp(-t/\tau)\,; \quad \tau = \mu_n/G_n$$

μ_n viscosity, τ relaxation time.

Using the formulae of the present paper we get after some calculations

$$w = w_0[1 + (1 + t/\tau)\sigma]$$

In dependence of the materials properties we can obtain the influence of anisotropy and the viscoelastic behaviour on the classical Kirchhoff solution. Other examples are given in [10], [15].

References

[1] Goldenveizer, A.L., *Methods of motivation and increasing accuracy of shell theories*, PMM **32**, (1968), 4, 684-695 (in Russian); (engl. transl. PMM **32**, (1969), 4, 704-718)

[2] Naghdi, P.M., *The Theory of Plates and Shells*, Flügge, S., (ed.), Handbuch der Physik VIa/2 (Festkörpermechanik), S. 425-640, Berlin, Heidelberg, New York, Springer 1972

[3] Wunderlich, W., *Vergleich verschiedener Approximationen der Theorie dünner Schalen (mit numerischen Beispielen)*, Tech.-Wiss. Mitt. d. Instituts für Konstructiven Ingenieurbau der Ruhruniversität Bochum, 73-1

[4] Reissner, E., *Reflections on the theory of elastic plates*, Appl. Mech. Rev. **38**, (1985), 11, 1453-1464

[5] Grigoriuk, E., Kogan, F.A., *Contemporary state of theories of multilayered shells*, Prik. mekh., VIII (1972), 6, 3-17 (in Russian); (engl. transl. Sov. Appl. Mech. **8**, (1974), 6, 583-595)

[6] Dudtchenko, A.A., Lurie, S.A., Obrazcov, I.F., *Anisotropic multilayered plates and shells*, Itogi nauki i tekhniki, Seria "Mekh. tverd. defor. tiela", **15**, 2-63, Moskva, WINITI 1983 (in Russian)

[7] Noor, A.K., Burton, W.S., *Assessment of computational models for multilayered composite shells*, Appl. Mech. Rev. **43**, (1990), 4, 87-97

[8] Zhilin, P.A., *Mechanics of Deformable Directed Surfaces*, Int. J. Solids Struct. **12**, (1976), 635-648

[9] Altenbach, H., *Eine direkt formulierte lineare Theorie für viskoelastische Platten und Schalen* Ing.-Arch. **58**, (1988), 215-228

[10] Altenbach, H., *Determination of given properties for multilayered viisco-plastic plates*, Mekh. komp. mat., (1988), 1, 58-66 (in Russian)

[11] Altenbach, H., Zilin, P.A., *The general theory of elastic simple shells*, Advances in Mechanics **11**, (1988), 4, 107-148 (in Russian)

[12] Christensen, R.M., *Theory of Viscoelasticity*, New York, Academic Press 1971

[13] Altenbach, H., *Dtermination of elastic moduli for plates made of anisotropic, nonhomogenous, in respect to the thickness, material*, Izw. AN SSSR Mekh. tv. tiela (1987), 1, 139-146 (in Russian)

[14] Altenbach, H., *Die Ermittlung effektiver Eigenschaften für viskoelastische ebene Flächentragwerke*, Technische Mechanik **8**, (1987), 1, 33-39

[15] Altenbach, H., *Eine direkt formulierte Biegetheorie für viskoelastische Ein- und Mehrschichtschalen*, Technische Mechanik **7**, (1986), 3, 38-43

The Verification of the Solution of Large Problems of Discontinuous Creeping Structures

R. G. ANDERSON

SIGMA EPSILON TAU
19 The Leylands,
Lytham, Lancashire, FY8 5QS
ENGLAND

Summary

Whilst computer solutions of large problems will be plentiful in the engineering of the 1990s, the establishment of their accuracy will not be easy. This paper draws attention to an exercise conducted in the 1980s to determine the accuracy of specialised codes developed to assess the interaction loads between many slowly deforming structures separated by hundreds of small gaps. Nine countries participated in the exercise and the set of results obtained provides an indication of the accuracy of each solution. The results suggest that the accuracy is not necessarily dependent on the size of the problem. If this conjecture can be supported for large, non-linear problems then benchmarking will become much easier. The effort required will still be large and often on an international scale. The IUTAM may wish to play a role.

Introduction

During the 1980s computer solutions of non-linear, three-dimensional structural mechanics problems became readily available to engineers. Given a large computing budget, three-dimensional solid mechanics problems with material, geometric and boundary non-linearities (such as gaps) can often be solved for a complete life history. As an illustration of a large problem, Squire-Taylor [1] reported modelling the heat transfer and stress analysis of a cylinder-head in three dimensions using 7364 elements associated with 122151 degrees of freedom and an RMS wavefront of 5440. The solution followed the component history and took about 7 hours using the full memory of a CRAY-2 (256 megawords).

Because of the large resources required it is unusual to repeat large calculations unless public safety or large investment risks are involved. The accuracy of the numbers

M. Zyczkowski (Ed.)
Creep in Structures
IUTAM Symposium Cracow/Poland 1990

provided at the end of a long computer run is thus usually unknown. The present paper addresses some of the problems of assessing this accuracy.

<u>What is currently done to assess accuracy for codes?</u>

At the level of single finite elements, the code developers do extensive work to check the output from their elements against hand calculations. Additional, independent tests produced by organisations such as the National Agency for Finite Element Methods and Standards (NAFEMS) in the UK (see [2]) and by individuals eg John Robinson (see [3]) are now accepted by many code developers. The problems considered are always small.

To cover larger problems (which will usually involve a combination of code features), code developers provide EXAMPLE problems. Major code developers will use several hundred EXAMPLE problems (some of which will have exact solutions) as standards against which results from each new release of the code are compared. When there is no exact solution the EXAMPLE problem can then only be used as a comparator. The EXAMPLE and VERIFICATION Manuals for the ABAQUS code [4] record the testing done by one of the major non-linear code developers.

<u>What is known about the accuracy of large problem solutions?</u>

The answer is "In general, very little". This paper puts forward some ideas which may prove useful. We first note that, because an exact solution will not be available, it is necessary to define a solution against which the accuracy can be assessed. This solution is defined as the average of a large number (eg 10) of solutions offered to a well posed problem. These solutions should, ideally, have been obtained by different people using different codes on different computers. Some solutions may be rejected to reduce the overall spread of the results. Such a process reduces the chances of input or code errors. The accuracy of any one solution offered (eg the maximum load) is then related to this average.

These ideas will now be illustrated using a set of solutions which have been obtained for the fast reactor core statics problem. The nature of the problem is outlined in the next section.

The fast reactor core statics problem

Fast reactor cores are made up of an array of hundreds of sub-assemblies which expand thermally as well as swell and creep under neutron irradiation. The boundary of each sub-assembly is a thin hexagonal shell of 0.1m-0.2m across, a few metres long and supported at its lower end. This shell is called a wrapper. The array is separated initially by small gaps between the wrappers but restrained against major movements during the slow deformations and seismic shocks by restraints around the edge of the core. The wrappers distort and take up the gaps between them. As the distortion continues, interactive loads develop between the wrappers and, to save money on fuel, it is very important to understand the mechanics of these interactions.

The fast reactor core statics problem is to understand the interactions in this discontinuous array. The analysis is difficult because, each time a gap closes, the stiffness matrix has to be recalculated. This is a severe non-linearity and, although progress was made in solving small arrays of sub-assemblies using commercial finite element codes, full cores of several hundred sub-assemblies needed specialised coding.

The code intercomparison exercise

By the early 1980s, thirteen organisations in ten countries which had an interest in fast reactors had developed specialised codes to deal with the non-linearity. Principally because of the cost of resourcing a duplicate calculation, most of them agreed to join a benchmarking exercise under the leadership of the International Atomic Energy Agency. The nine participating countries were Belgium, Czechoslovakia, France, the Federal Republic of Germany, India, Italy, Japan (with three separate organisations), the USSR and the UK. The benchmarking covered a range of problems from the calculation of the free bow of a hexagon (for which an analytical solution was available) to the solution of the interactions in a 127 sub-assembly core with no symmetry (for which no known solution was available).

The benchmark exercise began in 1985 and was organised as follows: A set of problems were proposed by RG Anderson and moderated at a Consultant Meeting of 5 participants. The solutions were discussed at a plenary session of all participants at the IAEA Headquarters in Vienna in March 1987. Particular solutions (eg maximum loads and deflections) were sought in order to facilitate discussion and reduce data handling. Five codes were unable to match all the benchmark problems set. A second round of problems were set in 1987, moderated by a Consultant Meeting and the results discussed at PNC-Oarai in Japan in May 1989. By that time the limitations of four of the codes had been substantially removed. The results of the exercise have been summarised by Anderson and Maeda [4].

Some results of the IAEA exercise

The IAEA exercise covered nineteen problems and, because of space limits in this paper, we discuss three problems from those dealing with the behaviour of a core due to the swelling of the wrappers in a neutron flux:

Problem 1: What is the *free bow* of a wrapper made of steel which will swell in a neutron flux gradient? An analtical solution of this problem was available.

Problem 2: What is the maximum *load* in an array of 127 wrappers with 30 degree symmetry? This problem could be

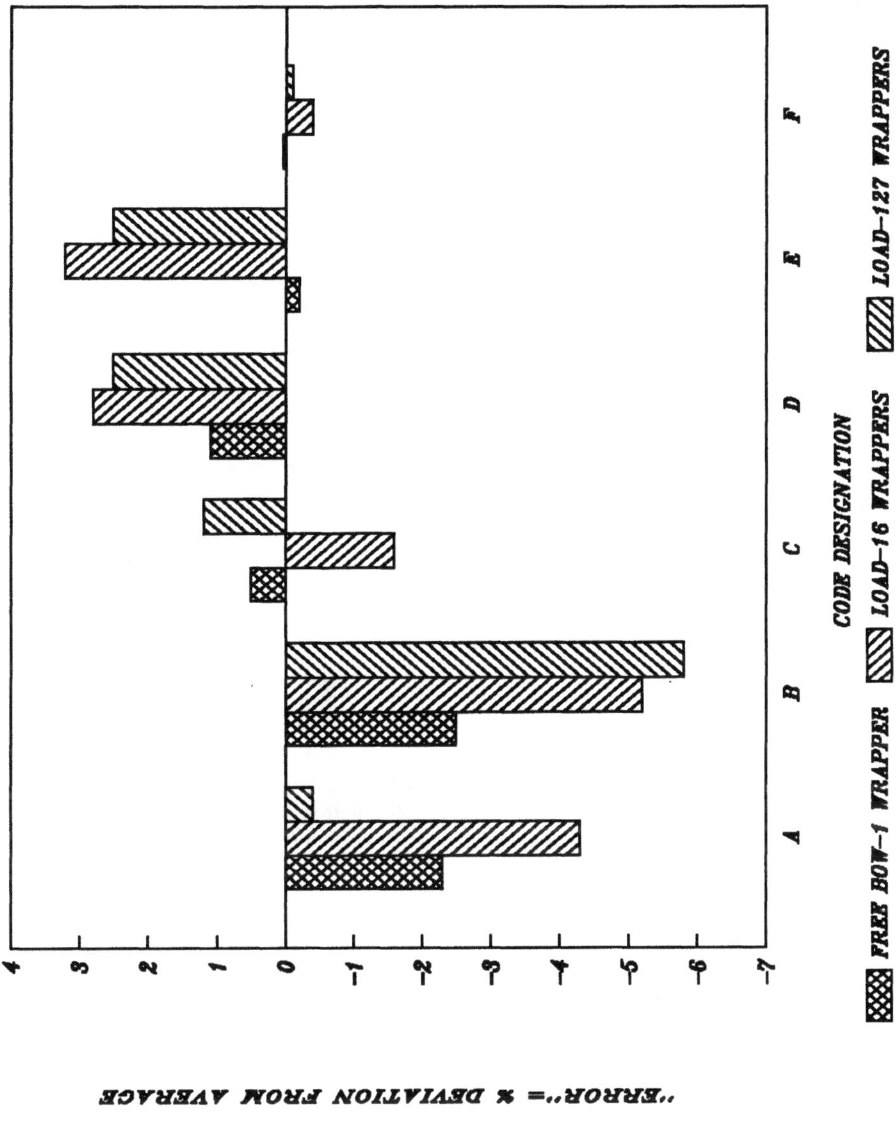
"ERROR"= % DEVIATION FROM AVERAGE
4
3
2
1
0
-1
-2
-3
-4
-5
-6
-7
A
B
C
D
E
F
CODE DESIGNATION
FREE BOW-1 WRAPPER
LOAD-16 WRAPPERS
LOAD-127 WRAPPERS

solved by some commercial finite element codes but hand calculation on the 16 wrappers involved was not attempted.

Problem 3: What is the maximum *load* in an array of 127 wrappers with no symmetry (a typical size of prototype core)? This problem is of the scale needed for calculations on reactors. It requires calculations on all 127 wrappers. No solution by a commercial code has been successful.

Problem 1 concerned the driving force producing the loads in Problems 2 and 3. The average results of the six codes and the corresponding errors were calculated for the three problems. The result is shown in the Figure for the six codes labelled A-F.

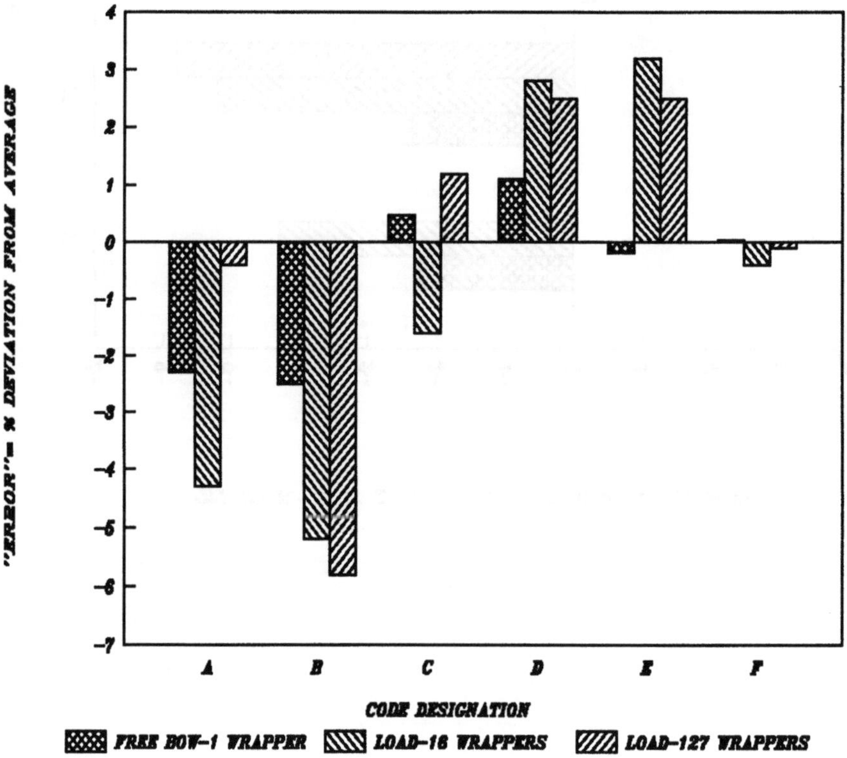

Fig. The errors of the free bow and load solutions.

<u>Some observations on the IAEA exercise</u>

The first observations are on the timescale and the effort. The complete exercise took about 4 years and involved people from nine countries. Two plenary meetings were organised for which the eleven participants had to prepare solutions to 19 problems (perhaps three months effort for each participant) as well as travel to Vienna and to Japan. In addition two Consultant Meetings of about one-half to one-third the size of a plenary session were called to moderate the problems. This scale of effort requires international organisation. The IAEA organise specialised benchmarking for nuclear problems (about one exercise/year) but there is a need for a further international organisation to cover the mechanics field more comprehensively. The IUTAM may be a suitable organisation.

We may turn now to the results from the exercise shown in the Figure. It is first noted that one code (F) obtained results very close to the average for all problems. The worst of the remaining 5 codes underestimated the loads by 6%.

Free-bow (Problem 1) physically produces the loads obtained in Problems 2 and 3 and it may be supposed that errors in free-bow will carry over to the errors in calculating the loads. The errors in the load calculations then range from fractions of a per-cent (code F) up to about three per-cent (codes B and E). Code B underestimates and code E overestimates.

We turn now to an associated observation which may have much more generality. Problem 2 possesses 30 degree symmetry and involves calculations on 16 wrappers. It is the type of calculation which would be done in large numbers to support exploratory design calculations. Problem 3 is a much larger problem involving 127 wrappers and would be useful for operational decisions on reactor handling and for detailed design. Comparison of the errors of Problems 2 and 3 in the Figure show that they are very comparable.

This suggests the extremely important possibility that errors associated with small problems may carry through to much larger problems. If this result can be established in some generality and its limitations explored, it will encourage many more benchmarking exercises to assess errors at the level of Problem 2 ie without a lot of the details which are frequently commercially protected.

Conclusions

1. Considerable work on benchmarking is needed to establish the accuracy of large computer calculations. In order to obtain the necessary resources and diversity of approach, many of the benchmark exercises may need to be conducted internationally- the IUTAM may be an appropiate body.

2. There is evidence that the errors of load prediction may not be dependent on problem size. This conjecture, if accepted, would reduce benchmarking effort substantially.

3. The calculation procedures for loads showed errors of less than three percent with code F showing an error of less than 0.5%.

References

1. Squire-Taylor, C.J. : Modelling and Analysis of a complex Three-dimensional Cylinder Head Structure. ABAQUS Users' Conference Proceedings, Newport RI, May 30-June 1, 1990.

2. BENCHmark : House magazine of NAFEMS, National Engineering Lboratory, East Kilbride, Glasgow G75 0QU, UK.

3. Robinson, J (editor): Finite Element News, Great Bidlake Manor, Bridestowe, Okehampton, Devon EX20 4NT, UK.

4. Hibbitt, Karlsson and Sorensen Inc. EXAMPLE, VERIFICATION Manuals for the ABAQUS code (1990 Version: 4.8), 100 Medway Street, Providence, Rhode Island 02906, USA.

5. Anderson, R.G. and Maeda, K : Verification and validation of LMFBR static core mechanics codes: report on an IWGFR co-ordinated research programme, Paper 25, BNES Conference on Fast Reactor Core and Fuel structural behaviour, Inverness, June 1990.

Creep Analysis in a Large Strain Context by Unified Constitutive Equations

S. CESCOTTO
Department M.S.M.
Université de Liège

Summary

When metals undergo inelastic strains at high temperature, it is well known that creep and plasticity interact and that unified constitutive equations, in which no distinction is made between plastic strains and creep strains are well suited to model such behaviour.
In this paper, we recall briefly the basic equations of unified models and their extension to the case of large strains. Then, we focus mainly on the following topics :
- How can we obtain the material parameters from experimental tests in the case of large strains ?
- What are the numerical problems to be solved when a unified constitutive equation is implemented in a large strain analysis computer code ?

Some numerical simulations illustrate the effectiveness of the proposed approach.

INTRODUCTION

Unified constitutive models for the elastic-visco-plastic behaviour of metal are characterized by the additive decomposition of the total strain rate $\dot{\underline{\epsilon}}$ into an elastic part $\dot{\underline{\epsilon}}^e$ and an inelastic part $\dot{\underline{\epsilon}}^n$:

$$\dot{\underline{\epsilon}} = \dot{\underline{\epsilon}}^e + \dot{\underline{\epsilon}}^n \tag{1}$$

with no distinction between plastic and viscous inelastic strains.
This approach has definite advantages over the traditional one, in which $\dot{\underline{\epsilon}}^n$ is considered as the sum of a plastic contribution $\dot{\underline{\epsilon}}^p$ and a creep contribution $\dot{\underline{\epsilon}}^c$

$$\dot{\underline{\epsilon}}^n = \dot{\underline{\epsilon}}^p + \dot{\underline{\epsilon}}^c \tag{2}$$

$\dot{\underline{\epsilon}}^p$ is computed according to a classical theory by choosing a flow rule and a hardening rule. $\dot{\underline{\epsilon}}^c$ is determined with the help of a creep flow law of the form

$$\dot{\underline{\epsilon}}^c = f(\underline{\sigma}, \underline{\epsilon}^c, T)$$

that can be established experimentally and is capable of representing primary as well as secondary (steady state) creep.

This classical approach suffers two main drawbacks: (1) physically, there is no fundamental difference between "plastic" and "creep" strains; at the microscopic level, both result from dislocation generation, motion or annihilation, so that their separation appears unnatural ; (2) phenomenologically, it has long been recognized that creep and plasticity interact.

For example, figure 1 taken from [1], shows the effect of prior creep on the yield strength of 2 1/2 Cr - 1 Mo steel at 510°C. In this figure, OAB is a

M. Zyczkowski (Ed.)
Creep in Structures
IUTAM Symposium Cracow/Poland 1990

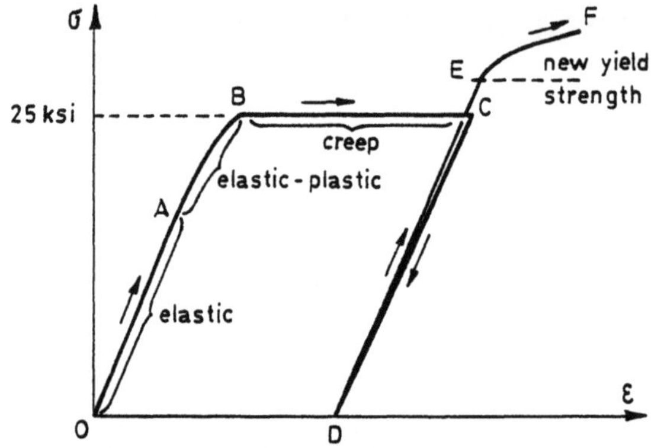

Figure 1

loading at constant strain rate $\dot{\epsilon} = 8{,}33\ 10^{-5}\ s^{-1}$; BC is a creep period of 105 h at a stress level $\sigma = 25$ ksi (172,3 MPa), CD is an unloading at $\dot{\epsilon} = 8{,}33\ 10^{-5}\ s^{-1}$; DEF is a reloading at $\dot{\epsilon} = 8{,}33\ 10^{-5}\ s^{-1}$. It is seen that the creep period has modified the yield strength.

UNIFIED CONSTITUTIVE EQUATIONS

In unified models, the inelastic strain rate is computed by a system of differential equations of the form :

$$\dot{\underline{\epsilon}}^n = f(\underline{\sigma}, q_1, q_2, \ldots, q_N, T) \qquad (3)$$

$$\dot{q}_i = h_i(\underline{\sigma}, q_1, q_2, \ldots, q_N, T) - r_i(\sigma, q_1, q_2, \ldots, q_N, T) \qquad (4)$$

where the q_i denote state variables and T the absolute temperature.
Then, the stresses are given by :

$$\dot{\underline{\sigma}} = \underline{H}\, \dot{\underline{\epsilon}}^e = \underline{H}\, (\dot{\underline{\epsilon}} - \dot{\underline{\epsilon}}^n) \qquad (5)$$

where $\underline{H}$ is Hooke's elastic tensor.
The choice of the state variables q_i and of the corresponding hardening (h_i) and recovery (r_i) functions determine the properties of the model. A comprehensive analysis of such functions can be found in [2] for monotonic and cyclic loadings.

Particular unified models proposed by Perzyna, Lagneborg, Ponter and Leckie, Hart, Miller, Krieg, Robinson, etc... have also been analysed in [3], [4]. Usually the number of material parameters is large, which renders their identification difficult. For the case of small strains, general guidelines are given in [2], and a practical example of experimental determination of material parameters for 304 stainless steel at 600°C under cyclic loading is described in [3], [5].

THE CASE OF LARGE STRAINS

A straightforward generalization of unified constitutive equations for large strains is the following.

Let $\underline{\sigma}$ be the Cauchy (true) stress tensor, $\underline{v}$ the material particle velocity and $\underline{x}$ its coordinate vector in the current deformed configuration of the solid. Let $\underline{L} = \partial \underline{v}/\partial \underline{x}$ be the velocity gradient.
Then, $\underline{L} = \underline{D} + \underline{W}$ with $\underline{D} = \mathrm{sym}(\underline{L})$ and $\underline{W} = \mathrm{skw}(\underline{L})$.
$\underline{D}$ is the strain rate tensor in the current configuration of the body and it can be decomposed it into an elastic and an inelastic parts

$$\underline{D} = \underline{D}^e + \underline{D}^n \qquad (6)$$

The generalization of (5) becomes :

$$\overset{o}{\underline{\sigma}} = \underline{H}\,\underline{D}^{e} = \underline{H}\,(\underline{D} - \underline{D}^{n}) \qquad (7)$$

where ^{o} denotes an objective time derivative.

Again, the inelastic strain rate is computed by differential equations :

$$\underline{D}^{n} = f\,(\underline{\sigma}, q_1, q_2, \ldots, q_N, T) \qquad (8)$$

$$\overset{o}{q_i} = h_i\,(\underline{\sigma}, q_1, q_2, \ldots, q_N, T) - r_i\,(\sigma, q_1, q_2, \ldots, q_N, T) \qquad (9)$$

Since state variables can be of tensorial nature, their time derivative must also be objective.

Jaumann's, Green-Naghdi's, Truesdell's, Oldroyd's or other objective time derivatives can be used but, in some instances, non-physical stress oscillations are obtained under monotonously increasing strains in simple shear [6, 7, 8]. Hence, the choice of an objective time derivative is still very controversial and will not be discussed here.

DETERMINATION OF MATERIAL PARAMETERS.

In the domain of large strains, classical tensile tests or creep tests in tension are not appropriate because of the striction of the specimen. The two best experiments are the upsetting of a cylinder (figure 2) and the torsion of a cylindrical bar (figure 3). In the first case, it is necessary to use a lubricant to reduce the barrelling effect which is responsible for a non uniform strain field in the cylinder. However, numerical simulations [9] of upsetting tests show that the friction coefficient between cylinder and plateau has little influence on the upsetting force up to 50 % height reduction. So, the practical strain range of this test is of $|\epsilon| \le 0.7$. The influence of strain rate can be studied with this test. Classical machines can reach strain rates up to 0.1 per s while cam plastometers reach 100 to 1000 per s. If the specimen can be heated, (furnace, Joule or electromagnetic effect) the influence of temperature can also be obtained. On the other hand the torsion test is usually performed in a furnace and is capable to produce equivalent strains up to $\epsilon_{eq} = 20$ in a strain rate range of $0 \le \epsilon_{eq} \le 1000$ per s.

Figure 2

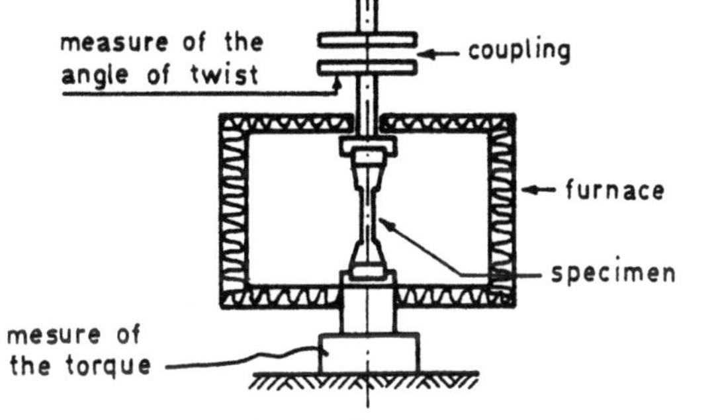

Figure 4

Figure 3

The main advantage of the upsetting test and of the torsion test is that the strain field in the specimen is very simple, so that an analytical formulation can be used to analyse the test results and to deduce the material parameters, at least for unified constitutive equations which are not too complex. When the complexity and the number of parameters increase, an optimization procedure must be designed to find the best fit of the model with the experiments. An example of such a procedure is described in [10].

On the other hand, it is also possible to use extrusion tests (figure 4) which produce very large strains (up to $\epsilon_{eq} = 1000$) and allow to study the influence of strain rate and temperature. Here, the analysis of experimental results is based on an approximate analytical solution proposed by Avitzur [11] which allows to calculate the yield limit of the material from the evolution of the extrusion pressure σ_o during the process [12]. From tests performed at different velocities and for different area reductions, it is then possible to determine the parameters of an unified constitutive equation. It is clear that upsetting tests and extrusion tests apply only in case of monotonic loading, while the torsion test could be used also for cyclic loading.

EXAMPLE OF TEMPERATURE DEPENDENT ELASTIC VISCO PLASTIC LAW.

As an example of unified constitutive equation, we give hereafter a material law valid at high temperature for steels submitted to large strains under monotonic loading. It is an extension of the material law presented in [10]. In this case, only one scalar state variable K is necessary. Then equation (7) becomes :

$$\overset{o}{\hat{\sigma}}_{ij} = 2G\,(\hat{D}_{ij} - \hat{D}^{n}_{ij})\ ;\quad \dot{\sigma}_m = 3\chi\, D_m \tag{10}$$

with

$$D_m = \frac{1}{3} D_{ii}\ ;\ \hat{D}_{ij} = D_{ij} - D_m\,\delta_{ij}\ ;\ \sigma_m = \frac{1}{3}\sigma_{ii};\ \hat{\sigma}_{ij} = \sigma_{ij} - \sigma_m\,\delta_{ij}$$

G = shear modulus, χ = bulk modulus.

The inelastic strain rate is

$$\hat{D}^{n}_{ij} = \Phi\, \hat{\sigma}_{ij}/\sqrt{J_2} \quad \text{with} \quad J_2 = \frac{1}{2}\hat{\sigma}_{ij}\,\hat{\sigma}_{ij} \tag{11}$$

$$\Phi = 0 \text{ if } J_2 \le K^2 \quad \text{or} \quad \Phi = B(\sqrt{J_2} - K)^n \text{ if } J_2 > K^2 \tag{12}$$

The evolution of the state variable is given by

$$\dot{K} = 2\, h_K\, \Phi/\sqrt{3} \qquad \text{with} \quad h_K = \theta(1 - K/K_S) \tag{13}$$

K_S is the hardening saturation level calculated by $K_S = C(\Phi)^{m'}$.

The material parameters of this model are temperature dependent. If T is the absolute temperature and r the gaz constant, we have :

Young's modulus	$E = E_o \exp(-b_E\, T)$
Poisson's ratio	$\nu = \nu_o \exp(+b_\nu\, T)$
Strain rate exponent	$n = n_o/T$
Strain rate coefficient	$B = B_o \exp(-Q/rT)$
Hardening saturation exponent	$m' = T/m_o$
Hardening saturation coefficient	$C = K_{So}\, [\frac{2/\sqrt{3}}{\gamma_o}]^{m'}$
Hardening coefficient	$\theta = \theta_o \exp(-b_\theta\, T)$
Initial yield limit	$K_o = K_{oo} \exp(-b_K\, T)$

The constants E_o, b_E, ν_o, b_ν, n_o, B_o, Q, m_o, K_{SO}, γ_o, θ_o, B_θ, K_{oo}, b_K are calibrated on torsion tests according to the optimization procedure described in [10]. Obviously, they depend on the steel grade.
To illustrate this constitutive law, a set of tensile tests are simulated. A specimen of initial length L = 20 is strained until its length becomes l = 30, that is up to a logarithmic strain $\epsilon = \ell n\ 1.5 = 0.41 = 41\ \%$.
Two temperatures (T = 800°C = 1073°K and T = 1200°C = 1473°K) and three test durations (t = 10s, t = 100s, t = 1000s) are considered.
The material parameters are (units: N, mm) : $E_o = 285900$:
$\nu_o = 0.255$; $n_o = 4768$; $B_o = 25125$;
$Q = 219274$; $m_o = 8691$: $K_{so} = 3090$;
$\theta_o = 10554$; $K_{oo} = 398$; $b_E = 0.76\ 10^{-3}$;
$b_\nu = 0.264\ 10^{-3}$; $\gamma_o = 0.108\ 10^{13}$
$b_\theta = 0.275\ 10^{-2}$: $b_K = 0.248\ 10^{-2}$.

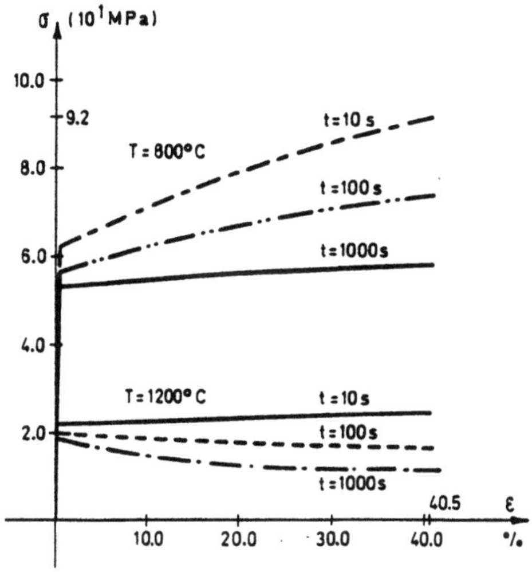

Figure 5

NUMERICAL ASPECTS.

The use of unified constitutive equations in non linear finite element requires the solution of some crucial numerical problems related to the stiff character of these equations and to the large strains. They are briefly examined hereafter.

Choice of a strain path between two neighbouring configurations.

If γ_A is a know configuration of the creeping solid at time t_A and γ_B is a trial configuration at time $t_B = t_A + \Delta t$ in the usual step by step procedure, the constitutive equation has to be integrated between γ_A and γ_B. If the coordinates $\underline{x}_A$ and $\underline{x}_B$ of a material particle are known, however, its path during the time interval Δt is unknown and many choices are possible [13]: constant velocity path, constant velocity gradient path, final rotation path. Here, an original approach is proposed. It will be called the "creep path". For simplicity, it will be presented for the case of a function Φ given by :

$$\Phi = B\ (\sqrt{J_2}/K)^n \tag{14}$$

with K constant but the extension to K variable is easy.
If $\underline{U}$ is the right stretch tensor obtained by the polar decomposition of the Jacobian tensor, the natural strain tensor is $\underline{\epsilon} = \ell n\ \underline{U}$.
Knowing $\underline{\sigma}_A$, $\underline{\epsilon}_A$ and $\underline{\epsilon}_B$, we try to calculate $\underline{\sigma}_B$. To this end, we assume that the strain path between $\underline{x}_A$ and $\underline{x}_B$ is such that :

$$\underline{\sigma}(\xi) = f_A(\xi)\ \underline{\sigma}_A + f_B(\xi)\ \underline{\sigma}_B \tag{15}$$

with $\xi = (t - t_A)/(t_B - t_A)$, $0 \le \xi \le 1$ and the conditions

$$f_A(0) = 1\ ;\ f_A(1) = 0\ ;\ f_B(0) = 0\ ;\ f_B(1) = 1$$

A straightforward calculation fives, for the hydrostatic stress

$$\sigma_{mB} = \sigma_{mA} + 3\chi\ (\epsilon_{mB} - \epsilon_{mA}) \tag{16}$$

while, for the deviatoric part of the equations, we have :

$$2G\ (\hat{\underline{\epsilon}}_B - \hat{\underline{\epsilon}}_A) = \hat{\underline{\sigma}}_B - \hat{\underline{\sigma}}_A + C \int_{t_A}^{t_B} [J_2(\xi)]^p\ \hat{\underline{\sigma}}_B\ dt \tag{17}$$

with $C = 2GB/K^n$; $p = (n - 1)/2$

Equation (17) can be simplified if $f_A(\xi)$ and $f_B(\xi)$ are chosen as shown on figure 6.

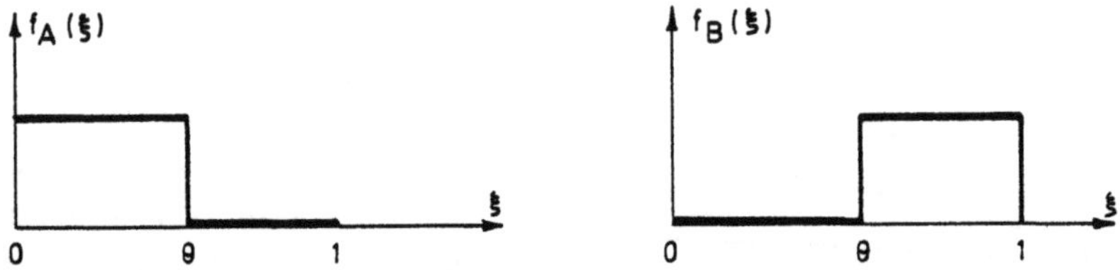

Figure 6

Then, we obtain :

$$2G\ (\hat{\underline{\epsilon}}_B - \hat{\underline{\epsilon}}_A) = \hat{\underline{\sigma}}_B - \hat{\underline{\sigma}}_A + C.\Delta t[\theta(J_{2A})^p\ \hat{\underline{\sigma}}_A + (1-\theta)(J_{2B})^p\ \hat{\underline{\sigma}}_B] \tag{18}$$

In particular, if $\theta = 1$, (18) gives

$$\hat{\underline{\sigma}}_B = \hat{\underline{\sigma}}_A\ [1 - C.\Delta t\ (J_{2A})^p] + 2G\ (\hat{\underline{\epsilon}}_B - \hat{\underline{\epsilon}}_A) \tag{19}$$

This result can be interpreted as follows : between t_A and t_B, the material particle is allowed to creep under constant stress $\underline{\sigma}_A$.
Hence (10) and (14) give :

$$\hat{\underline{D}} = \hat{\underline{D}}^n = B\ (\sqrt{J_{2A}}/K)^n\ .\ \hat{\underline{\sigma}}_A/\sqrt{J_{2A}} = \frac{B}{K^n}\ (J_{2A})^p\ \hat{\underline{\sigma}}_A \tag{20}$$

At the end of this creep period, the deviatoric strain is :

$$\hat{\underline{\epsilon}}^{*B} = \hat{\underline{\epsilon}}_A + \hat{\underline{D}}\ \Delta t = \hat{\underline{\epsilon}}_A + (C.\Delta t/2G)\ (J_{2A})^p\ \hat{\sigma}_A \tag{21}$$

Then, at time t_B, a stress $\Delta\hat{\underline{\sigma}} = 2G\ (\hat{\underline{\epsilon}}_B - \hat{\underline{\epsilon}}^*_B)$ is applied instantaneously (i.e elastically) to restore the strain $\hat{\underline{\epsilon}}_B$.
Finally :

$$\hat{\underline{\sigma}}_B = \hat{\underline{\sigma}}_A + \Delta\hat{\underline{\sigma}} = \hat{\underline{\sigma}}_A + 2G\ (\hat{\underline{\epsilon}}_B - \hat{\underline{\epsilon}}_A) - C.\Delta t\ (J_{2A})^p\ \hat{\underline{\sigma}}_A \tag{22}$$

which is equivalent to (19). This shows that the choice of the "creep path" gives an analytical solution to the integration of the unified constitutive equation during a time step.

Numerical integration of unified equations along a path.

If another path is chosen, the stresses and state variables are computed by numerical integration of (7), (8), (9) along this path. Classical integration schemes (explicit, implicit, sub-steps...) can be used. Since these equations are usually stiff, implicit scheme with or without sub-steps are required to avoid oscillations and divergence of the solution [13].

Incidence constitutive tensor.

The stresses $\underline{\sigma}_B$ depend on the numerical integration scheme and on the strain path between γ_A and γ_B. During the classical equilibrium iterations, the coordinates $\underline{x}_B$ are progressively modified until the out-of-balance forces in γ_B vanish. If $d\underline{x}_B$ is the correction at a given iteration, the corresponding stress correction $d\underline{\sigma}_B$ also depends on the integration scheme and the strain path. Let $d\underline{L}_B$ be the modification of the velocity gradient induced by $d\underline{x}_B$. The incidence constitutive tensor $\underline{C}_T$ is defined by :

$$d\underline{\sigma}_B = \underline{C}_T \, d\underline{L}_B \, \Delta t \tag{23}$$

This tensor can be calculated analytically or by numerical perturbation on $\underline{L}$. It indicates how stresses are modified by a perturbation of γ_B. It must be distinguished from the usual tangent constitutive tensor often used in small strains elasto-plastic computations $\underline{C}^{ep}$. For elastic-visco-plastic materials, the latter does simply not exist.
The introduction of $\underline{C}_T$ allows to extend the notion of tangent stiffness matrix to the case of unified constitutive equations and, consequently, to develop a unified approach of elasto-plastic and elastic-visco-plastic problems in non-linear computer codes [14].

APPLICATION.

The creep of a circular plate (radius = 10 mm; thickness = 1 mm) under uniform pressure is studied at two temperatures (600° C and 1000° C) with the help of the temperature dependent elastic-visco-plastic law described previously.

The pressure increases linearly from 0 to 1.85 N/mm^2 in 1,85 s and then is maintained constant. The data are such that, at 600°C, the deformations remain small and the bending effect is dominant while, at 1000°C, large deflections occur and membrane effects develops.

Figure 7 shows the plate deflection at 1000°C at the end of the loading (time = 1,85 s) and after creep (time = 39,7 s). The radial stresses σ_R at points 1 and 2 near the center are compared on figure 8. The creep and the membrane effects (at T = 1000°C) are clearly seen.

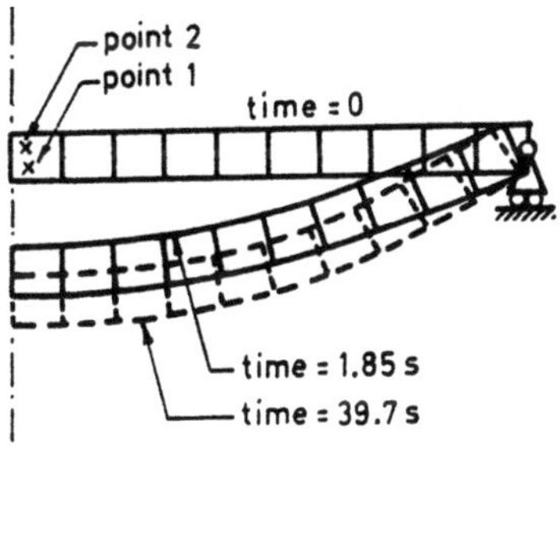

Figure 7

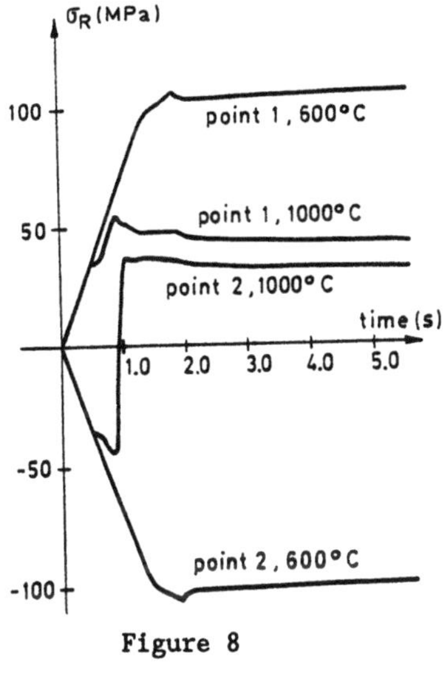

Figure 8

REFERENCES.

1. Onat, E.T.: Representation of inelastic behavior, ORNL-SUB-3863-2, Nov. 1976, Yale University.

2. Lemaitre, J.; Chaboche, J-L.: Mecanique des matériaux solides, Dunod, 1985.

3. Cescotto, S.; Leckie, F.; Abrahamson, E. : Unified constitutive models for creep and plasticity of metals at high temperature, T.A.M. report, University of Illinois at Urbana-Champaign (U.S.A.), nov. 1982.

4. Cescotto, S.; Leckie, F.: Analysis of unified constitutive equations for metals at high temperature, in Num. Meth. in Industrial Forming Processes, Ed. J.F.T. Pittman et al., pp. 461-470, Pineridge Press, Swansea, U.K., 1982.

5. Cescotto, S.; Leckie, F.: Deetermination of Unified Constitutive Equations for Metals at High Temperature, Proc. Int. Conf. on Constitutive Equations for Engineering Materials, Tucson, U.S.A., Jan. 1983.

6. Nagtegaal and De Jong : Some aspects on non isotropic workhardening in finite strain platicity, Proc. Workshop on Plasticity of Metals at Finite Strains, pp. 65 - 102, Lee and Mallett Ed., Standford University, 1981.

7. Dogui : Plasticité anisotrope en grandes déformations, Thèse de Doctorat d'Etat présentée devant l(Université Claude Bernard, Lyon 1, 1989.

8. Cescotto, S. and Habraken, A-M.: A note on the response to simple shear of elasto-plastic materials with isotropic hardening, to appear in European Journal of Mechanics, A / Solids.

9. Belkhiri, L.; Charlier, R.; Detraux, J.M.: On the large deformation of axisymmetric bodies : experimental and numerical studies, Proc. Int. Conf. NUMETA, University College Swansea, U.K., Jan. 1985.

10. Cescotto, S.; Grober, H.: Calibration and Application of an Elastic Viscoplastic Constitutive Equation for Steels in Hot-Rolling Conditions, Engineering Computations, Vol. 2, n° 2, June 1985.

11. Avitzur, B.: Analysis of Metal Extrusion, J. of Eng. for Ind., Feb.1965, pp. 57 sqq.

12. Cescotto, S.: Détermination expérimentale des paramètres d'une loi constitutive visco-plastique pour métaux dans le domaine de l'extrusion, rapport final, Convention R. et D. dans le cadre du contrat IRSIA n° 4398/1986/1988, janvier 1988.

13. Cescotto, S.: General Strategy for Nonlinear Finite Element Formulation, Proc. 5th Int. Symposium on Numerical Methods in Engineering, Ed. R. Gruber et al, Springer Verlag, 1989, Vol. 1, pp. 155-164.

14. Cescotto, S.; Charlier, R.: Numerical simulation of elasto-visco-plastic large strains of metals at high temperature, Proc. 8th Conf. on SMIRT, Brussels, 1985.

The Large Time Increment Method Applied to Cyclic Loadings

J.-Y. COGNARD, P. LADEVEZE

Laboratoire de Mécanique et Technologie
(E.N.S. de Cachan / C.N.R.S. / Université Paris 6)
61, avenue du Président Wilson - 94235 CACHAN CEDEX (France)
G.R.E.C.O.- G.I.S. "Calcul des Structures"

Summary
A new computational method is proposed to deal with non-linear mechanical behaviours for cyclic loadings, especially for behaviours described by internal variables (plasticity, viscoplasticity, ...). This approach contrast with the classical step by step method. It is an iterative procedure which accounts for the whole loading process in a single time increment, without any a priori hypothesis. To give an idea of the step length, several hundred cycles can be simulated in a single increment with the large time increment method. The CPU time high performances of this method are clearly evidenced when the number of degrees of freedom is large. Moreover, for this method, the use of parallel computers is easy and very efficient.

Introduction

Life prediction of structures submitted cyclic loadings is of great interest for high-tech industries (nuclear, space, ..). But for industrial structures, non-linear numerical simulations are today very hard to perform when complex loading histories are considered (several hundred cycles can be computed). Moreover, more and more accurate behaviour modellings, with numerous internal variables, are necessary to describe correctly cyclic behaviour including multiaxial and out-of-phase loading conditions, ratcheting effects, ...[5].
The classical way consists in splitting the loading path into small increments (step by step method). The history of variables being known up to a time "t", a loading time "Δt" is prescribed. The problem is then to complete the history of the variables on the time interval $[t, t + \Delta t]$, i.e. to compute them at a time "$t + \Delta t$". The last problem is time-independent and can be solved with Newton-type algorithms. Moreover, accurate numerical integration methods are used in order to ensure the stability [2], [4]. Strongly non-linear constitutive relations require a sufficiency fine discretization of the loading path, especially under cyclic conditions to get accurate numerical results. And so complex simulations cannot be kept between sensitive limits [4]. Even to obtain the limit response in strain and stress, simplified procedures are not easy to develop and to implement [1], [8], [10], [12].
The LArge Time INcrement method (LATIN method) introduced by P. Ladevèze [10] contrast with the step by step methods, and allows an important reduction of the numerical cost [3], [7]. It is an iterative procedure which accounts for the whole loading process in a single time

M. Zyczkowski (Ed.)
Creep in Structures
IUTAM Symposium Cracow/Poland 1990

increment, which is not a priori limited. To give an idea, several hundred cycles can be simulated in a single increment. Iterations are not associated with the time description of the loading process. One of the main idea is to treat the non-linearities locally so that the global problems are linear. The other key-point of the LATIN method is the choice of appropriate space-time representations of the various functions defined over space and time.
The structure of the new LATIN method version adapted to behaviours described by internal variables, which convergence has been proved, is described. The performances are presented on a simple example with to a large number of cycles in case of viscoplastic behaviour.

Presentation of the problem

We consider a structure which occupies a domain Ω. At each time $t \in [0, T]$. The displacement U_d is given on the part $\partial_1\Omega$ of the boundary of Ω. The surface force F_d is given on the complementary part $\partial_2\Omega$. A body force f_d is applied on Ω. For quasi-static response the structure evolution is described by the problem :

Find the couple $s = (\dot{\varepsilon}(U(M,t)), \sigma(M,t))$, $M \in \Omega$, $t \in [0,T]$, verifying $\forall\, t \in [0,T]$:

- kinematic equation $U|_{\partial_1\Omega} = U_d +$ "regularity" (U kinematically admissible, KA)
- equilibrium equation $\int_\Omega \mathrm{Tr}[\,\sigma \,.\, \varepsilon(U^*)\,]\, d\Omega = \int_\Omega f_d \,.\, U^*\, d\Omega + \int_{\partial_2\Omega} F_d . U^*\, dS$
  $\forall\, U^* \in \{\, U ,\, U|_{\partial_1\Omega} = 0 \,\}$ (σ statically admissible, SA)
- constitutive relations $\dot{\varepsilon}(t) = \mathcal{A}\,(\sigma\,(\zeta)\,,\, \zeta \le t)$ $\mathcal{A}$ characteristic of the material.

The LArge Time INcrement method

According to the properties of the mechanical equations, the difficulties are splited into two groups :

- linear equations which are possibly global in space,
- equations which are local in space and possibly non linear.

Classically we note $\mathbf{A_d}$ the linear subspace of couples s verifying the admissibility conditions (static and kinematic). The couples s checking the constitutive relations is a non-linear subspace noted Γ. So the problem is to find the intersection of $\mathbf{A_d}$ and Γ. The LATIN method start with an initial couple s_0 element of $\mathbf{A_d}$ (for instance, the elastic response of the structure over [0,T]), and build couples successively belonging to Γ and to $\mathbf{A_d}$, up to converge to the couple solution denoted by s_{ex} (Fig. 1). An iteration, i.e. the building of a new admissible couple s_{n+1} from a given admissible couple s_n, requires two stages :

- local stage : find $\hat{s} \in \Gamma$ such as $\hat{s} = (\Gamma) \cap (\, s_n + E^+ \,)$
- global stage : find $s_{n+1} \in A_d$ such as $s_{n+1} = (A_d) \cap (\, \hat{s} + E^- \,)$

E^+ is a given direction, linear and local in space. E^- is a given linear direction .

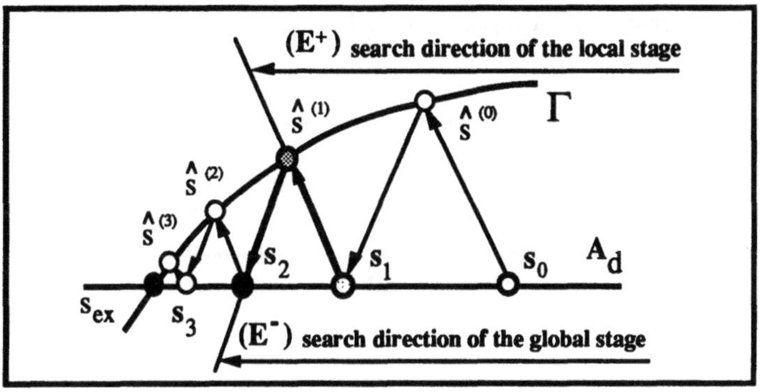

Fig. 1. Presentation of the LATIN method

For materials which verify the Drucker's inequality, the conditions about E^+ and E^- which ensure the convergence of the LATIN method are developed in [9]. The directions E^+ and E^- are parameters of the LATIN method and so we can define different versions [3], [6]. For example we can use the tangent rigidity tensor $\mathcal{R}_t$ associated with $\hat{s} \in \Gamma$ to define the search directions [3]. The different stages are as follows :

- local stage : find $\hat{s} \in \Gamma$ such as $\quad \dot{\hat{\varepsilon}} - \dot{\varepsilon}_n = - \mathcal{R}_t^{-1} \frac{d}{dt} (\hat{\sigma} - \sigma_n)$
- global stage : find $s_{n+1} \in A_d$ such as $\quad \dot{\varepsilon}_{n+1} - \dot{\hat{\varepsilon}} = + \mathcal{R}_t^{-1} \frac{d}{dt} (\sigma_{n+1} - \hat{\sigma})$

For the local stage, at each time t of [0, T] the problem to solve is local in space but non-linear. On the other hand for the global stages, associated to kinematic and static admissibilities, the problems are linear and global, but time parametered. So to solve them, at a low cost, a key-point of the method is to choose appropriate space-time representations of the various functions defined over [0, T] x Ω. For example, in the case of small strains and quasi-static approximations, the correction of the variables $x_n(t, M)$ can be approximated by :

$$x_{n+1}(t, M) - x_n(t, M) = \sum_{i=1}^{m} \alpha_i(t)\, a_i(M) \qquad (x = \sigma \text{ or } \varepsilon)$$

In practice, the number of the unknown functions $\alpha_i(t)$ and $a_i(M)$ can be limited to m = 1 or 2 at each global stage. These fields and time functions are entirely determined by the algorithm.

For structures submitted to complex loading (several cycles), and for large number of degrees of freedom the convergence has always been obtained after a few iterations (nearly five), in case of elastoplastic behaviour. Comparisons with classical step by step method (table 1) show the efficiency of the method. A structure with 18000 degrees of freedom submitted to an increasing loading was studied, and only one global matrix has been inversed for the different simulations (for this example the platified part of the structure is important). Other results of this method are presented in [3] and [7], particularly its performances on parallel computers are shown.

method	BFGS	LATIN
nb. of resolutions	650	70
CPU time (ALLIANT)	40 000 s	12 000 s

Table 1: comparison with incremental methods

Case of cyclic loadings

For cyclic loadings it is possible to increase the performances of the first versions of the LATIN method introducing new search direction $\mathbf{E}^+$ and $\mathbf{E}^-$. The idea is to use the real tangent subspace to Γ to define the search directions $\mathbf{E}^+$ and $\mathbf{E}^-$ (an element of Γ is defined over $[0, T] \times \Omega$). The two main difficulties, existence of this tangent subspace and convergence of the algorithm, have been solved by a modification of the mathematical frame [11]. The classical modeling of behaviours described by internal variables is changed in order to introduce the concept of "normal formulation" which is characterized by a quadratic free energy. It is shown that most of the present plasticity and visco-plasticity models admit such a formulation.

This new method is presented in case of Marquis-Chaboche's model with kinematic and isotropic hardening, well adapted to cycled processes. The free energy is such as :

$$\rho\Psi = \frac{1}{2}\mathrm{Tr}\left[\mathbf{K}_e\, \varepsilon^e\, \varepsilon^e\right] + \frac{1}{2}\mathrm{Tr}\left[\alpha\, \mathrm{C}\, \alpha\right] + \mathrm{G(p)}$$

where ε^e is the elastic strain tensor, $\mathbf{K}_e$ the elastic operator, p the isotropic hardening parameter and α the kinematic one (G and C are material characteristics). The associated forces are noted R and $\mathbf{x}$, the equations of state are :

$$\sigma = \frac{\partial\rho\Psi}{\partial\varepsilon^e} = \mathbf{K}_e\, \varepsilon^e \,; \qquad \mathbf{x} = \frac{\partial\rho\Psi}{\partial\alpha} = \mathrm{C}\,\alpha \;; \qquad \mathrm{R} = \frac{\partial\rho\Psi}{\partial p} = \mathrm{G'(p)}$$

The viscoplastic potential is taken as :

$$\varphi^* = \frac{k}{n+1}\langle z\rangle_+^{n+1}; \quad z = \|\sigma^D - \mathbf{x}\| + \frac{1}{2}\,a\,\mathrm{Tr}[\mathbf{x}\mathrm{C}^{-1}\mathbf{x}] - \mathrm{R} - \mathrm{R}_o \;; \quad \|y\|^2 = \mathrm{Tr}[y.y]$$

C, a , R_o, k and n are material parameters and $\langle y\rangle_+$ represent the positive part of y. If we note $\dot\varepsilon = \dot\varepsilon^e + \dot\varepsilon^p$, the evolutions laws are :

$$-\dot{p} = \frac{\partial\varphi^*}{\partial \mathrm{R}} = -\,k\,\langle z\rangle_+^n; \qquad \dot\varepsilon^p = \frac{\partial\varphi^*}{\partial\sigma} = \dot{p}\,\frac{\sigma^D - \mathbf{x}}{\|\sigma^D - \mathbf{x}\|} \;; \qquad -\dot\alpha = \frac{\partial\varphi^*}{\partial\mathbf{x}} = -\,\dot\varepsilon^p + \dot{p}\,a\,\mathrm{C}^{-1}\mathbf{x}$$

These evolution laws are strongly non-linear. A change of variables, which can be proposed for a large class of models, is introduced so that the previous state equations become linear [11]. For the proposed model the change of variables is such as :

$$\tilde{R} = A\,\tilde{p} \quad (A > 0)\,; \quad \tilde{p} = \int_0^p \left(\frac{\partial R}{\partial p}\frac{1}{A}\right)^{1/2} dp$$

This "normal formulation" allows us to introduce a new partition of the equations. The first group of equations is associated with free energy and the second one with dissipation. Let us note again **s** the set of variables :

$$\mathbf{s} = (\dot{\varepsilon}^p, \underline{\dot{X}}\,;\, \sigma, \underline{Y}) \qquad M \in \Omega \text{ and } t \in [0,T]$$

$$\underline{X} = (\alpha, \tilde{p})^T \text{ (internal variables)} \qquad \underline{Y} = (\mathbf{x}, \tilde{R})^T \text{ (associated thermodynamic forces)}$$

We note $\mathbf{A_d}$ the set of couples **s** such as :

$$\sigma\,SA\,;\; \dot{\varepsilon} = (\mathbf{K}_e^{-1}\,\dot{\sigma} + \dot{\varepsilon}^p)\,KA \qquad \text{admissibility conditions}$$

$$\underline{X} = \Lambda\,\underline{Y} \qquad \text{equations of state } (\Lambda \text{ linear operator})$$

And we note Γ the set of couples **s** such as :

$$\frac{d}{dt}\begin{bmatrix} \varepsilon^p \\ -\underline{X} \end{bmatrix} = \mathbf{B}\begin{bmatrix} \sigma^D \\ \underline{Y} \end{bmatrix} \qquad \text{evolution laws (}\mathbf{B}\text{ non-linear operator)}$$

The general framework of the LATIN method can still be used, only search directions have to be defined. The real tangent subspace to Γ is introduce, for $\mathbf{s} \in \Gamma$ we note **L** the operator verifying the relation :

$$\begin{bmatrix} \Delta\dot{\varepsilon}^p \\ -\Delta\underline{\dot{X}} \end{bmatrix} = \mathbf{L}\begin{bmatrix} \Delta\sigma^D \\ \Delta\underline{Y} \end{bmatrix}$$

ΔU means the differentiation of U. The different stages of the method are :

- local stage : find $\hat{\mathbf{s}} = (\hat{\dot{\varepsilon}}^p, \hat{\underline{\dot{X}}}\,;\, \hat{\sigma}, \hat{\underline{Y}}) \in \Gamma$ such as

$$\frac{d}{dt}\begin{bmatrix} \hat{\varepsilon}^p - \varepsilon^p_n \\ -(\hat{\underline{X}} - \underline{X}_n) \end{bmatrix} = \mathbf{L}^T\begin{bmatrix} \hat{\sigma}^D - \sigma^D_n \\ \hat{\underline{Y}} - \underline{Y}_n \end{bmatrix} \qquad \text{(search direction)}$$

- global stage : find $\mathbf{s}_{n+1} = (\dot{\varepsilon}^p_{n+1}, \underline{\dot{X}}_{n+1}\,;\, \sigma_{n+1}, \underline{Y}_{n+1}) \in \mathbf{A_d}$ such as

$$\frac{d}{dt}\begin{bmatrix} \varepsilon^p_{n+1} - \hat{\varepsilon}^p \\ -(\underline{X}_{n+1} - \hat{\underline{X}}) \end{bmatrix} = \mathbf{L}\begin{bmatrix} \sigma^D_{n+1} - \hat{\sigma}^D \\ \underline{Y}_{n+1} - \hat{\underline{Y}} \end{bmatrix} \qquad \text{(search direction)}$$

Remark : for plasticity the operator **L** does not exist, but plasticity can be considered as a limiting case of viscoplasticity.

At each iteration the local stage requires the resolution of non-linear and local space problems at each point of the structure. These problems can be reduced to simple non-linear equations where time t is a parameter over [0, T]. Similar technics to those defined for the previous versions ([3], [6]) can be used to solve the global stage at a relatively low cost. Here we must introduce representations of the corrections for the different variables (i.e. $\dot{\varepsilon}^p$, $\underline{\dot{X}}$, σ, $\underline{Y}$). Sum of products of space fields by scalar time functions gives very good results for quasi-static problems. The form of the time functions adapted to the problem can be chosen, for cyclic loadings functions described with two time scales can be used (a long and a short characteristic time) [8]. Efficient numerical methods, such as Fast Fourier Transform or wavelet technics [13] must be adapt to these time representations.

At present time some adjustments of these representations of the time functions are being carried out in order to develop an efficient computational method for cyclic loadings.

Numerical results

The first results of the new version of the LATIN method are presented on a simple example (fig.2 : uniaxial problem). As for complex structures, an important stress redistribution is observed. For this example we have not used representations of the corrections, the only goal of such examples is to test the method's performances. The computation was carried out over 200 cycles. An excellent accuracy is reached after only six iterations.

Results are presented in case of isotropic hardening and for only 25 cycles sine shaped loading in order to simplify the different figures. Similar numerical results are obtained for a loading path of 25 cycles and for 200 cycles. The elastic initialization is shown fig.3 (thin lines represent the short bar, thick lines the long one). The 6th iteration results are presented in fig.4 with the same scale. At this iteration, the convergence is reached, the two couples s_n and $\hat{s}$ are equal and represent the solution of the problem. Figure 5 present the 2nd local stage results for the short bar (full lines represent the value of $s_n \in A_d$, dashed lines represent the value of $\hat{s} \in \Gamma$). At each iteration of the method, corrections are carried out over the whole interval [0,T], and we can note important corrections during the two first iterations (for the elastic initialization we have $\dot{\varepsilon}^p = 0$ and $\dot{p} = 0$ over [0,T]).

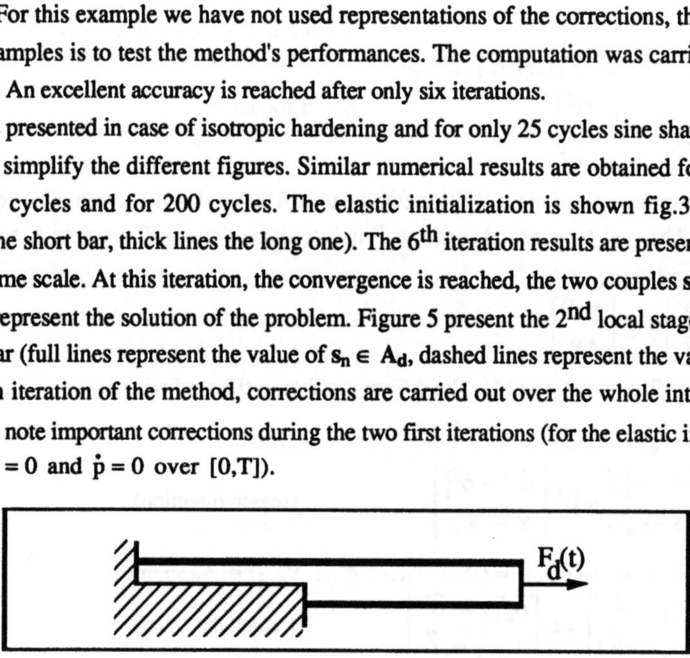

Fig. 2. Structure studied

The parameters of the problem studied are such as :

$G'(p) = A\,[1 - \exp(-\gamma\, p)]$ (isotropic hardening law)

$\tilde{p} = \frac{2}{\gamma^{1/2}}[1 - \exp(-\frac{\gamma\, p}{2})]$ $R = A\,(\frac{\tilde{R}}{A}\frac{\gamma^{1/2}}{2})[2 - \frac{\tilde{R}}{A}\frac{\gamma^{1/2}}{2}]$ (change of variables)

$E = 20\,000$ daN mm^{-2} $R_0 = 10$ daN mm^{-2} $A = 10$ daN mm^{-2}

$\gamma = 10$ $n = 10$ $k = 1.10^{-5}$

ratio of the bar lengths = 1.3 ratio of the sections = 1. T = 20 s

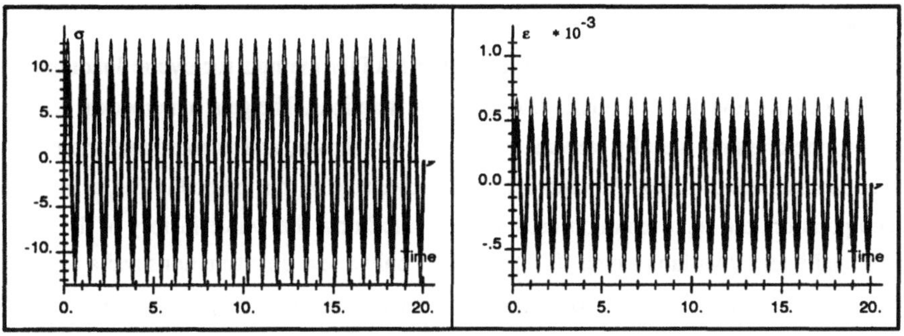

Fig. 3. Elastic initialization

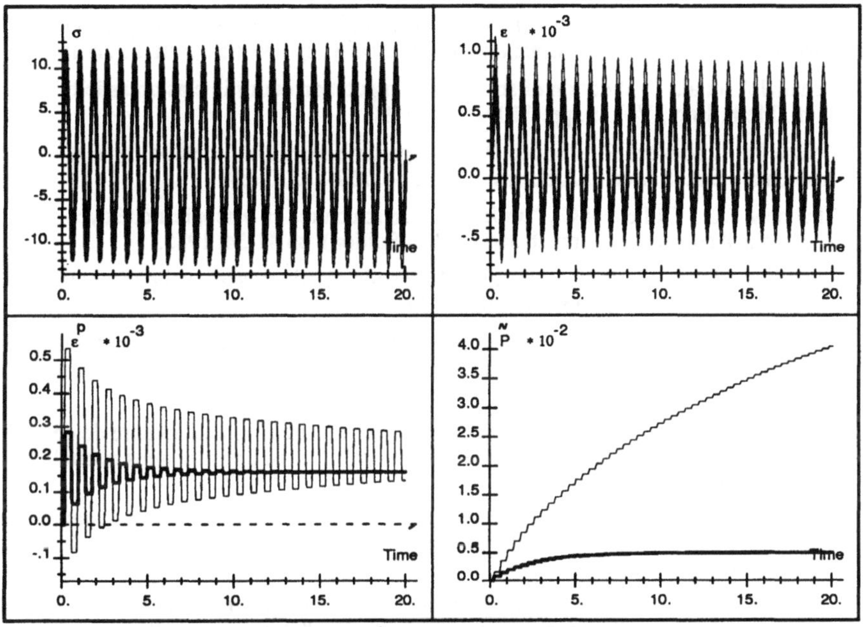

Fig. 4. Results of the 6th iteration (solution of the problem)

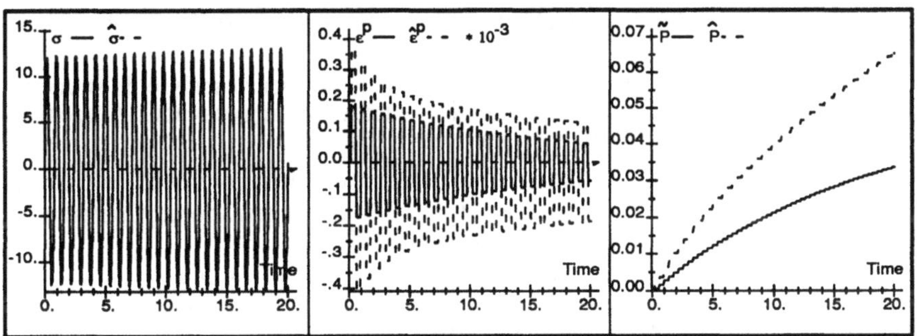

Fig. 5. Results of the 2th local stage (short bar)

Conclusion

The results presented show the feasibility of the LATIN method for cyclic loadings. The convergence have always been obtain after very few iterations. The gain in numerical cost with respect to the classical step by step methods has been shown, and is more important as the loading path is more complex. An other important advantage of this method is that, at the end of the computation, the representation used allows us to know the whole history of the strains, stresses and internal variables over [0, T] x Ω with only a few time functions and a few space fields. It is particularly interesting to analyse the results, especially for cyclic loadings.

References

1. Akel S., Nguyen Q.S., "Determination of the limit response in cyclic plasticity", 2nd International Conference on Computational Plasticity, pp 639-650, Barcelonna 1989.
2. Benallal A. et al., "Validation of structural computation codes in elastoplasticity", Int. J. Num. Meth. Eng., Vol 29, pp. 1109-1130, 1990.
3. Boisse Ph., Ladevèze P., Rougée P. "A large time increment method for elastoplastic problems", Eur. J. Mech., A/Solids, 8, n°4, pp. 257-275, 1989.
4. Chaboche J.L., Cailletaud G., "sur le calcul de structures en viscoplasticité cyclique", La Recherche Aérospaciale (French and English Edition), pp 41-54, 1985.
5. Chaboche J.L., "Constitutive equations for cyclic plasticity and cyclic viscoplasticity", Int. J. of Plasticity, n°5, pp. 245-302, 1989.
6. Cognard J. Y., "Une nouvelle approche des problèmes de plasticité et de viscoplasticité : la méthode à grand incrément de temps". Thèse de l'Université Pierre et Marie Curie, 1989.
7. Cognard J. Y., "Le traitement des problèmes non linéaires à grand nombre de degrés de liberté par la méthode à grand incrément de temps", Calcul des structures et intelligence artificielle, Vol 4, J.M. Fouet, P. Ladevèze, R. Ohayon, Pluralis 1990.
8. Guennouni T., "Sur une methode de calcul de structures soumises à des chargements cycliques : l'homogénéisation en temps", Mathematical Modelling and Numerical Analysis, n°3, Vol 22, pp. 417-455, 1988.
9. Ladevèze P., "Sur une famille d'algorithmes en mécanique des structures", Comptes-rendus, Académie des Sciences - Paris, 300, série II, n° 2, pp. 41-44, 1985.
10. Ladevèze P., Rougée P., "Plasticité et viscoplasticité sous chargement cyclique : propriétés du cycle limite", Comptes-rendus, Académie des Sciences - Paris, 301, série II, n°13, pp. 891-894, 1985.
11. Ladevèze P., "La méthode à grand incrément de temps pour l'analyse de structures à comportement non linéaire décrit par variables internes", Comptes-rendus, Académie des Sciences - Paris, 309, série II, n°11, pp. 1095-1099, 1989.
12. Lesne P. M., Savalle S., "An efficient cycles jump technique for viscoplastic structure calculations involving large number of cycles", 2nd International Conference on Computational Plasticity, pp. 591-602, Barcelonna 1989.
13. Perrier V., Basdevant C., "La décomposition en ondelettes périodiques, un outils pour l'analyse de champs inhomogènes. Théorie et algorithmes", La Recherche Aérospaciale (French and English Edition), pp. 53-67, 1989.

Some Analytic Results Pertaining to Stress Redistribution in a Class of Structures Undergoing Primary or Secondary Creep

Warren S. Edelstein
Illinois Institute of Technology, Chicago 60616

Summary

Upper and lower bounds are derived and various qualitative properties are deduced for solutions of a class of transient creep problems.

1. General Constitutive Relations

Infinitesimal strains ϵ_{ij} have the form

$$\epsilon_{ij} = \epsilon_{ij}^{(e)} + \epsilon_{ij}^{(c)}$$

where $\epsilon_{ij}^{(e)}$ and $\epsilon_{ij}^{(c)}$ denote elastic strains and creep strains respectively. These are related to the stresses σ_{ij} by the equations

$$\epsilon_{ij}^{(e)} = \frac{1}{E}[(1+\nu)\sigma_{ij} - \nu\delta_{ij}\sigma_{kk}]$$

$$\epsilon_{ij}^{(c)}|_{t=0} = 0$$

$$\dot{\epsilon}_{ij}^{(c)} = \frac{F(\sigma_e)}{[\epsilon_e^{(c)}]^m} \cdot s_{ij}, t > 0 \tag{1}$$

Here, s_{ij} are the deviatoric components of the stress given by $s_{ij} = \sigma_{ij} - \frac{\delta_{ij}}{3} \cdot \sigma_{kk}$ while σ_e and $\epsilon_e^{(c)}$ are the effective stress and effective creep strain respectively defined by $\sigma_e = \sqrt{\frac{3}{2} s_{ij} s_{ij}}$ and $\epsilon_e^{(c)} = \sqrt{\frac{2}{3}\epsilon_{ij}^{(c)}\epsilon_{ij}^{(c)}}$. It is assumed that

$$E > 0, -1 < \nu \leq \frac{1}{2}, m \geq 0,$$

that F is C^2, and that* for $z > 0$

$$F(z) > 0, \frac{d}{dz}[zF(z)] > 0,$$

$$zF'(z) \geq mF(z).$$

Special cases: When F takes the form

$$F(z) = \frac{3K}{2} z^{n-1} (K > 0),$$

(1) becomes the Odqvist-Hult primary creep power law. In this case, the last restriction on F reduces to

$$n \geq m + 1,$$

a condition which occurs repeatedly in the primary creep literature. When m=0, we get secondary creep.

*There is an additional, more complicated assumption on F, [1].

M. Zyczkowski (Ed.)
Creep in Structures
IUTAM Symposium Cracow/Poland 1990

2. Results for Primary Creep

Consider a right-cylindrical beam with a uniform cross-section represented by a region R in the x,y plane. Generators are parallel to the z-axis. Assume lateral boundaries are traction-free and that resultant loads and moments on the ends are all zero, except that

$$-\int_R x\sigma_{zz}dA = M(t).$$

R is symmetric with respect to the x and y axes. It is assumed that

$$M(0) \leq 0; M < 0, \dot{M} \leq 0 \quad (0 < t < \infty),$$

so that the region $x > 0$ is in a state of tension. As in the elastic case, one assumes a solution of the form

$$\sigma_{xx} = \sigma_{yy} = \sigma_{xy} = \sigma_{xz} = \sigma_{zy} = 0$$

$$\sigma_{zz} = \sigma(x,t).$$

On physical grounds, we consider only solutions for which

$$\sigma > 0, \epsilon_{zz}^{(c)} > 0$$

in the half-beam $0 < x < c$, where 2c is the x-diameter of the cross section. Then it can be shown [1] that σ satisfies an integral equation of the form

$$\frac{\sigma(x,t)}{E} = -\frac{M(t)x}{EI} + \frac{x}{I}\int_0^c [\int_0^t G(\sigma)d\tau]^{\frac{1}{m+1}} k(\xi)d\xi$$

$$-[\int_0^t G(\sigma)d\tau]^{\frac{1}{m+1}}.$$

Here,

$$G(\sigma) = \frac{2}{3}(m+1)F(\sigma)\sigma$$

$$I = \int_0^c xk(x)dx$$

where k(x) depends only on the geometry of R. With this integral equation and the assumption that the limit $M(\infty)$ exists one can *formally* derive the following expression for the limiting value of σ

$$\sigma(x,\infty) = \frac{-M(\infty)x^{\frac{m+1}{n}}}{\int_0^c \xi^{\frac{m+1}{n}} k(\xi)d\xi}.$$

The above integral equation can also be used to derive a priori bounds for σ. Once σ is bounded, one can then bound all quantities of physical interest. Bounds for σ are obtained using the following simple

Lemma. *Let $y(t)$ satisfy the integral equation*

$$y(t) + k_1(t)\int_0^t k_2(\tau)y(\tau)d\tau = f(t)(t > 0)$$

where k_1, k_2 and f are nonnegative on $(0,\infty)$. *Suppose that k_1 is nonincreasing and f is nondecreasing on* $(0,\infty)$. *Then y is nonnegative on* $(0,\infty)$.

Since $\sigma(x,t)$ is assumed to be continuous and odd in x, $\sigma(0,t) = 0$ for all t. This together with arguments based on the above Lemma, gives the bounds

$$0 \leq \sigma(x,t) \leq -\frac{cM(t)}{I}.$$

This type of reasoning also gives information about the geometry of the stress profile. In fact, if we define $\phi(x,t) = \int_0^t G(\sigma(x,\tau))d\tau$, then differentiation of the integral equation with respect to x yields

$$\frac{1}{E}\frac{\partial \sigma}{\partial x} + \frac{1}{m+1}\phi^{\frac{-m}{m+1}} \int_0^t G'(\sigma)\frac{\partial \sigma}{\partial x} d\tau =$$

$$\frac{1}{I}\Big(-\frac{M(t)}{E} + \int_0^c \phi^{\frac{1}{m+1}}(\xi,t)k(\xi)d\xi\Big).$$

Using the assumptions on G and M, and applying the Lemma to this equation, we can conclude that

$$\frac{\partial \sigma}{\partial x} \geq 0 \qquad (0 < x \leq c, t \geq 0).$$

This, in fact, is one of the inequalities used to obtain the above bound on σ. If one makes the additional assumptions $m = 0, G'' \geq 0$ one can also show that

$$\frac{\partial^2 \sigma}{\partial x^2} \leq 0 \qquad (0 \leq x \leq c, t > 0).$$

Analogous results [2] exist for other simple geometries. My first results of this type were, in fact, for internally loaded, hollow symmetric pressure vessels of inner radius a and outer radius b. In this case, the basic integral equation is

$$\sigma(r,t) = \frac{\beta}{r^j}\Big(P(t) + \mu \int_a^b \Big[\int_0^t \sigma^n(\xi,\tau)d\tau\Big]^{\frac{1}{m+1}} \frac{d\xi}{\xi}\Big)$$

$$-\mu\Big[\int_0^t \sigma^n(r,\tau)d\tau\Big]^{\frac{1}{m+1}}$$

where

$$\beta^{-1} = \int_a^b \frac{d\xi}{\xi^{j+1}}, \qquad P(t) = (\frac{1}{2})3^{\frac{(3-j)}{2}} \cdot p(t).$$

Here, $j = 2$ for hollow incompressible cylinders, $j = 3$ for hollow spheres, $p(t)$ is the internal pressure and σ stands for effective stress. Also, the constant μ is different for the two cases. P is assumed to be positive and nondecreasing. By virtue of the above-stated Lemma, one obtains the basic bounds

$$\frac{\beta P(t)}{b^j} \leq \sigma(r,t) \leq \frac{\beta P(t)}{a^j} \qquad (a \leq r \leq b, t \geq o).$$

From these, follow bounds on all relevant stresses, strains and displacements. For example, the paper gives explicit upper and lower bounds on $u(b,t)$.

Corresponding bounds have also been obtained for the problems of torsion in circular cylinders [3] and the torsion of thin strips [4].

3. Results for Secondary Creep

All of the above-mentioned results also hold for secondary creep, since it simply represents the special case $m = 0$. However, in this mathematically simpler case, additional and much stronger results have been obtained. The secondary creep problems corresponding to all of the geometries described above are contained in the following unified equation [3]

$$s(r,t) = \frac{r^l}{I}\left(N(t) + \int_0^t \int_a^b H(s)q(\xi)d\xi d\tau\right) - \int_0^t H(s)d\tau \qquad (2)$$

where

$$I = \int_a^b \xi^l q(\xi)d\xi. \qquad (a \le r \le b, t \ge 0)$$

In all cases, the function $s(r,t)$ is equal or proportional to the effective stress. Depending on the circumstances, N(t) is proportional to an internal pressure or an applied moment. It is assumed that

$$N > 0, \dot{N}(t) \ge 0 \qquad (0 < t < \infty)$$

$$q(r) > 0 \quad (a < r < b), \quad l \neq 0, 0 < I < \infty$$

The basic constitutive assumption is that

$$H(0) = 0, H'(z) > 0, H''(z) > 0 \qquad (z > 0)$$

Notice that this is **much more general** than simple power law creep. It follows from the assumptions on H that

$$H'(z)z - H(z) > 0 \qquad (0 < z < \infty).$$

It can be shown using (2) that two sequences of functions, $s_k(r,t)$ and $\sigma_k(r,t)$ exist such that s_k is monotone nonincreasing, σ_k is monotone nondecreasing, i.e.,

$$s_1 \ge s_2 \ge \dots, \qquad \sigma_1 \le \sigma_2 \le \dots$$

and both sequences converge uniformly to the exact solution s(r,t) of (2) on $[a,b] \times [0,\infty)$. The sequence of descending approximations is defined recursively by the equation

$$s_{k+1} + \int_0^t H'(s_k)s_{k+1}d\tau =$$

$$\frac{r^l}{I}\left(N + \int_0^t \int_a^b H(s_k)q(\xi)d\xi d\tau\right) + \int_0^t [H'(s_k)s_k - H(s_k)]d\tau.$$

Here, we have modified the last term in (2) by means of a Newton-type approximation

$$H(s) \simeq H(s_k) + H'(s_k)(s_{k+1} - s_k).$$

In order to complete the definition of s_k in such a way that the sequence will have the desired properties, we need an appropriate choice of s_0. One which works is

$$s_0(r,t) = J\left[\left(\frac{r}{m}\right)^l H\left(\frac{m^l N}{I}\right)\right], \qquad m^l = \max_{[a,b]} r^l$$

where J is the functional inverse of H on $[0,\infty)$.

It is desired to show that

$$s_k > 0, \qquad s_{k+1} \le s_k \qquad (k = 0, 1, 2, \ldots; t > 0)$$

on $[a, b] \times [0, \infty)$. Our basic tool is the fact that if v satisfies the equation

$$v(t) + \int_0^t Q(\tau)v(\tau)d\tau = f(t) \qquad (t \ge 0)$$

where $f \ge 0$ is continuous and nondecreasing on $[0, \infty)$ and $Q \ge 0$ and continuous, then $v \ge 0$ on $[0, \infty)$. If, in addition, $f > 0$ on $(0, \infty)$, then $v > 0$ on $(0, \infty)$. Applying this observation to the iteration equation, we see that, if $s_k > 0$, then so is s_{k+1}. But $s_0 > 0$ by construction. Therefore, $s_k > 0$ for all k by mathematical induction. To show that the sequence is descending, we write

$$s_k - s_{k+1} + \int_0^t H'(s_k)(s_k - s_{k+1})d\tau =$$

$$\frac{r^l}{I}\int_0^t \int_a^b [H(s_{k-1}) - H(s_k)]q(\xi)d\xi d\tau$$

$$+ \int_0^t [H'(s_{k-1})(s_{k-1} - s_k) - H(s_{k-1}) + H(s_k)]d\tau.$$

Suppose that $s_{k-1} \ge s_k$ for some $k \ge 1$. We already know that $H'(s_k) > 0$. Therefore, it suffices to show that the two integrands on the right-hand side are nonnegative. One sees this immediately in the case of the first integral, since H is monotone increasing. As for the second integral, it follows from the Mean Value Theorem that at points where $s_{k-1} > s_k$,

$$H'(s_{k-1})(s_{k-1} - s_k) - [H(s_{k-1}) - H(s_k)] =$$

$$[H'(s_{k-1}) - H'(\xi)](s_{k-1} - s_k)$$

for some $s_k < \xi < s_{k-1}$. But H' is an increasing function. It is now established that $s_k - s_{k+1} \ge 0$, provided that $s_{k-1} - s_k \ge 0$. One completes the proof by showing that $s_0 - s_1 \ge 0$, the details of which are here omitted. In order to construct a sequence σ_k of lower bounds, we define

$$\sigma_0(r, t) = J\Big[\Big(\frac{r}{\mu}\Big)^l H\Big(\frac{\mu^l N}{I}\Big)\Big], \qquad \mu^l = \min_{[a,b]} r^l;$$

$$\sigma_{k+1} + \int_0^t \Big[\frac{H(s_0) - H(\sigma_k)}{s_0 - \sigma_k}\Big]\sigma_{k+1} d\tau =$$

$$\frac{r^l}{I}\Big(N + \int_0^t \int_a^b H(\sigma_k)q(\xi)d\xi d\tau\Big)$$

$$+ \int_0^t \Big[\frac{H(s_0)\sigma_k - H(\sigma_k)s_0}{s_0 - \sigma_k}\Big]d\tau.$$

Here, the second integral on the right-hand side of (2) has been modified by means of the "regula falsi" approximation

$$H(s) \simeq H(\sigma_k) + \Big[\frac{H(s_0) - H(\sigma_k)}{s_0 - \sigma_k}\Big](\sigma_{k+1} - \sigma_k).$$

The construction of the upper bound and lower bound sequences is heuristically justified by the following simple diagram.

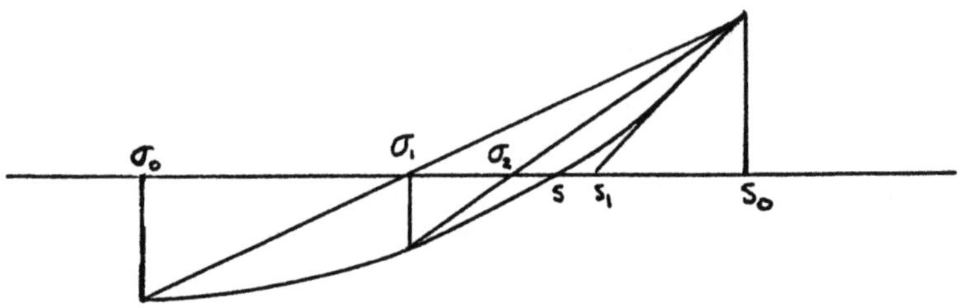

By means of the methods sketched above, one establishes the inequalities

$$\sigma_k < s_0, \quad \sigma_k \le \sigma_{k+1} \quad (k = 0, 1, 2, ...; t > 0).$$

Using the Monotone Convergence Theorem plus Dini's Theorem, one can show that the sequence s_k converges uniformly to a solution s of (2) while σ_k converges uniformly to a solution σ of (2). Since it can be shown that (2) has at most one continuous solution, it follows that $s = \sigma$ identically in $[a, b] \times [0, \infty)$. Denoting this unique solution by s, we have

$$\sigma_k \le s \le s_k \quad (k = 0, 1, 2, ...).$$

For example, let N be constant for $t \ge 0$ and set $k = 1$ in the above inequalities. There results the bounds

$$s(r,t) \le s_0(r) + \left[\frac{r'N}{I} - s_0(r)\right] exp[-tH'(s_0)],$$

$$s(r,t) \ge \sigma_0(r) + \left[\frac{r'N}{I} - \sigma_0(r)\right] \exp\left[-t\left(\frac{H(s_0) - H(\sigma_0)}{s_0 - \sigma_0}\right)\right],$$

where s_0 and σ_0 are as defined above.

Applications of Bounds and Related Results

1. Can be applied directly to provide bounds for a variety of simple problems.
2. Can be used to furnish rough bounds for more complicated geometries
3. Can be used to validate computer codes, not only by means of bounds but also using related geometric information on stress profiles.

4. Skeletal Points

Consider a creeping metal structure which is subject to a *time-independent* loading. As before, let $s(r,t)$ denote a one-dimensional stress profile (e.g. effective stress through the wall of a pressure vessel). In general, if one plots the initial elastic stress $s(r,0)$ and the steady-state creep stress $s(r,\infty)$ on the same graph, patterns like the following result:

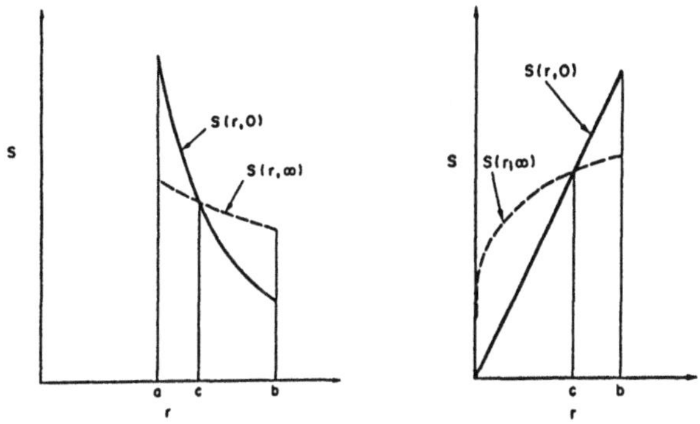

The point of cross-over, which occurs at $r = c$, was named the *skeletal point* by Marriott and Leckie in a 1964 paper [5]. An important engineering assumption is that $s(c,t)$ is constant for all time. It was discovered by Professor Jan Hult that this is not true. This result [6], which I helped to generalize, holds for power-law primary and secondary creep. The proof can be obtained by applying equation (2) in the secondary creep case or its primary creep generalization together with Hölder's inequality.

5. Relaxation Time

The *relaxation time* $\tau(r)$ estimates the time required for the stress at point r to realize its limiting value $s(r,\infty)$. It is defined by

$$\tau(r) \equiv \frac{s(r,\infty) - s(r,0)}{\dot{s}(r,0)}, \qquad (a \leq r \leq b).$$

This definition is motivated by the following picture:

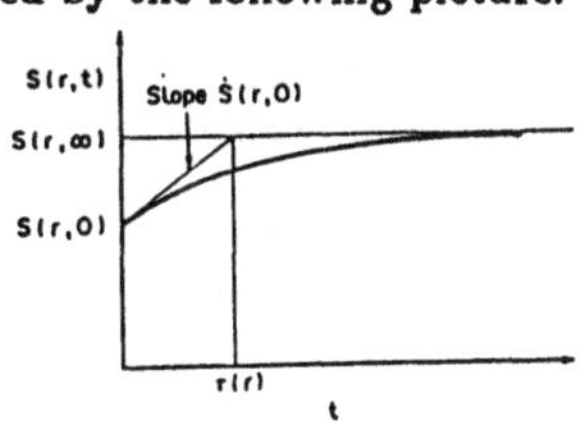

The problem which I considered [6], which was suggested by Professor Hult, concerns relaxation times in a one dimensional continuum consisting of an infinite number of infinitely thin vertical bars which occupy the unshaded portion of the following diagram:

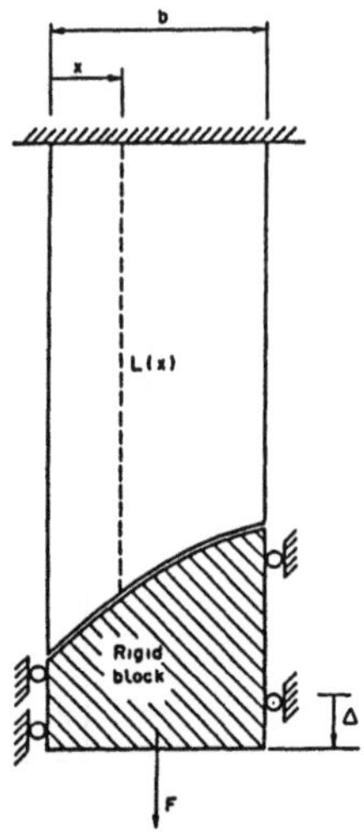

$L(x)$ is the initial length of the bar at position x. Its reciprocal $f(x) \equiv L^{-1}(x)$ is called the *shape function*. A question which naturally arises in the design of such a structure under creep conditions is whether a nonconstant shape function $f(x)$ exists such that all of the bars relax with the same relaxation time, i.e., such that $\tau(x) \equiv$ *constant*. It turns out that if $\tau(x)$ is identically constant, then $f(x)$ *cannot take three distinct values*. The proof assumes power law secondary creep. It can also be shown that if f is any integrable step function which takes two distinct values, then $\tau(x)$ is identically constant.

6. References

1. W.S. Edelstein, On Transient Creep Bounds for Saint-Venant Pure Bending Problems. *Int. J. Solids Structs.* **15**, 659 (1979).

2. W.S. Edelstein, On Bounds for Primary Creep in Symmetric Pressure Vessels. *Int. J. Solids Structs.* **12** 107 (1976).

3. W.S. Edelstein and P.G. Reichel, On Bounds and Approximate Solutions for a Class of Transient Creep Problems. *Int. J. Solids Structs.* **16**, 108 (1980).

4. W.S. Edelstein, On Transient Creep Bounds and Approximate Solutions for the Torsion of Thin Strips. *Int. J. Solids Structs.* **16**, 1053 (1980).

5. D.L. Marriott and F.A. Leckie, Some Observations on the Deflections of Structures During Creep. *Proc. I. Mech. E.* **178**, 115 (1964).

6. W.S. Edelstein and J. Hult, On Stress Redistribution in Structures During Creep. *Int. J. Solids Structs.* **19**, 915 (1983).

On Optimal Design of Disk with Respect to Creep Rupture

A. GANCZARSKI and J. SKRZYPEK

Institute of Mechanics and Machine Design,
Cracow University of Technology, Poland

Summary

The paper deals with optimal structural design in creep conditions, introduced for the first time by Życzkowski (1968) and Prager (1968). The objective of the present paper is the analysis of the effect of the form of damage law on the optimal shape of disk with respect to time to rupture under unsteady creep conditions. It is the extension of the authors' previous results (Ganczarski and Skrzypek (1990)) to the cases of both scalar and tensor descriptions of creep rupture with a propagation of failed zones admitted. The clamped annular disk is subject to combined loadings: the radial tension, the body forces due to rotation and the temperature field. The tensor formulation of the damage law results in slight overestimation of the time to rupture initiation τ_I of optimal disk when compared to a scalar one. More advantages are achieved if the propagation of the fronts of damage with respect to circumferential component of damage tensor is considered until a complete failure mechanism is reached, τ_{II}.

1 Formulation of the problem

Consider an axially-symmetric disk of jump-like variable thickness as shown in Fig.1. The annular disk, clamped at the inner edge, is subject to rotation about the axis of symmetry with a constant angular velocity ω .The disk is simultaneously loaded by the external normal tension and the nonhomogeneous field of temperature. The assumptions of the plane stress state and the small strains are applied. The basic equations, written for each constant thickness N^{th} disk portion, take the following form (cf. Penny and Marriott (1971))

$$\frac{\partial^2 u}{\partial r^2}+\frac{1}{r}\frac{\partial u}{\partial r}-\frac{u}{r^2}=-kr+(1+\nu)\alpha\frac{\partial t}{\partial r} \qquad for\ \tau=0\ ,$$
$$\frac{\partial^2 \dot{u}}{\partial r^2}+\frac{1}{r}\frac{\partial \dot{u}}{\partial r}-\frac{\dot{u}}{r^2}=-\frac{\dot{f}}{r}+\frac{\partial \dot{g}}{\partial r} \qquad for\ \tau>0\ , \tag{1}$$

where for a stationary field of temperature the functions $\dot{f}, \dot{g}$ and k are defined as follows

$$\dot{f}=(1-\nu)(\dot{\varepsilon}_r^c-\dot{\varepsilon}_\theta^c)\ , \quad \dot{g}=\dot{\varepsilon}_r^c+\nu\dot{\varepsilon}_\theta^c\ , \quad k=\frac{1-\nu^2}{E}\rho\omega^2\ , \tag{2}$$

and the total strains are decomposed into the elastic (superscript "e") and the creep (superscript "c") parts: $\varepsilon_r=\varepsilon_r^e+\varepsilon_r^c+\alpha t$, $\varepsilon_\theta=\varepsilon_\theta^e+\varepsilon_\theta^c+\alpha t$.

M. Zyczkowski (Ed.)
Creep in Structures
IUTAM Symposium Cracow/Poland 1990

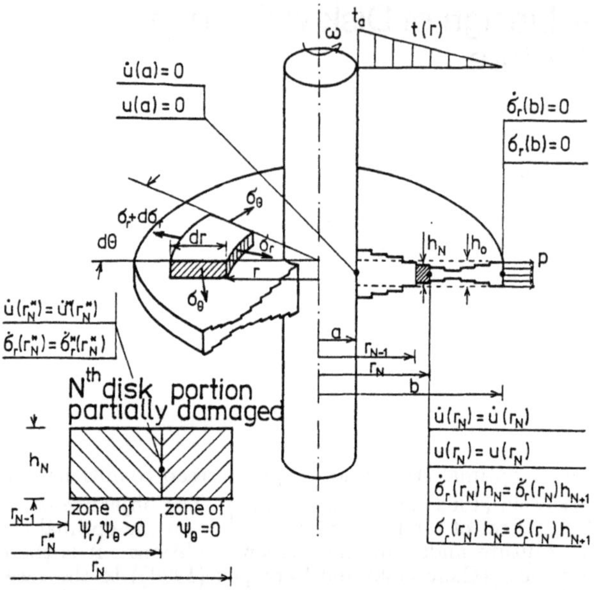

Fig.1. Axially-symmetric disk of a jump-like variable thickness subject to radial tension, rotation and temperature field

The similarity of deviators from the flow theory and the time hardening hypothesis associated with the Kachanov's brittle rupture theory are taken as the constitutive relationships for creep (cf. Boyle and Spence (1983), Kachanov (1986))

$$\dot{\varepsilon}_{ij}^{c} = \frac{3}{2}\frac{\dot{\varepsilon}_i^c}{\sigma_i} s_{ij} \, , \quad \dot{\varepsilon}_i^c = \left(\sigma_i^{net}\right)^m \dot{f}(\tau) \, , \quad \dot{\psi}_k = \begin{cases} -C\left(\dfrac{\sigma_k}{\psi_k}\right)^n & \text{if } \sigma_k > 0 \\ 0 & \text{if } \sigma_k \leq 0 \end{cases} \quad k = r, \theta \, . \tag{3}$$

For the plane stress state the intensities of the stress, the net stress and the strain rates are defined by the following formulae

$$\sigma_i = \sqrt{\sigma_r^2 + \sigma_\theta^2 - \sigma_r \sigma_\theta} \, , \quad \sigma_i^{net} = \sqrt{\left(\frac{\sigma_r}{\psi_r}\right)^2 + \left(\frac{\sigma_\theta}{\psi_\theta}\right)^2 - \frac{\sigma_r}{\psi_r}\frac{\sigma_\theta}{\psi_\theta}} \, ,$$
$$\dot{\varepsilon}_i^c = \frac{2}{\sqrt{3}}\sqrt{\left(\dot{\varepsilon}_r^c\right)^2 + \left(\dot{\varepsilon}_\theta^c\right)^2 + \dot{\varepsilon}_r^c \dot{\varepsilon}_\theta^c} \, , \tag{4}$$

respectively, where the symbols C, n, m, denote the temperature dependent material constants determined from experiments. Finally, assuming the creep incompressibility, the principal components of creep strain rates may be written as

$$\dot{\varepsilon}_r^c = \frac{\left(\sigma_i^{net}\right)^m}{\sigma_i}\left(\sigma_r - \frac{\sigma_\theta}{2}\right)\dot{f}(\tau) \, , \quad \dot{\varepsilon}_\theta^c = \frac{\left(\sigma_i^{net}\right)^m}{\sigma_i}\left(\sigma_\theta - \frac{\sigma_r}{2}\right)\dot{f}(\tau) \, ,$$
$$\dot{\varepsilon}_z^c = -\left(\dot{\varepsilon}_r^c + \dot{\varepsilon}_\theta^c\right) \, . \tag{5}$$

In the case of scalar damage the equations (3) and (4) take the simple form since $\psi = \psi_r = \psi_\theta$ and, hence, the intensity of the net stress reduces to $\sigma_i^{net} = \sigma_i/\psi$ (cf. Ganczarski and Skrzypek (1989)).

2 Basic equations

2.1 Solution of the elastic problem ($\tau = 0$)

The first of Eqs.(1) yields the formulae

$$
\begin{aligned}
u &= \frac{A_N}{2}r + \frac{B_N}{r} - \frac{1-\nu^2}{8E}\rho\omega^2 r^3 + (1+\nu)\frac{\alpha}{r}\int_{r_N}^{r} t\xi d\xi \, , \\
\sigma_r &= \frac{E}{1-\nu^2}\left[\frac{1+\nu}{2}A_N - (1-\nu)\frac{B_N}{r^2}\right] - \frac{3+\nu}{8}\rho\omega^2 r^2 - \frac{\alpha E}{r^2}\int_{r_N}^{r} t\xi d\xi \, , \\
\sigma_\theta &= \frac{E}{1-\nu^2}\left[\frac{1+\nu}{2}A_N + (1-\nu)\frac{B_N}{r^2}\right] - \frac{1+3\nu}{8}\rho\omega^2 r^2 + \frac{\alpha E}{r^2}\int_{r_N}^{r} t\xi d\xi - \alpha E t \, ,
\end{aligned}
\tag{6}
$$

where, for the linear field of temperature

$$
t(r) = t_a + \Delta t\frac{r-a}{b-a} \, , \tag{7}
$$

the integrals may be expressed in the form

$$
\int_{r_N}^{r} t\xi d\xi = \frac{\Delta t}{b-a}\frac{r^3 - r_N^3}{3} + \left(t_a - \frac{a}{b-a}\Delta t\right)\frac{r^2 - r_N^2}{2} \, . \tag{8}
$$

The symbol r_N denotes the outer radius of the Nth disk portion, whereas t_a denotes the temperature at the inner radius of the disk $r = a$ (see Fig.1).

2.2 Solution of the creep problem $(0 < \tau \leq \tau_I)$

The second of Eqs.(1) leads to the following expressions for the velocity and stress rates (cf. Ganczarski and Skrzypek (1989))

$$
\begin{aligned}
\dot{u} &= \frac{A_N'}{2}r + \frac{B_N'}{r} + \left(\frac{I_1}{1+\nu} - \frac{I_2}{1-\nu}\right)r \, , \\
\dot{\sigma}_r &= \frac{E}{1-\nu^2}\left[\frac{1+\nu}{2}A_N' - (1-\nu)\frac{B_N'}{r^2} + I_1 + I_2\right] \, , \\
\dot{\sigma}_\theta &= \frac{E}{1-\nu^2}\left[\frac{1+\nu}{2}A_N' + (1-\nu)\frac{B_N'}{r^2} + I_1 - I_2\right] - E\dot{\varepsilon}_\theta^c \, .
\end{aligned}
\tag{9}
$$

The integrals I_1, I_2 in (9) are defined as

$$
I_1 = \frac{1+\nu}{2}\int_{r_N}^{r}\frac{\dot{f}}{\xi}d\xi \, , \quad I_2 = -\frac{1-\nu}{2r^2}\int_{r_N}^{r}\left(2\dot{g} - \dot{f}\right)\xi d\xi \, , \tag{10}
$$

whereas $\dot{f}$ and $\dot{g}$ are expressed in terms of the creep strain rates (2) and (5) .

2.3 Solution of the creep problem when propagation of failed zones is admitted ($\tau_I \leq \tau \leq \tau_{II}$)

Admitting for the growth of the damaged zones with respect to the circumferential component of the damage tensor $\psi_\theta = 0$ the circumferential stress in the partially damaged zone of the disk becomes zero $\sigma_\theta = 0$ and, hence, the definitions (4) $\div$ (5) take the new form

$$\sigma_i = |\sigma_r| \;, \quad \sigma_i^{net} = \frac{|\sigma_r|}{\psi_r} \;, \quad \dot{\varepsilon}_r^c = \frac{|\sigma_r|^{m-1}}{\psi_r^m}\sigma_r \dot{f}(\tau) \;, \quad \dot{\varepsilon}_\theta^c = \dot{\varepsilon}_z^c = -\frac{1}{2}\dot{\varepsilon}_r^c \;. \tag{11}$$

Finally, the solution of the creep problem in the partially damaged zone reduces to

$$\dot{u}^* = C_N \ln r + D_N + I_3 \;, \quad \dot{\sigma}_r^* = E\frac{C_N}{r} \;, \quad \dot{\sigma}_\theta^* = 0 \;, \quad I_3 = \int_{r_N^*}^{r} \dot{\varepsilon}_r^c d\xi \;, \tag{12}$$

where r_N^* denotes the radius separating damaged and undamaged zones in N^{th} disk portion (Fig.1). The boundary and the continuity conditions and the additional continuity conditions for each of the partially damaged disk portions enable to evaluate the constants A_N, B_N, A_N', B_N' and C_N', D_N', respectively.

3 Optimal design of disks

The optimization problem is formulated as follows: the distribution of the jump-like variable disk thickness which maximizes the time to failure (τ_I or τ_{II} for the creep problems without or with a propagation of failed zones admitted, respectively) under the constant volume condition, is sought for

$$\tau_{I,II}(h_N) \longrightarrow \max \;, \quad \sum_{N=1}^{N_0} \delta h_N \left[(r_N)^2 - (r_{N-1})^2\right] = 0 \;, \tag{13}$$

where N_0 is the number of the constant thickness disk portions. The optimization requires the solution of the subsequent creep rupture problems. The numerical solution of the creep problems starts from the elastic solution of the constant thickness disk, with $\psi_r = \psi_\theta = 1$. Then, the "new" values of strains, stresses and damage parameters are computed for the subsequent time steps. The proposed procedure of optimization is based on the iterative corrections of the thickness in order to equalize the lower of damage parameters at a given level

$$(\psi_{min})_N = \inf\{\psi_r, \psi_\theta\}_N = D_i \;, \tag{14}$$

where D_i denote the constant values of the damage levels at the ith step of the thickness correction. The corrections of thicknesses are introduced proportionally to the current values of damage

$$\delta h_N - \delta\overline{h} = \mathcal{P}\delta\psi_N \;, \tag{15}$$

where the average value of the correction $\delta\overline{h}$ must satisfy the condition of constant volume (13) $\delta\psi_N = 1-(\psi_{min})_N$, whereas the step factor $\mathcal{P}$ should be chosen experimentally. Then,

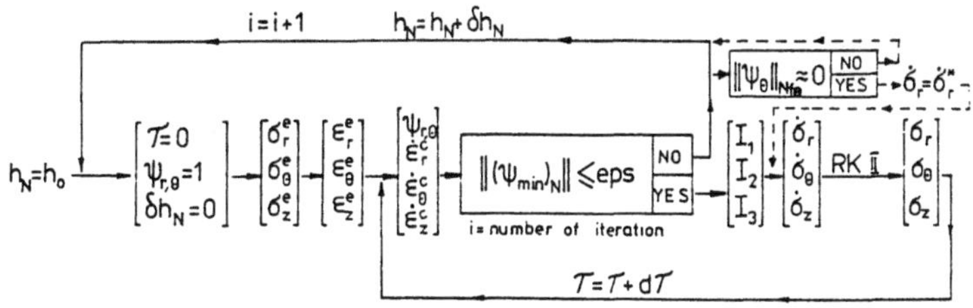

Fig. 2. Algorithm for the optimization procedure with respect to the brittle rupture time

the described procedure is repeated for the more advanced damage until, finally, the first macrocracks $(\psi_{min})_N$ will be initiated simultaneously at the all N_0 points. In the case when the propagation of the partially damaged zones of the circumferential microcracks, $[(\psi_\theta)_{min}]_{N_{f\theta}} = 0$ is admitted, the sequence of modified boundary value problems is solved and the appropriate corrections of the thickness are introduced until a simultaneous radial damage is reached in the remaining disk portions $[(\psi_r)_{min}]_{N_{fr}} = 0$. In all cases under consideration and on each step of time the corresponding integrals I_1, I_2, I_3 (10), (12) are numerically calculated, whereas for the time increments the Runge-Kutta II method is applied.

4 Results

The numerical examples are presented for the disks made of the ASTM 321 stainless steel: $E = 1.77 \times 10^5 MPa$, $\nu = 0.3$, $\rho = 7.9 \times 10^3 kg/m^3$, $a = 0.02m$, $b = 5 \times a$, $h_0 = 0.002m$, $p = 11.8MPa, \omega = 100s^{-1}, \alpha = 1.85 \times 10^{-5} 1/^o, t_a = 783K, \Delta t = -10K$; the temperature dependent material constants for creep rupture are shown in the Table 1 (cf. Odqvist (1966)).

absolute temperature	$C[Pa^{-n}/s]$	n	m
783 K	2.13×10^{-42}	3.90	5.60
773 K	1.35×10^{-41}	3.82	5.49

Table 1

The optimal profiles of the jump-like variable thickness are shown in Fig.3 for the following cases:

a) the scalar formulation of the damage law $\psi_r = \psi_\theta = \psi$ (dashed line),

b) the tensorial formulations of the damage law (solid line),

both compared to the disk of constant thickness of the same volume. The distribution of the scalar damage parameter ψ at the instant of rupture is presented in Fig.4a. The

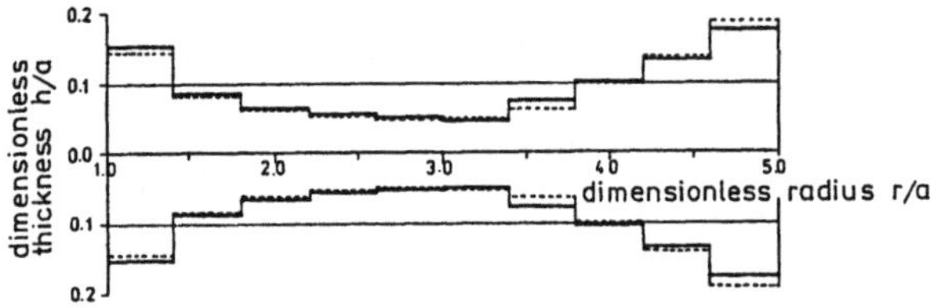

Fig. 3. Optimal profiles of disks: scalar formulation (dashed line) and tensorial formulation (solid line) of the damage law, vs the constant thickness disk

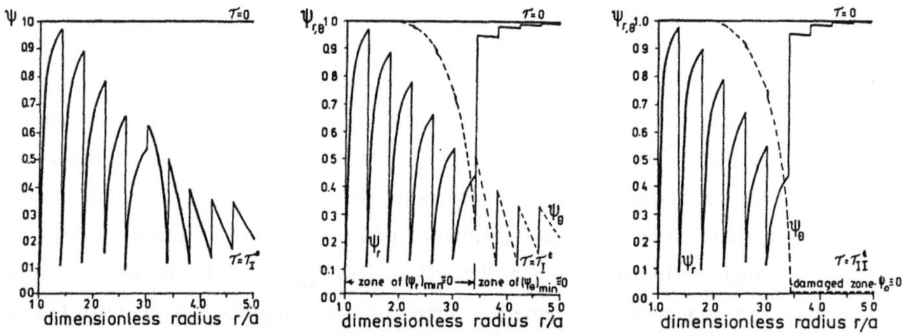

Fig. 4. Distributions of the damage parameters in the optimal disks at rupture: a) scalar formulation b) tensorial formulation c) propagation of the circumferential damage zone admitted

first macrocracks simultaneously appear in all jump-like variable thickness disk portions. Fig.4b shows the distribution of both the radial (ψ_r) and the circumferential (ψ_θ) components of the damage function at the instant of rupture. A mixed failure mechanism consists of two zones: of the radial damage ($\psi_\theta = 0$) in the inner part of the disk and of the circumferential damage ($\psi_\theta = 0$) in the outer one (cf. Fig.4b). The last distribution of damage components is obtained when the propagation mechanism is admitted. The damage zone occupies the outer part of the disk where the circumferential component of damage vanishes ($\psi_\theta = 0$) (Fig.4c).

The time to rupture for all previously discussed cases is compared in the Table 2.

	scalar damage	tensorial damage	tensorial damage with propagation zone
time to rupture	τ_I^s	$\tau_I^t = 1.04\tau_I^s$	$\tau_{II}^t = 1.05\tau_I^s$

Table 2

5 Conclusions

1. The tensorial formulation of the damage law allows for the 4% increase in the time to rupture initiation τ_I^t when compared to the scalar formulation τ_I^s.

2. The tensorial formulation with a propagation zone of circumferential damage $\psi_\theta = 0$ results in the 5% increase in the time to rupture τ^t_{II}.
3. The optimal shapes obtained from the applied approaches are closely similar each to the other. A scalar damage description results in a sharper change of thickness than a tensorial one. The propagation zone yields the negligible change in the optimal shape.

6 References

Boyle, J. T.; Spence, J. 1983: *Stress Analysis for Creep*, London -Boston, Butterworths

Ganczarski, A.; Skrzypek, J. 1989: *Optimal Prestressing and Design of Rotating Disks against Brittle-Rupture under Unsteady Creep Conditions*, Rozpr. Inż. (Eng. Trans.) **4**, pp. 49-71

Ganczarski, A.; Skrzypek, J. (1990): *Optimal Shape of Prestressed Disks against Brittle-Rupture under Unsteady Creep Conditions*, Struct. Opt. (to be published)

Kachanov, L. M. 1986: *Introduction to Continuum Damage Mechanics*, Martinus Nijhoff Publishers, Dortrecht, Boston, Lancaster

Odqvist, F. K. G. 1966: *Mathematical Theory of Creep and Creep Rupture*, Oxford Mathematical Monographs, Clarendon Press

Penny, R. K.; Marriott, D. L. 1971: *Design for Creep*, Maidenhead, Berkshire, England, McGraw, Hill

Prager, W. 1968: *Optimal Structural Design for Given Stiffness in Stationary Creep*, ZAMP, Vol. 19, pp. 252-256

Życzkowski, M. 1968: *Optimal Structural Design in Rheology*, 12th Int. Congress Appl. Mech., Stanford; also Trans. ASME, Vol. E38, 1971, pp. 39-46

Prediction of Macroscopic and Local Creep Behavior of Perforated Plates

T.IGARI*, K.SETOGUCHI*, S.NOMURA**

* Nagasaki R&D Center, Mitsubishi Heavy Industries, Ltd., Akunoura-machi 1-1, Nagasaki 850-91, Japan
** Nagasaki Shipyard and Engine Works, Mitsubishi Heavy Industries, Ltd., Akunoura-machi 1-1, Nagasaki 850-91, Japan

Summary

Prediction methods of both macroscopic and local creep behavior of perforated plates in a general case involving both primary and secondary creep are proposed in this paper. Predicted results are compared with both experimental results and numerical ones by FEM. Macroscopic creep behavior is predictable by use of the effective stress of perforated plates which is concretely defined by the predictive equation. As for stress-strain concentration around a hole, predictive equations which are basically based on Neuber's rule and which consider the effect of exponent of stress in creep equation give good results.

1. Introduction

In the design of high-temperature components containing perforated plates, such as heat exchangers for high-temperature nuclear reactors, creep analysis of perforated plates is required for estimating creep-fatigue damage reasonably. The simplified method based on the equivalent-solid-plate concept[1] is usually used for the creep analysis of perforated plates. In this case, the results of the analysis depend on the prediction methods of both equivalent-solid-plate properties and stress-strain concentration around a hole, that is, the prediction methods of macroscopic and local creep behavior of perforated plates. Uragami-Nakamura[2] and Porowski-O'Donnell[3] have proposed these prediction methods. Recently, Igari-Setoguchi[4] have proposed advanced prediction methods which are applicable to wider range of materials and temperatures than previous works. These methods were derived based on analytical approach assuming the Norton's creep law. In order to apply the proposed methods by the authors to actual structural materials, however, further work involving experimental study is

M. Zyczkowski (Ed.)
Creep in Structures
IUTAM Symposium Cracow/Poland 1990

necessary. This paper presents the results of study for extending the proposed methods to actual materials with the primary creep in addition to the secondary creep.

2. Prediction Method of Macroscopic and Local Creep Behavior

2.1 Prediction Method of Macroscopic Creep Behavior

The prediction method which was proposed by the authors[4] is summarized here. Macroscopic creep behavior of perforated plates with a triangular penetration pattern of circular holes is describable from two kinds of point of view, that is, a point of view from the equivalent-solid-plate concept and that from the effective-stress concept as shown in Fig.1. Assume that the creep property of the base material as

$$\dot{\varepsilon}_c = A\sigma^n \tag{1}$$

The equivalent solid plate is an imaginary plate which has the same cross section and which shows the same macroscopic deformation as those of the perforated plate. Stress σ^* and creep strain rate $\dot{\varepsilon}^*_c$ of the equivalent solid plate, which are the macroscopic stress and creep strain rate of the perforated plate, are related to each other as follows.

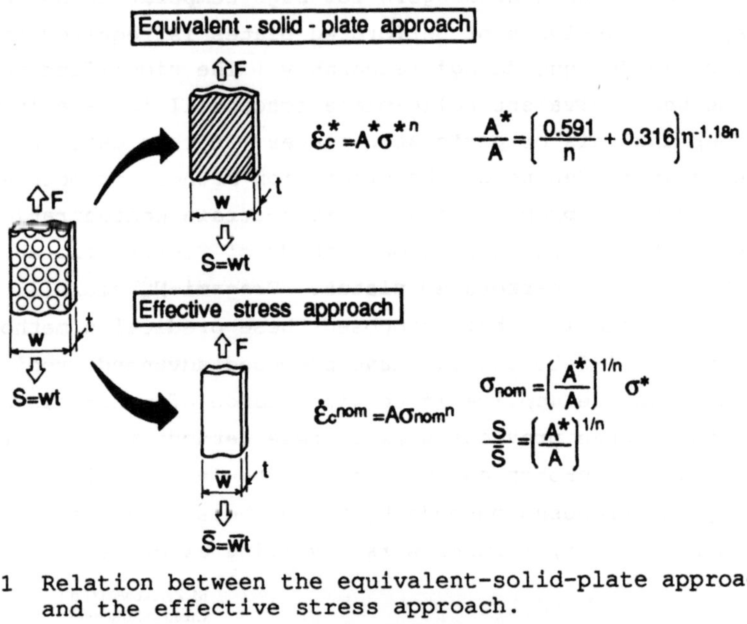

Fig.1 Relation between the equivalent-solid-plate approach and the effective stress approach.

$$\dot{\varepsilon}^*_c = A^* \sigma^{*n^*} \tag{2}$$

As for the properties of the equivalent solid plate, the coefficient A^* is expressed [4] by the eq.(3), and the exponent n^* is equal to the exponent n of the base material [2].

$$A^* = (0.591/n+0.316)\eta^{-1.18n} \bullet A \tag{3}$$

Where η is a ligament efficiency of perforated plate and is expressed as $\eta=h/p$ (p:pitch of hole, h:ligament width). The relation of eq.(3) is expressed in Fig.2 along with the inverse of creep exponent n which covers wide range of n from unity to infinite. Solid lines expressing the eq.(3) agree well with numerical results by FEM. An important point of this prediction method is a compatibility with the elastic case shown in ASME Boiler and Pressure Vessel Code A-8000. That is to say, the eq.(3) is determined by considering the relation "$A^*/A=E/E^*$" for the special case of n=1 by assuming the Hoff's analogy.

The other point of view for the macroscopic creep behavior of perforated plates is the effective-stress concept. This concept

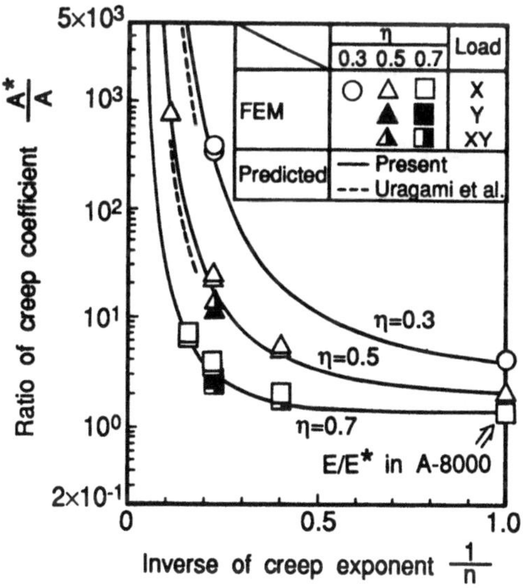

Fig.2 Prediction of A^*

assumes the solid plate with the same creep properties as the base material and with the smaller cross section than that the perforated plate so as to show the same creep deformation as the perforated plate. By using the effective stress σ_{nom} against this cross section, the concept of which is often used in damage mechanics, the macroscopic creep strain rate of this effective solid plate ε_c^{nom} can be expressed as follows.

$$\dot{\varepsilon}_c^{nom} = A\sigma_{nom}^{\ n} \tag{4}$$

From the definition that these three plates have the same creep strain rate against the same load, the concrete expressions of the effective stress σ_{nom} is obtained as

$$\sigma_{nom} = (A^*/A)^{1/n}\sigma^* \tag{5}$$

2.2 Prediction Method of Local Creep Behavior

The prediction method of stress-strain concentration was proposed [4] by assuming the Norton's creep law. By assuming the Neuber's rule, peak stress $\sigma_{\phi max}$ and peak creep strain rate $\dot{\varepsilon}_{\phi cmax}$ around the hole were determined as follows.

$$\sigma_{\phi max}/(\sigma^*/\eta) = \eta(CK_t)^{2/(n+1)}(A^*/A)^{1/n} \tag{6}$$

$$\dot{\varepsilon}_{\phi cmax}/\dot{\varepsilon}^*c = (CK_t)^{2n/(n+1)} \tag{7}$$

$$C = (-1.359 \bullet \log_{10}\eta - 0.551)(1/n-1)+1 \tag{8}$$

The elastic stress concentration factor K_t in these equation is expressed as "$K_t=(E^*Ymax)/(E\eta)$", where Ymax is the same stress multiplier as shown in ASME Boiler and Pressure Vessel Code Sec.III A-8000. Examples of comparison between predicted results and analytical ones by FEM are shown in Fig.3 and Fig.4. Predicted results coincide well with the prediction.

3. Application of the Prediction Methods to Actual Material

3.1 Macroscopic Creep Behavior

As a result of creep tests of perforated plates which were performed under constant loads at 800°C and 950°C by adopting Haste-

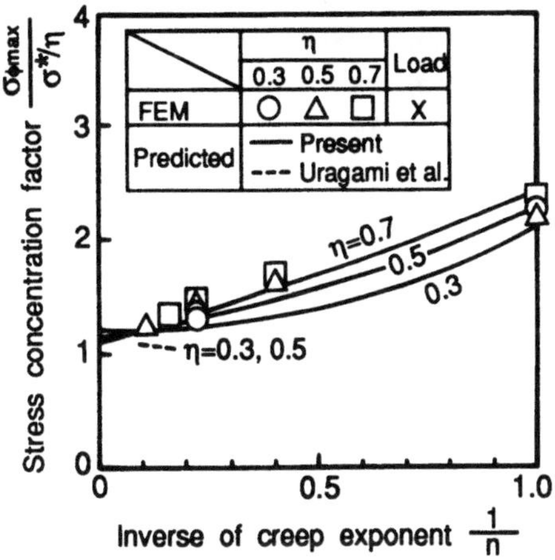

Fig.3 Stress concentration around a hole.

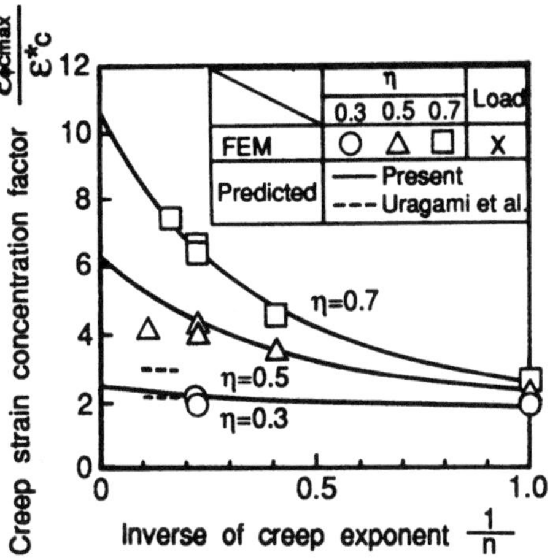

Fig.4 Creep strain concentration around a hole.

lloy X(ASTM B435-77) as a test material, macroscopic minimum creep rates of perforated plates, which correspond to the secondary creep and are directly related to the Norton's law, could be predicted[5] by using the eq.(3).

In order to describe the macroscopic creep behavior of perforated plate involving both primary and secondary creep, a method using the effective stress σ_{nom} of eq.(5) is proposed in this paper. The effective stress σ_{nom} is a representative stress like "reference stress" in a sense that this value can describe the macroscopic creep behavior of perforated plate. Let us assume the base material with the following creep property,

$$\varepsilon_c = f(\sigma,t) = B\sigma^m t^l + A\sigma^n t \tag{9}$$

where B, m, l are the material constants, and t is a time. The first term of the right side expresses the primary creep, and the second one corresponds to the secondary creep. The prediction of macroscopic creep behavior of perforated plates consisting of this material is made here by the following equation.

$$\varepsilon^*_c = f(\sigma_{nom},t) = B\sigma_{nom}{}^m t^l + A\sigma_{nom}{}^n t \tag{10}$$

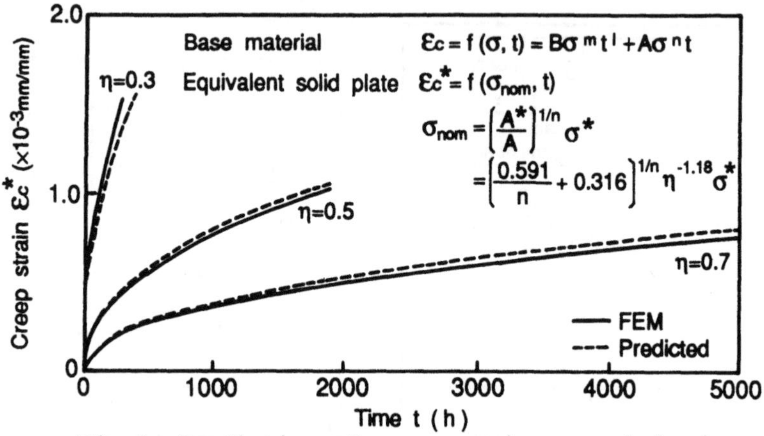

Fig.5 Prediction of macroscopic creep behavior of perforated plate.

Figure 5 shows an example of comparison between analytical results by FEM and predicted ones by the eq.(10). Analysis by FEM in this figure were performed for perforated plates with three kinds of ligament efficiency and with the material property of eq.(9). Predicted results coincided with those by FEM, and a possibility of predicting the macroscopic creep behavior of perforated plates in a case involving both primary and secondary creep was clarified.

3.2 Local Creep Behavior

In order to understand the local creep behavior of perforated plates consisting of actual material with both primary and secondary creep, both experimental and analytical studies were made. As for experimental study, creep strain measurement by Moire method was performed[6] at 800°C for the perforated plates made of the same material shown above. As for the analytical study by FEM, the perforated plates with the creep property of the eq.(9) were analyzed under the condition of constant load. In this situation, steady state can be obtained after the initial rapid change of both peak stress and peak creep strain rate which is caused by both material and structural transient creep[6].

Here prediction was made by using eqs.(6) and (7), and by considering that the stress distribution in the perforated plates was determined by the exponent of stress in the creep law. That is, consider the Bailey-Norton type creep law of eq.(9). If there is no primary creep, stress distribution in the perforated plate is determined by the exponent n. On the other hand, stress exponent m is a dominant factor for the stress distribution if there is no secondary creep. In this paper prediction of stress and strain concentration by eqs.(6) and (7) was made for the following two cases :

(1) Case 1 where exponent n in eqs.(6) and (7) is taken as exponent n of secondary creep in eq.(9).
(2) Case 2 where exponent n in eqs.(6) and (7) is taken as exponent m of primary creep in eq.(9).

Predicted results for stress concentration and creep strain concentration were compared with both experimental and analytical results at the above-mentioned steady state in Fig.6 and Fig.7 respectively. Regarding the stress concentration, results of case 2 show higher values than those for case 1. On the other hand, the opposite tendency is found for creep strain concentration. This tendency can be explained from Figs.3 and 4 by considering the fact that exponent n has larger value than exponent m for this material. As for the stress concentration, the analytical results were located near the band defined by two prediction lines, and the prediction based on secondary creep (Case1) seems to be the

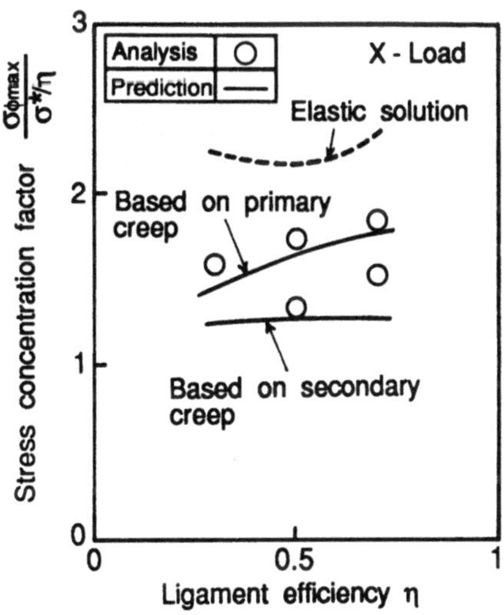

Fig.6 Prediction of stress concentration (X-Load)

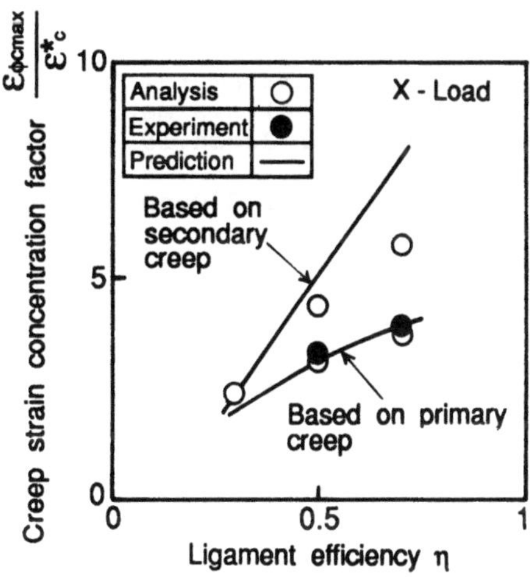

Fig.7 Prediction of creep strain concentration (X-Load)

lower found. The prediction based on stress exponent m of primary creep (Case2) seems to the first approximation of the stress concentration factor of actual structural material exhibiting both primary and secondary creep. As for the creep strain concentration, on the other hand, experimental and analytical results were located between two prediction lines. Prediction based on exponent of secondary creep seems to be an upper bound for the condition studied here.

4. Conclusion

Prediction methods of both macroscopic and local creep behavior of perforated plates in a case involving both primary and secondary creep were proposed.

(1) Macroscopic creep behavior of perforated plates could be predicted by use of the effective stress which is concretely defined by the prediction equation.

(2) Prediction method considering the effect of exponent of stress in creep equation gave good results for stress and creep strain concentration around a hole.

References

1. O'Donnell,W.J.; Porowski,J.S.: Design analysis of perforated plates and shells, Zamrik,S.Y. and Dietrich,D.(eds.) Pressure Vessels and Piping Design Technology 1982, A Decade of Progress. ASME 1982.

2. Uragami,K.; Nakamura,K.; Asada,K.; Kano,T.: Simplified inelastic analysis method of ligament plate. J.High Pressure Institute of Japan.19 (1981) 57-65.

3. Porowski,J.S.; O'Donnell,W.J.; Tanaka,T.; Badlani,M.: Stress and strain concentrations in perforated structures under steady creep conditions. J.Pressure Vessel Tech. 102 (1980) 419-429.

4. Igari,T.; Setoguchi,K.; Nomura,S.: Simplified creep analysis of perforated plates under steady creep condition. Nuclear Engineering and Design. 97 (1986) 161-166.

5. Igari,T.; Setoguchi,K.; Nomura,S.: Experimental study on macroscopic creep behavior of perforated plates. Trans.SMiRT-9 L4/4 (1987) 219-224.

6. Igari,T.; Setoguchi,K.; Matsumura,N.; Hamaue,Y.; Kawaguty,K.: Experimental study of macroscopic and local creep behavior of perforated plates. Proc. ICPVT-6 Beijing China (1988) 1093-1100.

A Viscoplastic Constitutive Relationship and the Finite Element Implementation of Stress/Strain Concentration in Notched Component

T. INOUE, S. IMATANI, and M. TANAKA

Department of Mechanical Engineering
Kyoto University, Kyoto, Japan

Summary

This paper deals with a unified inelastic constitutive model and its application to finite element implementation. In the first part, the validity of the constitutive model is evaluated under complex loading history. The finite element formulation is then obtained, based on the constitutive model proposed. The local strain response is analyzed for notched cylinder, being compared with the axial/torsional strain response near the notched root.

Introduction

In order to evaluate the safety assessment of high temperature machine components under complicated loading conditions, a suitable analytical procedure is preferably established, based on a unified inelastic constitutive relationship.

The aim of this paper is to develop a unified inelastic constitutive model and to apply it to finite element analysis for a circumferentially notched specimen. Experimental verification is carried out for a PCA material (primary candidate alloy for fusion reactor component) under a biaxial stress state of tension and torsion. Local strain trajectory is measured for notched specimens with an electric capacitance type strain gauge.

A Unified Inelastic Constitutive Model

General Frame for the Constitutive Relationship

Figure 1 represents the schematic view of the excess stress theory [1]. When the stress is beyond the static yield stress r, the inelastic strain is produced in proportion to the exceeding stress. Provided that the steady creep law of the material is described by the Norton's law, the inelastic strain rate is expressed as

$$\dot{\varepsilon}^{i}_{ij} = \langle \frac{F - r}{D} \rangle^{n} \frac{\partial F}{\partial \sigma_{ij}} = \omega p_{ij} \quad . \tag{1}$$

where $< \cdot >$ stands for the Macauley bracket, F indicates the yield function, D is the drag stress, and n means the power factor. The inelastic strain rate is decomposed into its magnitude ω and the direction p_{ij} for simplicity.

There are two kinds of internal variables in the basic equation; one is the kinematic back

M. Zyczkowski (Ed.)
Creep in Structures
IUTAM Symposium Cracow/Poland 1990

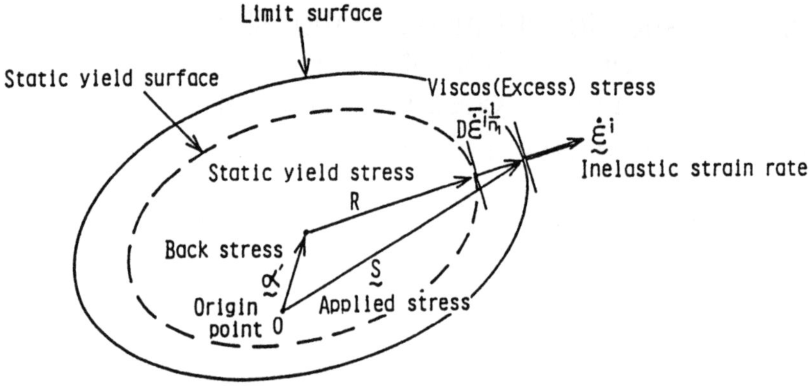

Fig.1. Schematic view of the excess stress hypothesis.

stress α_{ij} defined as a tensor variable and the other is the isotropic yield stress r with a scalar property. The evolution of the back stress is given by

$$\dot{\alpha}'_{ij} = C_1 \frac{2}{3} \dot{\varepsilon}^i_{ij} - C_2 (s_{kl} - \alpha'_{kl}) \dot{\varepsilon}^i_{kl} \alpha'_{ij} - C_3 \bar{\alpha}^{n-1} \alpha'_{ij} \quad . \tag{2}$$

in which C_1, C_2, and C_3 stand for material parameters. The parameter C_1 in the first term represents the work-hardening due to the inelastic deformation while the second parameter C_2 dominates the description of the rounding phenomenon of stress-strain curve. The parameter C_3 corresponds to the rate-dependent recovery appearing in the creep process. The isotropic yield stress r is decomposed into the temperature dependent term r_T and the deformation-induced term r_a. Hence, r is expressed as

$$r = r_T + r_a \quad , \tag{3}$$

in which r_a mainly governs the description of hardening phenomenon during cyclic processes. The variation of r_a can be represented by

$$\dot{r}_a = R_1 (R_2 - r_a)(s_{ij} - \alpha'_{ij}) \dot{\varepsilon}^i_{ij} \quad . \tag{4}$$

in which R_1 and R_2 are material parameters.

Inelastic Response of PCA material

We present typical examples of the validity of the model, predicting the inelastic behavior of a high temperature material under biaxial stress state. The tested material is an austenitic stainless steel, PCA material, which contains 14% Cr and 15% Ni in weight. The material is shaped into thin walled tubular specimen with 1mm in thickness to impose the combined stress of tension and torsion. The experiment was carried out at 700°C.

Hysteresis loops under a circular strain path are demonstrated in Fig.2 where the solid line indicates the hysteresis during early cycles. The shape of the hysteresis is entirely different from the uniaxial loop which we are used to see. The prediction agrees well with the experiment, particularly on the shape of the hysteresis loop. This is due to the second term of eq.(2) which is devoted to the description of the rounding phenomenon. Cyclic hardening behavior is evaluated as another example. Figure 3 implies the relationship between the stress amplitude and the number of cycles under the circular strain path with increasing strain range, together with the predictions by other constitutive models. The non-proportional straining induces much more hardening than the uniaxial cycling. The proposed model follows such a tendency, so that the validity of the model is now proved.

Finite Element Formulation

Alternative Representation of the Constitutive Model

The tangent modulus method [2] is developed to avoid an unexpected oscillation in numerical calculation. We first take an intermediate value of the objective function ω between the two steps at t and $t + \Delta t$ such that

$$\omega^* = (1-\theta)\omega_t + \theta\omega_{t+\Delta t} \quad . \tag{5}$$

Here, θ denotes a scalar representing the weight between the two steps. The Taylor expansion is adopted to get the value $\omega_{t+\Delta t}$ at the following time step. Applying this

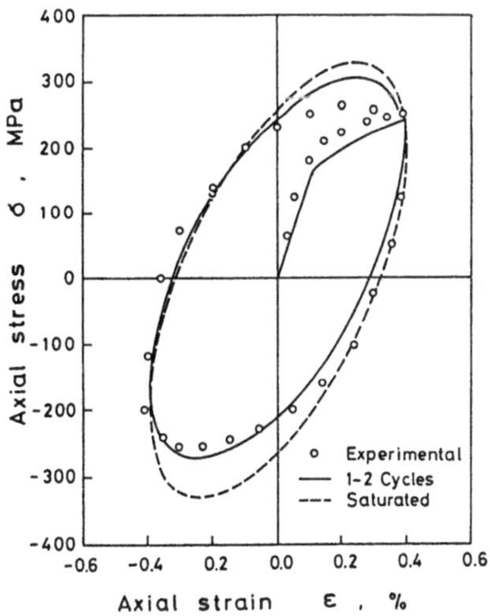

Fig.2. Hysteresis loop under the circular strain cycles.

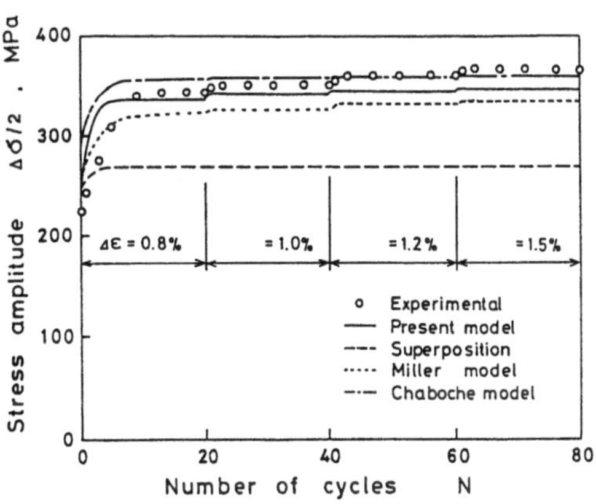

Fig.3. Cyclic hardening behavior under circular strain path.

procedure to the viscoplastic constitutive model, we get the following form for the stress rate as

$$\dot{\sigma}_{ij} = D^{tan}_{ijkl}\dot{\varepsilon}_{kl} - \frac{1}{1+\xi}(\omega_t + \frac{\xi}{h}C_3 p_{mn}\alpha'_{mn})E_{ijkl}p_{kl} \quad . \tag{6}$$

D_{ijkl}, ξ, and h are respectively expressed as

$$D^{tan}_{ijkl} = E_{ijkl} - \frac{\xi}{1+\xi}\frac{1}{h}E_{ijpq}p_{pq}p_{rs}E_{rskl} \quad , \tag{7}$$

$$\xi = \Delta t \cdot \theta \cdot \frac{n}{D}\langle\frac{F-r}{D}\rangle^{n-1} \cdot h \quad , \tag{8}$$

$$h = \frac{3E}{2(1+\nu)} + C_1 - C_2 F p_{ij}\alpha'_{ij} + R_1(R_2 - r_a)F \quad . \tag{9}$$

Note that the time step Δt explicitly appears in eq.(8). Therefore, the differential equation (6) is not an "exact" but an "approximated" form by use of the Taylor expansion.

Governing Equations and Matrix Representations

Fundamental equations to solve a boundary-value problem are given as follows;

balance equation for stress $\qquad \dot{\sigma}_{ij,j} + \rho\dot{b}_i = 0 \quad ,$ (10)

strain-displacement relation $\qquad \dot{\varepsilon}_{ij} = \frac{1}{2}(\dot{u}_{i,j} + \dot{u}_{j,i}) \quad ,$ (11)

constitutive relationship $\qquad \dot{\sigma}_{ij} = D^{tan}_{ijkl}\dot{\varepsilon}_{kl} - \Omega E_{ijkl}\, p_{kl} \quad ,$ (12)

boundary conditions $\qquad \dot{\sigma}_{ij}n_i = \dot{\bar{t}}_j \text{ on } S_\sigma, \quad \dot{u}_i = \dot{\bar{u}}_i \text{ on } S_u \quad .$ (13)

Here, n_i stands for a unit normal vector being outward to the surface S_σ on which the external force $\bar{t}_i$ is imposed. All these equations are expressed in rate form since the constitutive equation is represented in a rate type formulation. The principle of virtual work leads to

$$\int_V \dot{\sigma}_{ij}\delta\dot{\varepsilon}_{ij}dv = \int_{S_\sigma} \dot{\bar{t}}_i\delta\dot{u}_i ds + \int_V \dot{b}_i\delta\dot{u}_i dv \quad . \tag{14}$$

We now divide a continuous body into finite elements, each of which is connected with neighboring elements by jointing the nodes. The principle of virtual work is expressible as a sum of the quantities of each element. Then, eq.(14) is replaced by

$$\sum_e [\int_{V_e}\{\dot{\sigma}\}^T\delta\{\dot{\varepsilon}\}dv - \int_{S_{\sigma e}}\{\dot{\bar{t}}\}^T\delta\{\dot{u}\}ds - \int_{V_e}\{\rho\dot{b}\}^T\delta\{\dot{u}\}dv] = 0 \quad , \tag{15}$$

in which all the quantities are expressed in matrix representation. Since it must hold for any (arbitrary) admissible displacement rate, we finally obtain

$$[K]\{\dot{u}\} = \{\dot{L}\}_t + \{\dot{L}\}_b + \{\dot{L}\}_\omega \ . \tag{16}$$

The following nomenclatures are utilized for simplicity;

$$[K] = \sum_e \int_{V_e} [B]^T [D^{tan}][B] \, dv \ , \tag{17}$$

$$[\dot{L}]_t = \sum_e \int_{S_{\sigma e}} [N]^T \{\dot{\bar{t}}\} \, ds \ , \tag{18}$$

$$[\dot{L}]_b = \sum_e \int_{V_e} [N]^T \{\rho \dot{b}\} \, dv \ , \tag{19}$$

$$[\dot{L}]_\omega = \sum_e \int_{V_e} \Omega [B]^T [D^e]\{p\} \, dv \ . \tag{20}$$

$[N]$ and $[B]$ indicate the matrices for the shape function and the strain-displacement relation, respectively.

Inelastic Response of Notched Component

Experimental Procedure

The PCA material is shaped into notched specimens as shown in Fig.4. When measuring the local strain at the notched root, we utilize an electric capacitance type strain gauge "strain-pecker" [3]. The experiment was carried out in strain controlled condition with the gauge length of 20mm. Here, we measure the local strain with the gauge length of 1mm by the strain-pecker and compare it with the analytical result.

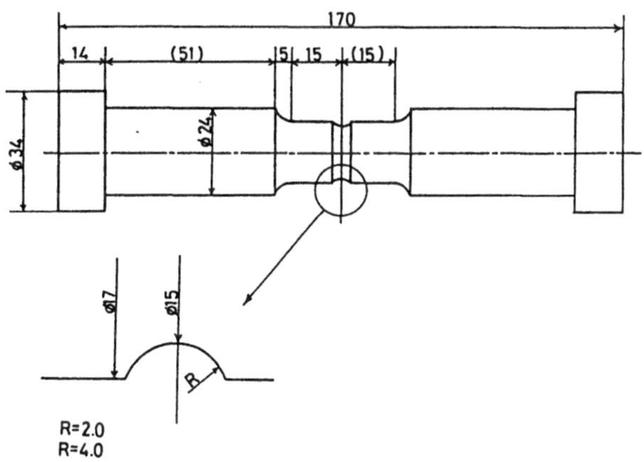

Fig.4. Shape and dimension of the notched specimens.

Local Stress/strain Response at the Notched Root

An axial cyclic straining is imposed to the notched cylinder at room temperature with the global strain range of 0.4%. Figure 5 represents the variation of both the global and the local strains with respect to the nondimensional time scale. The analytical result shows good agreement with the experimental data. If the whole region is in elastic state, the local strain should exhibit a straight line corresponding to the global strain. However, we observe a slightly curved response in the local strain. This means that a part of the region yields and that the strain concentration progresses.

In order to compare the loading pattern, a similar experiment was carried out under shear straining. The variation of shear strain is shown in Fig.6 where the lines and the symbols have similar meaning to the axial case. The shear strain exhibits a different aspect from the axial response. Since the prescribed displacement at the boundary is proportional to the radius from the center in this case, the elastic region always exists near the central axis, so the yield zone is limited around the outer surface. It implies that the strain concentration is accelerated in such a mixed mode of elastic and inelastic regions [4].

Figure 7 represents the shear stress distribution along the minimum cross section during a semi-cycle of the shear straining. When the stress is in the elastic state, the stress concentration is accelerated in proportion to the radius as depicted in the broken line. On the contrary, it is released, i.e. the slope is gentle, when a part of the region yields around the surface. In such a way, the stress trajectory at the notched root as well as its surroundings is so complicated that the inelastic analysis is prerequisitely carried out in real engineering field.

Rate-dependence of Notched Component

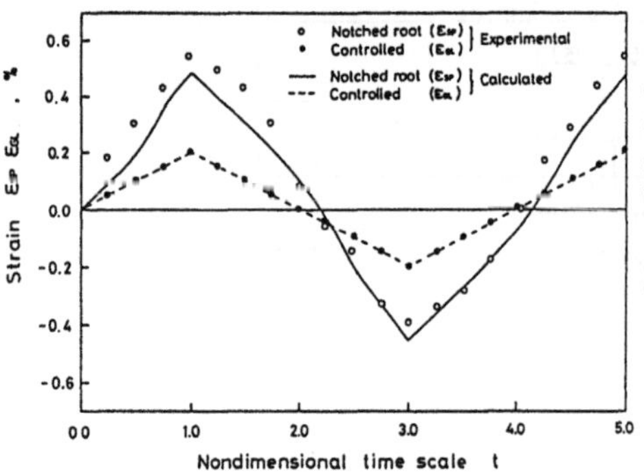

Fig.5. Strain response under the axial cyclic straining for the notched cylinder.

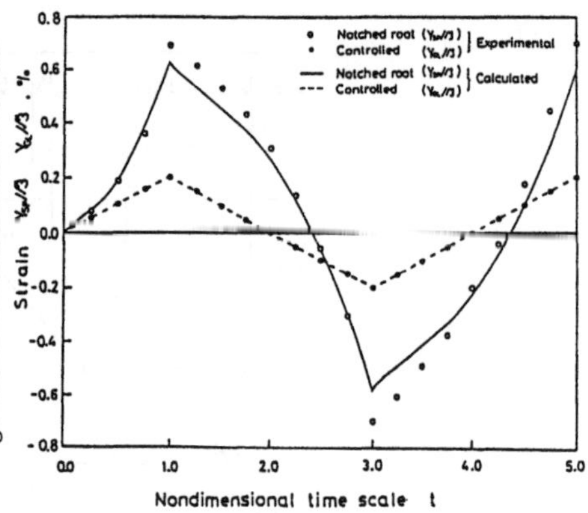

Fig.6. Local strain response subjected to the shear cycling.

The prescribed displacement is kept constant for 1000s at the global strain of 0.2%. This condition corresponds to a tension hold cyclic straining. The stress-strain diagram is presented in Fig.8. The nominal stress is defined as the global force divided by the minimum cross section. When the displacement is kept constant, the local strain is still progressing whereas the stress decreases. This means that the creep is induced in the localized region of the notched root.

A creep loading with the nominal stress of 120MPa is examined as the final simulation. Figure 9(a) indicates the stress distribution just after the preloading is terminated, and the stress distribution at 10hr is presented in Fig.9(b). The stress contour of higher level shifts towards the center of the cylinder. Once higher stress is imposed, more creep strain is induced in the localized region, and then a stress relaxation occurs due to the constraint by the surroundings.

Conclusion

An inelastic constitutive model is proposed in this paper with reference to the excess stress theory. The proposed model is expected to cover a wide range of material behavior under complicated loading conditions.

The second part deals with the application of the model to the stress analysis of notched bar under tension and torsion. The analytical results agree well with the experimental data for both tension and torsion. When a part of the notched root yields, the local displacement is elongated, which implies that the strain concentrates at the notched root.

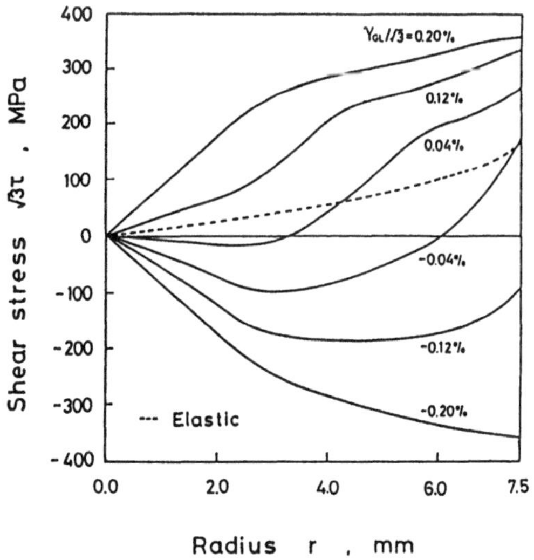

Fig.7. Stress history along the minimum cross section.

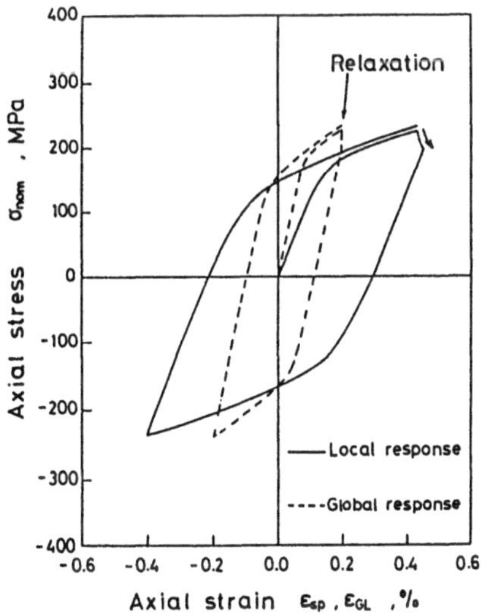

Fig.8. Stress-strain diagram under the strain hold test.

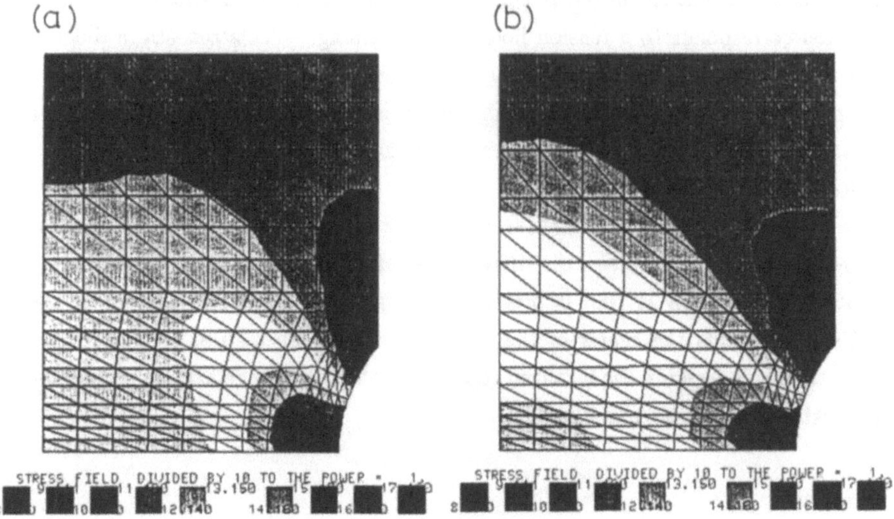

Fig.9. Axial stress distribution in the notched cylinder. Fig.(a) represents the initial state and Fig.(b) corresponds to the final stress distribution.

References

1. Perzyna, P: Thermodynamic Theory of Viscoplasticity. Adv. Appl. Mech., 11 (1971) 313-354.
2. Peirce, D.; Shih, D.F.; Needleman, A.: A Tangent Modulus Method for Rate Dependent Solids. Computer & Struct., 18 (1984) 875-887.
3. Subcommittee on Inelastic Analysis of High Temperature Materials, JSMS : Evaluation of Inelastic Constitutive Models: FEM Analysis of Notched Bars of 2·1/4Cr-1Mo steel at 600°C. Proc. 3rd Int. Conf. Constitutive Laws, (1991) To appear.
4. Neuber, H.: Theory of Stress Concentration for Shear-strained Prismatical Bodies with Arbitrary Nonlinear Stress-strain Law. J. Appl. Mech., Trans. ASME, 28 (1961) 544-550.

Creep Buckling of Inelastic Plates

M. PALEY and J. ABOUDI

Department of Solid Mechanics, Materials and Structures.
Faculty of Engineering, Tel-Aviv University,
Ramat-Aviv, Tel-Aviv 69978, Israel.

Summary

Creep buckling of inelastic plates is analysed by employing a unified theory of viscoplasticity. The analysis is based on a derivation of an instantaneous properties dependent buckling condition of the plate. Creep buckling times against applied stress levels are given for an aluminum plate at elevated temperature and the effect of initially applied strain rate is discussed.

Introduction

The subject of creep buckling of inelastic plates received considerable attention, see the review of Kurshin [1], for example. Recent investigation of creep buckling of elastic-viscoplastic materials are presented by Tvergaard [2].

In a recent paper [3], the authors analyzed the bifurcation buckling of elastic-viscoplastic homogeneous plates. The plastic buckling analysis relies on the instantaneous properties of the material which are load and history dependent. The modeling of the material behavior was given by the unified theory of viscoplasticity of Bodner and Partom [4]. This unified theory provides the creep behavior of rate-sensitive materials which are subjected to constant stress loading.

In the present investigation, the previous methodology is generalized for the creep buckling analysis of viscoplastic plates. This analysis is based on a derivation of a buckling condition of the plate, which involves the instantaneous properties of the material. The buckling condition is established from the governing equations of the plate and the type of boundary conditions involved. The determination of the creep buckling time of the plate is performed incrementally by applying an initial strain rate until the imposed stress level is achieved. Henceforth the stress level is kept constant, while the corresponding strain rate becomes time dependent.

M. Zyczkowski (Ed.)
Creep in Structures
IUTAM Symposium Cracow/Poland 1990

Creep buckling analysis

Consider an elastic-viscoplastic material. Due to the application of an external loading, plastic deformations are developed, modifying the stiffness of the material. The instantaneous constitutive relation of the material is given by

$$\dot{\sigma}_{ij} = C^{VP}_{ijkl}\,\dot{\epsilon}_{kl} \tag{1}$$

where $\dot{\sigma}_{ij}$ and $\dot{\epsilon}_{ij}$ are rates of stress and strain, respectively. It was shown by the authors [3] that the instantaneous stiffness tensor C^{VP}_{ijkl} is a function of the material elastic properties, the current stress level and loading history:

$$C^{VP}_{ijkl} = \lambda\delta_{ij}\delta_{kl} + \mu(\delta_{ik}\delta_{jl} + \delta_{il}\delta_{kj}) - \frac{\mu}{J_2}\frac{\dot{W}^P}{\dot{W}^\Gamma} s_{ij}\,s_{kl} \tag{2}$$

where μ and λ are Lamé constants of the material and δ_{ij} is the Kronecker delta; $s_{ij} = \sigma_{ij} - \delta_{ij}\sigma_{kk}/3$ is the deviatoric part of the stress tensor σ_{ij}, and $J_2 = \frac{1}{2}s_{ij}\,s_{ij}$ is its invariant. In eqn (2), $\dot{W}^P = \sigma_{ij}\,\dot{\epsilon}^P_{ij}$ is the plastic work rate, and $\dot{W}^\Gamma = s_{ij}\,\dot{\epsilon}_{ij}$ is the rate of work of the total deformation which does not include volume changes.

The elastic part $\dot{\epsilon}^E_{ij}$ of the total strain-rate tensor $\dot{\epsilon}_{ij}$ is related to the stress-rate tensor $\dot{\sigma}_{ij}$ by the derivative of the generalized Hook's law (the problem is associated with the small deformation theory)

$$\dot{\sigma}_{ij} = [\lambda\delta_{ij}\delta_{kl} + \mu(\delta_{ik}\delta_{jl} + \delta_{il}\delta_{kj})]\dot{\epsilon}^E_{kl}, \tag{3}$$

while the irreversible (plastic) components $\dot{\epsilon}^P_{ij}$ of the total strain-rate tensor are controlled by Prandtl-Reuss flow law

$$\dot{\epsilon}^P_{ij} = \Lambda s_{ij} \tag{4}$$

where Λ is the flow rule function of the adopted viscoplastic constitutive model. It should be noted that plastic incompressibility is implied from eqn (4),i.e., $\dot{\epsilon}^P_{kk} = 0$.

Consider a rectangular isotropic elastic-viscoplastic plate, the middle plane of which concides with the x-y plane of the Cartesian coordinate system (xyz). The x and y axes are parallel to the plate edges, and the origin of the system is taken at the lower left corner. The dimensions of the plate are: a and b in the x and y directions, respectively, and thickness h in the z-direction. When it is convenient, we will use the following alternative notation for the axes: $x_1 = x$, $x_2 = y$ and $x_3 = z$. The constitutive equation (1) can be rewritten using the plane-stress approximation, in the contracted form

$$\dot{\sigma}_i = Q^{VP}_{ij}\,\dot{\epsilon}_j \qquad (i,j = 1,2,6) \tag{5}$$

where $\dot{\sigma}_1 = \dot{\sigma}_{11}$, $\dot{\sigma}_2 = \dot{\sigma}_{22}$, $\dot{\sigma}_6 = \dot{\sigma}_{12}$ and $\dot{\epsilon}_1 = \dot{\epsilon}_{11}$, $\dot{\epsilon}_2 = \dot{\epsilon}_{22}$, $\dot{\epsilon}_6 = 2\dot{\epsilon}_{12}$. The elements of Q^{VP}_{ij} are obtained from C^{VP}_{ij} (where C^{VP}_{ij} is the instantaneous material stiffness tensor

C_{ijkl}^{VP}, eqn (1), which is rewritten in the contracted notation) as follows

$$Q_{ij}^{VP} = C_{ij}^{VP} - C_{i3}^{VP} C_{3j}^{VP} / C_{33}^{VP} \qquad (i,j = 1,2,6) \tag{6}$$

In framework of the classical plate theory, the governing equation of the plate, which is subjected to a combined normal and shear loading, is given by

$$\begin{aligned} D_{11}^{VP} \partial^4 \dot{w} / \partial x^4 + 2(D_{12}^{VP} + 2D_{66}^{VP}) \partial^4 \dot{w} / \partial x^2 \partial y^2 + D_{22}^{VP} \partial^4 \dot{w} / \partial y^4 &= \\ (\sigma_{xx} h) \partial^2 \dot{w} / \partial x^2 + 2(\sigma_{xy} h) \partial^2 \dot{w} / \partial x \partial y + (\sigma_{yy} h) \partial^2 \dot{w} / \partial y^2 & \end{aligned} \tag{7}$$

It can be shown that the following bifurcation buckling condition for a biaxially compressed, simply supported plate can be established

$$\sigma_{xx} h (m\pi/a)^2 + \sigma_{yy} h (n\pi/b)^2 = D_{11}^{VP} m^4 + 2(D_{12}^{VP} + 2D_{66}^{VP}) m^2 n^2 (a/b)^2 + D_{22}^{VP} n^4 (a/b)^4 \tag{8}$$

where σ_{xx} and σ_{yy} are the normal stresses at the instant of buckling of the plate, and mn is the buckling mode; the plate bending stiffnesses D_{ij}^{VP} are defined by

$$D_{ij}^{VP} = Q_{ij}^{VP} h^3 / 12 \tag{9}$$

In this paper, the inelastic response of the material is described by Bodner and Partom [4] unified viscoplasticity theory. This theory does not assume the existence of a yield condition, which eliminates the need to specify loading or unloading criteria, and the same equations can be directly used in all stages of loading and unloading. According to these equations plastic deformation always exists, but it is negligibly small when the material behavior should be essentially elastic.

The flow rule function Λ in eqn (4) is given, according to this theory, as follows

$$\Lambda = D_0 \exp\left(-\bar{n} \left[Z^2/(3J_2)\right]^n\right) / J_2^{1/2} \tag{10}$$

where $\bar{n} = \frac{1}{2}(\tilde{n}+1)/\tilde{n}$; D_0 and $\tilde{n}$ are inelastic material parameters, and Z is a state variable that represents the hardened state of the material with respect to resistance to plastic flow. In the case of isotropic hardening, the evolution law of this variable is given by

$$\dot{Z} = \tilde{m}(Z_1 - Z)\dot{W}^P / Z_0 \tag{11}$$

where Z_0, Z_1 and $\tilde{m}$ are additional inelastic parameters of the material. Note that the plastic work W^P, whose rate is $\dot{W}^P = \sigma_{ij} \dot{\epsilon}_{ij}^{(P)}$, is taken as the measure of hardening. The physical significance of the above inelastic constants is as follows. The parameter D_0 is the limiting strain rate, Z_0 is related to the "yield stress" of a uniaxial stress-strain curve, and Z_1 is proportional to the ultimate stress. The material constant $\tilde{m}$ determines the rate of work hardening, and the rate sensitivity is controlled by the temperature

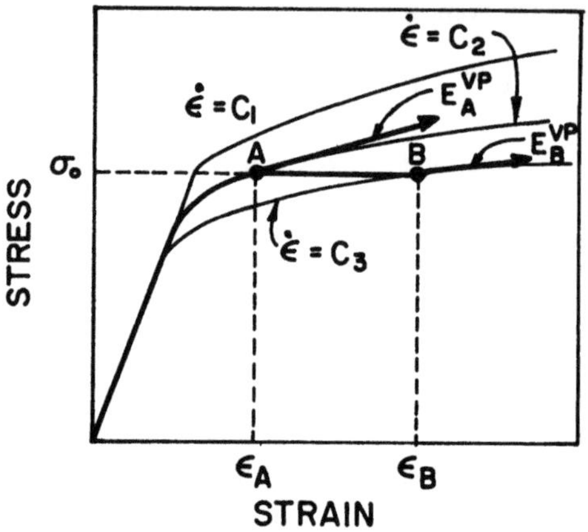

Figure 1: Material response to varios values of strain-rate and local instantaneous material properties during creep.

dependent parameter $\tilde{n}$. The unified theory of Bodner and Partom can be utilized to model primary creep behavior of the material which is subjected to a constant stress loading. In this situation, the total strain rate is equal to the plastic strain rate, since the elastic strain rate vanishes.

In the present paper, creep buckling of viscoplstic plates is investigated by employing a buckling condition which is given in terms of the instantaneous properties of the plate. To this end, consider, for example, the simply supported plate whose buckling condition is given by eqn (8). Suppose that the plate is subjected to a compressive normal loading $\sigma_{xx} = \sigma_0$. In order to determine the critical time of creep buckling of the plate, let us generate its uniaxial responses at various values of applied strain rates. In Fig.1, such a family of curves are depicted by fine lines. For a chosen applied strain rate, one can proceed with the computations along the appropriate stress-strain curve until the value of σ_0 is reached. At each increment the instantaneous properties of the plate D_{ij}^{VP} are examined to verify that buckling condition (8) is not fulfilled. At point A, where $\sigma_{xx} = \sigma_0$, the computitions are performed incromentally along the path AB. Here the applied stress is kept constant σ_0, whereas the strains and strain rates are time dependent. At each increment, the buckling condition is examined in conjunction with the current properties of the plate. These instantaneous properties are determined from eqn (9) together with (6) and (2). This process is continued until buckling of the plate occurs at point B which correspondes to creep buckling.

The present method of the determination of plastic bifurcation buckling can be applied for various types of inelastic platos, subjected to various boundary conditions. If

Table 1: Material constants of an aluminum alloy (2024-T4)

E (GPa)	ν	D_0 (sec^{-1})	Z_0 (MPa)	Z_1 (MPa)	$\tilde{m}$	$\tilde{n}$
65.7	0.33	10000	340	435	300	4.0

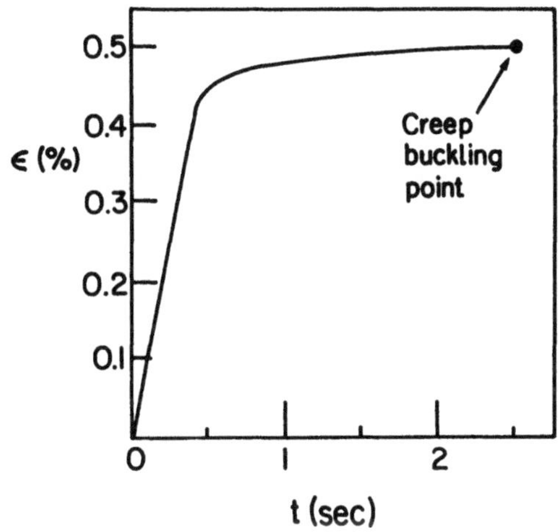

Figure 2: Creep curves and creep buckling point

an analytical, exact solution to the governing equations is not possible, approximate methods (e.g., numerical) can be employed in conjunction with the proposed methodology.

Application

The proposed method is illustrated by the prediction of the critical time of creep buckling of a square viscoplastic plate. The material of the plate is an alluminum alloy 2024-T4 that is rate sensitive at elevated temperatures. In Table 1, the elastic and inelastic properties of the material are given in framework of the Bodner-Partom unified viscoplastic theory at a temperature $T = 204.4°C$. The aluminum alloy is isotropic in the elastic region and isotropic work hardening in the plastic region.

In Fig.2, a creep curve (i.e., strain ϵ_{xx} versus time t) is given for stress level $\sigma_0 = 273MPa$. The curve was generated by applying an initial strain rate $\dot{\epsilon}_0 = 0.01/sec$

in the x-direction. Plastic buckling of the plate occurs at time $t = 2.5sec$ and strain $\epsilon_{xx} = .49\%$. For a different values of initial strain rate $\dot{\epsilon}_0 = 10^{-4}$ and $\dot{\epsilon}_0 = 10^{-6}$, buckling of the plate occurs at $\epsilon_0 = .51\%$ and $t = 56.5sec$, and $\epsilon_{xx} = .57\%$ and $t = 95.0\,min$, respectively. It should be noted that, in general the critical time of creep buckling of the plate exhibits a strong sensitivity to the values of the applied stress σ_0 and initial strain rate $\dot{\epsilon}_0$.

References

1. Kurshin, L.M.: Creep stability: a survey. *Mech. of Solids*, 13 (1978), 111–146.
2. Tvergaard, V.: Plasticity and creep at finite strains. *Proc. 17th Congress of Theoretical and Applied Mechanics* (edited by P.Germain, M.Pian and D.Caillerie), Ammsterdam: Elsevier Sci.Pub. (1989).
3. Paley, M. and Aboudi, J.: Viscoplastic bifurcation buckling of plates. *AIAA J*, (1991).
4. Bodner, S.R. and Partom, Y.: Constitutive equations for elastic-viscoplastic strain-hardening materials. *J.Appl.Mech.*, 42 (1975), 385–389.

Contributions to Robust Methods of Creep Analysis

R K Penny

Department of Mechanical Engineering, University of Cape Town, South Africa

Summary

Robust methods for the predictions of deformations and lifetimes of components operating in the creep range are presented. The ingredients used for this are well-tried numerical techniques combined with the concepts of continuum damage and so-called reference stresses. The methods described are derived in order to obtain the maximum benefit during the early stages of design where broad assessments of the influences of material choice, loadings and geometry need to be made quickly and with economical use of computers. It is also intended that the same methods will be of value during operation if estimates of damage or if exercises in life extension or inspection timing are required.

Introduction

The need for better creep design methods needs no justification : parts are getting hotter and more highly stressed and systems have to last longer and cost less. Present day codes of practice are usually conservative in their design rules and this conservatism not only leads to higher capital costs but also to unnecessarily early retirement of plant. Designers are therefore faced not only with the need to provide accurate, reliable predictions of operating lifetimes but also with the need to incorporate in their designs a sound basis for updating and extending lifetimes during operation. The operating plant has to become the designer's laboratory wherein in-situ, on-line, periodic and well-chosen measurements of creep and its interaction with other deteriorating mechanisms will be continuously monitored.

In addressing these issues some interesting and powerful techniques deriving from the bases of mechanics have emerged. Generally speaking, these include techniques for stress analysis and the use of concepts to account for progressive material degradation. The coupling of stress analysis with material degradation ("damage") is an attractive proposition and one that, if feasible, has promising capabilities for predicting reductions in strength, stiffness, toughness, residual and ultimate lives. More powerful computers and computational techniques have become available to help in these predictions but there is conflict - not only with prohibitive costs of

M. Zyczkowski (Ed.)
Creep in Structures
IUTAM Symposium Cracow/Poland 1990

computer processing time - between increased accuracy and insufficient relevant materials data. In any case though there is the increasing recognition that the designer needs to be able to *de-couple* the effects of geometry, loading and materials characteristics. This is needed not only to be able to determine which factors are important and which are not, but also to be able to make decisions quickly and economically in ways which are within the scope of available tools. The call therefore, and it appears to be a desperate one [1], is for *robust* methods which are rigorously based, can be used flexibly and which provide solutions commensurate with the quality of available data and the tolerances which usually have to be accepted in modeling geometries and loadings. The quest for such methods should also lead to carefully planned physical testing programmes.

The basic ingredients for robust methods in creep mechanics are available : well-tried numerical methods of stress analysis and the concepts of "continuum damage" and "reference stresses". The work of Kachanov [2] and of Rabotnov [3] in their early studies of macroscopic damage variables during creep has developed continuously and forms a framework for a broader approach to other damaging mechanisms such as fatigue and large scale plasticity [4]. In fact, so prolific has been the literature on "damage mechanics" that the stage of reviews of reviews [5] has already been reached. Far less spectacular has been the application of reference stress techniques [6], [7] which aim at the derivation of representative stress levels which characterise the creep behaviour of components from a minimum amount of data.

The aim of this paper is to contribute to the generation of robust methods which have the following characteristics:

- cater for several interpretations of failure which may be used by designers in assessing lifetimes,
- offer the means for parametric studies which enable the identification of essential features affecting component behaviour with the fewest number of iterations,
- allow graphical representations of results for commonly occurring components, and,
- contribute to the furtherance of Codes of Practice.

Normalised Arrangements of Constitutive Laws for Creep Including Damage Effects

In order to maximise the potential for assessing the effects of dependent variables (loading, material and temperature) on any calculated results of the creep response of a component, it seems sensible to cast the results in normalised form whenever possible. This can be accomplished by using the well-tried device of a distorted time measure [7] and converting the creep and damage relationships [3] accordingly. Inherent in such a procedure is also the use of the time-hardening hypothesis as follows for the uniaxial case during which long term, low stress conditions leading to brittle failure apply :

$$\varepsilon_c' = A\sigma^m f'(t) \quad ; \quad \varepsilon_{cd}' = A\,\sigma^m f'(t)(1-\omega)^{-p} \quad ; \quad \omega' = B\sigma^k f'(t)(1-\omega)^{-r} \qquad (1)$$

where σ is the nominal stress at time t which gives rise to creep strain ε_c in the absence of damage, ε_{cd} when damage occurs and at which time damage is ω. A, B, m, p, k and r, are material constants; also $( )' = \partial( )/\partial t$ This formulation seems to work quite well in the uniaxial case for calculating yield strength reduction caused by creep damage and residual and rupture lifetimes [9]. It is also worth noting that the same formulation is easily extended to include temperature dependence [10] but in this case the likelihood of sufficient relevant data is small.

Assuming that the von Mises yield criterion and the Prandtl-Reuss relations are valid during creep [7] and that an equivalent stress criterion holds for damage accumulation [8], the multi-axial stress creep and damage laws can be written:

$$\dot{\lambda}_{ij,c} = \Sigma^{*^m} \frac{d\Sigma^*}{d\Sigma_{ij}} \quad ; \quad \dot{\lambda}_{ij,cd} = \Sigma^{*m}\,(1-\omega)^{-p}\,\frac{d\Sigma}{d\Sigma_{ij}} \quad ; \quad \dot{\omega} = C\,\Sigma_\beta^k\,(1-\omega)^{-r} \qquad (2)$$

where Σ_β is an equivalent stress given by :

$$\Sigma_\beta = \beta\Sigma^I + (1-\beta)\,\Sigma^* \qquad (3)$$

Further definitions needed are :

$$\Sigma = \frac{\sigma}{\sigma_o} \quad ; \quad \lambda = \frac{\varepsilon}{\varepsilon_o} \quad ; \quad \sigma_o = E\,\varepsilon_o \quad ; \quad \tau = \int A\,E\,\sigma_o^{m-1} f'(t)dt \qquad (4)$$

where σ_o is some convenient normalising stress, whilst :

$$C = \frac{B\,\sigma_o^{1+k-m}}{A\,E} \qquad \text{and} \qquad (\dot{\ }) \equiv \frac{\partial(\ )}{\partial\tau} \tag{5}$$

In addition, Σ^* and Σ^I are the Mises effective and maximum principal stresses respectively whilst $0 \le \beta \le 1$. Defining an initial damage ω_o as the material damage caused by initial flaws such as inclusions, micro-cracks and the like, equation (6) can be integrated between the limits ω_o and ω whilst failure occurs at $\tau = \tau_R$ when $\omega = 1$; these conditions result in:

$$\tau_R = \frac{(1-\omega_o)^{1+r}}{C\,(1+r)\,\Sigma_\beta^k} \tag{6}$$

$$(1 - \omega) = (1 - \omega_o)\,(1 - \Gamma)^{1/(1+r)} \tag{7}$$

$$\lambda_{ij,c} = K\,(1 - \omega_o)^{1+r}\,\Gamma \tag{8}$$

$$\lambda_{ij,cd} = \eta\,K\,(1 - \omega_o)^{1+r-p}\left[1 - (1 - \Gamma)^{1/\eta}\right] \tag{9}$$

$$\frac{d\lambda_{ij,c}}{d\Gamma} = K\,(1 - \omega_o)^{1+r} \tag{10}$$

$$\frac{d\lambda_{ij,cd}}{d\Gamma} = \eta\,K\,(1 - \omega_o)^{1+r-p}\,(1 - \Gamma)^{(1-\eta)/\eta} \tag{11}$$

$$\text{where} \qquad \eta = \frac{1 + r}{1 + r - p} \quad \text{and} \quad K = \frac{1}{C\,(1+r)}\,\frac{\Sigma^{*m}}{\Sigma_\beta^k}\,\frac{d\Sigma^*}{d\Sigma_{ij}} = \tau_R\,\Sigma^{*m}\,\frac{d\Sigma^*}{d\Sigma_{ij}} \tag{12}$$

These equations form the basis for any stress analysis which aims to calculate the deformation and rupture characteristics of a component during creep. If the coupling between the progressive degradation of the material and its stiffness is to be taken into account then equations (9) and (11) must be used in stress analysis. When prohibitive computer costs make this unattractive, then material degradation is decoupled for the purpose of the stress analysis and equations (8) and (10) are used. If the latter course of action is taken then the analysis will not be capable of predicting the rupture characteristics directly. However, prudent manipulation of the results of the analysis, interpreted with the use of (9) and (11), is likely to provide a method of component assessment which is particularly useful at the early stages of design and in which economy in computer processing time makes it viable.

Criteria for Component Failure

An analysis which couples progressive degradation with stiffness changes will allow calculation of all characteristics including the rupture time. A simpler approach is to estimate strains and rupture times based upon stationary state stresses. This will underestimate the strains and, therefore, more conservative criteria need to be formulated; for example by adjustment of the calculated stationary state strains and strain rates in the component, such that the effects of damage are taken into account. In this process, the judgment and experience of the designer will be crucial and much will depend on the intended use of the component. For example, a pressure vessel with a lifetime expectancy of 20-30 years could be monitored carefully and estimates of the remaining life can be made and updated against predicted values in a systematic way. Short of emergencies, which might include overloading, it is likely that changes in strain rate will be reasonable predictors of micro-structural change and damage. On the other hand, in which life-times may be shorter and in which operating conditions contain more imponderables, predictions and measures of total strain will be more important ; e.g. turbine blade integrity assessment. Whichever the case, it is a relatively simple matter for stationary strain estimates to be adjusted to account for damage in current or forward predictions. So far as rupture estimates are concerned, stationary state stress values are likely to be too conservative in most cases because of the stress shedding from high stress regions to regions of lower stress that will happen, in the usual way, for statically indeterminate components. If it is accepted that stationary state stresses are too conservative then it is certain that initial elastic values are even more so, for creep conditions - a very restricting approach which has been suggested in some Codes of Practice. For uniaxial conditions, stress redistribution is constrained and net section stresses will most likely govern rupture. In this case, conservative life times can be calculated on the basis of limiting strain rates to realistic values.

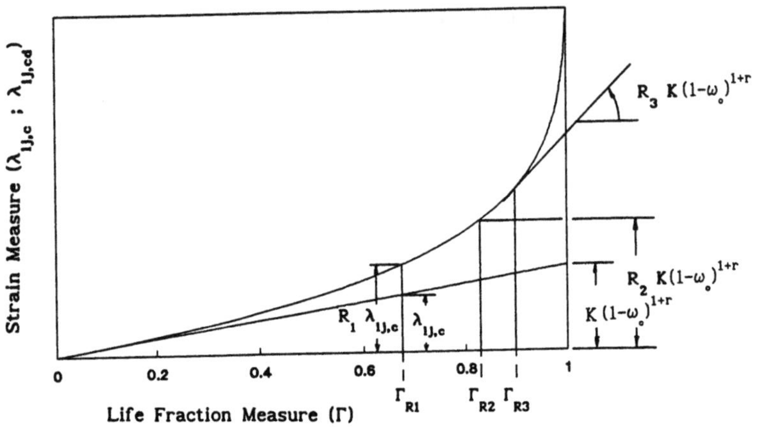

Figure 1. Schematic representation of failure criteria based on strain measures

Figure 1 is a generalised representation of the strain measures $\lambda_{ij,c}$ and $\lambda_{ij,cd}$ as a function of the life fraction measure Γ for a given value of $\eta_1 = \eta\ (1-\omega_o)^{1+r-p}$, together with several failure criteria which could be used to calculate likely component retirement times from the gross deformation or deformation rate. The choice here will depend on the degree of conservatism thought necessary. In the first case, an acceptable value of creep strain which includes contributions from the *hardened* stationary state and the *softened* (void accumulated) state is chosen as a proportion R_1' of the hardened state. The corresponding life fraction measure Γ_{R1} to achieve this state is easily calculated from (8) and (9) and is given by :

$$R_1'\ (1 - \omega_o)^p\ \Gamma_{R1} = \eta \left[1 - (1 - \Gamma_{R1})^{1/\eta}\right] \tag{13}$$

Another method based upon a strain measure would be in terms of a factored amount R_2' $(1 < R_2' < \eta\ (1-\omega_o)^{1+r-p})$ of the stationary state creep strain at rupture. The life fraction measure Γ_{R2} corresponding to the choices of R_2' is:

$$R_2'\ (1 - \omega_o)^p = \eta \left[1 - (1 - \Gamma_{R2})^{1/\eta}\right] \tag{14}$$

A third approach would be to regard the strain rate as a more critical measure of the imminence of failure. In this case, a proportion R_3' of the stationary state creep rate is used to guard against the effects of excessive void formation. The life fraction measure Γ_{R3} at which this occurs is given by (10) and (11) :

$$R_3' \ (1 - \omega_\circ)^p = \eta \left[1 - \Gamma_{R3}\right]^{(1-\eta)/\eta} \tag{15}$$

The factors $R_i = R_i' \ (1-\omega_\circ)^p$ etc. can all be represented graphically and by way of example, Figure 2 gives particular values for R_3.

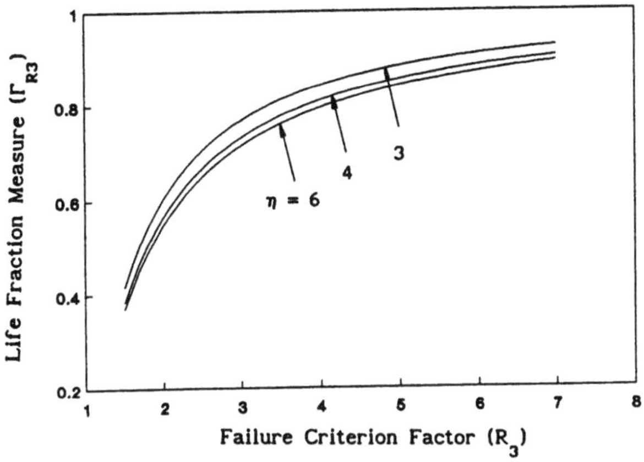

Figure 2. Variation of life fraction using third failure criterion.

<u>Stationary State Analyses and Use of Reference Stress Techniques</u>

Methods for calculating stationary state behaviour of components are well known [7]. In performing these calculations, the governing equations are all phrased in normalised terms as outlined earlier ; this not only makes for more efficient computational routines but also gives the results in a form that is convenient in design. Stresses and strains are calculated as proportions of nominal values (stress/strain concentration factors) and these and other quantities can be studied and presented parametrically so that the effects of geometry or material constants can be easily evaluated. For example, if a power law is used to represent creep then the results obtained from the stationary state calculations will quickly reveal the importance or otherwise of the creep exponent. Likewise, other material constants are embodied in the distorted time measure τ through equation (4) so that without performing new calculations every time the material is changed during early design studies, the effects can be judged from the amount of distortion that results in τ. In the same manner, if the component response is examined from a viewpoint of temperature changes, this can also be accommodated in the distorted time measure.

Another powerful concept in creep design is that of the so-called reference stress [5]. What this amounts to is that there exists for a given component, a unique stress level at which the stationary state creep response is relatively insensitive to the precise stress dependence of creep ; in our case the response does not vary much with the stress exponent m. The significance of this is manifold but, for the purpose of this paper, use of a reference stress as that value for normalising the results of creep calculations means that all data can be computed once and for all for a given component regardless of the material. The results for standard components can then be catalogued for use in design in ways similar to those in which elastic stress concentration factors have been collected together. Likewise, estimates of damage accumulation and rupture times can be tabulated for various life- fractions. This is particularly valuable in the early stages of design where design regions can be mapped.

References

1. C E Jaske "Predicting the Residual Life of Plant Equipment - why worry?". J Press. Vess. Tech., Trans. ASME vol 107, 214-217, 1985.

2. L M Kachanov "On Rupture Times Under Conditions of Creep". Izvestia Akademi Nauk SSR,Otd. Tekhn. Nauk, 8, 26-31, 1958.

3. Y N Rabotnov "**Creep Problems in Structural Members**". North-Holland Co., Amsterdam, 1969.

4. J L Chaboche "Continuum Damage Mechanics : Part 1 - General Concepts". J Appl. Mech., Trans. ASME, vol. 55, 59-64, 1988.

5. D Krajcinovic "Damage Mechanics", J Mechanics of Materials". Elsevier, 8, 117-197, 1989.

6. R G Sim "Creep of Structures". Ph.D. Dissertation, Cambridge University, 1968.

7. R K Penny and D L Marriott "**Design for Creep**". McGraw Hill Co., 1971

8. D Hayhurst "Creep Rupture Under Multi axial States of Stress". J Mech Phys Solids, 20, 381-390, 1972

9. R K Penny "The Usefulness of Engineering Damage Parameters During Creep". J Met and Materials, 8, 278-283, 1974

10. M A Weber "Study of Creep in Machine Elements Using Finite Element Methods". M.Sc(Eng) thesis, University of Cape Town, 1990.

On the Stability of Viscoelastic Beams Under a Stochastic Excitation

V.D. POTAPOV

Moscow Railway Engineering Institute, USSR

Summary

The experience acquired when observing real viscoelastic structures under operating conditions indicates that there is a great random scattering of values of control parameters that define strength, rigidity and stability of structures. This is explained primarily by the stochastic nature of external loads and viscoelastic material properties being described by random functions. Hence solving the stochastic viscoelastic problems using the technique of the theory of random functions seems to be important not only from the theoretical point of view but also from the practical one. The present work intends to investigate stability of viscoelastic beams under random effects.

The stability of the viscoelastic beams

The stability of viscoelastic beams is investigated using the model problem given in [1]. The elastic beam located in a viscous medium and compressed by the force of the form of stationary random process with rational spectral density is considered. The equations describing the motion of the beam are as follows

$$\frac{\partial W}{\partial t} = -A(EIW'''' + FW'')$$

$$\frac{d^i F^0}{dt^i} + a_1 \frac{d^{i-1} F^0}{dt^{i-1}} + \cdots + a_i F^0 = \frac{\pi^2 EI}{l^2} M\xi \tag{1}$$

where $F(t) = F_0 + F^0(t)$; F_0 =const., $F^0(t)$ are the mean value and random fluctuation of longitudinal force respectively; $\xi(t)$ is the Gaussian white noise; $A, a_1, \ldots, a_i, M$ are constants. The remaining designations are commonly used ones. The angle brackets denote the ensemble average. For a beam pivoted at the ends, the condition of asymptotical stability in the sense of Liapunov was obtained with reference to statistical moments of deflection $< W^n(t,x) >$ under perturbation of initial conditions. It is indicated that the condition of stability of beam as a system with infinite number of degrees of freedom

$$\alpha_0 < 1 - \frac{n}{2}\left(\frac{M}{a_i}\right)^2, \quad \alpha_0 = < \alpha > = \frac{|< F(t) >|^2}{\pi^2 EI} \tag{2}$$

coincides with the one for beam with given initial perturbation of the form $f(0)\sin \pi x/l$.

The solution to the problem with constantly acting excitations of the form of a transverse loading or initial bending of a beam is proved to be possible only in the case of

M. Zyczkowski (Ed.)
Creep in Structures
IUTAM Symposium Cracow/Poland 1990

the fluctuations of compressing force proportional to white noise [2]. The condition of the existence of moments $< W^n(t, x) >$ in this case coincides with the condition of the asymptotic stability of the beam with reference to statistical moments of the same order (2).

The considered model problems give an idea about the features of behaviour of viscoelastic structures under stochastic compressing forces. However, the problems related to the analysis of behaviour of the real structures may be solved only by use of numerical methods.

Let's consider the problem of longitudinal bending of viscoelastic beam pivoted at the ends and having initial sine bending of axis with amplitude f_0. The beam is subjected to longitudinal force of the form of normal stationary random process. Supposing that the force slowly vary in time considerations may be restricted to quasi-static problem. Then, amplitude $f(t)$ of the additional deflection of the beam $f(t) \sin \pi x/l$ is defined by the solution of stochastic nonlinear integral equation

$$f(t) - \alpha(t)[f(t) + f_0] - \int_0^t K(t-y)\alpha(y)[f(y) + f_0]dy = 0 \tag{3}$$

where kernel $K(t-y)$ characterizes the creep properties of the material. With a view to solve this equation, the Krylov-Bogolubov method [3] is used. To this end, the interval of integration $[0, t]$ is divided into n sufficiently small parts with duration $\Delta t = t/n$. Time values t_k $(t_0 = 0, \ldots, t_k = k\Delta t, \ldots, t_n = n\Delta t)$ correspond to the boundaries of the parts. Finally, integral equation (3) is replaced by the system of algebraic equations of the form

$$\begin{aligned} f(t_k) - \alpha(t_k)[f(t_k) + f_0] - 0.5 \sum_{i=0}^{k-1} \{\alpha(t_i)[f(t_i) + f_0] + \\ + \ \alpha(t_{i+1})[f(t_{i+1}) + f_0]\} \int_{t_i}^{t_{i+1}} K(t_k - y)\, dy = 0\,, \quad k = 0, 1, \ldots, n \end{aligned} \tag{4}$$

Simulation of Gaussian stationary random process $\alpha(t)$ may be carried out, for example, according to the known algorithm of autoregressive moving average based on formula [4]

$$\alpha_k = \sum_{j=0}^{m} a_j c_{k-j} - \sum_{j-1}^{n} b_j \alpha_{k-j} \tag{5}$$

where c_{k-j} is the sequence of independent Gaussian numbers with parameters $(0, 1)$; m, n, a_j, b_j, are the parameters determined for the given correlation function of stationary process $\alpha(t)$. Kernel of creep $K(t-y)$ is also a random function and any numerical algorithm can be used for simulation of its realization. However, due to the absence of experimental data, let's limit ourselves to discussing a special case, when this kernel is present by the deterministic function with random parameters. Finally, the amplitude of the beam initial bending f_0 (or in general case function $W_0(x)$) should be considered as a random one. To emphasize the influence of longitudinal force scattering and characteristics of material viscoelasticity on the statistical characteristics of function f_0, it is further on assumed that the function is equal to 1. Let's consider exemplary power kernel of creep of the form

$$K(t-y) = g\,E\,b(t-y)^{g-1}\,, \quad g<1 \tag{6}$$

Here $E\,,b$ are random values having logarithmically-normal and g-normal distribution of probabilities with parameters obtained from experiments for polymer specimens: $<\ln E>$=11.173; $<\ln b>$=10.740; $<g>$=0.230; mean-root-square scattering is $S_{\ln E}$=0.313; $S_{\ln b}$=0.500; S_g=0.059, while E is expressed in kgf/cm^2, b - in cm^2/(kgf·hour^g).

The correlation function of non-dimensional random fluctuation of longitudinal force

$$\alpha^0(t) = F^0(t)l^2/[\pi^2\exp(\ln E)I]$$

is assumed to be

$$K(t-y) = S^2\alpha\exp(-Q|t-y|)\,, \tag{7}$$

where $S^2\alpha$ is the dispersion of function $\alpha^0(t)$, Q is the parameter characterizing the scale of correlation of the force $F(t)$. For simulation of random process $\alpha^0(t)$ is described by correlation function (6) and Eq. (5) takes the following form

$$\alpha_k^0 = a_1c_k + b_1\alpha_{k-1}^0\,,$$

where

$$\alpha_0^0 = S\alpha C_0\,, \quad a_1 = S\alpha\sqrt{1-r^2}\,, \quad b_1 = r\,, \quad r = \exp(-Q\Delta t)$$

The values of random numbers: $E\,,b\,,g$ are received according to standard programs. Realization of nonstationary process $f(t)$ is determined from Eqs. (4) for each realization of stationary process $\alpha(t)$ and random values $E\,,b\,,\lambda$. The moments of unknown function $f(t)$ are determined by statistical analysis after sufficiently large number of these realizations. Further, let's limit ourselves to consideration of only two first statistical moments - $<f(t)>$ and dispersion $S^2(t)$ though computation of the moments of higher order does not cause any principal difficulties.

Figs. 1 and 2 show $<f>$ and S versus time charts which are the result of analysis of 200 realizations for α_0=0.25. Curves 1 are built for $S\alpha$=0.05, Q=0.1 1/hour and deterministic values of parameters $\ln E =<\ln E>$, $\ln b =<\ln b>$, $g=<g>$; curves 2 - for $S\alpha$ =0 and random values $e\,,b\,,g$ with distribution parameters specified above; curves 3 - for $S\alpha$=0.05, Q=0.1 1/hour and the same random values $E\,,b\,,g$. For comparison chart 4 presented in Fig. 1 corresponds to deterministic initial data.

The presented charts give an opportunity to estimate the influence of each type of initial random information on mean value and mean-root-square scattering of the function $f(t)$. For the sake of comparison the charts 1', 2', 3' built in conformity with the result of analysis of 100 realizations are shown. Comparison of charts 1, 2, 3 and 1', 2', 3' indicates a more precise definition of the unknown characteristics for the increasing number of analyzed realizations.

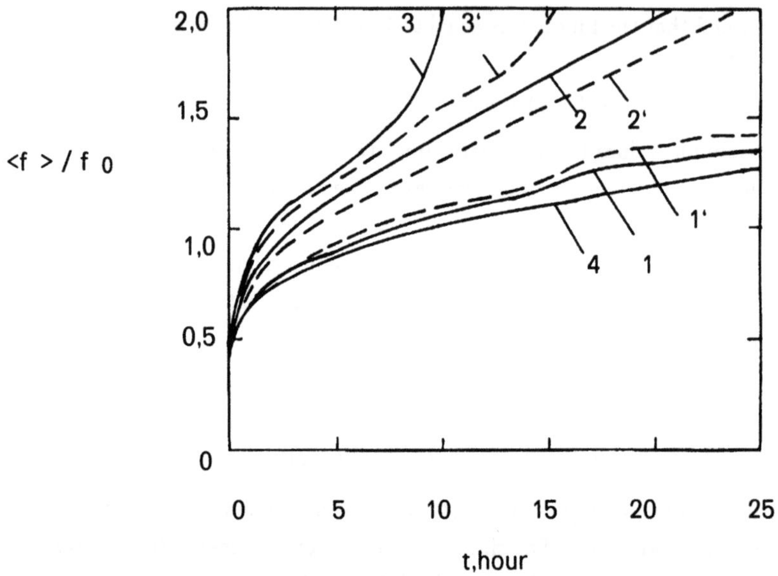

Figure 1: Mean value $< f(t) >$

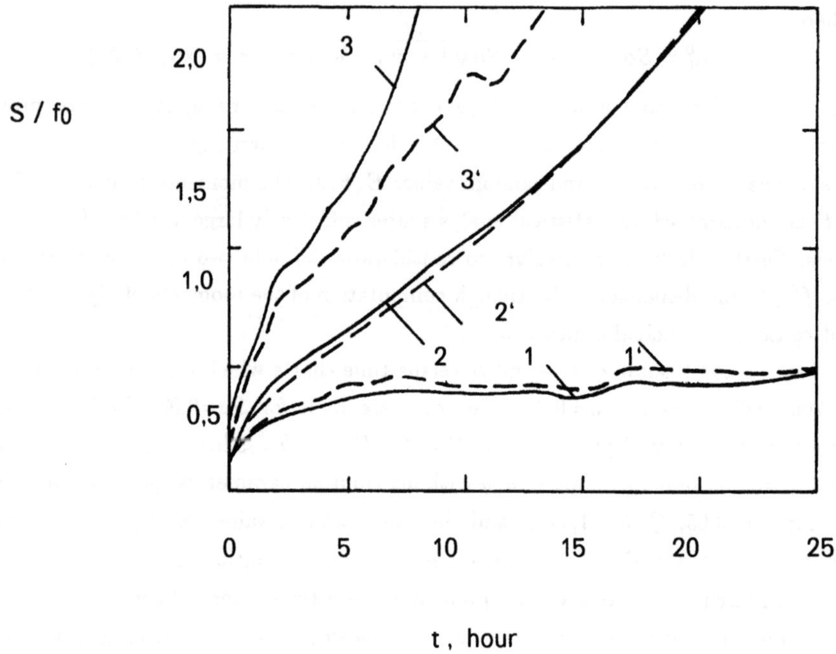

Figure 2: Mean-root-square scattering $S(t)$

Since $< f(t) >$ and $S(t)$ can unrestrictedly increase in time the stability of the beams in question may be considered only as the stability in final time interval. In case of stochastic formulation of the problem, it is connected with the problem of reaching the boundaries, i.e. a random process reaches any initially chosen value D.

The charts showing the variation of probability for cases for which function $f(t)$ reaches limit value D equal $2f_0$ are given in Fig. 3. Curves 1, 2, 3 are built for the same initial data as those given in Figs. 1 and 2.

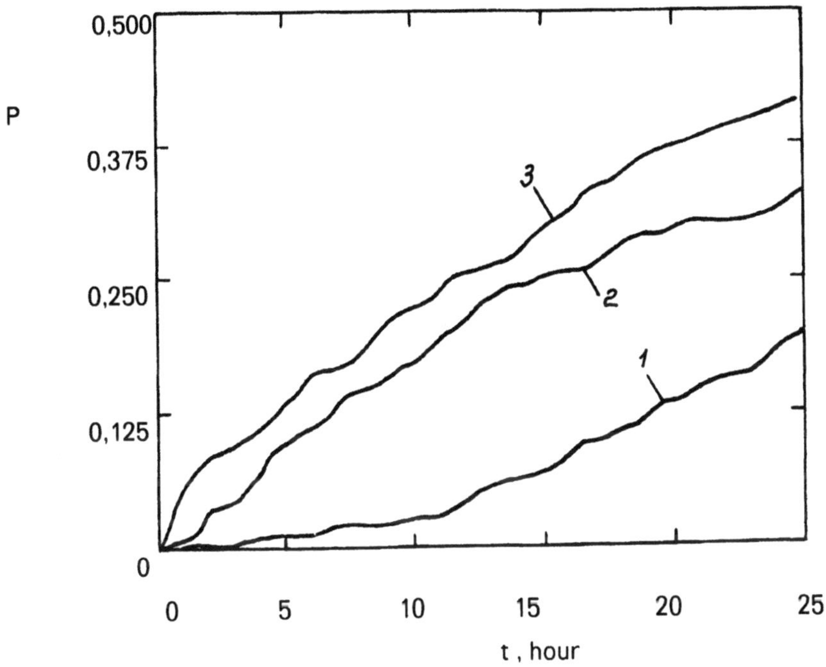

Figure 3: Probability of reaching by function $f(t)$ the limited value $D = 2F_0$

It follows from the given charts that the deflection of the beam in time interval $[0\,,t]$ with probability $1 - P(t)$ will be less then value D, which will indicate the stability of the beam at the given level of the probability.

The given results show the considerable influence of the longitudinal force and of random scattering material properties on the characteristics of the beam deflection and its stability.

References

[1] Potapov, V.D.: *On the stability of beams under stochastic excitation*, Appl. Math. and Mech., vol. 53, **6**, (1989), 1006-1013 (in Russian)

[2] Potapov, V.D.: *Stability of viscoelastic elements of structures*, Moscow, Stroyizdat, 1985 (in Russian)

[3] Kantorowich, L., Krylov, V.I.; *Approximate methods of higher analysis*, Moscow-Leningrad, GITTL, 1952 (in Russian)

[4] Shalygin, A.S., Palagin, Y.I.: *Applied methods of statistical simulation*, Leningrad, Mashinostroenie, Leningrad department, 1986 (in Russian)

On the Non-Linear Creep Flow Potential

S.A. SHESTERIKOV and M.A. YUMASHEVA

Institute of Mechanics,
University of Moscow, USSR

Summary

New relation of mechanical constitutive equation for non-linear creep is represented. This one describes non-linear flow and ideal plasticity simultaneously. It is shown that this new relation permits more adequate description of experimental data. There are some examples of using this relation for particular problems of creep in structures.

General relations

The concept of a mechanical equation of state is currently accepted in the theory of creep. The problem is that the defining equations should adequately describe the experimentally observed phenomena. There are many particular forms of these equations in the theory of creep. That is a result of broad spectrum of variety of the creep phenomenon. Theory of the mechanical equation of state is widely discussed in Rabotnov's fundamental monograph [1]. We propose one variant of equation of state which can describe some particular properties of metals at state of creep.

Let us consider an elementary case of creep, namely non-linear viscous flow. For the uniaxial stress state we have $\dot{p} = (\sigma, T)$ where $\dot{p}$ is the creep strain rate, σ is the stress, and T is the temperature. Usually power law of Bailey-Norton is taken in the form

$$\dot{p} = A\sigma^n , \tag{1}$$

or an exponential law, or more complex relations of the type of Garofalo's law (hyperbolic sine to the power n). It should be pointed out that when relation of type (1) is used, it is impossible (even for constant temperature) to select the only one value of n for entire range of stresses where creep must be taken into account. For instance, experimental data were obtained from duraluminum speciments, and it was necessary to take n=5.15 for $\sigma < 275$ MPa and n=29.07 for $\sigma > 275$ MPa. An analysis of experimental data performed by the authors for various metals indicate that for steady-state creep we take relation of the form

$$\dot{p} = A(\sigma - \sigma_0)^n / (\sigma_b - \sigma)^n \qquad (\sigma_0 < \sigma < \sigma_b) , \tag{2}$$

M. Zyczkowski (Ed.)
Creep in Structures
IUTAM Symposium Cracow/Poland 1990

where σ_0 - is the stress below which there is virtually no creep, σ_b - is the stress of the yield-point type. We should mention that for structural alloys n can be taken equal to one in many cases.

In case if the difference of properties in tension and compression must be taken in account we can postulate next equation

$$\begin{aligned} \dot{p} &= A\sigma\sigma_{1b}/(\sigma_{1b}-\sigma) \qquad (\sigma > 0)\,, \\ \dot{p} &= A\sigma\sigma_{2b}/(\sigma_{2b}-\sigma) \qquad (\sigma < 0)\,. \end{aligned} \tag{3}$$

Obviously, relation (3) is a continuous dependence for $\dot{p}$ upon passing through $\sigma = 0$. If $\sigma_{1b} = -\sigma_{2b}$ the material can be regarded as having different tensile and compressive strengths at high stress levels, and nearly the same rate deformation of creep at low levels. The existing experimental data confirm these notions. In case when $\sigma_{1b} = -\sigma_{2b}$ relation (3) can be represented in a form

$$\dot{p} = A\sigma\sigma_b/(\sigma_b^2 - \sigma^2) \quad or \quad \dot{p} = A\sigma/(\sigma_b^2 - \sigma^2)^{1/2}\,. \tag{4}$$

That form is good for writing a law of creep for complex stress state. In that case relations of creep can be represented in a form of flow rule with potential of creep Φ

$$\dot{p}_{ij} = \frac{\partial\Phi}{\partial s_{ij}}\,, \tag{5}$$

where $\dot{p}_{ij}$ - tensor of creep strain rates and s_{ij} - deviator of stress tensor. Thus

$$\Phi = -A(\sigma_b^2 - \sigma_i^2)^{1/2}\,, \tag{6}$$

where σ_i denotes stress intensity (the Huber-Mises effective stress). That relation is a direct generalization of the second expression in (4). In case (3) generalization can be taken in the form:

$$\Phi = -A(B + B_1\sigma_i^2 + b_1\sigma_i J_0^2)^{1/2}\,, \tag{7}$$

where $J_0 = \sigma_{kk}$. In case if σ_b - yield stress in pure tension and σ_c - yield stress in pure compression we use the following form instead of (7)

$$\Phi = -A\left[2\sigma_b^2\sigma_c^2 - (\sigma_b^2 + \sigma_c^2)\sigma_i^2 - (\sigma_c^2 - \sigma_b^2)\sigma_i J_0\right]^{1/2}\sigma_b^{-1}\,. \tag{8}$$

Let us point out some general properties of the equation proposed. Obviously, all particular forms of the potential have two essential features. The first, in case $\sigma_i << \sigma_b$ a practically linear relation between stress and strain rate is obtained. At $\sigma_i \rightarrow \sigma_b$ from equation (6) we see that state of stress tends in limit to a perfectly plastic stress distribution. Similarly, the expression under square root in (8) tends to zero in this case. Scheme of perfect plasticity is often used as a limit creep analysis with power law. In this case that scheme is the direct consequence of the potential form proposed.

Hardening and softening materials

In usual creep tests we observe three stages: hardening, non-linear flow and softening. In case when it is necessary to take into account not only steady-state flow but also the hardening region it is effective to accept the hypothesis of seperation of the total deformation into non-linear flow p_2 and bounded creep $p_1 (p = p_1 + p_2)$. For p_2 we employ relation (2) or (4), while for p_1 we employ the hardening theory relation

$$\dot{p}_1 = j(\sigma, p_1, T) . \tag{9}$$

In the equation (9) p_1 is responsible for hardening. In simplest form for $T = const$ we have

$$\dot{p}_1 = C(\sigma)(p_0 - p_1) . \tag{10}$$

Paper [2] gives in details a method of determination of all parameters and functions in (2),(3),(9) and (10). As an example, we can cite the case of processing the creep curves obtained by Rabotnov [1] (p. 203). For that example

$$\dot{p}_1 = C(p_0 - p_1) , \qquad p_0 = B(\sigma/\sigma_0 - 1) , \qquad \dot{p}_2 = A(\sigma - \sigma_0)/(\sigma_b - \sigma) ,$$

where $\sigma_0 = 117$ MPa, $\sigma_b = 297$ MPa, $C = 0.15$ h^{-1}, $A = 3.8 \times 10^{-5}$ h^{-1}, $B = 0.027$. The results of analysis have been shown in the figure. The solid lines in the figure are experimental creep curves. Results of approximation are indicated by open circles. The dashed lines represent approximation using expression (1) with usually taken form of hardening law. The solid circles show the refinement provided by (1), when parameters are assumed to be linear functions of the stress. The numbers indicate the stresses at which the creep curves were obtained.

When it is necessary to allow for the third segment of the creep curve and for the effect of creep damage and rupture, relation (9) can be written in the form

$$\dot{p}_2 = A(\sigma - \sigma_0)/[\sigma_b(1 - \omega) - \sigma] . \tag{11}$$

That expression can naturally be supplemented by a kinetic (evolution) equation for the damage parameter ω of type

$$\dot{\omega} = \dot{\omega}(\sigma, \omega, T) \quad or \quad \dot{\omega} = D[\sigma/(1 - \omega^r)]^n . \tag{12}$$

Detailed analysis and comparison with experimental data was carried out in paper [3]. In last investigations on creep rupture [4] it was shown that in the theory of creep as well as in approximate data of creep rupture an expression analogical to (3) holds.

Some examples of application of the relations proposed

Bending with tension of beams.

The use of the relations (4) for analysis of a rectangular beam under combined bending and tension leads to following two equations for rate of curvature $\dot{\kappa}$ and rate of axial strain $\dot{\varepsilon}_0$ in terms of longitudinal force N and bending moment M:

$$N\dot{\kappa}/b\sigma_b = 2\dot{\varepsilon}_0 - \ln(1+\dot{\varepsilon}_0+\dot{\kappa}h) + \ln(1-\dot{\varepsilon}_0+\dot{\kappa}h)$$

and

$$M\dot{\kappa}^2/b\sigma_b = -\dot{\varepsilon}_0^2 + \dot{\kappa}^2h^2 - 2\dot{\kappa}h + (1-\dot{\varepsilon}_0)\ln(1-\dot{\varepsilon}_0+\dot{\kappa}h) + (1+\dot{\varepsilon}_0)\ln(1+\dot{\varepsilon}_0+\dot{\kappa}h) .$$

These equations were derived by usual hypothesis of beams. In limits, when $\dot{\varepsilon}_0$ and $\dot{\kappa}h << 1$, we can easily obtain

$$N = 2bh\sigma_b\dot{\varepsilon}_0 \quad \text{and} \quad M = 2/3bh^3\sigma_b\dot{\kappa} .$$

When $\dot{\kappa} \to \infty$ (at $\dot{\varepsilon}_0/\dot{\kappa}h = C$) (limit state) we arrive at

$$N/hb\sigma_b = 2c \quad \text{and} \quad M/h^2b\sigma_b = 1 - c^2 .$$

These expressions are similar to those for perfect plasticity.

Buckling of a column under the action of an axial constant load P.

In this case we can employ the assumption of an idealized cross section and use the collocation method. Then, using the equlibrium conditions we can readily obtain expressions for the stresses in terms of the dimensionless deflection u and the mean value of the stresses σ_0

$$\sigma_1 = \sigma_0(1+u) \quad \text{and} \quad \sigma_2 = \sigma_0(1-u) .$$

The plane-section hypothesis and the assumption that the deflections are small yield the following expressions for the strains

$$\varepsilon_1 = \varepsilon_0 - \varrho(u-u_0) , \quad \varepsilon_2 = \varepsilon_0 + \varrho(u-u_0) , \quad \varrho = \pi^2h^2/L^2 .$$

Combining these expressions with (3), we obtain a differential equations for u which will be different for $u < 1$ (both flanges are compressed) and for $u > 1$. For $u > 1$ we have

$$\varrho\dot{u}/A\sigma_0 = \frac{(\sigma_{1b}+\sigma_{2b})\sigma_0(1-u^2) - 2\sigma_{1b}\sigma_{2b}u}{2[\sigma_{2b}+\sigma_0(1+u)][\sigma_{1b}-\sigma_0(1-u)]} ,$$
$$\sigma_{2b} < 0 .$$

The limit state is achieved when $u = u^*$, where $u^* = 1 - \sigma_{2b}/\sigma_0$ and $\dot{u} \to \infty$ at $u \to u^*$. We should note that, in contrast to a power law or exponential relationship, the use of linear-fractional relation leads to a qualitatively new results: loss of carrying capacity at small deflections.

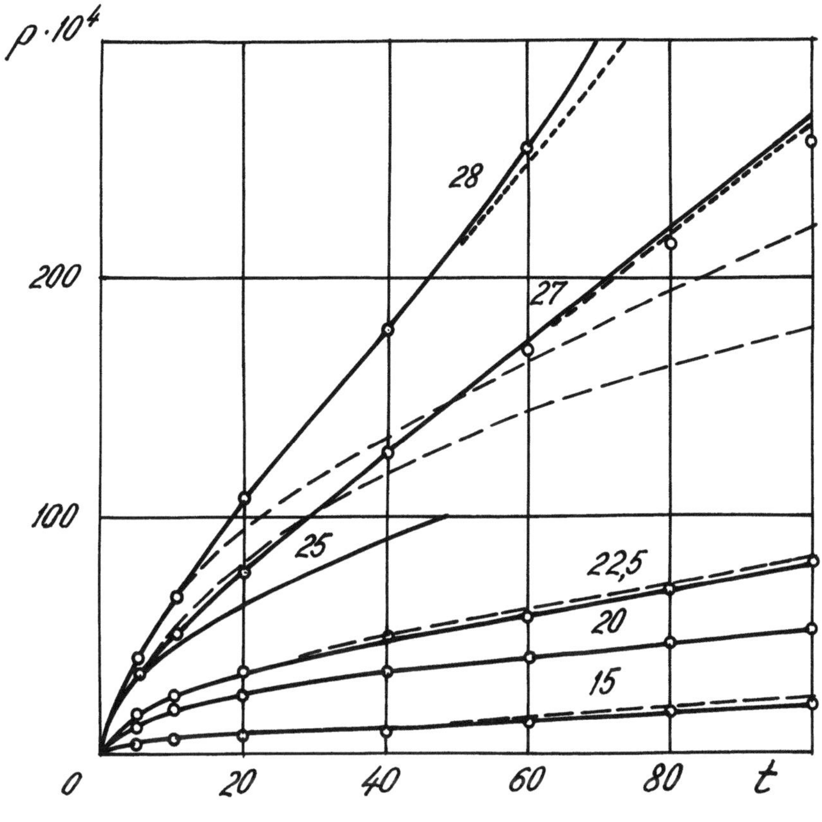

Figure 1: Creep curves

References

[1] Rabotnov, Yu.N., *Creep problems in structural members*, Amsterdam; North Holland 1969

[2] Shesterikov, S.A., Yumasheva, M.A., *More precise specification of the equation of state in creep theory*, Mekh. Tverd. Tela, **1**, (1983), 86-91 (in Russian)

[3] Lokoshtchenko, A.M., Shesterikov, S.A., *Method for description of creep and long-term strength with pure elongation*, Prikl. Mekhan. i Tekhn. Fiz., **3**, (1980), 155-159 (in Russian)

[4] Shesterikov, S.A., Arshackuny, A.L., Shtcheredeeva, L.V., *Method of temperature-loading prediction of creep rupture of metals*, Probl. Prochn., **9**, (1989), 6-9 (in Russian)

Bending of Orthotropic Viscoelastic Plates with Thermal Effects

Z. SOBOTKA

Mathematical Institute
Czechoslovak Academy of Sciences, Prague

Summary

The paper deals with the analysis of the orthotropic viscoelastic plates in non-stationary temperature fields. It is assumed that the temperature varies linearly over the thickness of the plate. The author introduces the equivalent stresses, strains, displacements, moments, loads and temperature in order to replace the time-varying relations by the equations of elasticity expressed in terms of equivalent variables. In this manner, he arrives at the separation of the solutions with respect to coordinates from those with respect to time. Such a procedure leads to considerable simplifications.

Definition of Equivalent Variables

The equivalent stress components are defined by

$$\bar{\sigma}_{ij} \equiv \tilde{A}\sigma_{ij} = a_0\sigma_{ij} + a_1\frac{\partial\sigma_{ij}}{\partial t} + a_2\frac{\partial^2\sigma_{ij}}{\partial t^2} + \ldots + a_m\frac{\partial^m\sigma_{ij}}{\partial t^m} + + a_{-1}\int_{t_0}^{t} K(t-\tau)\sigma_{ij}(\tau)\,d\tau\,, \quad (1)$$

while, for the equivalent strain, we have the relation [1]:

$$\bar{\varepsilon}_{ij} \equiv \tilde{B}\varepsilon_{ij} = b_0\varepsilon_{ij} + b_1\frac{\partial\varepsilon_{ij}}{\partial t} + b_2\frac{\partial^2\varepsilon_{ij}}{\partial t^2} + \ldots + b_n\frac{\partial^n\varepsilon_{ij}}{\partial t^n} + + b_{-1}\int_{t_0}^{t} L(t-\tau)\varepsilon_{ij}(\tau)\,d\tau\,, \quad (2)$$

where $\tilde{A}$ and $\tilde{B}$ are linear integro-differential operators which have in special cases the differential or integral forms, re-

M. Zyczkowski (Ed.)
Creep in Structures
IUTAM Symposium Cracow/Poland 1990

spectively. The coefficients a_k and b_ℓ in the foregoing relations may be constants or functions of time and temperature, respectively, but cannot depend in any way on the coordinates.

It should be emphasized that the definitions of equivalent stress and strain components given by Eqs. (1) and (2) also include among their integral terms all integral transforms possible after the substitution of appropriate kernels and appropriate limits of integration.

In an analogous manner, we can express the definition of the equivalent temperature:

$$\overline{T} \equiv \tilde{\mathcal{C}}T = c_0 T + c_1 \frac{\partial T}{\partial t} + c_2 \frac{\partial^2 T}{\partial t^2} + \ldots + c_1 \frac{\partial^p T}{\partial t^p} + + c_{-1} \int_{t_0}^{t} N(t - \tau) T(\tau)\, d\tau \, . \tag{3}$$

The temperature which varies linearly with the distance z from the middle surface of the plate is expressed by [2]

$$T = \frac{T_1 + T_2}{2} + \frac{T_1 - T_2}{h} z = \theta_0 + \theta z \, , \tag{4}$$

where T_1 and T_2 denote respectively the temperature on the top and bottom surface and h is the thickness of the plate.

Constitutive Relations

The equivalent strain-stress relations for the plane-stress state of the orthotropic viscoelastic plates at the non-stationary thermal effects have the following form [3]

$$\overline{\varepsilon}_x = A_{11}\overline{\sigma}_x + A_{12}\overline{\sigma}_y + \alpha_1 \overline{T} \, , \tag{5}$$

$$\overline{\varepsilon}_y = A_{21}\overline{\sigma}_x + A_{22}\overline{\sigma}_y + \alpha_2 \overline{T} \, , \tag{6}$$

$$\overline{\varepsilon}_{xy} = A_{33}\overline{\sigma}_{xy} + \alpha_3 \overline{T} \, . \tag{7}$$

In order to determine the relation between the coefficients α_1 ,

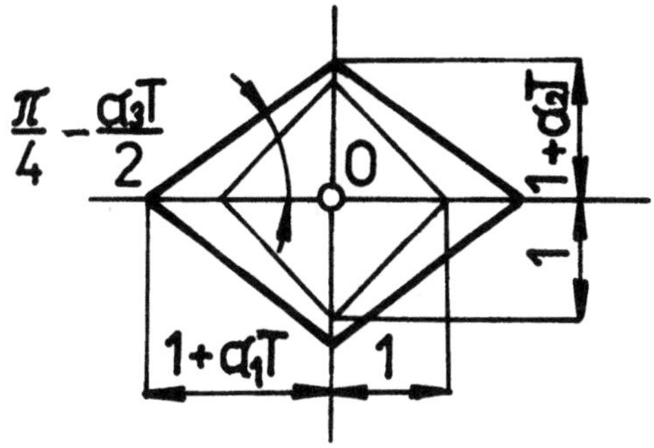

Fig. 1. Thermal deformation of the square cross-section

α_2 and α_3, let us consider the thermal deformation of a prismatic element with the square cross-section shown in Fig. 1. This prismatic element is bounded by the planes parallel to the z-axis and forming the angles 45° with the x and y-axes. In the course of thermal deformation, the right angles between the sides of the initially square cross-section become $\pi/2 - \alpha_3 T$ or $\pi/2 + \alpha_3 T$, respectively. From the geometrical relations in Fig. 1, we have

$$\tan\left(\frac{\pi}{4} + \frac{\alpha_3 T}{2}\right) = \frac{1 + \alpha_1 T}{1 + \alpha_2 T} . \tag{8}$$

Replacing the tangents by the arcs of unit circle in radians, we obtain after rearranging the terms the resultant relation:

$$\alpha_3 = \alpha_1 - \alpha_2 . \tag{9}$$

From Eqs. (5) to (7), we get the equivalent stress-strain relations:

$$\bar{\sigma}_x = \frac{1}{A_{11}A_{22} - A_{12}A_{21}} \left[A_{22}(\bar{\varepsilon}_x - \alpha_1\bar{T}) - A_{12}(\bar{\varepsilon}_y - \alpha_2\bar{T})\right], \tag{10}$$

$$\bar{\sigma}_y = \frac{1}{A_{11}A_{22} - A_{12}A_{21}} \left[A_{11}(\bar{\varepsilon}_y - \alpha_2\bar{T}) - A_{21}(\bar{\varepsilon}_x - \alpha_1\bar{T})\right], \tag{11}$$

$$\bar{\sigma}_{xy} = \frac{1}{A_{33}}(\bar{\varepsilon}_{xy} - \alpha_3\bar{T}) . \tag{12}$$

The foregoing relations can be rewritten as follows

$$\bar{\sigma}_x = B_{11}\bar{\varepsilon}_x + B_{12}\bar{\varepsilon}_y - \beta_1\bar{T} , \tag{13}$$

$$\bar{\sigma}_y = B_{21}\bar{\varepsilon}_x + B_{22}\bar{\varepsilon}_y - \beta_2\bar{T} , \tag{14}$$

$$\bar{\sigma}_{xy} = B_{33}\bar{\varepsilon}_{xy} - \beta_3\bar{T} . \tag{15}$$

Equivalent Deflection, Moments, Load and Equilibrium

According to the Kirchhoff-Love hypothesis, the strain components of a plate can be expressed in terms of derivatives of the deflection by means of the following tensor relation:

$$\varepsilon_{k\ell} = - z w_{,k\ell} . \tag{16}$$

Multiplying this relation by the rheological operator $\mathfrak{B}$, we obtain the equivalent relation:

$$\bar{\varepsilon}_{k\ell} = - z \bar{w}_{,k\ell} , \tag{17}$$

which can be rewritten in the form of three component equations between equivalent strains and derivatives of the equivalent deflection:

$$\bar{\varepsilon}_x = - z \frac{\partial^2 \bar{w}}{\partial x^2} , \quad \bar{\varepsilon}_y = - z \frac{\partial^2 \bar{w}}{\partial y^2} , \quad \bar{\varepsilon}_{xy} = - 2 \frac{\partial^2 \bar{w}}{\partial x \partial y} . \tag{18}$$

The equivalent deflection is defined by

$$\bar{w} \equiv \mathfrak{B} w = b_0 w + b_1 \frac{\partial w}{\partial t} + b_2 \frac{\partial^2 w}{\partial t^2} + \ldots + b_n \frac{\partial^n w}{\partial t^n} + \\ + b_{-1} \int_{t_0}^{t} L(t - \tau) w(\tau) \, d\tau . \tag{19}$$

The equivalent bending and twisting moments per unit length of the plate are in view of Eqs. (4), (13), (14), (15) and (18) given by

$$\bar{m}_x = \int_{-h/2}^{h/2} \bar{\sigma}_x \, z \, dz = - \frac{h^3}{12} \left(B_{11} \frac{\partial^2 \bar{w}}{\partial x^2} + B_{12} \frac{\partial^2 \bar{w}}{\partial y^2} + \beta_1 \bar{\theta} \right) , \qquad (20)$$

$$\bar{m}_y = \int_{-h/2}^{h/2} \bar{\sigma}_y \, z \, dz = - \frac{h^3}{12} \left(B_{21} \frac{\partial^2 \bar{w}}{\partial x^2} + B_{22} \frac{\partial^2 \bar{w}}{\partial y^2} + \beta_2 \bar{\theta} \right) , \qquad (21)$$

$$\bar{m}_{xy} = \int_{-h/2}^{h/2} \bar{\sigma}_{xy} \, z \, dz = - \frac{h^3}{6} \left(B_{33} \frac{\partial^2 \bar{w}}{\partial x \partial y} + \frac{1}{2} \beta_3 \bar{\theta} \right) . \qquad (22)$$

Multiplying the equation of moment equilibrium by the rheological operator $\tilde{A}$ yields the equation of equivalent equilibrium:

$$\frac{\partial^2 \bar{m}_x}{\partial x^2} + 2 \frac{\partial^2 \bar{m}_{xy}}{\partial x \partial y} + \frac{\partial^2 \bar{m}_y}{\partial y^2} = - \bar{q} , \qquad (23)$$

where

$$\bar{q} \equiv \tilde{A} q = a_0 q + a_1 \frac{\partial q}{\partial t} + a_2 \frac{\partial^2 q}{\partial t^2} + \ldots + a_m \frac{\partial^m q}{\partial t^m} + + a_{-1} \int_{t_0}^{t} K(t - \tau) q(\tau) \, d\tau \qquad (24)$$

is the equivalent quasi-static time-varying load.

Introducing Eqs. (20), (21) and (22) into Eq. (23) yields

$$B_{11} \frac{\partial^4 \bar{w}}{\partial x^4} + 2(B_{12} + 2B_{33}) \frac{\partial^4 \bar{w}}{\partial x^2 \partial y^2} + B_{22} \frac{\partial^4 \bar{w}}{\partial y^4} + \beta_1 \frac{\partial^2 \bar{\theta}}{\partial x^2} + + 2\beta_3 \frac{\partial^2 \bar{\theta}}{\partial x \partial y} + \beta_2 \frac{\partial^2 \bar{\theta}}{\partial y^2} = \frac{12 \bar{q}}{h^3} . \qquad (25)$$

Solution of the Boundary-Value Problem

The solution of the linear differential equation for equivalent deflection surface with respect to the coordinates x and y can be carried out by means of the procedures of linear theory of elasticity. Consider, for example, the simply supported orthotropic restangular viscoelastic plate with side lengths a and b shown in Fig. 2. For the solution in the form of a double sine series, the equivalent load should be expanded as follows

$$\bar{q}(x,y,t) = \sum_{k=1}^{\infty} \sum_{\ell=1}^{\infty} \bar{Q}_{k\ell}(t) \sin \lambda_k x \sin \mu_\ell y \, , \tag{26}$$

where

$$\bar{Q}_{k\ell}(t) = \frac{4}{ab} \int_0^a \int_0^b \bar{q}(x,y,t) \sin \lambda_k x \sin \mu_\ell y \, dx \, dy \, , \tag{27}$$

$$\lambda_k = \frac{k\pi}{a}, \quad \mu_\ell = \frac{\ell\pi}{b} . \tag{28}$$

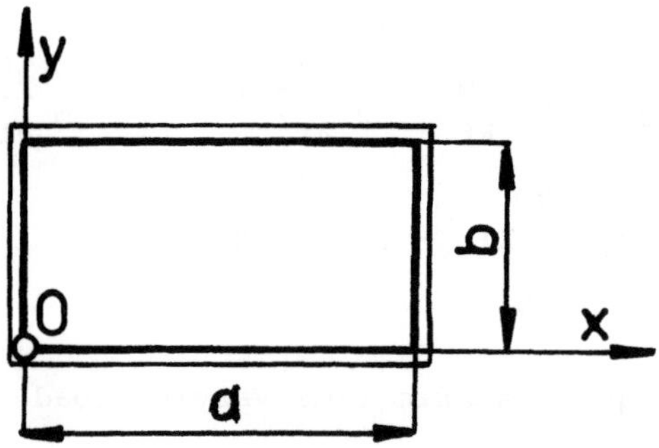

Fig. 2. Scheme of a simply supported rectangular plate

The function of the equivalent temperature will be expanded in an analogous form

$$\bar{\theta}(x,y,t) = \sum_{m=1}^{\infty} \sum_{n=1}^{\infty} \bar{\theta}_{mn}(t) \sin \lambda_m x \sin \mu_n y \, , \tag{29}$$

where

$$\bar{\theta}_{mn}(t) = \frac{4}{ab}\int_0^a\int_0^b \bar{\theta}(x,y,t)\ \sin\lambda_m x\ \sin\mu_n y\ dx\ dy\ . \tag{30}$$

The equivalent deflection will be sought in the following form

$$\bar{w}(x,y,t) = \sum_{k=1}^{\infty}\sum_{\ell=1}^{\infty} \bar{T}_{k\ell}(t)\ \sin\lambda_k x\ \sin\mu_\ell y\ . \tag{31}$$

Combining the effects of the external load and temperature according to Eq. (25), we find the formula for the coefficients $\bar{R}_{k\ell}(t)$:

$$\begin{aligned}\bar{R}_{k\ell}(t) &= \bar{Q}_{k\ell}(t) + \frac{h^3}{12}\,(\beta_1\lambda_{k\ell}^2 + \beta_2\mu_\ell^2)\bar{\theta}_{k\ell}(t) - \\ &- \frac{2h^3\beta_3}{3ab}\sum_{m=1}^{\infty}\sum_{n=1}^{\infty}\bar{\theta}_{mn}(t)\ \cdot \\ &\cdot \int_0^a\int_0^b \cos\lambda_m x\ \cos\mu_n y\ \sin\lambda_k x\ \sin\mu_\ell y\ dx\ dy\ .\end{aligned} \tag{32}$$

Introducing the foregoing expressions into Eq. (25) yields the formula for the general coefficient of the series (31):

$$\bar{T}_{k\ell}(t) = \frac{12\ \bar{R}_{k\ell}(t)}{h^3\left[B_{11}\lambda_k^4 + 2(B_{12} + 2B_{33})\lambda_k^2\mu_\ell^2 + B_{22}\mu_\ell^4\right]}\ . \tag{33}$$

The solution will be simplified if we neglect the considerably small effect of torsional temperature effect. In such a case, we obtain

$$\bar{T}_{k\ell}(t) = \frac{12\left[\bar{Q}_{k\ell} + \frac{h^3}{12}\,(\beta_1\lambda_k^2 + \beta_2\mu_\ell^2)\ \bar{\theta}_{k\ell}(t)\right]}{h^3\left[B_{11}\lambda_k^4 + 2(B_{12} + 2B_{33})\lambda_k^2\mu_\ell^2 + B_{22}\mu_\ell^4\right]}\ . \tag{34}$$

Determination of Actual Deflections

The actual deflection $w(x,y,t)$ can, in the general case, be determined from the integrodifferential equation (2). In many cases, the rheological relations are just differential ones. In the simplest case, we have for the actual deflection, the following first-order differential equation

$$\bar{w} = b_0(t) + b_1(t) \frac{\partial w}{\partial t} . \qquad (35)$$

Its solution is expressed by

$$w = \exp\left[-\int_{t_0}^{t} \frac{b_0(\tau)}{b_1(\tau)} d\tau\right] \left[\int_{t_0}^{t} \frac{\bar{w}(\tau_1)}{b_1(\tau_1)} \exp\left(\int_{t_0}^{\tau_1} \frac{b_0(\tau_2)}{b_1(\tau_2)} d\tau_2\right) d\tau_1 + \right.$$
$$\left. + w_0\right] , \qquad (36)$$

where w_0 is the initial deflection at time t_0 . In another typical case, the actual deflection follows from the linear differential equation with constant coefficients:

$$\bar{w} = b_0 w + b_1 \frac{\partial w}{\partial t} + b_2 \frac{\partial^2 w}{\partial t^2} , \qquad (37)$$

which has the following solution

$$w = \frac{e^{-\chi_1 t}}{\chi_2 - \chi_1} \int_{t_0}^{t} \bar{w}(\tau) e^{\chi_1 \tau} d\tau + \frac{e^{-\chi_2 t}}{\chi_1 - \chi_2} \int_{t_0}^{t} \bar{w}(\tau) e^{\chi_2 \tau} d\tau +$$
$$+ c_1 e^{-\chi_1 t} + c_2 e^{-\chi_2 t} , \qquad (38)$$

where χ_1 and χ_2 are the reciprocal times of relaxation represented by the roots of the characteristic euqation which are taken with opposite signs.

References

1. Sobotka, Z.: Rheology of Materials and Engineering Structures. Amsterdam, Oxford, New York, Tokyo: Elsevier 1984.

2. Nowacki, W.: Thermoelasticity. Warszawa: PWN-Polish Scientific Publishers; Oxford, New York, Toronto, Sydney, Paris, Braunschweig 1986.

3. Zyczkowski, M.: Combined Loadings in the Theory of Plasticity. Warszawa: PWN-Polish Scientific Publishers 1981.

Stress Concentrations Resulting from Longitudinal Butt-Welds in Piping at Elevated Temperatures

By: G.R. Stevick and I. Finnie
Department of Mechanical Engineering
University of California, Berkeley, CA 94720 USA

Abstract

It is shown that significant stress concentrations may develop in longitudinally welded pipe when the creep strain rate of the base metal and weld metal are different. The stress concentration increases from unity to a steady state value after several years. Implications for failure of this type of weld at elevated temperatures are discussed.

Introduction

Large diameter pipe is often manufactured by bending a flat plate into a cylinder which is closed by a longitudinal weld. Although this type of construction is not acceptable for elevated temperature service in certain European Design Codes, it is permitted in the United States.

The ASME/ANSI piping codes permit longitudinally welded pipe which has been heat treated in accordance with code specifications and 100% radiographed to be used at the same stress levels as seamless pipe. In recent years, two failures of longitudinally welded pipe in power plants have caused concern about the long-term reliability of this type of construction. In June 1985, failure of a 0.76 m diameter 1-1/4Cr-1/2Mo steel hot reheat line resulted in six fatalities and a number of injuries. Shortly later in January 1986, a 0.81 m diameter 2-1/4Cr-1Mo hot reheat line also failed catastrophically. In both cases, the cracks propagated for large distances in the axial direction, 5.5 m and 9 m respectively [1]. The failures initiated from "flaws" near the weld fusion line at the midwall of the pipes. It is an understatement to say that these failures were a cause of concern to the electric power industry. Longitudinally welded pipe was used extensively in older plants and 15% of the total power capacity of the United States as well as 75% of the reserve capacity is provided by units which are over 30 years old [1]. As a result, inspection of at least 10 km of steam piping has been carried out in the last 4 years and a number of flaws have been detected [2,3].

M. Zyczkowski (Ed.)
Creep in Structures
IUTAM Symposium Cracow/Poland 1990

Approximately two thirds of these flaws were located at the midwall region either along the fusion line or in weld metal.

The weld geometry of longitudinal pipe is not specified in the Piping Code but typically has the double-V configuration shown in Fig. 1 with a width near the midwall, much less than the width at the surfaces. Investigations of failures in welded pipe generally focus on the welding processes and metallurgical variables. A factor which has received little attention is the stress concentration which may develop in the weld region due to different creep properties of the base and weld metals. This is the topic of the present investigation.

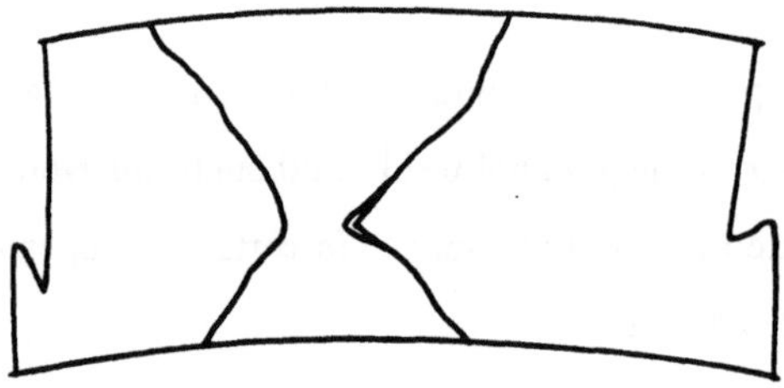

Fig. 1. Sketch of a typical longitudinal pipe weld cross-section with a midwall flaw [2].

Stress Concentrations Due to Weld Geometry

An idealized double-V weld geometry is shown in Fig. 2. For simplicity in presenting our preliminary conclusions, we neglect the curvature of the pipe and assume plane strain conditions.

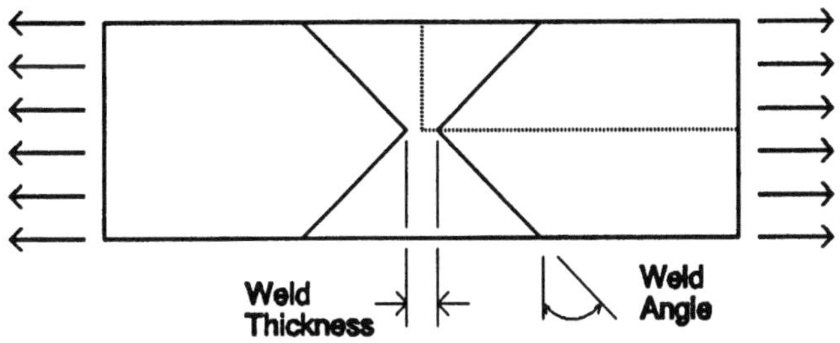

Fig. 2. Idealized model of a longitudinal pipe weld. Pipe curvature is neglected and plane strain conditions assumed. The dotted line bounds the finite element model.

Modeling the entire pipe circumference could be carried out but is unlikely to significantly modify the results. Under elastic conditions if the weld and base materials are similar there is, of course, no stress concentration. However, creep properties are notoriously sensitive to small changes in alloy content. The carbon content of the weld is often below that of the base metal which by itself would be expected to lead to higher creep rates in the weld metal. For example, Saxena [4] provides power law equations for the steady-state creep rate for 2-1/4Cr-1Mo base and weld metal. For the code allowable stress of 53.8 MPa at 538 °C, these equations predict a creep rate in the weld metal almost an order of magnitude greater than the base metal. It can be seen that such a difference in creep behavior will lead to a slowly increasing stress at the midwall as creep deformation occurs. Eventually when creep strains are large compared to elastic strains, a steady state stress distribution will be obtained. The final stress concentration factor should depend on the relative creep properties of weld and base materials as well as the geometry of the weld. Indiscriminately increasing the creep strength of the weld metal is not a solution to this problem. Significant stress concentrations will occur near the fusion line at the inside and outside surfaces of the pipe if the weld metal creeps significantly slower than the base metal.

In view of the potential importance of this problem, it is remarkable that it has received very little attention in the literature. Goodall and Walters [5] examined the influence of different creep properties in weld and base metal for circumferential welds. They considered steady-state creep behavior which may be written in the form:

$$\dot{e}_w = B_w S^{n_w} \; ; \quad \dot{e}_b = B_b S^{n_b} \tag{1}$$

In these equations, strain rates, $\dot{e}$, and stress, S, are related through the constants, B and n, with subscripts, w and b, referring to weld and base metal. In Goodall and Walters' work, it was assumed that $n_w = n_b$; we will adopt the same simplification. Their detailed analysis of circumferential welds led to the conclusion that it was preferable to select a weld metal which creeps more rapidly than the base metal.

The first and virtually the only reference in the literature to the problem we are considering is due to Wells [6] in 1986. He commented "superimposed on the nominal membrane stress there may be a concentration of stress resulting from differences between the weld metal and the plate". Well's comment appears to have been relatively ignored in a growing literature which deals extensively with such topics as non-destructive testing, welding procedures and creep-crack propagation.

Finite Element Formulation

To assess the effects of differing strain rates between the weld and base metals, a parametric study was performed using a finite element code which considers elastic deformation and steady-state creep [7]. The element size was decreased by factors of 2 until the maximum principal stress varied less than 1 percent from one size to the next. With the mesh size chosen, the maximum principal stresses correspond approximately to the average stress across 5 metallurgical grains, assuming a .05 mm (.002 inch) average grain diameter. The length of the model was determined by progressively increasing the length until convergence was reached. Models more than 2 thicknesses in length were found to give stresses differing by less than 1 percent.

The material constants used were obtained from the Electric Power Research Institute's Users Guide for their high temperature crack growth program [8] and the creep data given in References [9,10]. For a temperature of 538 °C the constant B in Equation 1 was taken as 2.2×10^{-16} for 1-1/4Cr-1/2Mo and 1.0×10^{-15} for 2-1/4Cr-1Mo in units of MPa and Hrs. For both alloys, the exponent n was assumed to equal 5 and the elastic constants were taken as E = 165.5 GPa and ν = 0.3.

The finite element analysis was used to obtain three results: 1) the time to reach a steady state stress distribution, 2) a parametric study of stress concentration as a function of weld angle and creep property differences between the weld and base metals, and 3) the effect of a crack on the stress distribution. The nominal stress level of 45.5 MPa used for all cases was based on the present ASME/ANSI B31.1 piping code allowable for 1-1/4Cr-1/2Mo at 538 °C.

Results of Analysis

The first result obtained from the finite element analysis was the maximum tensile stress as a function of time for both materials assuming a 45 degree, double-V weld and a strain rate constant for the weld metal 5 times that of the base metal. The results are shown in Fig. 3. Not surprisingly, 1-1/4Cr-1/2Mo reaches steady state equilibrium before 2-1/4Cr-1Mo and it takes several years for the stress concentration to approach its maximum value of approximately 1.5. Stress concentration is defined as the maximum tensile stress divided by the uniform stress far from the weld.

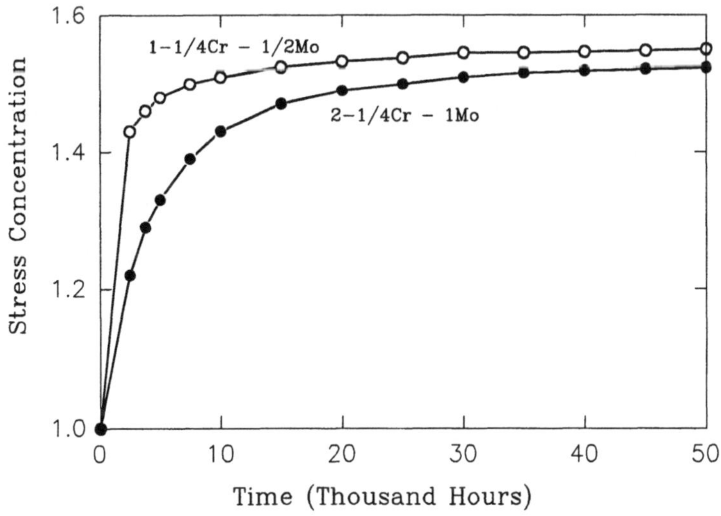

Fig. 3. Stress concentration as a function of time.

A parametric assessment of steady state creep stress concentration was then determined by carrying out the analyses to 100,000 hours and varying weld angle and ratio of weld metal to base metal creep constants. The results are shown in Fig. 4 below.

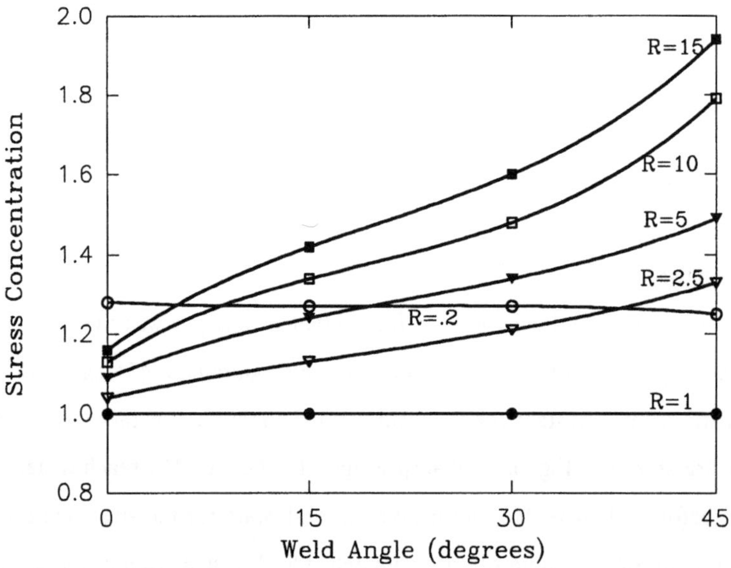

Fig. 4. Stress Concentration as a function of weld angle and strain rate constant ratio.

The stress concentration results shown in Fig. 4 clearly show that differing creep properties have a large effect on the steady state stresses at high temperatures. These results have several implications for the designer: 1) weld metal strengths should be as close as possible to that of the base metal, 2) lowering the weld angle when the base metal is stronger than the weld metal reduces the stress concentration, 3) weld metal stronger than the base metal can also result in large stress concentration as indicated by the curve in Fig. 4 for a creep constant ratio of .2, and 4) a 0 degree weld can still have a stress concentration if the properties of the weld and base metals differ. Weld root thickness is also a parameter which needs to be investigated. With high stress concentrations premature crack initiation would be expected. Following initiation from limited preliminary studies it appears that creep damage relieves some of the high local stresses. For example, Fig. 5 shows the stress distribution for creep strain ratios of 1 and 5 and a crack extending approximately 3 mm along the weld line on both sides of the root of a 45 degree weld as shown in Fig. 1.

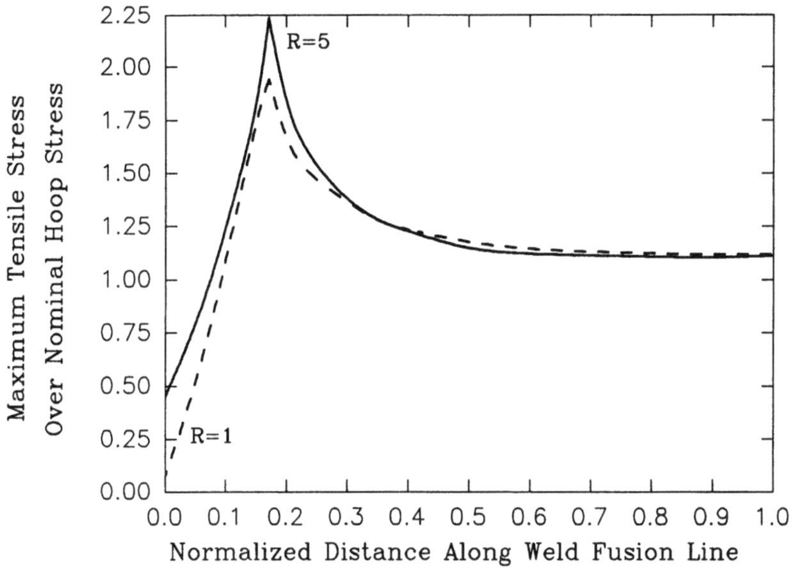

Fig. 5. Maximum tensile stress along the weld fusion line with a midwall crack (3.1 mm half length) as shown in Fig. 1.

Conclusions

This work is only a start in what we believe to have been a neglected area in the field of creep rupture. It shows that a stress concentration will occur if creep properties of the weld and base metals are different. To obtain predictive results, crack initiation and crack growth need to be studied. A major problem is the uncertain nature of creep properties. However, some typical solutions would be of value in establishing controls on weld chemistry and weld geometry. Such a study could also lead to a reappraisal of weld efficiency factors and possibly a decision as to whether this type of construction should even be permitted in future designs for elevated temperature operation.

<u>References</u>

1. Jaske, C.E., "Life Assessment of Hot Reheat Steam Pipe", Proceedings Session II of the International Conference on Life Extension and Assessment, June 13-15, 1988, Congress Center, The Hague, The Netherlands.

2. Becker, F.L., Walker, S.M., and Viswanathan, R., "Guidelines for the Evaluation of Seam-Welded Steam Pipes", Final Report on Research Project 2596-7, CS-4774, Electric Power Research Institute, Palo Alto, Calif., February, 1987.

3. Viswanathan, R., Dooley, R.B., Saxena, A., "A Methodology for Evaluating the Integrity of Longitudinally Seam Welded Pipe in Fossil Plants", ASME JOURNAL OF PRESSURE VESSEL TECHNOLOGY, Vol. 110, August 1988, pp. 283-290.

4. Users Guide for PCPIPE, A Computer Code for Integrity Analysis of Elevated Temperature Steam Pipes, Developed by A. Saxena, Professor, Georgia Institute of Technology and M. Saxena, Software Consultant, Marietta, GA, 1987.

5. Goodall, I.W., and Walters, D.J., "Creep Behavior of Butt-Welded Tubes", Institution of Mechanical Engineers, Conference Publication C190/73, May, 1973.

6. Wells, C.H., "On the Life Prediction of Longitudinal Seam Welds in Hot Reheat Piping", EPRI Conference on Life Extension and Assessment of Fossil Plants, Washington D.C., June 2-4, 1986.

7. Hallquist, J.O., "Nike2D: An Implicit, Finite Deformation, Finite Element Code for Analyzing the Static and Dynamic Response of Two-Dimensional Solids," University of California, Lawrence Livermore National Laboratory, Report UCRL-52678 (March, 1979).

8. Users Guide for SLIC2, Computer Code for Analysis of Steam Line Cracking, Prepared by Failure Analysis Associates, Palo Alto, CA for Electric Power Research Institute, Palo Alto, CA, EPRI Project 2595-7, March 1989.

9. Smith, G.V., "Supplemental Report on the Elevated Temperature Properties of Chromium Molybdenum Steels (An Evaluation of 2-1/4Cr-1Mo Steel)", ASTM Data Series Publication No. DS-681, 1966.

10. Van Echo, J.A., Simmons, W.F., "Supplemental Report on the Elevated Temperature Properties of Chromium Molybdenum Steels", ASTM Data Series Publication No. DS-681, 1966.

Bars of Uniform Strength vs. Optimal with Respects to Ductile Creep Rupture Time

K. SZUWALSKI

Institute of Mechanics and Machine Design,
Cracow University of Technology, Poland

Summary
The conditions under which the bar of uniform initial strength remains the bar of uniform strength throughout the whole creep process up to the ductile rupture, are formulated. The discussion whether such a bar is an optimal one, with numerical example is given.

1 Introduction

In problems of optimal design of structures in creep conditions many various criteria may be adopted. The review of papers on the optimization in creep conditions was given by M. Życzkowski [4]. Some of them are connected with life-time to rupture. In such cases usually the Kachanov's theory of brittle rupture is used, leading often for simple, statically determine structures with neglected geometrical changes to structures of uniform strength. The possibility of application of the ductile rupture theory was pointed out by M. Życzkowski [3] and results for the Hoff's theory may be different. Then geometrical changes must be taken into account, in order to follow the whole process from its beginning up to the ductile rupture (infinite strains). Details are presented by author [2].

In paper the problem of nonuniform tension of bars is discussed in details. We shall make use of some solutions obtained by author [1] in the first paper on the optimization with respect to ductile rupture time. Discussing the bars of uniform strength we must divide them into several groups:

- bars of uniform initial strength - in which the initial stress Σ in each cross section is proportional to the critical stress (for homogeneous bar - the same);
- bars of uniform strength with respect to ductile rupture time in which simultaneously all cross-sections diminish to zero;
- bars of uniform deformability - in which at each moment strains are the same in all cross-sections (various in time).

The relations between them are shown in in Fig.1. for homogeneous bar.

M. Zyczkowski (Ed.)
Creep in Structures
IUTAM Symposium Cracow/Poland 1990

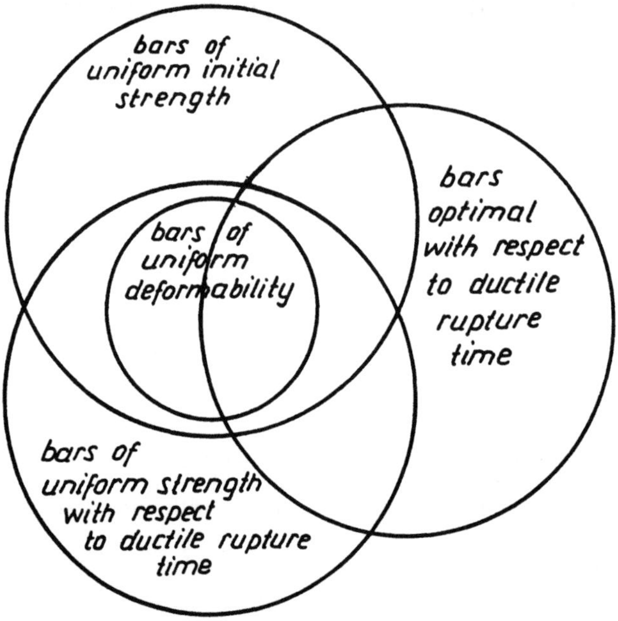

Figure 1: Bars of uniform strength and optimal

2 Bars of uniform deformability

Under assumptions:

1. incompressibility of material;
2. body force is a homogeneous function of any order m, of spatial coordinate x (arbitrary function of material coordinate X);
3. external force on the free end of the bar is also a homogeneous function of the same order m of material coordinate x;
4. homogeneity of material,

the bar of uniform initial strength is the bar of uniform deformability i.e. the bar of uniform strength with respect to ductile rupture time.

It can be proved that additional assumption of uniform deformability, leading to the relationship between material and spatial coordinates:

$$x = X\psi(t) \ , \tag{1}$$

makes it possible to satisfy all the governing equations, and causes no contradictions. From the assumption (1) we have

$$a = \frac{A}{x'} = \frac{A}{\psi(t)} \ , \tag{2}$$

where A is initial and a - current cross-section, and "prime" denotes the derivative with respect to the material coordinate X. Assumption (2) means that

$$b(x,X) = \psi^m B(X,X) , \tag{3}$$

where $B(X,X)$ stands for initial body force, while $b(x,X)$ for current value of it. When self-weight of the bar is taken into account - $m=0$, for a rotating bar - $m=1$. The bar of uniform initial strength is described (regardless to the physical law) by the equation

$$A(X) = \frac{P}{\Sigma} \exp\left(\frac{1}{\Sigma}\int_L^X B(\Xi,\Xi)\,d\Xi\right) , \tag{4}$$

where L is the initial length of the bar, P - initial value of external force at $X=L$, and Ξ - material coordinate as integration variable. The condition of internal equilibrium for an element of deformed bar, derived in [1], has a form:

$$\frac{\partial}{\partial x}(a\sigma) + b(x,X)\,a = 0 . \tag{5}$$

Putting into it conditions (2) and (3) and Eq.(4), we come to

$$\psi^{m+1}(t) = \frac{\sigma(X\,,t)}{\Sigma} - \frac{\sigma'(X\,,t)}{B(X\,,X)} , \tag{6}$$

To satisfy this equation, the function of material coordinate on the right hand side must vanish. It is possible if

$$\sigma(X\,,t) = \Sigma\psi^{m+1}(t) , \tag{7}$$

what means that stress field remains homogeneous through the whole creep process. Bar of uniform initial strength is then the bar of uniform deformability, provided the external force P changes in time in the same way, as body force does (according to assumption (3)).

3 Optimality of bars of uniform strength

The bar of uniform strength is simultaneously the optimal one in case of statically determine bars. By statically determine we understand here bars for which distribution of normal force is independent from physical law, i.e. it is function of only material coordinate (so must be body force). Then true stress is equal:

$$\sigma(X\,,t) = \frac{P + \int_X^L B(\Xi)A(\Xi)\,d\Xi}{A(X)} x'(X\,,t) = \Sigma(X)\,x'(X\,,t) . \tag{8}$$

For the physical law of type

$$\dot{\varepsilon} = F(\sigma\,,\varepsilon\,,X) , \tag{9}$$

where ε is a logarithmic strain, and $\dot{\varepsilon}$ its velocity

$$\varepsilon = \ln\frac{dx}{dX} = \ln(x') , \quad \dot{\varepsilon} = \frac{\dot{x}'}{x'} , \tag{10}$$

we can separate variables: time t and x'. This equation may be integrated with initial condition for $t = 0$ - $x' = 1$. For the time of ductile rupture t_*, x' tends to infinity.

Therefore

$$t_* = \inf_{(X)} \int_1^\infty \frac{d(x')}{x' F[\Sigma(X)x', \ln(x'), X]} \ . \tag{11}$$

The optimal solution is obtained for the same value of this integral in all cross-sections (bar of uniform strength with respect to ductile rupture). Any change of initial shape will cause increase of initial stress, at least in one cross- section, and consequently t_* would be there smaller. When body force depends on spatial coordinate (e.g. rotating bar), the variables cannot be separated, and Eq.(9) must be solved with respect to x, instead of x'. Then any change of initial shape will affect the ductile rupture time not only in this particular cross-section, but through integration it will influence the whole creep process of the bar. The shape of the bar of uniform strength can be improved, and therefore it will be not the optimal bar.

4 The bar of uniform strength as nonoptimal one

We shall discuss in details the solution for a rotating bar made of the material described by the Norton law

$$\dot{\varepsilon} = k\sigma^n \ , \tag{12}$$

here generalized for true stresses and logarithmic strains. As it is not statically determine bar, slight changes of the uniform strength shape may lead to longer times of work to ductile rupture. We shall make use of equations derived for such bar in [1], for defined there dimensionless quantities.

$$\begin{aligned} \frac{\partial \overline{x}}{\partial \overline{X}} &= y \ , \\ \frac{\partial \overline{x}}{\partial \overline{t}} &= \overline{z} \ , \\ \frac{\partial \overline{z}}{\partial \overline{X}} &= \frac{1}{n}\left(\frac{\overline{N}}{\overline{A}}\right)^n y^{n+1} \ , \\ \frac{\partial \overline{N}}{\partial \overline{X}} &= -\mu \overline{A}\overline{x} \ , \end{aligned} \tag{13}$$

with four unknowns: spatial coordinate x, its derivatives y and z, and normal force N.

By μ is here denoted ratio of of mass of the bar to the mass m placed on the free end:

$$\mu = \frac{V\gamma}{mg} \ , \tag{14}$$

where V is given volume of the bar, γ - the specific weight. The initial distribution of cross-section:

$$A(\overline{X}) = A_{us}(\overline{X}) + A_{cor}(\overline{X}) \ , \tag{15}$$

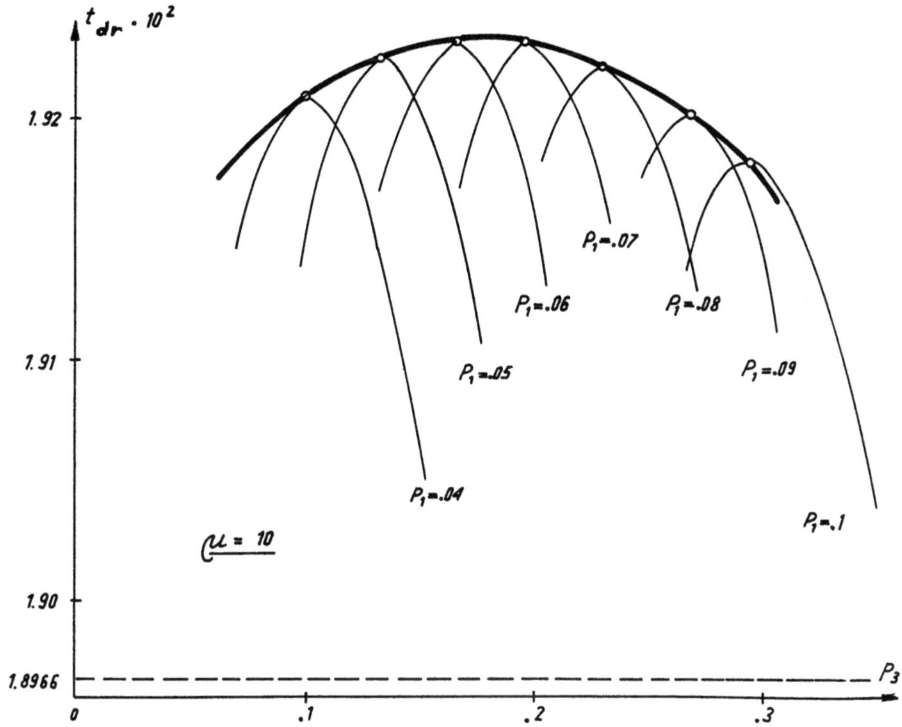

Figure 2: Ductile rupture time for corrected bars

consists of two parts, where the first expression describes the bar of uniform strength:

$$A_{us}(\overline{X}) = \frac{1}{\overline{\Sigma}} \exp\left[\frac{\mu}{2\overline{\Sigma}}\left(1 - \overline{X}^2\right)\right] , \tag{16}$$

with initial stress $\overline{\Sigma}$ calculated for the given volume from the transcendental equation

$$\overline{\Sigma} = \exp\frac{\mu}{2\overline{\Sigma}} \int_0^1 \left[\exp\left(-\frac{\mu}{2\overline{\Sigma}}\overline{X}^2\right)\right] d\overline{X} . \tag{17}$$

The second term in Eq.16., the correction we adopt in form:

$$A_{cor}(\overline{X}) = p_0 + p_1\overline{X} + p_2\overline{X}^2 , \tag{18}$$

with two independent parameters (the third results from the condition, that correction cannot change volume of the bar). The results of numerical calculations are presented in Fig.2. For given parameter p_0 we can find corresponding parameter p_1 ensuring the longest life-time to ductile rupture. Finally we look for maximum maximorum, obtaining the optimal correction. Note, that all results are above the dotted line determining the ductile rupture time for the bar of uniform strength. The optimal values of parameters

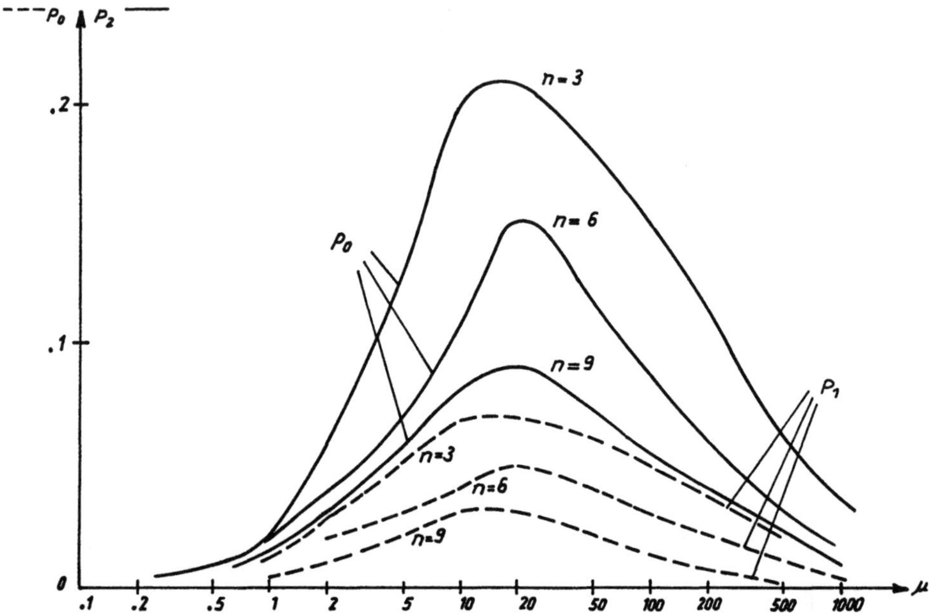

P_0
P_2
.2
.1
0
$n=3$
$n=6$
$n=9$
P_0
P_1
$n=3$
$n=6$
$n=9$
.1
.2
.5
1
2
5
10
20
50
100
200
500
1000
μ

Fig. 3

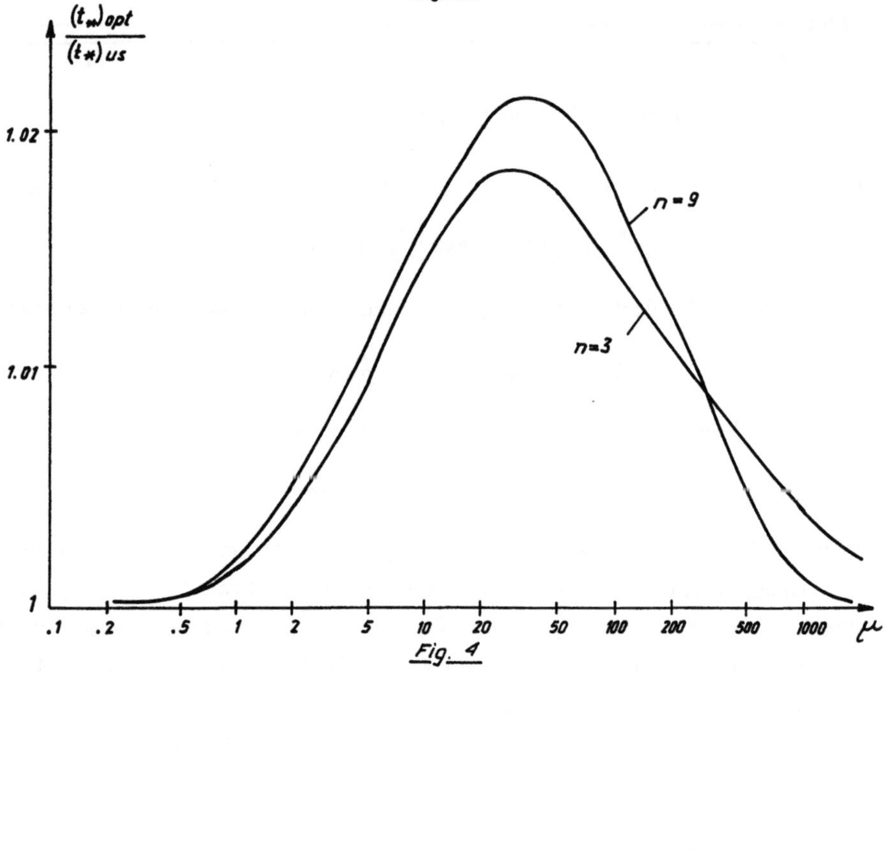

$\frac{(t_*)_{opt}}{(t_*)_{us}}$
1.02
1.01
1
$n=9$
$n=3$
.1
.2
.5
1
2
5
10
20
50
100
200
500
1000
μ

Fig. 4

in Eq.19. are for various exponents n in Norton's law presented in Fig.3. in terms of parameter μ (Eq.15). The ratio of ductile ruptures time for optimal bar and for bar of uniform strength is shown in Fig.4. The gain is not very large - only slightly exceeds 2% , but the most important are qualitative results. The shape of bar of uniform strength can be corrected in order to obtain longer life-time to ductile rupture.

References

[1] Szuwalski, K.: *Optimal design of bars under nonuniform tension with respect to ductile creep rupture*, Mech. Struct. & Mach. **17**, (3), 303-319, (1989).

[2] Szuwalski, K.: *Optimal design of structures with respect to ductile creep rupture time*, Zeszyty Naukowe Politechniki Krakowskiej, Kraków 1991. (in Polish)

[3] Życzkowski, M.: Optimal structural design in rheology, J. Appl. Mech. Trans ASME **38** (1971), 39-46.

[4] Życzkowski, M.: *A survey of optimal design in creep conditiions*, Mech. Teor. Stos. **3** (1986), 24, 243-258 (in Polish); Appl. Mech. Rev. **41** (1988) 453-461.

A Numerical Method of Nonlinear Viscoelastic Structural Analysis

W. SZYSZKOWSKI[1] AND P.G. GLOCKNER[2]

[1]DEPARTMENT OF MECHANICAL ENGINEERING
UNIVERSITY OF SASKATCHEWAN
SASKATOON, SK, CANADA

[2]DEPARTMENT OF MECHANICAL ENGINEERING
UNIVERSITY OF CALGARY
CALGARY, AB, CANADA

Summary

An engineering multimodal constitutive law modelling in a continuous manner the main features of tme-dependent materials including creep, strain hardening and softening, recovery and creep rupture is briefly reviewed. Material response, dependent on the stress history, is defined by means of single Volterra integrals. The method allowing the effective handling of these integrals is also discussed.

Introduction

Most plastics, composites and other synthetic materials due to their relatively low melting temperatures and long molecular chain structure exhibit strong, hereditary type, nonlinear viscoelatic behaviour. The analysis and design of structures made of such materials continues to present a serious challenge to the engineer despite of the recent progress in computer codes and computing equipment.

The general description of nonlinear viscoelastic is provided by the multiple integral formulation [1,2]. This approach, however, seems to be numerically intractable as well as it poses enormous experimental difficulties when attempting to establish the values of the large number of parameters required [3,4]. The single integral formulations, as for example, the Modified Superposition Principle [2,6,8], or the Schapery Thermodynamic Theory [5,7], though significantly simpler in form, have still been found mathematically too complex to be used in engineering analyses, due to presence of Volterra-type integrals.

M. Zyczkowski (Ed.)
Creep in Structures
IUTAM Symposium Cracow/Poland 1990

For the stresses vary with time the integrand of every Volterra type integral depends on the whole stress history and also changes with time. Any direct treatment of such integrals would require to repeat the whole integration process from the beginning for every time step. Since the time domain is continuously increasing, the number of operations and the storage requirements may become prohibitively large in a relatively short time. These calculations, when applying the Finite Element Method technique, should be performed for all strain components at every integration point of each element used in modelling the structure.

Perhaps that is why the complex integral material models have been used almost exclusively for creep samples with uniform stress distributions. Most analyses of real structures, presented in the literature, have been obtained assuming substantially simplified material descriptions, which neglected either hereditary or nonlinearity or both.

In this paper an approximate method which is able to handle the nonlinear constitutive equations of the single integral theories is discussed. The method is also acceptable for the FEM numerical analysis.

The time related functions in the Volterra integrals are approximated either by Prony's series or by equivalent creep compliance characteristics. Consequently, the method requires storing and continuously updating only the current stress and some internal strain components.

Various time-integration schemes are examined. A significant improvement in the numerical efficiency was achieved by calculating "exact" increments of the strain and linear increment of the stress during the assumed time step. This approach allows for relatively long time steps without loosing numerical stability.

The relations derived were coded to interact with the ANSYS finite element program and used for the analysis of materially and geometrically nonlinear problems.

The Material Model

Only the material models which can be included into the FEM technique, have a chance to be recognized by the engineers and used to the analysis of real structures. These models should be "calculable" by means of some reasonable effort and should incorporate the basic characteristics of time-dependent materials.

Such an engineering model is briefly presented here. The model is able to simulate the following features:

a) a linear instantaneous response typical of Hookean bodies,

b) material isotropy, extending for some range into the primary and secondary creep stages,

c) the stress anistropy related to the appearance of microcracking at some advanced stage of deformation.

Postulating the creep potential, P, be a function of the stress deviator, s_{ij}, one can formally obtain the constitutive relation in the form [9]:

$$\dot{\varepsilon}_{ij} = \frac{\partial P}{\partial s_{ij}} = \dot{\varepsilon}_{ij}^{e} + \dot{\varepsilon}_{ij}^{r} + \dot{\varepsilon}_{ij}^{p} \tag{1}$$

where: $\dot{\varepsilon}_{ij}^{e} = \frac{1}{E}\,[(1-\mu)\,\dot{\sigma}_{ij} - \mu\dot{\sigma}_{ij}\,\delta_{ij}]$; $\dot{\varepsilon}_{ij}^{r} = \frac{1}{v_1}\frac{d}{dt}\int_0^t [\,\tilde{\sigma}_{ij}(\tau)]^n\, j(t-\tau)\, d\tau$

$\dot{\varepsilon}_{ij}^{p} = \frac{1}{v_2}\,[\tilde{\sigma}_{ij}(t)]^n$; $\tilde{\sigma}_{ij} = s_{ij}\,[\frac{3}{2}\,|\frac{S}{s_{ij}}|^{n-1}]^{1/n}$

and where S is the second invariant of the stress deviator, $\tilde{\sigma}_{ij}$ denotes the "effective" viscous stress components. The total strain rate is expressed in terms of the usual elastic portion $\dot{\varepsilon}_{ij}^{e}$, the reversible viscous $\dot{\varepsilon}_{ij}^{r}$ and the permanent viscous $\dot{\varepsilon}_{ij}^{p}$ rates. The function j(t) in the reversible creep is controlling the transient (primary) creep while ε_{ij}^{p} can be looked at as the secondary creep (Fig. 1). In general, the consitutive relations (1) cover the deformations specified in (a) and (b).

For the advanced (tertiary) stage of creep the permanent creep rate is adjusted to:

$$\dot{\varepsilon}_{ij}^{p} = \frac{1}{v_2}\,[\frac{\tilde{\sigma}_{ij}(t)}{1-\omega(t)}]^n\,[1 + \alpha\,(\frac{\varepsilon_{eff}^{p}}{\varepsilon_0^{p}} - 1)^a] \tag{2}$$

where $\omega(t)$ is the skalar damage function satisfying the evolution law in the form:

$$\dot{\omega}(t) = R[\frac{\sigma_{max}^{+}}{1-\omega(t)}]^{\Gamma}\,[1-\omega(t)]^{\Gamma-\eta} \tag{3}$$

and where σ_{max}^{+} denotes the largest principal tensile stress. The function ε_{ef}^{p} is the square root of the second invariant of the permanent strain deviator.

The constants v_1, v_2, n and the function j(t) can be obtained from the primary and secondary phases of the creep test, the parameters R, η and Γ from the creep rupture tests in tension and the parameters α, ε_0^{p} and a from the tertiary phase of creep when tested in compression [10]. This model simulated reasonably well the behaviour of quite a wide class of real materials [10,11].

The heredity is included in the reversible creep, ε^r. Because of the Volterra integrals, the numerical treatment of this portion of creep is very challenging. For the reasons outlined earlier a direct handling of the integral is not feasible.

A method of approximate calculations of ε_r was presented in [12] where the Eqn. (1b) was replaced by the formula (the subscripts i,j are omitted for convenience):

$$\dot{\varepsilon}^r \cong (\sigma'')^n / v_1 \tag{4}$$

where σ'' is determined from a current magnitude of σ and ε^r using the relation:

$$(\sigma - \sigma'')^n + c\sigma^{n-1}(\sigma - \sigma'') = E_1\,\varepsilon^r \tag{4a}$$

Here the entire stress history is included in the "hidden" stress σ'' and c and E_1 are parameters defining the functions j(t) [13]. Eqn. (1) was derived assuming the creep potential in the form of the power function with one term only (Norton's power law). In this formulation the power n, the magnitude of which is selected to fit the experimental data from the creep tests in the secondary phase, in general, is not integer.

Assuming the creep potential as a power series with K terms, the formulation which is more general than given by Eqn. (1), can be obtained [2,6]. The reversible creep, for example, is written in the form:

$$\dot{\varepsilon}^r(t) = \frac{1}{v_1}\frac{d}{dt}\int_0^t \sum_{k=1}^{K} \frac{\partial(\sigma^k)}{\partial\tau} D_k(t-\tau)\,d\tau \tag{5}$$

Note that the functions $D_k(t)$ may be interpreted as similar to the creep curves at a constant stress (increasing in time up to same limit) while the function j(t) was similar to the creep rate at a constant stress (decreasing in time to zero).

Each functions $D_k(t)$ can be expanded into n "modes of creep" using the Prony series:

$$D_k(t) = \sum_{i=1}^{n} a_{ik} \ (1-\exp(-t/T_i)) \tag{5a}$$

The $(k+1)\cdot n$ parameters, a_{ik} and T_i, are to be determined from the creep tests. For example in [14] the best set of these parameters was selected by the optimization procedure minimizing the error:

$$E(a,T_t) = (\frac{1}{T_t} \int_0^{T_t} [\ \varepsilon^r_{exp} - \sum_{i=1}^{K} \sum_{i=1}^{n} a_{ik} \ (1-e^{-t/Ti})]^2 \ w_k(t)^2 \, dt)^{1/2} \tag{6}$$

where T_t is the total time of the experiment and $w_k(t)$ is a given set of weighting functions.

Under same assumptions discussed in [14] Eqn. (5) can be transformed to:

$$\dot{\varepsilon}^r = \sum_{i=1}^{n} \dot{\varepsilon}^r_i \tag{7}$$

where each i-th term is governed by the relation

$$\dot{\varepsilon}^r_i = \frac{1}{T_i} [g_i(\sigma) - \varepsilon^r_i] \tag{7a}$$

and where: $g_i(\sigma) = \sum_{i=1}^{K} a_{ik}\sigma^k$

As can be seen from Eqn. (7a) the parameters T_i and g_i can be interpreted as the retardation time and amplitudes of the i-th mode of creep.

Clearly, Eqn. (7), similarly as Eqn. (4), represents an alternative to Eqn. (1b). However, the Volterra integrals, present in Eqn. (1b) were effectively eliminated from Eqn. (7) or Eqn. (4).

The reversible creep definitions as given by these two equations have some common features as well as some differences. Mechanically, Eqn. (4) can be viewed as the governing relation for the Kelvin body with the nonlinear spring and dashpot characteristics [10] while Eqn. (7) defines the behaviour of a series of Kelvin bodies with somewhat simpler, nonlinear spring and dashpot characteristics [14].

Time-Step Integration Schemes

When inertial forces are neglected and when utilizing the FEM to solve the boundary values problem (which eliminates the spatial derivatives), the behaviour of the structure in time will be determined when solving the following initial values problem (in terms of the displacement vector, u):

$$\dot{u}(t) = f[\sigma(t),u(t),t] \tag{8}$$

or at the element level for the strain tensor:

$$\dot{\varepsilon}(t) = F[\sigma(t),\varepsilon(t),t] \tag{8a}$$

Both equations represent a typical parabolic initial values problem. When the explicit integration scheme is used the integration steps may need to be very small to avoid the numerical instability generating undesired oscillations. The implicit schemes, though numerically stable, are not convenient since they require corrections to the stiffness matrix at each iteration. Also, the accuracy in the time domains following rapid changes of the external loadings is poor. The efficiency of integration of the equations (8) can be substantially improved when the formulation (7) is used [14]. For example, if the stresses vary linearly over the assumed integration step Δt, the corresponding exact increments of the recoverable strain in the i-th mode, $\Delta\varepsilon_i^r$, can be determined explicitly as (the increment of the permanent creep are easily obtainable from (1c)):

$$\Delta\varepsilon_i^r = \xi_i\, \Delta t + \mu_i\, \Delta\sigma \tag{9}$$

where:

$$\xi_i = (g_i(\sigma) - \varepsilon_i)\, Y_i/T_i\,; \quad \eta_i = \frac{dg_i}{d\sigma}\,(1 - Y_i)$$

$$Y_i = (1 - \exp\,(-\Delta t/T_i))/(\Delta t/T_i)$$

The difference between the formula (9) and the standard explicit or implicit integration schemes is shown in Fig. 2a. Note that for constant stresses, which is the case for some statically determinate problems, only one step of integration is required to find the solution at an arbitrary time t. If the stresses vary, a recursive procedure has to be applied to eliminate the residual nodal forces (Fig. 2b). Any standard procedure handling numerically plasticity problems, after some modifications [14], can be used for this purpose.

Conclusions

An engineering multiaxial hereditary constitutive model for nonlinear time-dependent materials is presented. It depicts all the main characteristic features of the materials behaviour. The model was successfully used in the analysis of ice plates [10] and in the geometrically nonlinear polypropyline beams [14]. In the latter case the model was incorporated into ANSYS, the commerically available FEM software.

References

1. Sobotka, Z., Rheology of Materials and Engineering Structures, Elsevier, Amsterdam, 1984.
2. Findley, W.N., Lai, J.S.Y., Onaran, K., Creep and Relaxation of Nonlinear Viscoelastic Materials, North-Holland Publishing Company, Amsterdam - New York - Oxford, 1976.
3. Neis, V.V., Sackman, J.L., An Experimental Study of a Nonlinear Material with Memory, Trans. of the Society of Rheology, 11:3, 1967, pp. 307-333.
4. Gradowczyk, M.H., On the Accuracy of the Green-Rivlin Representation for Viscoelastic Materials, Int. J. Solids and Structures, 5, 1969, pp. 873-877.
5. Augl, J.M., Land, D.J., A Numerical Approach for Obtaining Nonlinear Viscoelastic Parameters of Polymeric Materials and Composites, J. of Appl. Polymer Sci., 30, 1985, pp. 4203-4233.
6. Smart, J., Williams, J.G., A Comparison of Single-Integral Non-Linear Viscoelasticity Theories, J. Mech. Psyc. Solids, Vol. 20, 1972, pp. 313-324.
7. Schapery, R.A., On the Characterization of Nonlinear Viscoelastic Materials, Polymer Engineering and Science, 1969, Vol. 9, pp. 295-310.
8. Pipkin, A.C., Rogers, T.G., A Nonlinear Integral Representation for Viscoelastic Behaviour, J. Mech. Phys. Solids, 1968, Vol. 16, p. 59.
9. Glockner, P.G., and Szyszkowski, W., An Engineering Multiaxial Constitutive Model for Nonlinear Time-Dependent Materials, Int. J. Solids and Structures, 26, 1990, pp. 73-82.
10. Szyszkowski, W. and Glockner, P.G., On a Multiaxial Constitutive Law for Ice, Mechanics of Materials, 5, 1986, pp. 49-71.
11. Mahrenholtz, O.H., Wu, Z., Koenig, J.A., Creep of Ice Under Multiaxial Lading: Analysis and Experiment, Proc. CSME Mech. Eng. Forum 1990, Toronto, 1990, pp. 359-365.
12. Szyszkowski, W., and Glockner, P.G., On a Multiaxial Non-Linear Hereditary Constitutive Law for Non-Ageing Materials with Fading Memory, Int. J. Solids Structures, 23, 1987, pp. 305-324.
13. Szyszkowski, W., Glockner, P.G., A Numerical Method for Creep Computations Including Hereditary, Finite Elements in Analysis and Design, 6, 1989, pp. 33-34.
14. Czyz, J., and Szyszkowski, W., An Effective Method of Nonlinear Viscoelastic Structural Analysis, Computers and Structures (in press).

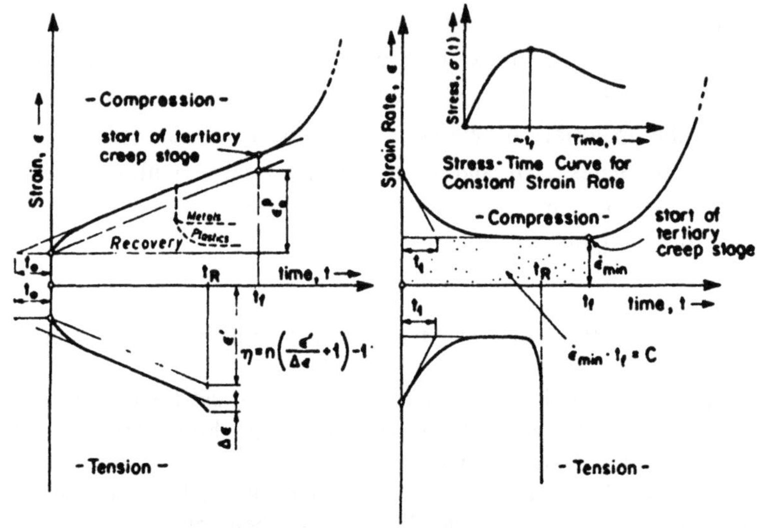

Fig. 1. Some characteristics of time-dependent materials

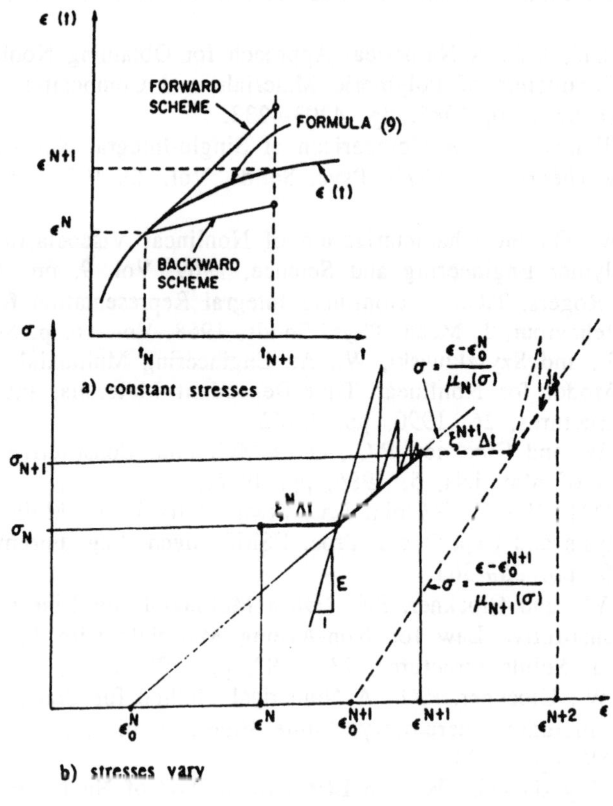

Fig. 2. Time-Step Integration Scheme

Stability and Bounds on Motion of Viscoelastic Columns with Imperfections and Time-Dependent Forces

A. TYLIKOWSKI

Institute of Mechanics
Warsaw University of Technology, Poland

Summary

The Voigt-Kelvin column with random initial geometrical imperfections subject to the axial force is considered. The derived motion bound is expressed in terms of energy-like function. Then the applicability of the Liapunov method is extended to the stability analysis of columns made of a linear standard material.

Introduction

It is well known that the load carrying capacity and stability of certain structures is greatly reduced by the presence of small geometric imperfections. The presence of time-dependent deterministic or stochastic loads and perturbations sometimes does not eliminate the existence of the equilibrium state, if it merely introduces time-varying coefficients in the equations of motion. Examples would be a perfectly straight column subjected to a time-dependent axial load or a perfectly plane plate subjected to in-plane membrane forces, for which the undeflected shape is a possible state irrespective of the nature of the load. In this case one may examine the stability of the equilibrium state. Such problems received considerable attention both in a linear [1,2] and a non-linear case [3]. The present paper is concerned with a second type of problem, in which the perturbations, imperfections lead to forcing terms in the equations of motion and no equilibrium state exist. First the problem was formulated by Plaut and Infante [4], who studied bounds on motions of some continuous system excited by time-dependent deterministic forces. For these problems one is interested in obtaining results concerning the motion which is caused by the perturbations. It is usually impossible to obtain closed form solutions of the equations of motion in these dynamic problems even in a linearized approach. The determination of the effect of perturbations on the solutions of partial differential equations has been the subject of numerous numerical investigations. The approach presented in the paper will be a qualitative one which will yield bound on the motion due to the initial imperfections. The imperfections considered are of a general nature and are not required to be small. Upper bounds on the motion is derived using a Liapunov type of approach and assuming that the column is made of the linear Voigt-Kelvin material. The derived motion is

M. Zyczkowski (Ed.)
Creep in Structures
IUTAM Symposium Cracow/Poland 1990

expressed in terms of energy-like functionals. The problem can be also treated in another way as a stability with permanently acting perturbations.

The second purpose of the present paper is to derive criteria for dynamic stability of viscoelastic columns subjected to time-dependent axial force. The problem of static buckling of viscoelastic columns under constant axial forces has been solved by De Leeuw [5]. One of the first analyses of the dynamic stability of viscoelastic continuous system with viscoelastic boundary conditions has been made by Genin and Maybee [6]. In the next significant study Plaut has used [7] the Liapunov method to determine the stability criteria of viscoelastic columns subjected to compressive axial force. The dynamic stability analysis of viscoelastic continuous system under time-dependent deterministic or stochastic forces was usually applied to the linear Voigt-Kelvin solid (cf. [8,9]). More sophisticated models of material damping such as a linear standard solid [1,2] or a non-linear standard model [3] were introduced to the stability analysis of structures but only constant forces and nonconservative time-independent loadings were assumed. In the present paper the applicability of the Liapunov method is extended to columns made of a linear standard model.

Bound on motion of thr Voigt-Kelvin column with initial imperfections

Let us consider the Voigt-Kelvin column with random initial geometrical imperfections $\eta(\gamma,\xi)$ subject to the axial force, where γ is an element of probability space $(\Gamma, \mathcal{B}, P)$. The field of imperfections is assumed to be two times differentiable with respect to ξ. The transverse displacement u of the column is measured from the initial state and is governed by the following dynamic equations for $\xi \in (0,1)$

$$A\rho u_{,\tau\tau} + F_o(u_{,\xi\xi} + \eta_{,\xi\xi}) + EIu_{,\xi\xi\xi\xi} + EI\lambda_1 u_{,\tau\xi\xi\xi\xi} = 0 . \tag{1}$$

where EI - bending stiffness of the column, ρ -the material density, A - cross section area, l - column length, F_0 - constant axial force.

Introducing dimensionless time $t = \tau/k_t$, $k_t = (A\rho/EI)^{1/2}$, $\lambda_1 = \lambda/2$, $f_0 = F_0/EI$, and dividing by $A\rho$ we obtain

$$v_{,t} + 2\lambda v_{,\xi\xi\xi\xi} + u_{,\xi\xi\xi\xi} + f_0 u_{,\xi\xi} + f_0 \eta_{,\xi\xi} = 0 . \tag{2}$$

where $v = u_{,t}$.

The purpose of the present paper is to derive criteria for solving the following problem: will the deviation of the deflection line from the initial state be bounded in some mathematical sense in the case when the initial imperfections exist. To estimate a perturbed solution of Eq. (2) we introduce a measure $||\cdot||$ of distance of the solution from the initial state. We are going to derive a bound on solutions for Eq. (2). The problem can also be treated in another way as a stability with permanently acting perturbations [10]. We construct the functional as a sum of modified kinetic energy and potential energy of the column

$$V = \int_0^l \left[\frac{1}{2}v^2 + \frac{1}{2}\left(v + 2\lambda u_{'\xi\xi\xi\xi}\right)^2 + u_{'\xi\xi}^2 - f_0 u_{'\xi}^2\right] d\xi \tag{3}$$

Boundary conditions corresponding to simply supported edges written in terms of the transverse displacement are as follows

$$u = 0, \quad u_{'\xi\xi} = 0 \quad \text{at } \xi = 0\,,1\,. \tag{4}$$

The functional is positive-definite as the first three terms of the integrand can be rearranged as a sum of squares and the rest is a potential energy of the column, positive if the constant axial force is sufficiently small and buckling does not occur. Therefore the square root of functional V can be chosen as the measure of distance of disturbed solutions from the straight line $||\cdot|| = V^{1/2}$. If the force acting in the column axis is constant the classic differentiation rule can be applied to calculate the time derivative of the functional (3)

$$\frac{dV}{dt} = 2\int_0^l \left[-2\lambda v_{'\xi\xi}^2 - \lambda u_{'\xi\xi\xi\xi}^2 + \lambda f_0 u_{'\xi\xi\xi}^2 - f_0\left(+\lambda u_{'\xi\xi\xi\xi}\right)\eta_{'\xi\xi}\right] d\xi\,. \tag{5}$$

The first three terms represent the time-derivative of functional for the perfectly straight column and it is possible by means of the calculus of variations to find a positive number κ satisfying inequality

$$\left(\frac{dV}{dt}\right)_p \leq -\kappa V$$

Therefore we can rewrite (5) in the form

$$\frac{dV}{dt} \leq -\kappa V - 2\int_0^l \left(V + \lambda u_{'\xi\xi\xi\xi}\right)\eta_{'\xi\xi}\, d\xi\,. \tag{6}$$

Our object now is to find a bound of the second term in inequality (6). Following Plaut and Infante [4] we use Schwartz's inequality and obtain the following estimation

$$\frac{dV}{dt} \leq -\kappa V + 2\mu\left[\int_0^l \left(V + \lambda u_{'\xi\xi\xi\xi}\right)^2 d\xi\right]^{1/2}\,. \tag{7}$$

where μ is a random number, depending on a random field of imperfections, defined as follows

$$\mu = |f_0|\left[\int_0^l \eta_{'\xi\xi}^2\, d\xi\right]^{1/2} \tag{8}$$

Using functional (3) we can obtain the differential inequality with respect to V

$$\frac{dV}{dt} \leq -\kappa V + 2\mu V^{1/2} \tag{9}$$

Upon substituting $U = V^{1/2}$ into nonlinear inequality (9) one obtains a linear one

$$\frac{dU}{dt} \leq -\frac{\kappa}{2}U + \mu\,. \tag{10}$$

Integration immediately yields

$$U(t) \leq U(0) \exp\left(-\frac{\kappa}{2} t\right) + 2\frac{\mu}{\kappa}\left[1 - \exp\left(-\frac{\kappa}{2} t\right)\right] . \tag{11}$$

Inequality (11) gives us an effective estimation of the measure growth in time. If the trivial solution of the uniform equation ($\eta = 0$) is Liapunov stable, which is equivalent to the fact that the axial force is sufficiently small in the case of perfect column, then the first term of the right hand side is nondecreasing. The asymptotic value of the second term in inequality (11) is equal to $2\frac{\mu}{\kappa}$. Thus, we see that if the sum of $U(0) + 2\frac{\mu}{\kappa}$ is sufficiently small in some probabilistic sense, then $U(t)$ is also bounded by this value

$$U(t) \leq U(0) + 2\frac{\mu}{\kappa} . \tag{12}$$

Applying Khasminski's theorem [10] one can easy prove that the initial state is stable for permanently acting excitations. As an example let us consider the random imperfections of the form

$$\eta(\gamma, \xi) = \eta_0 \sin\frac{\pi\xi}{l} .$$

It means that the displacement in the middle of the column is a random variable with a known probability density distribution. Then the probability density distribution of the random variable μ can be easy obtained by means of the well known relation between the probability distributions of the non-inertial transformations

$$\mu = |f_0\eta_0|\, \pi(2l)^{-1/2}$$

Therefore, the initial state is stable with permanently acting excitations (due to the initial imperfections) if the mean value of displacement in the middle is sufficiently small. The bound of motion can be obtained also for the stochastic axial force f_ξ, independent of the initial imperfections. If the axial force f_ξ is stationary and ergotic and $E\kappa \geq 0$ then one can finally obtain an asymptotic bound

$$\lim_{t\to\infty} U(t) \leq \frac{2\mu}{E\kappa}$$

Stochastic stability of standard material column

In the present point the applicability of Liapunov's method is extended to linear standard viscoelastic columns subject to the axial force. In the case when the axial force is a broad band, the Gaussian process is considered and the uniform stochastic stability of a straight form of columns is analyzed. Assuming the Euler-Bernoulli model of the column and taking into account the mathematical model of standard solid we describe the transverse displacement of column by the integro-partial equation

$$A\rho u_{,tt} + [F_0 + F(t,\gamma)]\, u_{,\xi\xi} + EI\left[u_{,\xi\xi\xi\xi} - \frac{E}{\eta}\int_0^t e^{-\lambda(t-\tau)} u_{,\xi\xi\xi\xi}\right] d\tau = 0 . \tag{13}$$

where $F(t,\gamma)$ - stochastic component of axial force, η, λ - constants of the standard solid. As previously, the beam is assumed to be simply supported at both ends. If the stochastic parametric excitation can be modeled as the Gaussian white noise with intensity σ_F equation (13) can be treated as the Ito integro-partial differential equation. Expanding the transverse displacement into series of functions $\sin\frac{i\pi\xi}{l}$ we obtain the infinite system of decoupled integro-differential equations with respect to modes. Omitting the index i we can write the governing dynamic equation of the i-th mode in the form

$$\ddot{x} + k^2 x + (f(\gamma,t))\,x - k^2\frac{E}{\eta}\int_0^l e^{-\lambda\tau}\,x(t-\tau)\,d\tau = 0\,. \tag{14}$$

where k^2 - natural frequency of i-th mode, f_0 and $f(\gamma,t)$ - constant and stochastic components of axial force divided by $A\rho$

$$k^2 = \frac{EI}{\rho A}\left(\frac{i\pi}{l}\right)^4,\quad f_0 = \frac{F_0}{\rho A}\left(\frac{i\pi}{l}\right)^2,\quad f = \frac{F}{\rho A}\left(\frac{i\pi}{l}\right)^2. \tag{15}$$

We rewrite equation (14) in the Ito differential form

$$\begin{aligned} dx &= y\,dt \\ dy &= -\left[\left(k^2 - f_0\right)x - k^2\frac{E}{\eta}z\right]dt + \sigma x\,dW\,, \\ dz &= (x - \lambda z)\,dt\,, \end{aligned} \tag{16}$$

where W denotes a standard Wiener process and $\sigma = \frac{\sigma_F}{\rho A}\left(\frac{i\pi}{l}\right)^2$. Using a method described by Barabashin [11] the Liapunov function should be chosen in the following way

$$\begin{aligned} V &= \left[k^2\left(1-\frac{E}{\lambda\eta}\right) - f_0\right]\left(k^2 - f_0 + \lambda^2\right)\lambda x^2 + \lambda k^4\left(\frac{E}{\eta}\right)z^2 + \\ &- 2k^2\frac{E}{\eta}\left[k^2\left(1-\frac{E}{\lambda\eta}\right) - f_0\right]xz + \lambda\left[\lambda^2 + k^2\left(1-\frac{E}{\lambda\eta}\right) - -f_0\right]v^2 + \\ &+ 2\lambda^2 k^2\frac{E}{\eta}vz\,. \end{aligned} \tag{17}$$

The function (17) is positive-definite if the static criterion of stability is fulfilled

$$f_0 < k^2\left(1 - \frac{E}{\lambda\eta}\right). \tag{18}$$

From inequality (18) we see that the critical axial force for the standard material is smaller then the Euler Force calculated for an elastic column. As we are going to apply the Liapunov method for the stability analysis a measure of distance of distributed solution of equations (16) from the trivial one can be chosen as the square root of function (17). Calculating the infinitesimal generator LV along the solution of equation (16) we have

$$LV = -\left\{2k^2\frac{E}{\eta}\lambda\left[k^2\left(1-\frac{E}{\lambda\eta}\right) - f_0\right] - \left[\lambda^2 + k^2\left(1-\frac{E}{\lambda\eta}\right) - f_0\right]\sigma_2\right\}x^2\,. \tag{19}$$

Therefore the trivial solution of equation (16) $x = y = z = 0$ is uniformly stochastically stable if

$$\sigma^2 < 2\frac{2k^2\frac{E}{\eta}\lambda\left[k^2\left(1-\frac{E}{\lambda\eta}\right)-f_0\right]}{\lambda^2+k^2\left(1-\frac{E}{\lambda\eta}\right)-f_0}. \tag{20}$$

Having the stability problem of the i-th mode solved we proceed to the stability analysis of viscoelastic column. Substituting relations (15) into the condition (20) we obtain the infinite system of inequalities for $i = 1, 2, \ldots$

$$\sigma_F^2 < \frac{EIA\rho\frac{E}{\eta}\left[\frac{EI\pi^4 i^4}{\rho A l^4}\left(1-\frac{E}{\lambda\eta}\right)-\frac{F_0\pi^2 i^2}{\rho A l^2}\right]}{\lambda^2+\frac{EI\pi^4 i^4}{\rho A l^4}\left(1-\frac{E}{\lambda\eta}\right)-\frac{F_0\pi^2 i^2}{\rho A l^2}}. \tag{21}$$

Maximizing the obtained result with respect to i we have the sufficient condition for the uniform stochastic stability of column in the form

$$\sigma_F^2 < \frac{EIA\rho\frac{E}{\eta}\left[\frac{EI\pi^4}{\rho A l^4}\left(1-\frac{E}{\lambda\eta}\right)-\frac{F_0\pi^2}{\rho A l^2}\right]}{\lambda^2+\frac{EI\pi^4}{\rho A l^4}\left(1-\frac{E}{\lambda\eta}\right)-\frac{F_0\pi^2}{\rho A l^2}}. \tag{22}$$

It is evident that the necessary condition (18) is implied by inequality (22).

References

[1] Potapow, W.D.: *Stability of viscoelastic construction elements*, Moscow: Stroizdat 1985 (in Russian).

[2] Bolotin, V.V.; Zhinger, N.I.: *Effects of damping on stability of elastic system subjected to nonconservative forces*, Int. J. Solids Structures **5** (1969) 965-985.

[3] Życzkowski, M.; Kowalski, A.: *Nonconservative stability problems for columns subject to nonlinear creep*, Euromech Coll. 190, Dynamical Stability of Inelastic Structures, Technische Universität Hamburg-Harbourg 1984 109-111.

[4] Plaut, P.H.; Infante, E.F.: *Bounds of motion of some lumped and continuous system*, J. Appl. Mech. **39** (1972), 251-256.

[5] De Leeuw: *Buckling criterion for linear viscoelastic column*, AIAA Journal **1** (1963) 2665-2666.

[6] Genin, J.; Maybee, J.S.: *On the asymptotic stability of solutions of a linear viscoelastic beam*, J. Franklin Inst. **293** (1972) 191-197.

[7] Plaut, R.H.: *Asymptotic stability and instability criteria for some elastic systems by Liapunov's direct method*, Quart. Appl. Math **29** (1973) 395-398.

[8] Tylikowski, A.: *Dynamic stability of viscoelastic continuous systems under time-dependent loadings*, Mech. Teoret. Stos. **24** (1986) 127-137

[9] Tylikowski, A.: *Dynamic stability of viscoelastic shells under time-dependent membrane loads*, Int. J. Mech. Sci. **31** (1989) 591-587.

[10] Khasminskii, R.Z.: *Stochastic stability of differential equations*, Moscow: Nauka 1969 (in Russian).

[11] Barabashin, E.A.: *Liapunov functions*, Moscow: Nauka 1970 (in Russian).

Nonlinear Instability Induced by Creep

A.M. VINOGRADOV

Department of Mechanical Engineering
University of Calgary
Calgary, Alberta, Canada

Summary

A nonlinear analysis of creep-induced instability phenomena associated with bifurcation and snap-through buckling is presented. The study concerns the response of the viscoelastic materials with fading memory. The obtained solutions afford a consistent examination of the time-dependent critical conditions and post-critical characteristics. The concept of the useful life of structures subjected to creep buckling is discussed.

Introduction

Creep-induced instability of structures has been studied extensively. The surveys by Hoff [1], Kempner [2], Kurshin [3], Odquist [4], and Samuelson [5] indicate that, despite the diversity of the problems, there are fundamental concepts and principles that must be clearly understood in order to develop a consistent view of the creep buckling phenomenon. In this regard, the studies by Hayman [6], Hult [7], and Souza and Bittencourt [8] demonstrate that the typical characteristics of creep buckling can be effectively examined through the analysis of simple structural systems. A similar approach is adopted in this paper with the objective to formulate the critical conditions and examine the time-dependent response associated with stable bifurcation, unstable bifurcation, and snap-through buckling of viscoelastic structures. The analysis deals with the materials defined by the constitutive equations of the linear hereditary viscoelasticity. The problems are solved by means of the quasi-elastic method and are examined in relation to limited and unlimited creep materials. In one special case, the quasi-elastic approximation is compared with the closed-form solution.

M. Zyczkowski (Ed.)
Creep in Structures
IUTAM Symposium Cracow/Poland 1990

Nonlinear Analysis

Three systems considered in the study are shown in schematic form in Fig. 1. The link model in Fig. 1a represents a rigid column laterally constrained by a flexible link AB. The Shanley model in Fig. 1b consists of a tee-shaped rigid member supported by two flexible bars, AB and CD. The shallow v. Mises truss in Fig. 1c is composed of two hinge connected flexible bars AB and BC, and is hinge supported at A and C. The models are subjected to a compressive force P which is applied instantaneously at the time t = 0 and sustained or monotonically increased at t > 0. In each case the deformed configuration of the system is defined in terms of the angular deflection, θ. Note that the Shanley model represents a two-degree-of-freedom system and, accordingly, its deformation is defined by an additional vertical displacement component, V.

The elastic buckling characteristics of the systems are also shown in Fig. 1. The responses of "ideal" structures and structures with imperfections are depicted, respectively, by solid and dashed lines. The diagrams describe stable symmetric bifurcation, unstable symmetric bifurcation and snapthrough buckling in terms of the loading parameter $\lambda = P/P_e$, where P_e denotes the elastic critical load.

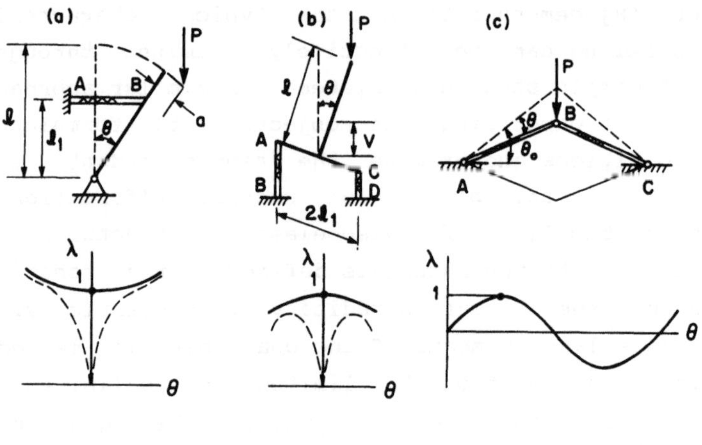

Fig. 1. Structural members and their elastic buckling characteristics

In this study, the material properties of the flexible members of the systems are defined by the linear viscoelastic constitutive law formulated in terms of integral operators of the Stieltjes convolution type. The one-dimensional constitutive equation is considered in the form

$$ku = (1 + \tilde{\Gamma}) * F \tag{1}$$

where k and u = u(t) denote, respectively, the instantaneous axial stiffness and axial displacement of the flexible members, and $\tilde{\Gamma}$ signifies an integral operator

$$\tilde{\Gamma} * F = \int_0^t G(t-\tau)\; F(\tau)\; d\tau \tag{2}$$

in which $G(t-\tau)$ describes the creep properties of the material.

Since the study deals with large deformations, the response of the systems is defined by geometrically nonlinear state equations comprising the kinematic relations and equations of quasi-static equilibrium. These equations combined with the constitutive law in the form of Eq. (1) lead to the nonlinear integral equations governing the behaviour of:

a) the link model:

$$\tan\theta = \lambda(1+\tilde{\Gamma})*[\cos\theta(\sin\theta + \frac{a}{\ell}\cos\theta)] \tag{3}$$

b) the Shanley model

$$2\sin\theta = (1+\tilde{\Gamma}_{AB})*[\lambda(\frac{\ell_1}{\ell}+\tan\theta)]-\frac{k_{CD}}{k_{AB}}(1+\tilde{\Gamma}_{CD})*[\lambda(\frac{\ell_1}{\ell}-\tan\theta)] \tag{4}$$

c) the v. Mises truss:

$$\lambda(1+\tilde{\Gamma})*[1/\sin(\theta_o-\theta)]=[1-\cos\theta_o/\cos(\theta_o-\theta)](1-\cos^{2/3}\theta_o)^{-3/2} \tag{5}$$

Note that, in the case of the Shanley model, it is assumed that the members AB and CD are made of different materials. Respectively, the imperfection of the system is created by asymmetry of the material properties.

The problems under consideration are solved by means of the quasi-elastic method suggested by Schapery [9] for linear viscoelastic stress analysis. The method stems from the observation that under static and quasi-static loading conditions the behaviour of the materials with fading memory can be simulated by the response of a fictitious elastic material the properties of which depend parametrically on time. The method implies that in the viscoelastic constitutive law the action of the integral operator $\tilde{\Gamma}$ can be replaced by the creep function defined as

$$\psi(t) = \tilde{\Gamma} * 1 = \int_0^t G(t-\tau)\, d\tau \qquad (6)$$

Respectively, Eqs. (3)-(5) are reduced to nonlinear algebraic equations that define the angular deflection, θ, as a function of two control parameters, λ and t, provided that the creep function $\psi(t)$ is specified.

Creep Buckling Characteristics

The response of the systems is examined in relation to two types of viscoelastic materials with limited and unlimited creep properties. The first type is defined by an asymptotically limited creep function, i.e. $\psi(t) \to \psi_\infty$ = const as $t \to \infty$. Respectively, in the case of unlimited creep, $\psi(\infty) \to \infty$.

Typical creep buckling characteristics of the link model with an imperfection are shown in Fig. 2. The diagrams indicate that the overall behaviour of the system directly depends on the value of λ. Under a sustained load the deformations increase in time. In the case of limited creep, θ is limited, alternatively, unlimited creep leads to an indefinite increase

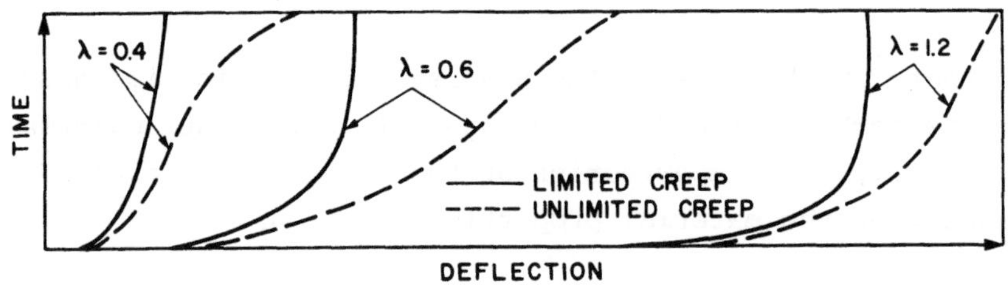

Fig. 2. Creep buckling characteristics of link model with imperfection

of deformations. As shown by Vinogradov [10], similar behaviour is exhibited by viscoelastic columns under an eccentrically applied compressive force.

In the analysis of the Shanley model it is assumed that the member AB is linearly elastic and the time-dependent behaviour of the system is governed only by the creep properties of CD. The respective buckling characteristics in the case of limited creep are depicted in Fig. 3. From these diagrams one observes that the deflection is limited as λ does not exceed a safe load limit λ_s, alternatively, at any $\lambda > \lambda_s$, the system eventually becomes unstable. The condition of instability defined in the form

$$\frac{dt}{d\theta} = 0 \tag{7}$$

yields a load-time relationship schematically shown in Fig. 4. This curve represents a locus of critical points which, for any given $\lambda > \lambda_s$, identifies the corresponding critical time, t_{cr}. One notes that at $\lambda = \lambda_s$, $t_{cr} \rightarrow \infty$, In the case of unlimited creep, $\lambda_s = 0$.

The creep buckling behaviour of the v. Mises truss depicted in Fig. 5 is qualitatively similar to the response of the Shanley model. The condition of instability in the form of Eq. (7) combined with the quasi-elastic solution of Eq. (5) results in the relation

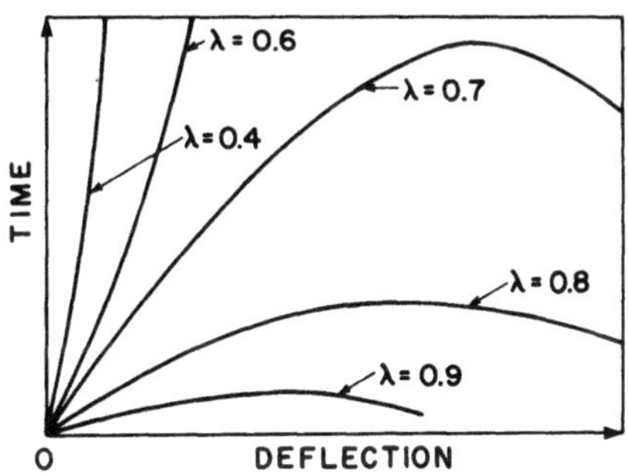

Fig. 3. Creep buckling characteristics of Shanley model (limited creep)

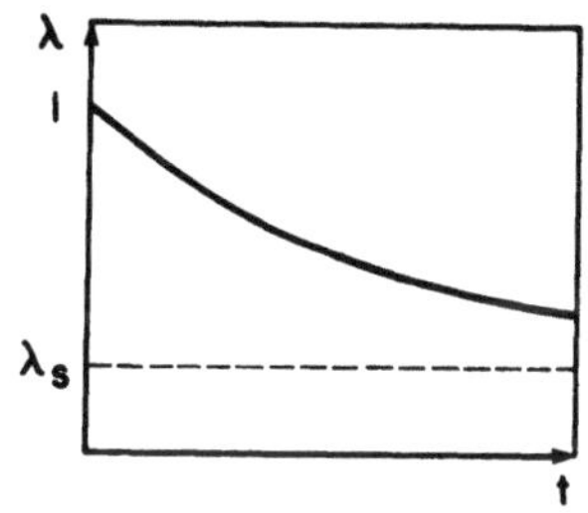

Fig. 4. Locus of critical points

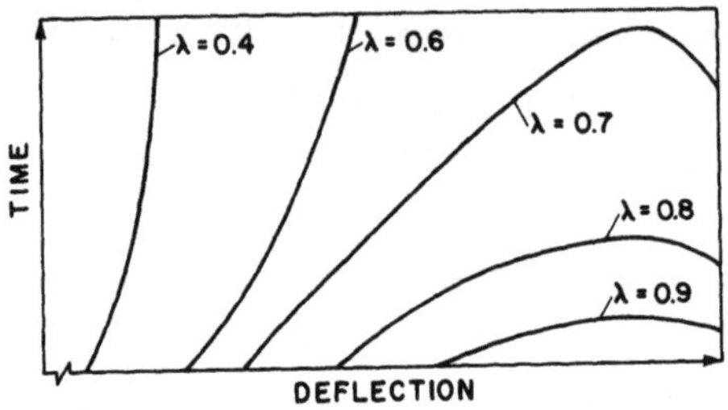

Fig. 5. Creep buckling characteristics of v. Mises truss (limited creep)

$$\lambda = 1/[1 + \psi(t)] \tag{8}$$

which defines the locus of the critical points for the v. Mises truss. It is clear from Eq. (8) that $\lambda_s = 1/(1+\psi_\infty)$ and $\lambda_s=0$ in the cases of limited and unlimited creep, respectively. At $\lambda=1$, $t_{cr}=0$ which indicates that the structure becomes instantaneously unstable at the time of the load application.

The quasi-elastic characteristics of the v. Mises truss are compared in Fig. 6 with the closed-form solution of a special case involving the Maxwell material model. The latter represents a steady state unlimited creep. The instability condition in the form of Eq. (7) applied to the exact analytical solution of the problem yields the load-time relation in the form

$$\lambda = \cos\theta_o \cdot (1 - \cos^{2/3}\theta_o)^{-3/2} \cdot \tan^3(\theta_o - \theta) \tag{9}$$

This equation is compared in Fig. 8 with the locus of the critical points defined by Eq. (8).

The diagrams in Figs. 6 and 7 demonstrate that the quasi-elastic solution provides a reasonably accurate description of the overall response of the system. The approximate values of the critical time tend to be more conservative than those obtained from the exact analysis. However, both curves in Fig. 7 are of similar shape, coincide at $t = 0$ and $t \to \infty$, and indicate that the safe load limit $\lambda_s = 0$.

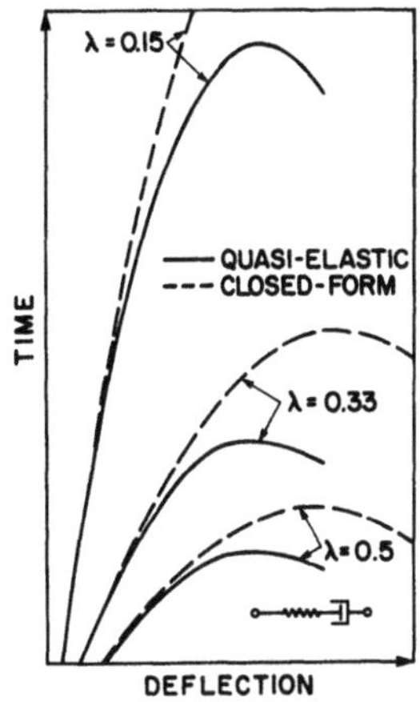

Fig. 6. Creep buckling characteristics of v. Mises truss (Maxwell material)

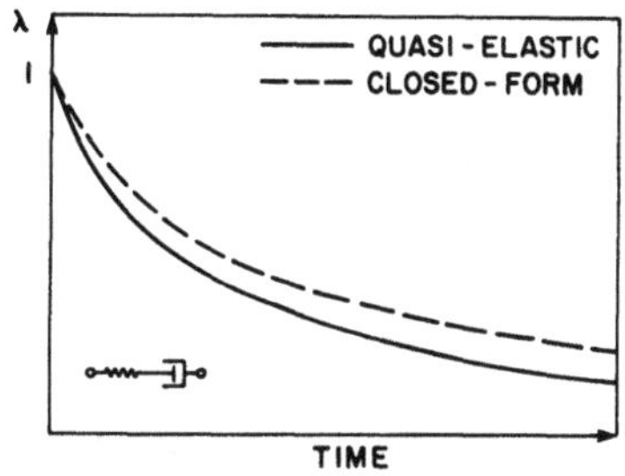

Fig. 7. Critical condition (Maxwell material)

Concluding Remarks

One of the main objectives of the creep stability analysis is to determine the critical conditions that result in termination of the useful life of structural members. This study illustrates that the concept of critical time in creep buckling is closely related to the type of the post-critical behaviour of the structures. In particular, stable bifurcation does not imply the loss of load carrying capacity as observed from the analysis of the link model. In this case, the critical time criterion depends on certain extraneous factors such as geometric constraints or specific design requirements. In the cases of unstable bifurcation and snapthrough buckling, the critical conditions are defined in terms of a functional relation between two destabilizing parameters, the applied load and time. However, for limited creep materials, there is a safe load limit below which the structure remains asymptotically stable. In general, one concludes that an assessment of the useful life of structures in creep buckling must necessarily rely on the geometrically nonlinear approach.

Acknowledgements

The financial support of this work provided by the Natural Sciences and Engineering Research Council of Canada is gratefully acknowledged.

References

1. Hoff, N.J.: Rules and methods of stress and stability calculations in the presence of creep. J. of Appl. Mech. 45, No. 3 (1978) 669-675.
2. Kempner, J.: Viscoelastic buckling. Handbook of Engineering Mechanics. Flugge, W. (ed.) New York: McGraw-Hill Book Co. 1962.
3. Kurshin, L.M.: Creep stability: a survey. Mech. of Solids. Translated M.T.T., Isv. A.N. S.S.S.R. 13, No. 3 (1978) 125-160.
4. Odquist, F.K.G.: Creep stability of bars, plates and shells. Stability. Leipholz, H.H.E. (ed.) Waterloo (1972) 199-242.
5. Samuelson, L.A.: Creep collapse of structures. Collapse analysis of structures. Sobel, L.H.; Thomas, E. (eds.) New York, A.S.M.E. (1984) 137-151.
6. Hayman, B.: Creep buckling - a general view of the phenomena. Creep in Structures. Ponter, A.R.S.; Hayhurst, D.R. (eds.) New York: Springer-Verlag (1981) 289-307.
7. Hult, J.: Creep buckling and instability. Creep of engineering materials and structures. Bernasconi, G.; Piatti, G. (eds.) London: Appl. Sci. Pub. (1978) 133-145.
8. Souza, M.A.; Bittencourt, T.N.: Visco-elastic effects on the vibration of structural elements lible to buckling. Recent developments in buckling of structures. Hui, D.; Birman, V.; Bushnell, D. (eds.), PVP - Vol. 18, New York: ASME (1989) 55-60.
9. Schapery, R.A.: A method of viscoelastic stress analysis using elastic solutions. J. Franklin Inst., 279, No. 4 (1965), 268-289.
10. Vinogradov, A.M.: Nonlinear effects in creep buckling analysis of columns. J. Engg. Mech. 111, No. 6 (1985) 757-767.

Finite Element-Implementation of Creep of Concrete for Thin-Shell Analysis Using Nonlinear Constitutive Relations and Creep Compliance Functions

H. WALTER and H.A. MANG

Institute for Strength of Materials
Technical University of Vienna, Austria

Abstract

A procedure for combining nonlinear short-time behavior of concrete with nonlinear creep compliance functions is presented. It is an important ingredient of a computer code for nonlinear finite element (FE) analysis of prestressed concrete shells, considering creep, shrinkage and ageing of concrete, and relaxation of the prestressing steel. The program was developed at the Institute for Strength of Materials of Technical University of Vienna, Austria.

The procedure has resulted from efforts to extend the range of application of a Finite Element program, abbreviated as FESIA [1], which originally was capable of modeling reinforced concrete in the context of thin-shell analysis, using nonlinear constitutive relations for both, concrete and steel. The extension encompasses the time-dependent behavior of concrete: creep, shrinkage and ageing. It is documented in detail in [2]. Creep is modeled with the help of creep compliance functions which may be nonlinear to conform with the short-time constitutive relations. Ageing causes an interdependence between long-time and short-time deformations.

The paper contains a description of the physical background of the procedure and hints on the implementation of the algorithm. The focus is on general aspects. Details of the aforementioned computer program are considered only where this is inevitable.

1 Basic Assumptions

The deformational behavior of concrete can be described mathematically as the sum of the short-time and the long-time behavior of the material. Accordingly, the strain vector $\boldsymbol{\epsilon}(t)$ at the time instant t can be decomposed as follows:

$$\boldsymbol{\epsilon}(t) = \boldsymbol{\epsilon}^I(t) + \boldsymbol{\epsilon}^C(t) + \boldsymbol{\epsilon}^0(t). \tag{1}$$

In Eq.(1),

$$\boldsymbol{\epsilon}^I(t) = \int_0^t \mathbf{D}^{-1}(t_0, \boldsymbol{\sigma}(t_0))\, d\boldsymbol{\sigma}(t_0) \tag{2}$$

M. Zyczkowski (Ed.)
Creep in Structures
IUTAM Symposium Cracow/Poland 1990

is the instantaneous part of the stress-dependent strains, which can be split into an elastic and a plastic portion, that is, $\boldsymbol{\epsilon}^I(t) = \boldsymbol{\epsilon}^E(t) + \boldsymbol{\epsilon}^P(t)$;

$$\boldsymbol{\epsilon}^C(t) = \mathbf{D}_0^{-1} \int_0^t C(t, t_0, d\boldsymbol{\sigma}(t_0), \boldsymbol{\sigma}(\tau), \boldsymbol{\epsilon}(\tau), 0 \leq \tau \leq t_0) d\boldsymbol{\sigma}(t_0) \tag{3}$$

are the creep strains and $\boldsymbol{\epsilon}^0(t)$ denotes stress-independent deformations such as shrinkage or temperature strains. In Eqs.(2) and (3), $\boldsymbol{\sigma}$ is the stress vector, t_0 is the time-instant at the application of a differential of stress, $d\boldsymbol{\sigma}$. $\mathbf{D}^{-1}(t_0, \boldsymbol{\sigma}(t_0))$ is the inverse of the material stiffness matrix for short-time behavior (which, in general, is path-dependent), and $C(t, t_0, d\boldsymbol{\sigma}(t_0), \boldsymbol{\sigma}(\tau), \boldsymbol{\epsilon}(\tau))$ denotes the creep compliance function. Eq.(3) implies the Boltzmann superposition principle which is commonly used in connection with creep of concrete [3]. Because of lack of information concerning creep under multiaxial states of stress, the one-dimensional creep compliance function C is generalized in the program FESIA for plane stress through multiplication by

$$\mathbf{D}_0^{-1} = \begin{bmatrix} 1 & -\nu_B & 0 \\ -\nu_B & 1 & 0 \\ 0 & 0 & 1+\nu_B \end{bmatrix}, \tag{4}$$

where ν_B is Poisson's ratio for concrete. The very general form of the creep compliance function according to Eq.(3), which might be useful to model special effects related to "delayed elastic deformations" [4], has been specialized within the framework of FESIA, for the purpose of computational efficiency, to the form

$$C = C(t, t_0, \boldsymbol{\sigma}(t_0)). \tag{5}$$

Two aspects of ageing of concrete are important: (a) the increase of the material stiffness with time, resulting in an increase of the (initial) modulus of elasticity $E_{B0} = E_{B0}(t)$, and (b) the increase of the concrete strength $\beta_P = \beta_P(t)$, leading to a change of the definition of the failure surface and the loading surfaces in the constitutive relations. In this paper, only general remarks on ageing will be made. Details depend on the material model used.

An incremental form of the principle of virtual displacements is used as the starting point for the derivation reported in this paper. Several terms of the mathematical formulation of this principle are omitted to focus just on the nonlinear and time-dependent part of the constitutive equations. Spatial discretization and use of numerical integration yields

$$\sum_{e=1}^{m} \Big\{ - \int_{S^{(e)}} \Big[\sum_{\ell=1}^{n_L} g^{(\ell)} (\delta\boldsymbol{\epsilon}^T \Delta\boldsymbol{\sigma} + \boldsymbol{\sigma}^T \delta\Delta\boldsymbol{\eta}) \Big] h dS + \int_{S^{(e)}} \delta \mathbf{u}^T \Delta \bar{p}_S dS \Big\} = 0. \tag{6}$$

In Eq.(6), the first sum extends over the number m of finite elements, whereas the second sum stretches over the number n_L of layers of FESIA's thin-shell element. $g^{(\ell)}$ denotes the weight coefficient of layer ℓ for the numerical integration over the shell thickness h. An analogous scheme is used to integrate over the shell surface $S^{(e)}$. The symbols δ and Δ stand for "variation" and "increment", respectively. $\boldsymbol{\epsilon}$ denotes the linear contribution to the (Green-Lagrange) strains in the middle surface of layer ℓ; $\Delta\boldsymbol{\eta}$ stands for an increment of the nonlinear contribution to the respective membrane strains. Eq.(6) also contains the contribution of an increment of the surface load, $\Delta\bar{\mathbf{p}}_S$, acting on virtual displacements $\delta\mathbf{u}$, to the virtual work of an increment of the forces. A full Newton-Raphson iteration scheme is used in FESIA to reduce the unbalance of forces (not contained in Eq.(6)), resulting from linearization of the incremental virtual work equation. This iteration scheme requires updating of the stiffness matrix at each iteration step.

In the following, the index q symbolizes a calculation step, independent of whether it is part of a load increment or a time increment. For the case of the second or of a following iteration step of a load increment, the increment of the external load is equal to zero. Taking the unbalance of forces after the calculation step $q-1$ into account, the mathematical formulation of the incremental form of the principle of virtual displacements for the analysis step q is obtained as

$$\sum_{e=1}^{m}\Big\{-\int_{S^{(e)}}\Big[\sum_{\ell=1}^{n_L} g^{(\ell)}(\delta\boldsymbol{\epsilon}^T_{(q-1)}\Delta\boldsymbol{\sigma}_{(q)}+\boldsymbol{\sigma}^T_{(q-1)}\delta\Delta\boldsymbol{\eta}_{(q)})\Big]\Big\} =$$

$$=\sum_{e=1}^{m}\Big[-\int_{S^{(e)}}\delta\mathbf{u}^T_{(q-1)}(\bar{\mathbf{p}}_{S(q-1)}+\Delta\bar{\mathbf{p}}_{S(q)})dS+\int_{S^{(e)}}(\sum_{\ell=1}^{n_L} g^{(\ell)}\delta\boldsymbol{\epsilon}^T_{(q-1)}\boldsymbol{\sigma}_{(q-1)})hdS\Big], \tag{7}$$

where

$$\boldsymbol{\sigma}_{(q-1)}=\sum_{k=1}^{q-1}\Delta\boldsymbol{\sigma}_{(k)}. \tag{8}$$

2 Temporal Discretization

Up to this point, nothing has changed in comparison with the solution strategy for short-time analyses. Considering long-time deformations requires a discretization in time. Fig.1b shows a stress history caused by five load increments applied at two different instants of time (Fig.1a). Full lines in the $\sigma - t$ diagram illustrate the actual stress history, whereas broken lines show its idealization within the framework of FE-analyses. Changes of stress are taken into account as stress jumps at a certain point within each time interval. In the time interval $[t_{i-1}, t_i]$, this point is defined as $t_{i-0.5} = \sqrt{t_i \cdot t_{i-1}}$ (see Fig.1b). Changes of stress caused by increments of the external load are considered as occurring within a time interval of zero length. Like the sizes of the load increments, the lengths of the time intervals are user-defined. It is recommended to start with a small increment after each load increment and to increase the size of the time steps continuously [3] (Fig.1b).

Using the temporal discretization according to Fig.1, renders

$$\boldsymbol{\epsilon}^C(t)=\mathbf{D}_0^{-1}\sum_{i=1}^{n} C\left(t,t_{i-0.5},\boldsymbol{\sigma}_{(i-0.5)}\right)\Delta\boldsymbol{\sigma}_{(i)} \tag{9}$$

from Eq.(3). In Eq.(9), n is the number of stress changes (in contrast to the index q in the Eqs.(7) and (8), denoting the actual calculation step). The term "stress change", either means the discrete change of stress "in the middle" of an actual time interval or the sum of stress changes produced by all load increments applied at a certain instant of time (Fig.1b). Noting that n is implicitly defined by q, one could write n_q instead of n. $\boldsymbol{\sigma}_{(i-0.5)} = (\boldsymbol{\sigma}_{(i)}+\boldsymbol{\sigma}_{(i-1)})/2$ is

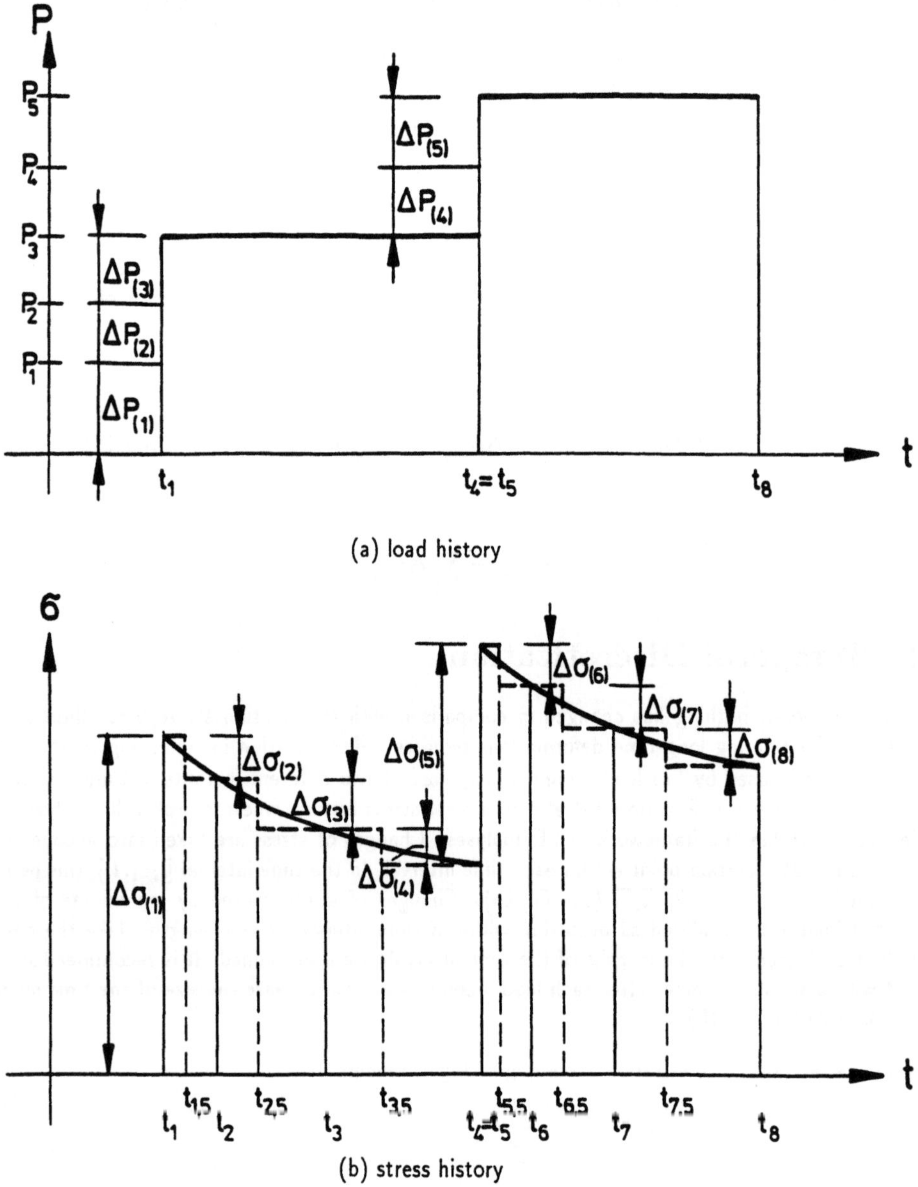

Fig. 1: Stress History (and its Discretization) Caused by Five Load Increments Applied at Two Different Instants of Time

an estimate of the mean value of the stresses within the time interval $[t_{i-1}, t_i]$. For $i = 1$, $t_{i-0.5}$, $\boldsymbol{\sigma}_{(i-1)}$ and $\Delta\boldsymbol{\sigma}_{(i)}$ are given as $t_{0.5} = t_1$, $\boldsymbol{\sigma}_{(0)} = \mathbf{0}$ and $\Delta\boldsymbol{\sigma}_{(1)} = \boldsymbol{\sigma}_{(1)}$, respectively. Eq.(9) shows that the proposed algorithm requires storage of the total stress history, which has the disadvantage of needing more computer resources and the advantage of greater flexibility than an algorithm using Dirichlet series.

3 Modification of Creep Strains and Creep Compliance due to Ageing

In order to satisfy the condition $\boldsymbol{\sigma} = \mathbf{E}(t)\boldsymbol{\epsilon}^E$ at any instant of time without making $\boldsymbol{\epsilon}^P$ time-dependent, $\boldsymbol{\epsilon}^C$ and C have to be substituted by

$$\bar{\boldsymbol{\epsilon}}^C(t) = \boldsymbol{\epsilon}^C(t) + \sum_{i=1}^{n} \left[\mathbf{E}^{-1}(t_{i-0.5}) - \mathbf{E}^{-1}(t)\right] \Delta\boldsymbol{\sigma}_{(i)} \tag{10}$$

and

$$\bar{C}(t, t_0, \boldsymbol{\sigma}(t_0)) = C(t, t_0, \boldsymbol{\sigma}(t_0)) + \frac{1}{E(t_0)} - \frac{1}{E(t)}. \tag{11}$$

For further details about these modifications see [2], [5].

4 Stress-Strain Relations for the Calculation Step q

4.1 Computation of $\boldsymbol{\sigma}_{(q-1)}$

The calculation of the strains $\boldsymbol{\sigma}_{(q-1)}$ is performed in two substeps:

- first, the strain increment $\Delta\boldsymbol{\epsilon}_{(q-1)} = \boldsymbol{\epsilon}_{(q-1)} - \boldsymbol{\epsilon}_{(q-2)}$ is split into the contribution from creep, $\Delta\bar{\boldsymbol{\epsilon}}^C_{(q-1)}$, and into the remaining parts, $\Delta\boldsymbol{\epsilon}^I_{(q-1)}$ and $\Delta\boldsymbol{\epsilon}^0_{(q-1)}$,
- then, the instantaneous part of the increment of the stress-dependent strains, $\Delta\boldsymbol{\epsilon}^I_{(q-1)}$, is used to determine the stress increment $\Delta\boldsymbol{\sigma}_{(q-1)}$:

$$\Delta\boldsymbol{\sigma}_{(q-1)} = \mathbf{D}_{IS}(t_{q-1}, \boldsymbol{\sigma}_{(q-1)}, \Delta\boldsymbol{\sigma}_{(q-1)}, \boldsymbol{\epsilon}^P_{(q-1)})\Delta\boldsymbol{\epsilon}^I_{(q-1)}, \tag{12}$$

  where $\mathbf{D}_{IS}$ symbolizes the incremental secant material stiffness matrix. $\Delta\boldsymbol{\sigma}_{(q-1)}$ is added to $\boldsymbol{\sigma}_{(q-2)}$, resulting in $\boldsymbol{\sigma}_{(q-1)}$.

The two substeps cannot be carried out in a straightforward manner because $\Delta\bar{\boldsymbol{\epsilon}}^C_{(q-1)}$ depends on the unknown $\Delta\boldsymbol{\sigma}_{(q-1)}$:

$$\Delta\bar{\boldsymbol{\epsilon}}^C_{(q-1)} = \mathbf{D}_0^{-1} \sum_{i=1}^{n_{q-2}} \bar{C}(t_{q-1}, t_{i-0.5}, (\boldsymbol{\sigma}_{(i-1)} + \boldsymbol{\sigma}_{(i)})/2)\Delta\boldsymbol{\sigma}_{(i)} +$$
$$+\mathbf{D}_0^{-1}\bar{C}(t_{q-1}, t_{n_{q-1}-0.5}, (\boldsymbol{\sigma}_{(n_{q-1}-1)} + \boldsymbol{\sigma}_{(q-1)})/2)\Delta\boldsymbol{\sigma}_{(q-1)} - \bar{\boldsymbol{\epsilon}}^C_{(q-2)}. \tag{13}$$

Care must be taken when performing the summation in Eq.(13). The index i is used to count time intervals, whereas the indes q denotes a calculation step. $\boldsymbol{\sigma}_{(i)}$ is the best approximation available for the stresses at the end of the time interval i. By "best approximation", for $i = n_{q-2}$, $\boldsymbol{\sigma}_{(q-2)}$ is meant, whereas for all other time steps the stresses at the end of the respective equilibrium iteration are meant. $\boldsymbol{\sigma}_{(q-2)}$ will only be equal to $\boldsymbol{\sigma}_{(n_{q-2})}$, if $n_{q-1} \neq n_{q-2}$. For the case of a nonlinear creep compliance function, Eq.(13) contains an approximation [2].

$\Delta\boldsymbol{\sigma}_{(q-1)}$ has started acting at $t_{n_{q-1}-0.5}$. At that time instant the modulus of elasticity was equal to $E(t_{n_{q-1}-0.5})$. Following from Eq.(11),

$$\bar{C}(t_{q-1}, t_{n_{q-1}-0.5}, \boldsymbol{\sigma}_{(n_{q-1}-0.5)}) = C(t_{q-1}, t_{n_{q-1}-0.5}, \boldsymbol{\sigma}_{(n_{q-1}-0.5)}). \tag{14}$$

Exact calculation of $\Delta\bar{\epsilon}^C_{(q-1)}$ would require an iterative procedure alternating between the two substeps. In order to reduce computer costs, $\Delta\bar{\boldsymbol{\epsilon}}^C_{(q-1)}$ will be approximated. For this purpose, the following approximations for C and $\mathbf{D}_{IS}$ will be used:

$$C(t_{q-1}, t_{q-1.5}, \boldsymbol{\sigma}_{(q-1.5)}) \approx C(t_{q-1}, t_{q-1.5}, \boldsymbol{\sigma}_{(q-2.5)}), \tag{15}$$

$$\mathbf{D}_{IS}(t_{q-1.5}, \boldsymbol{\sigma}_{(q-1)}, \Delta\boldsymbol{\sigma}_{(q-1)}, \boldsymbol{\epsilon}^P_{(q-1)}) \approx$$
$$\approx \mathbf{D}^*_{T(q-2)} = \mathbf{D}_T(t_{q-2.5}, \boldsymbol{\sigma}_{(q-2)}, \Delta\boldsymbol{\sigma}_{(q-2)}, \boldsymbol{\epsilon}^P_{(q-2)}) E(t_{q-1.5})/E(t_{q-2.5}), \tag{16}$$

where

$$t_{q-1.5} = \sqrt{t_{n_{q-1}-1} t_{n_{q-1}}}, \qquad t_{q-2.5} = \sqrt{t_{n_{q-2}-1} t_{n_{q-2}}}, \tag{17}$$

$$\boldsymbol{\sigma}_{q-1.5} = (\boldsymbol{\sigma}_{(n_{q-1}-1)} + \boldsymbol{\sigma}_{(q-1)})/2, \qquad \boldsymbol{\sigma}_{q-2.5} = (\boldsymbol{\sigma}_{(n_{q-2}-1)} + \boldsymbol{\sigma}_{(q-2)})/2, \tag{18}$$

and $\mathbf{D}^*_{T(q-2)}$ is the tangent material stiffness matrix of the previous calculation step, modified by the factor $E(t_{q-1.5})/E(t_{q-2.5})$ representing the increase of Young's modulus due to ageing.

The approximations (15) and (16) result in an approximation of the stresses,

$$\Delta\hat{\boldsymbol{\sigma}}_{(q-1)} = \mathbf{D}^*_{T(q-2)} \Delta\boldsymbol{\epsilon}^I_{(q-1)}, \tag{19}$$

where the increment of the instantaneous part of the stress-dependent strains, $\Delta\boldsymbol{\epsilon}^E_{(q-1)}$, is approximated as

$$\Delta\boldsymbol{\epsilon}^I_{(q-1)} = \Delta\boldsymbol{\epsilon}^E_{(q-1)}(t_{q-1.5}) + \Delta\boldsymbol{\epsilon}^P_{(q-1)} \approx$$
$$\approx \Delta\boldsymbol{\epsilon}_{(q-1)} - \Delta\hat{\bar{\boldsymbol{\epsilon}}}^C_{(q-1)} - \Delta\boldsymbol{\epsilon}^0_{(q-1)} - [\mathbf{E}^{-1}(t_{q-1.5}) - \mathbf{E}^{-1}(t_{q-2.5})]\boldsymbol{\sigma}_{(q-2)}. \tag{20}$$

$\Delta\hat{\bar{\boldsymbol{\epsilon}}}^C_{(q-1)}$ is the aforementioned approximation of $\Delta\bar{\boldsymbol{\epsilon}}^C_{(q-1)}$, given as

$$\Delta\hat{\bar{\boldsymbol{\epsilon}}}^C_{(q-1)} = \tilde{\bar{\boldsymbol{\epsilon}}}^C_{(q-1)} + \mathbf{D}_0^{-1} C(t_{q-1}, t_{q-1.5}, \boldsymbol{\sigma}_{(q-2.5)}) \Delta\hat{\boldsymbol{\sigma}}_{(q-1)} - \bar{\boldsymbol{\epsilon}}^C_{(q-2)}, \tag{21}$$

where

$$\tilde{\bar{\boldsymbol{\epsilon}}}^C_{(q-1)} = \mathbf{D}_0^{-1} \sum_{i=1}^{n_{q-2}} \bar{C}(t_{q-1}, t_{i-0.5}, \boldsymbol{\sigma}_{(i-0.5)}) \Delta\boldsymbol{\sigma}_{(i)} \tag{22}$$

is the first term of Eq.(13). The last term in Eq.(20) is equal to

$$\Delta\epsilon^E_{(q-1)} - \left[\epsilon^E_{(q-1)}(t_{q-1.5}) - \epsilon^E_{(q-2)}(t_{q-2.5})\right], \tag{23}$$

where

$$\Delta\epsilon^E_{(q-1)} = \mathbf{E}^{-1}(t_{q-1.5})\Delta\boldsymbol{\sigma}_{(q-1)}, \tag{24}$$

$$\epsilon^E_{(q-1)}(t_{q-1.5}) = \mathbf{E}^{-1}(t_{q-1.5})\boldsymbol{\sigma}_{(q-1)}, \quad \epsilon^E_{(q-2)}(t_{q-2.5}) = \mathbf{E}^{-1}(t_{q-2.5})\boldsymbol{\sigma}_{(q-2)}. \tag{25}$$

Substituting the approximation for $\Delta\epsilon^I_{(q-1)}$ according to Eq.(20) into Eq.(19), yields

$$\Delta\hat{\boldsymbol{\sigma}}_{(q-1)} = \mathbf{D}^*_{T(q-2)}\left[\Delta\epsilon_{(q-1)} - \Delta\epsilon^*_{(q-1)} - \mathbf{D}_0^{-1}C(t_{q-1}, t_{q-1.5}, \boldsymbol{\sigma}_{(q-2.5)})\Delta\hat{\boldsymbol{\sigma}}_{(q-1)}\right], \tag{26}$$

where

$$\Delta\epsilon^*_{(q-1)} = \tilde{\epsilon}^C_{(q-1)} - \bar{\epsilon}^C_{(q-2)} + \epsilon^0_{(q-1)} - \epsilon^0_{(q-2)} + \left[\mathbf{E}^{-1}(t_{q-1.5}) - \mathbf{E}^{-1}(t_{q-2.5})\right]\boldsymbol{\sigma}_{(q-2)}. \tag{27}$$

Solving Eq.(26) for $\Delta\hat{\boldsymbol{\sigma}}_{(q-1)}$, gives

$$\Delta\hat{\boldsymbol{\sigma}}_{(q-1)} = \left[\mathbf{I} + C(t_{q-1}, t_{q-1.5}, \boldsymbol{\sigma}_{(q-2.5)})\mathbf{D}^*_{T(q-2)}\mathbf{D}_0^{-1}\right]^{-1}\mathbf{D}^*_{T(q-2)}\left[\Delta\epsilon_{(q-1)} - \Delta\epsilon^*_{(q-1)}\right], \tag{28}$$

where $\mathbf{I}$ is a unit matrix. Inserting Eq.(28) into

$$\Delta\hat{\epsilon}^C_{(q-1)} = \mathbf{D}_0^{-1}C(t_{q-1}, t_{q-1.5}, \boldsymbol{\sigma}_{(q-2.5)})\Delta\hat{\boldsymbol{\sigma}}_{(q-1)}, \tag{29}$$

yields the approximation $\Delta\hat{\epsilon}^C_{(q-1)}$ of $\Delta\bar{\epsilon}^C_{(q-1)}$. Thus, $\epsilon^I_{(q-1)}$ can be approximated as

$$\epsilon^I_{(q-1)} = \epsilon^E_{(q-1)} + \epsilon^P_{(q-1)} = \epsilon_{(q-1)} - \bar{\epsilon}^C_{(q-1)} - \epsilon^0_{(q-1)}, \tag{30}$$

where

$$\bar{\epsilon}^C_{(q-1)} = \tilde{\epsilon}^C_{(q-1)} + \Delta\hat{\epsilon}^C_{(q-1)}. \tag{31}$$

The second substep (see Eq.(12)) can now be performed as in a conventional short-time analysis. Care has to be taken that Young's modulus and the compressive strength are evaluated at the time instant $t_{q-1.5}$. Depending on the kind of material model, the change of the concrete strength due to ageing may require special provisions concerning loading and failure surfaces.

Ths size of the error resulting from the above approximations depends, on the one hand, on the difference between the incremental secant material stiffness matrix and, on the other hand, on the amount of nonlinearity of the creep compliance function. Since $\Delta\bar{\epsilon}^C_{(q-1)}$ is only a part of $\Delta\epsilon_{(q-1)}$, it will decrease as $\Delta\epsilon_{(q-1)}$ decreases in the course of the iteration procedure. One can expect that the convergence of the iteration process will not be slowed down noticeably.

4.2 Computation of $\Delta\boldsymbol{\sigma}_{(q)}$

Analogous to Eq.(12), $\Delta\boldsymbol{\sigma}_{(q)}$ can be written as

$$\Delta\boldsymbol{\sigma}_{(q)} = \mathbf{D}_{IS}(t_{q-1.5}, \boldsymbol{\sigma}_{(q-0.5)}, \Delta\boldsymbol{\sigma}_{(q)}, \boldsymbol{\epsilon}^P_{(q)})\Delta\boldsymbol{\epsilon}^I_{(q)}, \tag{32}$$

where $t_{q-0.5}$ is an abbreviation for $t_{i-1.5}, i = n_q$, and $\boldsymbol{\sigma}_{(q-0.5)}$ stands for $(\boldsymbol{\sigma}_{(n_q-1)} + \boldsymbol{\sigma}_{(q)})/2$. Because of the stress-dependence of $\mathbf{D}_{IS}$, this matrix has to be replaced by the modified tangent material stiffness matrix $\mathbf{D}^*_{T(q-1)}$, given as

$$\mathbf{D}^*_{T(q-1)} = \mathbf{D}_T(t_{q-1.5}, \boldsymbol{\sigma}_{(q-1.5)}, \Delta\boldsymbol{\sigma}_{(q-1)}, \boldsymbol{\epsilon}^P_{(q-1)})E(t_{q-0.5})/E(t_{q-1.5}). \tag{33}$$

Analogous to Eq.(16), the factor $E(t_{q-0.5})/E(t_{q-1.5})$ represents the increase of Young's modulus due to ageing.

The strains $\boldsymbol{\epsilon}_{(q)}$ can be split as follows:

$$\boldsymbol{\epsilon}_{(q)} = \boldsymbol{\epsilon}^E_{(q-1)}(t_{q-0.5}) + \Delta\boldsymbol{\epsilon}^E_{(q)}(t_{q-0.5}) + \boldsymbol{\epsilon}^P_{(q-1)} + \Delta\boldsymbol{\epsilon}^P_{(q)} + \tilde{\boldsymbol{\epsilon}}^C_{(q)} + \mathbf{D}_0^{-1}C(t_q, t_{q-0.5}, \boldsymbol{\sigma}_{(q-0.5)})\Delta\boldsymbol{\sigma}_{(q)} + \boldsymbol{\epsilon}^0_{(q)}, \tag{34}$$

where, analogous to Eq.(22),

$$\tilde{\boldsymbol{\epsilon}}^C(q) = \mathbf{D}_0^{-1} \sum_{i=1}^{n_q-1} \bar{C}(t_q, t_{i-0.5}, \boldsymbol{\sigma}_{(i-0.5)})\Delta\boldsymbol{\sigma}_{(i)}. \tag{35}$$

Thus,

$$\begin{aligned}\Delta\boldsymbol{\epsilon}_{(q)} = \boldsymbol{\epsilon}_{(q)} - \boldsymbol{\epsilon}_{(q-1)} = \Delta\boldsymbol{\epsilon}^E_{(q)}(t_{q-0.5}) + \left[\mathbf{E}^{-1}(t_{q-0.5}) - \mathbf{E}^{-1}(t_{q-1.5})\right]\boldsymbol{\sigma}_{(q-1)} + \\ + \Delta\boldsymbol{\epsilon}^P_{(q)} + \tilde{\boldsymbol{\epsilon}}^C_{(q)} + \mathbf{D}_0^{-1}C(t_q, t_{q-0.5}, \boldsymbol{\sigma}_{(q-0.5)})\Delta\boldsymbol{\sigma}_{(q)} - \tilde{\boldsymbol{\epsilon}}^C_{(q-1)} + \Delta\boldsymbol{\epsilon}^0_{(q)}.\end{aligned} \tag{36}$$

By analogy to Eq.(15), $C(t_q, t_{q-0.5}, \boldsymbol{\sigma}_{(q-0.5)})$ is approximated as $C(t_q, t_{q-0.5}, \boldsymbol{\sigma}_{(q-1.5)})$. $\Delta\boldsymbol{\epsilon}^E_{(q)}(t_{q-0.5}) + \Delta\boldsymbol{\epsilon}^P_{(q)}$ is approximated as $\mathbf{D}^{*-1}_{T(q-1)}\Delta\boldsymbol{\sigma}_{(q)}$. With the help of these approximations, $\Delta\boldsymbol{\epsilon}_{(q)}$ can be written as

$$\Delta\boldsymbol{\epsilon}_{(q)} = \mathbf{D}^{*-1}_{T(q-1)}\Delta\boldsymbol{\sigma}_{(q)} + \mathbf{D}_0^{-1}C(t_q, t_{q-0.5}, \boldsymbol{\sigma}_{(q-1.5)})\Delta\boldsymbol{\sigma}_{(q)} + \Delta\boldsymbol{\epsilon}^*_{(q)}, \tag{37}$$

where $\Delta\boldsymbol{\epsilon}^*_{(q)}$ is analogous to $\Delta\boldsymbol{\epsilon}^*_{(q-1)}$ according to Eq.(27). From Eq.(37), $\Delta\boldsymbol{\sigma}_{(q)}$ is obtained as

$$\Delta\boldsymbol{\sigma}_{(q)} = \left[\mathbf{D}^{*-1}_{T(q-1)} + \mathbf{D}_0^{-1}C(t_q, t_{q-0.5}, \boldsymbol{\sigma}_{(q-1.5)})\right]^{-1}(\Delta\boldsymbol{\epsilon}_{(q)} - \Delta\boldsymbol{\epsilon}^*_{(q)}). \tag{38}$$

5 Implementation

Eqs.(12), (30) and (38) serve as the basis for rearranging the original computations for short-time analysis of reinforced concrete shells in a series of substeps in a manner that the time-dependent behavior of concrete can be considered. These substeps (for details see [2]) were grouped such that

- all modifications of the original short-time algorithm are concentrated at as few locations as possible to keep the code clear,

- there is a balance of the amount of computer storage required and the number of recomputations
- the amount of I/O operations to mass storage devices is minimized.

Furthermore, a few additional approximations and linearizations were performed in order to reduce computer costs at the price of a slower rate of convergence for the case of strongly nonlinear creep compliance functions. Ref. [2] contains a full description of both, the program structure and the storage organization.

6 Summary

The basis of a computer algorithm for consideration of the time-dependent behavior of surface structures made of concrete were presented. Main ingredients of this algorithm are the use of creep compliance functions which may depend nonlinearly on the stress history, the storage of the whole stress history, and consideration of the evolution of material parameters due to ageing. Care has been taken to balance the accuracy of the analysis and the rate of convergence of the iterations, on the one hand, and the amount of computer resources required, on the other hand. The algorithm was built on top of an existing computer code for nonlinear short-time analysis; keeping the code modular was a major concern.

Applications to the analysis of reinforced and prestressed concrete slabs and shells were reported elsewhere [5].

References

[1] Floegl, H.: Traglastermittlung dünner Stahlbetonschalen mittels der Methode der Finiten Elemente unter Berücksichtigung wirklichkeitsnahen Werkstoffverhaltens sowie geometrischer Nichtlinearität. Dissertation, Technical University of Vienna, Vienna, Austria, 1981.

[2] Walter, H.: Finite Elemente Berechnungen von Flächentragwerken aus Stahl- und Spannbeton unter Berücksichtigung von Langzeitverformungen und Zustand II. Dissertation, Technical University of Vienna, Vienna, Austria, 1988.

[3] Bažant, Z.P.: Mathematical models for creep and shrinkage of concrete. In: Creep and shrinkage in concrete structures (eds.: Bažant, Z.P.; Wittmann, F.H.), John Wiley & Sons, 1982.

[4] Rüsch, H.; Jungwirth, D.; Hilsdorf, H.K.: Creep and Shrinkage. Springer-Verlag, 1983.

[5] Hofstetter, G.; Walter, H.; Mang, H.A.: Finite Elemente Berechnungen von Flächentragwerken aus Spannbeton unter Berücksichtigung von Langzeitverformungen und Zustand II. Bauingenieur 64 (1989) 449-461.

Optimal Design of Annular Plates Against Creep Buckling

A. WRÓBLEWSKI

Institute of Mechanics and Machine Design
Cracow University of Technology, Poland

Summary

The paper deals with the problem of optimal design of annular plates against creep buckling. The design variable, i.e. the plate thickness function, is determined so as to minimize the total volume of a plate under given external load (radial pressure) and the critical time. A circularly symmetric thickness function is analysed but unsymmetric buckling modes are also admitted. The problem is solved by the use of Pontryagin's maximum principle.

1 Introduction

Optimal design of circular and annular plates against creep buckling is a new problem and therefore a through study of it has not been made so far. In a monograph by Gajewski & Życzkowski [1] that gives a broad review of papers on optimal design under stability constraints there are only five papers mentioned that deal with that problem. All of them consider plates in elastic range. The papers by Frauenthal [2] and Grinev & Philippov [3] undertake the problem of optimization of simply supported or clamped circular plates. Energy approach and the formalism of Pontryagin's maximum principle are applied respectively. The other three papers (Tarasov [4]; Rzegocińska-Pelech & Waszczyszyn [5]; Strzelczyk & Drobniewicz [6]) present numerical approach to the optimal design problem of annular plate. A plate is divided into many annular segments each of constant or linearly varying thickness. In practice the problem is reduced to the non-linear programming one.

Optimal design of structures under creep conditions is the youngest branch of structural optimization. A survey by Życzkowski [7] reviews about 120 papers where that problem has been investigated. The only paper that deals with optimization of circular plates with respect to creep buckling is by Wróblewski [8]. The loss of stability criterion is based on the theories of creep buckling of perfect structures formulated by Shanley [9], Gerard [10], and Rabotnov-Shesterikov [11]. The present paper is an extension of that paper to the case of annular plates. Optimization with one design variable, namely a circularly symmetric thickness function of solid cross-section is discussed in detail, but unsymmetric buckling modes are also admitted. With a view to solving this problem Pontryagin's maximum principle formalism is adopted.

M. Zyczkowski (Ed.)
Creep in Structures
IUTAM Symposium Cracow/Poland 1990

2 Basic equations

It is assumed, in general, that the stress and strain components of basic state (for example pure compression) are slightly varied due to bifurcation and it is possible to linearise creep law with respect to their variation so that equations of equilibrium or motion are obtained in variations. The analysis of either quasi-equilibrium (static approach) or stability of motion (kinetic approach) is then performed on the basis of these equations.

Let us consider the Rabotnov strain hardening creep theory as the physical law:

$$\Phi\left(\tilde{\sigma}_e, \varepsilon_e^C, \dot{\varepsilon}_e^C\right) = \dot{\varepsilon}_e^C \left(\varepsilon_e^C\right)^{\mu} - \Gamma \tilde{\sigma}_e^n = 0 \, ,$$
$$\varepsilon_e^C = \varepsilon_e - \frac{\tilde{\sigma}_e}{E} \, , \quad \varepsilon_e^2 = \tfrac{2}{3} e_{ij} e_{ij} \, , \quad \tilde{\sigma}_e^2 = \tfrac{3}{2} \tilde{s}_{ij} \tilde{s}_{ij} \, , \tag{1}$$

where ε_e^C denotes inelastic strain intensity, ε_e total strain intensity, $\tilde{\sigma}_e$ stress intensity, $\tilde{s}_{ij}, e_{ij}$ are the stress and strain deviators, E, Γ, n, μ are material constants dependent on temperature. The tilde '~' denotes dimensional quantities and the dot denotes differentiation with respect to time $\tilde{t}$.

The components of the stress and strain tensors will receive small increments (variations) $\delta\tilde{\sigma}_{ij}, \delta\varepsilon_{ij}$ due to buckling. Assuming similarity of stress and strain deviators and incompressibility of material ($\varepsilon_{ij} = e_{ij}$), one can derive the constitutive equations for postbuckling state in the incremental form:

$$\begin{aligned} \delta\tilde{s}_{ij} &= \frac{\tilde{s}_{ij}}{\tilde{\sigma}_e^2}\left(\frac{\delta\tilde{\sigma}_e}{\delta\varepsilon_e} - \frac{\tilde{\sigma}_e}{\varepsilon_e}\right)\tilde{s}_{kl}\delta\varepsilon_{kl} + \frac{2}{3}\frac{\tilde{\sigma}_e}{\varepsilon_e}\delta\varepsilon_{ij} = \\ &= \frac{\tilde{s}_{ij}}{\tilde{\sigma}_e^2}\left(\tilde{E}_t - \tilde{E}_s\right)\tilde{s}_{kl}\delta\varepsilon_{kl} + \frac{2}{3}\tilde{E}_s\delta\varepsilon_{ij} \, . \end{aligned} \tag{2}$$

The relation between stress and strain intensities (the secant modulus $\tilde{E}_s$) in the precritical state, for $\sigma_e = const(\tilde{t})$ is obtained from Eq. (1). The way of determining of the tangent modulus $\tilde{E}_t = \delta\tilde{\sigma}_e / \delta\varepsilon_e$ is different for the creep buckling theories considered. Shanley [9] determined the tangent modulus for an isochronic curve $\tilde{t} = const$. Gerard's hypothesis [10] of a critical strains is obtained if we formally repleace the tangent modulus $\tilde{E}_t$ by the secant modulus $\tilde{E}_s$. Rabotnov-Shesterikov [11] determined $\tilde{E}_t$ for constant rate of inelastic strain intensity ($\dot{\varepsilon}_e^C = const(\tilde{t})$). For the considered Rabotnov's law secant and tangent moduli take the following form in dependence on the creep buckling theory:

$$\frac{1}{\tilde{E}_t^S} = \frac{1}{E} + \frac{n}{\mu+1}\frac{1}{\tilde{\sigma}_e}\left[\Gamma(\mu+1)\tilde{\sigma}_e^n \tilde{t}\right]^{1/(\mu+1)} \, , \tag{3}$$

$$\frac{1}{\tilde{E}_t^G} = \frac{1}{\tilde{E}_s} = \frac{1}{E} + \frac{1}{\tilde{\sigma}_e}\left[\Gamma(\mu+1)\tilde{\sigma}_e^n \tilde{t}\right]^{1/(\mu+1)} \, , \tag{4}$$

$$\frac{1}{\tilde{E}_t^{RS}} = \frac{1}{E} + \frac{n}{\mu}\frac{1}{\tilde{\sigma}_e}\left[\Gamma(\mu+1)\tilde{\sigma}_e^n \tilde{t}\right]^{1/(\mu+1)} \, . \tag{5}$$

Let us consider an annular plate of variable thickness. The plate is loaded by a constant in time radial pressure that is uniformly distributed along the inner and outer contours (Fig. 1). We analyse circularly symmetric thickness distribution $\tilde{h} = \tilde{h}(\tilde{r})$, but unsymmetric buckling modes $\tilde{w} = \tilde{w}(\tilde{r}, \theta)$ are also admitted.

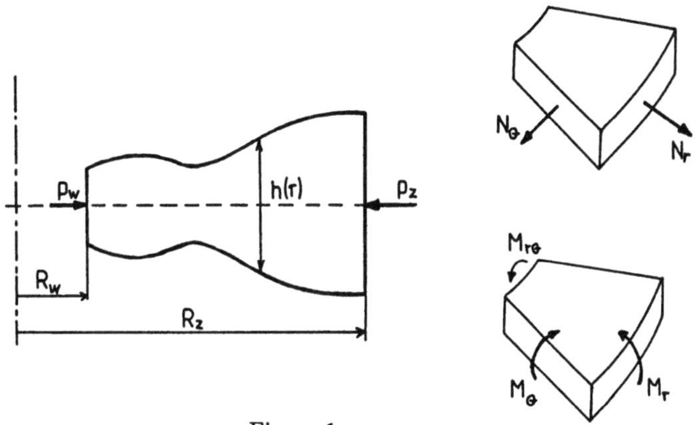

Figure 1:

The stability analysis of plates of variable stiffness requires at first, the solution of membrane stress distribution problem (precritical state), and just then application of stability criterion, making use of small superposed bending. The assumption of small deflections during buckling makes it possible to disregard mutual interaction between membrane state and bending state. The details of the state equations derivation are given in [8] and can be omitted here. We present only the final form for dimensionless variables and parameters:

1) the precritical state (membrane state)

$$\begin{aligned} N_\theta &= \frac{1}{2}\frac{N}{\xi} + E_s \frac{u}{\xi}\frac{h}{\chi}\,, \\ N' &= \frac{1}{2}\frac{N}{\xi} + E_s \frac{u}{\xi}\frac{h}{\chi}\,, \\ u' &= \frac{3}{4}\frac{1}{E_s}\frac{\chi}{h}\frac{N}{\xi} - \frac{1}{2}\frac{u}{\xi}\,, \end{aligned} \tag{6}$$

2) the postbuckling state

$$\begin{aligned} w' &= \varphi\,, \\ \varphi' &= -\frac{M}{E_s h^3 a_{11} \xi} - \frac{a_{12}}{a_{11}}\frac{\varphi}{\xi} + m^2 \frac{a_{12}}{a_{11}}\frac{w}{\xi^2}\,, \\ M' &= Q + \frac{a_{12}}{a_{11}}\frac{M}{\xi} + E_s h^3 \frac{\varphi}{\xi}\left(\frac{a_{12}^2}{a_{11}} - a_{22} - m^2\right) + E_s h^3 \frac{w}{\xi^2}\left(1 + a_{22} - \frac{a_{12}^2}{a_{11}}\right)\,, \\ Q' &= \frac{M}{\xi^2}\left(\frac{1}{E_s h^3 a_{11}}\frac{N}{\xi} + m^2 \frac{a_{12}}{a_{11}}\right) + E_s h^3 \frac{\varphi}{\xi^2}\left(\frac{a_{12}^2}{a_{11}} - a_{22} - 1\right) + \end{aligned}$$

$$+ \left[\frac{N}{\xi}\left(\frac{1}{2}-\frac{a_{12}}{a_{11}}\right)+E_s\frac{h}{\chi}\frac{u}{\xi}\right]\left(m^2\frac{w}{\xi}-\varphi\right)+E_s h^3\frac{w}{\xi^3}\left[1-m^2\left(\frac{a_{12}^2}{a_{11}}-a_{22}\right)\right] , \quad (7)$$

where:

$$a_{11}=\frac{\tilde{\sigma}_r^2}{\tilde{\sigma}_e^2}(a+1) , \quad a_{12}=a_{21}=\frac{\tilde{\sigma}_r\tilde{\sigma}_\theta}{\tilde{\sigma}_e^2}(a+\frac{1}{2}) ,$$
$$a_{22}=\frac{\tilde{\sigma}_\theta^2}{\tilde{\sigma}_e^2}(a+1) , \quad a_{33}=\frac{1}{2} , \qquad a=\frac{3}{4}\left(\frac{\tilde{E}_t}{\tilde{E}_s}-1\right) . \quad (8)$$

The state variables are defined as follows:

$$M=\xi M_r , \quad N=\xi N_r , \quad \varphi=\frac{dw}{d\xi} , \quad Q=\frac{dM}{d\xi}+2mM_{r\theta}-M_\theta , \quad (9)$$

and the dimensionless variables:

$$\xi=\frac{\tilde{r}}{\tilde{R}_2} , \quad h=\frac{\tilde{h}}{\tilde{h}^*} , \quad w=\frac{\tilde{w}}{\tilde{R}_2} , \quad u=\frac{\tilde{u}}{\tilde{R}_2} , \quad E_s=\frac{\tilde{E}_s}{E} , \quad E_t=\frac{\tilde{E}_t}{E} ,$$
$$N_i=\frac{9\tilde{R}_2^2\tilde{N}_i}{E\left(\tilde{h}^*\right)^3} , \quad M_j=\frac{9\tilde{R}_2\tilde{M}_j}{E\left(\tilde{h}^*\right)^3} , \quad \sigma_k=\frac{\tilde{\sigma}_k}{E} , \quad \begin{array}{l} i=r,\ \theta \\ j=r,\ \theta,\ r\theta \\ k=r,\ \theta,\ e \end{array} , \quad (10)$$

and parameters:

$$\eta=\frac{\tilde{R}_1}{\tilde{R}_2} , \quad p_1=\eta\frac{9\tilde{R}_2^2}{E\left(\tilde{h}^*\right)^3}\tilde{p}_1 , \quad p_2=\frac{9\tilde{R}_2^2}{E\left(\tilde{h}^*\right)^3}\tilde{p}_2 , \quad \chi=\frac{9\tilde{R}_2^2}{\left(\tilde{h}^*\right)^2} ,$$
$$T=\left[\Gamma(\mu+1)E^n t_{cr}\right]^{1/(\mu+1)}\left(\frac{1}{\chi}\right)^\nu , \quad \nu=\frac{n-\mu-1}{\mu+1} , \quad (11)$$

where $\tilde{h}^*=const(\tilde{r})$ denotes the thickness of the plate which has the same volume as the optimal plate:

$$\tilde{V}=\pi\left(\tilde{R}_2^2-\tilde{R}_1^2\right)\tilde{h}^*=2\pi\tilde{R}_2^2\tilde{h}^*\int_\eta^1\xi\ h(\xi)\ d\xi . \quad (12)$$

The boundary conditions for the clamped plate take the forms:

$$N(\eta)=-p_1 , \quad N(1)=-p_2 ,$$
$$w(\eta)=\varphi(\eta)=w(1)=\varphi(1)=0 . \quad (13)$$

3 The optimal design problem

The problem of optimization is to determine the design variable $h(\xi)$ (thickness function) that satisfies the state equations (6), (7), the conditions (13) and minimizes the total volume of the plate for given external load and the critical time:

$$V=\int_\eta^1\xi\ h(\xi)\ d\xi\rightarrow min ,$$
$$p,\ T=const . \quad (14)$$

The necessary optimality condition is derived by use of Pontryagin's maximum principle. The Hamiltonian H may be written in the form:

$$\mathcal{H} = \overline{u}u' + \overline{N}N' + \overline{w}w' + \overline{\varphi}\varphi' + \overline{M}M' + \overline{Q}Q' + \lambda\xi h \ , \tag{15}$$

where barred symbols denote the adjoint functions and λ is a constant which is determined by relation (12). The adjoint functions defined as:

$$\overline{f}' = -\frac{\partial\mathcal{H}}{\partial f} \ , \quad f = u, N, w, m, M, Q \ , \tag{16}$$

and we assumed here that the terms $\partial E_{s,t}/\partial u$ and $\partial E_{s,t}/\partial N$ may be neglected. It is shown in the paper [8] that the influence of these terms on the optimal solution is very small. The boundary conditions one can obtained from the transversality conditions.

Finally, the Hamiltonian takes the form:

$$\begin{aligned} \frac{1}{\xi}\mathcal{H} &= \frac{1}{E_s h^3 a_{11}} - \left(\frac{\overline{Q}}{\xi}\frac{N}{\xi}\frac{M}{\xi^2} - \frac{\overline{\varphi}}{\xi}\frac{M}{\xi}\right) + E_s h^3 \left(\frac{a_{12}^2}{a_{11}} - a_{22} - m^2\right)\frac{\overline{M}}{\xi}\frac{\varphi}{\xi} + \\ &+ E_s h^3 m^2 \frac{\overline{Q}}{\xi}\left\{\left[1 - m^2\left(\frac{a_{12}^2}{a_{11}} - a_{22}\right)\right]\frac{w}{\xi^3} + \left(\frac{a_{12}^2}{a_{11}} - a_{22} - 1\right)\frac{\varphi}{\xi^2}\right\} + \\ &+ E_s \frac{h}{\chi}\frac{u}{\xi}\left[\frac{\overline{Q}}{\xi}\left(m^2\frac{w}{\xi} - \varphi\right) + \frac{\overline{N}}{\xi}\right] + \frac{3}{4}\frac{1}{E_s}\frac{\chi}{h}\frac{\overline{u}}{\xi}\frac{N}{\xi} + \lambda h \ , \end{aligned} \tag{17}$$

where terms without design variable $h(\xi)$ are omitted. The necessary optimality condition:

$$\frac{\partial\mathcal{H}}{\partial h} = 0 \ , \tag{18}$$

determines the shape of plate.

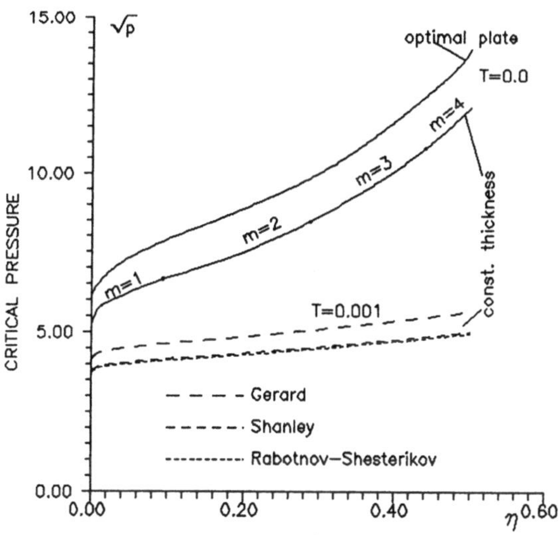

Fig. 2

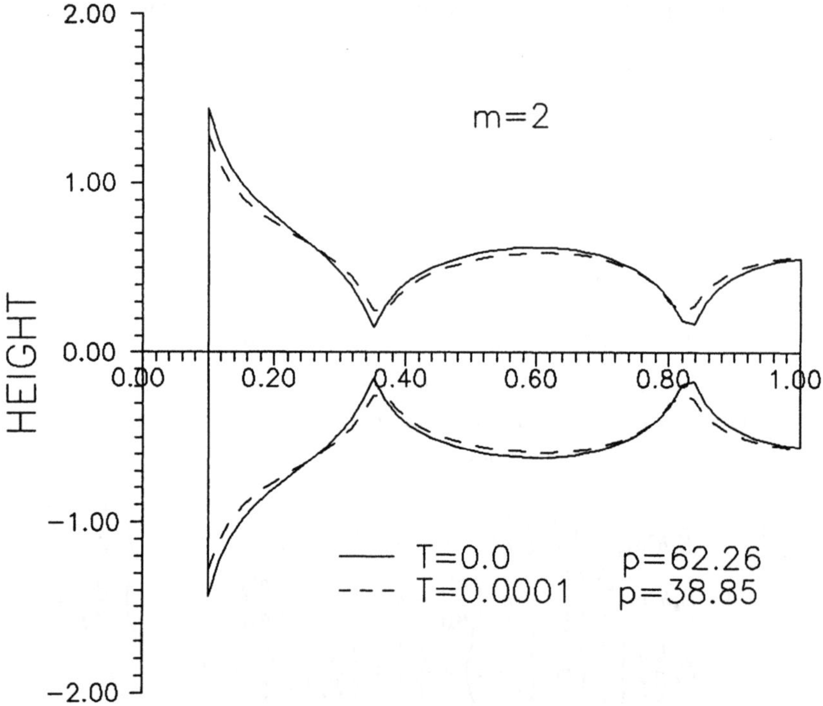
2.00
1.00
0.00
−1.00
−2.00
HEIGHT
0.00
0.20
0.40
0.60
0.80
1.00
m=2
T=0.0 p=62.26
T=0.0001 p=38.85

FIG. 3

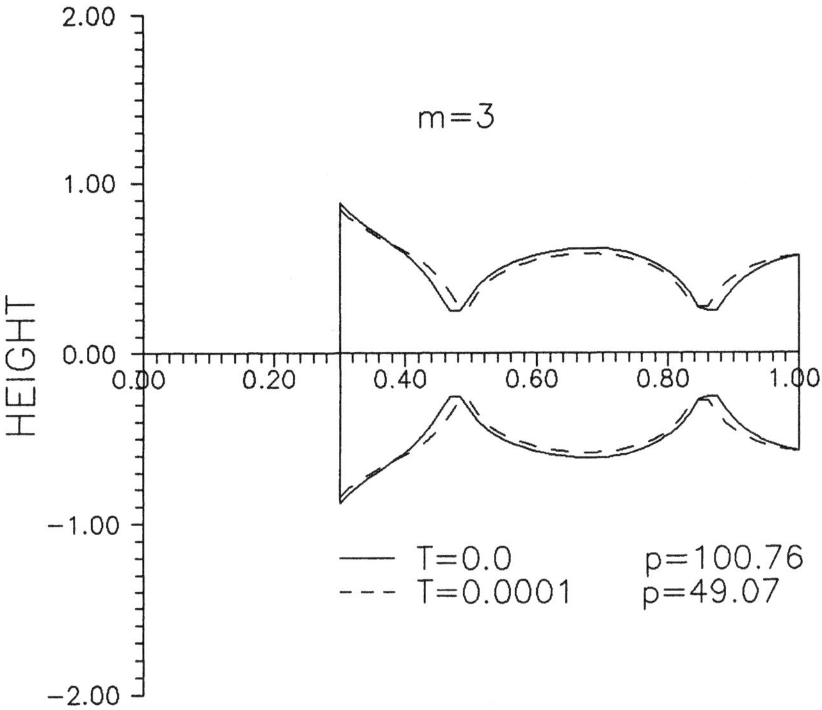

2.00
1.00
0.00
-1.00
-2.00
HEIGHT
0.00
0.20
0.40
0.60
0.80
1.00
m=3
T=0.0 p=100.76
T=0.0001 p=49.07

FIG. 4

4 Examples

The problem is solved by use of the iterative method presented in [3]. Fig. 2 shows changes in the critical load of plates of constant thickness in dependence on the inner radius η and given critical time $T = 0$ or $T = 0.001$ for the three creep buckling theories considered. The quantity m denotes the number of half-waves along the perimeter. In addition the dependence of the critical load on the plate radius η for optimal shapes is given. The necessity of multimodal formulation of optimization problem has not been observed. Two different forms of buckling may occur simultaneously only in point η^* for which the number of half-waves changes.

In Fig. 3 and 4 optimal shapes of elastic plates and plates for critical time $T = 0.001$ are given. Creep buckling has been analysed within the Rabotnov-Shesterikov theory approach.

References

[1] Gajewski, A., Życzkowski M., *Optimal structural design under stability constraints*, Dordrecht; Nijhoff 1988

[2] Frauenthal, J.C., *Constrained optimal design of circular plates against buckling*, J. Struct. Mech., **1**, (1972), 2, 159-186

[3] Grinev, V.B., Philippov, A.P., *Optimal design of circular plates against buckling*, Stroit. Meh., (1977), 2, 16-20, (in Russian)

[4] Tarasov, V.L., *Optimal design of annular plates under critical loadings*, Prikl. Problemy Proch. Plast. **7**, (1977), 97-103, (in Russian)

[5] Rzegocińska-Pełech, K., Waszczyszyn, Z., *Numerical optimum design of elastic annular plates with respect to buckling*, Computer & Struct., **18**, (1984), 2, 369-378

[6] Strzelczyk, A., Drobniewicz, M., *Elastic buckling of annular plates of variable thickness*, Arch. Bud. Maszyn, **28**, (1981), 2, 91-103

[7] Życzkowski, M., *A survey of optimal design in creep conditions*, Mech. Teor. Stos. **3**, (1986), 24, 243-258 (in Polish); Appl. Mech. Rev. **41**, (1988) 453-461

[8] Wróblewski, A., *Optimal design of circular plates with respect to creep buckling*, (in print)

[9] Shanley, F.R., *Weight-strength analysis of aircraft structures*, MacGraw-Hill, New York 1952

[10] Gerard, G., *A creep buckling hypothesis*, J. Areo. Sci., **23**, (1956), 9, 879-882

[11] Rabotnov, G.N., Shesterikov, S.A., *Creep stability of columns and plates*, J. Mech. Phys. Solids, **6** (1957), 27-34

Dynamic Response of a Viscoelastic Circular Plate on a Viscoelastic Half Space Foundation

Yang Ting–Qing (Huazhong University of Science & Technology, Wuhan, China)
Wang Ren (Peking University, Peking, China)
Yang Zheng–Wen (Shanghai University of Technology, Shanghai, China)

Abstract

In the present paper, a dynamic problem of a thin viscoelastic circular plate under axisymmetric loads and resting on a viscoelastic half–space is discussed . This problem is reduced to a Fredholm integral equation of the first kind in Laplace transform space. After solving the integral equation by numerical method, the dynamic response of the problem can be obtained by means of numerical Laplace transform inversion.

Introduction

In engineering practice, many problems, such as airport pavements, highways and bridge surfaces, can be considered as a plate resting on a foundation. In general, these problems can not be simply treated as pure elastic ones because real materials are viscous more or less.

For quasi–static problems, there were a lot of papers dealing with elastic plates on viscoelastic foundations [1,2,3] . Some literatures also dealt with viscoelastic plates [4,5,6] These problems could be easily analysed with the help of the elastic–viscoelastic correspondence principle. Under dynamic cases, however, few papers could be found. Lin [7] studied the response of elastic circular plates on a viscoelastic half–space under displacement excitations.

In this paper, the dynamic response of a thin viscoelastic circular plate subjected to transverse load and on a viscoelastic half–space foundation is discussed. The problem is reduced to a first kind Fredholm integral equation in the Laplace transfom space and solved by numerical procedures.

Fundamental Equations

The equation of motion of a thin viscoelastic plate can be written in the form [8,9]

$$D_1 \nabla^4 w + \rho h \partial^2 w / \partial t^2 = q \tag{1}$$

* The project is supported by the National Natural Science Foundation of China

M. Zyczkowski (Ed.)
Creep in Structures
IUTAM Symposium Cracow/Poland 1990

where w is the deflection, ρ is the mass density, h is the thickness, q is the intensity of the transverse load, and D_1 is a differential operator related to material functions.

If the initial values of material functions satisfy the smoothing conditions [10], and define the Laplace transform of function $f(t)$ as $\bar{f}(s)=\int_{0^-}^{\infty} f(t)e^{-st}dt$, from Eq. (1) it's found that

$$s\bar{D}\nabla^4\bar{w}+\rho hs^2\bar{w}=\bar{q} \tag{2}$$

where $$\bar{D}=\bar{E}h^3/12(1-s^2\bar{\mu}^2) \tag{3a}$$

$$\bar{E}=9\bar{G}\bar{K}/(3\bar{K}+\bar{G}), \qquad s\bar{\mu}=(3\bar{K}-2\bar{G})/2(3\bar{K}+\bar{G}) \tag{3b}$$

$G(t)$ and $K(t)$ are the shear and dilatation relaxation functions of the material, respectively, and $\mu(t)$ is the Poisson's ratio.

Assume the viscoelastic Poisson's ratio to be a constant, μ_0, and the loads on the plate as well as the boundary conditions can be separated in time and space, a dynamic problem of linear viscoelastic plate can be divided into a static elastic problem and an eigenvalue problem [11].

For a circular plate, introducing the following substitution

$$\bar{w}_A=\bar{w}\bar{D}/\bar{g} \tag{4}$$

where $g(t)$ is the time part of the known forcing boundary condition. Then w_A can be formulated as

$$w_A=U-\sum_{n=1}^{\infty}V_n\varphi_n(t) \tag{5}$$

in which U satisfies

$$\nabla^4U=0, \quad L_1(U)=F(r), \quad L_2(U)=0 \tag{6}$$

V_n satisfy

$$\nabla^4V_n-\rho hk_n^2V_n=0, \quad L_1(V_n)=0, \quad L_2(V_n)=0 \tag{7}$$

where L_1 and L_2 are linear differential opertors, $F(r)$ is the space part of the known forcing boundary condition, and

$$\bar{\varphi}_n=\frac{\int_0^R\left(\frac{\bar{q}}{\rho h}+sU\right)rV_n(r)dr}{(s\bar{D}k_n^2+s^2)\int_0^R rV_n^2(r)dr} \tag{8}$$

where k_n are eigenvalues (n = 1,2···), R is the radius of the plate.

From Eqs (4) and (5), we have

$$\bar{w}=\frac{\bar{g}}{\bar{D}}\left(\frac{U}{s}-\sum_{n=1}^{\infty}V_n\bar{\varphi}_n\right) \tag{9}$$

Now, we discuss a viscoelastic half–space foundation without inertia effect.

For an elastic half–space loaded axisymmetrically, the surface settlement is

$$W_f = \frac{2(1-\mu_f^2)}{E_f}\int_0^\infty \sigma_c^*(\zeta)J_0(r\zeta)\mathrm{d}\zeta \tag{10}$$

where E_f and μ_f are the elastic modulus and Poisson's ratio of the half–space, respectively, σ_c is the intensity of the axisymmetric load, σ_c^* is the Hankel transform of σ_c, namely

$$\sigma_c^*(\zeta) = \int_0^\infty \sigma_c(\alpha)\alpha J_0(\zeta\alpha)\mathrm{d}\alpha \tag{11}$$

Then the surface settlement of a viscoelastic half–space foundation in the Laplace transform space can be shown as

$$\overline{W}_f(r,s) = \frac{2(1-s^2\bar{\mu}_f^2)}{s\bar{E}_f}\int_0^\infty\int_0^\infty \alpha J_0(\zeta\alpha)J_0(\zeta r)\bar{\sigma}_c(s,\alpha)\mathrm{d}\alpha\mathrm{d}\zeta \tag{12}$$

Use Laplace transform inversion techniques, the surface settlement W_f can be obtained from Eq. (12) if σ_c is known.

Circular Plate On A Half–Space Foundation

The dynamic response, here, of a viscoelastic circular plate on a viscoelastic half–space foundation implies only the inertia of the plate being included, while that of the foundation being neglected. This is consistent with engineering practice circumstances.

Assume the plate and the foundation to remain in contact and no friction exist between them, i.e., their normal displacements and stresses at the contact points are continuous:

$$w(r,t) = W_f(r,t) \tag{13a}$$

$$q(r,t) = p(r,t) - \sigma_c(r,t) \tag{13b}$$

where $p(r,t)$ is the intensity of the external transverse load.

For the plate with edges being simply–supported or free, from Eqs. (8), (9), (12) and (13), we get the governing equation of the problem as

$$\int_0^R\Big[\frac{2(1-s^2\bar{\mu}_f^2)}{s\bar{E}_f}\int_0^\infty \alpha J_0(\alpha\zeta)J_0(r\zeta)\mathrm{d}\zeta - \frac{\bar{g}}{\rho h\bar{D}}\sum_{n=1}^\infty \frac{V_n(r)\alpha V_n(\alpha)}{(s\bar{D}k_n^2+s^2)\int_0^R \zeta V_n^2(\zeta)\mathrm{d}\zeta}\Big]\bar{\sigma}_c(s,\alpha)\mathrm{d}\alpha$$

$$= \frac{\bar{g}}{s\bar{D}}U(r) - \frac{\bar{g}}{\bar{D}}\sum_{n=1}^\infty \frac{V_n(r)\int_0^R\Big[\frac{\bar{p}}{\rho h}+sU(\alpha)\Big]\alpha V_n(\alpha)\mathrm{d}\alpha}{(s\bar{D}k_n^2+s^2)\int_0^R \alpha V_n^2(\alpha)\mathrm{d}\alpha} \tag{14}$$

For a Laplace transform parameter s, Eq.(14) is a Fredholm integral equation of the first kind with respect to $\bar{\sigma}_c$.

If the plate edge is free, all the boundary conditions mentioned above would be homogeneous and it's unnecessary to make substitution (4). Eq.(14) can be simplified to

$$\int_0^R \overline{N}(s,r,\alpha)\overline{\sigma}(s,\alpha)\mathrm{d}\alpha = \overline{B}(s,r) \tag{15}$$

where

$$\overline{N}(s,r,\alpha) = \int_0^\infty \alpha J_0(\alpha\zeta)J_0(r\zeta)\mathrm{d}\zeta + \sum_{n=1}^{\infty} \frac{s\overline{E}_f V_n(r)\alpha V_n(\alpha)}{2\rho h(1-s^2\overline{\mu}_f^2)(s\overline{D}k_n^2+s^2)\int_0^R \zeta V_n^2 \mathrm{d}\zeta} \tag{16a}$$

$$\overline{\sigma}(s,\alpha) = \frac{2(1-s^2\overline{\mu}_f^2)}{s\overline{E}_f}\overline{\sigma}_e(s,\alpha) \tag{16b}$$

$$\overline{B}(s,r) = \sum_{n=1}^{\infty} \frac{V_n(r)\int_0^R \zeta\overline{p}(\zeta)V_n(\zeta)\mathrm{d}\zeta}{\rho h(s\overline{D}k_n^2+s^2)\int_0^R \zeta V_n^2(\zeta)\mathrm{d}\zeta} \tag{16c}$$

To find $V_n(r)$, let $\rho hk_n^2 = \lambda_n^4$, from Eq. (7) we have the eigenvalue equation for the case of simply–supported boundary

$$(1-\mu_0)[J_1(\lambda_n R)I_0(\lambda_n R) + J_0(\lambda_n R)I_1(\lambda_n R)] - 2\lambda_n R J_0(\lambda_n R)I_0(\lambda_n R) = 0 \tag{17}$$

where μ_0 is constant. Similarly, for the case of free boundary we have

$$2(1-\mu_0)I_1(\lambda_n R)J_1(\lambda_n R) - \lambda_n R[J_0(\lambda_n R)I_1(\lambda_n R) + J_1(\lambda_n R)I_0(\lambda_n R)] = 0 \tag{18}$$

Numerical Procedures

Because of the complicated kernel function we have to use numerical method to solve Eq.(15). An accelerated iteration method is used here in order to give numerical results of $\overline{\sigma}_e$. Let

$$||x||^2 = \int_0^R x(\alpha)x^*(\alpha)\mathrm{d}\alpha \tag{19a}$$

$$\overline{N}x = \int_0^R \overline{N}(s,r,\alpha)x(\alpha)\mathrm{d}\alpha \tag{19b}$$

where x^* is the conjugate of x. Then, for Eq.(15), we have the following accelerated iteration formula [12]

$$\overline{\sigma}_{k+1}(s,r) = \overline{\sigma}_k(s,r) + \beta\overline{N}_0[\overline{B}(s,r) - \overline{N}\overline{\sigma}_k(s,r)] \tag{20}$$

where

$$\beta = \frac{||\overline{B} - \overline{N}\overline{\sigma}_k||^2}{||\overline{N}_0\overline{B} - \overline{N}_0\overline{N}\overline{\sigma}_k||^2}, \quad \overline{N}_0(s,r,\alpha) = \overline{N}(s,\alpha,r) \tag{21}$$

Now Laplace transform inversion will give the dynamic responses in the real space. Due to the impossibility to make an analytical Laplace transform inversion, a numerical one would be utilized [13]. If the Laplace transform of function $f(t)$ is known as $\overline{f}(s)$, then function $f(t)$ can be expanded in series

$$f(t) = \sum_{n=0}^{\infty} C_n P_n^{(0,\gamma)}(2e^{-s_0 t} - 1) \tag{22}$$

where $P_n^{(0,\gamma)}$ is the Jacobi polynomials of order n, coefficients C_n satisfy

$$s_0 \bar{f}[(\gamma + 1 + k)s_0] = \sum_{n=1}^{k} \frac{k(k-1)\cdots[k-(n-1)]}{(k+\gamma+1)(k+\gamma+2)\cdots(k+\gamma+1+n)} C_n \tag{23}$$

and s_0 and γ are small contants which can be adjusted to give satisfactory results.

Numerical Examples

Suppose the materials of the plate and foundation are standard linear solids in shear [14] :

$$2G(t) = q_0 + \left(\frac{q_1}{p_1} - q_0\right)e^{-t/p_1}, \quad 3K(t) = \frac{1+\mu_0}{1-2\mu_0} 2G(t) \tag{24}$$

$$2G_f(t) = q_{0f} + \left(\frac{q_{1f}}{p_{1f}} - q_{0f}\right)e^{-t/p_{1f}}, \quad 3K_f = 3K_0 \tag{25}$$

Consider a thin asphalt–concrete plate with free boundary on the loamy soils, the plate is subjected to an uniformly distributed load $p(t) = p_0 H(t)$ in the circular area with radius $R_0 \leqslant R$. Take $\rho = 2400$ kg / m^3, and [5]

$$p_1 = 0.05\text{sec}, \quad q_0 = 3000\text{MPa}, \quad q_1 = 500\text{MPa} - \text{sec}, \quad \mu_0 = 0.30;$$
$$p_{1f} = 0.2\text{sec}, \quad q_{0f} = 40\text{MPa}, \quad q_{1f} = 12\text{MPa} - \text{sec}, \quad K_0 = 70\text{MPa}.$$

Numerical calculation is carried out for several cases as follows

$$R = 1\text{m}, \quad h = 0.1\text{m and / or } 0.2\text{m}, \quad p_0 = 0.5\text{MPa}, \quad R_0 = 0.5\text{m and / or } 0.9\text{m}.$$

Some results are shown in Fig.1~4.

Both Fig.1 and Fig.2 are for the viscoelastic plate on the viscoelastic half–space foundation. It's found that the period of w becomes larger with time. Drastic reduction of w in amplitude can be seen due to the doubling thickness of the plate. Probably because the redistribution of the contact stress constraints the deformation, the deflection w decreases in amplitude as the loaded circular area increases.

From Fig.1~ 2, we also notice the deflections attenuate more quickly than those by Robertson [11] when the coupling effect between the plate and the foundation is considered.

The contact stress is shown in Fig.3. It varies with time in a similar way as the deflection does. Note that σ_c is negative near the plate edge, this phenomenon is the same as that in the static elastic case [15] .

Fig.4 depicts the shapes of the plate or the surface of the foundation in four instances: initially, the first peak, the first trough and the steady state. It is interesting to note that the highest bulging point is not on the plate edge due to the interaction between the plate and the foundation. For example, if $h/R = 1/10$ and $R_0/R = 1/2$, the highest point is at about $r/R = 9/10$.

The settlement of the foundation tends to zero at those points far from the center of the plate. Some numerical results are displayed in table 1.

Tab.1. Settlements of the foundation surface
(R=1m, h=0.1m, R_0=0.5m, p_0=0.5MPa)

w/m \ r/m ; t/sec	1.2	2.0	5.0	10.0	20.0
2	-6.69×10^{-4}	-6.87×10^{-4}	8.54×10^{-5}	2.74×10^{-4}	-1.39×10^{-5}
10	-5.54×10^{-4}	-5.67×10^{-4}	7.07×10^{-5}	2.26×10^{-4}	-1.15×10^{-5}
>502	-5.99×10^{-4}	-6.14×10^{-4}	7.65×10^{-5}	2.45×10^{-4}	-1.24×10^{-5}

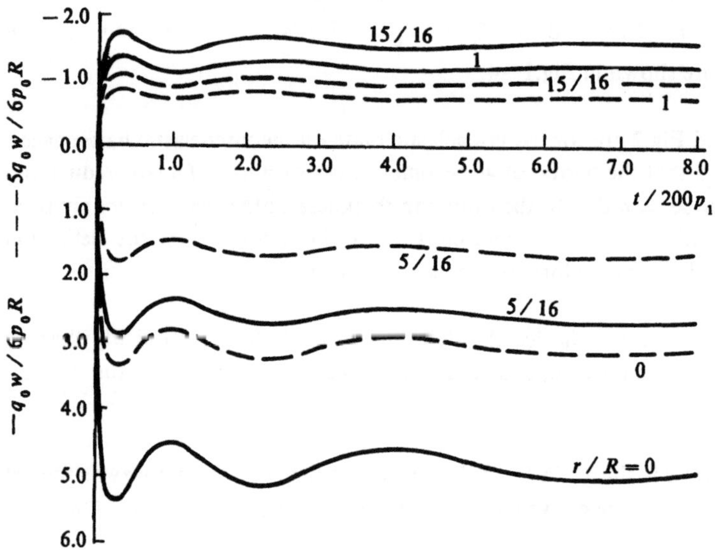

Fig.1. Deflections vs. times
— $h/R=1/10$, − − −$h/R=1/5$; $R_0/R=1/2$

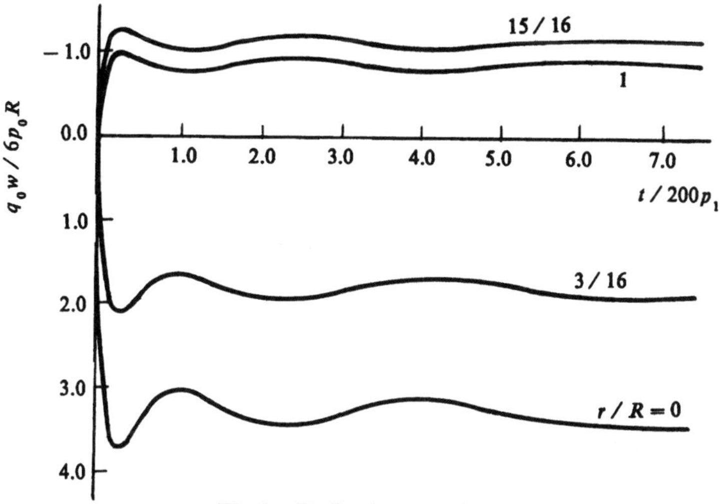

Fig.2. Deflections vs. times
$h/R=1/10$; $R_0/R=9/10$

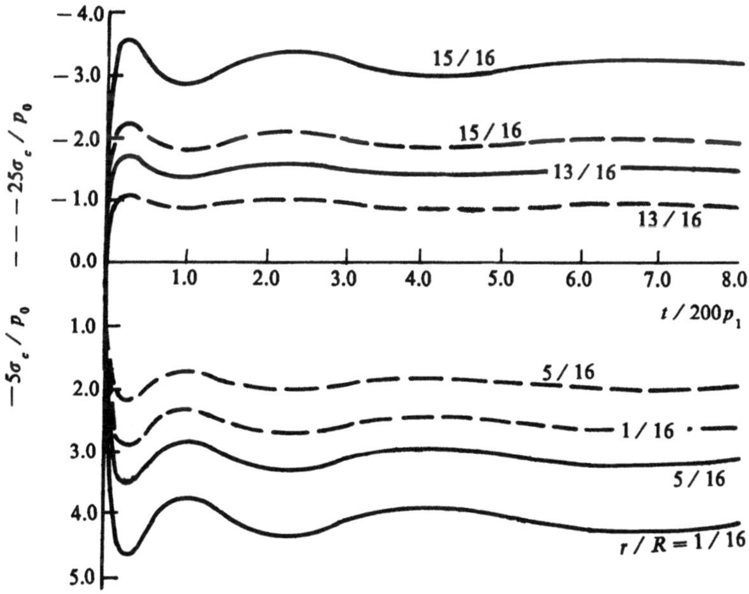

Fig.3. Contact stresses σ_c vs. times
— $h/R=1/10$, - - - $h/R=1/5$; $R_0/R=1/2$

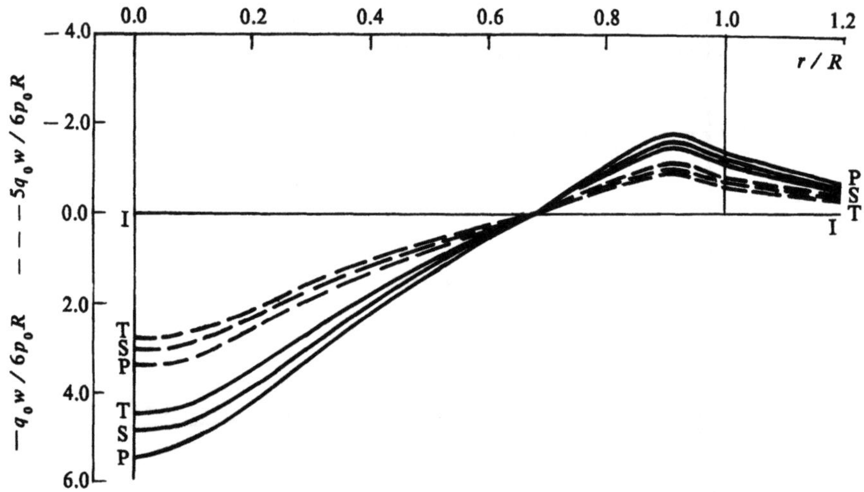

Fig.4. Settlements vs. times & radius

— $h/R=1/10$, - - - $h/R=1/5$; $R_0/R=1/2$

I—initially, P—the first peak, T—the first trough, S—the steady state

Reference

1. Pister K S. & Williams M L., J. Engng Mech. Div. ASCE 86 (1960) EM5, 31–44.
2. Sonoda K. & Kobayashi H., J. Engng Mech. Div. ASCE 106 (1980)EM2, 323–338.
3. Kobayashi H. & Sonoda K., Theo. & Appl. Mech., 31 (1981) 153–164.
4. Pister K S., J. Engng Mech. Div. ASCE 87 (1961)EM1, 43–54.
5. Radovskii B S., Soviet Appl. Mech., 15, 10 (1979) 940–946.
6. Yang Ting–Qing et al , J. Huazhong Univ. of Sci. & Tech., 15 (1987) 1–6 (in Chinese with English abstract).
7. Lin Y J., J. Appl. Mech. 45 (1978) 379–384.
8. Mase G E., J. Engng Mech. Div. ASCE 86 (1960) EM3, 25–39.
9. Yang Zheng–Wen & Yang Ting–Qing, to be published in J. Appl. Math. & Mech.
10. Gurtin M E. & Sternberg E., Arch. Rational Mech. Anal., 11(1962) 291–356.
11. Robertson S R., J. Sound Vib. 14, 3 (1971) 263–278.
12. Yun Tian–Quan, J.Huazhong Univ. of Sci. & Tech., 3 (1978) 94–98 (in Chinese with English abstract).
13. Miller M K. & Guy W T., SIAM J. Numer. Anal., 3, 4 (1966) 624–635.
14. Yang Ting–Qing, Theory of Viscoelasticity, HUST Press (1990) (in Chinese).
15. Barden L., Struct. Engng , 43 (1965) 153–154.

Low-Stress Level of a Compound Wedge Under Steady Creep

M.A. ZADOYAN

Institute of Mechanics,
Armenian Academy of Science, Yerevan

Summary
The paper considers steady creep in a wedge composed of two materials under longitudinal shear and under plane strain. It is shown that for certain values of the wedge angle and material constants the stress concentration at the angular point of the wedge may be eliminated (low-stress level).

Introduction

The publication of M. Williams and A. Zak [1], K.S. Chobanyan [2], D.B. Bodgy [3] and others are devoted to problems of stress concenteration at the angular point of a compound linearly elastic body. The fundamental research conducted in this area by K.S. Chobanyan [2] resulted in discovery of the low-stress level phenomenon which occurs at the angular point of a compound loaded elastic body. It is proved in a space of geometric and physical parameters certain areas of zero stress exist which are called by the author the areas of low-stress level. The edge of the contact surface which corresponds to points of this area remains unloaded and manifests itself as a monolithic body.

The present article investigates problems of low-stress level at the edge of the contact surface of a compound wedge under the conditions of steady creep according to the following low

$$\dot{\varepsilon}_0 = K\sigma_0^n \tag{1}$$

where $\dot{\varepsilon}_0$ denotes effective strain rate, σ_0 - effective tangential stress, K and n are mechanical parameters of materials. For both materials it is accepted that the exponents n are the same and the multipliers K may be different from each other.

The paper considers the state of longitudinal shear and plane strain of the wedge. We shall limit ourselves here to discussing boundary conditions of the first kind.

1 Longitudinal shear

The problem of longitudinal shear for homogeneous bodies with non-linear relation between stress and strain is studied by H. Neuber [4], V.V. Sokolovsky [5], G.P. Cherepanov [6,7] and J. Rice [8]. A detailed review of this study is given in publications [7,8].

M. Zyczkowski (Ed.)
Creep in Structures
IUTAM Symposium Cracow/Poland 1990

1.1 Formulation of the problem

Let us consider the strtess state under steady creep on a wedge projecting or entering edge of the contact surface of a cylindrical body under the conditions of longitudinal shear. Such stress state on this edge takes place also at torsion of the body under consideration.

The angular point is taken as the origin of a cylindrical coordinate system where a $\theta = 0$ axis coincides with the contact surface and a Z axis is directed along the longitudinal axis of the body (Fig. 1). In every area of the cross-section we have the known equilibrium

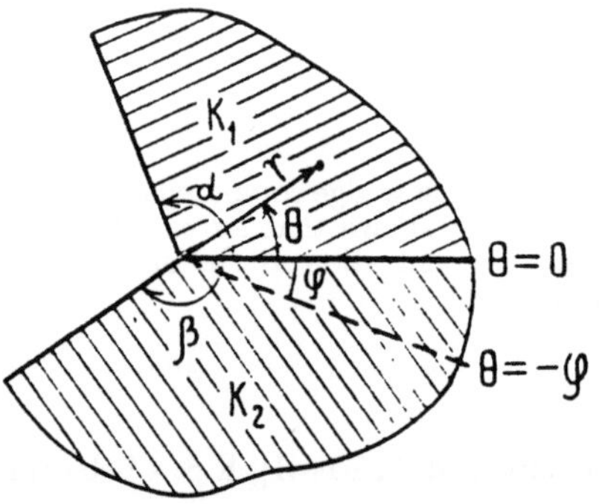

Figure 1:

equation, the relation between strain rate components and displacements

$$2\dot{\gamma}_{rz} = \frac{\partial \dot{w}}{\partial r}\,, \quad 2\dot{\gamma}_{\theta z} = \frac{1}{r}\frac{\partial \dot{w}}{\partial \theta}\,,$$

power law of creep (1), such that

$$\sigma_0 = \sqrt{\tau_{rz}^2 + \tau_{\theta z}^2}\,, \quad \dot{\varepsilon}_0 = 2\sqrt{\left(\frac{\partial \dot{w}}{\partial r}\right)^2 + \frac{1}{r^2}\left(\frac{\partial \dot{w}}{\partial \theta}\right)^2}\,,$$

relation between stress components and strain rate components

$$\tau_{iz} = 2k\dot{\varepsilon}_0^{m-1}\dot{\gamma}_{iz}\,,$$

where $m = 1/n$, $k = 1/K^m$ and i adopts symbols r and θ.

We adopt the following boundary condition on the lateral surface of the wedge

$$\tau_{\theta z} = 0 \quad for \quad \theta = \alpha\,, \ -\beta\,.$$

We also suppose that on a certain ray $\theta = \theta_*$ the velocity of displacement $\dot{w}$ changes its sign, say, from negative to positive.

1.2 Representation of the solution

Without loss of generality, we shall discuss the case $\beta > \alpha$ and suppose that $\theta_* = -\varphi$ where $\varphi > 0$ and is subject to be calculated. We shall solve the problem in three domains $0 \leq \theta \leq \alpha$, $-\varphi \leq \theta \leq 0$, $-\beta \leq \theta \leq -\varphi$.

We shall denote values in the domains by symbols $i = 1, 2, 3$, respectively.

We shall represent components of stress and velocity of displacements in the form

$$\tau_{rzi} = \lambda k_i r^{(\lambda-1)m} f_i \chi_i \ , \quad \tau_{\theta zi} = k_i r^{(\lambda-1)m} f_i' \chi_i \ ,$$
$$\dot{w}_i = r^\lambda f_i \ , \quad \chi_i = \left(\sqrt{f_i'^2 + \lambda^2 f_i^2}\right)^{m-1} \ .$$

Here $f_i = f_i(\theta; \lambda)$ and λ are the unknown characteristic function and characteristic value of the problem, $k_3 = k_2$.

Using the equlibrium equation we arrive at the differential equation

$$(f_i' \chi_i)' + \eta f_i \chi_i = 0 \ , \quad \eta = \lambda [1 + (\lambda - 1)m] \ . \tag{2}$$

We have boundary conditions

$$f_1'(\alpha) = f_2(-\varphi) = f_3(-\varphi) = f_3'(-\beta) = 0 \ , \quad f_2'(-\varphi) = f_3'(-\varphi) \ , \tag{3}$$

and the mating conditions on the contact surface

$$f_1' \chi_1 = \gamma f_2' \chi_2 \ , \quad f_1 = f_2 \quad \text{at} \quad \theta = 0 \ , \tag{4}$$

where $\gamma = k_2/k_1$. Introducing a new functions $\psi_i(\theta; \lambda)$ through the relations $f_i' = f_i \psi_i$ from (2) - (4) we arrive at the differential equation:

$$\psi_i' = -\frac{(\psi_i^2 + \lambda^2)(\psi_i^2 + \omega^2)}{\psi_i^2 + \lambda^2 n} \ , \tag{5}$$

where $\omega^2 = \lambda(\lambda + n - 1)$, with the boundary conditiions:

$$\psi_1(\alpha) = \psi_3(-\beta) = 0 \ , \quad \psi_2(-\varphi) = \psi_3(-\varphi) = \infty \ , \tag{6}$$

and the conditions on the contact surface

$$\mu_1 = \left(\sqrt{\mu_1^2 + \lambda^2}\right)^{m-1} = \gamma \mu_2 \left(\sqrt{\mu_2^2 + \lambda^2}\right)^{m-1} \ , \quad \mu_i = \psi_i(0) \ . \tag{7}$$

It is easy to see that equation (5) at $\lambda(\lambda + n - 1) \leq 0$ has no bounded solution.

The general solution of equation (5) will be

$$\arctan \frac{\psi_i}{\lambda} + \frac{1 - \lambda}{\omega} \arctan \frac{\psi_i}{\omega} = H_i - \theta \ , \tag{8}$$

where H_i is an arbitrary constant.

Using conditions (6) we find

$$\begin{aligned} H_1 &= \alpha\,, \qquad H_2 = \tfrac{\pi}{2}\left(1+\tfrac{1-\lambda}{\omega}\right)-\varphi\,, \\ H_3 &= -\beta\,, \qquad \varphi = \beta - \tfrac{\pi}{2}\left(1+\tfrac{1-\lambda}{\omega}\right)\,. \end{aligned} \tag{9}$$

Supposing in (8) $\theta = 0$ we obtain the equation ($i = 1,2$):

$$\arctan\frac{\mu_i}{\lambda} + \frac{1-\lambda}{\omega}\arctan\frac{\mu_i}{\omega} = H_i\,, \tag{10}$$

which with (7) determine $\lambda = \lambda(\alpha\,,\beta\,,\gamma\,,n)$.

1.3 Homogeneous material

In case when wedge is made of a homogeneous material suppose $\gamma = 1\,, \mu_1 = \mu_2$. Then, denoting $\alpha + \beta = 2\alpha_*$ and $\alpha_*/\pi = \nu$ from (7), (9), (10) we obtain $\lambda - 1 = (1-2\nu)\omega$. From here at $\nu = 1/2$ (half-plane) we have $\lambda = 1$ and at $\nu = 1$ i.e. at a semi-infinite slot, we have $(\lambda-1)/n = -1/(n+1)$. Such result for plane deformation was first obtained by Cherepanov [6] and Rice [8] through other methods. When $\nu \neq 1$ we have:

$$\lambda = \frac{2+(1-2\nu)^2(n-1) \pm \sqrt{[2+(1-2\nu)^2(\nu-1)]^2 - 16\nu(1-\nu)}}{8\nu(1-\nu)}$$

with the upper sign for $\nu \leq 1/2$ and the lower sign for $\nu \geq 1/2$ (Fig. 2).

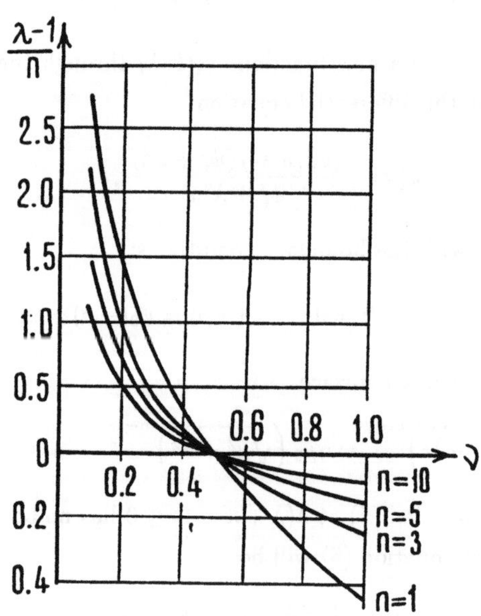

Figure 2:

1.4 Areas of low-stress level

In space of parameters α, β, γ, n the condition $\lambda = 1$ from (7), (10) determines an equation of hypersurface of limit stresses [9]

$$\tan\alpha\,|\cos\alpha|^{1-m} + \gamma\tan\beta\,|\cos\beta|^{1-m} = 0\,, \tag{11}$$

at $\alpha \neq \beta$. In case $\alpha = \beta$ we have $\alpha = \beta = \pi/2$.

Fig. 3 shows traces of this surface on the coordinate plane (α, β). The family of limit curves $\beta = \beta(\alpha, \beta, n)$ seperates the areas of low-stress level (below curves) and the areas of stress concentration (above curves). At $n = 1$ equation (11) and curves on Fig.3 tend to respective results [2].

2 Plane deformation

In this case in each area $0 \le \theta \le \alpha$ and $-\beta \le \theta \le 0$ we have differential equilibrium equations, relations between strain rates and velocities, power law of creep (1). Here

$$\sigma_0 = \frac{1}{2}\sqrt{(\sigma_r - \sigma_\theta)^2 + 4\tau_{r\theta}^2}\,, \qquad \dot{\varepsilon}_0 = \sqrt{(\dot{\varepsilon}_r - \dot{\varepsilon}_\theta)^2 + 4\dot{\gamma}_{r\theta}^2}\,,$$

relations between stress components and strain rates

$$\sigma_j - \sigma = 2k\dot{\varepsilon}_0^{m-1}\dot{\varepsilon}_j\,, \qquad \tau_{r\theta} = 2k\dot{\varepsilon}_0^{m-1}\dot{\gamma}_{r\theta}\,,$$

where $j = r, \theta$ and incompressibility condition $\dot{\varepsilon}_r + \dot{\varepsilon}_\theta = 0$ is adopted; σ stands for the mean stress.

Represent the solution in the way:

$$\begin{gathered}\sigma_{ri} = \sigma_{0i} + 4k_iB_i\psi_i\chi_i\,, \quad \tau_{r\theta i} = k_iB_i(1+\psi_i')\chi_i\,, \\ \sigma_{\theta i} = H_i - 2B_ik_i\textstyle\int_0^\theta(1+\psi_i')\chi_i\,d\theta\,, \quad B_i = C_i\,|C_i|^{m-1}\,, \\ u_i = C_ir\psi_i\,, \quad v_i = -2c_irf_i + C_ir\ln r\,,\end{gathered}$$

with $\psi_i = f_i', \chi_i = \left(\sqrt{(1+\psi_i')^2 + 4\psi_i^2}\right)^{m-1}$ where C_i and H_i are arbitrary constants, $f_i = f_i(\theta)$ is unknown function.

After satisfying to equilibrium equations, the boundary conditions and the conditions on the contact surface, we arrive at the differential equation for the function ψ_i

$$\psi_i'' = -4\psi\frac{4\psi_i^2 + (1+\psi_i')(1+m\psi_i')}{4\psi_i^2 + m(1+\psi_i')^2}\,, \tag{12}$$

at the boundary conditions

$$\psi_1(\alpha) = \psi_2(-\beta) = -1\,, \tag{13}$$

$$(1+\psi_1')\chi_1 = \gamma(1+\psi_2')\chi_2\,, \quad \psi_1 = \psi_2 \quad \text{at} \quad \theta = 0\,, \tag{14}$$

$$\int_0^\alpha (1+\psi_1')\chi_1\,d\theta + \gamma\int_{-\beta}^0 (1+\psi_2')\chi_2\,d\theta = 0\,. \tag{15}$$

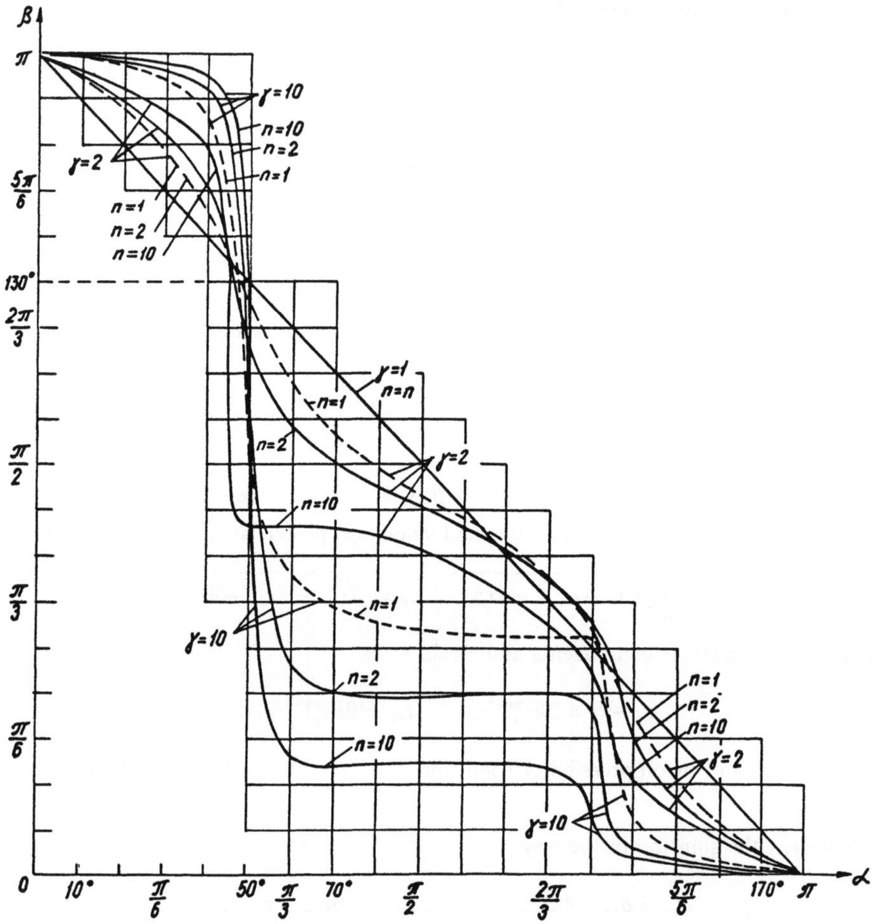
β
π
γ=10
n=10
n=2
n=1
γ=2
5π/6
n=1
n=2
n=10
130°
2π/3
γ=1
n=n
n=1
n=2
γ=2
π/2
n=10
π/3
n=1
γ=10
n=2
n=1
n=2
n=10
π/6
n=10
γ=2
γ=10
0
10°
π/6
50°
π/3
70°
π/2
2π/3
5π/6
170°
π
α

Figure 4:

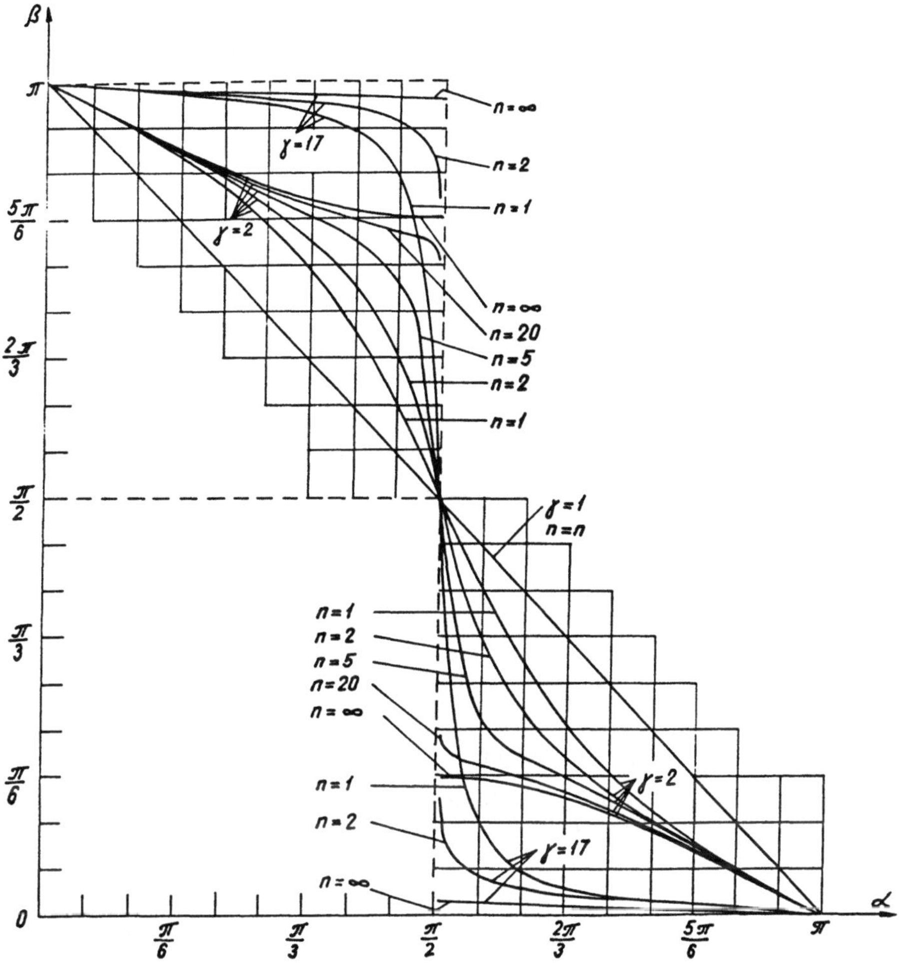
β
π
5π/6
2π/3
π/2
π/3
π/6
0
α
n=∞
n=2
n=1
γ=17
γ=2
n=∞
n=20
n=5
n=2
n=1
γ=1
n=n
n=1
n=2
n=5
n=20
n=∞
n=1
n=2
γ=2
γ=17
n=∞
π/6
π/3
π/2
2π/3
5π/6
π

Figure 3:

The boundary problem (12)-(15) determines the hypersurface of limit stress $F(\alpha, \beta, \gamma, n) = 0$ which separates the area of low-stress level from area of high-stress concentreation. At $n = 1$ from (12)-(15) we shall have [10]

$$\begin{aligned} F(\alpha,\beta,\gamma,1) &= (\gamma^2-\gamma-1)\cos 2\alpha\cos 2\beta+\gamma(\gamma-1)\cos 2\alpha+ \\ &+ (\gamma-1)\cos 2\beta-\gamma\sin 2\alpha\sin 2\beta+ \\ &+ (\alpha+\beta\gamma)(\sin 2\alpha\cos 2\beta+\gamma\cos 2\alpha\sin 2\beta)-\gamma=0 . \end{aligned} \quad (16)$$

This equation coincides with the respective equation of Chobanyan [2] for Poisson's ratios of materials equal to $1/2$.

The traces of surface (16) are shown on the coordinate plane (α, β) through the numerical method of system (12)-(15) in Fig.4. The family of these curves $\beta = \beta(\alpha, \gamma, n)$ separates the zone of low-stress level (below curves) from the zone of high-stress concentration (above curves). The curves for the linear elastic ($n = 1$) incompressiblle material are shown by dashed lines.

References

[1] Williams, M., Zak, A., *Crak Point Stress Singularities at a Bimaterial Interface*, Journ. of Appl. Mech., **30**, (1963), 1, 169-171.

[2] Chobanyan, K.S., *Stress State in Compound Elastic Bodies*, Armenian Academy of Sciences Press, Yerevan 1987, 338 pp.

[3] Bodgy, D.B., *Influence of Spatial Loads on a System of Two Joints Along a Side of Elastic Wedges Made of Different Materials and Having Arbitrary Spread Angles*, Appl. Mech., Trans. ASME, **2**, (1971), Series E, 2, 87-96.

[4] Neuber, H., *Theorie der Spannungskonzentration für beliebiges nichtlineares Spannungs-Dehnungsgesetz in Schubbeanspruchten Prismen*, Proc. 10th Congr. Appl. Mechanics, Stresea 1960; Mekhanika **4**, (1961), 116-130 (Russian version).

[5] Sokolovsky, V.V., *Concentration of Tangential Stresses at Non-linear Law of Deformation*, Engineering Journal, **2**, (1962), issue 2, 332-372.

[6] Cherepanov, G.P., *Elastic and Plastic Problem under Antiplane Deformation*, Appl. Math. and Mech., **26**, (1962), issue 4, 697-708.

[7] Cherepanov, G.P., *Mechanics of Brittle Failure*, Nauka, Moscow 1974, 840 pp.

[8] Rice, J., *Mathematical Methods in Mechanics of Failure*, Collection Failure, v.2, Mir, Moscow 1975, 204-335.

[9] Zadoyan, M.A., *Longitudinal Shear of Compound Wedge*, Papers of the USSR Academy of Sciences, **296**, (1987), 2, 297-302.

[10] Zadoyan, M.A., *Problem of Low-stress Level of Compound Bodies*, Papers of the Armenian Academy of Sciences, **87**, (1988), 4, 160-164.

Creep Instability of Spinning Viscoelastic Rod

Wen ZHANG,[2] Shing-Tung Yau, Ya-Yan Lu

Department of Mathematics, Harvard University

Cambridge, MA 02138, USA

Abstract

In this paper, A general theory for studying dynamic instability of viscoelastic spinning rod(shaft, arm etc.) is presented. The continuous and multi-degree-of-freedom(MDOF) discrete model are both considered. The Boltzmann equation is used to be the general linear viscoelastic constitutive equation. The equation of motions are derived for continuous as well as MDOF discrete models. The analytical dynamic instability criterion is deduced. The *creep instability* phenomenon for viscoelastic shafts is revealed. The results obtained in the paper cover all the results given in literature known to the authors.

1 Introduction

Internal damping in structure usually is a stabilization factor for any perturbed motion from its equilibrium position. However, if the system is in a steady spinning(or partly spinning) state, the internal damping may play a destablizing role. For example, in rotordynamics, viscoelastic property of high spinning shafts will lead to a creep instability when the spinning speed falls into certain regions[1][9]. The secular attitude drift of spinning gyrostats due to dissipative rotor or/and dissipative plateform is the another example[11].

Theoretically, these topics belong to the category of *Dynamics of Viscoelastic Flexible Systems* which is much more complex than the *Dynamics of Elastic Systems*. Mathematical difficulty increases greatly due to introducting viscoelastic constitutive equation instead of the Hooke's law. Up to now, in this field a few works have be done, mainly dealing with single degree of freedom rotating systems and simple viscoelastic models[1]-[10].

[1]Research Projects Supported by NFS/DARPA grant DMS 87-11394and Foundation for Developing Chinese Education and Science

[2]permanent address: Department of Applied Mechanics, Fudan University, Shanghai, 200433, P.R.China

M. Zyczkowski (Ed.)
Creep in Structures
IUTAM Symposium Cracow/Poland 1990

In this paper, A general theory for studying dynamic instability of viscoelastic spinning rod(shaft, arm etc.) is presented. The continuous and multi-degree-of-freedom(MDOF) discrete model are both considered. The Boltzmann equation is used to be the general linear viscoelastic constitutive equation. The equation of motions are derived for continuous as well as MDOF discrete models. The analytical dynamic instability criterion is deduced. The *creep instability* phenomenon for viscoelastic shafts is revealed. The results obtained in the paper cover all the results given in literature known to the authors.

2 Constitutive Equation

One-dimensional Boltzmann equation is[12]

$$\sigma(t) = \int_{-\infty}^{t} G(t-\tau)\frac{d\epsilon(\tau)}{d\tau}d\tau, \tag{1}$$

Where $G(t)$ is the relaxation function of the material under unit step strain history $H(t)$. The general figure of $G(t)$ is shown in Fig.1, which is a decreased function of t, decreasing from its initial value E_0, quite rapidly at first, then gradually, approaching its limiting value G_∞. For solid, $G_\infty > 0$. $G(t)$ could be expressed roughly by following formula:

$$G(t) = E_\infty + (E_0 - E_\infty)e^{-\frac{t}{\tau}} \tag{2}$$

where τ is the characteristic relaxation time, For many practical materials, the order of magnitude of τ is about a week, a month even a year. (2) is called three-parameter model because of three parameters in (2) included.

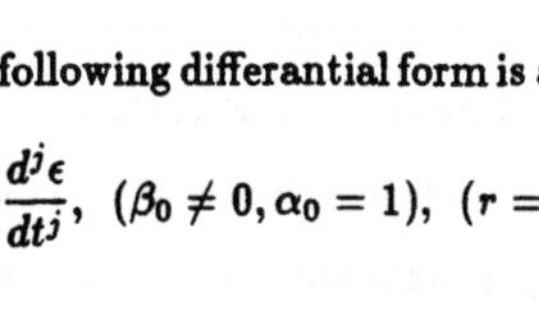

Fig.1

Besides the integral form of (1), following differantial form is also used frequently in literature [12][13]:

$$\sum_{j=0}^{r}\alpha_j\frac{d^j\sigma}{dt^j} = \sum_{j=0}^{p}\beta_j\frac{d^j\epsilon}{dt^j}, \quad (\beta_0 \neq 0, \alpha_0 = 1), \quad (r = p-1 \quad \text{or} \quad r = p). \tag{3}$$

The Laplace transform $\widehat{G}(s)$ of G(t) corresponding to (3) is a rational fraction form:

$$\widehat{G}(s) = \sum_{j=0}^{p}\beta_j s^j / s\sum_{j=0}^{r}\alpha_j s^j. \tag{4}$$

3 Equation of Motion of A Rotating Viscoelastic Rod

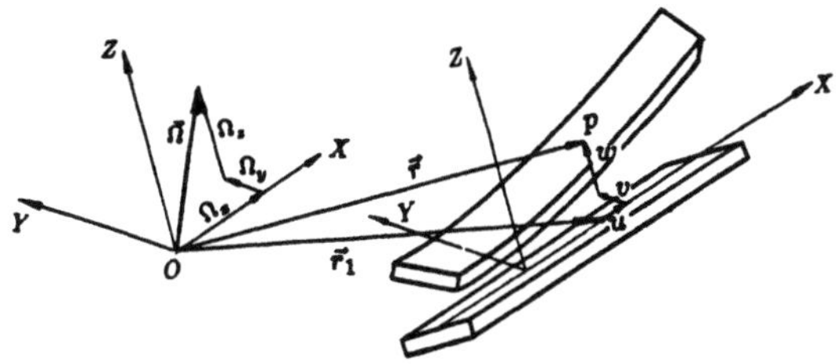

Fig.2

Assuming that the rod is rotating about fixed point O with a fixed angular velocity $\vec{\Omega}$(Fig.2), the absolute simultaneous displacement $\vec{r}(t)$ of a typical point p on the rod can be described by the rigid body motion $\vec{r}_1(t)$ and the deformation motion $\vec{r}_2(t)$ relative to the rigid body position:

$$\vec{r}(t) = \vec{r}_1(t) + \vec{r}_2(t). \tag{5}$$

Taking a moving coordinate system O-XYZ rotating with $\vec{\Omega}$ as shown in Fig.2, and letting $\vec{\Omega} = \Omega_x\mathbf{i} + \Omega_y\mathbf{j} + \Omega_z\mathbf{k}$, $\vec{r}_1 = x\mathbf{i} + y\mathbf{j} + z\mathbf{k}$ and $\vec{r}_2 = u\mathbf{i} + v\mathbf{j} + w\mathbf{k}$. While u, v and w are functions of t, the x, y, z and Ω_x, Ω_y, Ω_z are all independent of t. The absolute velocity $\vec{v} = v_x\mathbf{i} + v_y\mathbf{j} + v_z\mathbf{k}$ and absolute acceleration $\vec{a} = a_x\mathbf{i} + a_y\mathbf{j} + a_z\mathbf{k}$ are then obtained as

$$v_x = \dot{u} + \Omega_y(z + w) - \Omega_z(y + v) \tag{6}$$

$$v_y = \dot{v} + \Omega_z(x + u) - \Omega_x(z + w) \tag{7}$$

$$v_z = \dot{w} + \Omega_x(y + v) - \Omega_y(x + u) \tag{8}$$

$$a_x = \ddot{u} + 2(\Omega_y\dot{w} - \Omega_z\dot{v}) - (\Omega_y^2 + \Omega_z^2)(x + u) + \Omega_x\Omega_y(y + v) + \Omega_x\Omega_z(z + w) \tag{9}$$

$$a_y = \ddot{v} + 2(\Omega_z\dot{u} - \Omega_x\dot{w}) - \Omega_x\Omega_y(x + u) - (\Omega_x^2 + \Omega_z^2)(y + v) + \Omega_y\Omega_z(z + w) \tag{10}$$

$$a_z = \ddot{w} + 2(\Omega_x\dot{v} - \Omega_y\dot{u}) + \Omega_x\Omega_z(x + u) + \Omega_y\Omega_z(y + v) - (\Omega_x^2 + \Omega_y^2)(z + w) \tag{11}$$

<u>General Equation of Motion</u>

$$\frac{\partial^2 M_y}{\partial x^2} + m(x)a_y + c(x)v_y = \frac{\partial}{\partial x}(p_x v') \tag{12}$$

$$\frac{\partial^2 M_z}{\partial x^2} + m(x)a_z + c(x)v_z = \frac{\partial}{\partial x}(p_x w') \tag{13}$$

$$\frac{\partial p_x}{\partial x} = m(x)a_x + c(x)v_x \tag{14}$$

where M_y and M_z are both the bending moments of the rod[1]:

$$M_y = I_y \int_{-\infty}^{t} G(t-\tau)\frac{dv''(t-\tau)}{d\tau}d\tau, \quad M_z = I_z \int_{-\infty}^{t} G(t-\tau)\frac{dw''(t-\tau)}{d\tau}d\tau \tag{15}$$

p_x is the axial force. $m(x)$ and $c(x)$ are the mass and external damping coefficient per unit rod length respectively.

4 Creep Instability of Viscoelastic Shafts

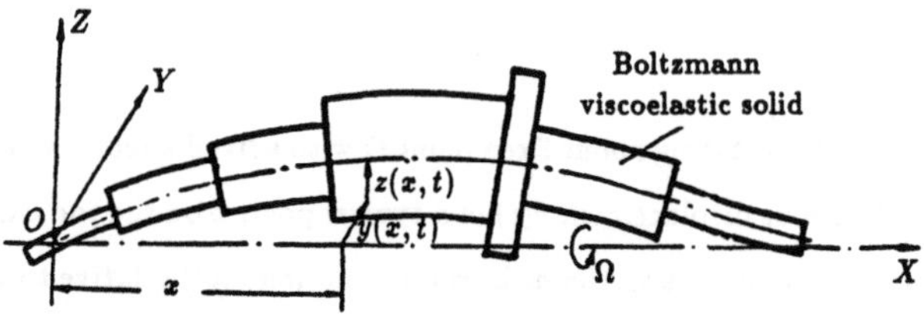

Fig.3

4.1 Equations of Motion for Continuous and Discrete Systems

In this case, the shaft is rotating around its longitudinal symmetric axis OX with $\Omega_x = \Omega$, $\Omega_y = \Omega_z = 0$, $I_y = I_z = I$ and $y = z = u = 0$. The equation of Motion is reduced to

$$m(x)(\ddot{q} + 2\Omega i\dot{q} - \Omega^2 q) + c(x)(\dot{q} + \Omega i q) + \left[I \int_{-\infty}^{t} G(t-\tau)\frac{dq''(x,\tau)}{d\tau}d\tau\right]'' = 0 \tag{16}$$

where $q = v + iw$. This equation is suitable for continuous shafts.

By following Rayleigh-Ritz procedure:

$$q(x,t) = \sum_{i=1}^{n} \phi_i(x) f(t) = \Phi(x)\mathbf{F}(t) \tag{17}$$

the equation of motion (16) can be discretized to the following matrix form:

$$\mathbf{M}(\ddot{\mathbf{F}} + 2\Omega i\dot{\mathbf{F}} - \Omega^2\mathbf{F}) + \mathbf{C}(\dot{\mathbf{F}} + \Omega i\mathbf{F}) + \mathbf{K}\left[\frac{1}{E}\int_{-\infty}^{t} G(t-\tau)\frac{dF(\tau)}{d\tau}d\tau\right] = \mathbf{0} \tag{18}$$

where E is the Young's modulus of the shaft, and

$$\mathbf{M} = \int_0^l m(x)\Phi^T\Phi dx, \quad \mathbf{K} = \int_0^l EI(x)\Phi''^T\Phi'' dx, \quad \mathbf{C} = \int_0^l c(x)\Phi^T\Phi dx. \tag{19}$$

4.2 Dynamic Instability Analysis

The Laplace transform of (18) is

$$\left[(s+i\Omega)^2\mathbf{M}+(s+i\Omega)\mathbf{C}+\frac{s\widehat{G}(s)}{E}\mathbf{K}\right]\widehat{\mathbf{F}}(s)=\mathbf{M}(s\mathbf{F}_0+\dot{\mathbf{F}}_0+2\Omega i\mathbf{F}_0)+\mathbf{C}\mathbf{F}_0 \tag{20}$$

where

$$\mathbf{F}_0=\mathbf{F}(0),\ \ \dot{\mathbf{F}}_0=\dot{\mathbf{F}}(0). \tag{21}$$

If $\widehat{G}(s)$ is a rational fraction of the form of (4), it can be verified that $\widehat{\mathbf{F}}$ can be expanded as

$$\widehat{\mathbf{F}}(s)=\sum_{k=1}^{(r+2)n}\frac{\mathbf{A}_k}{s-s_k}, \tag{22}$$

where $\mathbf{A}_k(k=1,...,(r+2)n)$ are the coefficient matrices. $s_k(k=1,...,(r+2)n)$ are the eigenvalues of the following characteristic equation:

$$\left|(s+i\Omega)^2\mathbf{M}+(s+i\Omega)\mathbf{C}+\frac{s\widehat{G}(s)}{E}\mathbf{K}\right|=0 \tag{23}$$

The inverse transform of (22) is

$$\mathbf{F}=\sum_{k=1}^{(r+2)n}\mathbf{A}_k e^{s_k t} \tag{24}$$

Stability Criterion:

If and only if the real parts of the complex eigenvalues of characteristic equation (23) are all negative, the system is stable.

If $\phi_i(x)(i=1,...,n)$ are the normal vibration modes of the shaft, and $c(x)$ belongs to Rayleigh proportional damping, $\mathbf{M},\mathbf{C}$ and $\mathbf{K}$ are all diagonal, (23) can be separated into following n independent equations:

$$(s+i\Omega)^2+(s+i\Omega)\frac{c_j}{m_j}+\frac{s\widehat{G}(s)}{E}\omega_j^2=0,\quad (j=1,...,n) \tag{25}$$

where m_j,c_j and k_j are the jth diagonal elements of $\mathbf{M},\mathbf{C}$ and $\mathbf{K}$ respectively, $\omega_j=\sqrt{\frac{k_j}{m_j}}$ is the jth natural frequency of the shaft with Young's modulus E.

Example 1: Voigt-Kelvin Model

$$\sigma_x=E(\epsilon_x+\mu\frac{d\epsilon_x}{dt}),\quad \widehat{G}(s)=\frac{E}{s}(1+\mu s) \tag{26}$$

Substituting it into (25), we get

$$(s+i\Omega)^2+(s+i\Omega)\frac{c_j}{m_j}+(1+\mu s)\omega_j^2=0,\ (j=1,...,n). \tag{27}$$

By Rough-Hurwitz criterion, the stable region of Ω is obtained as

$$\Omega>\Omega_0=\min_j\left(\frac{c_j}{\mu_j m_j\omega_j}+\omega_j\right). \tag{28}$$

If Ω greater than Ω_0, the spinning Voigt-Kelvin shaft will be unstable. Ω_0 is called *threshold speed* of the system.

Example 2: Three-Parameter Model and the Creep instability

From (2), we have

$$\widehat{G}(s) = \frac{E}{s}\frac{s+\frac{e}{\tau}}{s+\frac{1}{\tau}}, \tag{29}$$

where $E = E_0$, $0 < e = \frac{E_\infty}{E_0} < 1$. Eq.(25) now is

$$(\lambda - i\Omega)(\lambda^2 + \frac{c_j}{m_j}\lambda + \omega_j^2) + \frac{1}{\tau}(\lambda^2 + \frac{c_j}{m_j}\lambda + e\omega_j) = 0, \quad (j = 1, ..., n), \tag{30}$$

where $\lambda = s + i\Omega$. It is very cambersome to use Routh-Hurwitz criterion for eq.(30). However, because $\frac{1}{\tau}$ is very small in practice, the first order perturbation solutions of the three eigenvalues $(\lambda_j)_k (k = 1, 2, 3)$ of eq.(30) can be solved by the standard perturbation procedure. The real parts of the solutions are

$$Re(\lambda_j)_{1,2} = -\frac{c_j}{2m_j} \mp \frac{1}{\tau}\frac{(1-e)\omega_j^2(\Omega \pm \omega_{dj})}{2\omega_{dj}\left[\left(\frac{c_j}{2m_j}\right)^2 + (\Omega \pm \omega_{dj})^2\right]}$$

$$Re(\lambda_j)_3 = -\frac{1}{\tau}\frac{(\Omega^2 - e\omega_j^2)(\Omega^2 - \omega_j^2) + \left(\frac{c_j}{m_j}\Omega\right)^2}{(\omega_j^2 - \Omega^2)^2 + \left(\frac{c_j}{m_j}\Omega\right)^2} \tag{31}$$

where $\omega_{dj} = \sqrt{\omega_j^2 + (\frac{c_j}{2m_j})^2}$.

In fact, the decremental rate of damped free vibration, $\frac{c_j}{2m_j}$, is always much faster than $\frac{1}{\tau}$. So $Re(\lambda_j)_{1,2}$ keep negative still. For $Re(\lambda_j)_3$, however, if the numerator of (31) is nagative:

$$\left(\frac{\Omega}{\omega_j}\right)^4 + (\mu_j^2 - e - 1)\left(\frac{\Omega}{\omega_j}\right)^2 + e < 0, \quad \left(\mu_j = \frac{c_j}{m_j\omega_j}\right). \tag{32}$$

the $Re(\lambda_j)_3$ will be positive and the unstable spinning motion occurs.

From (32), the unstable regions of Ω are

$$\Omega_{1j} < \Omega < \Omega_{2j} \quad (\text{if } \mu_j < 1 - \sqrt{e}) \tag{33}$$

$$\Omega_{1j}^2 = \frac{\omega_j^2}{2}\left[e + 1 - \mu_j^2 - \sqrt{(e+1-\mu_j^2)^2 - 4e}\right]$$

$$\Omega_{1j}^2 = \frac{\omega_j^2}{2}\left[e + 1 - \mu_j^2 + \sqrt{(e+1-\mu_j^2)^2 - 4e}\right]$$

(33) is shown in Fig.4.

In undamped case($c_j = 0$), the unstable regions are simply reduced to

$$\sqrt{e}\omega_j \le \Omega \le \omega_j, \quad (j = 1, ..., n). \tag{34}$$

It should point out that $Re(\lambda_j)_3$ is the order of $\frac{1}{\tau}$. The instability developing process is very slow. The shaft spins spirally outward very slowly(over days, monthes even year) without bound. This is the typical *Creep Instability* of viscoelastic shafts[1][4][9] and can be realized by rotating a net noodle.

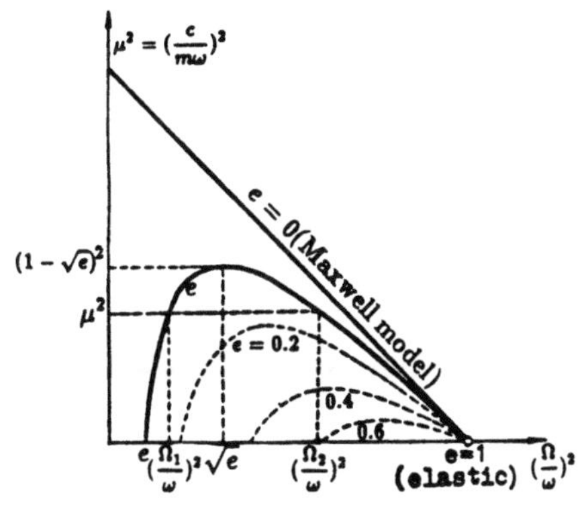

Fig.4

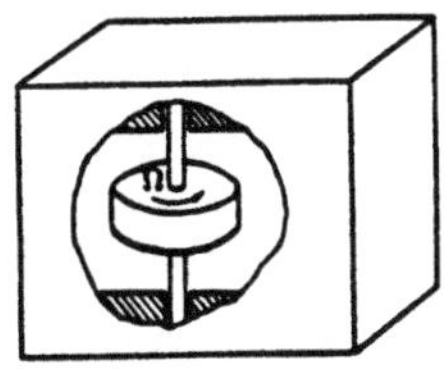

Fig.5

General viscoelastic material

For much general material, we can take

$$\widehat{G}(s) = \frac{E_0}{s}\left(1 - \frac{1}{r}\sum_{j=1}^{r}\frac{1-e_j}{\tau_j}\frac{1}{s+\frac{1}{\tau_j}}\right) \tag{35}$$

which corresponds to $r = p$ in (4) and is called *generalized three-parameter model*[12][13]. Substituting it into (23) and ignoring all the high order terms of $1/\tau_j (j = 1, ..., r)$, we get

$$\left|(\lambda - i\Omega)(\lambda^2\mathbf{M} + \lambda\mathbf{C} + \mathbf{K}) + \frac{1}{\tau}(\lambda^2\mathbf{M} + \lambda\mathbf{C} + e\mathbf{K})\right| = 0 \tag{36}$$

where

$$\frac{1}{\tau} = \sum_{j=1}^{r}\frac{1}{\tau_j}, \quad e = \frac{r-1}{r} + \frac{\tau}{r}\sum_{j=1}^{r}\frac{e_j}{\tau_j} \tag{37}$$

Thus, the conclusions obtaioned from three-parameter model in example 2 can be extended to general. The creep instability is a common phenomenon for most of voscoelastic shafts. In advanced special technology and industry, it is necessary to design high rotating units such as the centrifugal isotope saparator or gyrostat in satellites(Fig.5) to run continuously for a very long period, typically one year or even more. In these cases, the designers should pay much attention to this kind of instability. High-polymer adhesive used on some small-sized exquisitely constructed high-speed rotating components should be very carefully selected to avoid the creep instability which is hardly found out by normal instruments during a normal experimental period.

References

[1] Zhang, Wen, (1986). Dynamic Stability of the Rotating Shaft Made of Boltzmann Viscoelastic Solid, ASME *Journal of Applied Mechanics*, v.53, 1986, 424-429.

[2] Zhang, Wen, (1984). Dynamic Stability of Rotating Viscoelastic Shafts(in Chinese), *Fudan Journal(Natural Science)*, v.23, no.4, 425-433.

[3] Zhang, Wen, (1990). *Basic Theory of Modern Rotordynamics*(in Chinese), Chinese Science Press, Beijing.

[4] Rosenblum, V.I., (1970). Stability of the Rectilinear Mode of a Rotating Shaft under Conditions of Material Creep, *Problems of the Mechanics of Continuous Media*, Izd. Sudostroenie, Moscow(in Russian), 359-365.

[5] Tondl, A.,(1965). *Some Problems of Rotor Dynamics,* Publishing House of the Czechoslovak Academy of Sciences, Prague.

[6] Torby, J., (1979). The Effect of Structural Damping upon the Whirling of Rotors, ASME *Journal of Applied Mechanics*, v.46, 469-470.

[7] Kellenberger, W., (1963). Dir Stabilität Rotierender Wellen Infolge Innerer und Äusserer Dämpfung, *Ingenieur-Archir*, v.32, 323-340.

[8] Pyrz, R., (1982). Stability of Motion of a Viscoelastic Rotating Shaft, Report No.237, Danish Center for Applied Mathematics and Mechanics.

[9] Hendricks, S. L., (1986). The Effect of Viscoelasticity on the Vibration of a Rotor, ASME *Journal of Applied Mechanics*, v.53, 412-416.

[10] Ehrich, F. F., (1964). Shaft Whirl Induced by Rotor Internal Damping, ASME *Journal of Applied Mechanics*, v.31, 279-282.

[11] Rimrott,F.P.J., (1989), *Introductory Attitude Dynamics*, Springer-Verlage.

[12] Pipkin, A.C., (1972). *Lecture on Viscoelasticity Theory*, Applied Mathematical Sciences, v.7, Springer-Verlag, New York.

[13] Creus, G.J., (1986), *Viscoelasticity-Basic Theory and Application to Concrete Structures*, Springer-Verlag.

MIX
Papier aus verantwortungsvollen Quellen
Paper from responsible sources
FSC® C105338

If you have any concerns about our products,
you can contact us on
ProductSafety@springernature.com

In case Publisher is established outside the EU,
the EU authorized representative is:
Springer Nature Customer Service Center GmbH
Europaplatz 3, 69115 Heidelberg, Germany

Printed by Libri Plureos GmbH
in Hamburg, Germany